Trace Metals in Aquatic Systems

Trace Metals in Aquatic Systems

Robert P. Mason
Departments of Marine Sciences and Chemistry
University of Connecticut Avery Point, Groton, CT, USA

WILEY-BLACKWELL
A John Wiley & Sons, Ltd., Publication

Blackwell Publishing was acquired by John Wiley & Sons in February 2007. Blackwell's publishing program has been merged with Wiley's global Scientific, Technical and Medical business to form Wiley-Blackwell.

Registered office: John Wiley & Sons, Ltd, The Atrium, Southern Gate, Chichester, West Sussex, PO19 8SQ, UK

Editorial offices: 9600 Garsington Road, Oxford, OX4 2DQ, UK
The Atrium, Southern Gate, Chichester, West Sussex, PO19 8SQ, UK
111 River Street, Hoboken, NJ 07030-5774, USA

For details of our global editorial offices, for customer services and for information about how to apply for permission to reuse the copyright material in this book please see our website at www.wiley.com/wiley-blackwell.

Library of Congress Cataloging-in-Publication Data

Mason, Robert P. (Robert Peter), 1956–
Trace metals in aquatic systems / Robert P. Mason.
pages cm
Summary: "The textbooks that currently exist do not deal with this particular subject in a comprehensive fashion, and therefore this book is being proposed to full this gap" – Provided by publisher.
Includes bibliographical references and index.
ISBN 978-1-4051-6048-3 (hardback)
1. Trace elements–Analysis. 2. Water–Analysis. 3. Water chemistry. I. Title.
QD142.M316 2013
553.7–dc23

2012038420

A catalogue record for this book is available from the British Library.

Wiley also publishes its books in a variety of electronic formats. Some content that appears in print may not be available in electronic books.

Cover image: © Willyam Bradberry / Shutterstock.com
Cover design by Design Deluxe

Set in 9/12 pt Meridien by Toppan Best-set Premedia Limited
Printed and bound in Malaysia by Vivar Printing Sdn Bhd

1 2013

Contents

Preface

This book is an outgrowth of a number of courses I have taught over the last seven years at the University of Connecticut. While written as an upper level text, it should also be useful to environmental scientists interested in trace elements in aquatic systems. Much of the information is derivative from a class in Environmental Chemistry that I introduced and later taught with Ron Siefert within the University of Maryland Center for Environmental Science. Readers will see the influences of my graduate and post-doc mentors, Bill Fitzgerald, Francois Morel and Harry Hemond, in the book's focus, and my exposure to books such as Broecker and Peng *Tracers in the Sea*, Stumm and Morgan *Aquatic Chemistry* and Morel and Hering *Principles and Applications of Aquatic Chemistry*, and their related texts. These excellent primary texts have been supplemented in recent years by many other books, including compilations such as the *Treatise on Geochemistry*, but I always find myself constantly referring to the earlier works. In studies of trace metals and metalloids one must always reflect on the underlying principles which are easy to forget, especially if an element is studied in isolation.

Much of my graduate work focused on making precise and accurate measurements of mercury speciation in the ocean and freshwaters, which solidified my appreciation for the care and rigor needed for environmental research. These pursuits took me to the equatorial Pacific in late 1989, and important research findings. Many other journeys, including oceanic cruises in the Atlantic and Pacific, and studies in small and large lakes and in the atmosphere, are what delivered me to where I am today, and the contents of this book. I acknowledge the many students who have helped me refine my teaching approach and improve the delivery of the often complex messages relating to the biogeochemical cycling of elements in aquatic systems, which is reflected in the book. There is obviously more emphasis on certain elements over others, but I have endeavored to focus on a particular topic because of its biogeochemical importance and potential impact on humans and the environment, rather than because of personal bias. I hope that I have succeeded as it is not possible to cover all topics in detail within the page limits of the publisher, or in a book suitable as a one semester course. In this vein, I have chosen focus topics in the latter chapters as examples demonstrating chemical principles and focusing on problems of global importance. While the book has almost 1000 references this is clearly the tip of the publication "iceberg". I have referred to the primary literature except in cases when there was a high quality recent compilation chapter or review article. The reader is encouraged to examine the references within these reviews. I am sure that some readers may not understand why I did not cite their work, but I endeavored to remove bias in the choice of citations. The book was an undertaking of many years and so the citations reflect the order the chapters were written, which was not totally chronological. Where possible, I used examples from locations around the world.

I thank Stan Wakefield and Ian Francis for helping me get connected with the publisher about 6 years ago and for the help of many people at Blackwell during the process of publication. I thank the reviewers of the original book proposal for their comments that helped frame the content. I especially thank two anonymous reviewers for their comments on the first complete draft of the book. This was a tall order that they did with diligence and their comments were very useful. I also thank those who read sections of the book, or helped with the editing, especially Elsie Sunderland, Brian DiMento and Amina Schartup. Also, I acknowledge support from the University of Connecticut Small Grant Program for student support for the book compilation and also for funds to help cover the costs of obtaining permission to reuse figures. Obtaining permissions, and the associated costs, are an unfathomable detour along the road to publication even given the advances of the electronic age. I will endeavor in the future to make figures in my publications as clear as possible as even with electronic tools it is not always easy to reproduce figures already in the literature.

I never thought it would take this long, but it did. This lack of foresight is probably a good thing as I am sure my wife Joan would have been less enthusiastic had she known the truth. I thank her for her support through the process. The book writing occurred in stages and that may be evident to the reader, and was interrupted by the realities of academic life. For example, three of my PhD students have graduated since I started the book, and much new research has come to light, which was an ongoing challenge to try and keep the text as current as possible. I hope to have succeeded in this. Enjoy.

Robert P. Mason

About the companion website

This book is accompanied by a companion website:

www.wiley.com/go/mason/tracemetals

The website includes:

- Powerpoints of all figures from the book for downloading
- PDFS of tables from the book
- Answers to end of Chapter Problems

CHAPTER 1
Introduction

1.1 A historical background to metal aquatic chemistry

In terrestrial waters, much of the initial work by environmental engineers and scientists was aimed at understanding the processes of waste water treatment, and of the reactions and impact of human-released chemicals on the environment, and on the transport and fate of radioactive chemicals. This initial interest was driven by the need to understand industrial processes and the consequences of these activities, and the recognition of the potential toxicity and environmental impact of trace metals released during extraction from the Earth as well as during refining and use. Initial interest in the chemistry of the ocean was driven by the key question of why the sea is salty. Of course, it is now known that freshwaters, and even rainwater, contain a small amount of dissolved salts; and that the high salt content of the ocean, and some terrestrial lakes, is due to the buildup of salts as a result of continued input of material to a relatively enclosed system where the major loss of water is due to evaporation rather than outflow. This is succinctly stated by Broecker and Peng [1]: "The composition of sea salt reflects not only the relative abundance of the dissolved substances in river water but also the ease with which a given substance is entrapped in the sediments".

The constituents in rivers are derived from the dissolution of rocks and other terrestrial material by precipitation and more recently, through the addition of chemicals from anthropogenic activities, and the enhanced release of particulate through human activity. The dissolution of carbon dioxide (CO_2) in rainwater results in an acidic solution and this solution subsequently dissolves the mostly basic constituents that form the Earth's crust [2]. The natural acidity, and the recently enhanced acidity of precipitation due to human-derived atmospheric inputs, results in the dissolution (weathering) of the terrestrial crust and the flow of freshwater to the ocean transports these dissolved salts, as well as suspended particulate matter, entrained and resuspended during transport. However, the ratio of river to ocean concentration is not fixed for all dissolved ions [3, 4] as the concentration in the ocean is determined primarily by the ratio of the rate of input compared to the rate of removal, which will be equivalent at steady state. Thus, the ocean concentration of a constituent is related to its reactivity, solubility or other property that may control its rates of input or removal [1, 5].

Many of the early investigations looked at the aquatic ecosystem as an inorganic entity or reactor, primarily based on the assumption that aquatic chemistry was driven by abiotic chemical processes and reactions, and that the impact of organisms on their environment was relatively minor. In 1967, Sillen [6], a physical chemist, published a paper *The Ocean as a Chemical System*, in which he aimed at explaining the composition of the ocean in terms of various equilibrium processes, being an equilibrium mixture of so-called "volatile components" (e.g., H_2O and HCl) and "igneous rock" (primarily KOH, $Al(OH)_3$ and SiO_2), and other components, such as CO_2, NaOH, CaO and MgO. His analysis was a follow-up of the initial proposed weathering reaction of Goldschmidt in 1933:

$$0.6 \text{ kg igneous rock} + \sim 1 \text{ kg volatiles} = 1 \text{ L sea water} + \sim 0.6 \text{ kg sediments} + 3 \text{ L air} \quad (1.1)$$

Sillen stated that "the composition may . . . be given by well-defined equilibria, and that deviations from equilibrium may be explainable by well-defined processes", and proposed the box model shown in Fig. 1.1 [6]. However, he also stated that this did not "mean that I suggest that there would be true equilibrium in the real system". In reality, the ocean composition is a steady state system where the

Trace Metals in Aquatic Systems, First Edition. Robert P. Mason.

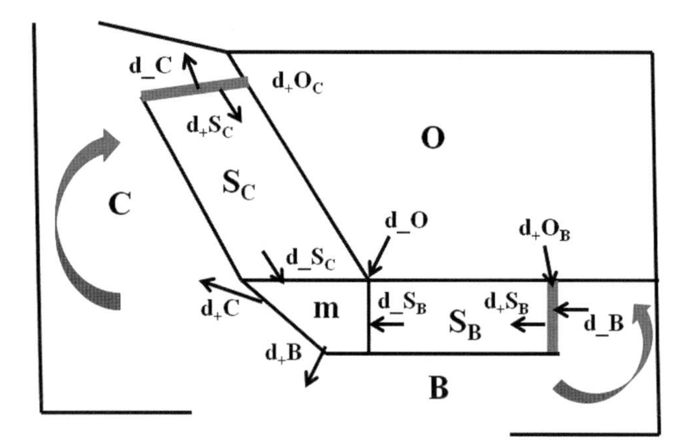

Fig. 1.1 A representation of the global material cycle. The reservoirs are indicated as symbols C = continental rock; B: basaltic rock; O: ocean; S_C: continental sediments; S_B: basaltic sediments. The fluxes into (+) and out of (−) each reservoir are indicated by appropriate symbols that represent the flow paths. Redrawn from Sillen (1967) *Science* **156:** 1189–97. Reprinted with permission of AAAS [6].

concentrations are determined primarily by the relative rates of addition and removal, and that reversible equilibrium reactions are not the primary control over ocean composition [1]; something Sillen [6] alluded to. The major difference between these two conceptualizations is that the equilibrium situation would lead to a constancy of composition over geological time while the steady state model accommodates variation due to changes in the rate of input of chemicals to the ocean. The Sillen paper established the idea that the ocean was a system that could be described as being at geological equilibrium in terms of the major reactions of the primary chemicals at the Earth's surface, and that the average composition of the overall system could be described. The composition of seawater, for example, was due to a series of reversible equilibrium reactions between the ocean waters, sediments and the atmosphere [6]. Similarly, it was proposed that the composition of the biosphere was determined by a complex series of equilibrium reactions – acid-base and oxidation-reduction reactions – by which the reduced volatile acids released from the depths of the Earth by volcanoes and other sources, reacted with the basic rocks of the Earth and with the oxygen in the atmosphere.

A follow-up paper in 1980 by McDuff and Morel [7] entitled *The Geochemical Control of Seawater (Sillen Revisited)*, discussed in detail these approaches and contrasted them in terms of explaining the composition of the major ions in seawater. This paper focused on the controls over alkalinity in the ocean and the recycled source of carbon that is required to balance the removal of $CaCO_3$ via precipitation in the ocean, and its burial in sediments. These authors concluded that "while seawater alkalinity is directly controlled by the formation of calcium carbonate as its major sedimentary sink, it is also controlled indirectly by carbonate metamorphism which buffers the CO_2 content of the atmosphere" [7]. They concluded that the "ocean composition [is] dominated by geophysical rather than geochemical processes. The acid-base chemistry reflects, however, a fundamental control by heterogeneous chemical processes."

McDuff and Morel [7] also focused attention on the importance of biological processes (photosynthesis and respiration) on the carbon balance, something that the earlier chemists did not consider [6]. These biological processes result in large fluxes via carbon fixation in the surface ocean and through organic matter degradation at depth, but overall most of this material is recycled within the ocean so that little organic carbon is removed from the system through sediment burial. Thus, the primary removal process for carbon is through burial of inorganic material, primarily as Ca and Mg carbonates. Recently, the short-term impact of increasing atmospheric CO_2 on ocean chemistry has been vigorously debated, and is a topic of recent research focus [8, 9] because of the resultant impact of pH change due to higher dissolved CO_2 in ocean waters on the formation of insoluble carbonate materials. Carbonate formation is either biotic (shell formation by phytoplankton and other organisms) or abiotic. Thus, there has been a transition from an initial conceptualization of the ocean and the biosphere in general, as a physical chemical system to one where the biogeochemical processes and cycles are all seen to be important in determining the overall composition. This is true for saline waters as well as large freshwaters, such as the Great Lakes of North America, and small freshwater ecosystems, and even more so for dynamic systems, such as rivers and the coastal zone [2, 4, 10].

It has also become apparent that most chemical (mostly redox) reactions in the environment that are a source of energy are used by microorganisms for their biochemical survival and that, for example, much of the environmental oxidation and reduction reactions of Fe and Mn are mediated by microbes, even in environments that were previously deemed unsuitable for life, because of high temperature and/or acidity [2, 10]. Overall, microorganisms specifically, and biology in general, impact aquatic trace metal(loid) concentrations and fate and their presence and reactivity play multiple roles in the biogeochemistry of aquatic systems. Understanding their environmental concentration, reactivity, bioavailability and mobility are therefore of high importance to environmental scientists and managers. Trace metals are essential to life as they form the basis of many important biochemicals, such as enzymes [10, 11], but they are also toxic and can cause both human and environmental damage. As an example, it is known that mercury (Hg) is an element that is toxic to organisms and bioaccumulates into aquatic food chains, while the other elements in Group 12 of the periodic table – zinc (Zn) and cadmium (Cd) – can play a biochemical role [11]. Until recently, Cd was thought to be a toxic metal only but the demonstration of its substitution for Zn in the important enzyme, carbonic anhydrase, in marine phytoplankton has demonstrated one important tenet of the trace metal(loid)s – that while they may be essential elements for organisms, at high concentrations they can also be toxic [11, 12]. Copper (Cu) is another element that is both required for some enzymes but can also be toxic to aquatic organisms, especially cyanobacteria, at higher concentrations. This dichotomy is illustrated in Fig. 1.2, and is true not only for metals but for non-metals and organic compounds, many of which also show a requirement at low concentration and a toxicity at high exposure.

This dual role is also apparent for the group of elements, termed the metalloids, which occupy the bottom right region of the periodic table (Groups 13–17, from gallium (Ga) to astatine (As); see Fig. 1.3) [12]. They do not fit neatly into the definition of "trace metals," as these elements are transitional between metals and non-metals in chemical character, but their behavior and importance to aquatic biogeochemistry argue for their inclusion in this book. Indeed, selenium (Se) is another example of an element that is essential to living organisms, being found in selenoproteins which have a vital biochemical role, but it can also be toxic if present at high concentrations [13]. As noted previously, the chemistry of both metals and microorganisms are strongly linked and many microbes can transform metals from a benign to a more toxic form, or vice versa; for example, bacteria convert inorganic Hg into a much more toxic and bioaccumulative methylmercury, while methylation of arsenic (As) leads to a less toxic product [14].

Therefore, in this book the term "trace metal(loid)" is used in a relatively expansive manner to encompass: (1) the transition metals, such as Fe and Mn, that are relatively abundant in the Earth's crust but are found in solution at relatively low concentrations due to their insolubility, and because they play an important role in biogeochemical cycles [2, 10]; (2) the so-called "heavy metals," such as Hg, Cd, lead (Pb) and Zn that are often the source of environmental concern, although some can have a biochemical role [4, 10]; (3) the metalloids, such as As and Se, which can be toxic and/or required elements in organisms; and (4) includes the lanthanides and actinides, which exist at low concentration and/or are radioactive [15]. Many of the lanthanides (so-called "rare earth elements") are now being

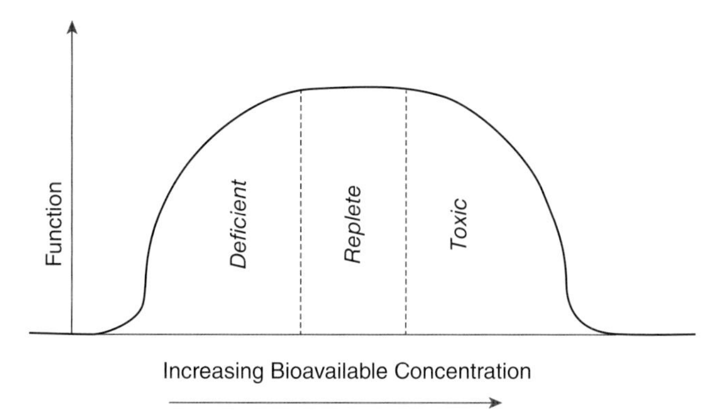

Fig. 1.2 Concentration-response curve for a trace element that is both required by organisms for survival but that can be toxic to the same organism at high concentration.

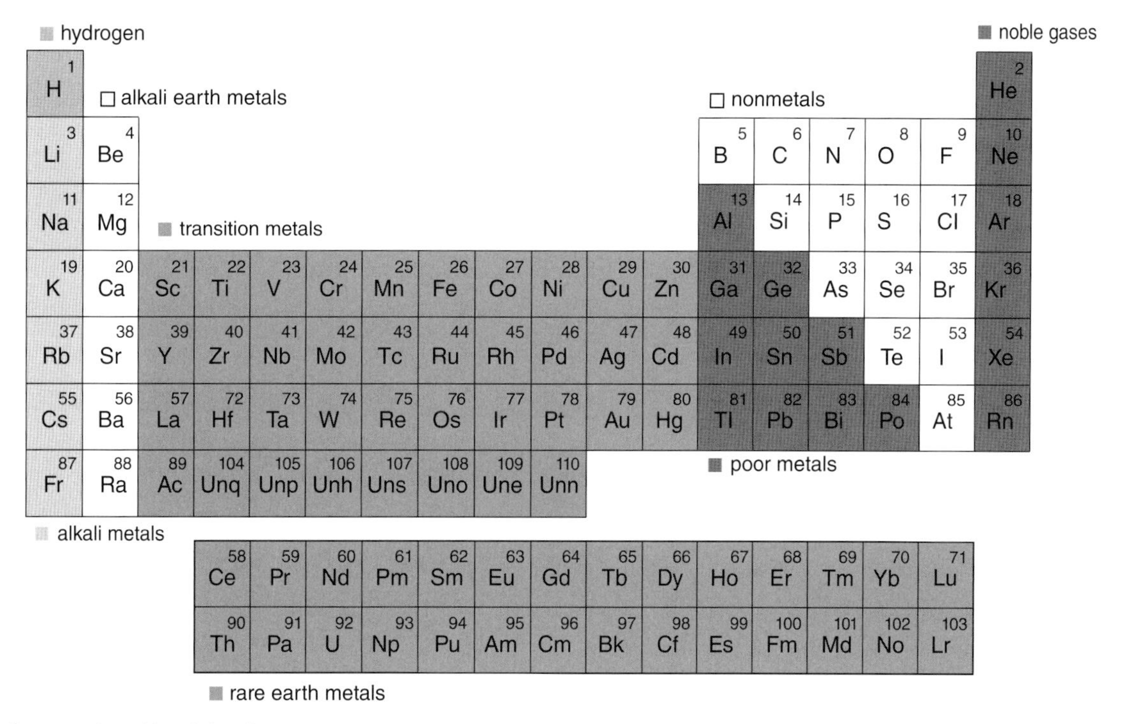

Fig. 1.3 The Periodic Table of the elements.

actively mined given their heightened use in modern technology [16]. Many of the actinides, including uranium (U), plutonium (Pu), are used or produced in the nuclear industry, while others, such as thorium (Th) and protactinium (Pa) are formed naturally from the decay of U in the environment and have been used as geochemical tracers. The elements that are the primary focus of the book are contained in Table 1.1, along with some key details about each element. The Periodic Table (Fig. 1.3) and the information in Table 1.1 provide an indication of the differences between these elements.

While Fe is not truly a trace element, being one of the most abundant elements on Earth, it is a metal that has been shown to be a limiting nutrient to oceanic phytoplankton [17]. This reason, and the fact that its chemistry plays a vital role in the overall fate and transport of many trace metals and metalloids argues for its inclusion in this book, along with Mn. The focus of the book will be on the metal(loid)s for which the most information is available, or which are the most environmentally relevant. While there will be a focus on metals of the first transition series, and all the metals of Groups 11 and 12, as they are either important

Table 1.1 Characteristics of the elements that will be principally discussed in this book. The average ocean and river concentrations for unpolluted waters are taken from various sources [4, 5, 17, 18].

Element	Symbol, Atomic Number	Atomic Mass	Abundance (μmol kg^{-1} or as noted)	Main Oxidation States	Average Ocean Concentration (nM)	Average River Concentration* (nM)
Aluminum	Al 13	26.982	Most	III	20	1850
Titanium	Ti 22	47.88	0.6%	IV	<20	
Vanadium	V 23	50.942	0.02%	II–V	30	
Chromium	Cr 24	51.996	2700	III, VI	4	19
Manganese	Mn 25	54.938	0.085%	II, IV, VII	0.3	145
Iron	Fe 26	55.847	2nd most	II, III	0.5	720
Cobalt	Co 27	58.933	500	II, III	0.02	3.4
Nickel	Ni 28	58.69	1500	II	8	8.5
Copper	Cu 29	63.546	1100	II, I	4	24
Zinc	Zn 30	65.39	1200	II	5	460
Gallium	Ga 31	69.72	270	III	0.3	1.3
Germanium	Ge 32	72.59	20	IV	0.07	0.1
Arsenic	As 33	74.913	28	III, V	23	23
Selenium	Se 34	78.96	0.6	–II, IV, VI	1.7	
Molybdenum	Mo 42	95.94	11	VI	110	5.2
Silver	Ag 47	107.868	0.7	I	0.025	2.8
Cadmium	Cd 48	112.41	1.3	II	0.6	0.2
Tin	Sn 50	118.71	18	II, IV	0.004	0.01
Antimony	Sb 51	121.75	1.7	III, V	1.6	8.2
Tellurium	Te 52	127.60	0.008	–II, IV, VI	<0.001	
Cerium	Ce 58	140.12	430	III, IV	0.02	
Other Lanthanides	59–71	140.91–174.97		II, III	La 0.03	La 0.4
Tungsten	W 74	183.85	0.6	IV–VI	0.06	
Rhenium	Re 75	186.207	0.016	–I, VI, VII	0.04	
Osmium	OS 76	190.2	0.011	II, III, IV	0.05	
Platinum	Pt 78	195.08	0.19	II, IV	0.003	
Gold	Au 79	196.967	0.015	I. III	<<0.001	0.01
Mercury	Hg 80	200.59	0.33	0, II	0.002	0.01
Lead	Pb 82	207.2	50	II	0.01	0.48
Bismuth	Bi 83	208.98	0.12	III, V	<0.025	
Polonium	Po 84	(209)	No data	II, IV, –II	<<0.001	
Thorium	Th 90	232.038	26	IV	<0.001	0.4
Protactinium	Pa 91	231.036				
Uranium	U 92	238.029	7.6	VI	14	1.0
Plutonium	Pu 94	244.064				

*Average for unpolluted rivers. For some elements, variability in river concentrations is high.

commercially, or are toxic/required elements, it is also necessary to discuss the radioactive and rare earth elements. Few of the elements from the second and third transition series have important biochemical or biogeochemical reactions, besides molybdenum (Mo), and possibly tungsten (W), but some of these elements, such as rhenium (Re) and osmium (Os) have been used as tracers of environmental processes. Overall, the elements of the second and third transition series have been little studied. The lanthanides, as a group, have been used by aquatic chemists as chemical tracers, and have similar chemistry, because, while the inner $4f$ orbital is being filled through the series, the electrons lost are the outer orbital electrons and they mostly behave as +3 cations in solution. For the actinides, which are often radioactive, the focus will be on the chemistry of U, Th, Pa, Pu as well as the products of nuclear fission reactions, which are often radioactive [4].

There are many texts that cover one or more aspects of the topic of this book, which are referred to throughout, such as Stumm and Morgan [2], Morel and Hering [10], Cotton and Wilkinson [18], Drever [3], Buffle [19], Langmuir [4] and Wilkinson and Leads [20]. A compilation by Holland and Turekian [21] and authors therein, *Treatise on Geochemistry*, synthesizes in the various volumes many of the topics that are contained in this book. Some books are specific to ocean metal chemistry [1, 5].

1.2 Historical problems with metal measurements in environmental media

While the concentrations of metals in solid phase media – rocks, sediments – and in biota are mostly sufficiently high that they can be analyzed using "typical" conditions found in a geochemical laboratory, this is not always the case, and it is certainly not true for the more trace constituents such as Cd, Hg, and Ag, and especially when making measurements in natural waters. Additionally, many other metals, such as Fe, Zn, and Cu, whose concentration in dust and in general laboratory materials (e.g., acids, plastics) is elevated, are difficult to accurately quantify without the proper scientific care. The potential for sample contamination during collection and analysis in many early environmental studies was not initially appreciated prior to the 1980s, and subsequently the techniques and approaches applied to the analysis of rocks were found not to be suitable for the analysis of natural water samples [22]. The difficulty in making measurements of the actual concentration of a metal in the water samples when collected, compared with the concentration in the sample bottle at the time of analysis, was not initially appreciated by many environmental scientists [23–25], and this still could cause erroneous results. Differences in concentration could result from any inadvertent addition of metal during sampling, handling, and analysis, and due to contaminants in the bottle plastics and in the reagents used. Losses, due to adsorption to container walls are a problem for some metals in water samples, and the subsequent leaching of these metals back into solution can produce sampling artifacts. The impact of all these factors on the measurement of the dilute metal concentrations in natural solutions (Table 1.1) was not fully appreciated in the early environmental studies of natural waters.

The first convincing demonstration of the errors in many of the measurements of trace metals in environmental waters was made by Patterson and Settle [22]. Their landmark paper, *The Reduction of Orders of Magnitude Errors in Lead Analyses of Biological Materials and Natural Waters by Evaluating and Controlling the Extent and Sources of Industrial Lead Contamination Introduced during Sample Collection and Analyses*, [22] fundamentally changed the approach to the collection of samples of natural waters for trace element analysis, and lead to the development of protocols, so-called "clean techniques", for the analysis of low level trace metal samples. These investigators, and the scientists who adopted their approaches, clearly demonstrated the potential for contamination of samples due to traces of metals in the plastic materials, the acids and other reagents used in storage and analysis, and the importance of excluding dust in laboratory settings and the potential for introduction of metals from humans during handling. They showed the extent of preparation and care that was needed at all stages of the process: in sample container cleaning and preparation, during sample collection and manipulation, and during analysis. Such precautions that are now routinely used by environmental scientists (acid cleaning of glassware and plastics, use of gloves, cleaning of sampling devices and sampling lines, use of specific materials for the contaminant prone metals (e.g., use of Teflon for Hg sampling), and use of non-metallic materials where possible) were not done routinely in the early studies.

Patterson and Settle [22] showed that their measurements of Pb in ocean waters were orders of magnitude lower than those reported by other investigators and furthermore, that their results were "environmentally and geochemically consistent". This realization of the magnitude and extent of the contamination of samples subsequently lead to a revolution in the way that samples were collected and analyzed, and, as a result, in the reported concentrations of metals in environmental waters. This is illustrated in Fig. 1.4 for a number of metals. In Fig. 1.4(a), a compilation of data from various sources shows the dramatic decrease in the reported concentration for the metals Cu, Fe, Zn, and Hg in open ocean seawater over time [49–52]. In all cases, the change between the reported concentrations in 1983 and 1990 are relatively small compared to the dramatic decrease between 1965 and 1983, especially for Fe, Hg, and Zn. For Cu, the reported concentration decrease is about a factor of 50, while is it more than two orders of magnitude for the other metals. Similar decreases in the reported metal concentrations for river water are illustrated by a comparison over time [23] (see Fig. 1.4b). Again, the use of clean approaches to

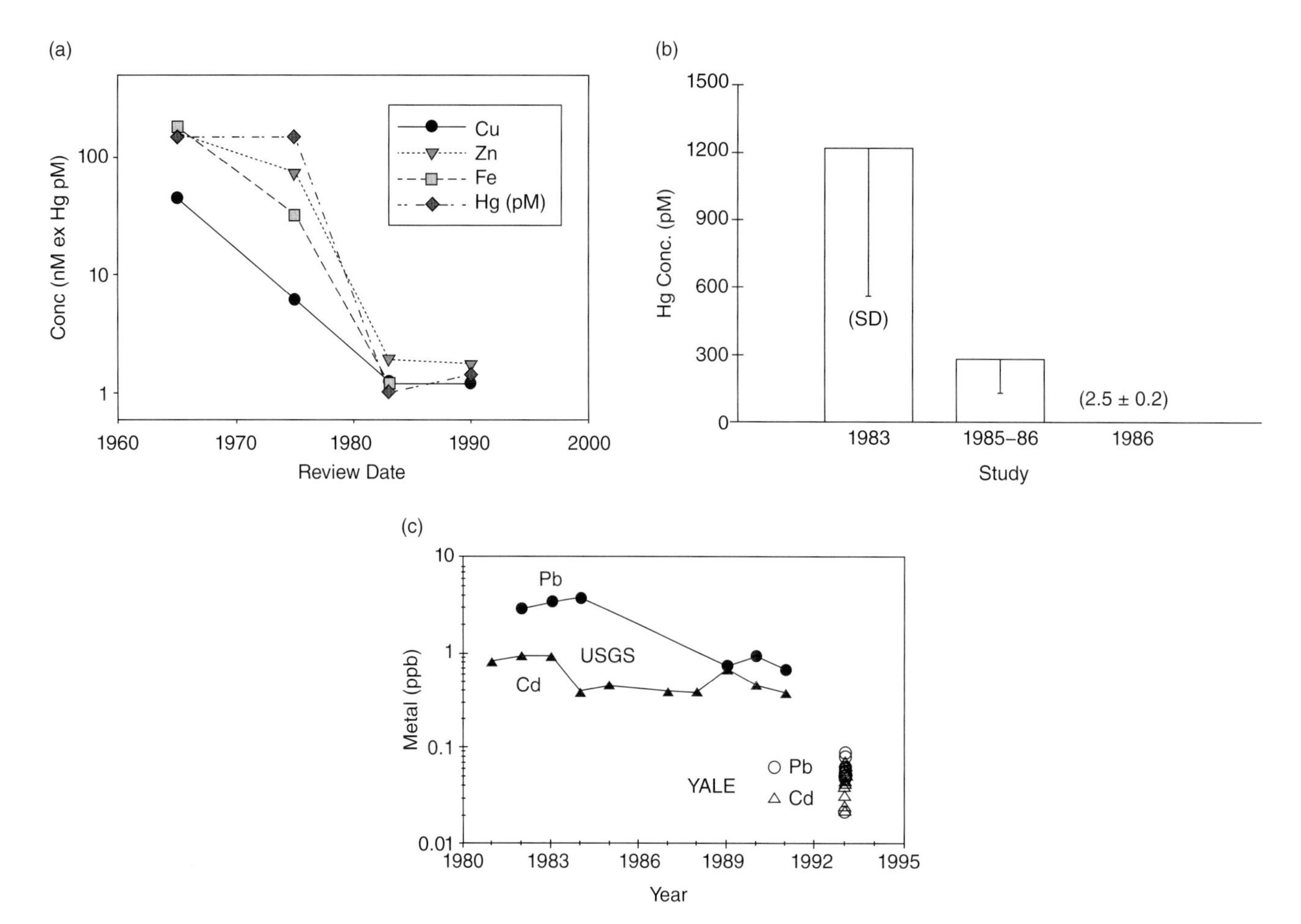

Fig. 1.4 (a) Graph showing the trend in the reported mean concentration of several metals in ocean waters. Early data are taken from reviews by Brewer [24]; Bruland [49]; Bruland et al. [40] and Goldberg [25]. The 1990 data are taken from Refs [50, 52] and are for the Pacific Ocean. Figure redrawn using this information and Mason [26]; (b) reported mercury concentrations for Vandercook Lake in Wisconsin over time. Figure reprinted from Fitzgerald, W. and Watras, C.J. (1989) *Science of the Total Environment* **87/88:** 223–32 with permission of Elsevier [27]. (c) Figure showing the difference between the reported concentrations of lead and cadmium for the Quinnipiac River. Figure reprinted with permission from Benoit, G. (1994) *Environmental Science and Technology* **28:** 1987–91, Copyright (1994), American Chemical Society [23].

sampling and analysis lead to more than an order of magnitude decrease in the measured concentrations of Cd and Pb in river water compared to the values reported previously. Finally, data from samples collected in a lake in Wisconsin show a similar trend for Hg [27] (Fig. 1.4c). The decrease in reported concentration was again more than two orders of magnitude between 1983–6. As noted earlier, the changes in the reported concentration in all these instances was not primarily the result of a "cleanup" of the environment, although there is evidence for changing concentrations over recent decades, but rather a realization and appreciation that the investigators themselves, and the plastics and other materials used for sampling contain significant traces of metals, and that these were being leached into the samples during handling and storage.

For the ocean, sampling methodologies that used metal wire cables led to contamination, as did the use of untreated plastic sampling bottles used to collect samples at depth (e.g., Niskin bottles). Metal inputs from the vessels themselves can also be a source of contamination. These sources of contamination were most severe for those waters that have low concentration due to their remoteness (e.g., the middle of the Pacific Ocean). A comparison by a reader of the reported concentrations of metals in the ocean in the recent literature would further illustrate the point. Such an examination indicates that while the reported concentrations of Al (~40 nM), Cr (40 nM) and other metals that are present in seawater at relatively high concentrations are similar to those in Table 1.1, many of the concentrations reported in various books are 2–100 times higher, especially for the heavy metals (e.g., Hg, Pb) and even for some of the transition metals (Ni, Co, Zn). So, there is still a need for caution when reading the literature and certainly much care and application is needed when doing field studies.

The extent of potential sample contamination is obviously, but not exclusively, a function of the concentration, with potentially more contamination likely for metals that are present at low concentrations. Cadmium provides one example where this is not true as, even though it exists in natural waters at low concentrations, there has been less evidence of its contamination in earlier studies. Sampling and analytical difficulties are most severe for those metals that are present in low concentration and which are routinely used in materials, or are part of typical sampling equipment and vessels, or occur in high abundance in atmospheric aerosols, such as Fe, Cu, Zn, Hg, and Pb. One example is the use of metal-containing dyes in plastics, such as in the coloring of pipette tips. Use of these tips with acidic solutions can lead to metal leaching and contamination. One recent example of contamination for Zn that has been highlighted in ocean studies is the use of Zn as "sacrificial anodes" on metal sampling devices to prevent corrosion – dissolution of Zn was found to be sufficient to contaminate samples during a recent international intercalibration study as part of the GEOTRACERS program [28].

The introduction of these approaches has led to a wealth of information on metal distributions in the ocean and in freshwaters that has resulted in an enhanced understanding of their toxicity to aquatic organisms, and their requirement as essential nutrients, and to the factors that control their cycling. Accurate measurements also led to a clearer and better understanding of the role of human inputs in environmental contamination, and the need to curb industrial and other emissions of metals to the atmosphere and aquatic systems. The importance of this development and the resultant explosion of high quality research can be illustrated by the papers presented at a special symposium held in honor of Claire Patterson on his retirement, in December 1992. The papers presented at the meeting were published in a special issue of *Geochimica et Cosmochimica Acta* in 1994 and many papers cover topics related to Pb and other metals, and their geochemical cycle, such as using ice core records and ocean measurements and Pb isotope ratios to examine recent changes and those over geological timescales [29–37]. The paper by Boutron et al. [29] illustrates the difficulty in making measurements of trace concentrations of metals in a situation where the ice core needed to be drilled and recovered using contaminating techniques. These authors measured the concentration of the metals in successive layers of ice shaved from the outside to the inside of the ice core and showed that while the outer layers were highly contaminated, the inner core was not (Fig. 1.5). This analysis further provided a clear record of the impact of human use and release of Pb, primarily from its use as a gasoline additive, on the concentrations of Pb in remote precipitation (Fig. 1.6) [29].

Other papers presented at the symposium, such as the review paper of Bruland et al. [40], detail measurements of reactive trace metals (Al, Fe, Mn, Zn and Cd) in ocean waters. These authors discussed the notion, first proposed by Whitefield and Turner [38], that the profiles of such elements reflect their biogeochemistry. Some metals such as Zn and Cd have so-called "nutrient (recycled) profiles" in the ocean as their concentration is depleted in surface waters and higher at depth, while others, such as Al, reflect their primary atmospheric input and their high association with

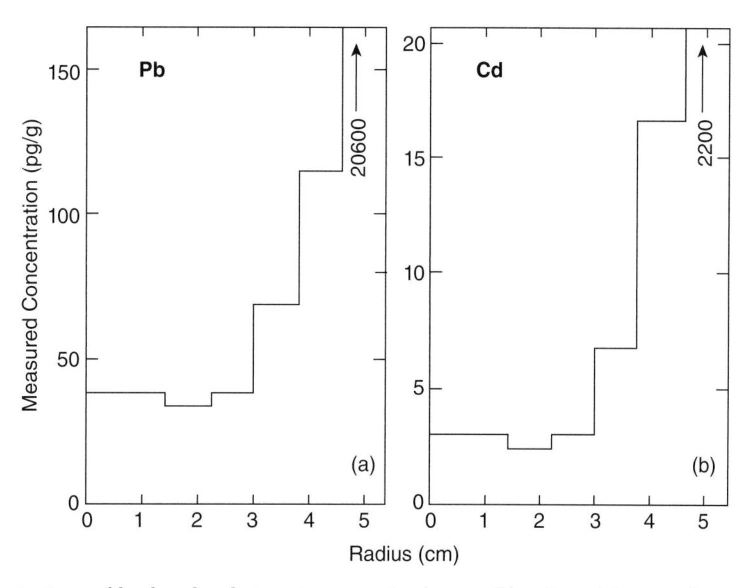

Fig. 1.5 The measured concentrations of lead and cadmium in successive layers of ice shaved from an ice core recovered from Vostok Station in East Antarctica. Figure reprinted from Boutron et al. (1994) *Geochimica et Cosmochimica Acta* **58:** 3217–25 with permission of Elsevier [29].

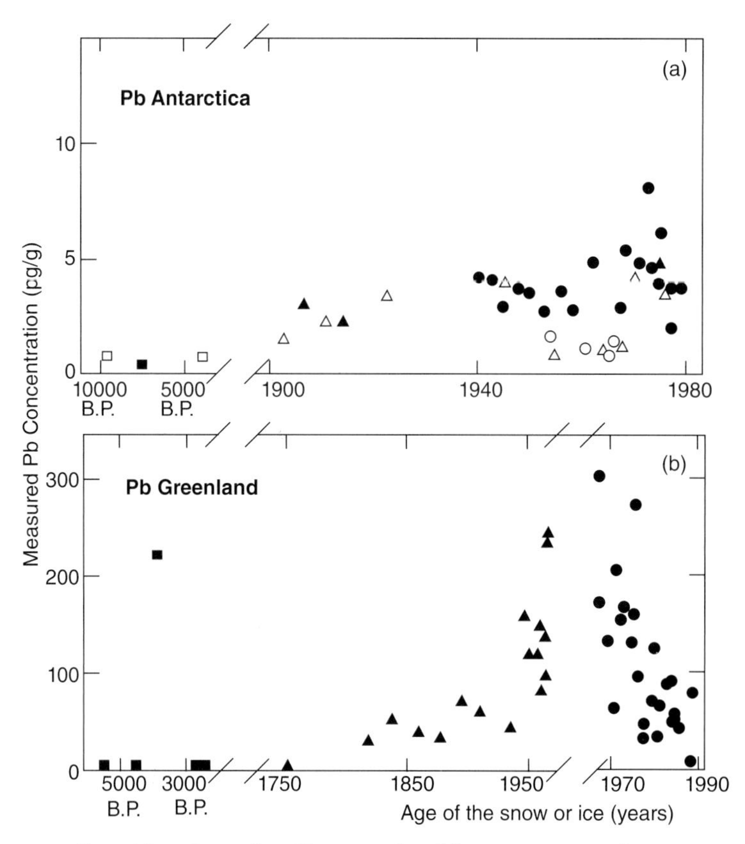

Fig. 1.6 Lead profiles for ice cores collected in polar region. Figure reprinted from Boutron et al. (1994) *Geochimica et Cosmochimica Acta* **58:** 3217–25 with permission of Elsevier [29].

the particulate phase (so-called "scavenged profiles"; high surface concentrations and low deep water values). Other metals, such as Hg, have a more complex ocean distribution due to their complex chemistry and cycling [39].

The notion that a metal, such as Zn, could have an ocean profile that is similar to that of the major nutrient elements [40] (i.e., nitrogen (N), phosphorous (P), and silica (Si) – required by diatoms), and that the concentration of Zn and Si appear closely correlated for open ocean waters [17] was reconciled with the knowledge that Zn has a biological role and forms part of many enzymes important to cellular biochemistry [11]. At the time, the correlation of Cd and P profiles was more of a "mystery" as Cd was considered a toxic heavy metal with no biological function and there was "no biological or chemical explanation for the (relationship)" [1]. It has since been shown that Cd can substitute for Zn in one important enzyme for ocean photosynthetic organisms (carbonic anhydrase) and therefore its role as a nutrient has been established [11].

It is worth noting that contamination is not only a problem in trace metal research where quantification of *in situ* concentrations is the principal focus. Fitzwater et al. [41] demonstrated, for example, that the introduction of metals into solutions from container vessels and reagents could lead to substantial under-estimation of the primary productivity when using ^{14}C-incubation approaches. They demonstrated the potential for leaching of Cu from glass bottles even after acid washing, and the converse, the adsorption of Cu onto glass containers in unacidified solutions. High metal levels were found in the standard ^{14}C solutions and in other solutions used in productivity studies. As a result, more recent techniques have been devised that use trace metal free "clean techniques" [42] for productivity and other studies with phytoplankton and other microbes, if experiments are performed under natural conditions.

The environmental scientist is therefore strongly cautioned to check the reported concentrations in any publication against other sources as, while there is obviously variability between ecosystems and especially for systems with high or variable TSS (total suspended solids) loads, there is still the possibility that reported concentrations in the literature are not reflective of actual concentrations. However, on the contrary, while it is often considered that the lowest concentration must be correct, this is not always the case, as losses and incomplete recovery of the metal during sample analysis is a real possibility, and therefore this metric should not be used exclusively to gauge whether a reported value is correct or not.

Table 1.2 Comparison of techniques for determining metal concentrations in water samples showing relative precision and other analytical metrics [46].

Technique	Linear Range	Precision	DL Cd (nM)	DL Cu (nM)	DL Zn (nM)	Interferences
GFAA	~10^2	0.5–5 %	0.02	0.2	0.3	Relatively few
ICP-AE	~10^5	<2%	0.9	6	3	Spectral; many lines overlap
ICP-MS	10^5–10^7	1–3%	10^{-3}	3×10^{-3}	4.5×10^{-3}	Many; isotopes, oxides and other molecular mass

1.3 Recent advances in aquatic metal analysis

While the previous section outlined the historical problems associated with sample contamination, this issue has been mostly addressed and most recent studies have demonstrated the ability to accurately quantify the total concentration of an element in environmental waters. Some recent intercalibration exercises in open ocean seawater indicate that contamination is always a possibility, and care and application of "clean techniques" is necessary as the concentrations are low (nM or pM; Table 1.1). Comparison and confirmation of accuracy can be achieved through the continued use and further development of low level reference materials. Standard (or Certified) Reference Materials (SRMs or CRMs) are environmental materials that have been repeatedly measured, often by more than one technique, so that their concentration is well-defined by an accepted value and a standard deviation which reflects the overall precision of the measurement. Such materials allow an analyst to confirm that the method and approach used is providing an accurate answer. The US National Institute for Standards and Technology (NIST), the European Institute for Reference Materials and Measures (IRMM), and other organizations in other countries around the world, provide an impressive array of SRMs in many different media. However, it is often difficult to provide adequate SRMs for the low levels of many metals in seawater or freshwaters (nM or lower concentrations), and the number of SRMs for water are limited. There are recent concerted efforts to develop such SRMs, and the international GEOTRACERS Program [43] is one avenue through which such reference materials are being developed.

While accurate measurement of the total concentration is a goal, there is an acknowledgement that the total concentration of an element in solution is not often an accurate predictor of the bioavailability, toxicity, or reactivity of the element [2–4, 10], as most metals do not exist in solution purely as the free metal ion, and are often associated with dissolved organic matter or can be in colloidal phases in the filtered fraction, as discussed in Chapter 2. Therefore, protocols have been developed to examine the distribution and *speciation* (chemical form) of the metal(loid) in solution, as well as isotope ratios and other more detailed analysis [2, 10, 40, 44, 45]. Analytical advances and improvements in sensitivity have also improved the amount of information that is collected. For example, graphite furnace atomic adsorption spectroscopy (GFAAS) was often the preferred method for trace metal analysis in the early 1980s but this method is slow as each element requires a different lamp, and the linear calibration range is small, often requiring reanalysis of samples [45]. The analysis is also time-consuming although there is the potential to use an autosampler. Additionally, extensive pretreatment is required to concentrate samples to levels sufficient for analysis, and to remove interferences from the matrix, although techniques, such as Zeeman correction, have evolved to reduce these issues. As shown in Table 1.2 [46], the detection limits (DL) of GFAAS for the metals shown are within an order of magnitude of their concentrations in seawater (Table 1.1), and so sample preconcentration and/or matrix elimination is often required. The development of ICP (inductively coupled plasma) as an atomization technique has allowed for the more rapid analysis of samples although when coupled with atomic emission quantification (ICP-AES) results in a higher detection limit which is unsuitable for many metals in environmental waters, or even in sediment extracts (Table 1.2). However, coupling to a mass spectrometer (ICP-MS) provides high throughput, and low DL with high precision. However, ICP-MS is prone to substantial interferences such as different elements with the same atomic mass, or interferences due to the formation of oxides and other molecules that form after atomization (e.g., $^{40}Ar^{16}O$ and ^{56}Fe; $^{40}Ar^{35}Cl$ and ^{75}As) but there are a variety of approaches to circumvent these interferences [47]. The severity of the interferences depends on the resolution of the MS, which is relatively low (0.8 amu) for a quadrupole ICP-MS, which are the most robust and stable instruments. Therefore, analysis using these instruments needs to contend with interferences. One approach that has been used to circumvent molecular interferences is the use of a reaction cell, which is designed to break up diatomic and polyatomic molecules prior to the MS, thereby removing the interferences [47]. High resolution instruments are also available that are capable of resolving these interferences because of their much higher resolving power, such as that with $^{40}Ar^{16}O$ and ^{56}Fe, and these are used in many instances. However, these instruments are more delicate and require more maintenance than a quadrupole [47]. More recently, multicollector ICP-MS instruments have been more widely used in aquatic metal analysis as these instruments are able to quantify the isotope ratios of specific elements and can be used to examine isotope fractionation and the factors causing such fractionation. Such isotope fraction can be used to examine environmental processes in a similar fashion to the more widely used fractionation of the major organic elements (C, N, S). Isotope analysis provides a new avenue for research and exploration as many abiotic and biotic processes can cause isotope fractionation in the environment. This is discussed further in Section 7.5. Additionally, ICP-MS is now being used to analyze radioisotopes in natural waters (e.g., Th, Pa, U, Pu) as this can be done with more accuracy and precision than can be obtained with conventional counting techniques.

Therefore, the amount of information has expanded rapidly with the development of more accurate and precise measurement techniques and with the ability to sample at higher resolution. Many of the factors that caused contamination in early studies have been overcome and sampling approaches that were not feasible 20 years ago, such as sampling for metals in seawater with an automated sampling system (rosette sampler), are now being used [43, 48], resulting in a higher throughput of sample collection. In freshwater systems, automated water samplers, such as Isco devices (www.isco.com), are now used to collect samples without contamination. In addition to intercalibration studies, which have allowed for more detailed understanding of sources of sampling and analytical error, methods have been developed to examine the distribution of metals between dissolved, colloidal and particulate fractions without contamination, using ultrafiltration and other approaches. Analytical class-100 clean rooms are now routinely used for most metal analysis and high precision instrumentation has also allowed for a better characterization of blanks and sources of error.

As concentrations in water samples are generally too low for direct analysis, sample preconcentration often occurs. This can additionally remove potential matrix effects and other interferences. Preconcentration techniques typically rely on [44]: (1) co-precipitation of insoluble (hydr)oxides (e.g., Fe, Mn, Mg); or (2) solvent or solid-phase extraction of the analyte from the matrix using organic ligands and organic solvents or onto chelating resins, with subsequent elution into a small volume. Analysis on board ship or in the field often relies on techniques other than ICP-MS as it is not transportable. Many of these analyses rely on flow injection approaches, which allow for preconcentration prior to detection using spectrometry of a metal-ligand complex, or after reaction of the metal with a complex that changes the signal, as occurs, for example, with the reaction of Fe^{II} or Mn^{II} with luminol [44]. Alternatively, as discussed in Chapter 6, many metals can be determined using electrochemical techniques using mercury drop electrodes.

One example of a detailed approach to standardizing methods and analysis is the GEOTRACERS Program intercalibration study [28, 43]. This program has provided a detailed manual on how to collect ocean water samples for the determination of a suite of metals, metalloids, radionuclides and other tracer compounds (TEIs). The manual that has been developed covers the sampling, filtration and storage with details on which type of equipment, filters and other methods to be used, and the best container and sample preservation protocols. These methods are for both radioactive elements as well as stable trace metal(loid)s and other elements, and their stable isotopes; see Section 6.4.2 for further details.

Finally, the development of *in situ* detection methods for the analysis of metal(loid)s in environmental waters is in its infancy but there is the promise of a further explosion of information as such techniques become available. It will be apparent through the discussions in the later chapters of the book, especially Chapters 6 and 7, that there has been enormous progress in recent decades in the understanding of the fate, transport and transformation of metal(loid)s in aquatic systems and the dynamics of the interactions between metals in solution and the solid phase, and with microorganisms that inhabit these waters. Many of the recent discoveries would not be possible without the development of new techniques and approaches, including the methods to quantify different fractions of an element in aqueous systems.

References

1. Broecker, W. and Peng, T.-H. (1982) *Tracers in the Sea*. Columbia University Press, Palisades, NY.
2. Stumm, W. and Morgan, J.J. (1996) *Aquatic Chemistry*. John Wiley & Sons, Inc., New York.
3. Drever, J. (1997) *The Geochemistry of Natural Waters*. Prentice Hall, Upper Saddle River.
4. Langmuir, D. (1997) *Aqueous Environmental Geochemistry*. Prentice Hall, Upper Saddle River.
5. Chester, R. (2003) *Marine Geochemistry*. Blackwell Science, Oxford.
6. Sillen, L. (1967) The ocean as a chemical system. *Science*, **156**, 1189–1197.
7. McDuff, R. and Morel, F.M.M. (1980) The geochemical control of seawater (Sillen revisited). *Environmental Science & Technology*, **14**, 1182–1186.
8. Doney, S.C. (2006) The dangers of ocean acidification. *Scientific American*, **294**, 58–65.
9. Orr, J.C., Fabry, V.J., Aumont, O. et al. (2005) Anthropogenic ocean acidification over the twenty-first century and its impact on calcifying organisms. *Nature*, **437**, 681–686.

10. Morel, F.M.M. and Hering, J.G. (1993) *Principles and Applications of Aquatic Chemistry*. John Wiley & Sons, New York.
11. Morel, F., Milligan, A.J. and Saito, M.A. (2004) Marine Bioinorganic chemistry: The role of trace metals in the oceanic cycles of major nutrients. In: Elderfield, H. (ed.) *The Oceans and Marine Geochemistry*. in Holland, H.D. and Turekian, K.K. (Exec eds) Treatise on Geochemistry. Elsevier Pergamon, Oxford, pp. 113–143.
12. Lindh, U. (2005) Biological functions of the elements. In: Selinus, O., Alloway, B., Centeno, J.A. et al. (eds) Chapter 5, *Essentials of Medical Geology*. Elsevier, Amsterdam, pp. 115–160.
13. Fordyce, F. (2005) Selenium deficiency and toxicity in the environment. In: Selinus, O., Alloway, B., Centeno, J.A. et al. (eds) Chapter 15, *Essentials of Medical Geology*. Elsevier, Amsterdam, pp. 373–416.
14. Craig, P.J. (ed.) (2003) *Organometallic Compounds in the Environment*. John Wiley & Sons, Ltd, Chichester.
15. Nozaki, Y. (2010) Rare earth elements and their isotopes in the ocean. In: Steele, J.H., Thorpe, S.A. and Turekian, K.K. (eds) *Marine Chemistry and Geochemistry*. Elsevier, Amsterdam, pp. 39–51.
16. USGS (2011) Rare earth elements – end use and recyclability, USGS Scientific Investigations Report 2011–5094. http://pubs.usgs.gov/sir/2011/5094/pdf/sir2011-5094.pdf (accessed April 16, 2011).
17. Bruland, K. and Lohan, M.C. (2004) Controls on trace metals in seawater. In: Elderfield, H. (ed.) *The Oceans and Marine Geochemistry*. in Holland, H.D. and Turekian, K.K. (Exec eds) Treatise on Geochemistry. Elsevier Pergamon, Oxford, pp. 23–47.
18. Cotton, F.A. and Wilkinson, G. (1972) *Advanced Inorganic Chemistry: A Comprehensive Treatise*. John Wiley & Sons, Inc., New York, 1355 pp.
19. Buffle, J. (1988) *Complexation Reactions in Aquatic Systems. An Analytical Approach*. Ellis Horwood Pub., Chichester, 700 pp. (downloadable at http://www.unige.ch/cabe/ (accessed April 16, 2011).
20. Wilkinson, K.J. and Lead, J.R. (eds) (2007) *Environmental Colloids and Particles. Behaviour, Separation and Characterization*. IUPAC Series in Analytical and Physical Chemistry of Environmental Systems, Buffle, J. and van Leeuwen, H.P., Series Editors. John Wiley & Sons, Ltd, Chichester.
21. Holland, H.D. and Turekian, K.K. (2004) *Treatise on Geochemistry, in 10 Volumes*. Elsevier, Oxford.
22. Patterson, C.C. and Settle, D.M. (1976) The reduction in orders of magnitude errors in lead analysis of biological materials and natural waters by evaluating and controlling the extent and sources of industrial lead contaminantion introduced during sample collection and analysis. In: LaFleur, P. (ed.) *Accuracy in Trace Analysis: Sampling, Sample Handling and Analysis*. US National Bureau of Standards, Washington, D.C., pp. 321–351.
23. Benoit, G. (1994) Clean technique measurement of Pb, Ag and Cd in freshwater: A redefinition of metal pollution. *Environmental Science and Technology*, **28**, 1987–1991.
24. Brewer, P. (1975) Minor elements in seawater. In: Riley, J. and Skirrow, G. (eds) *Chemical Oceanography*. Academic Press, London, pp. 415–496.
25. Goldberg, E. (1965) Minor elements in seawater. In: Riley, J. and Skirrow, G. (eds.) *Chemical Oceanography*. Academic Press, London, pp. 163–196.
26. Mason, R. (1991) *The Chemistry of Mercury in the Equatorial Pacific Ocean*. University of Connecticut, Storrs-Mansfield, CT.
27. Fitzgerald, W. and Watras, C.J. (1989) Mercury in surficial waters of rural Wisconsin lakes. *Science of the Total Environment*, **87/88**, 223–232.
28. Geotracers (2011) The GEOTRACERS Cookbook. http://www.geotraces.org/science/intercalibration/222-sampling-and-sample-handling-protocols-for-geotraces- (September 25, 2012).
29. Boutron, C.F., Candelone, J. and Hong, S.M. (1994) Past and recent changes in the large-scale tropospheric cycles of lead and other heavy-metals as documented in Antarctic and Greenland snow and ice – a review. *Geochimica et Cosmochimica Acta*, **58**, 3217–3225.
30. Rosman, K.J.R., Chisholm, W., Boutron, C.F., Candelone, J. and Hong, S. (1994) Isotopic evidence to account for changes in the concentration of lead in Greenland snow between 1960 and 1988. *Geochimica et Cosmochimica Acta*, **58**, 3265–3269.
31. Erel, Y. and Patterson, C.C. (1994) Leakage of industrial lead into the hydrocycle. *Geochimica et Cosmochimica Acta*, **58**, 3289–3296.
32. Hirao, Y., Matsumoto, A., Yamakawa, H., Maeda, M. and Kimura, K. (1994) Lead behavior in abalone shell. *Geochimica et Cosmochimica Acta*, **58**, 3183–3189.
33. Veron, A.J., Church, T.M., Patterson, C.C. and Flegal, A.R. (1994) Use of stable lead isotopes to characterize the sources of anthropogenic lead in North-Atlantic surface waters. *Geochimica et Cosmochimica Acta*, **58**, 3199–3206.
34. Boyle, E.A., Sherrell, R.M. and Bacon, M. (1994) Lead variability in the western North-Atlantic Ocean and central Greenland ice – implications for the search for decadal trends in anthropogenic emissions. *Geochimica et Cosmochimica Acta*, **58**, 3227–3238.
35. Ritson, P.I., Esser, B.K., Niemeyer, S. and Flegal, A.R. (1994) Lead isotopic determination of historical sources of lead to Lake Erie, North-America. *Geochimica et Cosmochimica Acta*, **58**, 3297–3305.
36. Rivera-Duarte, I. and Flegal, A.R. (1994) Benthic lead fluxes in San-Francisco Bay, California, USA. *Geochimica et Cosmochimica Acta*, **58**, 3307–3313.
37. Sanudo-Wilhelmy, S.A. and Flegal, A.R. (1994) Temporal variations in lead concentrations and isotopic composition in the southern California Bight. *Geochimica et Cosmochimica Acta*, **58**, 3315–3320.
38. Whitefield, M. and Turner, D.R. (1987) The role of particles in regulating the composition of seawater. In: Stumm, W. (ed.) *Aquatic Surface Chemistry*. John Wiley & Sons, Inc., New York, pp. 457–493.
39. Mason, R.P., Fitzgerald, W.F. and Morel, F.M.M. (1994) The biogeochemical cycling of elemental mercury – Anthropogenic influences. *Geochimica et Cosmochimica Acta*, **58**(15), 3191–3198.
40. Bruland, K.W., Orians, K.J. and Cowen, J. (1994) Reactive trace-metals in the stratified central North Pacific. *Geochimica et Cosmochimica Acta*, **58**, 3171–3182.

41. Fitzwater, S., Knauer, G.A. and Martin, J.H. (1982) Metal contamination and its effects on primary production measurements. *Limnology and Oceanography*, **27**, 544–551.
42. Price, N., Harrison, G.I., Hering, J.G., Hudson, R.J., Nirel, P., Palenik, B. and Morel, F.M.M. (1988) Preparation and chemistry of the artificial algal medium Aquil. *Biological Oceanography*, **5**, 43–46.
43. Henderson, G..M., Anderson, R.F., Adkins, J., et al. (SCOR Working Group) (2007) GEOTRACES – An international study of the global marine biogeochemical cycles of trace elements and their isotopes. *Chemie der Erde-Geochemistry*, **67**, 85–131.
44. Sohrin, Y. and Bruland, K.W. (2011) Global status of trace elements in the ocean. *Trends in Analytical Chemistry*, **30**, 1291–1307.
45. Harrington, C.F., Clough, R., Drennan-Harris, L.R., Hill, S.J. and Tyson, J.F. (2011) Atomic spectrometry update: Elemental speciation. *Journal of Analytical Atomic Spectrometry*, **26**, 1561–1595.
46. Thermo (2012) ICP or ICP-MS? Which technique should I use? An elementary overview of Elemental Aalysis. Available at http://www.thermo.com/eThermo/CMA/PDFs/Articles/articlesFile_18407.pdf (accessed October 29, 2012).
47. Linge, K.L. and Kym, J.K. (2009) Quadrupole ICP-MS: Introduction to instrumentation: Measurement techniques and analytical capabilities. *Geostandards and Geoanalytical Research*, **33**, 445–467.
48. Measures, C.I., Landing, W.M., Brown, M.T. and Buck, C.S. (2008) A commercially available rosette system for trace metal-clean sampling. *Limnology & Oceanography-Methods*, **6**, 384–394.
49. Bruland, K. (1983) Trace elements in seawater. In: Riley, J. and Chester, R. (eds) *Chemical Oceanography*. Academic Press, London, pp. 157–215.
50. Coale, K. and Bruland, K.H. (1990) Spatial and temporal variability in copper complexation in the North Pacific. *Deep-Sea Research Part I*, **37A**, 809–832.
51. Landing, W. and Bruland, K.H. (1987) The contrasting biogeochemistry of iron and manganese in the Pacific Ocean. *Geochimica et Cosmochimica Acta*, **51**, 29–43.
52. Mason, R.P. and Fitzgerald, W.F. (1990) Alkylmercury species in the equatorial Pacific. *Nature*, **347**, 457–459.

Problems

1.1. Discuss briefly the chemistry of the elements that form the "rare earths" and provide brief details of the use of four of these elements in industrial applications.

1.2. A chemist makes a standard solution for the analysis of metals by ICP-MS. She wants the final standards to cover the range from one tenth to 10 times the average concentration of the element in seawater. To do this she will make a composite stock solution that is 1 μM in the following metals (Co, Cu, Zn, Cd, Ag) and 100 nM in Pb and Hg. How much metal does she need to weigh out for each element, using the anhydrous chloride salt (e.g., $CuCl_2$) given that the final solution will be made in 1 l of acidified water?

1.3. Examine the concentrations of the elements in Table 1.1 and summarize any evident trends between the concentrations in seawater versus those in river water. Discuss why this may be so, and comment on the elements that do not fit the trend and why you may suspect they do not.

1.4.

a. If the pH of seawater is 8.1 and the pH is mostly controlled by the acid-base chemistry of the carbonate system, calculate the concentration of dissolved carbonate assuming the total carbonate concentration (all species) is 2 mM. The following acid-base constants are appropriate for this problem:

$$H_2CO_3 = H^+ + HCO_3^- \quad K = 10^{-6.1}$$

$$HCO_3^- = H^+ + CO_3^{2-} \quad K = 10^{-9.3}$$

Equilibrium constants have been adjusted to account for ionic strength effects.

(Hint: given the pH, it is reasonable to assume that the bicarbonate concentration is approximately equivalent to the total carbonate concentration.)

b. If the pH decreased to 7.9, how much would the carbonate concentration change?

1.5. It has been stated that the concentrations of Fe and Mn in seawater are controlled by the precipitation of their solid phases. Estimate the concentrations based on the precipitation of $Fe(OH)_3$ and MnO_2 and compare these values to those in Table 1.1. If the concentrations are different, provide a reasonable explanation for the differences.

CHAPTER 2
An introduction to the cycling of metals in the biosphere

2.1 The hydrologic cycle

As the focus of this book is aquatic systems, it is worthwhile briefly considering the movement and distribution of water at the Earth's surface (Table 2.1) [1–4]. Water is exchanged between the principal reservoirs (freshwater bodies and the ocean) and the atmosphere by precipitation and evaporation (Fig. 2.1). Freshwater refers to waters with relatively low total dissolved solids (TDS) ($<10^3$ mg l^{-1}) while seawater has 35×10^3 mg l^{-1} TDS on average. Brackish water has intermediate TDS [1]. The freshwater and ocean ecosystems are linked through the input of water to the ocean via surface continental runoff, groundwater input and wet deposition. The exchange volumes relative to the various reservoir sizes indicates the rapidity of recycling between each reservoir. Precipitation adds about 3.5×10^{17} l yr^{-1} to the ocean, for example, while evaporation removes about 3.8×10^{17} l yr^{-1}. The difference is due to terrestrial inputs (0.36×10^{17} l yr^{-1}) [1]. Note that the atmospheric (pluvial) flux is about 10 times the fluvial input. The inputs are small compared to the overall reservoir size. It can be seen from Table 2.1 that the majority of the water at the Earth's surface is in the ocean (97%) and the amount in the atmosphere is trivial.

The precipitation input to the terrestrial reservoir is about 1.1×10^{17} l yr^{-1} and evaporation removes 0.73×10^{17} l yr^{-1}. The difference is the fluvial input to the ocean. The inputs into the terrestrial environment per year are of the same order as the reservoir size (Table 2.1) and therefore the *residence time* of water in the terrestrial surface layer is about a year. In this context, the residence time (τ) is the average time a molecule of water spends in a particular reservoir and is estimated as:

$$\tau = \text{reservoir size (mass units)/Input flux (mass yr}^{-1}) \quad (2.1)$$

In contrast to the ocean, inputs into the atmosphere are much greater than the atmospheric reservoir (0.13×10^{17} l) and thus the residence time of water in the atmosphere is very short – about 10 days on average [1]. The cycling of water in the surface terrestrial environment is complex as water is lost from the surface to the atmosphere by both direct evaporation and by transpiration from plants. Losses to deeper layers result from infiltration of water into soils. Transport of water from its site of deposition to rivers is through both overland flow, which occurs mainly during high precipitation events, and through transport in the subsurface. The subsurface environment is divided into the *unsaturated* (*vadose*) zone, the region above the water table, and the *saturated* (*phreatic*) zone, which is below the water table. An *aquifer* is a saturated zone with enough water to allow for water extraction.

For the various ocean basins, based on the estimations of Budyko [5], the sources and cycling of water are different. For the Atlantic Ocean, because of the inputs from large rivers and its relatively small surface area, about 70% of the water input is from precipitation, while greater than 90% of water input to the Pacific and Indian Oceans is from precipitation. The Arctic Ocean, being small and also having large river inputs, has only about 50% of its water input from the atmosphere. For all the major oceans, loss to the atmosphere via evaporation is the dominant loss term (>90%) with water exchange between ocean basins being a minor component of the hydrological cycle; again, the Arctic Ocean is an exception as most of its water loss (~75%) is via exchange with the major oceans. Overall, ~36% of the freshwater input is to the Atlantic Ocean, with ~40% to the Arctic. Continental runoff to the Pacific and Indian Oceans are relatively minor components of the overall global water cycle.

Trace Metals in Aquatic Systems, First Edition. Robert P. Mason.

Table 2.1 The distribution and cycling of water at the Earth's surface. Data taken from various references [1–6].

Reservoir	Volume ($\times 10^{18}$ l)	Inputs ($\times 10^{18}$ l yr^{-1})	Outputs ($\times 10^{18}$ l/yr^{-1})	Exchange time (yrs)
Surface Ocean	338	Precip. 0.35	Evap. 0.38	800
Deep Ocean	1032	Runoff 0.036		
Total Ocean	*1370 (97%)*			*3240*
Lakes and Rivers	0.13	Precip. 0.110	Evap. 0.073	
Saline Lakes and Inland Seas	0.10		Runoff 0.036	
Ice	29 (2%)			
Groundwater	9.5			
Total Freshwater	*39*			*354*

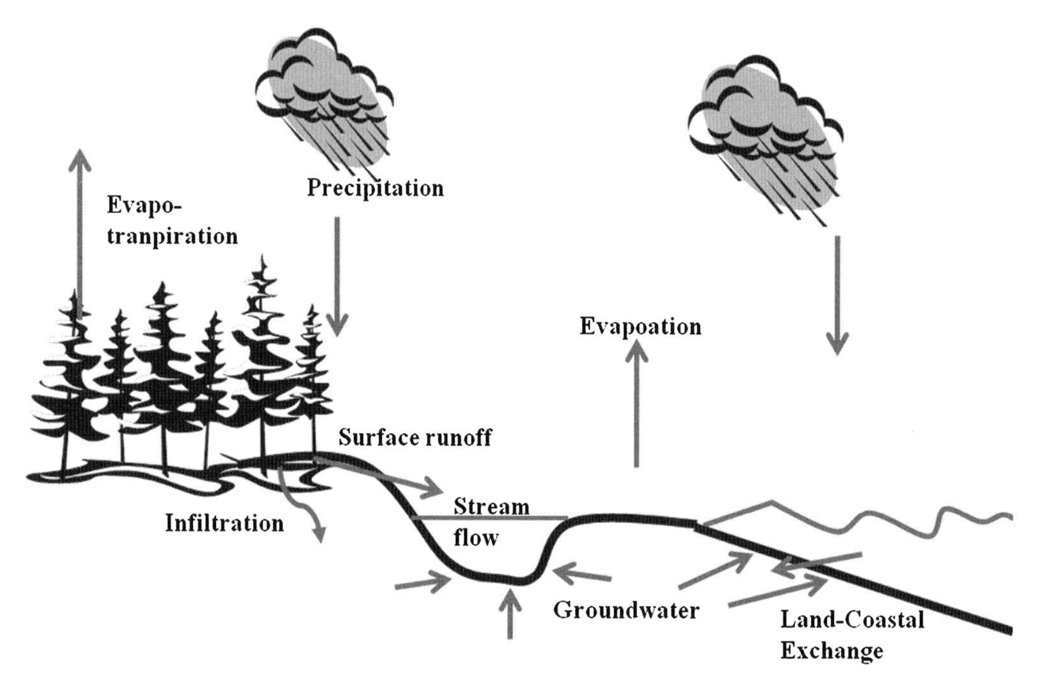

Fig. 2.1 The basic parts of the hydrologic cycle at the Earth's surface. Adapted and redrawn based on the understanding of the hydrologic cycle and various references [2–4, 10].

A simple, first-order estimate of the importance of the relative inputs for metal(loid)s can be made by assuming that these elements have the same concentration in precipitation as in riverine (estuarine) inputs. If so, then the dominant source of metal(loid)s to the ocean would be atmospheric deposition, based on the hydrologic fluxes. This simplistic view is illustrative of the potential importance of atmospheric deposition as a source of trace elements to the ocean and aquatic systems. However, while the water flux to the ocean is dominated by the atmosphere, the particulate flux is dominated by surface waters inputs. So, for those metals that reside primarily in the particulate phase, it may be suggested, based on these first-order considerations, that the riverine input will be the dominant source. This basic approach does not account for the large removal of particles in the estuarine mixing zone, and, with it, much of the strongly particulate bound metals, as discussed in Section 2.2.6. Thus, while there is a strong flux of Fe, Mn, and Al to coastal zones due to particulate transport in rivers, more than 90% of the particulate is removed in estuaries and only a small fraction is exported to the shelf and ocean reaches [7].

For freshwater environments, and particularly lakes, the relative importance of the fluvial versus pluvial flux is related to the watershed to waterbody : surface area ratio, and the degree of retention of metal(loid)s within the watershed [8, 9]. In the absence of point source inputs, the flux of an element in a river is some fraction of the overall atmospheric deposition to the watershed, assuming that atmospheric deposition is the major metal(loid) source. The validity of this assumption is discussed in Section 2.2.5. Thus, for a waterbody:

$$\begin{aligned}\text{Total Metal (loid) Input } (F_T) &= \text{Fluvial flux } (F_F) + \text{Pluvial Flux } (F_A) \\ &= F_A \cdot A_{WS} \cdot T_{WS} + F_A \cdot A_{WB} \\ &= F_A(A_{WS} \cdot T_{WS} + A_{WB})\end{aligned} \quad (2.2)$$

where F_A is the atmospheric deposition flux ($mol\,m^{-2}\,yr^{-1}$), A_{WS} is the watershed area (m^2), A_{WB} is the water body surface area (m^2) and T_{WS} is the *transmission factor* for the watershed. The transmission factor is *a measure of the propensity of the watershed to retain metal(loid)s* and is generally defined as the fluvial flux relative to that of watershed deposition, and is sometimes termed the *retention factor*. It should be noted that measurements of retention (transmission) factors are based on current atmospheric deposition and river fluxes, and are often yearly averages, while it is known and acknowledged that metal(loid)s are not transmitted from their deposition location to the river within a year, and that their residence time in the watershed could be relatively long. For example, during an experiment in the Experimental Lakes Area in Canada where different Hg isotopes were added directly to a lake and its watershed, the isotope added directly to the lake surface water was measurable in the fish within months while the isotope added to the watershed (via a crop-duster plane) had hardly migrated and a very small fraction of the isotope added had made it to the lake within five years [6]. Thus, there is an implicit assumption in these calculations that inputs have not changed dramatically over the relevant time period for transport of the metal(loid) from its deposition site to the river [8, 9]. This is not likely to be the case in many instances for metal(loid)s that interact and are strongly bound up in the solid phase, as changes in landuse will impact flux. This is also true for ecosystems close to local sources, or those changed by human activity (biomass clearing or burning).

It can be deduced from Equation 2.2, for example, that for a strongly retained metal(loid) (e.g., $T_{WS} < 0.1$), the fluvial flux will be the same as the pluvial flux when $A_{WS}/A_{WB} > 10$, and thus the fluvial flux will only dominate if the watershed is large relative to the water body. For a metal(loid) that is poorly retained in the watershed, this is not so, and the fluvial input is important for both relatively small and large watersheds. Retention and/or transmission factors have been estimated for various metals and metalloids by a number of investigators. Metal(loid)s such as Pb and Hg, and the crustal metals (Fe, Mn) are often strongly retained in the watershed (>80% retention) while metals such as Cu and Cd (50–70% retained) and the metalloids As (<10% retained) and Se (60–70% retained) appear to be more mobile [8, 9].

Overall, the global cycle of each metal(loid) is different due to the differences in sources, the relative importance of anthropogenic point source inputs to the atmosphere, and the propensity for the element to remain in solution or be taken up by biotic or abiotic particles. It is therefore useful to examine in some detail the sources and cycling of the more important metals and metalloids, and to understand the factors that influence their concentration in solution. This will be briefly discussed in this chapter and dealt with in more detail throughout the book.

2.2 An introduction to the global cycling of trace metal(loid)s

2.2.1 The sources and cycling of metal(loid)s in the biosphere

In considering the cycling of metal(loid)s at the Earth's surface, it is worthwhile discussing briefly the main processes involved in their global cycling. Beginning in the terrestrial realm, these cycles involve processes within the surface layers as well as the extraction of metal(loid)s from deep reservoirs in the Earth, and their transport to the atmosphere or surface waters. The extraction from deep reservoirs in the pre-industrial world was primarily due to volcanoes and oceanic hydrothermal inputs. However, more recently, mining activity (e.g., metal ore extraction and processing) and extraction of coal and hydrocarbon production have greatly exacerbated the inputs of metal(loid)s to the atmosphere and to aquatic ecosystems. Additionally, metal(loid)s are also reduced during the consumption of these products, especially through coal and hydrocarbon burning for energy and for transportation. The sources of metal(loid)s to freshwaters (rivers and lakes) are the erosion of surface terrestrial material, the input of metal(loid)s associated with groundwater and runoff, and deposition of metal(loid)s from the atmosphere. Again, these processes have a natural component but the input has been increased by human activity. Direct inputs from point source emissions are obviously an important local component in many ecosystems. Changes resulting from forest clearing and other landscape perturbations have also lead to metal(loid) mobilization.

For most metal(loid)s, transport at the Earth's surface predominantly occurs through the particulate phase: suspended solids in rivers and the coastal zone, and aerosols in the atmosphere. This is primarily due to the fact that many metal(loid)s are strongly associated with particles due to: (1) their dominance in terms of abundance in crustal material; (2) their strong partitioning to inorganic and organic solid phases; and (3) their accumulation in the particulate residue of high temperature combustion for industrial and energy extraction (small diameter atmospheric aerosols). Some elements are volatile enough, or form volatile compounds (e.g., Hg, Se, and As), so that their atmospheric transport may occur in the gas phase. Clearly, the major metal constituents of the crust, Fe, Mn, and Al, and many other metals, are transported and associated with particulate material primarily because they are incorporated into crustal material and their oxidized forms are highly insoluble. Other elements, such as Cr, and some of the other transition metals, are also relatively abundant in crustal material but are more mobile because of their speciation in solution (Table 1.1). In only a few cases are metal(loid) elements less soluble in freshwaters compared to the ocean. Of the elements listed in Table 1.1, it is only Mo that has a much higher concentration in seawater than in rivers, and this is primarily due to the fact that it exists as a highly stable and soluble oxyanion in solution. Arsenic, which is also found as an oxyanion in solution, also has a relatively high ocean concentration compared to that of freshwater. Of all the first row transition metals, Ni appears to be the element that is least reactive, having an ocean concentration similar to that of rivers (Table 1.1).

The transport of metal(loid)s from the terrestrial environment to the ocean occurs through atmospheric transport and deposition of metal(loid)s, through transport in riverine discharge, as well as through hydrothermal sources. There is a substantial removal of metal(loid)s in the coastal zone in combination with particulate deposition that occurs during the mixing of river water and seawater in the estuary, and results in their trapping in the sediment. Deep ocean sedimentation and burial is the long-term sink for metal(loid)s and would have balanced the net surface erosion and volcanic inputs in a pre-industrial steady state world.

2.2.2 Metal(loid) partitioning and solubility in natural waters

The extent that trace metals and metalloids cycle through the various pathways in the biosphere depends on their chemistry, their abundance, and their usefulness to humans. Many metals (e.g., Fe, Cu, Zn, and Au) are purposefully extracted from the Earth's interior while other elements are often added as a result of human activity, such as the burning of fossil fuels (e.g., V, Se, and Hg). Some metals fit both descriptions (e.g., Hg, Zn, and Pb). In aqueous systems, the factors that determine the overall concentration of the metal(loid) in solution are the strength of metal(loid) complexation, the solubility of the hydroxide and carbonate phases, and the propensity of the metal(loid)s to adsorb onto inorganic and organic solids and form complexes with dissolved natural organic matter (NOM) [2, 10]. A measure of the propensity of an element to be associated with the particulate phase can be defined by the partition coefficient (K_D), which is the ratio of the concentration of an element in the solid phase (on a mass basis) to that in solution:

$$K_D = C_P/C_W \qquad (2.3)$$

If the solid concentration (C_P) is expressed in moles per kilogram of solid, and the dissolved concentration (C_W) is in molar units, then the K_D is essentially unitless (assuming that 1 l of solution is 1 kg). This measure is widely used as an overall measure of partitioning. It does not distinguish between sorption to or binding to the solid phase, or uptake into living organisms, but it can be a useful relative measure of the tendency of a particular element to remain in solution. The K_Ds for metals range widely in the water column, from values that are $>10^5$ (e.g., Ag, Hg and Pb) to other metals where values are $<10^4$ (e.g., Cd) [2, 4, 8, 10]. However, values vary for a particular metal as well across ecosystems due to differences in the characteristics of the suspended material and the presence or absence of NOM in solution. In addition to passive sorption, the metal(loid)s that are required nutrients for algal growth can be actively assimilated and this can have a dramatic impact on their aquatic concentration. This is especially true in remote locations, such as the open ocean, or in remote temperate lakes.

Additionally, the distribution of metal(loid)s between the dissolved and particulate phases is not simple as there is a continuum of particle sizes, ranging from microorganisms and dust, which are all typically $>0.4\,\mu m$ (the typical size cutoff size of filters used by environmental scientists to separate the filtered fraction from the particulate fraction), to colloidal material, which are typically $<500\,nm$ ($0.5\,\mu m$) in size. The continuum of sizes for various types of particles is shown in Fig. 2.2 [11]. Colloids, both inorganic and organic in nature, are produced biotically or abiotically and include small organisms such as viruses and some small bacteria, and are often operationally defined as the >0.02 and $<0.4\,\mu m$ fraction. Their presence in the filtered fraction can distort estimates of partitioning as they are considered within the dissolved fraction in Equation 2.3 [10–12].

The partition coefficient, K_D, should be independent of particulate load as it is a normalized value, but, as shown in Fig. 2.3, the measured values often show a strong relationship with total suspended solids (TSS). For all metals shown (Zn, Pb, Cu, and Ag), there is a strong decrease in the value of K_D with increasing TSS and all have a similar slope [12–14]. The accepted explanation for this trend is that the experimentally separated dissolved fraction (typically $0.4\,\mu m$ filtered) contains both truly dissolved and "colloidal" material and that the relative amount of colloidal material is related to the TSS. Therefore, at high TSS, the error is enhanced and the value measured experimentally (K_D) is lower than the true value (K_P). The overall mechanism used to explain this phenomenon has been termed *colloidal pumping* [12]. Overall, it stipulates that the partitioning between the particulate and truly dissolved metal(loid) fractions involves the intermediate stage of colloidal material, and that there is a dynamic steady state between these three fractions. This further suggests that particles or colloids are not discreet entities but can be aggregates and can be formed by the combination and coagulation of smaller particles and colloids, especially in coastal waters and freshwaters. In the open ocean, it is generally accepted that in addition to colloids, there are two distinct groups of particles: the suspended smaller particles, which consist of phytoplankton

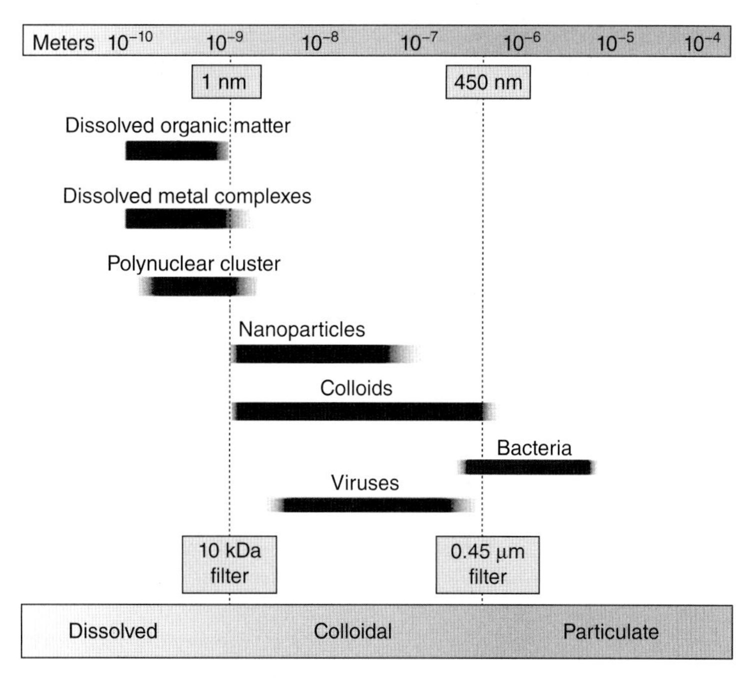

Fig. 2.2 Range in sizes for various fractions existing in a natural water sample and the typical constituents of these fractions. Typical filtration is using either a 0.2 or 0.45 µm pore size. Reprinted with permission from Aiken et al. (2011) *Environmental Science & Technology* **45**: 3196–201, Copyright (2011), American Chemical Society [11].

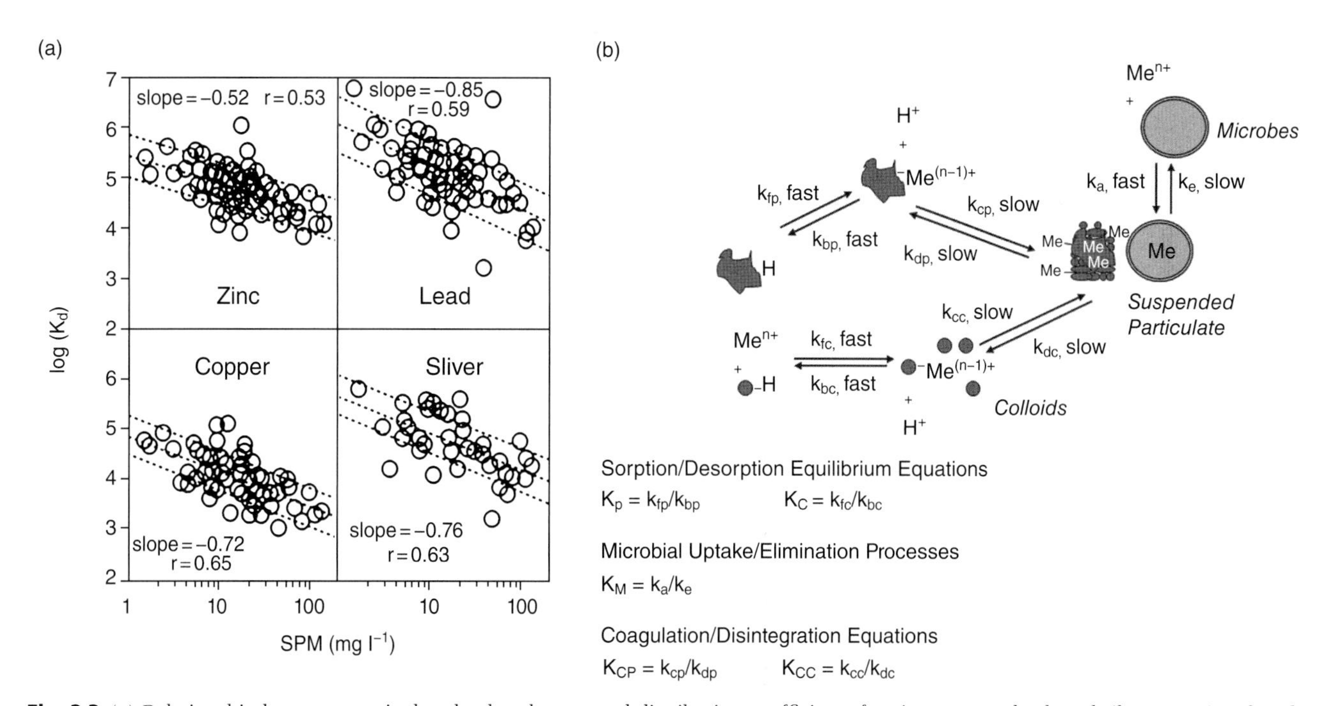

Fig. 2.3 (a) Relationship between particulate load and measured distribution coefficients for zinc, copper, lead, and silver. Reprinted with permission from Honeyman and Santschi (1988) *Environmental Science & Technology* **22**: 862–71, Copyright (1988), American Chemical Society [12]; (b) Diagram representing the interactions that can occur between microorganisms, particles, colloids and dissolved species and how this impacts metal cycling in aquatic systems. Note that SPM and TSS both refer to the same suspended solids.

and other microbes, and small detrital material, and larger sinking particles, which comprise zooplankton fecal pellets, skeletal material, and other biogenic debris. The larger fraction also includes agglomerations of smaller particles due to coagulation and adherence to the surfaces of "sticky" biological material, and this fraction forms the bulk of the sinking particulate matter [15, 16].

It is possible to define partition coefficients in terms of both the particulate (>0.4 μm material) and colloidal fractions if measurements of the colloidal fraction are made using ultrafiltration or other techniques [14]:

$$K_D = C_P/C_W \tag{2.4}$$

$$K_P = C_P/C_D \tag{2.5}$$

$$K_C = C_C/C_D \tag{2.6}$$

where C_P and C_C are particulate and colloidal concentrations (mol kg^{-1}) in suspended matter or colloidal material, and C_D is the truly dissolved concentration ([M] – mol l^{-1}). Given the definition of K_D, C_W is the concentration of dissolved and colloidal material (i.e., the total filtered fraction) on a molar basis, so:

$$C_W = C_D + C_C \cdot M_C \tag{2.7}$$

where M_C is the mass of colloidal material (in kg l^{-1}). Substituting Equation 2.7 into Equation 2.4, rearranging, and using Equation 2.5, it is apparent that:

$$K_D = K_P/(1 + K_C M_C) \tag{2.8}$$

Therefore, K_D approaches the true value of K_P at low colloidal mass concentrations or if the K_C value is much smaller than that of K_P.

The adsorption of metal(loid)s to particulate and colloidal material is often considered to be through surface complexation with active acidic sites (see Section 3.4). The interaction is often considered to be reversible and essentially at instantaneous equilibrium given changes in concentration; however, this is not always the case. As an example, metal adsorption to particulates from the Changjiang River, the fourth largest river in the world and the largest in China, showed that the adsorption followed typical adsorption isotherms but that the maximum adsorption was achieved only after five or more hours, and more than a day for Cd [17]. A number of studies of both dissolved–particulate interactions and interactions in sediments between the solid phase and porewater also suggest that while adsorption may be rapid, desorption is often a much slower process [10, 18]; therefore, it is not always appropriate to model the system as a reversible equilibrium system. One reason for differences in the rate of adsorption versus desorption is that metals may become incorporated into the inner matrix of the particulate, which in the environment may not be a well-ordered, spherical, non-porous solid, as many models assume. The particulate phase is much more likely a porous material due to the coagulation processes discussed previously, the heterogeneity of the particles, and the likely presence of pores in terrestrial material.

On encountering a solid phase, dissolved metal(loid)s first bind to the more abundant but relatively weaker sites. This rate is rapid compared to desorption, where the metal(loid)s are being released from the less abundant but stronger binding sites where they have migrated over time since adsorption. Studies have shown that while metal(loid)s reach a steady state in terms of K_D after addition of a metal(loid) to a particle-containing solution relatively rapidly, this initial K_D value is often lower than that measured using environmental samples. Reasons for this include the fact that some of the metal(loid) in natural solids is likely unavailable for exchange with the dissolved phase due to its incorporation into the internal structure of the particle. This has led to the notion that metal(loid)s exist in particulate in two phases – an *exchangeable* fraction and a *non-exchangeable* fraction. As an example, in the Changjiang Estuary in China, <25% of the metals measured were in the exchangeable fraction, likely reflecting the dominance of eroded terrestrial material in the TSS [17]. In contrast, in Galveston Bay, >50% was in the exchangeable pool [19]. Differences in fractionation reflect both differences in the solid phase composition, and in the quantity and strength of binding ligands in solution.

Therefore, the partitioning between the dissolved and particulate phase is not a simple reversible equilibrium situation. One approach to modeling the differences in exchange dynamics is to assume the K_D (or more appropriately the K_P) is related to the ratio of the adsorption and desorption constants, which are assumed to be first order:

$$K_P = k_{ad}/k_{ds} \tag{2.9}$$

Where k_{ad} is the adsorption rate constant and k_{ds} is the desorption rate constant. If k_{ds} can be determined, then $k_{ad} = K_P/k_{ds}$. Then, the half-life to steady state is $t^{1}\!/\!_{2} = (k_{ad} + k_{ds})^{-1}$ and is mostly determined by the rate of desorption. Another approach is to use a binary or multi-component model with different rates of binding of metal(loid)s to the various phases [20].

A number of studies examining the adsorption and desorption of metals from sediments are detailed in a publication edited by Allen [18]. The rates of adsorption and desorption estimated in the various studies of a number of metals (transition metals, Zn, Cd, and Pb) allow a number of conclusions or tenets to be proposed. It appears in many instances that adsorption and desorption are multiphasic and that the process can often be represented by a two-phase system with a "fast" and a "slow" process [21]. This reflects the fact that there are many different binding sites on the surfaces of marine solids (carboxylic acidic groups, N-containing groups, thiols and both organic and inorganic binding sites). Additionally, given that desorption is slower than adsorption, some studies have likely not reached viable conclusions as a result of insufficient study time, resulting in non-steady state conditions in the experiments.

Experimental evidence and thermodynamic theory suggest that there is an initial rapid adsorption from solution to readily available binding sites, followed by a re-equilibration over time with the resultant strength of the metal bond increasing with time. Such an idea makes sense in terms of complexation as it is known that there are many more weakly binding sites on the surfaces of aquatic solids than strongly binding sites, as will be discussed in Chapters 3 and 4; therefore, as the metal(loid) initially binds to the first site encountered, it is more likely to be a weaker site. However, over time, the metal(loid) will migrate from weaker sites to stronger sites.

Additionally, there could be penetration of the metal(loid) into the particle, or re-association of particles into flocs and other conglomerates that would result in the metal(loid) adsorbed being less available for desorption. In envisioning adsorption and desorption, the particle is often thought of, or parameterized, as a sphere of uniform shape and density when in reality, the particles are a conglomerate of colloids and particles derived from both inorganic and biological sources. Electron microscope studies confirm the heterogeneous nature of aquatic particles, but scientists studying processes do not often acknowledge this. Thus, given an understanding of the likely fate of a metal(loid) ion after adsorption, it is not difficult to understand that desorption is slower, and that desorption of recently adsorbed metal(loid) is much faster than that of metal(loid) that has been associated with the solid phase for an extended period.

If one considers that adsorption is due to chemical binding to sites within the solid material, then it is reasonable to conclude that the amount of adsorption will also depend on the amount of the metal(loid) already present in the solid phase. Additionally, it is possible that exchange reactions may occur and one metal(loid) could out-compete another for a particular site; therefore, the results of adsorption studies with multiple metal(loid)s in solution could be different depending on the actual conditions used. As a result, it is difficult to extrapolate from the results of studies done at high metal(loid) solution concentrations to the natural environment. Studies should endeavor to examine changes in partitioning at levels as close to natural levels as possible.

Differences in the rate of adsorption and desorption, and for the changing value of K_D with time after addition of a metal(loid) to solution, fits with the colloidal pumping mechanism described earlier [12]. As the colloidal fraction is measured with the filtered fraction, and given that colloidal material can coagulate and form particles, and vice versa, such processes can explain changing K_D values over time. This is especially true in environments where there is substantial organic matter in the filtered fraction, or colloidal inorganic oxides and other material. The colloidal pumping model (Fig. 2.3b) [11, 12] postulates that there are rapid adsorption and desorption reactions occurring between the truly dissolved species and the colloidal material in solution. The particulate phase is formed by the relatively slow coagulation of this colloidal material, and assumes there is also disintegration of particles to produces colloids. These processes are continually occurring but reach a steady state so that particles are being continuously created and destroyed, and adsorption/desorption involving surface complexation and other interactions is continually occurring. Overall, this is a purely physical interpretation of the environment in the absence of living microbes, as these "particles" will actively accumulate metal(loid)s from the dissolved phase and not from colloidal material. Additionally, it is possible for direct adsorption and desorption to occur from the dissolved phase to the particulate material and therefore the model shown in Fig. 2.3b is likely a simplification of a more complex set of interactions. However, it does provide a clear conceptual framework of the processes involved in the interactions of metal(loid)s with particles.

An estimation of the relative concentration of the free metal ion ($[M^{n+}]$) to the total concentration in natural waters (considering both dissolved and particulate species) ($[M_T]$), based on thermodynamic calculations discussed in Chapter 3, demonstrates the relative importance of these processes in controlling trace metal concentrations in solution. Ignoring binding with NOM, for Fe, Al, Hg, U, and Bi, the estimated ratio of M^{n+}/M_T is very small ($<10^{-5}$) in both freshwater and seawater, while for transition metals such as Ni, Co, and Zn, the ratio is relatively high (0.1–0.6) in both media [11]. Some of the other metals (e.g., Cd and Ag) have lower free ion concentrations in seawater because of strong complexation by the chloride ion, and this is also the case for Hg. Metals such as Pb and Cu have relatively low free ion concentrations in both media because of their tendency to form strong complexes with hydroxide, chloride, and carbonate species in solution. These estimations are not completely accurate for many metals because of their strong binding to organic matter. This binding could also enhance the association with the solid phase or increase the amount in solution depending on the relative circumstances in a particular ecosystem, and whether the particles are primarily living organisms or detrital particles, or primarily inorganic solids. Furthermore, as noted previously, some metals and many of the metalloids exist in solution as oxyanions that are negatively charged and therefore have very different aquatic chemistry to that of metal cations. This will be discussed further in this chapter and throughout the book.

2.2.3 Human influence over metal(loid) fate and transport

While considerations of solubility and complexation can give an indication of relative metal concentration, it is also necessary to consider the impact of human activity: the remobilization of metal(loid)s from deep reservoirs (as, for example, in ores, coal, and crude oil) to the Earth's surface and their enhanced cycling through the biosphere [22–25]. Many of the metals discussed in this book are economically important to society and therefore have been extracted from the Earth and remobilized from their primary phases by mining and elemental extraction from ores [12]. Many metal(loid)s are also trace substances in ores, or in materials that are combusted for energy (e.g., coal and oil), and are released to the environment as a result of industrial processing [22, 23]. The major metals are those of industry and chemical application and include Fe, Al, Cr, Cu, Zn, and the noble metals, such as Au, Ag, Pt, and others. Many metals and metalloids, such as Hg, Pb, As, and Se, are also released as by-products of industry [22, 23]. Major non-ferrous metal ores are sulfides, and while there is a dominant metal being targeted for extraction (such as Cu or Zn), there are often traces of the other heavy metal(loid)s. These contaminants form highly insoluble sulfide phases, or adsorb/partition strongly to these phases (e.g., Pb, Hg, Cd, and As), and are released into the environment during mining or through ore processing. The concentrations of some metal(loid)s in crustal material are shown in Table 2.2 [24, 25] and the concentrations of some of the metal(loid)s in sulfide ores are shown in Table 2.3 [22, 23]. Levels of various metal(loid)s in representative coals are also shown in Table 2.3. The sulfur content of various coals varies from <0.5–5%, and this obviously has an impact of the extent of resultant acid mine drainage that can occur in waters interacting with mining deposits, as discussed next.

Early in the Industrial Revolution, and ever earlier in human history, extraction of metals for usage in products was done without any concern for the impact of the trace contaminants on the environment. As a result, the legacy of these early methods is found in many locations. Additionally, the use of metal(loid)s as chemicals and reactants/catalysts in industry has also lead to large scale contamination of some watersheds. The legacy of these past excesses is still with us. One branch of environmental chemistry is concerned with the assessment of the continuing impact of these legacy inputs, and their persistence in the environment in locations where they can be bioaccumulated or impact aquatic organisms.

Acid mine drainage, for example, is a pervasive legacy of mining activity worldwide [22]. It is an important process whereby metal(loid)s, in addition to being transported with solid material, are also solubilized and transported large distances from abandoned mines. The process of metal release during the creation of acid mine drainage is related to the dissolution of sulfide ores which contain the traces of these elements (e.g., Table 2.3). The same processes can occur in relation to coal mining activity. The primary reactions are dissolution of the ores due to sulfide oxidation, and subsequent further oxidation of reduced Fe by oxygen [1, 8, 22]:

$$2FeS_2 + 7O_2 + 2H_2O = 2Fe^{2+} + 4SO_4^{2-} + 4H^+$$

$$2Fe^{2+} + \frac{1}{2}O_2 + 5H_2O = 2Fe(OH)_3 + 4H^+$$

Table 2.2 The average concentrations of metals in various terrestrial compartments. All concentrations are in μmol kg^{-1}. Data taken from Garrett [24] and Li [25].

Element	Earth's crust	Upper Continental crust	Igneous rock	Sedimentary rocks
Hg	0.4	0.4	0.02–0.l	0.25–1
Pb	~50	50–100	5–100	50–5000
Cd	1.8	1–2	1–2	0.1–30
Cr	2000	500–1500	70–3 × 10^4	10^2–10^4
Ni	~1500	500–1000	500–4 × 10^4	50–5 × 10^3
As	~25	10–20	10–20	10–300
Cu	~10^3	500–1000	200–1500	50–3 × 10^3
Zn	1–2 × 10^3	1–2 × 10^3	1–3 × 10^3	10–3 × 10^4

Table 2.3 Concentrations of some of the more common contaminant metal(loid)s in ores that are mined for their primary metal. All concentrations are in mmol kg^{-1}. Values in brackets represent the maximum concentrations. Data taken from Refs [22, 23].

Element	Galena (PbS)	Sphalerite (ZnS)	Chalcopyrite ($CuFeS_2$)	Pyrite (FeS_2)	Coal
Ag	5–50 (300)		0.1–10 (23)		0.004
As	3–70 (150)	3–5 (150)		7–15 (750)	0.01–1.0
Cd	–	10–50 (440)			1–25 × 10^{-3}
Cu	0.2–4 (60)	20–100 (1000)	Major	0.2–2000 (12,000)	0.01–1.0
Hg		0.05–0.3 (50)			0.1–5 × 10^{-3}
Co, Ni	–		0.2–1 (40)	0.2–10 (50)	0.008–0.75
Pb	Major			1–3 (30)	0.01–0.4
Sb	2–50 (300)			1–2 (7)	0.4–80 × 10^{-3}
Sn	–	1–2 (7)	0.1–2 (8)		0.008–0.08
Zn	–	Major		20–100 (900)	0.08–5.0

Notes: bituminous coal, NIST 1632e.

Thus, for every mole of pyrite (FeS_2) oxidized, 4 mol of H^+ are produced. The Fe oxidizing catalytic reactions can lead to the buildup of high acidity in streams. The reactions are often microbially mediated as there are a number of organisms that can use the redox reactions as an energy source. Most of the organisms are either Fe or S oxidizing bacteria (e.g., *Thiobacillus, Acidithiobacillus,* and *Leptosprillum*), with the S-oxidizing bacteria using S in various oxidation states (S(-II), S(0) and S(II) (as $S_2O_3^{2-}$)). The fate of the metals depends on the degree to which the Fe(III) produced by oxidation is precipitated from solution. This is a function of pH (discussed in more detail in Chapter 3), as most metals will be co-precipitated or adsorbed onto the Fe (hydr)oxide surfaces. Other solid metal sulfides are mostly dissolved due to the oxidation of sulfide by oxygen, although they can also react with Fe(III) [22]:

$$PbS(s) + 2O_2 = Pb^{2+} + SO_4^{2-}$$

or

$$PbS(s) + 8Fe^{3+} + 4H_2O = Pb^{2+} + SO_4^{2-} + 8Fe^{2+} + 8H^+$$

The acid mine waters are neutralized over time due to the dissolution of basic rocks (carbonates, hydroxides, and aluminosilicates), and the resulting increasing pH leads to oxide precipitation, especially Fe(III) phases. Acid mine drainage waters can have very low pHs (<0), and very high concentrations of sulfate (>1 M), Fe (>0.2 M), trace metals (e.g., Zn > 60 mM and Cu > 10 mM), and metalloids (e.g., As – up to 1 mM). Elevated concentrations have been found in waters of mine workings and in pit lakes that are often the legacy of such activity [22].

Impacts on the environment due to past or current mining activities continue. It takes, for example, about 100 tons of rock to produce 1 ton of Cu, and thus the waste generated from hard rock mining is large. There are many instances in the literature of current substantial inputs of metal(loid)s into the environment from such human activity, and many of these cases are in developing nations where concern for industrial development often trumps environmental concern, as it did in western nations in the last century. Additionally, there are many other industrial processes that have contributed to substantial contamination of local waters. For example, the Upper Mystic Lake, in Massachusetts, USA, has a legacy of metal and metalloid contamination [26] that reflects usage of sulfide ores in the manufacture of sulfuric acid (Fig. 2.4a), and also the use of Cr and other metals in the tanning industry and other processes upriver (Fig. 2.4b) [26]. It is clear from the profiles of As and Cr over time, converted from a sediment depth into time based on ^{210}Pb and ^{137}Cs sediment distributions, that there is a dramatic increase in sediment As, Cr and other metal content that coincides with specific industrial activities in the watershed of the lake. The profiles of As and Cr are not concurrent, reflecting the different industrial sources for the elements – As release due to the oxidation of sulfide ores in the manufacture of sulfuric acid and Cr release as a result of its use in the tanning industry. The remaining legacy of

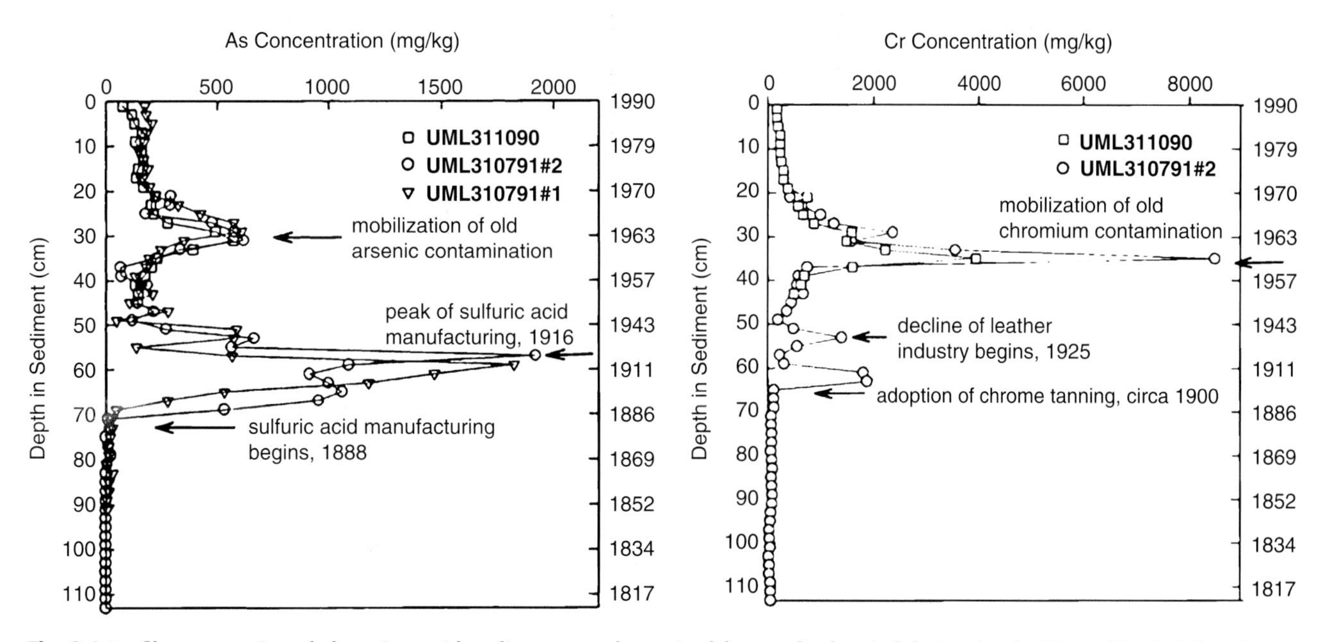

Fig. 2.4 Profiles or arsenic and chromium with sediment age, determined from radiochemical dating, for the Upper Mystic Lake, MA, USA, which is within the Abejona watershed where historical industrial activities occurred. Concentrations are given in mg kg^{-1}. For As, 1000 mg kg^{-1} = 13.35 mM; for Cr, 1000 mg kg^{-1} = 19.23 mM. Reprinted with permission from Spliethoff and Hemond (1996) *Environmental Science & Technology* **30**: 121–8, Copyright (2011), American Chemical Society [26].

metals in the watershed is evident from the recent peak due to sediment remobilization, resulting from a substantial development within the contaminated section of the watershed in the early 1960s.

While it appears clear from the profiles in Fig. 2.4 that there has been intermittent and large inputs of metal(loid)s to the Aberjona watershed, it is often not as easy to assign concentrations in sediments to anthropogenic versus natural sources. There is generally much variability in the inputs over time, especially for rivers and other water bodies subject to seasonal and annual variation in terrestrial inputs. Thus, to clearly demonstrate the legacy of past inputs, it is often necessary to normalize the metal(loid) concentration in the sediment to an element that is not subject to anthropogenic input, providing a measure of the amount of material being added to the sediment. Suitable candidate elements are those that are present in high concentrations in crustal material or have a dominant crustal signal, such as Al, Fe, Ti, and Li. Another approach is to calculate an enrichment factor (EF) for the metal(loid) in the sediment by normalizing the relative concentration to that of an unpolluted site:

$$EF = (Me/X)_{site}/(Me/X)_{ref} \quad (2.10)$$

Here, Me refers to the concentration of the metal(loid) of interest, X is the concentration of the normalizing element, and ref refers to the ratio for the reference (unpolluted) site.

An example of such normalization is shown with sediment data from the Lot River, a river that flows into the Gironde estuary outside of Bordeaux, France [27]. The river has received inputs as a result of Zn mining and smelting from 1842–1987. The legacy of this contamination for Cd is evident in the sediment profiles for the Lot River (Fig. 2.5b) when compared to a reference site (Fig. 2.5a), and when displayed as an enrichment factor (Fig. 2.5c). The metals Cd, Zn, and, to a lesser extent, Pb are substantially enriched while the enrichment of Cu is small relative to the reference site. In this case, the normalization is to the scandium (Sc) content of the sediment which is assumed to be conservative as it is a component of Fe-containing minerals. The sediments were dated based on ^{137}Cs and ^{210}Pb, and the profiles show that there is still substantial mobilization and contamination of the sediments even after the cessation of activity at the mine in the 1987. However, there is evidence of a decrease in relative metal concentrations in the sediment since then, illustrating the additional input that was occurring during the active operation of the mine.

Much of the early studies of environmental contamination were confined to the study of watersheds associated with mining activity, industrial activity, or urban inputs, and were focused on the local environment. In many countries, enhanced environmental regulation and monitoring has led to a decrease in such direct inputs to the environment. However, while these recent practices may have led to an overall decrease in the direct release of contaminants to the aquatic environment, this cannot necessarily be concluded for releases to the atmosphere.

The extent of the atmospheric transport of metal(loid)s was not appreciated until the middle of the 20th century. Since then, there has been much study of the fate and transport of metal(loid)s from the continents, and of their importance and impact on the ocean. While there has been an obvious focus on anthropogenic inputs of metal(loid)s, and their transport, deposition, and impact on the environment, it is also well-known that the atmospheric transport of particulate material to open ocean surface waters is vital in sustaining the phytoplankton in these environments. Phytoplankton, like all living organisms, require Fe and other trace metals for inclusion into vital enzymes [4, 26, 29]. This is especially true for photosynthetic organisms as the most important enzymes in the electron transfer process associated with photosynthesis are Fe-containing [29].

Metal(loid)s released to the atmosphere from natural and anthropogenic sources have a global impact. For example, many of the metals and metalloids released from coal combustion and other high temperature combustion are associated with the fine particulate fraction generated through high temperature combustion and therefore can be transported large distances prior to their deposition, through either wet (rain, snow, dew) or dry deposition. For large particles, the impact is more local as the rate of removal of particles from the atmosphere is a complex function of particulate size. This is discussed in more detail later in the book (Chapter 5).

Furthermore, the metal(loid) concentrations in the atmosphere have increased over time, specifically in the last century, for a number of elements. These changes in concentration, and resultant deposition to the Earth's surface, can be ascertained from the analysis of "historical archives" such as ice cores, sediment cores, coral, and other sedimentary phases that are accumulated over time, if they are not impacted by local direct or indirect inputs. For example, it was shown by the analysis of ice cores collected from Greenland, which were handled in a rigorous way to avoid contamination during sampling, transport and analysis (as discussed in Chapter 1, Fig. 1.5), that the Pb content had increased dramatically in the 20th century and the rate and extent of the increase coincided with the increase in Pb use in gasoline and other products (Fig. 1.6) [30, 31]. Studies in the Midwestern USA in the 1990s also showed the impact of human-derived emissions on the levels of Hg in the environment [32], with relative increases in deposition over historical times of up to a factor of 10 for lakes near sources (Minneapolis) to about a factor of two for remote Alaskan lakes (Fig. 2.6). Note that there is evidence in the more impacted sites of decreases in deposition in the last 20 years, especially for the urban locations. Such decreases are not evident at the remote sites. There are many other examples

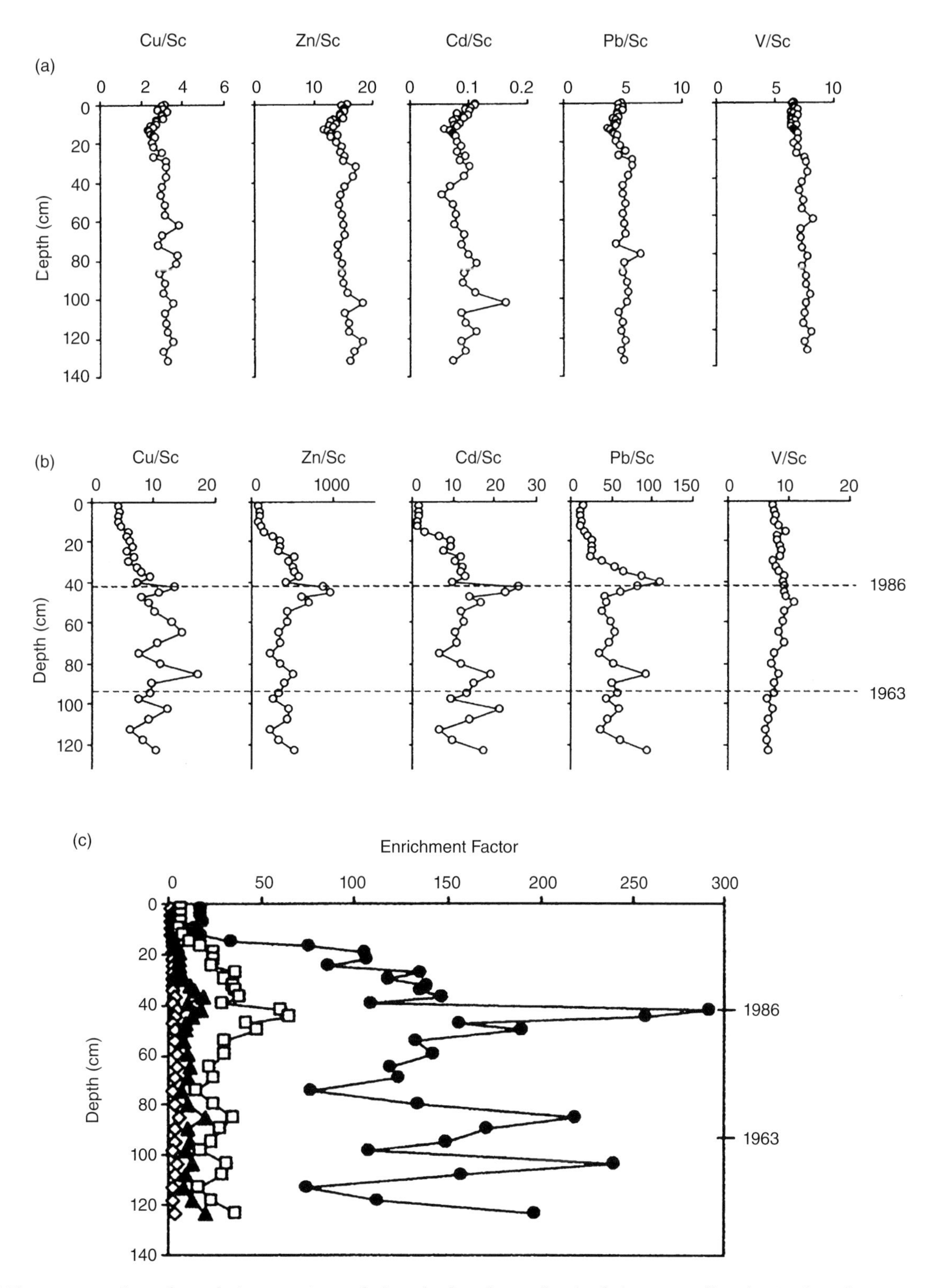

Fig. 2.5 The concentrations of metals (copper, zinc, cadmium, lead, and vanadium) relative to scandium for (a) the reference site (Marcenac), and (b) the contaminated site (Cajarc) on the Lot River in southern France. The enrichment factor for these metals is shown in (c): Cu (open circle), Cd (closed circle), Pb (triangle) and Zn (square). Reprinted from Audry et al. (2004) *Environmental Pollution* **132**: 413–26 [27], with permission from Elsevier.

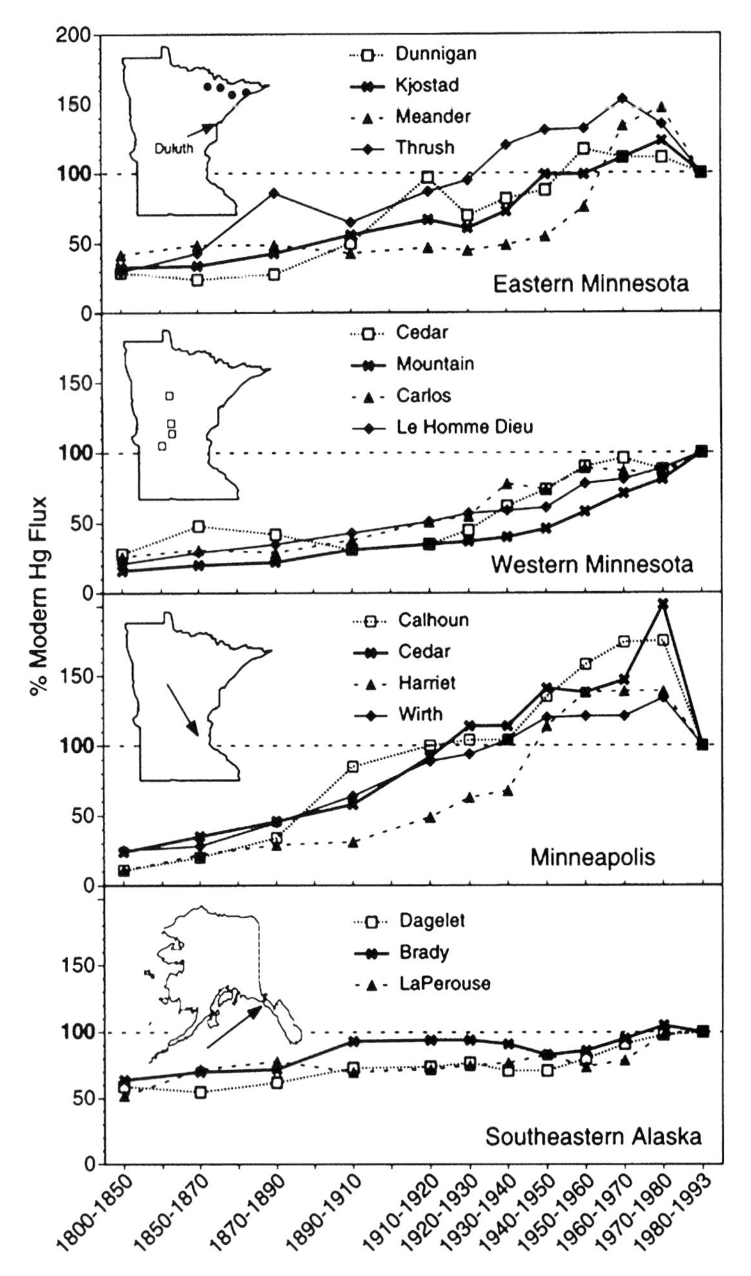

Fig. 2.6 Relative changes in the concentrations of mercury accumulating in sediment cores collected from lakes in various parts of Minnesota and Alaska, USA. The data are plotted as a relative concentration to that found in the upper section of the sediment core (denoted as 100%). Reprinted with permission from Engstrom and Swain (1997) *Environmental Science & Technology* **31**(4): 960–7 [32], Copyright (1997), American Chemical Society.

of similar historical records that document the contamination of local or remote environments with metal(loid)s as a result of human activity [33], and these will be discussed in more detail in later chapters.

The remainder of this chapter will discuss the overall cycling of metal(loid)s in the biosphere, concentrating on the upper regions of the terrestrial environment including the atmosphere and surface waters. Discussions of this cycling will provide information and detail that can then be used to present and discuss the global cycling of some of the trace metal(loid)s of most importance from a biogeochemical, environmental or socioeconomic viewpoint, which will give a focus to understanding the role of aquatic systems in trace metal(loid) biogeochemistry.

2.2.4 Trace metal(loid) inputs to the atmosphere

Atmospheric transport and dispersion of metals and metalloids is an important mechanism for their redistribution at the Earth's surface. For those elements that are relatively volatile or can exist as relatively stable volatile species, such as methylated compounds or hydrides, dispersion can occur through gaseous atmospheric transport. However, for many metals, dispersion through the atmosphere is mostly in the particulate phase. Atmospheric particulate matter is a mix of aerosols derived from natural sources (e.g., dust) and from anthropogenic activities. Many of the elements released from anthropogenic sources, such as from high temperature combustion, are the more volatile metal(loid)s. Given these sources, there is a relationship between the relative enrichment of a particular element in aerosols compared to Al (a tracer of natural sources), expressed as an enrichment ratio relative to terrestrial rocks, and the boiling point (Fig. 2.7). The more volatile elements are enriched to the greatest extent [2]. This is likely due to their volatility and partitioning between the gaseous and particulate phase in high temperature gases (e.g., combustion flue gases from coal burning and ore processing, and in volcanic emissions). Thus, the metalloids and Hg are the most highly enriched relative to the crustal element Al.

The release of metal(loid)s into the atmosphere as a result of human activity has increased their concentration and atmospheric burden in remote locations. For many of the metal(loid)s discussed in this book, natural sources, such as dust and volcanic eruptions, do not dominate over anthropogenic sources [34–37] (Table 2.4). Human activity has exacerbated the concentration of most metal(loid)s in the atmosphere (gaseous and particulate) and in atmospheric deposition, especially in locations close to sources. Additionally, close to sources, dry deposition of metal(loid)s likely dominates over wet deposition, while in remote locations, wet deposition is likely the more important flux.

The obvious major natural source for the "crustal elements" is windblown dust and crustal particulate (Table 2.4) Volcanoes and related emissions are important sources for the more volatile elements such as Hg, As, Cd, Cu, Ni, and, to a lesser extent, Zn [34, 37]. Biogenic emissions are important for those elements that can be released as gaseous species, either in their elemental form or as methylated species or hydrides (As, Hg, Se). Biomass burning is a source for many elements but it is not the primary natural source for any of the elements listed in Table 2.4. The case of biomass burning highlights one of the problems with categorizing emissions to the atmosphere as either natural or anthropogenic. While biomass burning is a natural phenomenon in many parts of the world, there are locations where the desire for land for cultivation and other uses has led to the intentional burning of forests, trees, and grassland, and it is thought that most of the biomass burning globally is human initiated [38, 39]. Currently, such activities are occurring in the Amazon and similar regions of South America, Asia, and Africa. Of course, the forests of Europe and North America were similarly destroyed in centuries past. Biomass burning is therefore not a new issue although its extent is accelerating with the rapid population increases of the 20th century. Estimates suggest that biomass burning has increased by about 50% since 1850, and that much of it is human-induced. Thus, it is somewhat a misnomer to list biomass burning emissions as a truly natural source [38].

Various metals and metalloids that are emitted from fossil fuel combustion have been used as "tracers" of anthropogenic sources as their source profile is relatively simple (Table 2.4). For example, essentially all the anthropogenic Se emissions are from the combustion of coal while emissions of V and, to a large extent, Ni are from the burning of fuel oils [34]. Coal burning is also the major anthropogenic source of Cr, Hg, Mn, Sb, and Sn. In some instances, while not obvious from Table 2.4, Cd and Cu can be used as indicators of waste incineration emissions. Most of the non-ferrous metal production emissions are related to the processing of Cu, which is often recovered from sulfide ores. These ores have an array of other metals and metalloids co-associated with it (Table 2.3). Thus, this industrial process is an important atmospheric source of many heavy and toxic metal(loid)s.

The inputs of Al to the atmosphere are dominated by natural sources (>90% natural). A number of other metals and metalloids have an overall dominant natural source. These can be broken down into those elements that have a relatively high crustal abundance (i.e., Fe, Mn, Cr, and Co), and those that are volatile elements and have a strong source from volcanoes and other emissions (e.g., As and Se). However, both As and Se have important anthropogenic sources (Table 2.4). For As, fossil fuel burning is an important source, but so is the refining of non-ferrous metals as these metalloids are an important trace constituent in many sulfide ores, as noted earlier (Table 2.3). Other metals released during non-ferrous metal refining and production are Cd, Cu, Sn, Hg, Ni, Pb, Sb, and Se [34]. Ferrous metal production can be an important source for As, Cr, and Mn, Pb, and Zn. Waste incineration is an important source of the more volatile elements (Sb, As, and Hg) as well as some metals used in various products that end up in the municipal waste stream, such as Zn, Cu, and Sn. It should be noted that the relative source strengths in Table 2.4 are global averages and there is likely to be strong regional differences depending on the relative sources within a particular location. From an environmental geochemical point of view, the terms "local" refer to deposition within a few hundred kilometers of the source, while "regional" reflects the next scale (100–1000 km). Thus, given typical wind speeds, local deposition reflects the removal of a contaminant within a day or two of its emissions while regional deposition likely will occur for highly soluble constituents (recall the average residence time of water in the atmosphere is 10 days).

Most of the emissions of metal(loid)s are in the particulate form and therefore dry deposition is often the most important sink from the atmosphere for these elements. For the crustal metals, dry deposition is the dominant process for relatively large dust and other natural particles (e.g., sea salt), but wet deposition of these elements, due to scavenging of the particulate material by precipitation, is not insignificant. For example, it was concluded based on measurements on the mid-Atlantic seaboard of the USA [40] that dry deposition of Al from natural sources dominated (~80%) over wet deposition to the Mid-Atlantic Bight, and that anthropogenic sources were insignificant (Fig. 2.8), in line with the data in Table 2.4. In complete juxtaposition, wet deposition of anthropogenically-derived Zn dominated over dry deposition, and natural (crustal) sources were insignificant. Anthropogenic sources are the dominant source for most metals to this region, even for an element with a strong crustal source such as Mn and Cr. Only Al and Fe have a predominantly crustal source. In comparing Fig. 2.8 and Table 2.4, it is apparent that the source signal for the mid-Atlantic Bight is different from the global average for most metals, reflecting the relative proximity of this region to major anthropogenic sources within the eastern USA. Given that these measurements were made about 15 years ago, and inputs to the atmosphere in North America have decreased substantially over that time, this observation may no longer be entirely valid.

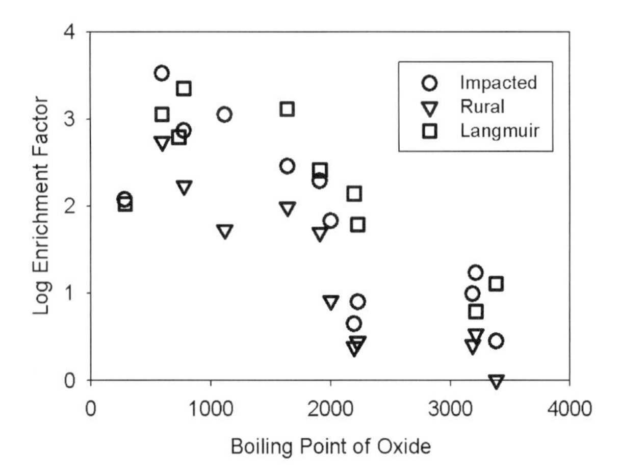

Fig. 2.7 Graph showing the relationship between the enrichment of metals and metalloids in atmospheric particulate matter, relative to that of terrestrial rocks, and the boiling point of their respective oxides. Figure created using information in a figure from *Langmuir* [2] as well as information from various tables in Chester [47].

2.2.5 Metal(loid)s in the terrestrial environment and freshwater ecosystems

The soils of the terrestrial environment, through which water moves on its journey from the atmosphere to fresh-

Table 2.4 The major natural and anthropogenic sources of metals and metalloids to the atmosphere. Taken from various sources [34–37], and based on the model discussed in Chapter 5, Section 5.6. Fluxes in Gmol yr^{-1}.

Metal	Natural Input*	Dominant Sources#	Anthro Input	Dominant Sources*	Total Input	Fraction Natural
Al	4320	WM	776	FFC (50%)	5100	0.85
As	0.16	VC, WM, BO	0.067	NFM (69%), FFC, FM	0.23	0.70
Cd	0.011	VC	0.067	NFM (73%), FFC	0.078	0.14
Co	0.51	WM, BO	0.051	NFM, FFC, OT	0.56	0.91
Cr	3.0	WM, VC	0.59	FFC (69%), FM, OT	3.6	0.87
Cu	0.44	VC, WM, SA	0.95	NFM (70%), FFC	1.4	0.32
Fe	1020	WM, VC, BO	109	FFC (50%), FM	1130	0.90
Hg	0.011	BO, VC**	0.022	FFC (66%), NFM, WI	0.033	0.33
Mn	26	WM, VC	0.70	FFC (85%), FM, WI	26.7	0.97
Ni	1.5	VC, WM	1.7	FFC (90%), NFM	3.2	0.47
Pb	0.06	WM, VC	2.3	MV (74%), FFC, NFM,	2.4	0.25
Sb	0.021	WM, VC, SA	0.029	FFC (47%), NFM, WI	0.050	0.43
Se	0.12	BO	0.049	FFC (89%), NFM	0.17	0.71
V	0.88	WM, VC	1.56	FFC (100%)	2.4	0.36
Zn	0.69	WM, VC, FF	2.0	NFM (72%), FFC, OT	2.7	0.26

Notes: #VC: volcanoes; WM: windblown crustal materials; BO: biogenic emissions; FF: biomass burning; SA: sea salt aerosol; FFC: fossil fuel combustion; MV: motor vehicles; FM: ferrous metal production; NFM: non-ferrous metal production; WI: waste incineration; OT: other (e.g., cement manufacture; gold production using Hg).

*Only sources accounting for greater than 5% of the total are listed. Emissions are those from Pacyna et al. (2006) [70], Nriagu and Pacyna (1988) [37] or from other compilations in the literature[19, 25] and represent emissions in the 1990s.

**Includes evasion of elemental Hg from water surfaces.

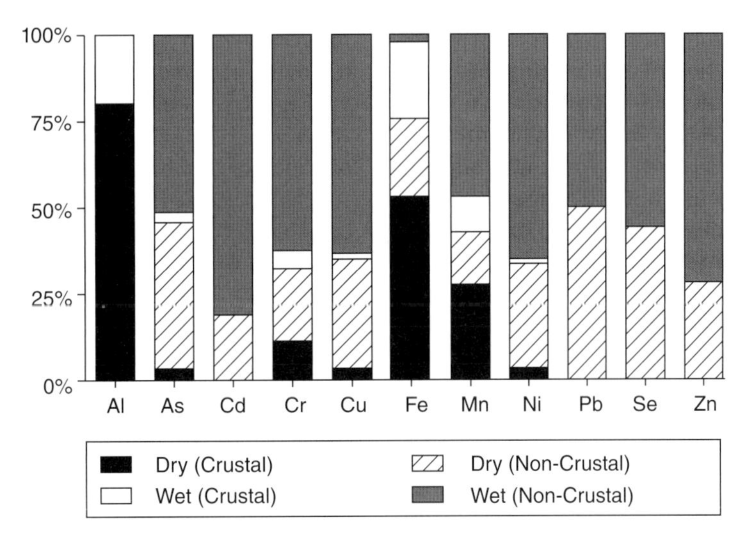

Fig. 2.8 Relative sources of atmospheric deposition to the mid-Atlantic Bight. Reprinted with permission from Scudlark and Church (1997) [40], *Atmospheric Deposition of Contaminants to the Great Lakes and Coastal Waters*, copyright 1997, Society of Environmental Toxicology and Chemistry (SETAC), Pensacola, FL, USA, ISBN 978-1-880611-10-4.

waters, have been classified in terms of different soils horizons [1]. In one soil classification scheme, the top layer, at the surface, is taken as the high organic matter layer (O Layer). This region is important in the trapping and retention of trace metal(loid)s which bind strongly to NOM. Below the O Layer is a coarse grained layer which is depleted in the major metals and in trace metal(loid)s due to the leaching of this layer by water percolation (the A or E Layer). The leached species are precipitated deeper in the soil, in the B Layer which contains precipitated crystalline oxides and hydroxides and clay minerals. The deepest layer (C Layer) consists of partially altered bedrock. The extent of these general strata depend on the location (tropical versus temperate) and the geology of the region. In humid temperatures, the oxides are removed from the A Layer primarily by dissolution and transport with organic material. They are re-precipitated deeper in the soils as the organic matter is consumed by biotic and abiotic decomposition. In more temperate environments, the extent of weathering is less. Trace metals are mostly present as trace constituents in the various primary phases, such as oxides, carbonates, and sulfates, and are released in conjunction with dissolution or alteration of these phases. Soil type depends on the bedrock composition, the local topography, the vegetation, and the time since its major alteration.

The concentration of solutes in the groundwater, and in the streams attached to a particular terrestrial environment, are related to the concentrations in the precipitation and dry deposition, the extent of weathering, and the extent of the exchange reactions that occur within the soil. In vegetated environments, the concentrations of metal(loid)s in precipitation reaching the soil surface are derived both from scavenging of particles and gases from the atmosphere, and from washoff of metal(loid)s from vegetation (so-called throughfall deposition) (Fig. 2.9) [41]. The metal(loid)s attached to leaves that are washed off in throughfall include dry deposited particles and leaf material, and for Hg, dry gaseous deposition. It is also possible that metal(loid)s are exuded from plant leaves.

Clearly, different plant types and leaf shapes impact the degree of metal(loid) accumulation via dry deposition. These differences relate to location and vegetation type and whether or not deciduous trees are present. Overall, the metal(loid) concentrations found in precipitation falling in open areas are lower than that of throughfall deposition (Table 2.5). The ratio of the metal(loid) concentration in throughfall relative to wet deposition in open areas ranges from high values (>2) for those metals that are mostly dry deposited (the crustal metals) such as Fe and Mn, to values between one and two for most of the anthropogenic metals and metalloids that have substantial wet and dry components to their deposition. Clearly, there are similarities between the ordering of the metal(loid)s in Fig. 2.8 and the ordering of those in Table 2.5. The ordering fits with the understanding of the processes contributing to metal(loid)s in throughfall deposition. Interestingly, the comparison of two sites in Spain, a more impacted site and a more remote site showed that there was a relative increase in the throughfall component at the more impacted site, likely reflecting the local input of particulate to the atmosphere [44]. In some cases, and in climates where deciduous trees dominate, inputs of metals from litterfall can also be a very important flux, especially if metal(loid)s in the gas phase are taken up through plant stomata. For example, it has been determined that leaves can take up elemental Hg from the atmosphere [41], and the flux of Hg due to litterfall can

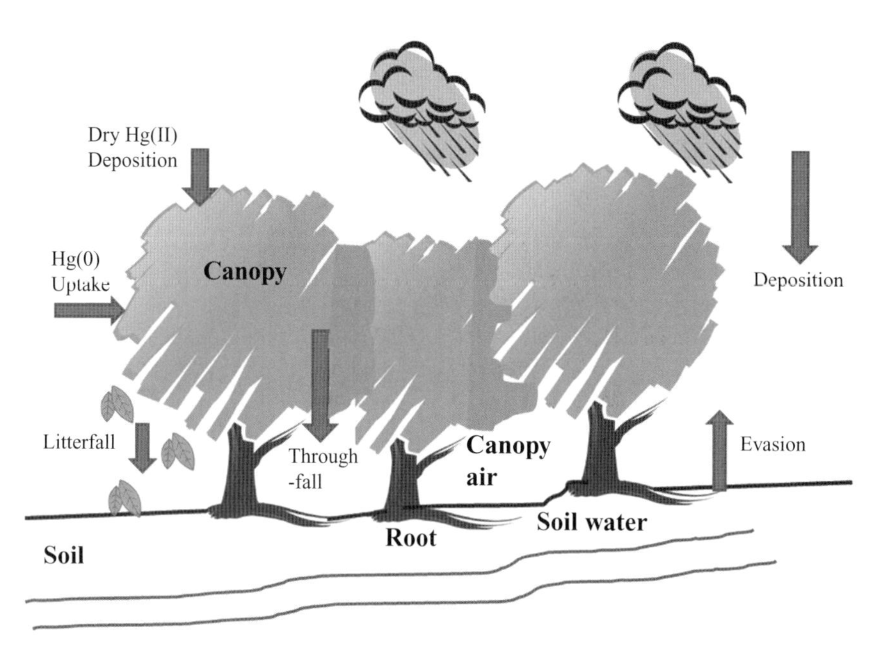

Fig. 2.9 Schematic of the processes involved in exchange at the terrestrial-air interface. Redrawn using information from various references.

Table 2.5 Estimates of enhancement of metal concentration in throughfall deposition (F_T) related to deposition to an open field (F_W). Relative ratios are given in the table as F_T/F_W. East Coast, USA data is from Lawson and Mason [8] and Scudlark et al. [9]; Japan data is from Itoh et al. [42] and Hou et al. [43]; Spain data from Avila and Rodrigo [44].

Metal	East Coast, USA	Japan	Spain (un-impacted)	Spain (impacted)
Hg	1.5–2	–	–	–
Pb	1–2	≤1	~1.0	1.3
Cd	~1	2.5±1.1	0.8	0.9
Zn	1.8	–	0.7	0.75
As	1–2	–	–	–
Se	1.5–2	–	–	–
Fe	1.6	4.5±0.9	–	–
Mn	>5	5.2±0.8	11	18

dominate the overall deposition to the forest floor in remote temperate regions.

For freshwater ecosystems, the input of trace metal(loid)s is from both the atmosphere and from erosion and transport of terrestrial materials in rivers. The importance of the atmospheric source relative to that of erosion and runoff was discussed briefly earlier. In addition, the relative importance of each source depends on the extent of rainfall. As noted above, under high rainfall, a larger fraction of the water is passed to freshwater environments via surface flow, which is rapid and immediate, rather than through infiltration and transport via groundwater to streams, rivers, or lakes, which is a much slower process (years to decades). Thus, during high rainfall, and at high river flows, much of the water is derived directly from surface flow. This water contains primarily metal(loid)s originating in precipitation and eroded suspended material, both from the terrestrial surface and the sediment in the stream or river. On the contrary, under dry conditions, when baseflow occurs, most of the metal(loid)s (and water) are derived from groundwater inputs. Thus, there are often substantial differences in the concentrations of metal(loid)s under different flow regimes. The changes in concentration are often directly related to the flow regime, or the hydrograph during a storm event, especially for those metal(loid)s that are strongly particle reactive. An example of changes in metal(loid) concentration during a storm, for two small streams in western Maryland, USA is shown in Fig. 2.10 [8]. As can be seen, the suspended particulate load increases by one to two orders of magnitude as a result of the increased flow, and the metals that are strongly particulate-associated (Hg, Pb) increase accordingly. Meanwhile, there is much less change in the concentrations of Cd and the metalloids, As and Se. Overall, the concentrations follow the hydrograph.

It is reasonable, as a first order approximation, to assume that the concentration of metal(loid)s in rivers will be some function of their concentration in the terrestrial environment (Earth's crust) as the particulate phase is the dominant vehicle for the transport of most of the trace elements. This is the case to an extent, as is shown by the correlation plotted in Fig. 2.11. The most particle reactive metals, such as Pb and Hg, have a low concentration relative to their abundance, while the transition metals (Fe, Mn, Zn, Ni) all

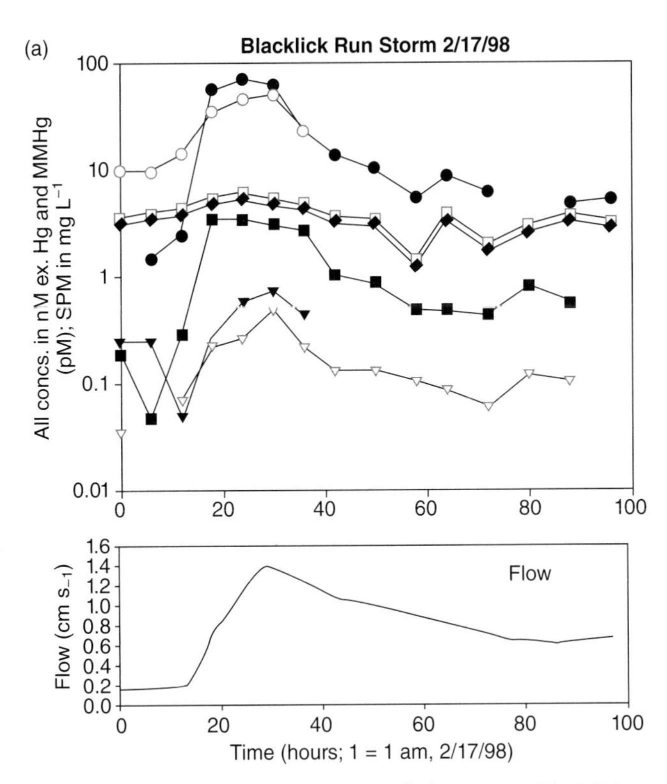

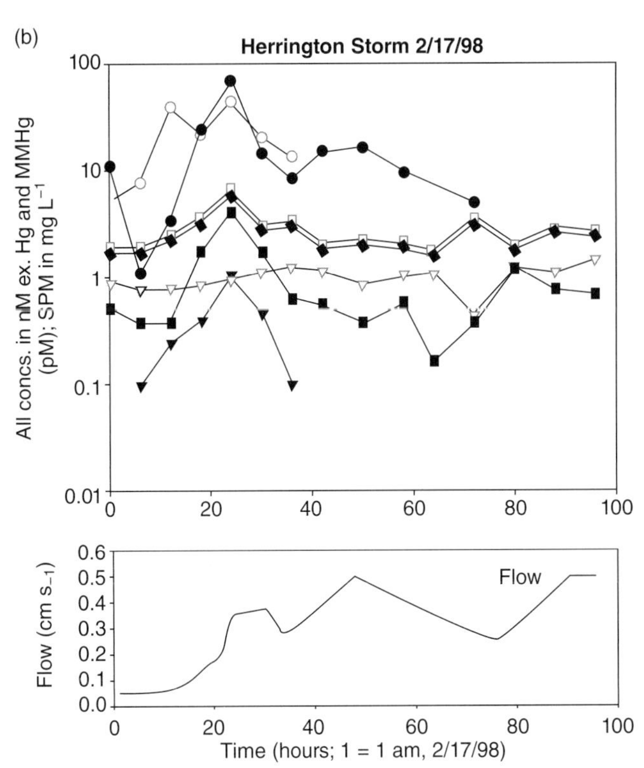

Fig. 2.10 Concentrations of total suspended matter (solid circle), mercury (open circle) methylmercury (solid triangle), cadmium (open triangle), lead (solid square), arsenic (open square), and selenium (solid diamond) for two small second order rural streams (a and b) in western Maryland during a storm event in February 1998. The flow regime (storm hydrograph) is also shown. All concentrations are in nanamoles except for mercury and methylmercury which are given in picomoles. Reprinted from Lawson and Mason (2001) *Water Research*, **35**(17): 4039–52 [8], with permission of Elsevier.

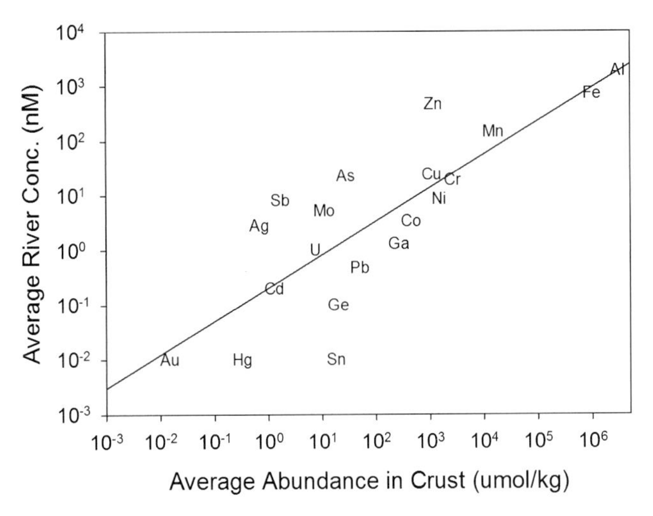

Fig. 2.11 Plot of the relationship between the average concentration of a metal(loid) in the Earth's crust and the average concentration in river water. Data extracted from various sources and the information in Tables 1.1 and 2.2.

have a similar river concentration to crustal abundance ratio. Finally, the metalloids, such as As and Sb, are present in relatively high abundance compared to the trace metals, likely a function of these elements being oxyanions in natural waters. Similarly, Mo and U, which also form oxyanions in solution, are relatively abundant in river water. The relative abundances and the degree of particle reactivity determines to a large extent the net export of metal(loid)s from the terrestrial system to the ocean. The processes that occur at the land-ocean interface – estuarine mixing and particulate matter deposition – lead to the relatively high removal of the particle reactive elements, such as Fe, Mn, and other first row transition metals and the heavy metals such as Pb and Hg. The more soluble metal(loid)s are transported to a greater degree, as discussed previously.

The export of metal(loid)s from a watershed can be characterized in terms of a watershed yield or transmission factor (T_{WS}), which is defined as the ratio of the metal(loid) flux from the atmosphere over the watershed to that in stream or river flow (Equation 2.2). The definition and relationship between pluvial inputs and fluvial export was discussed above and it was noted, for the particle reactive metals, that most of the metal deposited from the atmosphere is retained in the watershed. Many metal(loid)s are transported from the freshwater reaches to the coastal and open ocean mostly in the particulate phase. There is a further trapping of metal(loid)s in the coastal zone as a result of the interaction of the fresh and marine environments, and due to factors such as changes in topography, flow rate and other factors. For some metal(loid)s, inputs from the atmosphere dominate over inputs from rivers, while for others, the atmospheric input is minor.

The cycling of metal(loid)s in lakes is governed by the biogeochemistry of these systems and the relative importance of atmospheric versus fluvial inputs. For most metal(loid)s, it is the fluvial inputs that are the dominant source. Additionally, much of the metal(loid) input is associated with particulate matter. The release of metal(loid)s from the solid phase in such systems is governed by the chemistry of the ecosystem. Many lakes, especially in temperate climates, become stratified in summer. This stratification results in oxygen depletion in the deeper waters of the lake. This water column oxygen depletion, and a more general depletion of oxygen within the upper layers of the sediment, leads to low oxygen (hypoxic) or anoxic conditions. These conditions promote the reductive dissolution of oxide phases, and the release of metals and metalloids into solution. This cycling of metal(loid)s is discussed in detail in Chapter 7.

2.2.6 The transport of metal(loid)s to the ocean

The export of material from rivers to the coastal zone is not evenly distributed globally as a result of spatial differences in terrain and because of the dominance of export from large rivers such as the Amazon, and the large Asian rivers (Fig. 2.12) [45, 46]. In addition, material export is higher for rivers draining steep topography compared to those rivers draining regions with a large coastal plain. The export of particulate material from rivers to the ocean has been estimated by various groups. Estimates from Milliman [45] are shown in Fig. 2.12(a). There are three regions of high particulate discharge based on the data in the figure: the Amazon River region (~14% of the total input), southern Asia (34%), and eastern Asia (36%). These regions all have large rivers and periods of high rainfall that drive the high particulate discharge. A comparison of the net export of particulate material and water further illustrates the differences for the different regions in terms of discharge to the coastal zone (Fig. 2.12b). While there is a large water discharge for the South American continent, this is relatively less that the particulate flux when compared to other regions (note the different scales on Fig. 2.12b) [46]. The Asian continent is opposite in that there is a much higher particulate discharge per volume of water. This is likely both a function of the terrain in this region and the increased discharge of solids as a result of intense human activity (e.g., deforestation) on these watersheds. Besides these two outliers, the other continents appear to have a relatively similar ratio of particulate to water volume in the discharges. It is likely that the metal(loid) discharges will follow a similar pattern for those metal(loid)s that are strongly particle associated. This is also the case for NOM [47] as the distribution of organic material between the dissolved and particulate fractions appears to follow similar trends to that of water and particulate material as shown in Fig. 2.12(b).

The concentration of NOM varies across the range of aquatic ecosystems. This has an impact on the concentration of the metals in solution as high organic matter content tends to enhance the concentration of many metals in solution. Typical ranges in concentrations are given in Table 2.6 [10, 14]. The lowest concentrations are found in the deep open ocean and can be an order of magnitude smaller than those found in coastal waters, which are typically lower than those of estuaries and rivers. Of course, as indicated in the table, there is much variability between actual ecosystems, but these general trends are applicable overall. The highest concentrations are found in some effluents, especially the discharges of unregulated sewage treatment plants. Modern facilities can have dissolved and particulate organic matter (DOM and POM) concentrations of $<10\,mg\,l^{-1}$, similar to those of the receiving waters.

The difference in NOM concentrations across the land-sea interface illustrates the impact of mixing of fresh and saline waters in estuaries. Because of differences in ionic strength, pH, and the concentrations of dissolved solids, there is a general coagulation and removal of organic material in the mixing zone. In addition, in many estuaries, the flow rate of the water decreases, and this further enhances coagulation and particle settling and removal. The continents with the widest coastal plain therefore tend to be locations where particulate removal in estuaries is enhanced. Thus, particulate concentrations decrease dramatically through an estuary, leading to a substantial removal of metals and metalloids in most circumstances.

Most metallic elements will be removed from the water column during estuarine mixing. However, it is possible for them to pass through without their concentration changing dramatically (so-called *conservative* elements) or they may even be added to the water column as a result of estuarine processing [47]. These different scenarios are illustrated in Fig. 2.13. If there is a linear change in the concentration of a particular element relative to salinity through an estuary, then it is assumed that the concentration is the result of dilution of the riverine signal with seawater (it is almost entirely the case that the seawater concentration is less than that of the riverine end member, but this does not have to be the case) (Fig. 2.13a). In estuaries, salinity changes are not spatially linear with most of the salinity change occurring in a small mixing zone where the freshwater and saline waters meet. This region is also a region of high turbidity and particle settling. Metal(loid)s are removed from the water column by precipitation, adsorption to suspended matter and resultant sinking to the sediment, or added due to the release from particles and other interactions. As noted earlier, if an element is not involved in such interactions, or is minimally so, then its concentration is considered to be "conserved" during its transport through the estuary and its

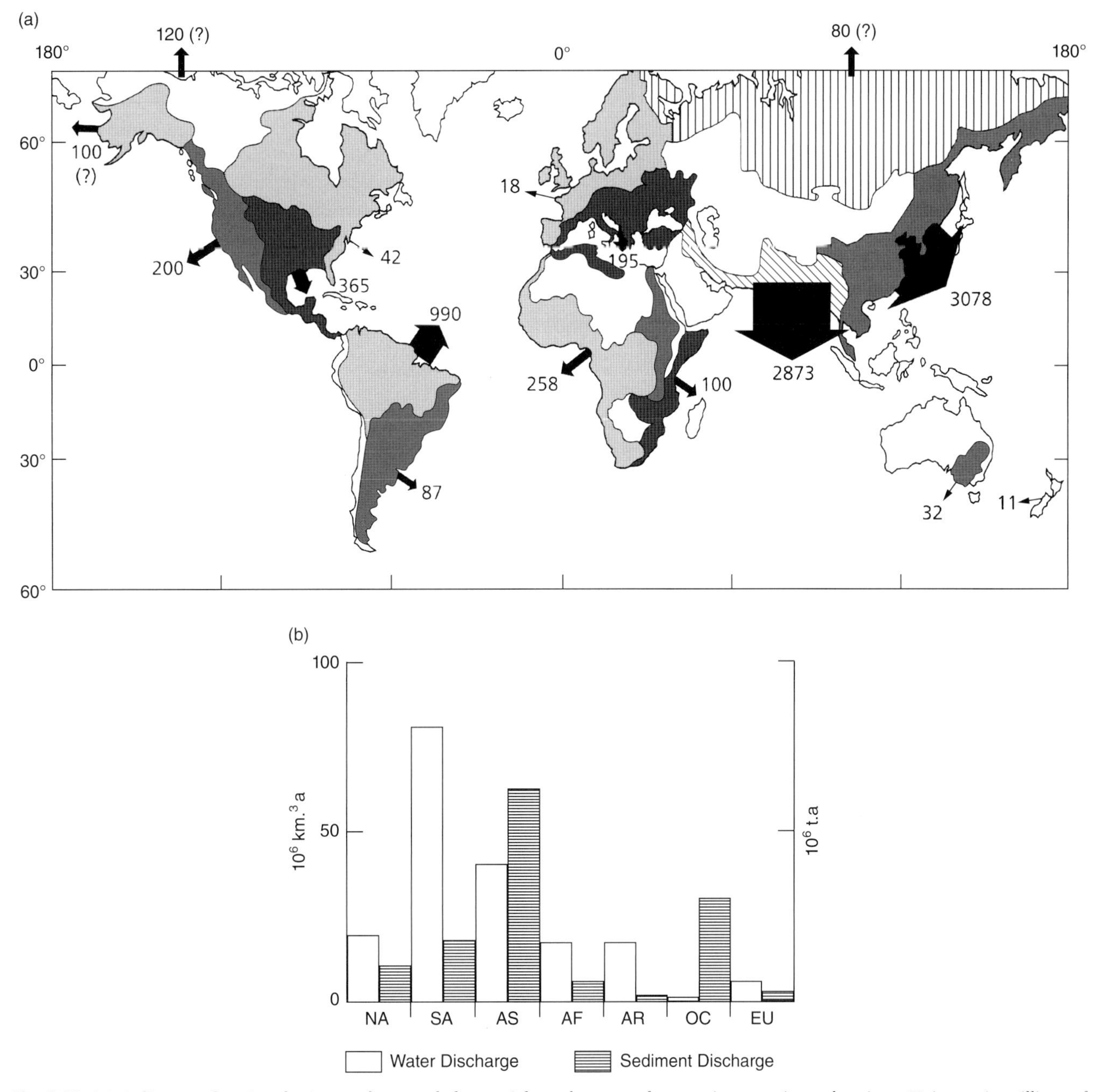

Fig. 2.12 (a) A diagram showing the input of suspended materials to the ocean from various continental regions. Units are in millions of metric tons/yr. Figure reprinted in Chester [46] but originally from Milliman [45] and reprinted with permission of the University of Chicago Press; (b) Relative fluxes of water and sediment by continental region. Data originally from Degens [46] and reprinted with permission from Chester, *Marine Geochemistry* [47] from John Wiley & Sons, Inc.

concentration change will be linear when plotted against salinity [47, 48]. For other elements, their profiles may show either the addition or removal through the estuary. The theoretical dilution lines for such profiles are shown in Fig. 2.13(a) for an element or compound which has a higher river concentration.

Many of the metals are retained in estuaries as they are strongly attached to particles. The model estimates below, by Yeats and Bewers (Table 2.7) [7], suggest that most (~95%) of the suspended matter entering from rivers is not exported out of the coastal zone. Note that their earlier estimate of the particulate flux in their Table 1.8 is about a factor of two higher than the estimated values in Fig. 2.12. It is possible that the inputs of particulate material to the ocean has decreased as a result of reservoir building on large river systems, which has accelerated in the last 3–5 decades,

Table 2.6 Range in concentrations of organic matter in the dissolved and particulate phase in various water types. Compiled from the literature as described in the text.

Conc (mg l^{-1})	River	Estuary	Coastal	Open Ocean	Deep Sea	Sewage*
Dissolved	10–20	1–20	1–10	1–2	0.5–0.8	<100
Particulate	5–10	0.5–10	0.1–1	0.01–1.00	<0.01	<100
Total	15–30	1–30	1–10	1–3	<1	<200

*High values represent plants without advanced treatment facilities. For plants with advance treatment, NOM can be <10 mg l^{-1}.

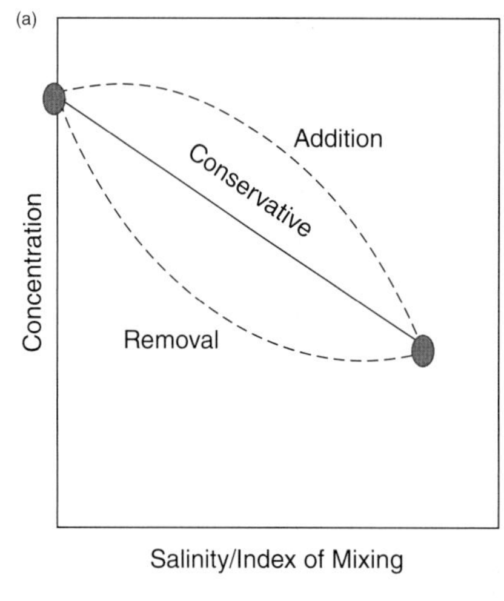

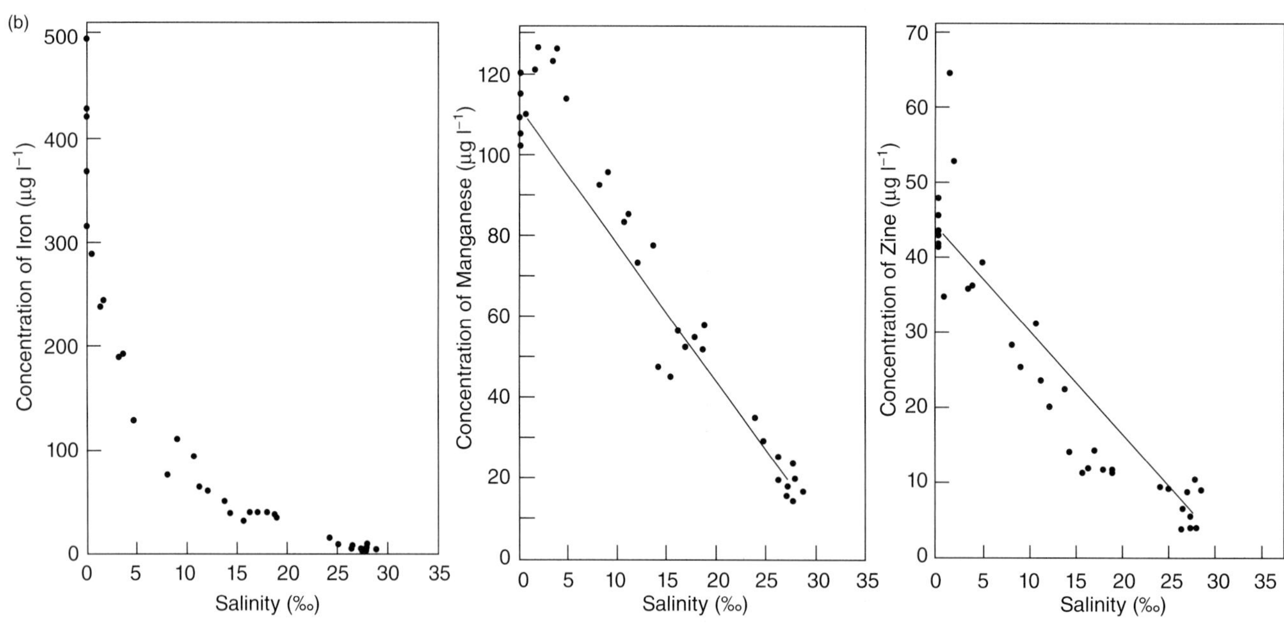

Fig. 2.13 (a) Theoretical dilution relationships for mixing for a constituent with a higher concentration in rivers (typical case) with sewater. (b) Examples of the various types of interactions that occur for various elements. For Fe, 1 µg l^{-1} = 17.92 nM; for Mn, 18.21 nM; for Zn, 15.29 nM. Taken from Holliday and Liss (1976), reprinted with permission of Elsevier.

Table 2.7 Estimated global river discharge of sediment and metals and the associated estimate of efflux to the coastal zone. The predicted concentrations are those derived from the mass balance model in Fig. 2.14. Recalculated from Yeats and Bewers [7].

Component	River flux*	Net efflux from coastal zone*	% Retention	Observed sediment content*	Model predicted content*
TSS	1.8	0.1	–	–	–
Fe	1650	100	99.8	5.3%	8.0%
Mn	40	14	69	5600	12,500
Co	0.51	0.13	79	60	115
Ni	2.6	0.39	92	180	290
Cu	3.9	0.91	66	410	540
Zn	8.3	4.3	54	1900	130
Cd	0.03	0.01	69	8.5	0.5

Notes: *Fluxes: $\times 10^7$ kmol yr^{-1} except $\times 10^{13}$ kg yr^{-1} for SPM; sediment concentrations in mol kg^{-1}.

although such building has now decreased in some countries. It is estimated that about a quarter of the particulate matter that was previously delivered to the ocean is now being trapped [49]. However, such trapping is somewhat counteracted by the simultaneous increased erosion that is likely occurring as human development on rivers and estuarine ecosystems increases, and the amount of impervious surfaces increases.

Iron, which is strongly particulate-associated, is trapped in the coastal zone, and typically shows such a mixing profile (Fig. 2.13b). The other metals (e.g., Cd, Cu, and Zn) which are less strongly particle associated are retained to a lesser extent, but are also mostly removed during estuarine mixing (see references in [47], for example). The results in Table 2.7 [7] were compiled from the literature (observed) as well as derived from their simple box model of estuarine mixing (predicted) (Fig. 2.14). The agreement between the predicted and observed concentrations is remarkable given the simple assumptions and the model framework.

The role of factors, such as biological productivity in surface waters and microbial degradation at depth, in metal(loid) cycling also need to be considered. While many of the redox processes that occur in estuaries can occur abiotically, it is now known that many of these processes are biotically-mediated. Phytoplankton productivity is driven by the supply of nutrients, which are typically in higher concentration in river waters than in coastal waters; thus, the input of freshwater and the estuarine mixing regime determines to a large degree the location where maximum productivity occurs. In many stratified estuaries, the mixing of the water masses and the resultant precipitation and coagulation of organic matter and other material leads to a high turbidity region, where nutrients are high but light is limited. In such instances, the peak in productivity and in biotic particles is somewhat downstream from the maximum mixing zone. As many metals are required co-factors in enzymes, they are actively accumulated by algae. Other metal(loid)s bind strongly to the plankton, or are taken up inadvertently. All of these processes lead to the removal of metal(loid)s in the high turbidity zone and in the maximum region of primary production, therefore potentially decreasing the dissolved concentration in surface waters. The ultimate fate of these metal(loid)s is sinking into the deeper waters or accumulation at the sediment-water interface. Therefore, there are multiple mechanisms for the removal of metal(loid)s in the low salinity mixing regime of estuaries.

Chemical, physical and biological transformations (so-called *diagenetic* processes) at the sediment-water interface lead to the release of metal(loid)s from particles. This is certainly true for elements that have important redox cycles, such as Fe and Mn, which are more soluble as reduced species (Fe^{II} and Mn^{II}) than in their common oxidized forms. The dissolution of Fe, Mn, and other oxide phases in low oxygen environments therefore can be an important mechanism for the release of metal(loid)s back to the water column, as discussed above for freshwater systems. Complexation of metals with NOM can enhance their mobility. In addition, decomposition of organic matter via the chain of respiratory processes involving different terminal electron acceptors can also provide a mechanism for metal(loid) release. These processes will be discussed in detail later in the book (Chapters 6 and 7), but the return of metal(loid)s from the sediment back to the water column cannot be ignored in studies of metal(loid) cycling in estuaries, and in the construction of mass balance budgets for such systems.

2.2.7 Trace metal(loid)s in ocean waters

The distributions of trace metal(loid)s in the oceans have been characterized in terms of their vertical profiles, which are typically divided into three basic categories based on their assumed or known behavior: conservative, nutrient-like, and scavenged. More recently, two other categories have been noted for metal(loid)s that do not fit these types:

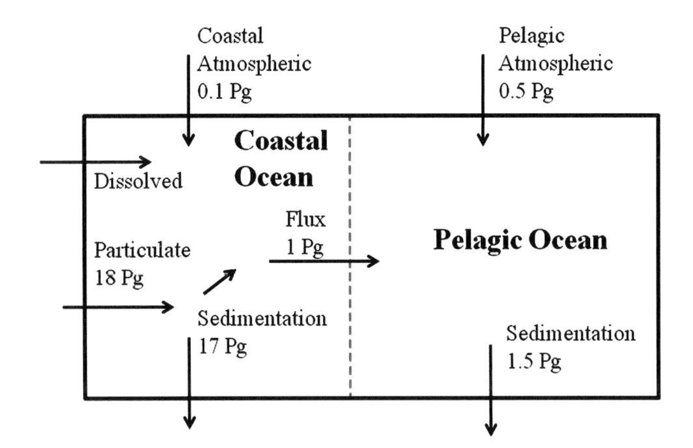

Fig. 2.14 A generic coastal zone model showing particulate fluxes as estimated by the model described by Yeats and Bewers (1982) [7]. Redrawn but original figure is Copyright (2008), Canadian Science Publishing or its licensors. Reproduced with permission.

mixed and hybrid distributions (Fig. 2.15) [9]. As more elemental distributions are examined in detail in the ocean, oceanographers have obtained an ever increasing complexity of knowledge about the factors controlling the speciation and fate of trace elements in ocean waters. The vertical distribution of an element in the ocean is determined by its solubility and reactivity. The simplest distribution is that for a "conservative" element whose concentration is not affected by short-term changes in inputs or biogeochemistry and which has a long residence time in ocean waters. These elements have an overall uniform distribution in all ocean basins. Such an element is being added and removed from the ocean very slowly relative to its *in situ* concentration, and therefore its residence time (τ) is long (10^3 – 10^5 yrs or more), much longer than the mixing time of the ocean (~1000 yrs) [50]. Such elements are not particle reactive, or if they are, these interactions do not impact their ocean distribution and concentration. The major cations is seawater (e.g., Na^+, K^+, Mg^{2+}, and Ca^{2+}) and most other alkali and alkali earth metals fall into this category ($\tau > 10^5$ yrs), although there is evidence for their local depletion in some environments. Such departures from the average are more likely on the continental shelf and in shallow environments where strong interactions with the sediment (supply or removal of an element) could lead to a local change in concentration. For example, precipitation of $CaCO_3$ in some regions could lead to depletion of the Ca^{2+} concentration relative to the ocean average. Such depletions could be driven, for example, by local increases in pH due to high primary productivity under conditions of low gas exchange (i.e., low CO_2 replenishment).

Molybdenum (Mo), which is present in seawater as an oxyanion, is another conservative element even though it is a required element for some enzymes. Its requirement by microorganisms is small compared to its concentration (~110 nM) and inventory, which is much greater than that of most trace metal(loid)s, whose concentrations typically range from nM to pM. It has a long residence time (τ ~8.2×10^5 yrs) as its river concentration is much less (~5 nM) than that in the ocean [50]. Other trace metals and metalloids that exhibit conservative behavior in the ocean are Sb ($\tau \sim 5.7 \times 10^3$ yrs), W, U ($\tau \sim 5 \times 10^5$ yrs) and Re [28, 50]. Most of these also exist as oxyanions in seawater in relatively high concentration (nM–μM); therefore, it is the relative unreactive nature of such compounds that leads to their long residence time. For the oxyanions, the lack of reactivity may be partially due to the fact that most environmental particles are negatively charged at ocean pH [11]. The interaction of anions with such particles is limited. Some of the oxyanions can be taken up by microbes either as they are required (e.g., sulfate) or because they are inadvertently taken up through channels designed for nutrient uptake (e.g., arsenate uptake through channels designed for phosphate uptake). In seawater, the relatively high As concentration has led to the evolution of mechanisms for dealing with its inadvertent uptake. Arsenate reduction and methylation are relatively ubiquitous mechanisms for converting and exporting As from cells, as discussed further in Chapter 8.

Alternatively, many elements have a defined biological as they are part of enzymes and biochemicals and are therefore needed by phytoplankton. If the concentration of these elements is low enough to be impacted by their uptake into primary producers in surface waters, the element will exhibit a "nutrient-like" distribution (Fig. 2.15) [6, 9]. Their concentration will be depleted in surface waters by uptake into phytoplankton or other microorganisms, and their deep water concentrations will be enhanced through their release into solution due to organic matter decomposition. If these metal(loid)s are relatively unreactive toward detrital and inorganic particles, then their distribution in the deep ocean will not be influenced by deep ocean scavenging processes, which can influence the distribution of the more particle-reactive metals. Examples of metals that demonstrate a nutrient-type distribution are Zn (Fig. 2.16) [28, 51, 52] and Cd [28, 29].

Zn has multiple roles in cellular metabolism and it is part of an important enzyme, carbonic anhydrase, in marine phytoplankton, which is required to interconvert CO_2 and bicarbonate (HCO_3^-) within cells [29]. Thus, it is actively accumulated into plankton in surface waters and is released back into the water at depths when the organic matter decomposes. It has a similar profile with depth to the nutrient, Si. The deep water concentrations of both Si and Zn are higher in the Pacific Ocean compared to the Atlantic Ocean, reflecting the continued buildup of these elements in the deep waters during the circulation of the "deep conveyor belt" through which surface waters sink in the North Atlantic and around Antarctica, and are ultimately transported to the North Pacific where they are returned to the surface via deep water upwelling [50].

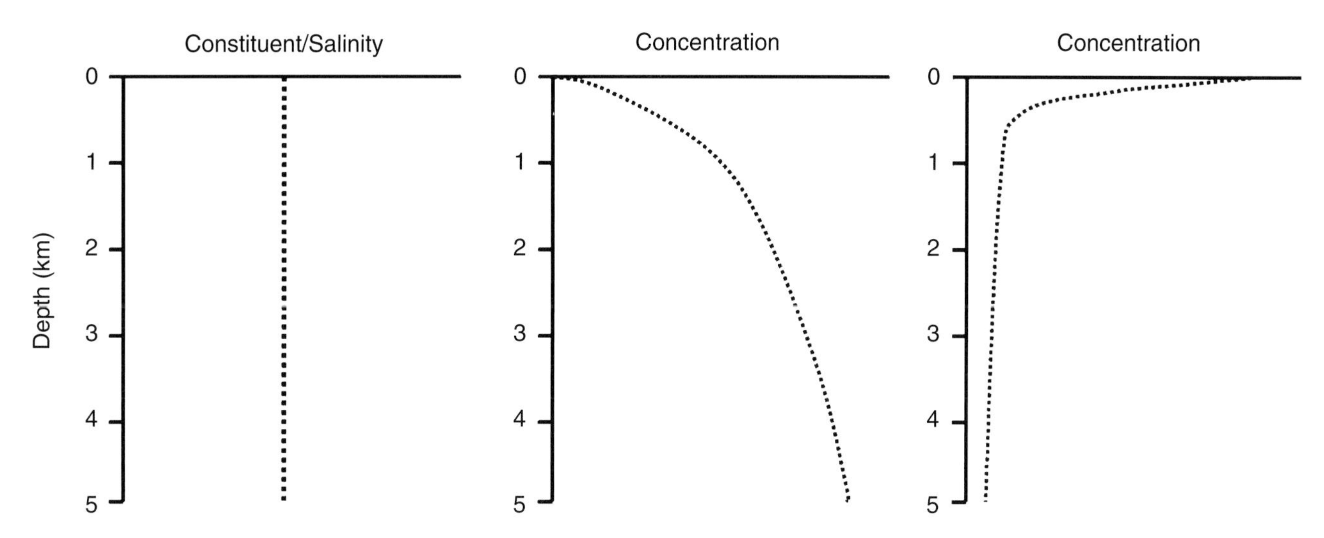

Fig. 2.15 Idealized profiles of metal distributions in open ocean waters. The graphs represent, from left, the distribution of an element that is conservative, that of an element with a nutrient type of profile and that of an element that has a strong atmospheric input but is scavenged in deep waters by particles.

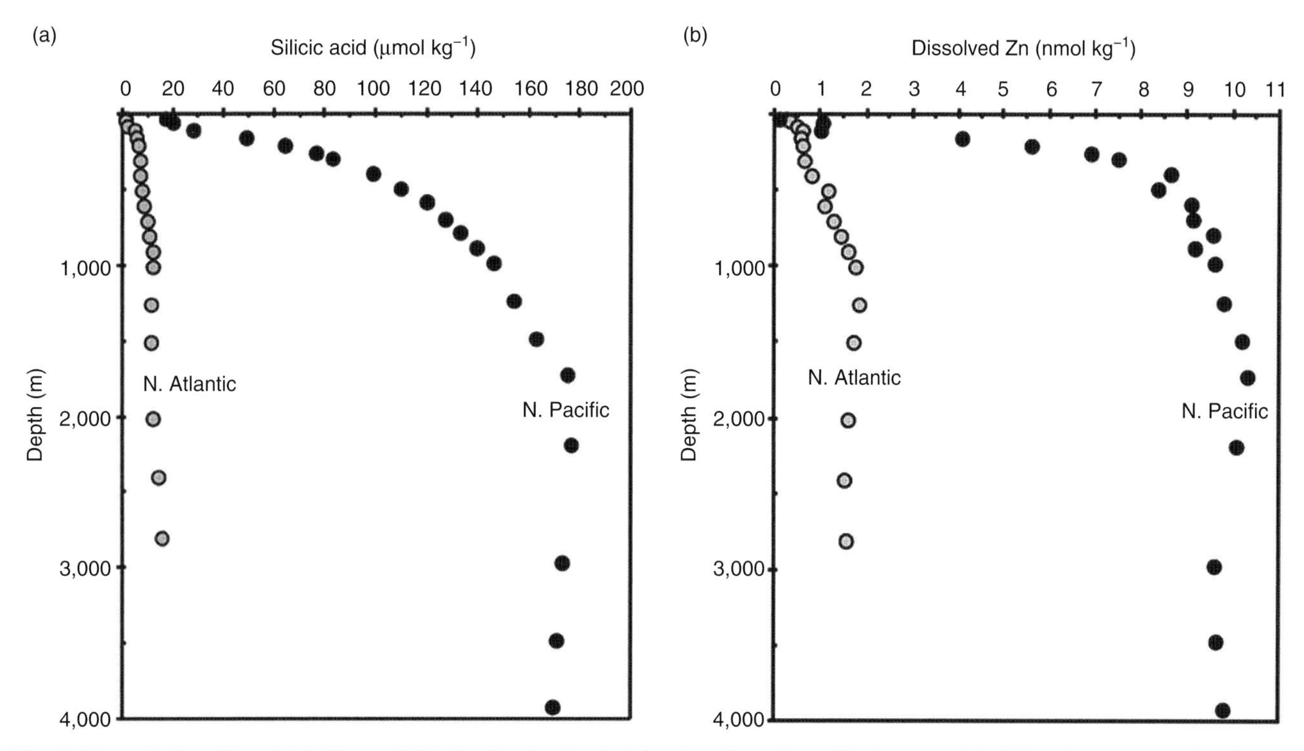

Fig. 2.16 Vertical profiles of (a) silica and (b) zinc for the North Atlantic and North Pacific Ocean. Original data from [51, 52] but figure redrawn by Bruland and Lohan (2004) In: H. Elderfield (ed.), *The Oceans and Marine Geochemistry*, vol. 6 of Treatise on Geochemistry: Holland, H.D. and Turekian, K.K. (eds), Elsevier, Amsterdam, pp. 23–47 [28] and reprinted with permission of Elsevier.

Another Group 12 metal, Cd was initially thought to be a toxic metal with no biochemical role, even though it exhibits a nutrient profile in the ocean. Its biochemical role was not known until the demonstration of its substitution for Zn in carbonic anhydrase in some marine phytoplankton [28]. Copper (Cu) is another element that is both required for some enzymes and toxic at high levels, but it has a more complex profile given its relatively high particle reactivity compared to Zn and Cd. An important micronutrient metal is Fe, and it also has depleted concentrations in the surface ocean (Fig. 2.17). However, the relative enhancement of deep Pacific Ocean water concentrations seen for Zn and Si is not apparent for Fe [28]. This is due to the high particle reactivity of Fe which results in its continual scavenging

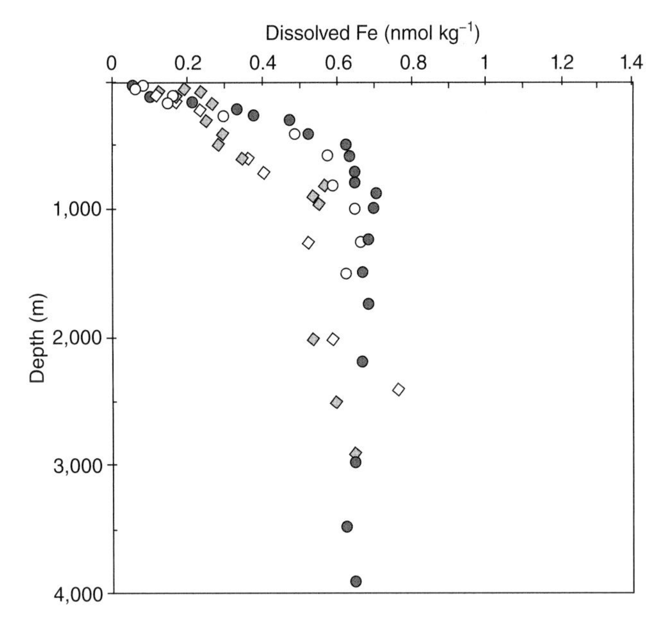

Fig. 2.17 Vertical profiles for dissolved iron from the North Atlantic and North Pacific. Original data from [51, 52] but figure redrawn by Bruland and Lohan (2004) In: H. Elderfield (ed.), *The Oceans and Marine Geochemistry*, vol. 6 of Treatise on Geochemistry: Holland, H.D. and Turekian, K.K. (eds), Elsevier, Amsterdam, pp. 23–47 [28] and reprinted with permission of Elsevier.

from the deep waters over time leading to its deep water concentration reflecting more the short-term history of particulate input to deep waters rather than the longer-term cycling of these deep waters [28]. The Pacific/Atlantic deep water ratio is much greater for Si than for Zn, suggesting that there is also scavenging of Zn in deep waters. On the other hand, ocean waters are undersaturated with respect to solid silica (opal) [50] and thus after the decomposition of the diatoms upon sinking, which are the main organisms incorporating Si into biogenic material, the Si is released into solution and is not further scavenged from the water column.

The distribution of Pb in ocean waters (Fig. 2.18) is illustrative of a scavenged profile and a metal with a strong atmospheric signal [53, 83]. As discussed next, most of the Pb entering the ocean is from the atmosphere and this leaves a signal of this input (higher concentrations) in surface ocean waters [36]. However, Pb is highly particle reactive and is taken up by biotic and abiotic particles in surface waters. When these particles decompose, the Pb repartitions to the remaining particulate matter in deep waters, which is a small fraction (<10%) of that in surface waters and is continually removed from the deep waters. The overall difference between the concentration of Pb in Atlantic Ocean waters compared to the Pacific is not just a consequence of ocean chemistry and physics [28, 84]. The input of Pb to the environment has been greatly perturbed by human activity, and the ocean profile reflects the magnitude of the inputs to the different ocean basins over time. The historical industrialization and use of Pb in gasoline in North America and Europe is reflected in the higher inputs of Pb to the North Atlantic and the resulting higher concentrations in these waters [85, 86]. Additionally, as Pb is a highly particle reactive metal it is not accumulated in the deep waters, in a similar fashion to Fe, as discussed previously. Another metal with a similar ocean chemistry and profile is Al [54], which also has a strong atmospheric signal, but from natural sources (dust), and which is actively removed from deep waters via particle scavenging [28]. Many metal(loid)s have a complex profile that is a mixture of the classic profiles illustrated in Fig. 2.15. As suggested earlier, this may be due to the modifying impact of deep water scavenging on a typical nutrient profile and this appears to be the reason that Cu, for example, does not have a typical nutrient profile even though it is a required metal by microorganisms. The cycling of metals in the ocean is discussed in detail in Chapter 6.

2.2.8 Trace metal(loid) inputs from hydrothermal vents

The discovery of hydrothermal vents on the deep ocean floor near the mid-ocean ridges and spreading centers in the late 1970s [55, 56] and the substantial input of high temperature fluids from these systems has transformed our understanding of the sources and sinks of major and minor elements to the ocean. The percolation of seawater through the ocean crust leads to its heating within the crust, followed by the addition or subtraction of chemicals due to reactions of the heated water with the crustal material. Temperatures can be as high as 350–400°C. At such high temperatures, the water becomes less dense and rises back to the surface where it is emitted and mixed with the surrounding seawater [57, 58]. Since their initial discovery, hydrothermal systems have been found on all of the major ocean ridge spreading centers, and associated with back-arc systems (crustal subduction zones). The total number of sites found exceeds 100. The biogeochemical processes that occur at crustal spreading centers are very important in the overall oceanic chemistry and fate of the major ions. Detailed chapters on hydrothermal vents and their importance to ocean geochemistry have been recently published [57, 58]. These provide much more information than can be detailed here, but this information is briefly summarized as the focus here is on trace metal(loid)s inputs and cycling in hydrothermal systems.

Magnesium, and to a lesser degree Na and K, is removed during hydrothermal activity, while Ca, Fe, Mn, and other metals are added to the fluids. Sulfate is removed by precipitation or by its conversion to sulfide in the low oxygen environment and HCO_3^- is converted into methane. Sulfide is extracted from the rock through which the water is moving. The low pH (<4), generated in part by the precipitation of Mg oxide minerals [58], leads to the conversion of

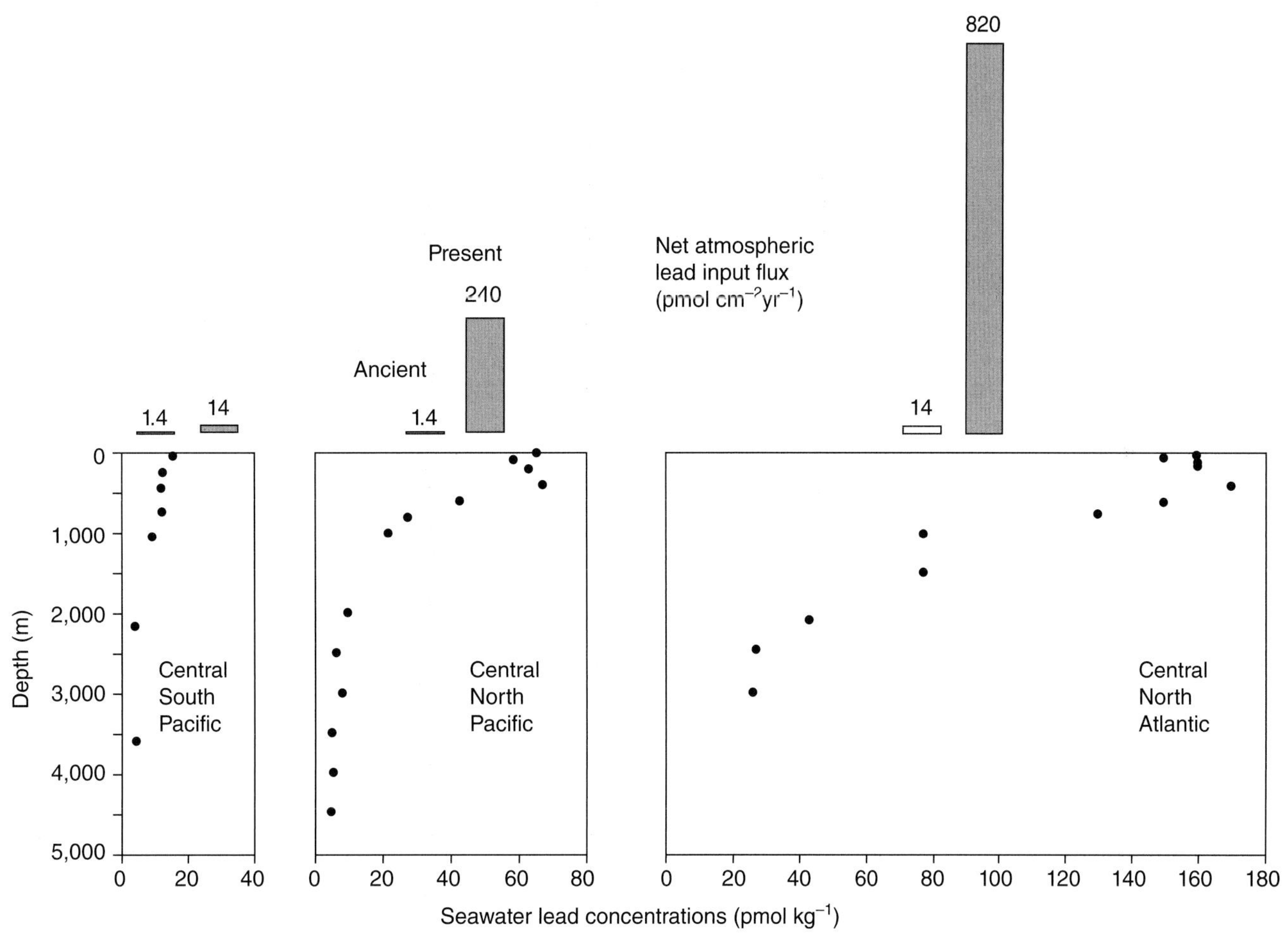

Fig. 2.18 Vertical profiles of lead in the North Atlantic and Pacific. Also included (bars) are estimates of the anthropogenic inputs of lead to the sites at the time of sampling. Original data taken from [53–55] but figure redrawn by Bruland and Lohan (2004) In: H. Elderfield (ed.), *The Oceans and Marine Geochemistry*, vol. 6 of Treatise on Geochemistry Holland, H.D. and Turekian, K.K. (eds), Elsevier, Amsterdam, pp. 23–47 [28] and reprinted with permission of Elsevier.

dissolved carbonate species (primarily HCO_3^-) into CO_2 [11, 57]. The enriched elements in hot hydrothermal fluids are subsequently precipitated as the fluids cool or as these sulfidic, hot solutions are emitted into the oxygenated ocean bottom waters. Sulfides of Fe and other metals are precipitated, forming precipitate chimneys through which the fluids continue to vent into the ocean waters, forming buoyant plumes where any remaining Fe and Mn will precipitate as their (hydr)oxides.

Overall, there is a wide range in both the temperature of the emitted fluids and in their chemical composition, with some locations emitting fluids with elevated trace metal(loid) concentrations and others with concentrations near or below seawater levels [57]. The lower temperature, more extensive, and more diffuse percolation off the mid-ocean ridges may potentially be a more important source of metal(loid)s to the deep ocean than the less extensive, higher temperature activity. Given the low solubility of many metal sulfides and/or their propensity to bind to and be scavenged by Fe-sulfide minerals and metal oxides precipitates, it is likely that most trace metal(loid)s in hydrothermal solutions are removed to the solid phase close to the emission source.

Sulfide, oxide, silicate and sometimes carbonate deposits (mostly $CaSO_4$ and $BaSO_4$) are formed as a result of precipitation from the cooling solutions. Initially sulfide minerals, mostly FeS and pyrite (FeS_2), form, and subsequently any remaining dissolved reduced Fe and Mn are oxidized and precipitated. The sulfide phases contain substantial amounts of trace metals such as Cu, Zn, and Pb. Given the slower rate of oxidation and precipitation of Mn oxides, the formation of these solids tend to lag behind the Fe (hydr)oxide formation. Evidence suggests that the oxidation of Mn is biologically-mediated. These initial deposits are further altered over time by various diagenetic processes and it is also possible that trace metal(loid)s can be released into

Table 2.8 Estimated inputs of metals and metalloids to the ocean from the atmosphere, riverine inputs and from hydrothermal vents. The relative importance of each source is indicated for each metal. Data taken from the literature as described in the text. All fluxes are in Gmol yr^{-1}.

Element	Net Fluvial Dissolved Flux	Net Fluvial Particulate. Flux	Pluvial Flux	Hydrothermal	%Pluvial*	%Hydrothermal
Al	35	5410	2156	0.4	28	<<1
V	0.73	5.1	1.1	–	16	–
Cr	0.35	2.9	1.5	–	32	–
Mn	5.4	29	11	25	16	36
Fe	5.4	1340	469	75	25	4
Co	0.13	0.51	0.23	0.070	24	7
Ni	0.51	2.4	1.5	–	34	–
Cu	0.95	2.4	0.65	0.75	14	16
Zn	0.38	6.3	1.3	2.0	13	20
As	4.0	1.0	0.08	–	2	–
Mo	1.0	0.47	0.10	–	6	–
Ag	0.56	0.0090	0.050	0.001	8	<1
Cd	0.028	0.014	0.04	0.007	45	8
Hg	0.001	0.002	0.017	0.001	81	5
Pb	0.02	0.76	1.3	0.001	62	<1

*Estimates of percentage are the fraction of the total flux.

solution during these processes. Many of the redox transformations are carried out by the extensive suite of chemolithoautotrophic microbes that exist in the vent environments. These processes are not only confined to the proximity of the vents, but are also occurring in the buoyant plumes formed by the expelled fluid water.

The extent to which trace metal(loid)s are added to the ocean from vents is difficult to accurately assess [57] because of artifacts in sampling and analysis, and because there is still limited measurement of the concentrations of trace metal(loid)s in such plumes. The measurements that have been made have found a large variation in concentration, reflecting the heterogeneous nature of this overall process. Average values for the concentration of trace metal(loid)s in hydrothermal fluids are given in Table 2.8.

Overall, it is apparent that the concentrations of some metals, such as Fe, can be enriched by many orders of magnitude in vent fluids over their concentration in seawater (Mn from 10 μM to 71 mM; Fe 7 μM to 25 mM; Al up to 20 μM) [57]. Most transition metals, and heavy metals such as Ag, Tl, Pb, and Hg, and the metalloids (As, Se, and Sb), are enriched in vent fluids (e.g., Cu up to 160 μM; Zn up to 3 mM; Co up to 14 μM; Pb up to 4 μM; and others (Cd, Ag) up to several 100 nM) relative to their concentration in seawater (Table 1.1) [25, 57, 58]. The level of sulfide in the fluids is likely one dominant control over the resultant concentration. For low temperature fluids there is little data, but it is speculated that the concentrations of some trace metal(loid)s in such fluids are similar or perhaps even lower than those in seawater. Overall, few of the first row transitions metals (e.g., V, Cr, Co, Ni, and Zn) and the heavy metals (Ag, Cd, Sn) of primary consideration in this book have a strong net hydrothermal source given their high particle affinity and their ability to bind to sulfide and oxide mineral phases. Even for Fe, the net input from hydrothermal activity is small relative to its other sources, but this is not the case for Mn (Table 2.8). Conversely, it appears that there may be net removal of some of the "oxyanions" such as As, Mo, and U by hydrothermal processes.

2.3 Global cycles of some important trace metals

Generally, for the ocean, it is possible to estimate the importance of atmospheric versus riverine inputs based on the estimates of the net fluvial flux, the pluvial (atmospheric inputs) inputs, and with consideration of the potential importance of hydrothermal sources, (Table 2.8). In making these estimates, different values in the literature need to be reconciled, and in some cases there are substantial differences in values from different sources [11, 30–34, 36, 40, 47, 57, 59–66]. In making these calculations it was assumed that 90% of the riverine suspended material is trapped in estuaries and does not get transported to the open ocean. The basis for this estimation was discussed above and will be covered further in Chapter 6. Additionally, it was assumed that if there was not data available for hydrothermal inputs for a particular element that it was not significant. Clearly, this assumption may not be valid. Also, there is little information on the potential importance of groundwater inputs or the release of metals from coastal sediments and their transport offshore, which has recently been suggested to be an important source of open ocean Fe. While these inputs potentially occur, it is likely that their magnitude is not sufficient to alter the relative importance of the various sources as outlined in Table 2.8. It is evident from Table 2.8 that riverine inputs are the dominant source for most trace metals to the ocean. Based on these values, Hg has the highest relative atmospheric input at 85% of the total flux. The average values listed in Table 2.8 are compiled from numerous recent global Hg cycle estimates and other sources [64, 67–71] and these all vary in some degree in the relative magnitude of the sources, although they all agree that the atmospheric signal is 5–10 times the riverine input. The input from riverine sources is higher than the hydrothermal input, but this flux is not well-known.

For Pb, the atmosphere is also the dominant source, with the estimate being 63% of the total (Table 2.8; Fig. 2.18). Again, the global cycle of Pb has been substantially perturbed by human-derived emissions of Pb to the atmosphere and this is likely the reason for the importance of the atmosphere as a source [36]. In addition, Pb is highly particle reactive and is therefore removed during transport through the estuary. Cadmium is another metal with a relatively high atmospheric signal, and hydrothermal sources are also relatively important for Cd. Additionally, hydrothermal sources are somewhat important for Co, Cu, and Zn. The input for Mn from hydrothermal sources, accounting for about 40% of the total input, is almost as much as is entering the ocean from riverine sources. Atmospheric inputs are not important for Mn, or any of the other crustal metals, especially Al, Fe, and Co. For most metal(loid)s, removal to deep ocean sediments is the major and dominant sink [47]. For Hg, which can exist in surface waters as a dissolved gas (elemental Hg), loss to the atmosphere via gas exchange is the dominant loss term [64]. This is not the case for any other trace metal(loid) except perhaps Se.

Overall, for many metals, the major fluxes at the Earth's surface, such as the terrestrial fluxes to the ocean, and the deposition from the atmosphere to land, are relatively well-known. Additionally, the anthropogenic inputs to the atmosphere are relatively well constrained for most metals [34, 37, 64]. Thus, it is possible to construct global mass budgets for these elements. Such budgets have been done to a greater degree for the important toxic metals, such as Hg and Pb, and to a lesser extent for the other trace metals. In the following section, the global cycling of a few important trace metals is discussed to illustrate the differences in the global cycling of different trace metal(loid)s. The global budgets for five metals (Cd, Zn, Cu, Pb, and Hg) will therefore be discussed. In constructing the budgets for the various metals, the concentration of the metals in the major reservoirs was also estimated as this can provide some important insights into the rate at which various reservoirs are changing in concentration as a result of an exacerbation of inputs due to human activity.

2.3.1 The global cycles of cadmium, copper, and zinc

The inputs of Cd, Cu, and Zn to the atmosphere have been substantially increased by human activity, so much that the natural sources are now a minor fraction of the total inputs to the atmosphere (14% for Cd, 35% for Cu, 25% for Zn) [34, 37, 72]. However, the net impact of these inputs on the metal concentration in the biosphere is not equivalent, and it depends on the relative magnitude of the inventory of the metals in each reservoir relative to current inputs. The overall global budget for Cd, Cu, and Zn is shown in Fig. 2.19. The amount of each metal in the atmosphere is small, and given the inputs, the average residence time is estimated at less than a month for Cd to two months for Cu and Zn. These are reasonable values given that these metals are essentially particulate-bound in the atmosphere and thus are generally removed on a relatively short timespan due to dry deposition and scavenging of particulate matter by wet deposition. As noted in Section 2.1, the average atmospheric residence time of water is 10 days. The longer residence time of metals represents their lack of complete particle scavenging during a precipitation event. The relatively shorter residence time for Cd could be due to higher amount of emissions from anthropogenic activities or it may reflect differences in the location and particle size of the emitted fraction. Overall, however, the average residence time of these metals in the atmosphere is consistent with the effectiveness of their removal by wet and dry deposition processes, as discussed further in Chapter 5.

Interestingly, the flux of the metals in rivers is larger than the terrestrial atmospheric inputs, suggesting that the metals are not strongly removed in estuaries (this is somewhat true) but also that all have substantial inputs to rivers besides atmospheric deposition (terrestrial and anthropogenic inputs). This discrepancy is greatest for Cu (river flow ~4 times greater than deposition) which has higher relative concentrations in soils and terrestrial material compared to Cd (Table 2.2). The estimated reservoirs of Cd, Cu, and Zn in the upper 10 cm of the soil layer are very large compared to the estimated annual atmospheric input. Therefore, the anthropogenic disturbance, while large in terms of inputs to the atmosphere, does not have a substantial impact on soil metal concentrations, even when the entire period of the heightened industrial revolution (last 150 years) is considered. Thus, the impact of enhanced deposition due to human-derived inputs is likely small. The same conclusion is essentially valid for the ocean as the net inputs are relatively small compared to the ocean inventory, representing an addition of <1% per year. Of course, there may be more of an impact in the surface layers where the metals are

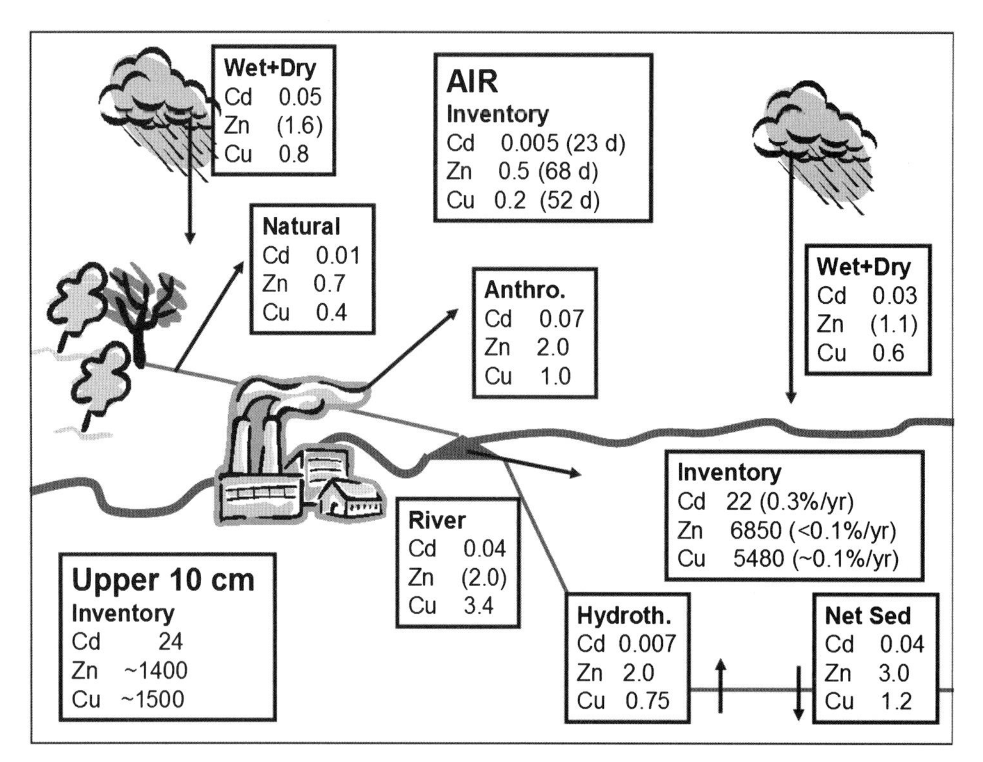

Fig. 2.19 The global cycling of copper, zinc and cadmium showing the magnitude of the fluxes for each principal exchange between the main reservoirs, as well as the magnitude of the reservoir inventory. For the atmosphere, the residence time of each metal is estimated (in brackets) while for the ocean, the rate of increase in concentration is estimated (in brackets). Fluxes for Zn given in brackets are uncertain and are constrained by mass balance. Fluxes are in Gmol yr^{-1} while reservoir inventories are given in Gmol. Information obtained from [47, 50, 73].

primarily being added. All the metals are considered to be nutrients and are therefore actively taken up in biological material in surface waters and transported to depth via the sinking of particulate material, and thus are relatively rapidly transported from surface waters to the deeper ocean.

The magnitude of the net sediment burial in ocean sediments for all the elements is relatively uncertain because the values in the literature vary greatly for different ocean sediments. The estimates of sedimentation rates by different authors also vary by more than an order of magnitude for each metal [47]. In contrasting the balances, it is also assumed, based on the predicted fate of these metals after addition to deep waters via hydrothermal activity [57], that 90% of the metal is removed to the sediment. To constrain the burial estimates further, the flux associated with sinking of biologically-related material from the surface ocean is estimated based on the burial of organic carbon in sediments (~1% of surface C) [50, 73] and a value for the metal : C ratio in biogenic material [29]. Using this approach, a burial flux of 0.04 Gmol yr^{-1} is estimated for Cd, a value of 3.0 Gmol yr^{-1} for Zn, and a value of 1.2 Gmol yr^{-1} for Cu. Thus, only about 50% of the Cd currently being added to the ocean is being removed by net burial in deep sediments, but as noted, the net input to the ocean is not changing the ocean concentration at a substantial rate either (~0.3% increase per year). As most of the Cd being added to the atmosphere is from anthropogenic sources, this lack of balance and steady state is not surprising. Clearly, a more detailed evaluation is required to determine the rate of increase in the surface water concentrations where most of the Cd is being added.

For Zn, there is also net input to the ocean but the actual extent of this input is difficult to ascertain, specifically given the uncertainty in the atmospheric flux estimate. The values used in Fig. 2.19 are not the same as those contained in Table 2.8, but are scaled to fit with the estimates of anthropogenic Zn inputs from Table 2.4, used in constructing the atmospheric inputs in Fig. 2.19. Clearly, it is not reasonable for the atmospheric deposition flux to be substantially greater than the total inputs to the atmosphere, given the fact that Zn is mostly particulate-associated in the atmosphere. Even given these budgeting constraints, there is a net input of Zn to the ocean, but the rate of change in concentration is again low. For Zn, the burial flux is about 20% of the current inputs. Overall, the fluxes and relative magnitude of the changes are similar for Cu and Zn. Hydrothermal inputs are of minor importance for Cu and the burial in deep ocean sediments accounts for about 30% of current inputs to the surface ocean. For all of the metals, there is a similar distribution between atmospheric deposition to the land compared to the ocean, with more relative input to the terrestrial environment. This contrasts to a degree their relative surface area (~75% of the Earth's surface being covered by water), but reflects the fact that essentially all the atmospheric inputs (both natural and anthropogenic) to the atmosphere are from the terrestrial environment for these metals.

The mass balance estimations highlight the fact that there is still uncertainty in many of these flux estimates. The reason for this uncertainty is primarily the lack of data for many aspects of the cycle. For example, while there is data on the concentration of metals in various sediment types, it is difficult to scale from these values to the long-term burial flux of the metals in deep sediments. In contrast, the wet deposition flux of many of the metals is relatively well constrained as there have been measurements made in both impacted and remote locations. However, these data span many decades in some cases and it is known that the global input of metals to the atmosphere is continually changing as the sources, location, and magnitude of the releases from human activity change. There is a critical need to carefully examine many of these biogeochemical cycles that have not been actively studied in the recent decade.

The previous comment does not suggest that the biogeochemistry of these metals has not been studied, either in ocean or freshwater ecosystems. On the contrary, the metals that are known to have a biological role are actively studied, especially in open ocean environments where there is the possibility that their low concentration may limit biological productivity, and as a result, influence carbon sequestration and removal from the biosphere and the associated global warming. The role of Fe as a potential limiting nutrient is relatively well-known, but it is possible that Zn could also be a limiting nutrient in some instances [29]. Given the global budget outlined above, inadvertent human releases to the atmosphere have not substantially altered this situation compared to the pre-industrial ocean. Thus, phytoplankton have evolved to deal with the low metal levels, such as through the acquisition of forms of carbonic anhydrase where there is a substitution for Zn in this enzyme by other metals (Cd or Co) [29]. It is interesting that ocean phytoplankton are more easily limited by lack of Fe inputs, which reflects the fact that Fe inputs and ocean concentrations may have been higher in the geological past due to higher winds and erosion, and higher inputs due to anoxia that persisted in the ancient ocean [74].

2.3.2 The global cycle of mercury and lead

Of all the potentially toxic and bioaccumulative metals, it is probable that the global cycle of Hg and Pb have been affected most by human activities. As noted in Table 2.4, the natural input of Pb to the atmosphere is about 3% of the current inputs, although the relative magnitude of these inputs is changing, especially as more countries phase out the use of Pb in gasoline. Again, this raises the issue of the timescale over which values are compared as the inputs to various parts of the globe have changed in recent decades, as shown, for example, in Fig. 1.6. While the data in the figure show a dramatic recent decrease in Pb concentration for the Arctic region, this may not be the case globally.

The global budget for Pb is shown in Fig. 2.19 [31, 37, 47, 83, 84–86]. The average residence time in the atmosphere of 14 days is less than that estimated for the other transition metals mentioned, possibly reflecting the differences in source. Much of the Pb added to the atmosphere, especially historically, was from automobile exhaust, and would have been removed by dry deposition faster than that added higher to the atmosphere from the stacks of combustion and other sources. Thus, the low residence time estimate appears reasonable. Alternatively, a low estimated residence time could indicate an averaging error in one of the flux estimates. In contrast to the metals discussed in Section 2.2.1, Pb inputs to the ocean are dominated by atmospheric inputs, with river inputs being about 40% of the total input. In addition, it is clear that the inventory of Pb in the ocean has been substantially changed by anthropogenic inputs. The estimated average rate of increase in the ocean inventory is 14% yr^{-1}, which is a large change. It is relatively obvious that this change is not the same for all oceans (Fig. 2.18). The value in the figure reflects the averaging involved in extrapolating from data collected in different locations over time. There is now relatively clear evidence that there has been an overall decrease in Pb concentration since the early 1980s for the North Atlantic Ocean (Fig. 2.21), which can be mostly attributed to the decrease in deposition of Pb as a result of the phasing out of Pb in gasoline in Europe and North America (Fig. 2.21) [75, 85, 86]. However, while input from Pb use in gasoline has declined, anthropogenic sources of Pb to the atmosphere still continue due to inputs from fossil fuel burning and metal refining. Such regional differences in response of the ocean to changing anthropogenic inputs will be discussed further in Chapter 5.

Hydrothermal inputs are insignificant for Pb and the burial in deep ocean sediments removes a small fraction of the estimated current input (<10%). The profiles for Pb in the ocean (Figs. 2.18 and 2.21) show an enrichment in the surface that is reflective of the importance of atmospheric input. Additionally, the rapid removal of the element from ocean waters via scavenging to particles and sedimentation to the deeper regions of the ocean depletes the deep water concentrations. The estimated overall residence time based on the inputs is not realistically valid given their transient nature. A more reasonable estimate can be made based on the removal of Pb, which would give an estimated residence time of about 100 yrs. This value is consistent with the known chemistry of Pb and the residence time of particulate matter and particulate-reactive elements.

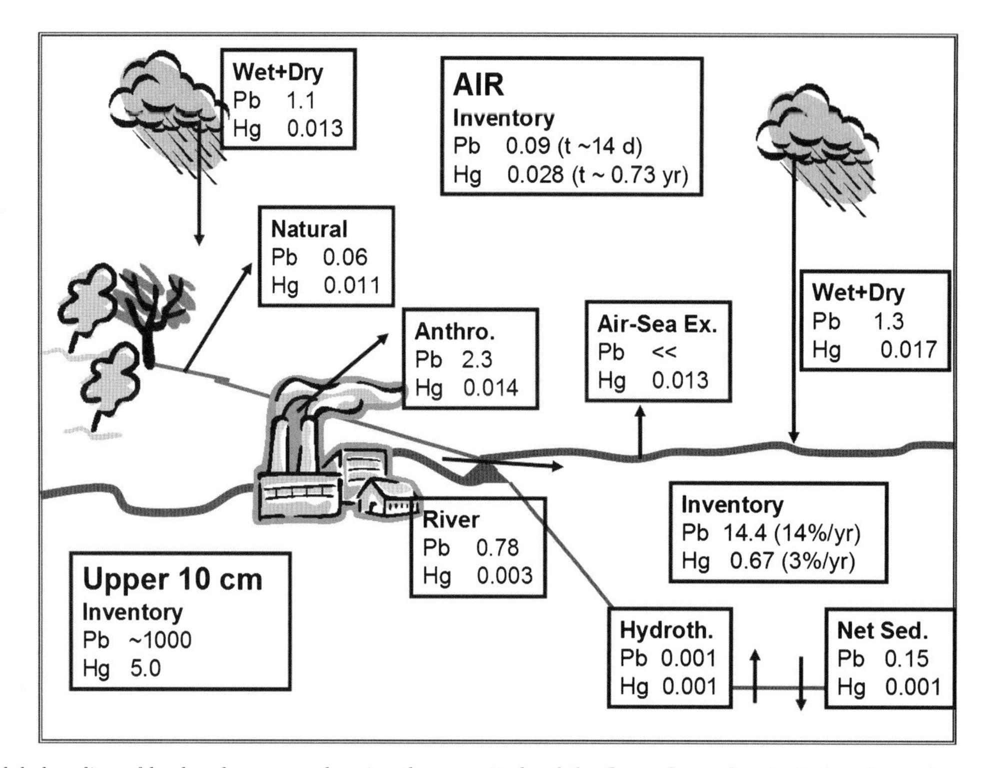

Fig. 2.20 The global cycling of lead and mercury showing the magnitude of the fluxes for each principal exchange between the main reservoirs, as well as the magnitude of the reservoir inventory. For the atmosphere, the residence time of each metal is estimated (in brackets) while for the ocean, the rate of increase in concentration is estimated (in brackets). Fluxes are in Gmol yr^{-1} while reservoir inventories are given in Gmol. Information obtained from [47, 50].

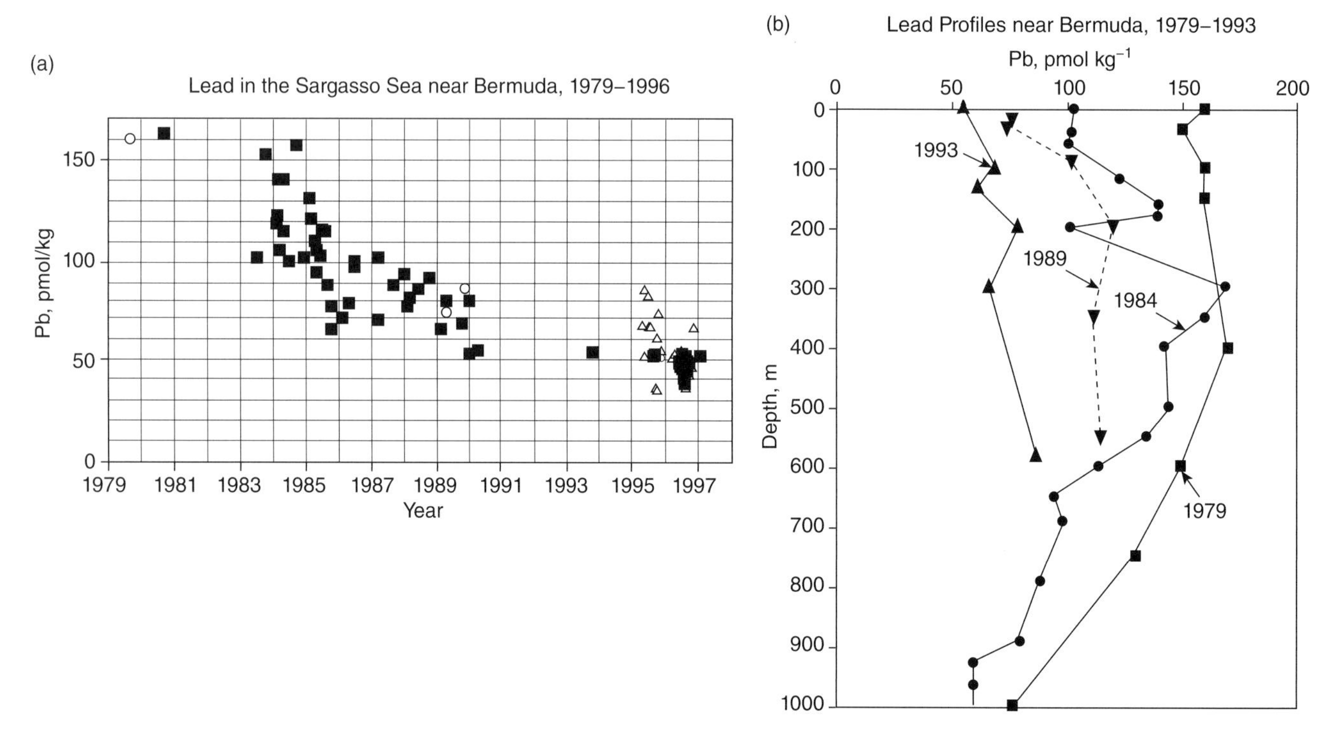

Fig. 2.21 Lead concentration in ocean waters in the vicinity of Bermuda. Figures compiled from data from [76, 82–86]. Reprinted from Wu and Boyle (1997) *Geochimica et Cosmochimica Acta* **61**(15): 3279–83 [75] with permission of Elsevier.

The global cycle of Hg (Fig. 2.20) has received much attention lately because of the heightened concern about the health risks of consuming methylmercury (CH_3Hg) laden fish and the likely impact that anthropogenic emissions have had on the ocean Hg burden. A number of recent papers have discussed the global cycle [8, 63, 63, 67, 71, 76] and while there is relative agreement in the magnitude of the estimates, there are also substantial differences in some aspects which reflect the complexity of the global Hg cycle and Hg biogeochemistry. For example, while most metals exist in the atmosphere only attached to particles, Hg can also exist in the gaseous phase, either as elemental Hg (Hg^0) or as ionic Hg species (e.g., $HgCl_2$, $HgBr_2$; collectively termed reactive gaseous Hg (RGHg)) [64, 66]. Thus, dry deposition of both gaseous and particulate species can occur, and all forms of Hg can also be scavenged by wet deposition to differing degrees. Additionally, Hg can exist as Hg^0 in ocean and freshwaters and therefore can be lost to the atmosphere via gas exchange. These complexities are not effectively captured in Fig. 2.21 and are discussed in detail later in the book (Chapter 6).

The global budget shows the relative importance of anthropogenic sources to the atmosphere. The residence time of Hg in the atmosphere is substantially longer than that of the other metals. This reflects the complexity of its atmospheric chemistry and the fact that most (>90%) of the Hg in the atmosphere is in the gaseous phase as Hg^0, which is removed from the atmosphere relatively inefficiently compared to that of particulate Hg or RGHg. Additionally, there is the potential for recycling of Hg between the surface reservoirs as ionic Hg deposited to surface waters can be reduced and then re-emitted to the atmosphere as Hg^0. This flux is a very important part of the global Hg cycle as it decreases the relative impact of anthropogenic emissions on ocean Hg concentration [64, 77, 78].

Atmospheric Hg^0 can be oxidized and deposited rapidly to surfaces and such processes appear to be enhanced in regions of ozone depletion due to photochemistry. The two processes are likely related [79, 80]. The atmospheric chemistry of Hg is discussed in much more detail later in the book. Locations where atmospheric oxidation is enhanced and deposition is increased include Polar Regions during polar sunrise, and over the ocean in the marine boundary layer [79–82].

Finally, recent modeling of the ocean cycling suggests that for Hg concentrations are not at steady state [77] (Fig. 2.22). This results from the relatively low inputs, and the dynamic nature of water movement and biogeochemical cycles in the upper ocean. Overall, the rate of change in ocean Hg concentration is lagging that of the atmosphere. These authors predict further that different ocean basins are changing in concentration at different rates due to the non-homogeneous distribution of anthropogenic Hg emission sources globally. For example, while North America and Europe have decreased emissions in the last 10 years, Asian emissions are increasing and now account for a large fraction of the total anthropogenic emissions [66]. This relative rate of change mirrors to some degree that of Pb whose concentration has decreased in the last decades (Fig. 2.21). There has probably been a similar decrease in Hg concentration, and the model

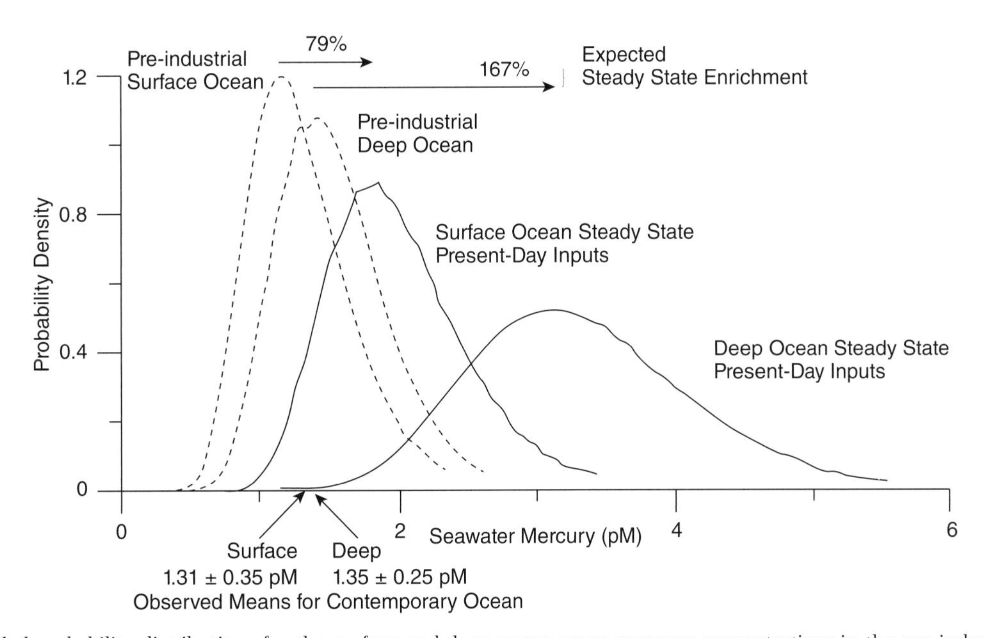

Fig. 2.22 Modeled probability distributions for the surface and deep water ocean mercury concentrations in the pre-industrial and current oceans. Figure reprinted from Sunderland and Mason (2007), *Global Biogeochemical Cycles* **21**, Article # GB4022 [76], with permission of the (AGU) American Geophysical Union (2007).

predicts such a decrease [71, 77], but the data to support this notion is lacking. These ideas are discussed further in Chapter 6.

Overall, the mass budgets, while they may not be able to provide details of the cycling of metals at the Earth's surface, are still insightful as they provide an indication of the major processes involved. These budgets highlight areas where more research is needed in the global cycling of metals and also the extent that human activities have impacted these cycles. The ideas and notions put forward in this chapter will be expanded and discussed in detail in the remainder of the book.

2.4 Chapter summary

1. Water fluxes within the surface biosphere play an important role in the transport of metal(loid)s between reservoirs. The residence time of metal(loid)s in a particular system is often different from that of water. For lakes with relatively large watersheds, fluvial inputs are likely more important than atmospheric inputs.
2. Human activities have exacerbated the input of many metal(loid)s to the biosphere. Many metal(loid)s are strongly retained in terrestrial soils and removed within the estuarine zone and therefore are not transported to the open ocean.
3. The partition coefficient is a parameter often used to describe the distribution of metal(loid)s in water between the filtered fraction and particles. The presence of colloids in the filtered fraction influences the measured degree of partitioning.
4. There is a dynamic steady state between the truly dissolved, colloidal and particulate fractions and these dynamics influence the fate and transport of many metal(loid)s in natural waters. The rate of adsorption or desorption from particles and colloids is, however, not instantaneous and therefore these kinetics can be important in some environments.
5. Acid mine drainage is one example of the impact of mining on the environment. Often metals can be transported large distances from their sources. The history of inputs can be inferred from the record of metal(loid) content in aquatic sediments.
6. While crustal sources (dust and sea salt) are vectors for trace element input to the atmosphere, they are also added by anthropogenic activities. Most metal(loid)s are associated with the particulate phase in the atmosphere. Mercury is a notable exception as it is mostly in the gas phase. The extent of transport of an element through the atmosphere depends on its volatility and the mechanism of input.
7. Metal(loid)s added to surface soils by atmospheric deposition are slowly transported through the soil layers to groundwater and to surface waters. Metals can also be scavenged from the atmosphere by vegetation.
8. For many metal(loid)s their degree of transport is associated with their tendency to bind to suspended particulate matter and to be complexed by inorganic and organic ligands, including NOM. Their solubility in natural waters is often controlled by the degree of complexation.
9. The distributions of elements in the ocean reflect their relative reactivity and their residence time. Elements with a residence time substantially greater than that of water tend to have a uniform distribution across ocean basins. Elements accumulated into phytoplankton (e.g., Fe, Zn, and Cu) can be depleted in surface waters, and higher in deep waters due to particle remineralization.
10. Metals whose concentrations in the biosphere have been strongly enriched due to anthropogenic inputs (e.g., Hg and Pb) often show large differences in concentration between surface and deep waters and between the North Atlantic and the North Pacific Ocean.
11. Hydrothermal vents release many metal(loid)s at relatively high concentrations into ocean waters. However, the relative importance of this source compared to atmospheric and coastal inputs is difficult to assess. It appears to be an important source for metals such as Mn.
12. For Pb, Hg, and Cd, inputs to the ocean from the atmosphere are similar or greater than the inputs from the coastal zone. Box models can be developed to examine the importance of the various inputs and can also be used to assess the extent of changes over time due to anthropogenic inputs.

References

1. Drever, J. (1997) *The Geochemistry of Natural Waters*. Prentice Hall, Upper Saddle River.
2. Langmuir, D. (1997) *Aqueous Environmental Geochemistry*. Prentice Hall, Upper Saddle River.
3. Berner, E.K. and Berner, R.A. (1996) *Global Environment: Water, Air and Geochemical Cycles*. Prentice Hall, Upper Saddle River, NJ.
4. Morel, F.M.M. and Hering, J.G. (1993) *Principles and Applications of Aquatic Chemistry*. John Wiley & Sons, Inc., New York.
5. Budyko, M.I. (1958) *The Heat Balance of the Earth's Surface*, Translated by Nina A. Stepanova, US Department of Commerce, Washington, D.C., 259 p.
6. Harris, R.C., Rudd, J.W.M., Amyot, M., et al. (2007) Whole-ecosystem study shows rapid fish-mercury response to changes in mercury deposition. *Proceedings of the National Academy of Sciences of the United States of America*, **104**, 16586–16591.
7. Yeats, P. and Bewers, J.M. (1982) Discharge of metals from the St Lawrence River. *Canadian Journal of Earth Sciences*, **19**, 982–992.
8. Lawson, N.M. and Mason, R. (2001) Concentration of mercury, methylmercury, cadmium, lead, arsenic and selenium in the rain and stream water of two contrasting watersheds in Western Maryland. *Water Research*, **35**(17), 4039–4052.
9. Scudlark, J., Rice, K.C., Conko, K.M., Bricker, O. and Church, T.M. (2005) Transmission of atmospherically derived trace ele-

ments through an undeveloped, forested Maryland watershed. *Water, Air, and Soil Pollution*, **163**, 53–79.

10. Stumm, W. and Morgan, J.J. (1996) *Aquatic Chemistry*. John Wiley & Sons, Inc., New York.
11. Aiken, G.R., Hsu-Kim, H. and Ryan, J.N. (2011) Influence of dissolved organic matter on the environmental fate of metals, nanoparticles, and colloids. *Environmental Science & Technology*, **45**, 3196–3201.
12. Honeyman, B.D. and Santschi, P.H. (1988) Metals in aquatic systems. *Environmental Science & Technology*, **22**, 862–871.
13. Stordal, M.C., Santschi, P.H. and Gill, G.A. (1996) Colloidal pumping: Evidence for the coagulation process using natural colloids tagged with Hg-203. *Environmental Science & Technology*, **30**, 3335–3340.
14. Wen, L.S., Santschi, P.H. and Tang, D. (1997) Interactions between radioactively labeled colloids and natural particles: Evidence for colloidal pumping. *Geochimica et Cosmochimica Acta*, **61**(14), 2867–2878.
15. Clegg, S. and Whitfield, M. (1990) A generalized model for the scavenging of trace metals in the open ocean–I. Particle cycling. *Deep-Sea Research*, **37**, 809–832.
16. Valiela, I. (1995) *Marine Ecological Processes*. Springer-Verlag, New York.
17. Zhang, Y.Y., Zhang, E.R. and Zhang, J. (2008) Modeling on adsorption-desorption of trace metals to suspended particle matter in the Changjiang Estuary. *Environmental Geology*, **53**(8), 1751–1766.
18. Allen, H.E. (1995) *Metal Contaminated Aquatic Sediments*. Ann Arbor Press, Chelsea, Michigan.
19. Tang, D.G., Warnken, K.W. and Santschi, P.H. (2002) Distribution and partitioning of trace metals (Cd, Cu, Ni, Pb, Zn) in Galveston Bay waters. *Marine Chemistry*, **78**(1), 29–45.
20. Turner, A., Millward, G.E. and Le, S.M. (2004) Significance of oxides and particulate organic matter in controlling trace metal partitioning in a contaminated estuary. *Marine Chemistry*, **88**(3–4), 179–192.
21. Jenne, E.A. (1995) Metal adsorption onto and desorption from sediments. I. Rates. In: Allen, H.E. (ed.) *Metal Contaminated Aquatic Sediments*. Ann Arbor Press, Chelsea, Michigan, pp. 81–110.
22. Blowes, D., Ptacek, C.J., Jambor, J.L. and Weisener, C.G. (2005) The geochemistry of acid mine drainage. In: Lollar, B. (ed.) *Environmental Geochemistry*. Volume 5, Treatise on Geochemistry, Holland, H.D. and Turekian, K.K. (Exec. eds) Elsevier, Amsterdam, pp. 149–203.
23. Fuge, R. (2005) Anthropogenic sources. In: Seleinus, O. (ed.) *Essentials of Medical Geology*. Elsevier, Amsterdam, pp. 43–59.
24. Garrett, R. (2005) Natural distribution and abundance of elements. In: Seleinus, O. (ed.) *Essentials of Medical Geology*. Elsevier, Amsterdam, pp. 17–41.
25. Li, Y.-H. (2011) Partitioning of elements between the solid and liquid pahses in aqueous environments. *Aquatic Geology*, **17**, 697–725.
26. Spliethoff, H. and Hemond, H.F. (1996) History of toxic metal discharge to surface waters of the Aberjona Watershed. *Environmental Science & Technology*, **30**, 121–128.
27. Audry, S.J., Blanc, G. and Jouanneau, J.-M. (2004) Fifty-year sedimentary record of heavy metal pollution (Cd, Zn, Cu, Pb) in the Lot River reservoirs (France). *Environmental Pollution*, **132**, 413–426.
28. Bruland, K. and Lohan, M.C. (2004) Controls of trace metals in seawater. In: Elderfield, H. (ed.) *The Oceans and Marine Geochemistry*. Volume 6 of Treatise on Geochemistry Holland, H.D. and Turekian, K.K. (Exec. eds) Elsevier, Amsterdam, pp. 23–47.
29. Morel, F.M.M., Milligan, A.J. and Saito, M.A. (2004) Marine bioinorganic chemistry: The role of trace metals in the oceanic cycles of major nutrients. In: Elderfield, H. (ed.) *The Oceans and Marine Geochemistry*. in Holland, H.D. and Turekian, K.K. (Exec. eds), Treatise on Geochemistry, Elsevier, Amsterdam, pp. 113–143.
30. Boutron, C., Rosman, K., Barbante, C., Bolshov, M., Adams, F., Hong, S.M. and Ferrari, C. (2004) Anthropogenic lead in polar snow and ice archives. *Comptes Rendus Geoscience*, **336**(10), 847–867.
31. Boutron, C.F., Candelone, J. and Hong, S.M. (1994) Past and recent changes in the large-scale tropospheric cycles of lead and other heavy-metals as documented in Antarctic and Greenland snow and ice – A review. *Geochimica et Cosmochimica Acta*, **58**(15), 3217–3225.
32. Engstrom, D.R. and Swain, E.B. (1997) Recent declines in atmospheric mercury deposition in the upper Midwest. *Environmental Science & Technology*, **31**(4), 960–967.
33. Callender, E. (2005) Heavy metals in the environment – historical trends. In: Lollar, B. (ed.) *Environmental Geochemistry*. Volume 5 Treatise on Geochemistry Holland, H.D. and Turekian, K.K. (Exec. eds) Elsevier, Amsterdam, pp. 67–105.
34. Berg, T. and Steinnes, E. (2005) Atmospheric transport of metals. In: Sigel, A., Sigel, H. and Sigel, R.K.O. (eds) *Biogeochemistry, Availability and Transport of Metals in the Environment*. vol. 44, Metal Ions in Biological Systems, Taylor & Francis, Boca Raton, pp. 1–19.
35. Duce, R., Liss, P.S., Merrill, J.T., et al. (1991) The atmospheric input of trace species to the World Ocean. *Global Biogeochemical Cycles*, **5**, 193–259.
36. Erel, Y. and Patterson, C.C. (1994) Leakage of industrial lead into the hydrocycle. *Geochimica et Cosmochimica Acta*, **58**(15), 3289–3296.
37. Nriagu, J. and Pacyna, J.M. (1988) Quantitative assessment of worldwide contamination of air, water and soils with trace metals. *Nature*, **333**, 134–139.
38. Van Der Werf, G.R., Randerson, J.T., Giglio, L., Collatz, G.J., Kasibhatla, P.S. and Arellano, A.F. (2006) Interannual variability in global biomass burning emissions from 1997 to 2004. *Atmospheric Chemistry and Physics*, **6**, 3423–3441.
39. Andreae, M.O. and Merlet, P. (2001) Emissions of trace gases and aerosols from biomass burning. *Global Biogeochemical Cycles*, **4**, 955–966.
40. Scudlark, J. and Church, T.M. (1997) Atmospheric deposition of trace metals to the mid-Atlantic Bight. In: Baker, J. (ed.) *Atmospheric Deposition of Contaminants to the Great Lakes and Coastal Waters*. SETAC Press, Pensacola, pp. 195–208.
41. St Louis, V., Rudd, J.W.M., Kelly, C.A., Hall, B.D., Rolfhus, K.R., Scott, K.J., Lindberg, J.E. and Dong, W. (2001) Importance of the forest canopy to fluxes of methyl mercury and total mercury to boreal ecosystems. *Environmental Science & Technology*, **35**, 3089–3098.

42. Itoh, Y., Miura, S. and Yoshinaga, S. (2006) Atmospheric lead and cadmium deposition within forests in the Kanto district, Japan. *Journal of Forest Research*, **11**(2), 137–142.
43. Hou, H., Takamatsu, T., Koshikawa, M.K. and Hosomi, M. (2005) Trace metals in bulk precipitation and throughfall in a suburban area of Japan. *Atmospheric Environment*, **39**, 3583–3595.
44. Avila, A. and Rodrigo, A. (2004) Trace metal fluxes in bulk deposition, throughfall and stemflow at two evergreen oak stands in NE Spain subject to different exposure to the industrial environment. *Atmospheric Environment*, **38**, 171–180.
45. Milliman, J.M. and Meade, R.H. (1983) World-wide delivery of river sediment to the oceans. *Journal of Geology*, **91**, 1– 21.
46. Degens, E. and Ittekot, V. (1989) Monitoring of carbon in world rivers. *Environment*, **26**, 29–33.
47. Chester, R. (2003) *Marine Geochemistry*. Blackwell Science, Malden.
48. Elbaz-Poulichet, F., Garnier, J.M., Guan, D.M., Martin, J.M. and Thomas, A.J. (1996) The conservative behavior of trace metals (Cd, Cu, Ni and Pb) and arsenic in the surface plume of stratified estuaries: Example of the Rhone River (France). *Estuarine, Coastal and Shelf Science*, **42**, 289–310.
49. Vorosmarty, C.J., Meybeck, M., Fekete, B., Sharma, K., Green, P. and Syvitski, J.P.M. (2003) Anthropogenic sediment retention: Major global impact from registered river impoundments. *Global and Planetary Change*, **39**(1–2), 169–190.
50. Broecker, W. and Peng, T.-H. (1982) *Tracers in the Sea*. Columbia University Press, Palisades, NY.
51. Martin, J.H., Gordon, R.M., Fitzwater, S. and Broenkow, W.W. (1989) VERTEX: Phytoplankton/iron studies in the Gulf of Alaska. *Deep-Sea Research*, **36**, 649–680.
52. Martin, J.H., Fitzwater, S.E., Gordon, R.M., Hunter, C.N. and Tanner, S.J. (1993) Iron, primary production and carbon–nitrogen flux studies during the JGOFS North Atlantic Bloom Experiment. *Deep-Sea Research*, **40**, 115–134.
53. Schaule, B.K. and Patterson, C.C. (1981) Lead concentrations in the northeast Pacific: Evidence for global anthropogenic perturbations. *Earth and Planetary Science Letters*, **54**, 97–116.
54. Han, Q., Moore, J.K., Zender, C., Measures, C. and Hydes, D. (2008) Constraining oceanic dust deposition using surface ocean dissolved Al. *Global Biogeochemical Cycles*, **22**. Article # GB2003.
55. Jannasch, H.W. and Mottl, M.J. (1985) Geomicrobiology of deep-sea hydrothermal vents. *Science*, **229**, 717–725.
56. Speiss, F.N., Ken, C.M., Atwater, T., et al. (1980) East Pacific Rise: Hot springs and geophysical experiments. *Science*, **207**, 1421–1433.
57. German, C.R. and von Damm, K.L. (2004) Hydrothermal processes. In: Elderfield, H. (ed.) *The Oceans and Marine Geochemistry*. Volume 6 of Treatise on Geochemistry, Holland, H.D. and Turekian, K.K. (Exec. eds) Elsevier, Amsterdam, pp. 181–222.
58. Herzig, P.M. and Hannington, M.D. (2000) Input from the deep: Hot vents and cold seeps. In: Schultz, H. and Zabel, M. (eds) *Marine Geochemistry*. Springer, Heidelberg, pp. 397–416.
59. Barbante, C., Schwikowski, M., Doring, T., Gaggeler, H.W., Schotterer, U., Tobler, L., Van De Velde, K., Ferrari, C., Cozzi, G., Turetta, A., Rosman, K., Bolshov, M., Capodaglio, G., Cescon, P. and Boutron, C. (2004) Historical record of European emissions of heavy metals to the atmosphere since the 1650s from Alpine snow/ice cores drilled near Monte Rosa. *Environmental Science & Technology*, **38**(15), 4085–4090.
60. Hong, S.M., Candelone, J.P., Soutif, M. and Boutron, C.F. (1996) A reconstruction of changes in copper production and copper emissions to the atmosphere during the past 7000 years. *Science of the Total Environment*, **188**(2–3), 183–193.
61. Mason, R.P., Fitzgerald, W.F. and Morel, F.M.M. (1994) The biogeochemical cycling of elemental mercury – Anthropogenic influences. *Geochimica et Cosmochimica Acta*, **58**(15), 3191–3198.
62. Mason, R.P. and Sheu, G.R. (2002) Role of the ocean in the global mercury cycle. *Global Biogeochemical Cycles*, **16**(4). Article # 1093.
63. Nriagu, J. (1978) Properties and the biogeochemical cycle of lead. In: Nriagu, J. (ed.) *The Biogeochemistry of Lead in the Environment*. Elsevier, Amsterdam, pp. 1–14.
64. Pirrone, N., Costa, P., Pacyna, J.M. and Ferrara, R. (2001) Mercury emissions to the atmosphere from natural and anthropogenic sources in the Mediterranean region. *Atmospheric Environment*, **35**(17), 2997–3006.
65. Sillman, S., Marsik, F.J., Al-Wali, K.I., Keeler, G.J. and Landis, M.S. (2007) Reactive mercury in the troposphere: Model formation and results for Florida, the northeastern United States and the Atlantic Ocean. *Journal of Geophysical Research*, **112**. Article # D23305.
66. Landis, M.S., Stevens, R.K., Schaedlich, F. and Prestbo, E.M. (2002) Development and characterization of an annular denuder for the measurement of divalent inorganic reactive gaseous mercury in ambient air. *Environmental Science and Technology*, **36**, 3000–3009.
67. Bergan, T., Gallardo, L. and Rodhe, H. (1999) Mercury in the global troposphere: A three-dimensional model study. *Atmospheric Environment*, **33**(10), 1575–1585.
68. Lamborg, C.H., Fitzgerald, W.F., Damman, A.W.H., Benoit, J.M., Balcom, P.H. and Engstrom, D.R. (2002) Modern and historic atmospheric mercury fluxes in both hemispheres: Global and regional mercury cycling implications. *Global Biogeochemical Cycles*, **16**(4), 51.1–51.11.
69. Lamborg, C.H., Fitzgerald, W.F., O'Donnell, J. and Torgersen, T. (2002) A non-steady-state compartmental model of global-scale mercury biogeochemistry with interhemispheric atmospheric gradients. *Geochimica et Cosmochimica Acta*, **66**, 1105–1118.
70. Pacyna, E.G., Pacyna, J.M., Steenhuisen, F. and Wilson, S. (2006) Global anthropogenic mercury emission inventory for 2000. *Atmospheric Environment*, **40**(22), 4048–4063.
71. Selin, N.E.D.J., Jacob, R.J., Park, R.M., Yantosca, S., Strode, L. and Jaegle, D. (2007) Chemical cycling and deposition of atmospheric mercury: Global constraints from observations. *Journal of Geophysical Research-Atmospheres*, **112**(D2), doi: 10.1029/2006JD007450.
72. Nriagu, J.O. (1989) A global assessment of natural sources of atmospheric trace metals. *Nature*, **338**, 47–49.
73. Karl, D. and Knauer, G.A. (1984) Vertical distribution, transport and exchange if carbon in the northeast Pacific Ocean: Evidence for multiple zones of biological activity. *Deep-Sea Research*, **31**, 221–243.

74. Saito, M., Sigman, D.M. and Morel, F.M.M. (2003) The bioinorganic chemistry of the ancient ocean: The co-evolution of cyanobacterial metal requirements and biogeochemical cycles at the Archean/Proterozoic boundary? *Inorganica Chimica Acta*, **356**, 308–318.
75. Wu, J.F. and Boyle, E.A. (1997) Lead in the western North Atlantic Ocean: Completed response to leaded gasoline phaseout. *Geochimica et Cosmochimica Acta*, **61**(15), 3279–3283.
76. Fitzgerald, W. and Lamborg, C.H. (2005) Geochemsitry of mercury in the environment. In: Lollar, B. (ed.) *Environmental Geochemistry*. Volume 5 of Treatise on Geochemistry, Holland, H.D. and Turekian, K.K. (Exec. eds), Elsevier, Amsterdam, pp. 107–147.
77. Sunderland, E.M. and Mason, R. (2007) Human impact on ocean mercury concentration. *Global Biogeochemical Cycles*, **21**. Article # GB4022.
78. Strode, S.A., Jaeglé, L., Selin, N.E. et al. (2007) Air-sea exchange in the global mercury cycle. *Global Biogeochemical Cycles*, **21**(1), doi: 10.1029/2006GB002766.
79. Laurier, F. and Mason, R. (2007) Mercury concentration and speciation in the coastal and open ocean boundary layer. *Journal of Geophysical Research- Atmospheres*, **112**(D6). Art. # D06302.
80. Lindberg, S.E. Brooks, S., Lin, C.J. et al. (2002) Dynamic oxidation of gaseous mercury in the Arctic troposphere at polar sunrise. *Environmental Science & Technology*, **36**(6), 1245–1256.
81. Hedgecock, I.M. and Pirrone, N. (2001) Mercury and photochemistry in the marine boundary layer-modelling studies suggest the in situ production of reactive gas phase mercury. *Atmospheric Environment*, **35**(17), 3055–3062.
82. Schroeder, W., Anlauf, K.G., Barrie, L.A., Lu, J.Y., Steffen, A., Schneeberger, D.R. and Berg, T. (1998) Arctic springtime depletion of mercury. *Nature*, **394**, 331–332.
83. Schaule, B. and Patterson, C.C. (1983) Perturbations of the natural lead depth profile in the Sargasso Sea by industrial lead. In: Wong, C., Boyle, E., Bruland, K.W., Burton, J.D. and Goldberg, E.D. (eds) *Trace Metals in Seawater*. Plenum Press, New York, pp. 487–503.
84. Flegal, A.R. and Patterson, C.C. (1983) Vertical concentration profiles of lead in the central Pacific at 158 N and 208 S. *Earth and Planetary Science Letters*, **64**, 19–32.
85. Vernon, A.J., Church, T.M., Patterson, C.C. and Flegal, A.R. (1994) Use of stable lead isotopes to characterize the sources of anthropogenic lead in North-Atlantic surface waters. *Geochimica et Cosmochimica Acta*, **58**(15), 3199–3206.
86. Boyle, E.A., Chapnik, S., Shen, G.T. and Bacon, M.P. (1986) Temporal variability of lead in the western North Atlantic. *Journal of Geophysical Research*, **91**, 8573–8593.

Problems

2.1. Two lakes, Lake A (area $10^4\,m^2$, depth 12 m) and Lake B (area $10^6\,m^2$, depth 50 m) are exposed to atmospheric wet deposition of $1\,m\,yr^{-1}$ of mercury (75 pM), arsenic (10 nM), and Zn (50 nM) to their surface and to the surrounding watershed (Watershed A, $5 \times 10^5\,m^2$; Watershed B, $5 \times 10^8\,m^2$). Assume dry deposition is not important. The river flow into Lake A is $1.4{\times}10^3\,m^3\,d^{-1}$ and that into Lake B is $1.4 \times 10^3\,m^6\,d^{-1}$, and the river concentrations are as follows: River A: 10 pM Hg, 7.5 nM As, and 25 nM Zn; River B: 15 pM Hg, 9 nM As, and 20 nM Zn. What is:

a. The residence time of water in the lakes?
b. The transmission factors for Hg, As and Zn for each watershed?
c. The total flux of each metal(loid) to each lake?
d. The percentage of the total flux that is directly to the surface in each lake for each metal(loid)?

2.2. In the rivers in Problem 2.1, the concentration of total suspended solids (TSS) is $5\,mg\,l^{-1}$ (River A) and $10\,mg\,l^{-1}$ (River B). If the dissolved concentrations are, respectively, for Hg, As, and Zn: 6 pM, 7 nM, and 22 nM for River A and 7.5 pM, 8 nM, and 19 nM for River B, what are the K_Ds for each metal(loid) in each river?

2.3. A pond containing $100\,m^3$ of water with a total carbonate concentration of 10^{-3} M and a pH of 8.0 is impacted by acid mine drainage (AMD) from a nearby mine site. The acid mine input is 100 l of water and the acidity is generated by the oxidation of 10 moles of FeS_2. Using the reactions in the chapter, estimate:

a. The acidity of the acid mine drainage prior to it impacting the pond, assuming no precipitation.
b. The change in pH and alkalinity of the pond immediately after the addition of the 100 l of AMD.
c. The pH and alkalinity of the pond after it has equilibrated with the atmosphere.
d. The metal concentrations in the AMD using the lower concentrations in Table 2.3. Would any of the metals, including Fe, precipitate in the AMD?
e. If any of the metals redissolve after equilibration of the AMD-impacted pond with the atmosphere.
f. How the fate of the metals would change the final pH of the system.

2.4. Consider the ratio of natural to anthropogenic inputs to the atmosphere, as detailed in Table 2.4. How would the total input and the fraction that is natural change if the input due to fossil fuel combustion (FFC) was doubled? Perform the calculation for Hg, Sb, Se, and Fe.

2.5. Using the information in Figs. 2.17 and 2.18, estimate the global budget for the preindustrial biosphere (i.e., anthropogenic emissions = 0). To perform this calculation, assume that 20% of the current river input occurred in the preindustrial area, and that the ratio of the atmospheric deposition to the land and the ocean was the same in the preindustrial era, and that total atmospheric deposition is equal to total inputs. Assume that the hydrothermal flux has not changed. To balance the overall budget, assume the ocean is in balance (i.e., input = outputs) in pre-industrial times. Focus on Cu, Cd, Zn, and Pb and:

a. Comment on the budgets and things that appear inconsistent between the different metals.

b. Estimate the factor by which the input to the ocean has changed between then and now.

c. Calculate the preindustrial ocean average concentration (recall, residence time = ocean inventory/total input), assuming a residence time for each metal in the preindustrial ocean of 500 years. Assume an ocean volume of 1.3×10^{21} l. Are the values reasonable? If not, give any suggestions to why this may not be so. Which metal has the largest enrichment?

2.6. In Fig. 2.19, the decrease in concentration of Pb in the surface ocean and in the upper 1000 m over the last 20 years is shown. The data in the table given (Table P.2.1) was extracted from the graph.

a. The decrease appears exponential. Does this mean that the changes in inputs, which are mostly atmospheric at this location, have also decreased exponentially? Explain your answer.

b. Calculate the rate of change in concentration over time. Predict what the concentration would be in 2010 and what it was in 1975, assuming the same rate of change.

Table P.2.1

Year	Pb conc. (pM)
1980	160
1985	110
1990	70
1995	60
1998	50

c. Assume that the top 1000 m of the ocean in this region mixes on a yearly timescale and that the input from the coastal zone to this region is equivalent to 4 μmol m^{-2} yr^{-1}. Further, assume that the coastal input has remained constant over the considered timeframe. Further, assume the net export of Pb at 1000 m is equivalent to 10% of the total inventory in this layer. Calculate the required atmospheric deposition in 1985 and 1995 needed to keep the system in steady state. How much did the atmospheric deposition change over that 10 year period? Is the change in atmospheric deposition of the same magnitude as the change in upper ocean concentration? If not, why not?

CHAPTER 3
Chemical thermodynamics and metal(loid) complexation in natural waters

3.1 Thermodynamic background for understanding trace metal(loid) complexation

This chapter provides an overview of factors affecting chemical speciation and reactivity of trace metal(loid)s in aquatic systems. Most students who have taken undergraduate chemistry classes have been exposed to the notion of free energy and the basic concepts of chemical equilibrium and kinetics, and the approaches to solving simple equilibrium calculations. This chapter therefore focuses primarily on systems in which trace metal(loid)s are present in solution and/or in the solid phase, while briefly discussing the theoretical chemical basis for thermodynamic or equilibrium relationships. The chapter will also discuss the impact of redox chemistry as the chemistry, reactivity and solubility of the important trace metals and metalloids is a function of their oxidation state.

Chemists use two approaches to understanding the interaction of metals and other chemicals in solution, which can be simply described as: (1) the equilibrium (*thermodynamic*) approach, which is aimed at estimating the final composition of the solution when all reactions have occurred and the system is in its lowest energy state; and (2) the *kinetic* approach, where the rate at which the composition of the solution is changing is followed. These two approaches are not mutually exclusive and are linked by relationships that will be discussed next. For example, envision a simple system where compound A is converted into compound B through some chemical process, and B can also be converted into A through a reverse reaction:

$$A \Leftrightarrow B \quad (3.1)$$

the equilibrium constant (K) for this reaction specifies the concentration ratio of A to B at equilibrium, which is a constant, and is given by the following expression:

$$K = [B]/[A] \quad (3.2)$$

where the brackets indicate the concentration of each species in solution, in molar units. In any situation, the equilibrium constant is given as the ratio of the concentration of the products, raised to the power of the stoichiometric coefficient for the reaction, over that of the reactants. Because of ionic strength effects in solution, this expression is valid for dilute solutions [1]. The kinetics of the situation are defined by the two rate constants for the forward (k_f) and reverse (k_b) reactions, and if they are reversible, first order reactions (which will be defined next), then the following relationship is valid:

$$K = k_f/k_b \quad (3.3)$$

To properly understand the equilibrium state of an aquatic system, it is necessary to characterize differences in energy between the different chemical states of the constituents within the system. The primary objective of chemical thermodynamics is to establish a criterion for determining the feasibility (*spontaneity*) of a particular chemical reaction or transformation. In an aquatic system, chemical thermodynamic calculations are useful for determining the overall state of the system and its likelihood to change. Note that such equilibrium calculations do not provide an understanding of the rates of change, but they do provide a good approximation of the composition of most natural waters. In most situations important to aquatic chemistry, the system of interest is not at *true equilibrium*, which requires the

Trace Metals in Aquatic Systems, First Edition. Robert P. Mason.

system be closed to outside inputs. In many instances, the system may not even be at *steady state* (i.e., a situation where the concentration in the bulk solution of the aquatic system remains constant over time even though there is continual input and output of constituents) [1–3]. However, if the rates of reaction are relatively rapid compared to the rate of input and output of the constituents then the system approaches steady state. Under such conditions, the steady state system becomes analogous to that of the true equilibrium state. Thus, the thermodynamic approach, which is the most widely used approach to understanding the fate, transport and reactivity of metal(loid)s in solution, gives an idea of whether a particular system is at steady state, and if not, allows an estimation of the direction of spontaneous change. Therefore it provides information on the energy available or required for a particular transformation to occur.

The chemical equilibrium approach is based on the *Laws of Thermodynamics* [4], which will be briefly discussed here. These laws are based on experience and experiment and are considered *fundamental* (primitive) postulates. One postulate is that there is an absolute temperature scale (*Zero Law*). The *First Law of Thermodynamics* is based on a series of experiments conducted by Joule, an English physicist (and brewer), born in Lancashire, England. Joule studied the nature of heat, and discovered its relationship to mechanical work and energy. Joule concluded that for a given amount of *work* (w) done on a system there was the same overall change in state regardless of how the work was done. Here the system is defined as the object of study, and it is separated from the surroundings by a boundary of some form. Such change, for an *adiabatic system* (i.e., a closed system where no *heat* (q) enters or leaves the system during the change), was independent of the path by which the change was achieved. From these considerations, the *energy* of the system (E) was defined. It is a *state function* (i.e., path independent) and the first law can be written as the following expressions:

$$dE = dq - dw \tag{3.4}$$

and

$$dE = dq - PdV \tag{3.5}$$

for a closed system, where w is the work done (work done by the system is negative, work done on the system by the surroundings is positive); q is the heat transferred to the system from the surroundings; P is the pressure; and V is the volume. The related thermodynamic quantity, *enthalpy* (H), is defined as:

$$H = E + PV \tag{3.6}$$

This is a useful concept and is a measure of the heat change that occurs during chemical transformations, and is derived from the first law of thermodynamics.

The *Second Law of Thermodynamics* can be stated in many ways and one definition is given here. It is impossible to construct a machine that, operating in a cycle, will take heat from a reservoir at constant temperature and convert it into work without accompanying changes in the reservoir or its surroundings. The concept of *entropy* (S) [4] needs to be defined to give the second law a mathematical formulation, and it can be thought of as "an index of exhaustion" of the reaction or a measure of the disorder of the system. Spontaneous changes proceed in the direction of increased entropy and approach minimum energy and a state of no further spontaneous change (equilibrium).

For a reversible process, the second law can be stated as follows:

$$dS_{sys} \geq dq/T \tag{3.7}$$

where dS_{sys} is the change in entropy of the system and T is the temperature on the Kelvin (absolute) scale. At equilibrium and fixed composition:

$$dq = TdS_{sys} \tag{3.8}$$

and, overall, for any possible process:

$$dS_{sys} \geq 0 \tag{3.9}$$

The internal entropy change for a closed system is zero at equilibrium and positive for a spontaneous process. Combining the first and second laws, the following expression is obtained:

$$dE = TdS_{sys} - PdV \tag{3.10}$$

J. Willard Gibbs, a physical chemist at Yale University working in the 1870s, defined the concept of the *free energy* of the system in his landmark paper *On the Equilibrium of Heterogeneous Substances*, a 323 page treatise, published in the *Transactions of the Connecticut Academy of Sciences* [5]. Gibbs was well-known for his abstract reasoning and apparently his paper was not clearly grasped by the journal reviewers as one member stated: "We knew Gibbs and took his contributions on faith". The importance of this work was not immediately appreciated but subsequently the concept of the *Gibbs Free Energy* became widely used.

The *Gibbs Free Energy*, G, of a system is given by the following expression:

$$G = H - TS \tag{3.11}$$

The free energy of the system can be defined as the amount of energy needed to recreate it from some arbitrary reference state.

At constant composition, it can be shown that [4]:

$$dG = -SdT + PdV \tag{3.12}$$

and

$dG < 0$ for an irreversible (spontaneous) change

$dG = 0$ for a reversible change.

Therefore, the parameter G provides an indication of whether a reaction may proceed or not. At constant T and P, for a finite change of state:

$$\Delta G - \Delta II - T\Delta S \tag{3.13}$$

As the state variables described above are defined in terms of changes in another particular variable, they provide information relative to some reference state that must be defined, that is, their values are relative and not absolute.

3.1.1 The relationship between free energy and the equilibrium constant

The overall free energy for a reaction is the sum of all the free energies of the chemical species involved in the reaction, and can be written as [1]:

$$G_{rxn} = \sum n_i \cdot \mu_i \tag{3.14}$$

where n is the number of moles of each species and μ is the *molar free energy* of that species. The molar free energy can be defined for an *ideal system* and consists of: (1) the energy of formation; and (2) the energy associated with obtaining the concentration of the species within the system. The energy associated with interactions between solutes within the medium is what leads to *non-ideality*, and this will be dealt with later in the chapter. For now, consider an ideal system. The ideal system assumption is that the energy is independent of other species in solution and therefore that there is no interaction between chemical species in solution other than that involved in the reaction of interest. This is clearly not the case in reality, especially for concentrated solutions, such as seawater, and the impact of these interactions will be discussed below. The molar free energy, μ, is further defined and consists of two parts consistent with the two contributions to the overall energy, and the equation can be written the following form:

$$\mu = \mu^o + RT \cdot \ln X \tag{3.15}$$

where μ^o is the *standard free energy of formation* for the species and RT.lnX is the free energy corresponding to the particular concentration of the species, defined in terms of X, the *mole fraction* ($X_i = n_i$/total n). Here R is the *universal gas constant* ($8.314\,J\,K^{-1}\,mol^{-1}$) and T is the temperature on the Kelvin temperature scale.

As noted above, the thermodynamic constants are defined relative to a *reference state* and thus the magnitude of the standard molar free energy depends on the choice of this reference state [1]. The *standard state* is 1 atmosphere of pressure (101.33 kP) and 25°C (298.15 K), and dilute solutions (i.e. an ideal solution). For the pure elements, the chemical standard state is defined as that of its most stable form. The following definitions are also applicable:

1. for *pure solids*: As $X = 1$, $\mu = \mu^o$

2. for *solutes*: $\mu = \mu^o + RT.\ln(n/n_w)$ as $X = n/n_w$ where n_w is the number of moles of water. The *molar concentration* of a constituent, [S], is n/V, where V is the volume of the solution (water). For dilute solutions, n_w/V is constant and can therefore be incorporated into the definition of the standard free energy and the expression becomes:

$$\mu = \mu^o + RT \cdot \ln[S]$$

3. for gases: $X = P_i/P$, where P_i is the *partial pressure* and P the total pressure. Again, at a constant pressure: $\mu = \mu^o + RT.\ln P_i$.

Overall, the *free energy change of a reaction* is given by:

$$\Delta G = \sum \upsilon_i(\mu_i^o + RT \cdot \ln X_i) \tag{3.16}$$

where υ_i is the stochiometric coefficient for each constituent in the reaction. This equation can be rewritten as:

$$\Delta G = \Delta G^o + RT \cdot \ln Q \tag{3.17}$$

where ΔG^o is the *standard free energy change* for the particular reaction, and is defined in Equation 3.18. The *reaction quotient*, Q, as defined in Equation 3.19, is a measure of the extent of reaction at any point in time and can indicate the direction of future change. At equilibrium Q = K, the *equilibrium constant*. Therefore, ΔG^o is equivalent to the sum for all species of the product of the molar free energies and their associated stoichiometric coefficients (υ_i) for the reaction for all species:

$$\Delta G^o = \sum \upsilon_i \mu_i^o \tag{3.18}$$

At equilibrium, as the system is in its lowest energy state:

$$\Delta G = 0 \tag{3.19a}$$

and

$$Q = \exp[-\Delta G^o/RT] = K \tag{3.19b}$$

K is therefore the specific value of the reaction quotient at equilibrium. Alternatively, Equation 3.19(b) can be written as:

$$\Delta G^o = -RT \cdot \ln K \tag{3.20}$$

Thus, the equilibrium constant is directly related to the standard Gibbs Free Energy of the reaction [1, 2, 4]. As noted earlier, to make the expression for K and Q more useful for environmental studies, it is most useful to use concentration units [2]. The concentration expression in molar units is equivalent to the expression in terms of mole fractions. Generally, the *mass law expression* can be written for a general reaction as, for the following reaction:

$$aA + bB = cC + dD \tag{3.21}$$

where A, B, C, D are the reacting chemicals and a, b, c, d are the stoichiometric coefficients for the reaction. The equilibrium expression is given as:

$$K = \frac{[C]^c \cdot [D]^d}{[A]^a \cdot [B]^b} \tag{3.22}$$

The magnitude of the equilibrium constant is the product of the concentration of the reaction products, as the reaction is written, raised to the power of the stoichiometric coefficients, over that of the reactants. As noted earlier, the equations so far defined are for ideal solutions. For non-ideal systems, it is necessary to introduce the additional concept of *activity*, which conceptually can be thought of as the active or reactive concentration rather than the actual total concentration. Non-ideal effects can lead to the activity being either larger or smaller than the actual concentration. In most instances, the activity is lower than the concentration due to the impact of interactions with other constituents in solution [1, 2].

The reactivity of a species in non-ideal solutions will be influenced by non-ideal effects so that this reactivity cannot be defined by the concentration alone in the equilibrium equation (Equation 3.22). However, reactivity can be represented by a parameter that is related to the concentration in a defined manner and this is what is defined as the *activity* of the species in solution. The activity of a species is not equivalent to its total dissolved concentration except when the system is behaving ideally. Therefore, when considering interactions in solution it is fundamentally more correct to talk in terms of activities rather than concentration. However, in most situations it is intuitively easier to consider and measure concentration rather than activity. However, it should be noted that some of the measurements commonly made, such as that of pH, which is typically made with an ion-selective electrode, is actually a measurement of the activity of the hydrogen ion in solution rather than the $[H^+]$ concentration. To examine more closely the relationship, we will define the relationship between concentration and activity. This also allows us to express reactions and equilibrium constants in concentration terms, if desired.

3.1.2 Ionic strength effects

As noted previously, the assumption of the ideal system is that the composition of the rest of the solution has no impact on the value of the molar free energy, μ_i, for the chemical of interest. Recall that this energy incorporates the energy of formation and the energy required to acquire a certain concentration in solution. However, in a non-ideal system, there is an interaction with other species in solution, and the energy associated with this interaction needs to be taken into account. Long range *columbic* interactions, due to the presence of charge in the solution, are one example of potential interactions that can occur between the species of interest and the major ions in solution, and this will be discussed in more detail in Section 3.2. As an example, consider the interaction of the major ions in solution due to the high concentrations of salt in seawater with that of a dissolved metal(loid) ion at a much lower concentration [2]. In addition to long-range columbic interactions, there are also dipole-related interactions due to *van der Waals, hydrogen bonding* and other interactions between molecules that have an asymmetrical electron density. Because water is dipolar and forms strong hydrogen bonds in solution, there is an interaction between the water and ions in solution that can become important if the ions are present in high concentrations. Ions in solution impact both the water molecules in their immediate vicinity, and can also interact with those at greater distances (see Section 3.2). The phenomenon whereby all ions or polar molecules interact with the polar water molecules to some degree is known as the *solvation effect* in aqueous solutions.

The importance of these effects is a function of the size and charge of the species of interest in aqueous solution. Additionally, the concentration of the species in solution will affect the extent of these interactions. In addition to the electrostatic and dipole interactions, at very high concentrations, *volume exclusion* effects are possible when molecules "crowd" each other in the solution. For example, Morel and Hering [2] calculate that Na^+ and Cl^- ions must "touch" each other in solutions of higher than 2.5 M NaCl. This being the case, one can expect that there will be significant ionic strengths effects at some fraction of this concentration. If we consider that seawater has an ionic strength of around 0.5 M, it is clearly apparent that non-ideal effects will be important in such solutions.

The interactions between dipoles do not occur in more dilute solutions when the molecules are sufficiently separated from each other. As a rule of thumb, dipole interactions are not important in solutions of <0.1 M ionic strength [2]. As dipole interactions are attractive, they influence the ability of the species to react and lead to a decrease in the activity of the species in solution. Coulombic interactions between charged species are most prevalent because these interactions occur over the broadest range of conditions. The magnitude of Coulombic interactions depends on the charge on the ion and the *dielectric constant* of water. Such interactions become important at relatively low ionic strengths ($>10^{-3}$ M). Hence, these interactions operationally define the range of ionic strengths where non-ideal conditions are significant enough to invalidate the assumption of ideality. Coulombic attraction or repulsion of ions in solution create the most important deviations from ideal behavior. As discussed in Section 3.2, the extent of interaction can be large, causing a different ordering of the water molecules around the ion compared to the bulk solution where the water molecules are ordered due to hydrogen bonding interactions. Such effects tend to occur when the *hydration number*, or number of water molecules interacting with a charged cation in solution, is large.

Non-ideal behavior of neutral (non-polar) molecules is likely to be much less important than that of charged species [2]. The most important interaction of neutral molecules is manifested as a decrease in solubility in a concentrated (>0.1 M) solution that is known as the *salting out effect* (named after the observed decrease in solubility of an organic molecule in a concentrated salt solution). High concentrations of a neutral species in solution decrease the overall dielectric constant of the solution. This causes an overall increase in the free energy of the charged ions, and the activity to be greater than the concentration.

We can account for non-deal effects by including an additional term in the definition of the molar free energy. Based on Equation 3.15, a new relationship for the non-ideal system can be represented as follows:

$$\mu_i = \mu_i^o + RT \cdot \ln[i] + RT \ln \gamma_i \quad (3.23)$$

where the parameter, γ represents the *activity coefficient* that takes into account the energy in the system due to non-ideal effects [1, 2]. When the two terms in Equation 3.23 are combined, by convention this is denoted by {i}:

$$\mu_i = \mu_i^o + RT \cdot \ln\{i\} \quad (3.24)$$

where {i} is *the activity* of species i, and $\{i\} = \gamma_i[i]$

The values of γ can be determined for pure solutions based on first principles, as shown for NaCl solutions by Stumm and Morgan [1]. However, γ is more difficult to determine for complex solutions. The *Debye–Huckel Theory* is a well-known relationship that defines the impact of Coulombic interactions on the activity of ions in solution [1, 2]. A number of other relationships have been developed to extend and refine this theory. Details of the derivation of the Debye–Huckel relationship are beyond the scope of this book, and the interested reader is referred to Morel and Hering [2], for a discussion of this derivation from the Poisson–Boltzman equation. However, it is useful to note the approach and some of the assumptions in the Debye–Huckel Theory to understand situations when the theory does not provide reliable predictions (such as those with concentrated solutions or complex mixtures).The relationship assumes that ions have a spherical surface with uniform charge in a medium of continuously changing charge density. Thus, the theory accounts for only the Coulombic terms and neglects others, such as dipole interactions and the effects of "crowding" in concentrated solutions [2]. Solving the Poisson–Boltzman equation, using the conditions as defined, results in the Debye–Huckel Relationship:

$$\ln \gamma = -AZ^2 \cdot I^{1/2}\left(1 + BR_oI^{1/2}\right)^{-1} \quad (3.25)$$

where Z is the ionic charge on the ion under consideration, I is the ionic strength of the solution, and A, B and R_o are constants. The constants A and B depend on the dielectric constant of the medium and the absolute temperature. For water at 25°C, $A = 1.17\,mol^{-1/2}\,l^{1/2}$ while $B = 0.329\,Å^{-1}\,mol^{-1/2}\,l^{1/2}$. The value of the constant R_o, is a fitting parameter equivalent to the effective ionic radius in Å. Thus the Debye–Huckel relationship results in a series of curves for each value of R_o, and the impact of both R_o and Z on the value of γ is apparent. The charge on the ion is very important [2]. Figure 3.1 shows the relationship derived from the Debye–Huckel equation as well as experimental data for a number of simple solutions. It can be seen from this equation that when I is small, $\ln\gamma$ is small and negative, and γ is therefore close to 1. For example, for a singly charged ion with $R_o = 4\,Å$, γ is 0.9 when I = 0.01 M and γ is 0.99 when $I = 7.5 \times 10^{-5}$ M. At 10^{-3} M, γ is 0.965 and the error in not making the ionic strength corrections in calculations is a few percent.

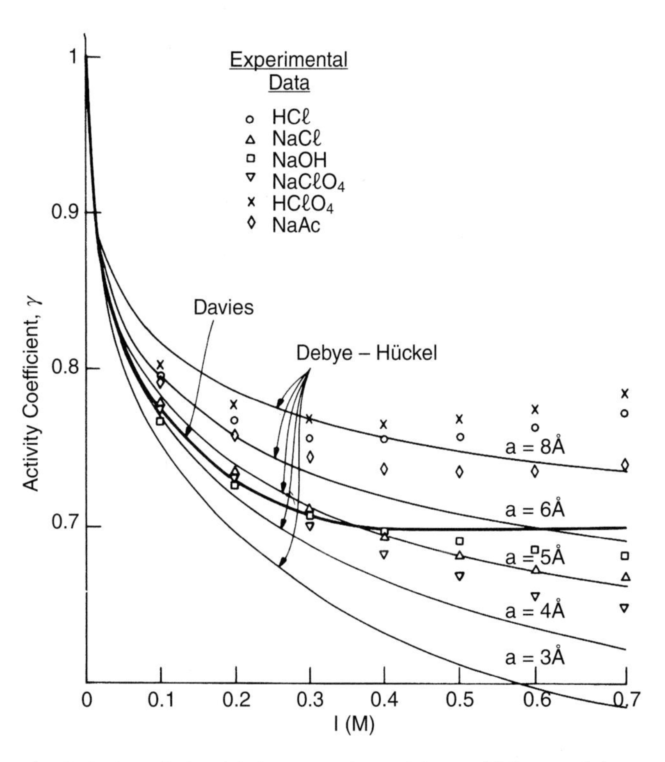

Fig. 3.1 The relationship between the activity coefficient and the ionic strength of the solution, as determined by the Davies Equation and the Debye–Huckel Theory. Figure taken from Morel and Hering (1993) *Principles and Applications of Aquatic Chemistry* [2], and reprinted with permission from John Wiley & Sons, Inc.

In addition to the Debye–Huckel relationship, the other commonly used equation for defining interactions in nonideal solutions is the *Davies Equation* (Fig. 3.1). This equation eliminates the size parameter and adds a term to the Debye–Huckel relationship based on fitting of empirical data [2]:

$$\ln \gamma = -AZ^2 \cdot [(I^{1/2}/1 + I^{1/2}) - bI] \quad (3.26)$$

where the value of b is 0.3. The results from both formulations are plotted in Fig. 3.1 and compared to actual data (symbols in the figure). The difference between the theory and the measurements is apparent.

The deviations shown between the theoretical lines in Fig. 3.1 and the experimental data are due to the fact that dipole interactions, crowding and other interactions in solution discussed above can be as important as the Coulombic effects. The theoretical equations and the experimental data agree best at the lower ionic strengths which correspond to most situations encountered in the freshwater environment, where the ionic strength is typically less than 0.01 M. Under these conditions the comparisons are reasonable and from the vantage point of most environmental chemists the errors are less of a concern than they might first appear. Errors between theory and reality are greatest at high ionic strength, and therefore this must be taken into account when dealing with metal(loid)s in seawater and other high ionic strength matrices. However, in the overall context of making equilibrium predictions, the error associated with the value of the activity coefficient is often smaller than the error associated with the value of the equilibrium constant, which can vary by an order of magnitude or more in some instances, if the results of different investigations are compared. These relative errors are discussed further next.

3.1.3 Thermodynamic equilibrium, kinetics and steady state

For reversible reactions in a closed system, there is no net change in concentration at equilibrium and the system has reached its lowest energy state.

Again, consider the simple example:

$$A \Leftrightarrow B$$

and

$$K_{AB} = C_B/C_A$$

where C represents the concentration of each component. This representation is used for clarity in the following equations. The magnitude of K determines the final ratio of products to reactants and also indicates the likelihood of the reaction proceeding in a particular direction; for example, large values of K suggest that the forward reaction is more favorable. In contrast, kinetics is the measurement of the rate at which the reaction proceeds. For a closed system, the rate of the forward and the reverse reactions are equal at equilibrium.

The following calculations and discussions are dealt with in detail in other texts [1, 2], and will be summarized here for situations where a thermodynamic approach provides a reasonable approximation of what is occurring in an open, steady state environment. Consider a system where initially there is only compound A (with concentration C_{A0}) reacting to form B (k_f is the forward rate constant; k_b is the back reaction rate constant):

$$\text{At } t = 0, C_A = C_{AO}; C_B = 0 \quad (3.27)$$

Then, at any point, for a reversible reaction:

$$dC_A/dt = -k_f C_A + k_b C_B \quad (3.28a)$$

and

$$dC_B/dt = k_f C_A - k_b C_B \quad (3.28b)$$

and, as:

$$C_{AO} - C_A = C_B \text{ and } C_A = C_{AO} - C_B \quad (3.29)$$

for this example, the expressions are solved as follows:

$$dC_A/dt = -k_f C_A + k_b(C_{AO} - C_A) \quad (3.30)$$

$$dC_B/dt = k_f(C_{A0} - C_B) - k_b C_B \quad (3.31)$$

At equilibrium, there is no net change in concentration and:

$$dC_A/dt = dC_B/dt = 0 \quad (3.32)$$

and therefore:

$$C_{A,eq} = k_b C_{AO}/(k_f + k_b) \quad (3.33)$$

Similarly:

$$C_{B,eq} = k_f C_{A0}/(k_f + k_b) \quad (3.34)$$

So:

$$\mathbf{C_{B,eq}/C_{A,eq} = k_f/k_b = K_{AB}} \quad (3.35)$$

Equation 3.35 represents the relationship between the equilibrium constant and reaction rates for a reversible first order reaction and is the fundamental relationship between equilibrium and kinetics for the criteria as defined previously [1].

Finally, we need to consider why the equilibrium approach is useful, and under what conditions it provides a reasonable estimate of concentrations that are present in an open system at steady state. Consider the same reversible reaction ($A \Leftrightarrow B$) in an open system with inflow concentrations of C_{AO} and C_{BO} and a flow rate of Q (volume per time), and a reservoir of volume, V (Fig. 3.2). The parameter r, the fluid transfer rate constant (units of per time), is defined as

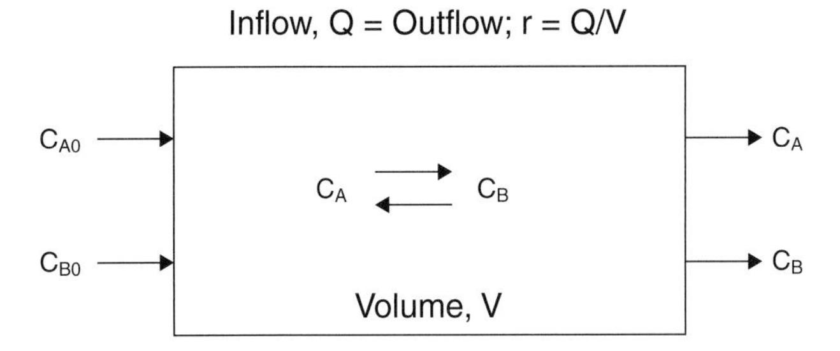

Fig. 3.2 Schematic diagram of the flow-through reactor system.

$r = Q/V$. At steady state, when there is no longer a concentration change in the system [1]:

$$dC_A/dt = -k_fC_A + k_bC_B + rC_{AO} - rC_A = 0 \quad (3.36)$$

and

$$dC_B/dt = k_fC_A - k_bC_B + rC_{BO} - rC_B = 0 \quad (3.37)$$

and, as

$$C_A + C_B = C_{AO} + C_{BO} \quad (3.38)$$

Then it can be shown that:

$$C_A = (rC_{AO} + k_b(rC_{AO} + C_{BO}))/(k_f + k_b + r) \quad (3.39)$$

and

$$C_B = (rC_{BO} + k_b(rC_{AO} + C_{BO}))/(k_f + k_b + r) \quad (3.40)$$

and

$$C_B/C_A = (rC_{BO} + k_f(C_{AO} + C_{BO}))/(rC_{AO} + k_b(C_{AO} + C_{BO})) \quad (3.41)$$

We can deduce from this expression that when r is relatively small, the values of the other expressions will be substantially greater and:

$$C_{B/}C_A \sim k_f/k_b \quad (3.42)$$

Similarly, if the concentrations of C_{A0} and C_{B0} are small compared to those in the system, then rC_{A0} and rC_{B0} may be small compared to the other expressions and the assumption of the steady state will also be valid, being essentially equivalent to the equilibrium condition. These are the conditions under which a thermodynamic equilibrium approach provides us with useful information about a steady state natural system. In the following chapters that discuss application of equilibrium calculations to evaluate the chemistry of natural waters, we assume that inputs are small compared to those within the system of study and that the system is at steady state.

Finally, for the open system described earlier, if one of the inputs is zero, for example, $C_{BO} = 0$, as in the example above for the closed system, then:

$$C_{B/}C_A = k_f/r + k_b \quad (3.43)$$

And again this expression will approach the equilibrium expression when $k_b >> r$. Remember, in a closed system, there is a direct relationship between the rate constants and the equilibrium constant. At steady state, an open system approaches the equilibrium condition at relatively low input concentrations or at low r, or when the rates of reaction are relatively rapid.

The thermodynamic chemical basis of the equilibrium constants reviewed here will be used throughout this chapter and the following chapters of the book. We turn our attention now to a brief examination of inorganic chemical concepts that are a useful introduction to understanding why metal(loid)s in solution and in the solid phase react in a certain way, and why some metals prefer to bind with compounds containing sulfur groups while others bind more strongly to oxygen-containing compounds. Traditionally, differences among metals have been characterized as "hard" and "soft" metals [1–3], depending on their electronic structure. The alkali and alkali earth metals are considered hard as they have a "noble gas" configuration when ionized and there is no electronic shielding due to the presence of filled or partially filled *d* or *f* orbitals. Aluminum also falls into this category, as do some other metals in various oxidation states. These metals form relatively simple complexes with anions and complexes are typically relatively weak, except perhaps the hydrolysis species for example, Al (Appendix 3.1). The larger atomic mass metals, for example, Hg and Pb have electron clouds that are readily polarized and these metals are considered "soft" in the sense that their clouds can be readily distorted – and the outer electrons are effectively shielded by the electronic cloud from the positive attraction of the central nucleus. The bonds of these metals are more "covalent" in nature, and these metals have a larger tendency to form complexes.

3.2 Bonding, electronic configuration, and complex formation

Metal(loid)s do not exist merely as charged ions in solution because water is a polar liquid. Molecules of water associate themselves around the ion in a fashion to neutralize the charge of the ion [2]. This is the case for any charged ion in aqueous solution and the interaction can be conceptualized as shown in Fig. 3.3. The area immediately around the central ion is termed the *primary solvation (hydration) shell* and this refers to the region where the water molecules are directly interacting with the central ion. This interaction can occur through the formation of a chemical bond, or involve a weaker interaction. The strength of this interaction varies considerably for the metal(loid)s of interest here. Most transition metal ions in solution are surrounded by six water

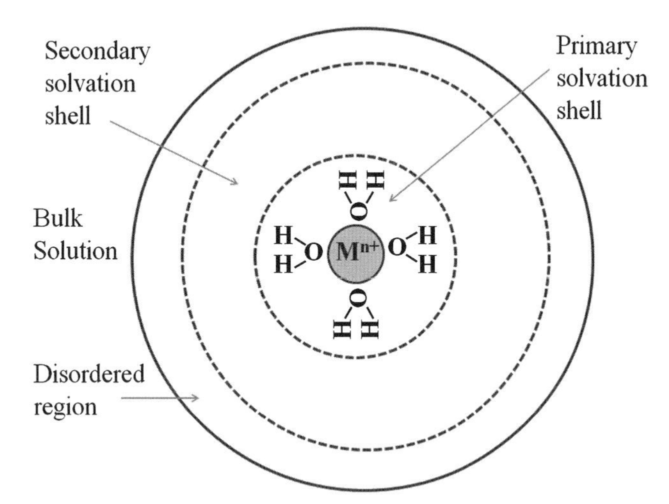

Fig. 3.3 Diagram showing the arrangement of water molecules around a central cation in dilute solution. Figure modified from Burgess (1999) *Ions in Solution* [6], and used with permission from Horwood Publishing.

molecules, complexed in an *octahedral* conformation to the central ion, and thus metals ions in solution should typically be written as $M(H_2O)_6^{n+}$ where M represents the metal ion and n the charge on the metal ion. Note that the term *octahedral*, as used to refer to the situation where six ligands are bonded to the central atom, may appear to be a contradiction. However, the term refers to the overall three dimensional shape of the resultant complex.

In most texts and papers, and in most of the discussions in this book, the presence of the water molecules is implicitly assumed and the ions are written as M^{n+}. While not all metals of interest form octahedral complexes in solution, for the transition and heavy metals being considered in this text, this is mostly the case. The ability to form complexes in solution generally requires the presence of unfilled *d* or *f* orbitals and thus it is the transition metals, lanthanides and actinides, and the heavy metals that are mostly involved in complex formation in solution. Metals at the end of the transition series such as Hg, Ag, and Au also form complexes in solution although they do not always form octahedral complexes. For example, Hg and Ag prefer to form either linear or tetrahedral complexes.

There is also a *secondary solvation (hydration) shell* around the central ion in which there is electronic interaction but no bonding between the central ion and the water molecules (Fig. 3.3) [2]. In this case, the water molecules are ordered based on the electrostatic influence of the central ion. For the alkali metals, which mostly form ion pairs in solution, the cations are not hydrated through complex formation with water like the transition metals. For K^+, there are four water molecules in the primary hydration shell, but the central ion influences the water molecules in the secondary shell, and the approximate hydration number for K^+ is given as 10.5 [7]. The extent of influence depends on the charge on the central ion and decreases with increased size of the central ion. Thus, the approximate hydration number decreases from 25.3 for Li^+ to 9.9. for Cs^+ for the Group I elements. In contrast, the Group 2 metals tend to be hydrated by six water molecules in solution. Group 2 metals also influence the water molecules in the second salvation shell but the number of molecules involved in this interaction has not been determined [7].

There is an additional region around the central ion where water molecules are still influenced by the central charge (Fig. 3.3) and thus are not ordered as in the bulk solution where hydrogen bonding between water molecules leads to an ordered solution. This transition zone is known as the *disordered solvent* region, where water molecules are less ordered than either in the bulk solution or in the secondary solvation shell [2]. In the *bulk solution*, forces between water molecules are not affected by that of the dissolved ions.

The association of the water molecules with the central ion in solution should in many instances be considered as a complex with the ion; that is, the water molecules are forming a bond with the central ion. The strength of this bond is related to the thermodynamics associated with the solvation of the metal ion, and the *energy of hydration* is a critical determinant of the strength of this interaction [7]. Most transition metals have high *enthalpies of hydration*, $\Delta H^0_{hydration}$ (i.e., for the reaction):

$$M^{n+}(g) + 6H_2O \rightarrow M(H_2O)_6^{n+}(aq) \tag{3.44}$$

Example enthalpies of hydration are shown in Table 3.1, and are large and comparable to the energy needed to remove the ion from a crystal lattice of the salt or due to the ionization of the metal (see Table 3.1). The large negative enthalpy associated with the formation of the hydrated ion, and the fact that in most cases the ions are being solubilized at low concentrations (recall that standard conditions are for 1 M solutions) are both sufficient for the solution of the ion in water to be favored.

Metal ions in solution are complexed to water molecules as free metal ions and to other *ligands* when in the presence of inorganic and organic compounds. Molecules or ions associated with the central metal ion are termed *ligands* and are referred to as *mono-, di- or poly- dentate* depending on the number of donor atoms that are available on the ligand to associate with the central atom. For example, OH^- is a monodentate ligand, ethylene diamine ($H_2NCH_2CH_2NH_2$) is a bidentate ligand, which binds to metal ions through both of the amine groups, and EDTA (ethylene diamine tetraacetic acid) is an example of a polydentate ligand. The structure

Table 3.1 Enthalpies of hydration and enthalpies of ionization for some ions. All values are in kJ mol^{-1}. Taken from various sources [1–3, 7, 8].

Metal	ΔH^0 (hydration)	ΔH^0 (ionization)	Metal	ΔH^0 (hydration)	ΔH^0 (ionization)
K^+	−322	419	Zn^{2+}	−2046	2639
Mg^{2+}	−1921	219	Ni^{2+}	−2105	2490
Ca^{2+}	−1577	1735	Cu^{2+}	−2100	2703
Ba^{2+}	−1305	1468	Cd^{2+}	−1807	2492
Cr^{2+}	−1904	2245	Hg^{2+}	−1824	2817
Mn^{2+}	−1841	2226	Pb^{2+}	−1481	2166
Fe^{2+}	−1946	2326	Fe^{3+}	−4430	5280

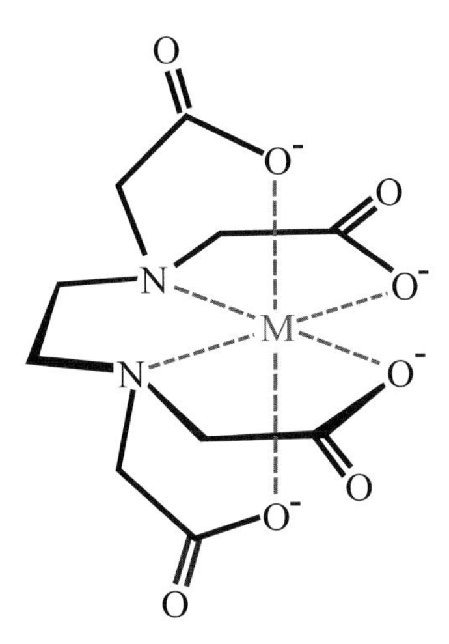

Fig. 3.4 Example of a metal-chelate complex. Here the central cobalt (Co) ion is bound to the polyprotic acid, ethylenediamine tetraacetic acid (EDTA).

of one such metal complex is shown in Fig. 3.4. Another term used for polydentate ligands is *chelate*, a term which derives from the Greek word for claw and refers to the ability of ligands, such as EDTA, to surround the central atom. The *chelating effect* of polydentate ligand enhances the stability of the complex compared to that of monodentate ligands [1]. This stability is demonstrated, for example, by comparing the degree of complexation of Cu in dilution solution with NH_3, ethylene diamine and EDTA [1]. The reasons for the differences in bonding strength depends on the type of ligand and the chemistry of the metal to which it is complexed.

The interaction and binding of a ligand to a metal ion results in the displacement of one of the water molecules. Therefore, when describing complex formation in solution with inorganic or organic complexes, this should in most cases be considered as a *ligand exchange reaction* [2]. The incoming ligand displaces a water molecule in forming the complex as illustrated by the following example:

$$Cu(H_2O)_6^{2+} + 4NH_3 = Cu(H_2O)_2(NH_3)_4^{2+} + 4H_2O$$

However, this reaction would usually be given in the form:

$$Cu^{2+} + 4NH_3 = Cu(NH_3)_4^{2+}$$

In both cases, given that, by definition, water has an activity of 1, the resultant equilibrium constant, K is the same:

$$K = [\text{Cu-complex}]/[Cu^{2+}][NH_3]^4$$

The displacement of the water molecules leads to an increase in the entropy of the system, due to the formation of a non-symmetrical complex with a mixture of ligands (water and the binding ligands). In many cases, this change in entropy is enough to drive the reaction because the change in enthalpy associated with complex formation is usually positive (see Table 3.2). The change in entropy associated with complexation is always greatest for the binding of the first ligand and decreases with the addition of more ligands (Table 3.2). Thus, for the reaction to proceed, the change in entropy must be sufficient to counter the change in enthalpy associated with the reaction. Overall, if there is a negative change in ΔG^0 ($\Delta G^0 = \Delta H^0 - T\Delta S^0$), then the reaction proceeds as reflected by the equilibrium constant. Recall that the magnitude of the equilibrium constant (K) for the reaction is related to ΔG^0 as:

$$\Delta G^0 = \Delta H^0 - T\Delta S^0 = -RT \ln K \tag{3.45}$$

Metal ions in solution are clearly not present without a counter anion or complex to balance the charge, even if the counter ion is the hydroxide ion. Thus, there is the further consideration of the association of the metal ion and its counter charged associate. Metal ions in the first and second period of the periodic table generally do not form complexes of high stability but form ion pairs in solution, especially the

Table 3.2 The enthalpy and entropy associated with the formation of various complexes and metals and the associated equilibrium constants for the reactions. Compiled from various sources [1–3, 7, 8]. Values are in kJ mol^{-1} for ΔH^0, J mol^{-1} K^{-1} for ΔS^0. I: ionic strength of the solution.

Ligand and Metal	ΔH^0	ΔS^0	Log K	I
Ca^{2+} and OH^-	8.4	51.5	1.2	0
Fe^{3+} and OH^-	5.0	209	11.7	0
Fe and L (malonate)	11.3	181.4	7.5	1
FeL and L	3.1	117	5.54	1
FeL_2 and L	–4.6	53	3.56	1

alkali ions [1, 7]. The strength of the ion pairs determines the degree to which the metal and its counter ion remain associated when in solution. Some metal salts, such as NaCl, are completely dissociated in water and there is little direct interaction between the ions. An equilibrium constant cannot be defined for the NaCl complex as it essentially does not exist in seawater. For other anions, such as sulfate, Na^+ forms a weak complex (i.e., the log K for $NaSO_4^-$ is 1.06) [2]. For example, if the concentrations of Na and sulfate are equivalent in dilute solution due to the dissolution of the salt to 10^{-4} M, the concentration of the complex is estimated to be 1.1×10^{-7} M, or ~0.1% of the complexed Na^+. Thus, in dilute solution Na^+ essentially dissolves completely in water. In contrast, most transition metal ions do not predominantly exist in solution as the free metal ion and are mostly in the form of complexes. A variety of terms are used to describe the type of associations that occur ranging from those that are relatively ionic in nature to those that have substantial covalent characteristics. Mercury, for example, forms linear complexes with some ligands, such as chloride, where the bond is essentially covalent.

For many of the trace metals, including transition metals, their empty *d* and *f* orbitals within the electronic shell have a dramatic impact on the extent of complex formation and their characteristics such as their paramagnetism and propensity to form colored solutions [7]. For metals and metalloids that have filled *d* and *f* orbitals, understanding their behavior in terms of complex formation is somewhat simpler. Therefore, a brief discussion of the inorganic chemistry theory related to complex formation by transition metals is warranted.

3.2.1 Ligand Field Theory

The discussion here will be focused on the *Ligand Field Theory* to describe transition metal interactions in solution [7]. Here a simplified description will be given, focusing on examples of 3*d* orbital interactions (for the first row transition metals). The following designations will be used: oxidation states will be denoted by Roman numerals (e.g., +I, +II, –IV) while the charge on a particular ion or species will be represented by numerals as a superscript (e.g., Cu^{2+}, Cl^-). This is done to avoid confusion, for example, between the discussion of the +II oxidation state for Cu (Cu(II) or Cu^{II}) and that of the free metal ion, Cu^{2+}. Given the order at which orbitals are filled, the atoms of the metals of the first row of the transition series have the electron configuration of $Ar4s^23d^x$, where x differs for the different elements of the series. There are two exceptions and these are due to the enhanced stability of the electronic configuration for d^5, which occurs when there is one electron in each *d* orbital, or d^{10}, where the orbitals are full. Given the enhanced stability of these configurations, the atoms of Cr have an electronic structure of $Ar4s^13d^5$, and those of Cu, $Ar4s^13d^{10}$, as opposed to structures with two 4*s* electrons [7]. This also explains, for example, the fact that Cu is found as a +1 cation to a greater degree in the environment than other transition metals, and the same is true for one of its counterparts in Group 11, Ag, which does not exist in the +II oxidation state to any significant degree. Gold, the final member of this group, is mostly found in the +III oxidation state, or the +I state, and does not exist in solution as a free ion. Besides these elements, it is the outer electrons from the 4*s* orbital that are first lost for most of the transition metal elements and therefore most elements of the transition series form +2 cations, or ions of higher charge, or they exist in higher oxidation states in complexes.

The *d* orbitals around the central atom occupy different locations in space, and these are shown schematically in Fig. 3.5 [7]. It is noted that three of the orbitals – d_{xy}, d_{yz} and d_{xz} – have their dominant electron density off the main axes while two of the orbitals – d_{z2} and $d_{x2\text{-}y2}$ – have most of their electron density on axis. These differences become important when the metal ion interacts with ligands in solution. To understand this better, consider the theoretical situation of a central positively charge transition metal ion with one electron in a d orbital (e.g., equivalent to Ti^{3+}) being approached by six negatively charged ions (e.g., Cl^-) which would be equidistantly placed in space and therefore would approach the cation along the x, y and z axes [7]. These incoming negatively charged ions would interact directly with the electron if it was in the two on-axis *d* orbitals (d_{z2} and $d_{x2\text{-}y2}$) but not if it was in the other orbitals. Such a situation would involve more interaction between the negative charges of the incoming ion and the electron, and would be a situation of higher energy than if the electron was in one of the off-axis orbitals. Thus, under this scenario, the *d* orbitals no longer all have equivalent energy, and the three off-axis orbitals have lower energy than the on-axis orbitals, and this energy difference would be sufficient to determine where the electron would likely reside. Such differences in *d* orbital energy have been termed *orbital splitting* in the pres-

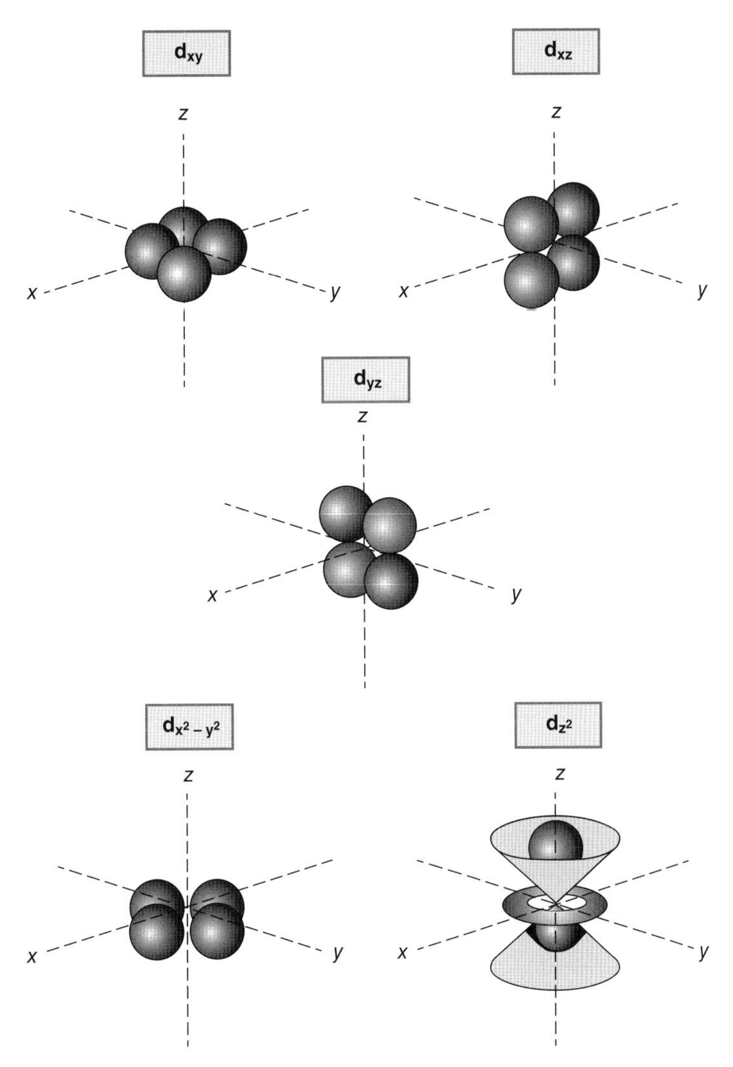

Fig. 3.5 Depiction of the *d* electron orbital density around a central nucleus.

ence of an external ion field. This is depicted in Fig. 3.6 for the octahedral environment (six equidistant ligands approaching and bonding to the central atom) described previously [7].

An alternative scenario occurs when four negatively charged species approach the cation as, in this case, the approaching ions would be tetrahedrally distributed in space and this would therefore result in the opposite situation – an electron in the three off-axis orbitals would interact with the incoming ions to a greater degree than an electron in the on-axis orbitals. Thus, the orbital splitting would be exactly the opposite of that which occurs in the octahedral situation, as also shown in Fig. 3.6.

Overall, the total energy of the five orbitals is unchanged and thus, for the octahedral environment, the increase in energy of the two orbitals of higher energy (denoted as e_g) is balanced by a decrease in energy of the other three orbitals (t_{2g}) of lower energy (Fig. 3.6) [7]. The total energy difference between the two sets of orbitals is termed the *orbital splitting energy* (Δ_o). For octahedral complexes, it can be seen that the energy increase of the two orbitals is $3/5\Delta_o$, while the decrease in energy of the three orbitals is $2/5\Delta_o$ that is, overall there is no change in total energy of the system. An analogous line of reasoning describes splitting in a tetrahedral environment, as shown in Fig. 3.6, where the three higher energy orbitals in this case are designated as t_2, and the two lower energy orbitals, e, and are separated by energy of Δ_t and the respective differences are an increase of $2/5\Delta_t$ in energy and a decrease of $3/5\Delta_t$.

This description is for a system where anions are assumed to be point charges, which is clearly a simplification, but the theory also holds for polar neutral ligands, such as water or

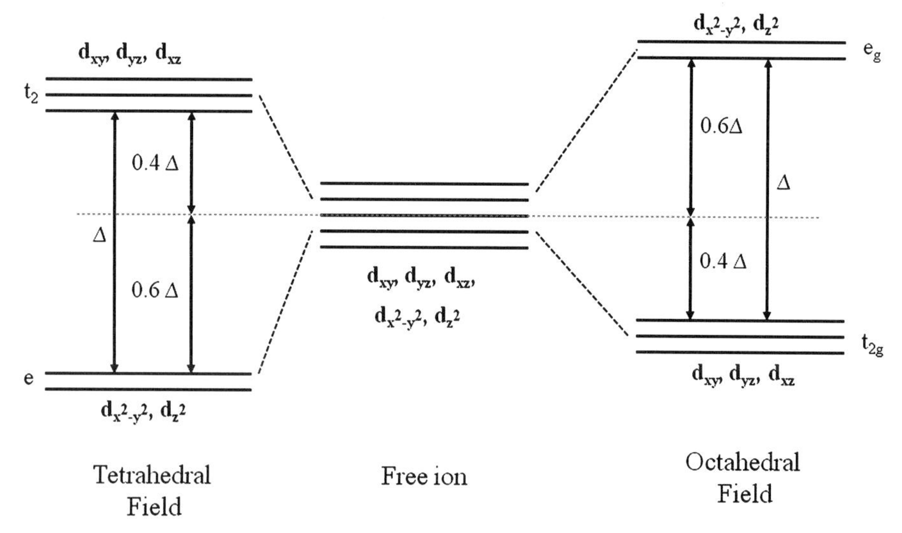

Fig. 3.6 Depiction of the splitting of the *d* orbitals under the influence of different electrostatic crystal fields.

NH_3. The splitting that occurs in the presence of other different complex configurations, such as with square planar complexes, will not be discussed here and interested readers are referred to Cotton and Wilkinson [7] or a similar text. Additionally, readers are referred to *Molecular Orbital Theory* for an explanation of differences among orbital energies as this is a more complete theory [9] of chemical bonding than the Ligand Field Theory. This more comprehensive theory of the impact of ligands on the electronic distribution around the central ion leads to similar results to that described above for the Ligand Field Theory.

The result of the orbital splitting is that filling of the d orbitals with electrons may no longer follow *Hund's Rule* which states that the electrons will not pair in an orbital until all the orbitals of the same energy contain one electron. The degree to which the orbitals are separated by the influence of the binding ligands (i.e., in the magnitude of Δ_o or Δ_t) determines whether all the *d* orbitals will fill with one electron first [7], as expected from Hund's Rule. Considering the octahedral case, if the value of Δ_o is large, the orbitals can be considered to form separate groups and therefore the orbitals will fill as though they are separate, and the three low energy orbitals will be completely filled prior to electrons occupying the two higher energy orbitals (i.e., they fill as expected but there are essentially two different energy levels of *d* orbitals). This is called the *low spin* state, for obvious reasons (i.e., the minimum amount of unpaired electrons), and is depicted in Fig. 3.7 for *d* orbitals with 4–7 electrons. If the d orbitals have 1–3 or 8–10 electrons, there is only one option available for their distribution and the value of Δ_o is immaterial to the outcome. If the orbital filling follows Hund's Rule, then all the *d* orbitals would be filled with one electron prior to filling orbitals as paired electrons, and this situation is referred to as the *high spin* state (Fig. 3.7) [7].

From this discussion, it can be determined that the number of free (unpaired) electrons in the d shells, which impacts such properties as paramagnetism, is dependent on whether the electron distribution is high spin or low spin. For example, under low spin, and with six electrons, there are no unpaired electrons, while under the high spin state, there are four unpaired electrons. The state that exists is ultimately determined by the relative magnitude of two competing energies: the energy associated with pairing electrons (P) and the energy difference (Δ_o for the octahedral case). The state which has the lowest energy clearly depends on the relative magnitude of P and Δ_o. The magnitude of P varies with the element and the ion charge while the magnitude of Δ_o clearly depends on the ligand structure and charge. Returning to the six-electron situation, the low spin state has energy, $E = 12/5\Delta_o + 3P$ (i.e., six t_{2g} electrons in three electron pairs) while the high spin case has an energy, $E = -2/5\Delta_o + P$ (Fig. 3.6) [7]. The more favorable energy therefore depends on the relative values of P and Δ_o, which varies for different elements with ions having the same electronic outer shell configuration (e.g., Fe^{2+} and Co^{3+}; Fe^{3+} and Mn^{2+}).

For ligands that induce strong orbital splitting, the resultant configuration is likely to be low spin, and vice versa. The relative capacity of common ligands to cause orbital splitting is:

$$I^- < Br^- < Cl^- < F^- < OH^- < H_2O < NH_3 < en < CN^- \quad (3.46)$$

where en represents ethylenediamine. Thus, for example, Fe^{2+}, which has six *d* electrons, forms a high spin octahedral

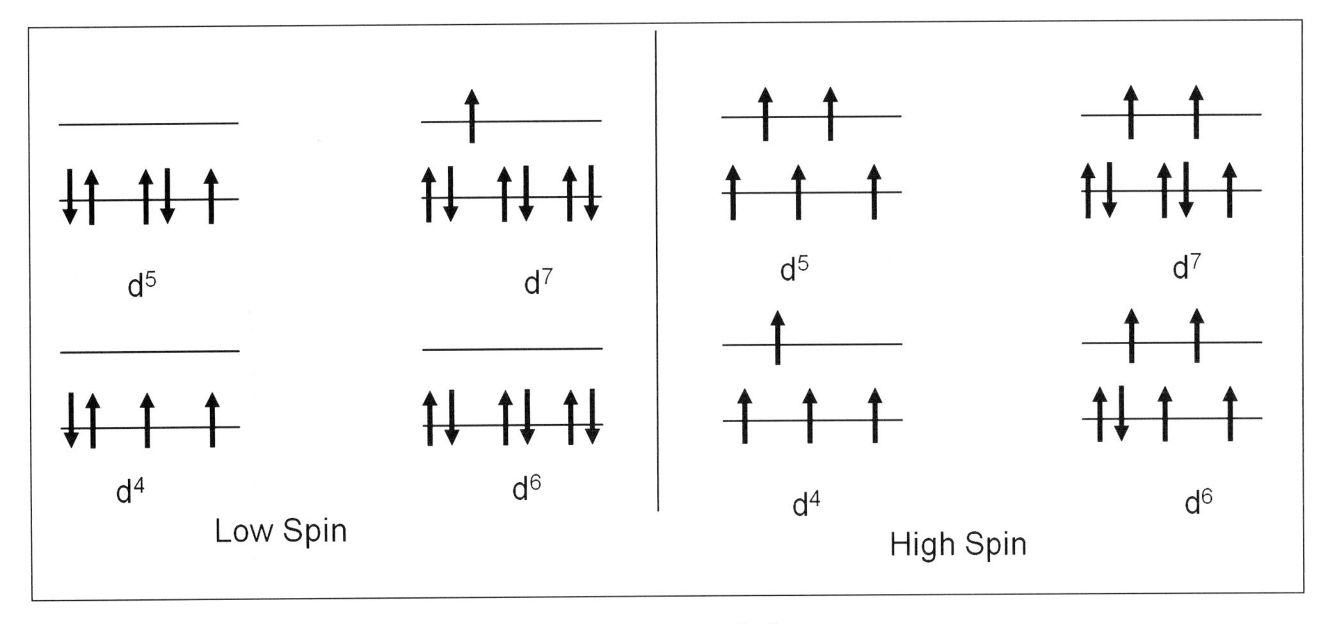

Fig. 3.7 Orbital diagrams for high and low spin electron configurations for d^4–d^7.

complex with water and a low spin complex with CN^-; Co^{3+} (also six *d* electrons) forms a high spin octahedral complex with F^- and a low spin complex with NH_3. A similar discussion pertains to the tetrahedral complexes and other configurations.

3.2.2 Thermodynamic effects of orbital splitting

The impact of d orbital splitting can be seen in the various properties of the transition metals which will be discussed briefly here. Firstly, as seen generally across the periodic table, there is an overall contraction in *atomic radii* along the series, and overall this is true for the transition metals. There is the expected decrease in ionic radii for the free ions in solution from Ca^{2+} (no *d* electrons) and Zn^{2+} (10 *d* electrons) (Table 3.3; radii calculated for an octahedral ligand environment) due to the increasing nuclear charge and the imperfect shielding of the *d* orbital electrons. Depending on the orbital the electrons are occupying, the impact of shielding is different and the changes in ionic radius are not uniform across the series with some ions having a smaller radius than would be predicted. Mn^{2+} fits the trend between Ca^{2+} and Zn^{2+} as it has one electron in each *d* orbital so it has a spherical electron distribution like that of Ca^{2+} and Zn^{2+} (Table 3.3). It can be shown that there is increased stability in the high-spin d^5 configuration [7] and this is therefore the preferred configuration for Mn^{2+} with an electron cloud that is evenly distributed around the ion, as is the case for Ca^{2+} and Zn^{2+}. However, the other ions have a smaller radius than would be predicted because of the orbital splitting and the order of filling of the *d* orbitals. With Sc^{2+}, Ti^{2+}, and V^{2+}, the electrons are in the t_{2g} orbitals, and are therefore not within the realm of the metal-ligand interaction (recall that there is an octahedral complex formed with water for the free metal ion), and the contraction of the ionic radius is greater than expected. However, Cr^{2+} has four *d* electrons and therefore one of these occupies an on-axis orbital in the high spin state, which leads to a small increase in radius for Cr^{2+} compared to V^{2+}. This increasing radius between ions continues from Cr^{2+} to Mn^{2+} (Table 3.3) [7]. For Fe^{2+}, Co^{2+} and Ni^{2+}, the electrons are forming pairs in the t_{2g} orbitals and again this leads to a decrease in radius compared to Mn^{2+}. Again, with additional electrons filling the e_g orbitals, there is an increase in atomic radius for Cu^{2+}. Note that there are additional effects on the overall orbital structure due to the interaction of the central ion and ligands involving the d^4 and d^9 electron ions (e.g., Cr^{2+} and Cu^{2+}) as the orbitals d_{z2} and $d_{x2\text{-}y2}$ are not analogous and have different structures (Fig. 3.5). There is a distortion of these electron clouds as a result of the presence of ligands and their uneven interaction with the electron clouds for these electron configurations. This is termed the *Jahn-Teller Effect* [7].

A similar relative change can be seen in the *electronegativity* (EN) (the power of an ion or atom to attract electrons) for the first transition series (Table 3.3). Here the values increase with atomic mass from Ca to Cr, with a subsequent decrease for Mn, followed by an increase in EN for Fe. An increasing EN with increasing atomic mass is apparent from Mn to Ni, but there is a slight decrease for Cu, and a large decrease for Zn. Thus, the changes in EN form a similar pattern to those for ionic radii.

Table 3.3 Relative values for various characteristics of the first series transition metals and their +2 cations. Taken primarily from various sources [3, 7, 8]. Electronegativity values are based on Pauling.

Cation	Ionic Radii (pm)	Relative Hydration Energy	Relative Hydration Energy (corrected)	Σ(IonizationEnergy) for ion (kJ mol^{-1})	Electronegativity
Ca^{2+}	106	0.15	0.15	1735	1.00
Sr^{2+}	127	–	–	1614	0.95
Ti^{2+}	80	0.62	–	1968	1.54
V^{2+}	72	0.72	0.40	2064	1.63
Cr^{2+}	84	0.74	0.56	2245	1.66
Mn^{2+}	91	0.64	0.64	2226	1.55
Fe^{2+}	82	0.85	0.72	2320	1.83
Co^{2+}	82	0.97	0.80	2406	1.88
Ni^{2+}	78	1.12	0.91	2490	1.91
Cu^{2+}	72	1.13	0.97	2703	1.90
Zn^{2+}	83	1.0	1.0	2369	1.65

The orbital splitting that occurs as a result of the ligand field can lead to a change in the overall energy of the system compared to what it would be if there was no splitting (i.e., five *d* orbitals of similar energy). So, there is a decrease in overall energy for the situation, considering again an octahedral ligand field, where the electrons are filling the t_{2g} orbitals. This decrease in energy is known as the *Ligand Field Stabilization Energy* (LFSE) [7]. These energies are of similar magnitude to those of other chemical changes and can influence the thermodynamic properties of the transition metals. For example, comparing the hydration energies of Cr^{2+}, Mn^{2+} and Fe^{2+} (Tables 3.1 and 3.3) reveals that the energies are higher for the ions where ligand field stabilization is present compared to Mn^{2+}, which has one electron in each orbital and therefore has no LSFE overall. As found for the atomic radii, the hydration energies of Ca^{2+}, Mn^{2+}, and Zn^{2+} all fall on a theoretical line while those of the other ions are higher than predicted due to their LSFE. When the LSFE is taken into account, all the hydration energies fall on a curve of increasing hydration energy from Ca to Zn. The impact of LFSE can also be seen if one compares the lattice energies of the ions, such as those of MCl_2, or the gas phase dissociation energies [7].

Another observation, which has been termed the *Irving-Williams Series*, can also be linked to the phenomenon of orbital splitting [7]. Generally, and as illustrated by the formation constants for the ions with N-containing donor atoms, equilibrium constant strengths are:

$$Mn^{2+} < Fe^{2+} < Co^{2+} < Cu^{2+} > Zn^{2+} \quad (3.47)$$

Recall that the magnitude of K is related to the magnitude of ΔG^0, which in this case depends mostly on differences in ΔH^0, as the magnitude of ΔS^0 is relatively constant for all these interactions. As N-containing compounds cause more orbital spitting that water (Equation 3.46), the effect of forming a complex with a N-containing molecule will increase the orbital splitting. This leads to LSFE effects for certain ions (recall Mn^{2+} and Zn^{2+} have no LFSE), which enhances the stability of their complexes. Thus, the complexes of Zn^{2+} are weaker than those of Cu^{2+}. For example, considering complexes with ethylenediamine, the formation constant for $Zn(en)_2$ of 3.98×10^{10} is much smaller than that of $Cu(en)_2$ (3.98×10^{19}) but larger than that of $Mn(en)_2$ (5×10^3) [1]. Thus, the degree to which these different metal ions form complexes in solution is determined to a large degree by their electronic structure. The degree of complexation has important implication as many metals are mostly bioavailable to organisms as the free metal ion, and the relative concentration of the free ion is strongly dependent on their propensity to form complexes in solution.

3.2.3 Inorganic chemistry and complexation of transition metals

For the transition metals of the first series, the maximum oxidation state is equivalent to the total number of s and d electrons, although the stability of the higher oxidation states decreases from Ti^{IV} to Mn^{VII}. For the remainder of the metals, the higher oxidation states are difficult to find in the natural environment [1, 7]. Overall, the oxides are more acidic for higher oxidation states and the halide complexes are more covalent. In most cases, the presence of higher oxidation states is often dependent on both stereochemistry as well as experimental conditions and many complexes will exist in solution but are not stable in the solid phase. All of the elements of the first transition series are found in the +II oxidation state and a number of well-defined binary compounds are formed with halides and oxides. These compounds are essentially ionic in nature [7]. In addition, a large

Table 3.4 The concentrations of the major metals in aquatic systems and their most important oxidations states. Taken from a number of sources [1–3, 10, 11].

Element	Atomic Number	Main Oxidation States	Average Ocean Concentration (nM)	Average River Concentration (nM)*	Ratio River/Ocean
Al	13	III	20	1850	93
Cr	24	III, VI	4	19	5
Mn	25	II, IV, VII	0.3	145	490
Fe	26	II, III	0.5	720	1430
Co	27	II, III	0.02	3.4	170
Ni	28	II	8	8.5	1.1
Cu	29	II, I	4	24	5.9
Zn	30	II	5	460	92
As	33	III, V	23	23	1
Se	34	–II, IV, VI	1.7	–	–
Mo	42	VI	110	5.2	0.05
Ag	47	I	0.025	2.8	110
Cd	48	II	0.6	0.2	0.3
Sn	50	II, IV	0.004	0.01	2.5
Sb	51	III, V	1.6	8.2	5.1
Ce	58	II	–	–	–
Other Lanthanides	59–71	II	–	–	–
Hg	80	0, II	0.002	0.01	5
Pb	82	II	0.01	0.48	48
Th	90	IV	<0.001	–	–
U	92	VI	14	–	–

**Range of values found in a particular system vary by as much as one order of magnitude.*

number of complexes are also formed with the metals in the +II oxidation state. For V, the +II oxidation state is of little importance and for Cr, it is unstable to oxidation in solution. These metals tend to form oxyanions in solution which impacts their reactivity and fate and transport in the environment. For example, Cr is much more mobile in groundwater as the chromate anion (Cr^{VI}) as compared to its mobility as Cr^{III}, its more reduced state (Table 3.4).

For Mn, the +II oxidation state is relatively stable and the metal exists as an aqua ion in solution except at high pH where the hydroxide will precipitate at higher concentration. The hydroxide is easily oxidized by air and this fact is exploited in the wet chemical (Winkler) method used to measure oxygen in natural waters. Because of the insolubility of the hydroxides and oxides of Mn^{III}, its concentration in oxidized natural waters is generally much lower than that of reduced environments. Similarly, Fe^{II} is relatively easily oxidized in oxic waters, resulting in the formation of insoluble Fe^{III} (hydr)oxides. Furthermore, the rate of oxidation of Fe^{II} is kinetically faster than that of Mn^{II} [1]. The initial precipitate formed is a mixed oxide-hydroxide and is referred to as hydrous ferric oxide (HFO) [2]. Additionally, in solution, Fe^{III} is strongly hydrolyzed forming a number of complexes in solution, some of which involve Fe-Fe bonds. In contrast, complexes formed by Mn^{II} in solution have relatively low binding constants, as noted above, because of the lack of LSFE stabilization. Finally, Fe forms many highly colored complexes in solution due to the number of unpaired electrons in the high spin state (Fe^{III} is $Ar4s^0 3d^4$) [7].

For Co, and the remaining members of the transition series, the +II oxidation state is the most important although the stability of the Co^{II} ion in solution depends on the type of ligands present, as these can enhance the oxidation by forming stable Co^{III} complexes. In solution, Co^{II} is relatively insoluble due to the formation of the hydroxide. In contrast to most other transition metals, it tends to form tetrahedral complexes. Co has important chemistry as Co^{III} in solution and the complexes formed are relatively kinetically inert to ligand exchange reactions. Co^{III} forms strong complexes with N-containing ligands, such as NH_3 and ethylenediamine [1].

Nickel's chemistry is almost entirely confined to the +II state and this is especially true for natural waters. Ni exists as an aqua ion in solution and salts of the various complexes can be isolated from solution. Ni forms complexes in a range of coordination numbers (4–6) and in all the main structural types, making it different from the other transition metals. Copper also has most of its chemistry in the +II oxidation state but, in contrast to the other transition metals, can be found in the +I oxidation state. This oxidation state is stabilized by the presence of a filled *d* orbital. In solution, Cu^I is only stable at low concentration and it can disproportionate to Cu^0 and Cu^{II}. The stability depends on the presence of

Table 3.5 Relative values for various characteristics of important metals of the second and third transition series and their most important cations. Taken primarily from [3, 7, 8]. Electronegativity values are based on Pauling.

Element/Ion	Atomic Radii (pm)	Ionic Radii (pm)	Σ(Ionization Potential.) for ion (kJ mol^{-1})	Electronegativity
Mo^{2+}	136	92	2243	2.16
Pd^{2+}	138	86	2680	2.20
Pt^{2+}	138	85	2661	2.28
Ag^{+}	144	113	731	1.93
Au^{+}	144	137	890	2.54

Cu^{II}-complexing ligands in solution. Also, there is the potential for some Cu^{II} complexes to decompose to Cu^{I} through an internal electron transfer reaction involving the ligand. In solution, Cu^{II} exists as an aqua ion but there is some distortion due to Jahn-Teller effects [7], and often only tetra ligand complexes are formed (e.g., $[Cu(NH_3)_4]^{2+}$).

The +III oxidation state is also important for most metals of the first transition series, being of the least importance for Cu (Table 3.4). For V, the aqua V^{III} ion is readily oxidized in solution to V^{IV}. In contrast to other transition metals (Fe, Mn), Cr is relatively insoluble in its more reduced (Cr^{III}) state but is much more soluble in the +VI state, where it exists in solution primarily as an oxyanion. Most of the transition metal compounds formed in the +III state are ionic, except perhaps for the chloride complexes which have more covalent character (e.g., $FeCl_3$). In solution, most metals, in both the +II and +III oxidation states are found as aqua ions, which can hydrolyze and therefore form acidic solutions. The halides are also acidic in solution. For Cr, the +III complexes in solution are relatively inert kinetically to ligand exchange reactions and therefore the complexes are relatively stable compared to those of other transition metals [7]. For Mn, the aqua ions tend to disproportionate to Mn^{II} and Mn^{IV}. The oxides of Fe^{III} are relatively insoluble and these are the primary forms in which Fe is found in the environment. As a result, Fe concentrations in oxic waters are generally low compared to what may be expected for such an abundant metal. When reduced to Fe^{II}, it is much more soluble, except in the presence of sulfide where Fe-S solid phases form due to their low solubility. In the environment, many of the oxide have Fe in both the +II and +II oxidation states.

The +IV oxidation state is the most important oxidation state for Ti and for V, and to a lesser extent for the other metals (Table 3.4) [7]. In this state, Ti has no d electrons. In nature, Ti is often found as TiO_2, a compound with many industrial applications. In solutions, Ti^{IV} is present as hydrolyzed species. The halides are important compounds in many applications. In addition, the Ti^{III} oxidation state has extensive chemistry with the aqua ion, which is weakly acidic, existing in solution. For V, the +IV oxidation state is the most stable, and is found in many complexes, and again the oxo species dominate in solution, as VO^{2+} which contains a V=O bond [7]. Complexes are formed by the addition of ligands to the VO unit. Higher oxidations states occur for Cr, Mn, and Fe and these compounds are mostly in the form of oxyanions such as permanganate (MnO_4^-) and chromate (CrO_4^{2-}) and these compounds are strong oxidizing agents.

The chemistry of metals of the second and third row of the transition series do not often closely follow their counterparts in the first row. This is primarily due to differences in their ionization potentials, electronegativity and other characteristics (compare Tables 3.3 and 3.5). For many second and third row metals, lower oxidation states are much less stable than for the first row metals and they do not have significant chemistry in the +II oxidation state. For the metals of most interest to aquatic chemists, the differences that occur down a group can be illustrated by comparing the importance of Cr^{III} chemistry to the lack of importance of Mo^{III} and W^{III} chemistry [7]. Also, while Cr^{VI} oxyanions are strongly oxidizing, those of Mo and W are stable in solution. For these metals, the oxyanions are the most environmentally important species. Overall, among heavier transition metals, higher oxidation states are more stable. Many heavy transition metals do not exist in solution as aqua ions, rather they are found as oxyanions and are more likely to form metal-metal bonds than the metals of the first transition series. Generally, these metals form complexes in low spin configurations.

Many of the transition metals form bonds with carbon and these have a variety of applications. Organo-Ti compounds are important catalysts, for example, in organic chemical reactions, and V forms a variety of organic compounds with a range of oxidation states. The role of Fe in biochemistry is well documented [12, 13] and the main function of these compounds are oxygen transport and electron transfer. The ferredoxins in plants have a role in photosynthesis and contain Fe-S clusters that are involved in the electron transfer reactions, as discussed further in Chapter 8. The most well-known biological role for Co is in enzyme, vitamin B_{12}

(cobalamin). Copper is another transition metal with an important biochemical role which is linked to the Cu^{I}–Cu^{II} redox transitions, and are mostly involved in oxygen transport or in oxidases. Of the heavier transition metals, Mo is known to have an important biochemical role (see Chapter 8).

3.2.4 Inorganic chemistry and complexation of non-transition metals and metalloids

In the discussion previously, the elements of Group 12 have been implicitly included in the discussion. However, as they have a complete *d* orbital in almost all their chemistry, in both their elemental state and as ions, they are not truly transition metals. They cannot form complexes that involve LSFE, although they can still form higher order complexes. Their chemistry does not involve oxidation states greater than +2. There are no simple +1 cations in this group although Hg can form metal-metal bonds as in Hg_2^{2+}, and thus exists in the +1 state in this form [7]. However, Hg in the +1 state disproportionates into Hg^{2+} and Hg^0, given the equilibrium constant for $Hg^{I} - Hg^{II}$ ($K = [Hg^{2+}]/[Hg_2^{2+}] = 6 \times 10^{-3}$). Additionally, in most environmental media the free ion concentration of Hg^{2+} is a small fraction of the total Hg^{II} due to complexation, so there are few environments where Hg_2^{2+} will be thermodynamically stable as Hg_2^{2+} forms few complexes in solution [7].

Even though the Group 12 metals are not transition metals by definition, they do form complexes with CN^-, the halides, NH_3, amines and reduced sulfur compounds, but not with larger organic molecules in the same way as transition metals. However, there are large differences in their ability to form complexes, and the nature of the complexes varies from being mostly ionic for Zn to being relatively covalent for Hg, and therefore the Hg complexes are much stronger (Table 3.6). The chemistries of Cd and Zn are relatively similar, as can be noted by the similarities in the binding constants [7]. Mercury is different from the others. For example, the hydroxide of Zn is amphoteric, that of Cd basic, while $Hg(OH)_2$ is an extremely weak base. Also, Hg forms such strong complexes with ligands so that for all intent and purposes the Hg^{2+} cation to be irrelevant in its aqueous chemistry as its concentration is very low in most situations [1, 2]. This is not the case for Cd^{2+} and Zn^{2+}, which form weaker complexes in solution. Therefore, the uptake and toxicity of Zn and Cd are based on similar mechanisms and depend on the concentration of the free metal ion in solution while for Hg it has been demonstrated that neutrally-charged complexes are the most important in terms of toxicity in many situations [14]. Furthermore, the chloride complexes of Zn and Cd are ionic while those of Hg have strong covalent character, which can be seen by the large differences in the value of the binding constants (Table 3.6) [2]. However, all three metals form covalent compounds in certain cases [7]. Finally, Hg is more likely to form linear two-coordination compounds than Zn or Cd. Overall, as shown in Tables 3.3 and 3.7, their electronegativities are less than those of many of the transition metals. The "covalent" nature of the bonds is due to the shielding by the filled d orbitals and the lack of unfilled d orbitals which can be involved in complex formation with electron-donating ligands.

Table 3.6 Formation constants (log K values) for the various metal (M) complexes for the reaction: $M + nX^{y-} = MX_n^{2-ny}$. From Morel and Hering [2].

Metal	MCl_2	$M(OH)_2$	MS (s)	$M(EDTA)^{2-}$	$M(Ac)_2$
Zn	0.2	11.1	24.7	18.3	1.8
Cd	2.6	7.6	27.0	18.2	3.2
Hg	14.0	21.8	52.7	26.8	10.1

Note: Ac = acetate ion.

The elements of the next group in the Periodic Table, Group 13, of importance in aquatic chemistry are Al, Ga, and to a lesser extent, In and Tl. All these elements have two *s* electrons and one *p* electron in their outer shells (e.g., Al is $[Ne]3s^2 3p$). Thus, their chemistry is mostly confined to the +3 state as this affords the outer shell configuration of the noble gas. The bonds formed are predominantly ionic, and for the larger members of this group (In and Tl), the +1 oxidation state is also important. The +1 ion of Tl is stable in solution and has similar chemistry to other +1 ions such as Ag^+, and forms complexes with halides and other ligands in solution. As a result, Tl^+ can be toxic, especially as it can be taken up from solution as a neutral chloride complex (TlCl), in a similar manner to Ag (see Chapter 8).

The ions exist in solution as $M(H_2O)_6^{3+}$ and can accept donor molecules and anions to form tetrahedral complexes under some circumstances, or to even dimerize. In addition, the free ions are acidic and can donate a proton:

$$M(H_2O)_6^{3+} = M(H_2O)_5(OH)^{2+} + H^+ \text{ and } K_a(Al) = 1.12 \times 10^{-5} \quad (3.48)$$

The other elements of the group are more acidic. This tendency to hydrolyze accounts for the toxicity of Al to fish as low pH environments tend to result in the release of Al from the soils and sediments, where it is present as alumina (Al_2O_3) or other solids (e.g., Al-silicates) in high concentrations. Therefore, there is often a link between freshwater acidification, due to acid rain inputs or acid mine drainage, and Al toxicity. All the elements of the group form complexes in solution through the interaction of the aqua ion with halides, oxygen containing compounds and similar ligands. Alumina exists in many forms and is an important binding phase for cations and anions because its surface

Table 3.7 Relative values for various characteristics of important heavy metals and the metalloids that are important for environmental chemistry, and their most important cations. Taken primarily from [3, 7, 8]. Electronegativity values are based on Pauling.

Element (Major Ion)	Atomic Radii (pm)	Ionic Radii (pm)	Σ(Ionization Potential for ion (kJ mol^{-1})	Electronegativity
Zn^{2+}	133	83	2369	1.65
Cd^{2+}	149	103	2492	1.69
Hg^{2+}	160	112	2817	2.00
Al^{3+}	143	57	5119	1.61
Ga^{3+}	122	62	5521	1.81
Tl^{3+}	170	105	5438	2.04
Ge^{2+}	123	90	2299	2.01
Sn^{2+}	141	93	2120	1.96
Pb^{2+}	175	132	2166	2.33
As^{5+}	125	69	5480	2.18
Sb^{3+}	182	89	5071	2.05
Se^{2-}	215	191	–	2.55

contains amphoteric groups. However, the binding strength of cations to Al_2O_3 is generally less than that to Fe-oxide/hydroxide phases.

The metal(loid) elements in Group 14 (Ge, Sn and Pb) have one more electron in the p orbitals in their atomic state and it is more typical for these elements to be found in the +2 state, especially for the heavier elements, such as Pb ([Xe] $4f^{14}5d^{10}6s^2 6p^2$), as the filled s orbital acts somewhat as an "inert pair". The cationic +2 state is the most important for Pb, while for Sn ([Kr]$4d^{10}5s^2 5p^2$), the +4 oxidation state is also important. While Sn forms SnX_4 complexes with the halides, these are not found for Pb. The bonds formed by elements in this group increase in ionic character down the series, and thus are more ionic for Sn and Pb, and therefore they differ markedly in character from the elements at the top of the group (C and Si), which are considered non-metallic elements. Ge is clearly intermediate in character between the two extremes in this group of the periodic table. Similarly to C and Si, Ge has a tendency to form compounds with Ge-Ge bonds, but this occurs to a lesser extent with Sn, and even less so with Pb. This is primarily due to the fact that the strength of the X-X bond for the element decreases relative to the strength of bonds with other elements [7]. The strength of covalent bonding decreases down the group and Pb forms a variety of cationic complexes although many of the complexes are highly insoluble in solution.

The Pb^{2+} ion is partially hydrolyzed in water to form the hydroxide (log K ~ –7.9) and therefore forms an acidic solution in the absence of other complexes. Higher forms of the hydroxide also exist in solution, as discussed in more detail further on. As shown in Table 3.4, Pb is highly enriched in freshwaters compared to ocean waters and this is related to its strong anthropogenic signal and the fact that it is highly particle-reactive. Sn^{2+} is also hydrolyzed in water but forms primarily a trimeric $Sn_3(OH)_4^{2+}$ species rather than a simple hydroxide. The Sn^{II} state is unstable to oxidation by air and is a mild reducing agent that is used in analytical applications.

The elements of this group can form a series of organometallic compounds in a similar fashion to carbon. In these compounds, Sn and Pb are the +IV oxidation state. For Pb, the environmental chemistry and impacts of anthropogenically manufactured alkyl-Pb compounds ($(CH_3)_4Pb$ and $(C_2H_5)_4Pb$) have been well-studied, as discussed earlier in the book. Additionally, alkyl Sn compounds are also well-known and studied as they have been used as anti-fouling agents on ships, as discussed further in Chapter 8.

The metalloids of most environmental interest are in Groups 15 and 16. Arsenic, and its associated elements, Sb and Bi, all have three electrons in the outer *p* orbital and therefore need to gain or share three electrons to obtain a noble gas configuration. No cationic chemistry is found for As and there is little for Sb, but Bi forms a hydrated ion in solution. Again, the "metallic" character increases down the group. For these elements the oxyanions dominate their solution chemistry and As is thermodynamically favored in the +5 state in the oxic environment but can also be found in the +3 oxidation state. Arsenic acid (H_3AsO_4) is the main form of As in solution although there is often measurable quantities of As^{III} and methylated As compounds in environmental waters. This is primarily due to the reduction and methylation of As^{IV} by microorganisms, as discussed in Chapter 8. The concentration of As in seawater is similar to its average freshwater value (Table 3.4) and this is different from most of the other trace metals and metalloids. This is one reason for the relative toxicity of As in seawater as it can be taken up instead of phosphate, given the similarity in the structure and speciation of their anions, especially from low phosphate open ocean seawater (see Chapter 8). Antimony does have some cationic chemistry in solution in the trivalent state, but mostly as antimonyl compounds (with the SbO^+ ion). Of environmental importance are compounds formed between As and S (As_xS_y) and Sb also forms similar complexes. Both can also be found in the environmental as oxides.

Selenium is another metalloid of environmental interest. Because Se is from Group 16, and has four outer *p* electrons ([Ar]$3d^{10}4s^24p^4$), it can exist in the –II oxidation state (e.g., H_2Se). Its chemistry is somewhat analogous to that of S (from the same group) and it is also acidic in solution and volatile as H_2Se. Overall, Se has similar chemistry to S in that it also forms oxyanions in the +IV and +VI oxidations states, mainly in solution, and that it can bind with metals when in its reduced –II oxidations state (e.g., forming compounds such as HgSe). The chemistry of Se is important as it is both a required element for organisms but is can also be toxic, most likely due to the fact that it can substitute for S in important biochemical pathways (see Chapter 8). Of all the elements that are both required by organisms but can also be toxic at higher concentration, Se is the element with the smallest "window" between these two states (i.e., as shown in Fig. 1.2) [13] and therefore relatively small changes in concentration and speciation can have a large impact on its toxicity to organisms.

3.3 Complexation of metals in solution

3.3.1 Inorganic complexation

In solution, the equilibrium distribution of the various complexes of a trace metal can be determined by considering the relative strength of the complexes formed, if the concentrations of the ligands in solution that can interact with the metal are known. Complexation is not important for major elements, but is very important for minor elements like trace metals. Some of the major ions are found as free anions and cations in solution and this is because they are derived from strong acids and bases (e.g., Na^+, Cl^- from NaOH and HCl) that are fully dissociated in water and do not significantly interact with water or other dissolved species (so-called *spectator ions*). Minor elements form complexes with dissolved ligands and also react with water. Many trace metals can be considered to be weak acids and bases, as they tend to hydrolyze in water. The equilibrium (*formation*) constant gives an indication of the strength of the complex formed. Values are low for electrostatic binding (e.g., for $NaSO_4^-$, apparent log K = 0.34 in seawater; for KSO_4^-, apparent log K = 0.13). For metals, binding constants (given here as log K values, I = 0) are much larger and variable between metals. This is shown by the data in Table 3.6, for example, and in the compendium of data of the more important equilibrium constants for trace metals in solution gathered in Appendix 3.1. Generally, aquatic chemists focus on reactions occurring in dilute solutions of low ionic strength as this mostly reflects the environmental reality. This allows approximation of interactions as those of "ideal solutions" and concentrations become a reasonable surrogate for the actual activity of the metal or ion in solution. For most freshwater environments, the overall ionic strength is low and it is not necessary to consider ionic strength effects. However, this is not so for estuarine (brackish) waters and seawater (Section 3.1.2).

Attempts to classify and contrast the complexation of metals in solution have led to different conceptual approaches to classification. One is to classify metals and ligands as either *hard* or *soft* acids or bases [1–3], with the cations being considered as Lewis acids. Soft refers to those cations where the electron cloud is easily deformed or polarizable, and these cations would be able to participate in "covalent" bonding by donating an electron while the hard cations would not. Furthermore, hard acid cations would preferably bond to hard bases (ligands), such as in the formation of KOH. Overall, the alkali and alkali earth metals, the elements at the beginning of the transition series (e.g., Ti, Cr, V, Fe, Mn in their most oxidized form (e.g., Fe^{3+}, Co^{3+}), the lanthanides and actinides are hard acids, while elements to the right of the periodic table are more likely to be soft acids (e.g., Hg, Cd, Ag, Au, Cu^+) [3]. Borderline acids would be some of the transition metals in their more reduced state (e.g., Fe^{2+}, Co^{2+}, Ni^{2+}) and metal ions such as Cu^{2+}, Zn^{2+}, Pb^{2+}, Sn^{2+} and Bi^{2+}, which can act as soft acids with particular ligands. Similarly, ligands from strong acids are hard, such as OH^- and most oxyanions (e.g., CO_3^{2-}, SO_4^{2-}) while soft bases include ligands containing S (e.g., HS^-, $S_2O_3^{2-}$) and N entities (e.g., CN^-, SCN^-).

Another approach has been to classify metals in groups (A, B and C) where the classification is based on the electronic configuration. Class A cations have a noble gas configuration and a complete outer shell. Class A therefore includes the alkali and alkali earth metals, Al^{3+} and some early transition metals in their more oxidized state (Sc^{3+}, Ti^{4+}, Y^{3+}, Zr^{4+}). Class B metals are those with an electronic shell equivalent to Ni^0, Pd^0 or Pt^0 and therefore includes most of the elements in Groups 11 and 12 as cations (e.g., Cu^+, Ag^+, Au^+, Zn^{2+}, Cd^{2+}, Hg^{2+}) and some other heavy metals in their higher oxidation states (e.g., Sn^{4+}). Cations such as Pb^{2+} are borderline Group B. Class C cations are intermediate and have partially filled inner *d* shells (e.g., Mn^{2+}, Fe^{3+}, Fe^{2+}, Co^{2+}, Cu^{2+}).

It can be seen that the two classifications are generally comparably but there are cations that do not fall either neatly into one of the other classification or are not correspondingly classified. For example, hard acids would mostly likely be Class A cations but this is not always the case, and Class C metals are both hard and borderline acids. Most Class

B metals are soft acids, except for Zn^{2+}. These classifications do allow for understanding and prediction of the potential of a specific cation to form a complex with a particular ligand. For example, Hg^{2+} forms strong complexes with S and N-containing ligands and weaker complexes with OH^-, although overall Hg^{2+} forms much stronger complexes with all ligands compared to alkali earth metal cations, for example. On the contrary, Fe^{3+} forms strong complexes with hard acid groups and weaker complexes with S-containing ligands.

In the environment, the concentration of a particular species is determined to a large degree by its reactivity and the source strength of its inputs. Concentrations of many metal(loid)s that are reactive and bind strongly to particles, or are taken up by organisms are generally low in aquatic systems. This is also true of any metal(loid) that is relatively insoluble in water, and for some metal(loid)s, solubility is low for seawater due to its generally higher pH relative to freshwater ecosystems. For elements whose input to the ocean is dominated by terrestrial runoff (river and groundwater inputs), the same processes that control their ocean concentration likely also control their input strength from the land. Overall, trace metal(loid) concentrations and distributions in aquatic systems are generally controlled by: (1) their degree of complexation as this leads to an enhancement of their dissolved concentration relative to the controlling solid phases; (2) the extent of sorption to suspended and other solid surfaces; (3) their propensity for precipitation; and (4) their degree of biological uptake. We will consider each of these interactions in detail in this and following chapters. Typical concentrations of dissolved metal(loid)s in environmental waters were given in Table 1.1, and the concentrations of some of the more important elements in seawater and freshwater are given again (Table 3.4). Many of the transition metals that are not abundant in crustal material have a ratio of river/seawater concentration that is relatively close to 1 (0.1–10). The exceptions are Zn and Co which appear to have relatively high river concentrations compared to their concentration in the ocean. One potential reason for this, as discussed in later chapters, is the fact that both of these elements are actively taken up by phytoplankton as they are required in essential enzymes (Zn in carbonic anhydrase and Co in cobalamin) [12]. Alternatively, these elements, as well as elements such as Pb and Ag, are likely effectively removed from surface seawater by particulate matter as they are known to be relatively particle reactive.

It can be clearly seen that some metal(loid)s exist in much higher concentrations in freshwater than seawater and these are mostly the crustal metal(loid)s that are highly insoluble in the high pH, saline waters of the ocean (Fe, Al, and Mn), and which have a strong terrestrial source due to their release from crustal material by weathering reactions. Both Fe and Mn are much more insoluble in their oxidized state compared to their reduced state and the concentrations given are those for oxic surface waters. In low oxygen waters, the concentration of these elements can be higher due to the dissolution of hydroxide or oxide solids. Also, for example, for Fe, the concentration of Fe^{3+} ion in the presence of the solid phase hydroxide is much less than the measured dissolved concentration in seawater, suggesting that this higher concentration is due to the formation of dissolved complexes. As discussed next, these complexes may not all be truly dissolved species as Fe can form strong associations with dissolved natural organic matter (NOM). Also, Fe could exist as colloidal particles in a 0.4 μm filtered water sample, used by most investigators to separate the dissolved and particulate phases.

3.3.2 An approach to determining metal(loid) speciation in solution

Before we proceed further with the discussion of metal(loid)s in aquatic systems, and the role of speciation, we must discuss approaches to estimating the speciation of the element of interest in natural waters. As noted earlier, we will be using the approach of considering the equilibrium speciation as a surrogate for steady state that pertains in these ecosystems. In discussing the approach to analytically calculating the metal(loid) speciation, we need to first consider the difference between the equilibrium constants for consecutive reactions (K) and the overall formation constants (β) as many trace metal(loid)s in solution form more than one complex with a ligand, with differing number of ligands bound to the central metal(loid). An example will suffice to explain the difference. Consider the interaction of the mercury ion with chloride ions in solution. The various reactions are given by:

$$Hg^{2+} + Cl^- = HgCl^+ \quad K_1 = [HgCl^+]/[Hg^{2+}][Cl^-]$$

$$HgCl^+ + Cl^- = HgCl_2 \quad K_1 = [HgCl_2]/[HgCl^+][Cl^-]$$

$$HgCl_2 + Cl^- = HgCl_3^- \quad K_3 = [HgCl_3^-]/[HgCl_2][Cl^-]$$

$$HgCl_3^- + Cl^- = HgCl_4^{2-} \quad K_4 = [HgCl_4^{2-}]/[HgCl_3^-][Cl^-]$$

The overall reaction is:

$$Hg^{2+} + 4Cl^- = HgCl_4^{2-} \quad \beta_4 = [HgCl_4^{2-}]/[Hg^{2+}][Cl^-]^4$$

So, what is the relationship between β_4 and K_4?

$$K_4 = [HgCl_4^{2-}]/[HgCl_3^-][Cl^-] \qquad (3.49)$$

and as:

$$[HgCl_3^-] = K_3[HgCl_2][Cl^-] \qquad (3.50)$$

and with a similar substitution for the other complexes, it can be shown that:

$$K_4 = [HgCl_4^{2-}]/K_1K_2K_3[Hg^{2+}][Cl^-]^4 \qquad (3.51)$$

or

$$\beta_4 = K_1K_2K_3K_4 \text{ or } \beta_i = \prod_1^i K_i \quad (3.52)$$

Thus, there is a simple relationship between the equilibrium constants for the consecutive reactions and the formation constants for the overall reactions. It is more useful in the solving of simultaneous equilibrium equations to write the reactions in terms of their formation constants rather than as equilibrium constants. From the above, it is obvious that $K_1 = \beta_1$ and that, in general, for the complexes:

$$\beta_i = [ML_i]/[M][L]^i \quad (3.53)$$

where the charges have been omitted for simplicity in the expression.

Complexation is predictive, if based on equilibrium theory, but it is difficult to verify these predictions as few suitable methods exist for determining the actual chemical species in solution. Most real situations are not truly at equilibrium but at steady state, under conditions where the chemical distribution amongst species approaches that of an equilibrium solution. Also, equilibrium constants are normally determined using simple solutions, often with the assistance of computer modeling to generate equilibrium constants. Thus, it is possible that other species exist in more complex solutions that are not accounted for by the values given in the literature (i.e., a complex is often only known to exist or not after someone has determined the relevant equilibrium constant). For example, in the reactions previously, it is possible to have, in addition to the reactions with chloride, reactions with water:

$$Hg^{2+} + OH^- = HgOH^+$$

$$HgOH^+ + OH^- = Hg(OH)_2$$

and what about?

$$Hg^{2+} + OH^- + Cl^- = HgOHCl$$

How could one experimentally determine if the complex HgOHCl existed? Experiments that vary both pH and chloride concentration could be used to fit the data to a set of constants and not realize that there was the mixed constant in solution. There are many instances in the literature of studies that have re-examined the equilibrium predictions and constants of earlier studies. Many of the constants found in thermodynamic databases were measured years ago and it is possible that these constants are inaccurate due to limitations in the technology available when the measurements were made. Students should always be aware that calculations and modeled estimates of chemical equilibria are only as good as the data used. For example, the reported equilibrium constants for the binding of Hg to sulfide complexes vary by several orders of magnitude, depending on the source of the information [15]. Overall, if the equilibrium constant for the metal-ligand interaction is not known or has not been measured, it is not possible to determine its potential for formation in solution using the methods discussed next. This is one limitation of the thermodynamic approach to determining the aquatic chemistry of the species of interest. However, comparing thermodynamic predictions to laboratory and/or field results has led to the discovery of interactions not known to exist. These examples illustrate that much can be gained from the use of thermodynamic equilibrium models.

As equilibrium equations are mathematical expressions, they can be manipulated accordingly. For example, for the following two reactions (the dissociation of the water and the reaction of carbon dioxide with the hydroxide ion) the overall reaction and equilibrium constant are as follows:

$$H_2O = H^+ + OH^- \quad \text{Log } K = -14$$

and

$$CO_2 + OH^- = HCO_3^- \quad \text{Log } K = 7.7$$

Adding the expressions for the overall reaction:

$$CO_2 + H_2O = H^+ + HCO_3^- \quad \text{Log } K = -6.3$$

Or, alternatively:

$$H_2O = H^+ + OH^- \quad \text{Log } K = -14$$

and

$$HCO_3^- = CO_2 + OH^- \quad \text{Log } K = -7.7$$

Subtracting

$$H_2O - HCO_3^- = H^+ - CO_2 \quad \text{Log } K = -6.3$$

or

$$CO_2 + H_2O = H^+ + HCO_3^-$$

Here the equilibrium constants are given as log values to the base 10 and are written as log in the text in contrast to the natural log which will always be written as ln.

3.3.2.1 A method for solving problems

Analytical solutions to problems concerning the speciation and complexation of an element in a relatively simple solution are possible. In more complex situations, spreadsheet calculations can be used to obtain a solution. Alternatively, the problem can be simplified if assumptions can be made about which are the dominant or controlling reactions. Another approach is an outgrowth of the Tableau method for solving equilibrium problems put forward by Morel and Hering [2] and which is also the structure and method of solutions of equilibrium problems by computer programs such as MINEQL+ (see Chapter 4). The Tableau method will be discussed in more detail later in the section describing methods for determining the concentration of the free anions for polyprotic acids and other ligands. An approach for determining metal speciation in solution is best demonstrated through the use of a specific example. While such a problem can be solved using the Tableau method, it can also be readily solved without using such an approach.

Example 3.1

Consider the situation of Zn^{2+} added to a freshwater solution. For now, the Zn concentration is not important except that it is low enough to prevent precipitation of solid phases such as $ZnCO_3$ (s) and $Zn(OH)_2$ (s). The pH is 6 ($[OH^-] = 10^{-8}$ M) and the total anion concentration, of importance for this calculation, is that of "typical freshwater" (5×10^{-4} M carbonate, 2×10^{-4} M chloride, 10^{-4} M sulfate). The speciation at equilibrium will be a mixture of all the possible complexes that can form between Zn and the anions in solution. This can be written in the form that follows, where the appropriate complexes are determined by examination of thermodynamic databases, such as that in Appendix 3.1:

$$\text{Total Zn} = [Zn^{2+}] + [ZnCl^+] + [ZnCl_2] + [ZnCl_3^-] + [ZnSO_4] + [ZnOH^+] + [Zn(OH)_2] + [Zn(OH)_3^-] + [Zn(OH)_4^{2-}]$$

Note that Zn does not form a dissolved carbonate or bicarbonate complex although there is the possibility of precipitation of the solid phase. For each complex in this equation, the concentration can be expressed in terms of an expression involving the appropriate β constant. For example:

$$[ZnCl^+] = \beta_{1Cl}[Zn^{2+}][Cl^-]$$

where the β_{ij} constant refers to the particular equilibrium constant for i number of the j ligand bound to the central metal ion. After these substitutions are done for all complexes and written in terms of the β expressions, all expressions on the right hand side of the equation contain the concentration of the free metal ion and therefore the expression can be simplified to:

$$\text{Total Zn} = [Zn^{2+}](1 + \beta_{1Cl}[Cl^-] + \beta_{2Cl}[Cl^-]^2 + \beta_{3Cl}[Cl^-]^3 + \beta_{1SO4}[SO_4^{2-}] + \beta_{1OH}[OH^-] + \beta_{2OH}[OH^-]^2 + \beta_{3OH}[OH^-]^3 + \beta_{4OH}[OH^-]^4)$$

All the expressions inside the bracket can be solved if the overall anion composition of the solution is known and using values for the formation constants from the literature. The most dominant complexes will therefore be those with the largest numerical value for the corresponding expression and therefore the equation can typically be solved by inspection. If the concentrations in solution are written in terms of the log values (to the base 10) as the equilibrium constants are typically given in the literature (Appendix 3.1), the expression can be simply solved. For this example, the solution for total Zn (Zn_T), for the above conditions, is:

$$\begin{aligned} Zn_T &= [Zn^{2+}](1 + 10^{0.4}10^{-3.7} + 10^{0.2}(10^{-3.7})^2 \\ &\quad + 10^{0.5}(10^{-3.7})^3 + 10^{2.1} \cdot 10^{-4} + 10^{5}10^{-8} \\ &\quad + 10^{11.1}(10^{-8})^2 + 10^{13.6}(10^{-8})^3 + 10^{15.5}(10^{-8})^4) \\ &= [Zn^{2+}](1 + 10^{-3.3} + 10^{-7.2} + 10^{-10.6} + 10^{-1.9} \\ &\quad + 10^{-3.0} + 10^{-4.9} + 10^{-10.4} + 10^{-16.5}) \end{aligned}$$

The largest numerical value is 1 in this expression, and so the free metal ion concentration is the most dominant form of Zn in freshwater in the presence of the chosen ligands. Of the complexes in solution, the sulfate, hydroxide and chloride are present as a small fraction of the total. If the Zn total concentration is known, the concentration of each complex can be determined. Their relative concentration can however be determined based on the relative importance to the expression in brackets. Adding all the values in the brackets, we get:

$$Zn_T = 1.014[Zn^{2+}]$$

and

$[Zn^{2+}]$ = 99% of the Zn in solution and $[ZnSO_4]$ = 1.2%. All others, <1%

It is necessary to check that none of the Zn would precipitate under the conditions of the experiment. For example, if the total Zn concentration was 500 nM (the average river concentration; Table 3.4) would the solids precipitate? The solid phases that could form are the hydroxide and carbonate, and their solubility is given by their respectively *solubility product constants* (K_{sp}) (Appendix 3.1). Note that given a pH of 6 and a total carbonate concentration of 5×10^{-4} M, the concentration of $[CO_3^{2-}]$ is 2.5×10^{-8} M (the details of how this value was achieved is outlined later in Section 3.3.3). The $[Zn^{2+}]$ concentration is 0.99×500 nM = 495 nM. The solids will precipitate if the *ion activity product* (IAP) of the complex exceeds that of the value of the K_{sp} or alternatively if the ratio IAP/K_{sp} >1. The solubility expressions are, for 0.5 μM Zn^{2+}:

$$K_{sp}(ZnCO_3) = 10^{-10}$$

and

$$IAP = [Zn^{2+}][CO_3^{2-}] = 0.495 \times 10^{-6} \cdot 2.5 \times 10^{-8} << 10^{-10}$$

Also

$$K_{sp}(Zn(OH)_2) = 10^{-15.5}$$

and

$$[Zn^{2+}][OH^-]^2 = 0.495 \times 10^{-6} \cdot 10^{-16} << 10^{-15.5}$$

So, neither of these solids would precipitate under the given conditions. It is interesting to investigate under what conditions precipitation would occur. As the formation of the solid occurs when the solubility product is exceeded, at a pH of 6 there would need to be unrealistically high Zn or ligand concentration for this to happen. Even at a higher pH, precipitation would not be likely at typical environmental $[OH^-]$ and $[CO_3^{2-}]$ concentrations, even though these increase dramatically with pH.

At what pH would the precipitation occur? This question can be solved by equating the ratio IAP/K_{sp} to 1 and therefore determining the minimum concentration that would cause precipitation. Therefore, for the hydroxide solid:

$$[OH^-]^2 = 10^{-15.5}/0.5 \times 10^{-6} = 10^{-9.2} \text{ or } [OH^-] = 10^{-4.1} \text{ (pH} = 9.9)$$

Of course, this calculation is assuming that no other complexes of Zn become important at the higher pH. At a higher pH, the carbonate will become the dominant form in solution and as shown in detail below in this chapter, above a pH equivalent to the acid dissociation constants, the acid anion (i.e., carbonate) is the dominant form of the acid. For the carbonate system, this occurs at a pH above 10.3 (pK_a = 10.3 for $HCO_3^- = H^+ + CO_3^{2-}$). Then $[CO_3^{2-}] = 5 \times 10^{-4}$ M for this problem. However, the system would exceed saturation when:

$$[CO_3^{2-}] = 10^{-10}/[Zn^{2+}] = 10^{-4.7}\ (1.9 \times 10^{-5}\ M)$$

which would occur at a pH slightly less than the pK_a.

Thus, at the low environmental concentrations typically found for Zn in freshwater, it appears that the precipitation of one of these solid phases will not be a controlling factor at low and typical pH, but the carbonate could precipitate under conditions of high dissolved carbonate or in very high pH media. Other processes must be controlling the dissolved Zn concentration in environmental media at low pH, as discussed further next.

Similar speciation calculations can be done for other metals. In Table 3.8, the calculated speciation is given for a number of +2 cations and for Ag^+ at pH 6 and 8, assuming 50 nM of each metal and the concentrations of anions as above. The calculations were done using the MINEQL+ computer program [17] and therefore relied on the equilibrium constants in that database. Solving complex equilibrium calculations using computer programs is the topic of the next chapter in this book. The equilibrium constants in that database may differ slightly from those in Appendix 3.1, as constants reported in the literature are not always consistent among sources. However, the general trends illustrate something about the speciation and chemistry of most of the trace metals in freshwater, in the absence of organic matter. As discussed later in this chapter, the complexation of these metals by organic matter can have an important impact on their dissolved speciation. However, under the conditions prescribed here, where the anion concentrations are as previously, and the metals are present at environmentally-relevant concentrations, the trends are illustrative of similarities and differences between the various trace metals. Note here that the potential complexation of the anions by the major ions in solution has not been considered. Major ions are relatively abundant compared to trace metals in most natural waters, and especially in seawater. The alkali and alkaline earth metals form relatively weak complexes, but because of their elevated concentration in seawater, they may still bind with anions to the extent that the free anion concentration is decreased. Binding of major ions to anions in high ionic strength media is discussed in detail in Section 3.3.4.

In most instances, the metals are primarily in solution as the free ion at the lower pH, especially those of the first transition series, where the free metal ion is >90% of the total metal in solution (Table 3.8). The fraction of Ag present as a free metal ion is lower (71%). Silver and Cd form stronger complexes with Cl and these complexes are present at low concentrations in freshwater. For the transition metals, the sulfate complexes are present as a low fraction of the total. For some of the transition elements and for Pb, the carbonate and bicarbonate complexes are present as a relatively low fraction of the total. Calculations were also

Table 3.8 The calculated speciation of metals in solution in freshwater (5×10^{-4} M carbonate, 2×10^{-4} M chloride, 10^{-4} M sulfate) at pH 6 and 8, for a 50 nM solution of each cation. The values in the table represent the percentages as each complex. Percentages of <1 are not shown and thus the lack of an entry in the table does not mean that the complex does not exist but that it is a minor component of the metal in solution. The speciation was calculated using the MINEQL+ computer program [17].

Metal	Free ion	$M(OH)^{n-1}$	$M(OH)_2^{n-2}$	MSO_4^{n-2}	MCO_3^{n-2}	$MHCO_3^{n-1}$	MCl^{n-1}
pH = 6							
Ag^+	70.5	–					28.8
Cd^{2+}	95.9	–	–	2.2	–	–	1.8
Cu^{2+}	90.3	2.9	–	2.1	3.8	–	–
Ni^{2+}	96.2	–	–	1.9	–	1.8	–
Pb^{2+}	82.1	2.1	–	4.0	1.8	9.4	–
Zn^{2+}	97.2	–	–	2.1	–	–	–
pH = 8							
Ag^+	70.5						28.8
Cd^{2+}	90.5	–	–	2.1	4.7	–	1.7
Cu^{2+}	5.4	17.3	3.5	–	73	–	–
Ni^{2+}	84.8	1.1	–	1.7	7.2	5.1	–
Pb^{2+}	9.1	23	21	–	62.7	3.3	–
Zn^{2+}	77.7	7.8	1.2	–	10.2	1.2	–

run for Hg, an important environmental trace contaminant, under these conditions, but the results are not included in the table as they are much different from those of the other metals. Importantly, under these conditions, and with 5 pM total Hg, $[Hg^{2+}] = 6.8 \times 10^{-19}$ M (0.68 aM), and is a very small fraction of the total metal. Accordingly, in most instances, the Hg free metal ion is a trivial fraction of the total Hg concentration in environmental solutions. The major species in solution at pH 6 is the mixed complex, HgOHCl, which accounts for 43.7% of the total; $Hg(OH)_2$ accounts for 8%; and $HgCl_2$ 48.2%. Interestingly, all these complexes have an overall neutral charge as compared to the other metals where the major species in solution are charged.

There are two major mechanisms for metal uptake from solution by microorganisms, such as phytoplankton and bacteria, as discussed in detail in Chapter 8. The most common mechanism is uptake of the free metal ion through a channel especially designed for these metals, some of which are required nutrients as they are present in enzymes [12, 13]. Alternatively, metals can be accumulated through the uptake of neutrally-charged complexes by passive diffusion [2, 14]. As can be seen in the calculations earlier, at low pH and in the absence of organic complexes, most of the transition metals are present as the free ion, with little of the metal in solution as a neutral complex, such as M^{2+}-carbonate complexes. Because the fraction as the carbonate is higher for Pb than for the transition metals, passive transport is a potential mechanism for Pb accumulation by microbes, in addition to Pb uptake through a channel designed for a cationic species. While Zn and, to some degree, Cu and possibly Cd, are required elements for phytoplankton growth, Pb has no biochemical role and its accumulation as a free metal ion is likely through channels designed for the uptake of transitions metal ions [13]. The interactions between metals and microorganisms are discussed in more detail in Chapter 8.

At a higher pH, there is an increase in the importance of the hydroxide complexes (Table 3.8). With increasing pH, carbonate complexes also become very important for Cu and Pb due to the increase in the CO_3^{2-} concentration, even at the same total carbonate concentration. For Hg, increasing pH results in dominance of hydroxide complexes ($Hg(OH)_2$ 94.8%; HgOHCl 5.2%). There is little influence of pH on Ag and Cd speciation because the Cl concentration is not pH dependent. Note that at this pH very little of Pb and Cu (<10%) are in solution as a free metal ion, in contrast to the situation at pH 6. In contrast, for Ni and Zn, the majority of the metal is still present in solution as the free metal. This example demonstrates important differences in speciation dependent reactivity and processes among metals, highlighting the need to consider metal speciation in solution if a scientist is to properly understand factors controlling metal(loid) fate, transport and bioaccumulation.

The example given previously and the related discussion provide an approach for determining the speciation of a metal of interest in an environmental sample. The example also illustrates that most transition metals in the first series are present as the free metal ion in relatively dilute solutions at low pH in the absence of organic molecules. Over the last 20 years research has revealed that for many metals in environmental matrices complexation to the major anions in solution are of secondary importance to the relatively strong bonds formed with natural organic matter. Natural organic matter is a "chemical soup" derived from release of organic compounds from biotic and abiotic sources, and by their continual degradation and transformation in natural waters. The binding of metals to natural organic matter is discussed further in Section 3.3.5.

3.3.3 The chemistry and speciation of metal-binding ligands

Many of the inorganic ligands in solution that are important for trace metal complexation in natural waters are acids and bases. Some of the ligands are the anions of strong acids and are therefore essentially fully dissociated in water (i.e., their free ion concentration is equivalent to their total concentration). Such anions include the halides (Cl^-, Br^-, F^-) and nitrate. The concentration of the anions of many of the other somewhat weaker acids are a function of pH but there is often a dominant form at the pH of typical natural waters (>5–8.5). This is true for sulfate (SO_4^{2-}) and acetate (CH_3COO^-), and most monoprotic low molecular weight carboxylic acids [1]. However, for many of the polyprotic acids, the pH has an impact on the concentration of the anion in solution and a particular free ion concentration will not be equivalent to the total concentration [2]. Additionally, some of these ligands form relatively strong bonds with the major ions in freshwater and seawater and therefore it is necessary to take the concentration of these cation complexes into account. Again, an example will suffice to demonstrate the need to consider the anion speciation and complexation in many instances.

Example 3.2

Consider the example previously, but instead of Zn as the metal, consider Cu which forms relatively strong complexes with the carbonate ion (Table 3.8). Also, assume that the major ions in freshwater are present at "typical" concentrations, and for this exercise, these will be set as:

$Na^+ = 0.3$ mM; $K^+ = 0.06$ mM; $Mg^{2+} = 0.15$ mM; $Ca^{2+} = 0.4$ mM.

The anion concentrations are those of Example 3.1 and for simplicity in the example, the pH is 8. For the calculation, the total Cu concentration is taken as 50 nM.

Comment: Because the concentration of Cu is much lower than that of any of the anions, any complexation of Cu with one of the anions cannot impact the overall anion concentration. Therefore only the major ions, which have concentrations similar or greater than the anions, can impact their free ion concentration through complexation.

For this example, we will estimate the carbonate speciation. Examining the formation constants in Appendix 3.1 shows that there is the potential for the following complexes to form, with the log values of the formation constants given next:

$NaCO_3^-$	$NaHCO_3$	$CaCO_3$	$CaHCO_3^+$	$MgCO_3$	$MgHCO_3^+$
1.27	10.08	3.2	11.59	3.4	11.49

For the cation carbonate species, the equilibrium constant refers to the following expression:

$$M^{n+} + CO_3^{2-} = MCO_3^{n-2}$$

For the bicarbonate species, the equilibrium constant refers to the following expression:

$$M^{n+} + H^+ + CO_3^{2-} = MHCO_3^{n-1}$$

As evaluated here, the following expression can be written:

$$\text{Total carbonate} = [H_2CO_3] + [HCO_3^-] + [CO_3^{2-}] + [NaCO_3^-] + [NaHCO_3] + [CaCO_3] + [CaHCO_3^+] + [MgCO_3] + [MgHCO_3^+]$$

There are obviously two issues associated with solving this problem. As carbonic acid is a diprotic acid, the relative amount of each form is dependent on the pH of the system. So, the concentration is not constant if pH changes. Also, there is the potential for the metal complexes to reduce the overall concentration of the uncomplexed species. This is explained next. The full details, fundamental background, and alternative methods of solving the equations for the total concentration of each of the carbonate forms are beyond the scope of the book, but will be summarized in the following. The acid–base chemistry of natural waters is comprehensively covered in Stumm and Morgan [1] and Morel and Hering [2], and many other texts. The acid dissociation constants for the carbonate system are:

$$K_1 = 10^{-6.3} = [H^+][HCO_3^-]/[H_2CO_3] \quad (3.54)$$

and

$$K_2 = 10^{-10.3} = [H^+][CO_3^{2-}]/[HCO_3^-] \quad (3.55)$$

Dissociation of carbonic acid is represented by a combined constant due to the fact that CO_2 in water can exist as both a dissolved gas and as carbonic acid (H_2CO_3). With this information, we can construct an equilibrium expression determining the relative concentration of these two species. However, in the discussions that follow and generally in the book, these two forms of dissolved CO_2 will not be treated separately and the expressions will be written as H_2CO_3 (i.e., that both forms react similarly in solution, with the implicit assumption that the notation used refers to H_2CO_3 plus dissolved CO_2). In other texts [1, 2], this is denoted as $H_2CO_3^*$.

To solve the sub-problem one can use "chemical intuition" (something professors espouse but students may not comprehend) to deduce that at a pH of 8, the concentration of the bicarbonate species is essentially equivalent to that of the total carbonate concentration (See Fig. 3.8). Alternatively, as the pH is known, the ratio of the components can be determined from the two expressions, or the two expressions can be combined to eliminate one variable. It is also possible to write both expressions in terms of two of the components. At a pH of 8, the following ratios result from the previous expressions:

$$[HCO_3^-]/[H_2CO_3] = 10^{1.7}$$

and

$$[CO_3^{2-}]/[HCO_3^-] = 10^{-2.3}$$

Therefore, the carbonate concentration is small compared to the bicarbonate concentration, and the bicarbonate concentration is large compared to the carbonic acid concentration. Although it is reasonable to assume that the bicarbonate concentration is essentially the same as the total carbonate concentration, a more rigorous calculation can be made by combining the two equilibrium expressions previously. This reveals that the bicarbonate concentration is 97.6% of the total (here the total is the total amount of carbonic species not complexed to any of the cations). From this, it follows that the carbonate concentration is 0.49% of the total and the carbonic acid concentration is 2% of the total uncomplexed species. To simplify the solution of this problem, we assume that the only form of the non-complexed anion of importance is HCO_3^-. Then, the expression can be written in terms of the formation constants for the different species, with the $[CO_3^{2-}]$ concentration being the common variable in all the equations:

$$\begin{aligned}
C_T &= [HCO_3^-] + [NaCO_3] + [NaHCO_3^-] + [CaCO_3] \\
&\quad + [CaHCO_3^+] + [MgCO_3] + [MgHCO_3^+] \\
&= [CO_3^{2-}]([H^+]/10^{-10.3} + [Na^+]10^{1.27} + [Na^+][H^+]10^{10.08} + [Ca^{2+}]10^{3.2} \\
&\quad + [Ca^{2+}][H^+]10^{11.59} + [Mg^{2+}]10^{3.4} + [Mg^{2+}][H^+]10^{11.49}) \\
&= [CO_3^{2-}](10^{2.3} + 10^{-2.23} + 10^{-1.44} + 10^{-0.2} + 10^{0.19} + 10^{-0.42} + 10^{-0.33}) \\
&= [CO_3^{2-}](202.5) \text{ or } 10^{2.307}[CO_3^{2-}]
\end{aligned}$$

This analysis shows that the dominant species in solution under these conditions is the bicarbonate ion. For the metal ions, the calcium complexes are the most important, followed by the Mg complexes, and the Na complexes are very minor components.

So, given that $C_T = 5 \times 10^{-4}$ M:

$$[CO_3^{2-}] = 10^{-2.307} C_T = 2.47 \times 10^{-6} \text{ M } (0.5\%)$$

$$[HCO_3^-] = 4.93 \times 10^{-4} \text{ M } (98.6\%)$$

$$[CaHCO_3^+] = 3.8 \times 10^{-6} \text{ M } (0.76\%)$$

$[CaCO_3] = 1.56 \times 10^{-6}$ M (0.32%) and so on.

Note that the concentrations and relative amount of the carbonate species is slightly different to those determined initially earlier. However, since the carbonic acid concentration is actually higher than the concentration of the major ion complexes, our above assumption ignoring this complex was inaccurate. When possible, it is always useful to estimate the error associated with such assumptions, particularly in situations where certain species are ignored in the calculations. The recalculated system, with the inclusion of H_2CO_3 provides a useful additional problem and exercise for the reader to perform. Mostly, for the application of speciation calculations in the context of this book, errors that are less than a few percent are generally considered acceptable.

The differences are clearly because carbonic acid was ignored as a species in the calculation and its concentration is not insignificant. However, the calculation does illustrate that, under these conditions, the complexation of the carbonate species by the major ions is relatively minor, accounting for less than 1% of the total carbonate in solution. A similar calculation could be done for the other anions in the example, using the same approach. Note that the changing speciation with pH that occurs with the anions that are bases of weak acids does not occur with the anions of strong bases.

Aside: A more rigorous general set of equations can be derived to define the acid-base chemistry of acid systems. For a monoprotic acid ($HB = H^+ + B^-$), the following equations can be used to analytically obtain a solution. For example, to calculate the anion concentration at a specific pH, recall:

$$K_a = [H^+][B^-]/[HB]$$

$$K_W = [H^+][OH^-]$$

and, for the charge balance equation:

$$[H^+] = [B^-] + [OH^-] \quad (3.56)$$

and, for the mass balance equation:

$$C_T = [HB] + [B^-] \quad (3.57)$$

Thus, for example, as $[HB] = C_T - [B^-]$:

$$[B^-] = K_a C_T([H^+] + K_a)^{-1} \quad (3.58)$$

If the pH is not known, but the total concentration of acid is given, then the analytical solution can be derived by substitution in the previous equations (Equations 3.56–3.58), which have four unknowns. This can be done, for example, by eliminating [OH-] from the charge balance Equation 3.56

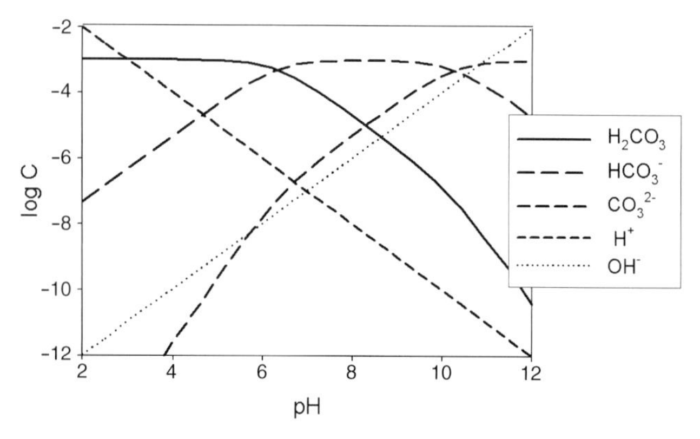

Fig. 3.8 Log C-pH diagram for the carbonate system. The relative concentration of each species is shown across the pH range. Total carbonate concentration is 10^{-3} M. The various lines overlap at the pH of the pK_as of the acid–base pairs. Figure created using MINEQL+, a computer program.

and then substituting for [B-] (Equation 3.58). This results in the following expression:

$$[H^+] = K_a C_T([H^+] + K_a)^{-1} + K_W/[H^+] \quad (3.59)$$

which can be rearranged and written as a cubic equation:

$$[H^+]^3 + K_a[H^+]^2 - (K_a C_T + K_W)[H^+] - K_a K_W = 0 \quad (3.60)$$

This equation can be solved by trial and error, or by using scientifically valid assumptions to yield a simpler expression. For example, one such assumption is that, for most weak acids, even though the dissociation of the acid is small, the resulting solution will be acidic and therefore $[H^+] >> [OH^-]$ and then, from Equation 3.60:

$$[H^+] = K_a C_T([H^+] + K_a)^{-1}$$

Or

$$[H^+]^2 + K_a[H^+] - K_a C_T = 0 \quad (3.61)$$

The resultant estimation of $[H^+]$ could then be substituted into Equation 3.60, and the other equations, to determine if the approximation is reasonable. Another approximation for a weak acid is that generally at equilibrium $[HB] >> [B^-]$ because the acid is only slightly dissociated, and therefore $[HB] \sim C_T$. When combined with the previous approximation, $[H^+] = [B^-]$, yields:

$$[H^+]^2 = K_a C_T \quad (3.62)$$

These expressions allow us to determine the pH and concentrations of various species in an acidic solution [1]. In the majority of cases dealt with in this book, problems will be solved for trace metal speciation in solutions where the pH is known and the concentration of the acid anion can be readily determined using Equation 3.59.

For diprotic acids and polyprotic acids, the analytical solution for determining the pH or $[H^+]$ concentration of the solution is beyond the scope of this book. However, in most cases it is possible to simplify the problem so that equations similar to those used for the monoprotic acid earlier can be applied [1, 2]. For a diprotic acid, generally only two of the forms of the acid will be dominant in solution at a particular time because of the pH. This is shown for the carbonate system in Fig. 3.8. For a diprotic acid, the following expressions can be used to solve the equations:

$$H_2A = H^+ + HA^- \quad K_{a1} = [H^+][HA^-]/[H_2A] \quad (3.63a)$$

$$HA^- = H^+ + A^{2-} \quad K_{a2} = [H^+][A^{2-}]/[HA^-] \quad (3.63b)$$

$$\text{Total concentration, } C_T = H_2A + HA^- + A^{2-} \quad (3.64)$$

It is possible through substitution to derive relationships for each species in terms of constants and the $[H^+]$ concentration:

$$[H_2A] = C_T \cdot (1 + K_{a1}/[H^+] + K_{a1} \cdot K_{a2}/[H^+]^2)^{-1} \quad (3.65a)$$

$$[HA^-] = C_T \cdot (1 + [H^+]/K_{a1} + K_{a2}/[H^+])^{-1} \quad (3.65b)$$

$$[A^{2-}] = C_T \cdot (1 + [H^+]/K_{a2} + [H^+]^2/K_{a1} \cdot K_{a2})^{-1} \quad (3.65c)$$

These expressions can be represented in terms of log concentrations and pH, and can be approximated as linear relationships for pHs that are not close to the pK_as. Firstly, when pH = pK, then the two species are equivalent in concentration, and this is equivalent to half the total concentration. For example, for Equation 3.64, when the pH = pK_{a1}, then $[H_2A]$ = $[HA^-]$ and $[A^{2-}]$ is very small in concentration so $C_T = [H_2A] + [HA^-]$. Considering Equation 3.65(a), when the pH << pK_{a1} << pK_{a2}, then the denominator approaches 1 and $[H_2A] \sim C_T$ or as shown in Fig. 3.8, $\log[H_2A] = \log C_T$. When the pH is between the pK_as, $K_{a1}/[H^+]$ is much greater than the other terms in the denominator and the expression simplifies to: $[H_2A] \sim C_T.(K_{a1}/[H^+])^{-1}$ or $\log[H_2A] = \log C_T + pK_{a1} - pH$. This is a linear line with a negative slope of −1. For the situation where pH > pK_{a2}, the value of $K_{a1}.K_{a2}/[H^+]^2$ in the denominator is much greater than the other terms and the expression becomes: $\log[H_2A] = \log C_T + pK_{a1} + pK_{a2} - 2pH$. This is a line with a negative slope of −2, as shown in Fig. 3.8. Similar expressions can be derived for the other acid-base species using the relationships above (Equations 3.65b and 3.65c)

Figure 3.8 is generic in terms of the total concentration, C_T. As discussed, at a pH << pK_{a1}, the dominant form of the acid is $[H_2CO_3]$ and essentially $[H_2CO_3] = C_T$. Similarly, when pH >> pK_{a2}, $[CO_3^{2-}] = C_T$. Finally, in the pH region numerically between the two pK_a values, $C_T = [HCO_3^-]$. It is only when the pH is close to a pK_a that more than one species in solution needs to be considered. While these approximations will need to be checked in any instance, and their validity depends on the accuracy required in the calculation, they can be used in most speciation calculations to simplify the problem. For most speciation calculations, given the errors inherent in the determination of the equilibrium constants, an error of <5% is more than acceptable. Thus, if the concentration of the minor acid species is two orders of magnitude less, then its impact on the overall speciation can be mostly discounted in solving the acid-base equations.

Based on these discussions, the estimates of concentration are most complex in the region of the pK_as. Consider the situation at pHs around the pK_{a1} in our example (Fig. 3.8). Here it is possible to simplify the solution by ignoring the very small relative concentration of $[CO_3^{2-}]$ that is present, when doing the calculations. Thus:

$$K_{a1} = [H^+][HCO_3^-]/[H_2CO_3]$$

and

$$C_T = [HCO_3^-] + [H_2CO_3]$$

and

$$[H^+] = [HCO_3^-] + [OH^-]$$

if the contribution of the carbonate species is ignored. These equations are essentially equivalent to those for a monoprotic acid discussed above (Equations 3.55–3.60), and can be solved similarly. Note that for the carbonate system, in this pH range, it cannot be assumed that $[H^+] >> [OH^-]$.

While this approach can be used to solve equations for simple solutions of one acid-base system, for equations with more than one set of ligands and pH dependent speciation, another approach is required. One solution is to set up a set of simultaneous equations in such a manner that they are more easily solved. For example, the results for metal speciation above were obtained by writing all the species in an equation so that each expression had the same component, which could then be factored out of the expressions to aid in their solution. A similar approach has been defined for the solution of complex problems, especially those where the pH is not known *a priori*, and this approach has been used and adapted for the computer programs that are available for solving equilibrium problems.

The approach is to set up a matrix, in which all the species in the solution of interest are written in terms of a minimum number of *principal components*, which for ease of solution should be the dominant forms in solution. Such a construction, and the approach, has been termed the *Tableau Method*. The details of how to develop and construct a Tableau is discussed in detail in Morel and Hering [2] and here it is sufficient to define the primary factors that need to be considered in using this approach, and in its application. Overall, it is an extension of the approaches used above to solve acid–base systems analytically, and is not the only way to solve the problems, but does have some efficiency. The basic approach is to choose a set of *principal components* (in essence, these should be the major components for each group of species in solution) that can be used to define the overall system and all the species in solution.

Firstly, because of the nature of the systems being examined, it is always chosen that [H+] and H_2O are principal components [2], and the reason for this will become more apparent through the examples discussed next. The other components are chosen so that *all species can be expressed stoichiometrically as a function of the components chosen* [2], and that there is only *one unique way of expressing each species* in terms of the chosen components that is, the number of components is minimized. Overall, the number of components is equal to the number of species in the system minus the number of independent reactions that take place in the system. For example, for water, there are three species ($[H^+]$, $[OH^-]$ and H_2O) and one reaction, the dissociation of water, and therefore the system can be described by the two components, $[H^+]$ and H_2O.

In using the Tableau approach to solving problems, it is advantageous in the solution that the principal components chosen are those that are the dominant species in the system (Example 3.3). This leads to equations containing mainly the minor species, which provides the easiest solution to the problem, and allows the resultant equations to be simplified. However, in theory, it is possible to solve the problem regardless of the choice of principal components, as demonstrated in detail in Morel and Hering [2]. This is the approach taken in the computer programs where the principal components are chosen and fixed for all problems, and computing power is used to solve the problems. While the Tableau approach has been used and can provide mechanisms for solving complex problems, this is obviously not the only approach and there are other mechanisms that can be used to solve the complex interactions between cations and anions, which often have acid–base chemistry in solution. For example, it is possible to solve a problem using an iterative approach, using hand calculations or a computer spreadsheet.

Example 3.3

To illustrate the Tableau approach, consider a system containing Cu and carbonate at a pH of 8. Given the results in Table 3.8, we will consider only the major complexes of Cu at this pH. As noted previously, at this pH, bicarbonate is the dominant species and therefore this is chosen as a component, along with $[Cu^{2+}]$, $[H^+]$ and H_2O (Table 3.9). The Tableau as set up has the individual equilibrium equations written in each row, in terms of the components. The equilibrium constant given for the specific reaction, as written in the Tableau, may be different from that given, for example, in Appendix 3.1. For example, the reaction for $CuCO_3$ (Table 3.9) is, given the principal components:

$$CuCO_3 = HCO_3^- - H^+ + Cu^{2+}$$

which is the combination of the following two equations:

$$CuCO_3 = Cu^{2+} + CO_3^{2-} \quad \log K = -6.7$$

$$H^+ + CO_3^{2-} = HCO_3^- \quad \log K = 10.3$$

Therefore, the overall log K is a combination of the two reactions, and is equivalent to 3.6 for the reaction as written. All the values for the equilibrium constants in the Tableau are written in this manner in terms of the actual reaction that is, for the equation in terms of principal components.

At the bottom of the Tableau is written the total concentration as the system is defined, that is, what was added to constitute the system. In this case, the problem is 50 nM Cu in a solution at pH 8, with a total carbonate concentration of 5×10^{-4} M. The column for the hydrogen ion is often the most useful equation for solving the Tableau and does not have a concentration associated with it in this example, as no strong acid was added to the solution. This equation, termed the TOTH equation [2], can be written as:

$$[H^+] - [OH^-] - [CO_3^{2-}] + [H_2CO_3] - [CuCO_3] - [Cu(OH)^+] - 2[Cu(OH)_2] = 0 \tag{3.66}$$

The approach to solving the equation is to eliminate minor species to the extent possible, and then solve the remaining equation. As noted previously, the Cu complexes are relatively minor and can be ignored in terms of the acid-base chemistry. Also, at a pH of 8, it can be assumed that $[H^+] << [OH^-]$ and so the simplified equation is:

$$[OH^-] + [CO_3^{2-}] = [H_2CO_3] \tag{3.67}$$

This equation can be solved given that the pH is known and by assuming that the bicarbonate concentration is essentially equivalent to the total carbonate concentration, as discussed above for this pH. Once the concentration of the carbonate species is known, the concentration of the various Cu species can then be determined. It is worth noting again that there are other ways to solve this problem and Equation 3.67 could have been reached from consideration of the charge balance of the system, and the mass balance for the carbonate species, without constructing a Tableau.

Table 3.9 The Tableau for the system of copper in water containing a fixed concentration of carbonate and at a fixed pH.

Species	$[H^+]$	$[HCO_3^-]$	$[Cu^{2+}]$	H_2O	Log K
$[H^+]$	1				
$[OH^-]$	−1			1	14
H_2O				1	
$[HCO_3^-]$		1			
$[CO_3^{2-}]$	−1	1			−10.3
$[H_2CO_3]$	1	1			+6.3
$[Cu^{2+}]$			1		
$[CuCO_3]$	−1	1	1		3.6
$[Cu(OH)^+]$	−1		1	1	−7.7
$[Cu(OH)_2]$	−2		1	2	−16.2
Total Concentration (M)	0	10^{-4}	5×10^{-8}	55.4	

3.3.4 The complexation of the major ions in solution

It is possible for the association between the anions and the major ions in solution to be important even though these species do not form strong complexes in solution, due to their relatively high concentration in seawater and some other natural waters. To estimate and demonstrate their importance, a number of different situations are considered. The conditions described earlier in Example 3.2 were used with the MINEQL+ program to determine the speciation of all the major anions in the presence of the major elements both at a pH of 6 and 8. The results are presented in Table 3.10. In freshwater, the cations are essentially present as free ions. For both Ca and Mg, there is a small fraction of the total cation as the sulfate complex. For understanding trace metal(loid) speciation, it is the speciation of the anions that are of most interest. We can see that Cl^- forms no complexes with major ions. For sulfate, a large fraction of the sulfate (92.2%) is present as the free ion, but there is some significant fraction as the Ca and Mg complexes. Thus, if sulfate complexation is to be determined in freshwater, one should take into account the reduction in the free anion concentration due to the complexation of sulfate by Ca and Mg. For the carbonate system, none of the major cation complexes are important species, but clearly, the pH of the system has an impact on the relative amounts of carbonic acid, bicarbonate and carbonate, and also on the absolute concentration of the associated complexes. Thus, in considering the complexation of trace metals with carbonate and bicarbonate, it is important to know the overall pH and/or alkalinity of the system to solve the problem. In the examples previously, the pH does not however have a dramatic impact on the relative speciation in terms of the cation complexes. At pH 6, the dominant species are H_2CO_3 and HCO_3^- and this is also the case at pH of 8, except that H_2CO_3 is a much more minor component.

The speciation of the anions and cations was also determined under conditions representative of seawater (Table 3.10). Again, these calculations were done with the MINEQL+ program [17] which made the appropriate ionic strength corrections to the equilibrium constants based on the overall ionic strength. Values for γ taken from the

Table 3.10 Calculated speciation for major ions in freshwater at two different pHs and at typical concentrations of the major cations and anions in freshwater. A similar calculation at the higher ion concentrations in seawater (pH 8.1) is also shown. Calculations were done using the computer program MINEQL+ [17].

Species	Log C (pH 8)	% total anion	% total cation	Log C (pH 6)	% total anion	Log C Seawater	% total anion	% total cation
Na^+	−3.52	–	100	−3.52	–	−0.31	–	98.2
K^+	−4.22	–	100	−4.22	–	−2.01	–	97.7
Ca^{2+}	−3.4	–	97.6	−3.4	–	−2.08	–	83.1
Mg^{2+}	−3.83	–	98.3	−3.83	–	−1.36	–	86.3
Cl^-	−3.7	100	–	−3.7	100	−2.6	100	–
SO_4^{2-}	−4.04	92.2	–	−4.02	92.2	−1.89	43	–
HCO_3^-	−3.32	96.4	–	−3.8	31.7	−2.75	70.8	–
CO_3^{2-}	−5.58	–	–	−8.07	–	−4.54	1.2	–
H_2CO_3	−4.99	2.1	–	−3.47	68.1	−4.65	–	–
HSO_4^-	−10.1	–	–	−8.1	–	−8.44	–	–
OH^-	−6.0	–	–	−8.0	–	−5.9	–	–
$NaCO_3^-$	−7.92	–	–	−10.4	–	−3.61	2.7	–
$NaHCO_3$	−7.13	–	–	−7.61	–	−3.61	9.9	–
$CaCO_3$	−5.96	–	–	−8.43	–	−4.6	–	–
$CaHCO_3^+$	−5.54	–	–	−8.01	–	−4.16	2.8	–
$MgCO_3$	−6.66	–	–	−9.14	–	−4.17	2.7	–
$MgHCO_3^+$	−6.22	–	–	−6.7	–	−3.7	8	–
$NaSO_4^-$	−6.91	–	–	−6.91	–	−2.06	28.9	1.7
KSO_4^-	−7.48	–	–	−7.48	–	−3.64	–	2.3
$CaSO_4$	−5.25	5.8	1.4	−5.23	5.8	−2.8	5.3	16
$MgSO_4$	−5.77	1.7	1.1	−5.76	1.7	−2.18	22	13.2
$CaOH^+$	−8.17	–	–	−10.2	–	−7.12	–	–
$MgOH^+$	−7.28	–	–	−9.29	–	−5.11	–	–

literature, calculated using the Davies equation, are given in Table 3.11. Note that the effect of ion size (parameter A in Equation 3.26) is relatively small. The pH was 8.1, and the program was set so that there was no precipitation under the conditions specified. The total concentrations of the cations were: $Na_T = 0.5\,M$; $K_T = Ca_T = 10^{-2}\,M$; $Mg_T = 5 \times 10^{-2}\,M$; and the cations: $Cl_T = 0.55\,M$; $SO_{4T} = 3 \times 10^{-2}\,M$; $CO_{3T} = 2.5 \times 10^{-3}\,M$.

Overall, under the much higher major cation concentrations of seawater one cannot completely ignore complexation of the major cations with the major anions. There is no change for Cl, but both Na and K have a small fraction of their total as sulfate complexes. Again, as we are concerned more here with determining the free anion concentrations available for complexation with trace metals, it is the effect of complexation on their concentration that is more important. The majority of the sulfate is present as complexes (57%), with $NaSO_4^-$ dominating (29% of the total sulfate) compared to $MgSO_4$ (22%) and $CaSO_4$ (5.3%), but all being important. If one is to estimate complexation of trace elements to sulfate in seawater, the impact of the major ions on the sulfate ion concentration needs to be considered. This is also the case for the carbonate system as complexes form an important fraction of the total: $NaHCO_3$ 10%, $MgHCO_3^+$ 8%, with $NaCO_3^-$, $CaHCO_3^+$ and $MgCO_3$ all being present at close to 3%. In seawater, because of the high pH, the bicarbonate ion is the major form of the uncomplexed acid, and accounts for 71% of the total carbonate.

There are two approaches to solving equilibrium calculations in seawater, or other matrices where the complexation of the anions by major ions is important. One approach is to initially determine the speciation for the system in terms of the major ions, and to then use this free anion estimation for calculating the speciation of the anions with the trace elements. For most of the major anions, their concentration in seawater is relatively constant and does not change significantly from location, and so this is a valid approach. For example, this adjustment of the anion concentration can be done by assuming that $[SO_4^{2-}] = 0.43SO_{4T}$, based on the results in Table 3.10. Recall, as well, that the equilibrium constants need to be adjusted for ionic strength effects, as discussed in Section 3.1.2. An example will suffice to examine these complications (Example 3.4).

Table 3.11 Values for the activity coefficients for ions, both positive and negative, as determined by the application of the Davies equation, discussed previously [1, 2].

I (M)	γ, Z = 1	γ, Z = 2	γ, Z = 3
0.001	0.95	0.87	0.72
0.01	0.89	0.66	0.40
0.05	0.81	0.45	0.17
0.1	0.78	0.36	0.10
0.3	0.74	0.30	0.068
0.5	0.71	0.25	0.045

Example 3.4

Consider the speciation of Cd in seawater. The total Cd concentration in seawater is typically around 1 nM, and the major anion and cation concentrations will be those of Table 3.10, with a pH of 8.1. From Appendix 3.1, considering only those complexes that are likely to be important, the speciation of Cd is given by:

$$Cd_T = Cd^{2+} + CdOH^- + Cd(OH)_2 + CdSO_4 + Cd(SO4)_2^{2-} + Cd(SO4)_3^{4-} + CdCl^+ + CdCl_2 + CdCl_3^- + CdCl_4^{2-}$$

It is worthy to examine the potential complexes and simplify the problem, if possible. For example, the concentration of $[OH^-]$ is $10^{-5.9}\,M$ and $[Cl^-]$ is 0.55 M. So, the hydroxide complexes would only be more important than the chloride complexes if the equilibrium constants for the hydroxide complexes were substantially larger than those of the chloride complexes. They are not, and by inspection one can eliminate these as unimportant species. For the other species, the equilibrium constants need to be adjusted for ionic strength through the use of activity coefficients (Section 3.1.2), or the activity coefficients need to be included in the relevant expressions. Thus, for the tri-sulfate complex:

$$[Cd(SO4)_3^{4-}] = \beta_3[Cd^{2+}][SO_4^{2-}]^3 \cdot \gamma_{Cd} \cdot (\gamma_{SO4})^3/\gamma_{CdSO43} \qquad (3.68)$$

where the free sulfate ion concentration is $0.43SO4_T$. If the problem is to be solved analytically, it may be expedient to determine which are the dominant complexes based on the total concentrations and the unadjusted equilibrium constants prior to doing all the ion strength calculations. Or the system could be relatively easily solved through the use of a spreadsheet program if more than one metal is to be analyzed for, or if a range of concentrations are to be considered. Recall from Table 3.5 that the dominant complexes in freshwater were the free ion, the sulfate complex and the chloride complex. Considering the concentration of free sulfate = $0.43C_T = 0.43.3 \times 10^{-2} = 10^{-1.89}$, and the general activity coefficients (Table 3.11) (I = 0.5), $\gamma(Z = 0) = 1.12$; $\gamma(Z = 1) = 0.71$; $\gamma(Z = 2) = 0.25$; $\gamma(Z = 3) = 0.045$; $\gamma(Z = 3) = 0.004$; and the relative values of the equilibrium constants, the expression above becomes, considering only the sulfate and chloride complexes, and with the γ values expressed as exponents:

$$\begin{aligned} Cd_T = {} & [Cd^{2+}](1 + 10^{2.3} \cdot 10^{-1.9} \cdot (10^{-0.6})^2/10^{0.05} + 10^{3.2} \cdot (10^{-1.9})^2 \cdot (10^{-0.6})^3/10^{-0.6} \\ & + \cdot 10^{2.7} \cdot (10^{-1.9})^3 \cdot (10^{-0.6})^4/10^{-2.4} + 10^{2.0} \cdot 10^{-0.26} \cdot 10^{-0.6} \cdot 10^{-0.15}/10^{-0.15} \\ & + 10^{2.6} \cdot (10^{-0.26})^2 \cdot 10^{-0.6} \cdot (10^{-0.15})^2/10^{0.05} \\ & + 10^{2.4} \cdot (10^{-0.26})^3 \cdot 10^{-0.6} \cdot (10^{-0.15})^3/10^{-0.15} \\ & + 10^{1.7} \cdot (10^{-0.26})^4 \cdot 10^{-0.6} \cdot (10^{-0.15})^4/10^{-0.6}) \\ = {} & [Cd^{2+}](1 + 10^{-0.85} + 10^{-1.8} + 10^{-3} + 10^{1.14} + 10^{1.13} + 10^{0.72} + 10^{0.06}) \\ = {} & [Cd^{2+}] \cdot 10^{1.54} \end{aligned}$$

Given a total Cd concentration of 1 nM, $[Cd^{2+}]$ = 2.9% of the total or 29 pM. The major species are:

$[CdCl^{+}]$ = 39.8% or 0.4 nM

$[CdCl_2]$ = 38.9% or 0.4 nM

$[CdCl_3^{-}]$ = 15.1% or 0.15 pM

$[CdCl_4^{2-}]$ = 3.3% or 33 pM

$[Cd^{2+}]$ = 2.9% or 29 pM

If a number of problems are to be solved for a specific metal, it could be useful to define apparent equilibrium constants that are relevant to the ionic strength of the solutions. For example, for the tri-sulfate complex earlier:

$$K^{C} = \beta_3 \cdot \gamma_{Ca} \cdot (\gamma_{SO4})^3 / \gamma_{CdSO43} \quad (3.69)$$

The calculations of the speciation of metals in seawater are therefore a fairly straightforward but rather tedious task. As will be discussed in the next chapter, these calculations are more conveniently done using an equilibrium computer program.

To follow on from the calculations previously, the speciation of the metals in seawater, at the concentrations given in Table 3.4, were estimated using the MINEQL+ program (Table 3.12) [17] and with the assumption that there were no strong organic complexing agents present. The results for Cd are very similar to those calculated earlier analytically. Essentially all the Cd is bound up in chloride complexes. Note that the MINEQL+ database has a constant for the complex CdOHCl which is of some importance. Thus, Cd also forms mixed complexes, as had been noted earlier for Hg. In the high chloride environment of seawater, complexation to Cl is much more important than in freshwater (compare Tables 3.12 and 3.8). Although not shown in the table, in all cases, it is the higher order chloride complexes that are the most important. For Ag and Cd, the free metal ion concentration is a small fraction of the total but this is not so for Zn and Ni where the free ion is the largest fraction of the total. Given the high pH of seawater and the relatively high carbonate concentration, these species are important for those metals that form strong associations with O-containing ligands.

3.3.5 Metal complexation with low molecular weight organic ligands

In addition to the interaction with inorganic ligands, ions can also interact with organic molecules. There are many organic molecules in the environment that are charged in solutions and many of these are the ions of weak acids and bases. For example, low molecular weight carboxylic acid ions such as acetate and oxalate are found in natural waters. In addition, amino acids released from microbes and other biochemicals can also interact with metal ions. In addition to low molecular weight compounds there are also complex organic molecules that are formed through the degradation of organic matter, such as humic and fulvic acids, than can have a large impact on metal speciation. We will first consider the interaction with simple molecules and then the complexities of the interactions with large molecular weight organic matter.

The complexation of metals with simple organic ligands can be dealt with in a similar fashion to that used for the inorganic ligands described in the previous sections [1–3]. Thus, if the binding ligand is acetate, or another simple organic acid, or a chelate such as EDTA or ethylenediamine, then it is sufficient to consider the complexation by these ligands in a similar fashion to that for the inorganic anions. As shown in Appendix 3.1, many of these organic ligands bind strongly to metals and thus can be important in environmental systems. Additionally, most of these ligands are weak polyprotic acids and their speciation needs to be considered, as discussed in Section 3.3.3. An example will be sufficient to demonstrate the manner for considering complexation of metals by organic ligands and there are many other examples in other texts [1, 2].

Table 3.12 Speciation of metals in seawater (percentage of the total) estimated using MINEQL+ and the concentrations for the metals as given in Table 3.4. The concentrations of the major ions are as in Table 3.10.

Metal	Free ion	$\sum M(OH)_x^{n-x}$	MCO_3^{n-2}	$\sum MCl_x^{n-x}$	Others
pH = <<8.1				99.7	
Ag^{+}					
Cd^{2+}	2.8	–	–	95.8	CdOHCl 1.4
Cu^{2+}	26.6	45.1	21.5	5.9	
Ni^{2+}	72.8	–	–	26.1	
Pb^{2+}	6.4	7.2	62.7	19.5	
Zn^{2+}	48	2.2	–	29.3	ZnOHCl 19.9

Example 3.5

For this example, recall the freshwater problem (Example 3.1, pH = 6) and consider only the +2 cations. We will estimate the speciation in the presence of citric acid (10^{-6} M) plus the inorganic ligands previously considered. Citric acid (Cit) is a triprotic acid with the following β values:

$$\beta_1 = 6.40 \text{ (i.e., for } Cit^{3-} + H^+ = HCit^{2-}\text{)}; \beta_2 = 11.16; \beta_3 = 14.29$$

The associated pK_a values can be determined from these formation constants as:

$$pK_{a1} = 3.13; pK_{a2} = 4.76; pK_{a3} = 6.40$$

Given the pH of 6, it is possible to determine by inspection of the pK_a values that the dominant uncomplexed species will be Cit^{3-} and $HCit^{2-}$. Note again that as the total citrate concentration is three orders of magnitude greater than the metal concentrations, their complexation has no impact on the major speciation of citrate. The results of the calculations below are that Cit^{3-} constitutes 31.3% of the total uncomplexed acid (3.1×10^{-7} M or $10^{-6.5}$ M) ; $HCit^{2-}$ is 65.1% ($10^{-6.19}$ M) and H_2Cit^- is 3.4% ($10^{-7.47}$ M). Thus, the likely dominant reactions are those that involve the dominant species. As done previously, the speciation of the various metals can be estimated if the concentrations of the ligands are known for the complexes under consideration. From the MINEQL+ database [17], the following complexes for Zn are found with citrate: $ZnCit^-$, $Zn(Cit)_2^{4-}$, $ZnHCit$ and ZnH_2Cit^+. The speciation calculation is therefore given as:

$$\begin{aligned}\text{Total Zn} = &[Zn^{2+}] + [ZnCl^+] + [ZnCl_2] + [ZnCl_3^-] + [ZnSO_4] + [ZnOH^+] \\ &+ [Zn(OH)_2] + [Zn(OH)_3^-] + [Zn(OH)_4^{2-}] + [ZnCit^-] + [Zn(Cit)_2^{4-}] \\ &+ [ZnHCit] + [ZnH_2Cit^+]\end{aligned}$$

The formation constants for the various metal-citrate species are given in Table 3.13, and following Example 3.1, we can write all the complexes in terms of their formation constants, the $[H^+]$ concentration and the concentration of the various inorganic ligands and $[Cit^{3-}]$. Recall that the magnitude of all the inorganic species is equivalent to 1.014 $[Zn^{2+}]$ from the earlier example. So therefore:

$$\begin{aligned}\text{Total Zn} = &[Zn^{2+}](1.014 + 10^{6.21} \cdot 10^{-6.5} + 10^{7.4} \cdot (10^{-6.5})^2 + 10^{10.2} \cdot 10^{-6} \cdot 10^{-6.5} \\ &+ 10^{12.84} \cdot (10^{-6})^2 \cdot 10^{-6.5})\end{aligned}$$

It can be seen from inspection that the only important Zn complex with citrate is $ZnCit^-$ and so:

$$\text{Total Zn} = [Zn^{2+}](1.014 + 10^{-0.29}) = [Zn^{2+}] \cdot 1.53$$

and Zn^{2+} = 65%, inorganic complexes = 0.9% and $ZnCit^-$ = 34% of the total Zn.

Table 3.13 Equilibrium (formation) constants for the various complexes formed between citrate and some transition metals. Data are taken from Appendix 3.1, which was compiled from various sources.

Metal Reactions	Cd	Cu	Ni	Pb	Zn
$M^{2+} + Cit^{3-} = [MCit]^-$	4.98	7.57	6.59	7.27	6.21
$M^{2+} + 2Cit^{3-} = [M(Cit)_2]^{4-}$	5.90	8.90	8.77	6.53	7.40
$M^{2+} + 2H^+ + Cit^{3-} = [MH_2Cit]^+$	12.9	13.23	13.30	–	12.84
$M^{2+} + H^+ + Cit^{3-} = [MHCit]^0$	9.44	10.8	10.5	–	10.20
$M^{2+} + H^+ + 2Cit^{3-} = [MH(Cit)_2]^{3-}$	–	–	14.9	–	–

The computer program MINEQL+ was used to generate results for the various metals under the conditions described previously (also see Table 3.14). As these calculations include all metals and all species, the results are slightly different from that obtained with the analytical solution outlined earlier. In addition, slight differences arise from the simplifying assumptions made in solving the problem analytically. In addition, to estimate the impact of ionic strength corrections in this instance, the model was rerun assuming that the ionic strength was 10^{-3} M (Table 3.15). This resulted in further small changes in the overall speciation, which is due to the impact of ionic strength corrections on the relative magnitude of the various equilibrium constants.

Overall, when ionic strength corrections were included, the importance of the free metal ion increased in all cases and this makes sense given the impact of the activity coefficients on the magnitude of the interaction for the M-citrate complexes: mostly, it will decrease the binding constant in this specific example, that is:

$$K^C \text{ (the adjusted constant)} = K \cdot \gamma_{2+} \cdot \gamma_{3-}/\gamma_- = 0.66K \quad (3.70)$$

These results provide another example of the impact of ignoring the ionic strength effects on the results of equilibrium calculations, even at a relatively low ionic strength.

The calculations show, in most instances, that the presence of 10^{-6} M citrate would have an important impact on the solution speciation. This is especially so for Cu, Pb and to a lesser extent for Ni as the organic complex is the mostly the dominant form of the metal in solution. This example is illustrative of the potential impact of complexation by organic acids on the speciation of metals in solution. It is unlikely, however, that the concentration of citrate in environmental waters is as high as in this problem. Concentrations of the simple organic acids vary depending on the type of water and can be relatively high in precipitation in urban environments (see Chapter 5). The lower molecular weight organic acids (formic and acetic acids) are likely to be present in higher concentration than the high molecular weight, polyprotic acids. It can, however, be seen from a quick inspection of Appendix 3.1 that the binding constants for citric acid are larger than the corresponding constants for acetic acid. Thus, even though the higher molecular weight ligands may be in lower concentration, their complexation of metals may be more important than those of the smaller ligands.

3.3.6 Complexation to large molecular weight organic matter

It has been demonstrated in many cases that trace metals bind strongly to the natural organic matter found in environmental waters. This organic carbon can be thought of as essentially a complex polyprotic organic acid mixture, with carboxylic, phenolic and other acid groups, and the extent of the interaction depends on the metal and the relative concentration of organic matter compared to other binding ligands. The amount of organic matter in solution varies widely between ecosystems being high in some natural freshwater environments, such as wetlands, and high in wastewaters, such as sewage treatment plant effluents, and being lowest in the deepest ocean waters (Table 3.16) [1, 2]. There are numerous acronyms that have been used to represent this fraction in solution: dissolved organic carbon (DOC), dissolved organic matter (DOM) and natural organic matter (NOM). In this book, the two acronyms that will be used for the filtered fraction (that which passes through a 0.4 μm filter; the typical definition for the dissolved fraction in environmental studies) are NOM and DOC, and DOC will be used to refer to measurements of this specific parameter. When referring to the particulate phase, the comparable acronyms will be POC and POM. The molecular weight range for NOM compounds is <5000 kDa to 50,000 kDa or greater, and thus it is probably a misnomer to refer to all these compounds as "truly dissolved" as they are likely present as colloidal phases suspended in the water. Additionally, many of the compounds that form part of the NOM, while containing charged surface ligands, such as carboxylic acid, amine and thiol groups, have a hydrophobic center, and are probably best considered as small charged particles (colloids, nanoparticles). Therefore, it is not sufficient

Table 3.14 Estimated speciation for metals in freshwater at pH 6 in the presence of 10^{-6} M citrate. Values given are the percentage of the total metal as each complex. M_I refers to all inorganic complexes. Values were estimated using the MINEQL+ program [17]. No ionic strength corrections were made.

Metal	$\%M^{2+}$	$\%M_I$	%M–citrate complexes
Cd	93.4	4.0	2.5 $CdCit^-$
Cu	8.8	–	90.0 $CuCit^-$
Ni	47.2	–	50.5 $NiCit^-$
Pb	15.8	1.8	80.8 $PbCit^-$
Zn	67.6	1.5	30.1 $ZnCit^-$

Table 3.15 Estimated speciation for metals in freshwater at pH 6 in the presence of 10^{-6} M citrate. Values are the percentage of the total metal as each complex. M_I refers to all inorganic complexes. Values were estimated using the MINEQL+ program [17]. Ionic strength corrections were made assuming an overall ionic strength of 10^{-3} M.

Metal	$\%M^{2+}$	$\%M_I$	%M–citrate complexes
Cd	94.8	3.3	1.8 $CdCit^-$
Cu	11.5	–	87.2 $CuCit^-$
Ni	54.6	–	49.3 $NiCit^-$
Pb	20.1	–	76.2 $PbCit^-$
Zn	73.8	1.2	24.4 $ZnCit^-$

Table 3.16 Dissolved organic carbon content of various waters. Compiled from [1, 2].

Water Body	DOC ($mg\,l^{-1}$)	DOC (μM)	Water Body	DOC ($mg\,l^{-1}$)	DOC (μM)
High latitude rivers	1–5	80–400	Sewage Effluent	<5–100	400–8000
Temperate rivers	2–15	160–1250	Groundwater	1–2	80–160
Tropical Rivers	2–15	160–1250	Estuarine Waters	5–20	400–1600
Rivers draining wetlands	5–50	400–4000	Surface ocean	1–10	80–800
Wetlands, bogs	10–50	800–4000	Deep ocean	1	~100
Lakes	2–10	160–800	Rainwater	1–5	80–400

to consider metal binding to NOM without considering the potential influences of surface charges on the extent of interaction, which will be discussed further next and in Chapter 4.

Overall, the composition of NOM in natural waters is only poorly known, and therefore it is difficult to characterize. For example, in ocean waters, more than 50% of the DOC is not identifiable by current methods of measurement [2]. A fraction of what can be characterized are biochemical compounds, such as lipids (fatty acids), proteins (amino acids) and saccharides (typically <20%), with an additional small fraction as chemicals such as chlorophyll. In addition, many of the initial degradation products of these chemicals can be identified [2]. In these cases, the concentrations of the compounds can be determined as the free entity (mono-compounds) and as the polymers (e.g., as free amino acids, polypeptides and as proteins). Other classes of chemicals, such as petroleum hydrocarbons, can be identified but are present in low concentrations.

Most of the remaining compounds are comprised of material that has not been identified in detail. This fraction is often designated as being either *humic* substances, humic or *fulvic* acids, which identifies the organic matter as classes of compounds based on the overall acidity of the functional groups, which is used in their separation and classification [1]. This terminology has been derived from studies initially done with soil extracts, based on the difference between the fractions that are isolated by separation methods involving XAD and similar type resins, and by the solubility of the compounds at different pHs. Humic and fulvic acids are hydrophobic compounds isolated from soils, sediments or solutions using these approaches (i.e., they are retained on the resins), while there is also a *hydrophilic* fraction that passes through these resins that is also part of the unidentified fraction of material in natural environments [1]. For example, fulvic acids are those compounds that remain soluble at low pH (pH <1) while humic acids precipitate at this pH. Humic acids are generally of higher molecular weight and contain less phenolic groups. All these compounds are derived from both terrestrial sources (so-called *allochthonous* material) or from *in situ* production by primary producers and bacteria, and are usually present as degraded compounds from these biochemicals (so-called *autochthonous* material) [1, 2]. In rivers, especially, the organic matter pool is mostly derived from the degradation of allochthonous material and these compounds are derived from the degradation of lignin and other plant material. Potential degradation pathways for these materials have been presented in various texts [1–3] and typically include many steps and condensation reactions. One pathway is illustrated in Fig. 3.9(a).

All fractions isolated using the techniques above contain compounds which are polyprotic acids, and contain a mixture of different acid groups that determine their chemistry in natural waters. The most dominant acid groups are carboxylic acids and phenols, which have different pK_a ranges, and their relative abundance can therefore impact the overall solubility of the organic compound. The typical pK_a ranges for the functional groups in humic substances range from 1–5 for carboxylic acids; from 1–11 for phenols and above 9 for saturated thiols. These are often the most important acids groups involved in metal complexation in solution although the complexation of metals by amines and other N-containing acids groups, which have a range of pK_as, but are generally more acidic than basic, can also be important. Acid groups attached to aromatic rings have a lower pK_a than the corresponding saturated compounds. The titration curve for a solution of a humic acid is compared to that of a mixture of 10^{-4} M acetic acid ($pK_a = 4.8$) and 10^{-4} M phenol ($pK_a = 10$) in Fig. 3.9(b). The similarity in the titration curves suggests that the humic substance has a range of acidic groups with pK_as that span those of these two end-member compounds.

Thus, it is most likely that any solution of humic material has a variety of metal binding ligands of various acidity and complexing ability, and it would be most accurate, but clearly difficult, to model complexation of metals to these compounds on this basis. As will be discussed in more detail in Chapter 4, the complexation of trace metals in solution by organic matter can be very important for many metals and it is necessary to understand and be able to characterize and model the extent of this interaction. While estimation of the extent of interaction of metals with small, well-defined organic molecules, such as oxalic acid, EDTA and cysteine can be accomplished using the equilibrium approaches already discussed, this is not the case for metal interactions with complex NOM. However, much of the earlier work on characterizing metal interactions with NOM used the notion that these interactions could be modeled without consideration of the surface charge on large NOM molecules. These studies considered that binding could be done using conditional stability constants determined from empirical studies that could be applied without modification across a pH gradient. If the surface charges are an important aspect of binding and lead to non-linearity in the interactions, such an approach is not valid, as discussed further next.

However, modeling all possible interactions is not usually possible or practical, or requires very detailed modeling studies and complex calculations, which will be discussed in Chapter 4. One approach that has been used to simulate the organic matter complexation is to define a model mixture of a number of well-characterized organic acidic compounds to simulate the range and variability in the acid groups present [2]. If the binding constants of metals to these simple ligands are known, then it would be possible to simulate their complexation to the complex mixture of NOM, taking into account surface effects. Another alternative is to

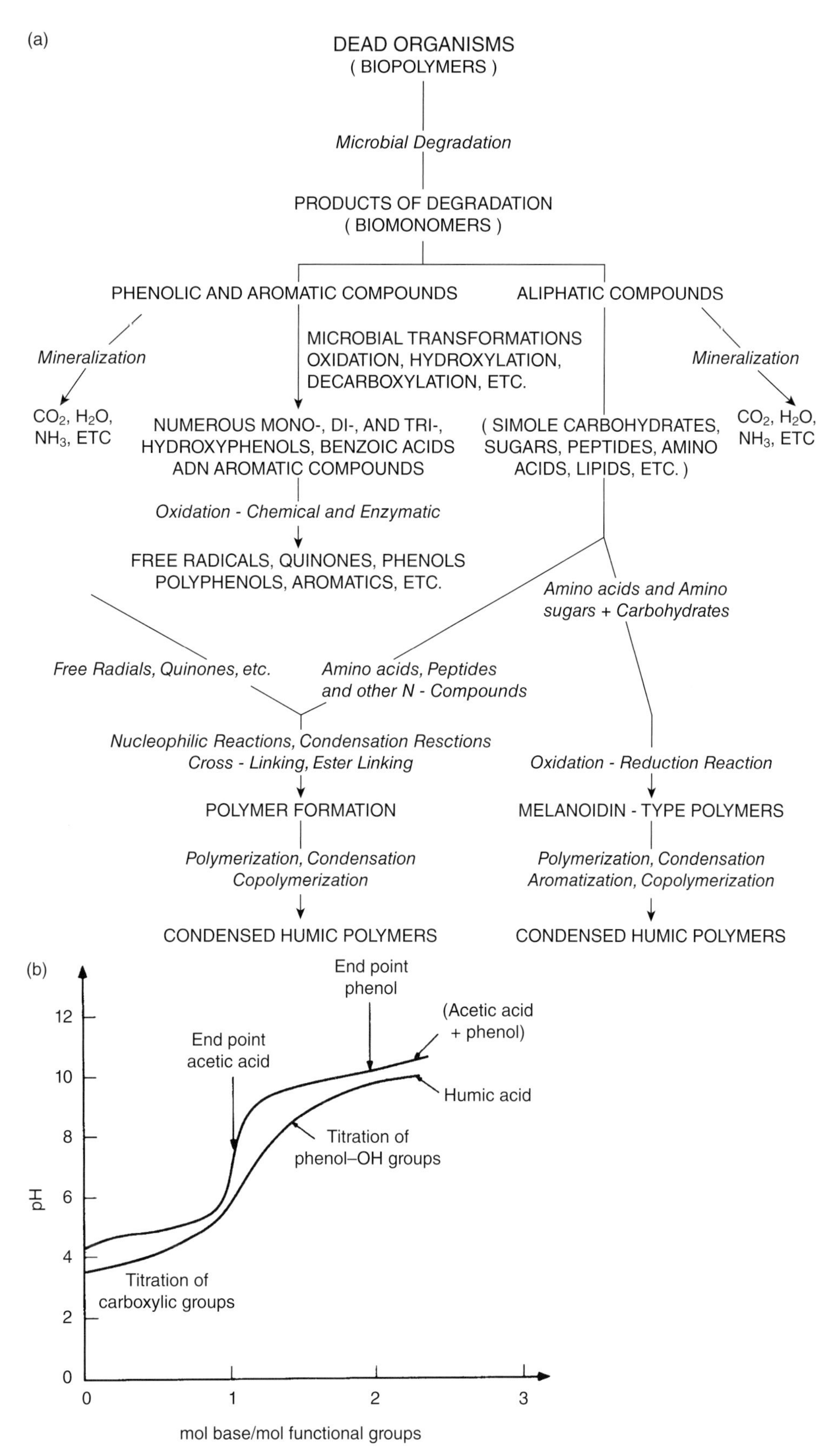

Fig. 3.9 (a) A representation of the degradation reactions that occur during the transformation of terrestrial organic material into degraded humic acid and other compounds. Figure taken from Rashid (1985) *Geochemistry of Marine Humic Compounds* [16], and reprinted with kind permission of Springer Science+Business Media. (b) Titration curves comparing the acid-base distribution of a humic acid mixture with a solution that is an equimolar (10^{-4} M) mixture of acetic acid and phenol. Taken from Stumm and Morgan (1996) *Aquatic Chemistry* [1] and reprinted with permission of John Wiley & Sons, Inc.

Table 3.17 The estimated degree of the metal complexed to ligands, which are presumed to be natural organic matter, and a brief description of the methods used to make these estimations. Information was gathered from a variety of sources, as indicated below.

Metal	% as organic complexes	Medium	Methods	Ref.
Al	~40	Fresh	Ion-exchange chromatography	Morel and Hering [2]
Fe	>99	Ocean	Electrochemical	Morel and Hering [2], Bruland and Lohan [18], Rue and Bruland [19]
Mn	<10	Ocean	Radiotracer/chromatography	Morel and Hering [2]
Cu	~100	Ocean	Electrochemical	Morel and Hering [2], Lalah and Wandiga [20]
	>95	Coastal	Electrochemical; Complex ligand; Bioassay	
	>85	Lake	Electrochemical	
Co	>90	Ocean	Electrochemical/Complex ligand	Saito and Moffett [23]
Zn	>95	Ocean	Electrochemical	Morel and Hering [2], Bruland and Lohan [18], Lohan *et al.* [21], Sander *et al.* [22]
	60–95	Estuarine	Electrochemical	
	10–20	River	Sorption	
	>85	Lake	Electrochemical	
Cd	<80	Ocean	Electrochemical	Morel and Hering [2], Bruland and Lohan [18], Sander *et al.* [22]
	~40	Lake	Electrochemical	
Hg	<50	Ocean	Titration; Ultrafiltration	Lamborg *et al.* [24], , Guentzel *et al.* [25]
Pb	50–70	Ocean	Electrochemical	Morel and Hering [2]

NB: Morel and Hering [2] and References in their Table 6.10.

determine empirical constants for a number of different solutions of humic material from a range of sources [1] and use the range in these constants, and knowledge on the number of acidic functional groups in the material to derive a range of values that is applicable to most environmental solutions. Both approaches have advantages and disadvantages.

As noted earlier, many trace metals and heavy metals, and other elements of environmental interest, bind strongly to DOC. The data in Table 3.17 gives the estimated relative fraction of various metals that are complexed to natural organic matter (NOM) in seawater and freshwater [2]. For most of the metals of importance, a large fraction of the dissolved metal is present in natural waters as NOM complexes. Some of the methods used to determine the extent of NOM complexation (e.g., electrochemical approaches and competitive ligand experiments) can provide some information about the strength of the association – as reflected in the apparent stability constant – and of the concentration of the ligand in solution, but they do not provide any details about the type of ligand, that is, the type of binding site, or the size of the complex [2, 18]. A number of studies using a variety of approaches tend to suggest that the ligands are molecules of relatively large molecular weight. It is often presumed that these ligands are derived from humic material, and that most of the sources for aquatic systems are terrestrial. This may be the case for many freshwater systems but there is little evidence to support this notion for ocean waters. If the organic material is not terrestrially-derived, then its most likely source is biochemicals released into the environment (e.g., proteins, carbohydrates) and their degraded products [18]. It is also possible that these "ligands" are not entirely organic in nature and there is the potential that they could be inorganic entities, such as metal-sulfide clusters [26], which would have a high affinity for other metal ions in solution, or be some complex colloidal mixture containing both organic and inorganic ligands.

Other methods, such as ultrafiltration, can determine the distribution of the metal between different size fractions, but do not also provide any notion of the type of bond formed. In the ocean, much of the trace metals present are complexed to ill-defined "organic ligands", and it has been shown that the complexation of metals in surface ocean waters is very important with the majority of the metal being associated with these ligands [12, 18]. Results for Zn and Cu are shown in Fig. 3.10 [18]. In both cases, it is apparent that much of the metal is complexed and not present as the free ion, especially in the surface waters. At depth, the fraction as the free ion is higher. As shown in Table 3.12, the free ion concentration of Zn in the absence of organic complexation is 48%, that of Cu, 27%. Thus, an examination of the data in Fig. 3.10 for Zn suggests that there is little ligand complexation at depth and most of the difference between [Zn^{2+}] and the total concentration (Zn_T) is due to the presence of other inorganic complexes. This is not the case for Cu, as there is more than an order of magnitude difference in the concentrations of [Cu^{2+}] and Cu_T, even at depth suggesting that there is still substantial complexation of Cu by these "ligands" in the deeper waters.

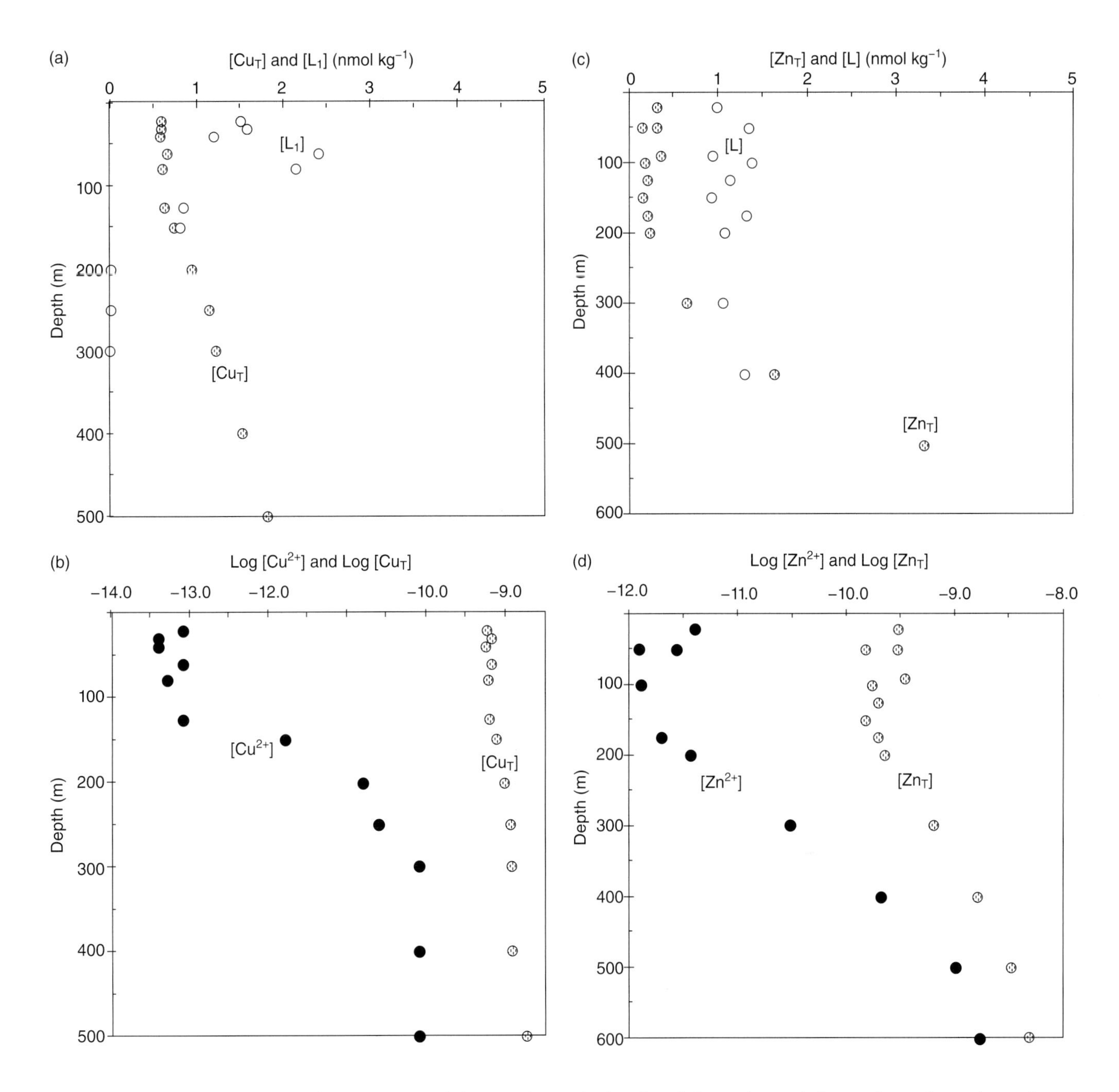

Fig. 3.10 The vertical distribution of copper (Cu) and zinc (Zn) in ocean waters showing the total concentration of each metal ($[X_T]$) and the calculated free ion concentration ($[X^{2+}]$) estimated from calculations of metal binding in solution using electrochemical techniques. These approaches also allow the determination of the concentration of organic binding ligands in solution ($[L_T]$), which is also plotted. For each metal, the top figure shows the total metal concentration and the estimated total ligand concentration while the bottom figure shows the estimated free ion content. Taken from Bruland and Lohan (2004) in *Treatise of Geochemistry*, vol. 6 [18] and reprinted with permission of Elsevier.

The distribution of the estimated ligand concentration with depth for the ocean tracks that of primary productivity and overall microbial activity suggesting that these ligands are derived directly by or indirectly from these organisms [12, 18]. It is known that some bacteria release specific metal binding chemicals into the environment to capture essential trace metals such as Fe and Zn, or to bind up potentially toxic metals, such as Cu, which can inhibit the growth and uptake of metals by some bacteria. Such chemicals would have a clear and recent biochemical source. Additionally, the concentration of the ligand appears to be of the correct order of magnitude required to complex most of the metal in solution, as shown in Table 3.17. This further suggests that such ligands are purposely released. This is discussed further in Chapter 4 in terms of modeling, and further in Chapter 8 in terms of the bioavailability and bioaccumulation and/or toxicity of the metals to organisms. Furthermore, the details of the role of organic matter in the fate and transport of metals in atmospheric waters, freshwaters and in the ocean are further discussed in the following chapters that are focused on examining the chemistry of metals in each media in more detail. The brief discussion earlier provides strong evidence that NOM has an important role in the chemistry of metals in natural waters.

The dissolved speciation to some degree determines the concentration of the particular metal(loid)s in solution. However, in understanding their distribution and fate in environmental waters, the interaction of metal(loid)s with the particulate phase cannot be ignored. In many instances it is these interactions that are the overall determining factor of the magnitude of the dissolved concentration. While most metals will precipitate from solution at high concentration, this is not often the case for most of the elements that are the primary consideration of this book, and therefore not the primary controlling mechanism. Rather it is the interaction of the dissolved metal ions and complexes with the inorganic and organic suspended material that is the major controlling factor, with complexation often controlling the partitioning between the dissolved and particulate fractions, as discussed in the following section. However, it is still necessary and pertinent to discuss the chemical principles behind the dissolution and precipitation of metal-containing solid phases in solution.

3.4 Trace metal interactions with the solid phase

Most of the discussion to this point has focused on metals in the dissolved phase but most metals are actually found in environmental solutions at relatively low dissolved concentrations and this is a function of various interactions: (1) their relative insolubility as oxide and carbonate phases in oxic waters, or as sulfides in anoxic waters; (2) their propensity to adsorb onto inorganic particles, or co-precipitate during their formation; and (3) their accumulation into living organisms, or their interaction with organic detritus. We will firstly consider the solubility of trace metals in natural waters due to their precipitation and in the presence of the most important inorganic solid phases. Then, we shall consider the interactions of metals with solid surfaces. The uptake of metals by living organisms will be dealt with more generally in Chapter 8.

3.4.1 Precipitation and dissolution

Precipitation and dissolution reactions have a major control over the movement of materials at the surface of the Earth, but these mostly involve the major elements and major ions in solution, and therefore are mostly beyond the scope of the book. However, there are some metals whose precipitation reactions are important and there are some specific waters where precipitation occurs because of elevated concentrations and as a result of very high or low pH. Thus, it is pertinent to review briefly the approach to determining whether a solid will precipitate or not, and this was briefly discussed in Example 3.1. For the purposes of this chapter, which is primarily involved in examining and discussing the underlying principles, it is not necessary to delve deeply into this discussion. It is assumed that the reader is versed in the principles of precipitation and dissolution that form part of most introductory undergraduate chemical curricula. The basic construct is that pure solids have unit activity (see Section 3.1) and therefore the reaction for the dissolution of a solid into the metal and associated ligand(s) (for illustrative purposes, a single charged ligand is shown) results in the following equilibrium constant, which is termed the *solubility product (K_{SP})* [1, 2]:

For

$$\begin{aligned} ML_n(s) &= M^{n+} + nL^- \\ K_{SP} &= [M^{n+}][L^-]^n \end{aligned} \tag{3.71}$$

In assessing the degree of dissociation of a solid in solution, it is also necessary to consider the concept of the *common ion effect*. This follows from the definition of the solubility product and relates to the fact that the presence of either the metal ion or the anion in solution will hinder the dissolution of any solid containing them that is added to the solution. Alternatively, the presence of other ligands in solution can enhance the dissolution of the solid by reducing the free metal ion concentration through complexation as it is the free ion concentration, rather than the total dissolved ion concentration, that determines the extent of dissolution of the solid (Equation 3.71).

A few notions are worth reiterating, that: (1) the formation of such solids from solution is a complex process, which can often be kinetically hindered, and more factors are involved than for the dissolved equilibrium reactions that occur in homogeneous solution; (2) the solid phase composition has an impact on its reactivity and solubility and thus solid phases of the same overall chemical structure, but with different morphologies, can have differing solubility; and (3) there is the potential that species can be overlooked in defining and quantifying the overall solubility of a solid [1]. It should also be noted that most methods of determining solubility are based on measurement of the concentration in solution in the presence of a solid, and it is assumed that the metal in solution is in a truly dissolved phase. With the emergence of evidence for metal "cluster" compounds [26], expanding evidence for metal-metal bonds and the potential for the formation of other nanoparticles [27] that would pass through most typically-used filtration devices, it is possible that some of the measured solubilities are higher than reality due to the inclusion of these types of compounds in the "dissolved" fraction. Specific examples of this are discussed in Chapter 6.

The mechanism of "precipitation" encompasses a variety of forms and processes. Often the first phase to form from a saturated solution is a metastable phase that is either amorphous or of a disordered crystalline nature, and this phase will initially be in equilibrium with the dissolved phase. As time passes, this phase will transform ("age") into a more stable crystalline form which may have a somewhat different solubility than the original phase. Thus, the equilibrium concentration would change with time for such a situation. The final crystalline form will likely be less "reactive" than the initial phase formed.

Table 3.18 lists the formation constants for the more important metals and their more important solid phases in natural waters. The constants are given for the formation reaction so that the larger the value in Table 3.18, the more insoluble the solid. These values are therefore the inverse of the solubility product (K_{SP}) as the constants in Table 3.18 are:

$$\beta_{MX} = ([M^{n+}][X^{m-}])^{-1} = (K_{SP})^{-1} \quad (3.72)$$

Given equilibrium conditions, there will be only one phase that will control the dissolved concentration, in the presence of excess solid phase, and therefore it is possible to estimate the equilibrium dissolved concentration if the anion concentration is known. For example, considering seawater at a pH of 8.1, it is possible to estimate the concentration of metals in equilibrium with the hydroxide ions. For example:

For iron, the free metal ion concentration at equilibrium is:

$$[Fe^{3+}]_{equil} = ([OH^-]^3 \cdot \beta_{Fe(OH)3})^{-1} = ((10^{-5.9})^3(10^{42}))^{-1} = 10^{-24.3} \quad (3.73)$$

and if this was the only form of the metal in solution, the concentration of Fe in seawater would be very small. Of course, the solubility of the hydroxide will increase with decreasing pH (decreasing $[OH^-]$) as the overall solubility is given by the product of the hydroxide and metal ion concentration. This is illustrated in Fig. 3.11 for a number of metal ions.

Table 3.18 Formation constants (log β) for various metal solid phases for free metals ions. Taken from [1–3]. Logβ is defined for the following system: $M^{n+} + L^{m-} = M_mL_n$ (s).

Metal	Hydroxide	Carbonate	Sulfide
+III ions			
Cr^{3+}	30.0	–	–
Fe^{3+}	38–42*	–	Reduced
+II ions			
Mn^{2+}	12.8	9.3	10–14
Fe^{2+}	15.1	10.7	18.1
Co^{2+}	15.7	10.0	21–26
Cd^{2+}	14.3	13.7	27.0
Cu^{2+}	19–21	9.6	36.1
Zn^{2+}	15–17	10.0	24.7
Hg^{2+}	25.4	16.1	52–54
Pb^{2+}	15.3	13.1	27.5
+I ions			
Ag^{+}	7.7	11.1	50.1

*A range of values are given as there are different forms of the solid phase that have slightly different equilibrium constants.

However, to determine the overall solubility, it is necessary to also take into account the fact that Fe^{3+} forms a number of dissolved complexes with the hydroxide ion (it is extensively hydrolyzed) and the presence of these complexes enhance the solubility above that given by the solubility product. For example, the free ion can act as an acid:

$$Fe(H_2O)_6^{3+} = Fe(H_2O)_5OH^{2+} + H^+$$

which can be rewritten in the form consistent with the constants given in Appendix 3.1:

$$Fe^{3+} + OH^- = FeOH^{2+}$$

The reader should be comfortable that these two relationships are equivalent, although, of course, the magnitude of the equilibrium constant for them differs by 10^{14}.

As the solubility product is determined by the free metal ion concentration rather than the total dissolved ion concentration, the actual concentration in solution in equilibrium with the solid is often much greater than the concentration of the free metal ion, determined by the solubility product. This notion is very often ignored by scientists when they discuss the relative solubility of various metals in solution in the presence of specific ligands: either the dominant

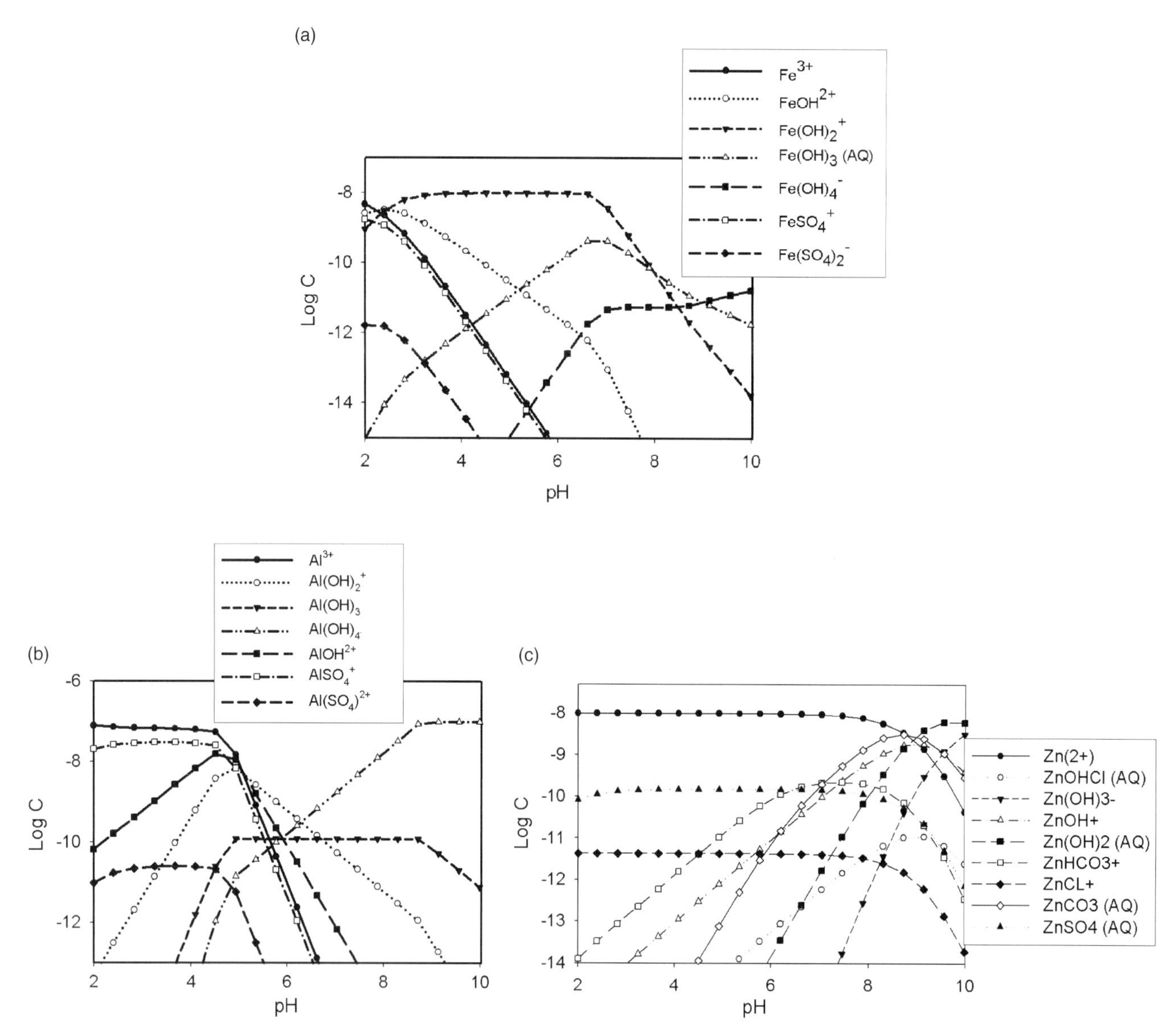

Fig. 3.11 The speciation of iron (Fe – part a) aliuminum (Al – part b), and zinc (Zn – part c) across a pH gradient showing the impact of pH on the dominance of the various dissolved hydroxyl species for each metal. Figure created using MINEQL+, a computer program. Metal concentration is 10^{-8} M for Fe and Zn, 10^{-7} M for Al.

oxic ligands of hydroxide and carbonate species or the sulfide ion in anoxic waters. Overall, there is not a linear relationship for a group of metals between the soluble concentration and the K_{SP} as the relationship depends on the potential for the metal to form strong complexes in solution. Such hydrolysis or complexation of the metal ion in solution has an important impact on its solubility, as does the complexation with other inorganic ions and organic ligands, and especially those associated with organic matter. This is explored further in the next example. Note from Table 3.16 that some metals, such as Fe, are found, of course, in different oxidation states depending on the system redox and this impacts the concentration as the ions of the different oxidation states have different solubility – for example, Fe^{III} oxides and hydroxides are much more insoluble than those of Fe^{II}.

The approach to solving such a problem analytically, where there is both the need to consider the potential precipitation of a solid, which requires determination of $[M^{n+}]$ and the calculation of the speciation, which needs to be completed to estimate $[M^{n+}]$, could be approached in an interactive fashion. However, the best approach is to first determine the dominant speciation, and as a result, the fraction of the total dissolved species that is present as $[M^{n+}]$. With this information in hand, it is possible to determine whether or not the solid will precipitate. If the answer is positive, then the presence of the solid phase "fixes" $[M^{n+}]$ and thus allows the determination of the concentration of each species as done above. If no precipitation occurs, the calculations are as described earlier. Again, an example will elucidate the approach.

Example 3.6

The solubility of Fe^{3+} will be determined in seawater (pH 8.1, total carbonate 3×10^{-3} M, 0.5 M Cl, 3×10^{-2} M sulfate; 0.5 M Na, 10^{-2} M Ca and K, 5×10^{-2} M Mg). The total Fe concentration is not important at this point in the calculations. To determine an analytical solution, we will make the following assumptions: (1) that the higher order complexes, which contain multiple Fe ions, are not formed (see Appendix 3.1) and (2) that the only solid phase formed is $Fe(OH)_3$ with a log K_{SP} of –42. Also, an inspection of the magnitude of the complexation constants should convince the reader that chloride and sulfate complexation is insignificant, and complexes with carbonate either do not exist, or their constant has not yet been determined (the former is likely the case). Ionic strength corrections need to be made and the values in Table 3.11 are used. Given this, the following equation is derived:

$$\begin{aligned}\text{Total } Fe_T &= [Fe^{3+}] + [FeOH^{2+}] + [Fe(OH)_2^+] + [Fe(OH)_4^{2-}] \\ &= [Fe^{3+}](1 + (\gamma_{3+} \cdot \gamma_- / \gamma_{2+}) \cdot 10^{11.8} \cdot 10^{-5.9} + (\gamma_{3+} \cdot (\gamma_-)^2 / \gamma_+) \\ &\quad \cdot 10^{22.3} \cdot (10^{-5.9})^2 + (\gamma_{3+} \cdot (\gamma_-)^4 / \gamma_{2-}) \cdot 10^{34.4} \cdot (10^{-5.9})^4\end{aligned}$$

By inspection, it can be determined that the two higher-order dissolved species will be dominant:

$$\text{Total } Fe_T = [Fe^{3+}](10^{9.4} + 10^{10.6})$$

So

$$[Fe^{3+}] = 2.36 \times 10^{-11} \cdot Fe_T$$

In other words, the free metal ion concentration is a very small fraction of the total dissolved Fe under these conditions because of the formation of the hydroxide complexes which dramatically enhances the solubility of Fe in seawater. It is possible to calculate the concentration at which the hydroxide would precipitate. Recall that:

$$[Fe^{3+}] = K_{SP}^C / [OH^-]^3 = 2.36 \times 10^{-11} \cdot Fe_T$$

where the equilibrium constant (K_{SP}^C) is the value corrected for ionic strength effects. In this case, $\log K_{SP}^C = -40.2$, and so:

$$Fe_T = 1.3 \times 10^{-12} \text{ M or } 1.3 \text{ pM.}$$

Comparison of this value with that of Table 3.6 shows that this value is much lower than the value reported for dissolved Fe in seawater (500 pM). There are at least two explanations for this discrepancy. One explanation could be that most of the Fe measured in a filtered seawater sample is actually not in a "truly dissolved" phase but is present as colloidal material which passes through the 0.4 μm filters typically used by oceanographers to separate the dissolved (filtered) and particulate fractions. It is likely that some fraction of the Fe is in colloidal material, but the nature of this material may relate to the second explanation. As discussed previously, much of the Fe (>95%) in surface ocean waters is complexed to organic ligands. Thus, the relatively high dissolved concentration is a function of the fact that much of the Fe is complexed to organic ligands that further substantially enhances its solubility above that which would occur in the absence of such ligands. Also, recall that the most insoluble form of $Fe(OH)_3$ was considered in these calculations and it is possible that such a form does not occur in seawater, or is kinetically hindered. It is known that Fe initially precipitates from environmental solutions as an amorphous phase, often called hydrous ferric oxide (HFO) which has a formula of FeOOH. The solubility of this form is different from that of the pure hydroxide. Finally, of course, it is worth noting that in many environmental waters the so-called NOM or DOC fraction contains both organic ligands that are truly dissolved and those that are colloidal material, and thus the explanation of the solubility of Fe in seawater probably incorporates both explanations, especially as some of the colloidal material is likely Fe-oxide containing entities.

The impact of organic complexation can be illustrated by the consideration of the impact of EDTA on the dissolved concentration of Fe in the presence of the hydroxide solid phase. Such a calculation is done for a freshwater environment by Stumm and Morgan [1] (10^{-8} M EDTA, 10^{-9} M Fe^{III}) and shows that the solubility is substantially increased (1–3 orders of magnitude) over a pH range of ~5.5–9. Again, this reinforces the notion that the solubility of Fe, and other sparing soluble cations, is likely enhanced greatly be the presence of NOM in natural waters, and this also likely explains the higher concentration of Fe in freshwaters, which have higher NOM, compared to ocean waters.

This example provides a demonstration of how to account for the precipitation of a trace metal solid from solution and how to incorporate such calculations into the overall scheme of determining the speciation and dissolved concentration of a trace metal in an environmental solution. There are many aspects of the process of precipitation, and the reverse, dissolution of the solid phase if the overlying solution becomes undersaturated, and the detailed mechanisms are still being studied [1]. These processes have enormous importance in the overall material cycle at the Earth's surface, though the dominant reactions are not those of trace metals. For example, the dissolution of rocks and terrestrial material by atmospheric precipitation provides the dissolved solids found in unimpacted remote rivers and streams. Transport of this dissolved material, in concert with the transport of solid material, is the main vector for the transfer of many of the crustal metals from land to the ocean. In the ocean, precipitation reactions, both those mediated by biota, and those that occur due to metals exceeding the solubility of various solid phases, and the sinking of organic material, lead to the removal of trace metals to the sediments, which is the primary long-term sink for most trace metals. On geological timescales, the process is repeated and the buried ocean material is returned to the surface of the Earth.

3.4.2 Adsorption of metals to aqueous solids

While the dissolution and precipitation of solids is of interest and important to many aspects of aquatic chemistry, it is mostly of lesser importance in terms of the cycling and fate of trace metals in natural waters. Most metals exist in solution in natural waters at concentrations that are below their intrinsic solubility and it is mostly the interaction of the metals with suspended materials found in environmental waters that control the dissolved concentration of a metal in solution, rather than those of precipitation or dissolution. Therefore, it is more pertinent and important to understand the interaction of metals with these suspended solids and sediment/soils, and this is mainly due to the abiotic adsorption/chemical interaction of metals with the solid phase for inorganic particles, and, to a degree, for biotic (microbial) particles. In addition, as discussed later in Chapter 8, there is the potential for the uptake, via both active and passive processes, of metals by living microorganisms. Some metals are required nutrients and therefore are actively accumulated while others are taken up inadvertently. Other metals are toxic if accumulated to high concentrations. Finally, it is evident that many particles suspended in solution do not consist of a single phase or entity but are a conglomeration of smaller particles and colloids. It is likely that many particles consist of a mixture or inorganic and organic materials, such as would be found with an organic coating of a oxide phase, or the precipitation of an oxide around an organic core particle. Furthermore, there is no distinct boundary between the dissolved and the particulate phases as there is a continuum of particles sizes from macromolecules, such as biochemicals and humic material, to colloidal material, both organic and inorganic, to sub-micron particles and larger material (Fig. 3.12), as discussed in Chapter 2. Microorganisms typically range in size from 0.1 μm upwards and there are an abundance of sub-micron clay particles (colloids) and other inorganic materials suspended in natural waters. In recent research, there is much focus on *nanoparticles* [28]. These are ultra-fine dispersed, solid particles in the lower size range of colloids (with typical sizes <100 nm (<0.1 μm)) which possess a very high specific surface to volume ratio. Much current research is focused on these very small particles and their role in the environment is not clearly understood. Given the separation techniques used by most environmental chemists, these particles will be in the so-called dissolved (filtered) fraction under most conditions unless specific ultrafiltration methods are used to separate the sub-micron fraction.

The adsorption/interaction of metals with inorganic and organic abiotic particles is due to a number of processes that can be mostly characterized as a direct interaction involving chemical bonding or another electrostatic association [1–3, 29]. The degree of association with the solid phase will

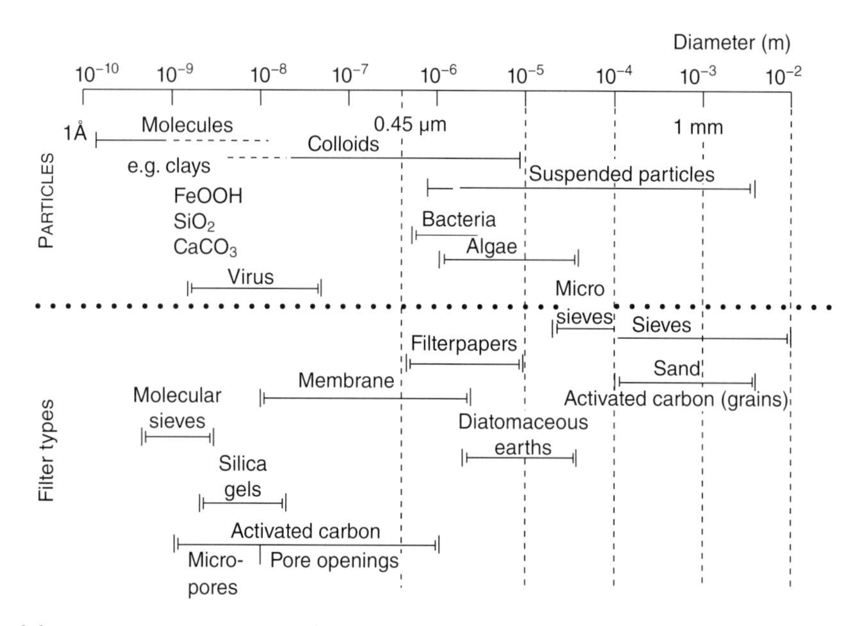

Fig. 3.12 A representation of the ranges in concentrations of different fractions of dissolved, colloidal and particulate material in environmental waters, as well as the typical size ranges of common filters and filtration devices. Figure redrawn by Langmuir (1997) *Aqueous Environmental Geochemistry* [3] but original figure reprinted with permission from Stumm (1977) [25] and, Copyright (1977), American Chemical Society.

clearly impact the rate of movement of a metal through the biosphere and especially in terrestrial waters where there is often a large movement of solid material due to river flow, storm events and other hydrologic processes. In the ocean, most of the particulate material is of biological origin and the affinity of a metal for these particles will determine to a large degree its fate and residence time, and concentration, in ocean waters. Many metals such as Fe and Mn, abundant elements in the biosphere, are found in the ocean at relatively low concentration due to their insolubility in general and their affinity for association with solid material.

3.4.3 Dissolved-particulate partition coefficients

In many instances in environmental application and determination, it is not necessary, possible or desirable to determine the detailed concentration of each species in solution if the overall partitioning of the metal between the dissolved (filtered) and particulate phase is the parameter of interest. Also, if the binding constants for the metal to the various phases are not known, or the composition of the solid phase is not well characterized, or the particulate material consists of both abiotic and biotic particles, then there is often a need to take a more empirical approach to examining the distribution and fate of metals in the environment. As discussed in Chapter 2, such an approach is the basis for the definition and use of *adsorption (partition) distribution coefficients*. This parameter is simply defined as the ratio of the amount of metal in the particulate phase (on a mass basis) to that in the filtered fraction, or:

$$K_d = C_p/C_W = C_p^W/(TSS \cdot C_W) \tag{3.74}$$

where C_p is the concentration of the chemical of interest on particles in mol kg^{-1}; C_w is the concentration in the dissolved phase in mol l^{-1}; C_p^W is the concentration of the chemical on particles in mol l^{-1} and TSS is the total suspended particulate load in kg l^{-1}. As noted in Chapter 2, if it is assumed that 1 l has a mass of 1 kg, the K_d is dimensionless. These values are relatively easy to obtain from field data but they are empirical values and don't provide any information about the binding phase, the type of interaction, or the impact of pH and other primary factors on the extent of binding. The K_d is often found to change between locations and there is evidence that there is a "particle effect" in that the K_d appears to decrease as the TSS increases, because of the presence of colloidal material in the filtered fraction (see Section 2.2.2).

The degree of association of various metal(loid)s with natural suspended matter differs greatly. Many of the transition metals are relatively weakly bound compared to the heavy metals which tend to form strong bonds with particulate material. The data in Table 3.19 [30] provide a comparison and demonstrates that the differences in the log K_d is several orders of magnitude for the different metals of importance in natural waters. A number of conclusions can be made based on this compilation of data. Firstly, for each metal, there is a large range in the values of the log K_d indicating that there are many different types of particles with different capacities to absorb or adsorb metals. The binding phases in sediments include inorganic oxide phases (which have acid–base surface sites), organic matter, and both inorganic sulfide phases and organic thiols. For most metals, the strongest binding constants are associated with reduced sulfur phases, with the weakest associations being with the oxide and mostly mineral phases. Secondly, the log K_d for sediments is most often lower than that for the water column/suspended sediments and this mostly reflects the colloidal effects noted previously. Additionally, there is the potential that, for some metals, uptake into living (microbial) organic matter enhances the particulate concentration in the water column. Moreover, the K_d values for soils are typically lower than for sediments (data not shown) and this likely reflects the lower organic content of most soils compared to sediments.

Table 3.19 A compilation of data from the literature of the median and range in values for the dissolved–particulate partition coefficient for a variety of metals and metalloids. Taken from [30].

Element	Median Log K_d TSS*	Range in values	Median Log K_d Sediment*	Range in values
Ag	4.9	4.4–6.3	3.6	2.1–5.8
As	4.0	2.0–6.0	2.5	1.6–4.3
Cd	4.7	2.8–6.3	3.6	0.5–7.3
Co	4.7	3.2–6.3	3.3	2.9–3.6
Cu	4.7	3.1–6.1	4.2	0.7–6.2
Hg	5.3	4.2–6.9	4.9	3.8–6.0
CH_3Hg	5.4	4.2–6.2	3.6	2.8–5.0
Ni	4.6	3.5–5.7	4.0	–
Pb	5.6	3.4–6.5	5.1	2.0–7.0
Sn	5.6	4.9–6.3	4.7	–
Zn	5.1	3.5–6.9	3.7	1.5–6.2

Finally, it is evident that some metals are very strongly particle-reactive (e.g., Pb and Hg) while others, and the metalloids (As being a representative example), are not strongly bound. The overall relative strength of binding for the water column particulate (i.e., to TSS), based on Table 3.19, is:

$$Pb > Hg > Zn > Ag > Cd, Cu, Co > Ni > As$$

The most strongly bound metals are the so-called "soft" metals which tend to associate strongly with reduced sulfur ligands and to a lesser degree with oxide phases. Overall, most of the transition metals behave similarly, with similar values for their K_ds. The Group 12 elements have a non-obvious distribution in that the log K_ds are: Hg > Zn > Cd and this again reflects their relative chemistry and binding capacities to the dominant ligands on particles. Of all the metals, where there is sufficient information available, it can be concluded that Hg and Pb have the strongest tendency to be associated with the solid phase.

In summary, the distribution coefficient is a parameter that defines the overall degree of association with the solid phase but does not give any indication or information about the type of association, and does not provide any information about the number and type of binding sites. There are, however, a number of approaches and models for examining the association of metals with the solid phase in more detail and these will be discussed in the following sections.

3.4.4 Adsorption isotherms

In many early studies of the interaction of metal(loid)s with solids surfaces it was not known what type on interaction was occurring and so rather than model the interaction in terms of a surface chemical reaction, the concept of an adsorption isotherm was used. This is a mathematical relationship that describes the adsorption of a dissolved species onto a solid surface and how this interaction changes as a function of concentration. Thus, given a surface with a fixed number of adsorption sites, S, then, for an adsorbing chemical, A, the adsorption reaction can be written as [1–3]:

$$S + A = SA \tag{3.75}$$

And, the associated distribution constant is therefore:

$$K_{ads} = [SA]/[S][A] \tag{3.76}$$

Now, as the total number of sites, $[S_T] = [SA] + [S]$, and by substitution it can be shown that:

$$[SA] = [S_T] \cdot K_{ads}[A]/(1 + K_{ads}[A]) \tag{3.77}$$

To make the relationship more useful for calculation, it is better to relate the adsorption to a surface concentration rather than the number of binding sites. Therefore, defining the surface concentration, Γ, as:

$$\Gamma = [SA]/\text{mass adsorbent} \tag{3.78a}$$

and

$$\Gamma_{max} = [S_T]/\text{mass adsorbent} \tag{3.78b}$$

then

$$\Gamma = \Gamma_{max} \cdot K_{ads}[A]/(1 + K_{ads}[A]) \tag{3.79}$$

This is the *Langmuir adsorption isotherm equation* and it is valid if there is only a monolayer formed on the surface, which is the basic inherent assumption in the formulation, and if the energy of adsorption is independent of the number of sites bound; that is, there are no local non-ideal interac-

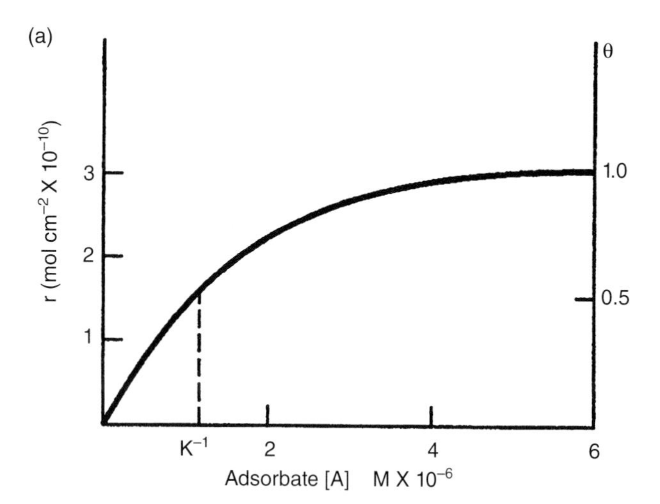

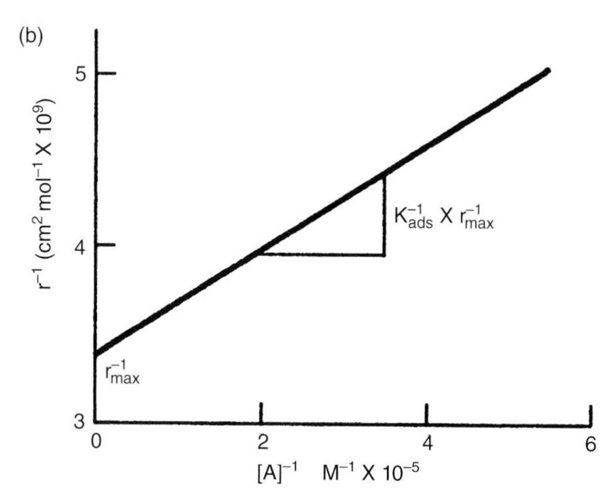

Fig. 3.13 (a) Demonstration of the relationship between the surface concentration of adsorbed material and the solution concentration of the adsorbant; (b) the linearized equations showing the relationship between the inverse of these parameters. Taken from Stumm (1992) *Chemistry of the Solid-Water Interface* [29] and reproduced with permission of John Wiley & Sons, Inc.

tions. Figure 3.13(a) shows the relationship, represented by Equation 3.79 while Fig. 3.13(b) shows the relationship for the linearized version of the same relationship, which is more typically used to estimate the various parameters (Equation 3.79):

$$\Gamma^{-1} = \Gamma^{-1}_{max} + K^{-1}_{ads} \cdot [A]^{-1} \cdot \Gamma^{-1}_{max} \tag{3.80}$$

From this equation it is apparent that a plot of Γ^{-1} versus $[A]^{-1}$ has a slope of $K^{-1}_{ads} \cdot \Gamma^{-1}_{max}$, and an intercept of Γ^{-1}_{max}. This equation therefore allows for the determination of the relevant parameters from experiments where the adsorption of a species is measured at a number of different concentrations.

A more general form of the Langmuir equation is derived if it is assumed that there is the potential for more than a monolayer of adsorption on the surface of the solid. This relationship is referred to as a *Freundlich Isotherm* relationship and is defined by the following expression:

$$\Gamma = m[A]^n \text{ or } C_s = mC_w^n \quad (3.81)$$

where m is the *Freundlich constant* and n is the measure of the nonlinearity involved. This equation does not have the limitation of monolayer adsorption – that is, the equation assumes unlimited adsorption – and it applies well to solids that are heterogeneous. This equation can be thought of as a generalization of a multi-site Langmuir isotherm equation.

Both equations have application in environmental studies of trace metals. However, most reports in the literature use either the concept of the distribution coefficient, which requires the least characterization of the waters – filtration, and determination of the concentrations of the metals in the dissolved and particulate fractions, as well as the TSS concentration. The approach that is mostly used to formulate the interaction of the metals with the surface of the particles in terms of a chemical interaction involving a surface site and the metal ion is through use of equilibrium interactions. As most inorganic phases important in the binding of trace metals in natural waters are oxide phases, the surfaces of these particles have protonated groups that can act as acids/bases, and therefore can interact with the metal ion [1, 2]. The interaction of a metal cation with the surface involves a substitution for the proton attached to the surface site. Similarly, anions will replace a hydroxyl ion on the surface. This approach was developed initially to consider the binding of metals to iron, manganese and other oxide phases, and the same approach has been also used to examine the interaction with organic particulate matter, where surface sites are likely a combination of carboxylic acid and other acid/base functional groups. The approach of this formulation is detailed in the following sections.

3.4.5 A complexation-based model for adsorption

The inorganic solids found in natural environments that interact with the dissolved trace metal(loid)s are mostly the various oxide surfaces (e.g., Fe-oxides; alumina, silica). The oxide surfaces can be parameterized as acid–base sites, which are protonated when the surface is neutral (overall surface charge is zero), deprotonated at higher pH, and positively charged at lower pH. The association of a metal oxide surface with species in solution is represented by Fig. 3.14, where the interaction of water molecules with the surface is also represented.

The possible acid base species, for an oxide surface, and the associated reactions, where the surface is represented by ≡X, are [1–3, 29, 31]:

$$\equiv X{-}O^- \quad \equiv X{-}OH \quad \equiv X{-}OH_2^+$$

$$\equiv XOH \Leftrightarrow H^+ + \equiv X{-}O^- \quad \text{dissociation of the surface acid group} \quad (3.82)$$

$$\equiv XOH + H^+ \Leftrightarrow \equiv XOH_2^+ \quad \text{protonation of the surface acid group} \quad (3.83)$$

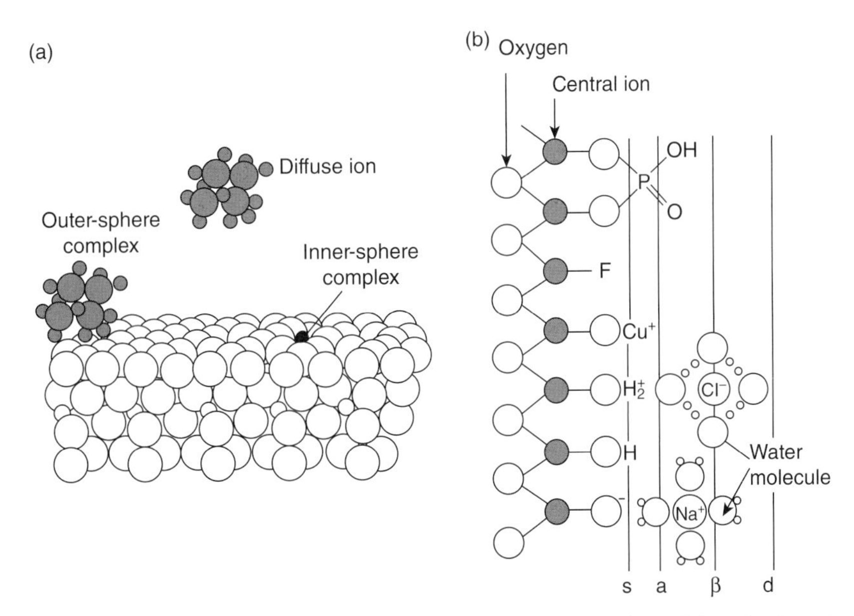

Fig. 3.14 Schematic showing the association and interaction of dissolved constituents with a solid oxide phase in solution. Taken from Stumm (1992) *Chemistry of the Solid-Water Interface* [29] and reproduced with permission of John Wiley & Sons, Inc.

In describing the charge on the surface of the solid, the concept of *point of zero charge (pzc)* is defined as the pH at which the overall charge on the solid surface is zero [1, 29]. This pH is not the same for all solids and can vary from being acidic to basic. In pure solution, the relative pH is a function of the strength of the acid/base groups as the degree of dissociation or protonation will be a function of the strength of the binding of the proton to the surface. In more complex solutions, the pzc depends on the composition of the medium as well as the composition of the solids. For the main oxide phases, the pzc pH is around 2 for SiO_2, around 5 for MnO_2, between 7 and 8 for HFO and between 9 and 10 for Al_2O_3 [1, 29]. At pH values above the pzc, the surface is overall negatively charged, and *vice versa*. Overall, cations or positively charged species will bind more strongly at pH values above the pzc, and often there is a so-called *adsorption edge*, which represents the pH range where the adsorption increases dramatically. This is shown in Fig. 3.15(b), for

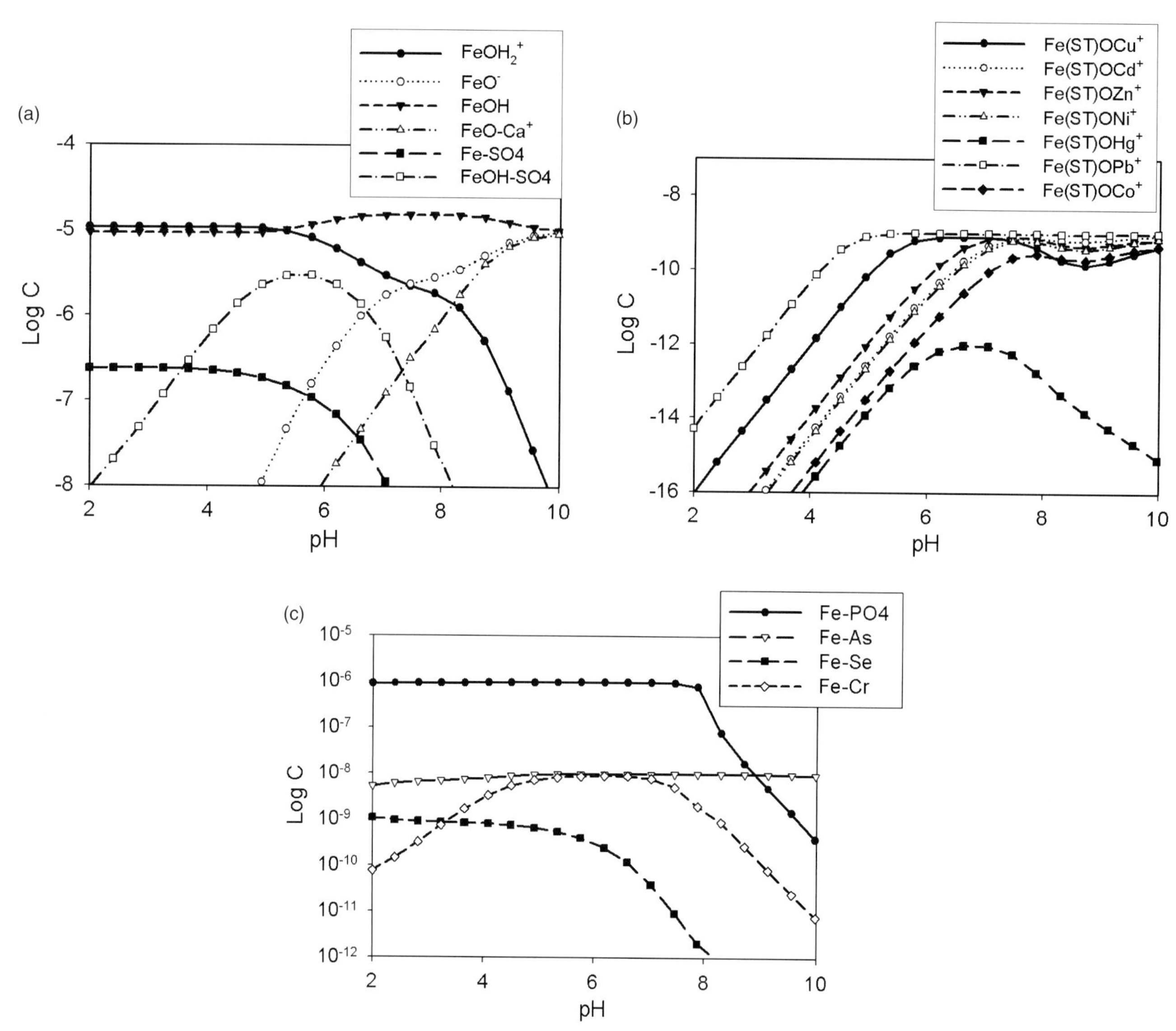

Fig. 3.15 Adsorption isotherms for hydrous ferric oxide (HFO) for cations and anions adsorbing onto the surface. Data generated using MINEQL+ and the incorporated HFO adsorption program from [27] under the following conditions: 10^{-4} M Na, K, sulfate; 2×10^{-4} M Mg, Cl; 10^{-3} M total carbonate; 4×10^{-4} M Ca; 10^{-9} M for all metals except Hg, 10^{-11} M; 10^{-8} for anions except as noted; phosphate 10^{-6} M. The total Fe was 10^{-4} M, which equates to 5×10^{-7} M strong sites (which the metals bind to) and 2×10^{-5} M weak sites (anions bind to preferentially). Plots show: (a) speciation of the surface sites on HFO with pH, as well as the total adsorption of calcium (Ca) and sulfate (SO_4^{2-}); (b) adsorption of +2 cations (Co, Ni, Cu, Zn, Cd, Hg and Pb); (c) adsorption of anions (PO_4^{3-}, AsO_4^{3-}, SeO_4^{2-} CrO_4^{2-}).

adsorption of +2 cations to an iron oxide surface [1–3]. The data in this figure was generated with MINEQL+ using conditions relevant to the environment (10^{-9} M for cations except Hg (10^{-11} M); 10^{-4} M Fe, "typical freshwater" anion concentrations). The model is described further next. Adsorption was calculated in the presence of major ions in the solution (0.1–1 mM; see figure caption for details) and therefore the results are somewhat different to the figures in the literature. As can be seen, however, for most metals, the adsorption edge occurs over about 1–2 pH units. Note that the strength of the binding influences the pH at which adsorption begins, with the more strongly bound species (e.g., Pb and Cu) being adsorbed at lower pH values, and the least-strongly bound, such as Co only being adsorbed when the surface is strongly negatively charged (Fig. 3.15b). The speciation of the surface as pH changes and the adsorption of the major ions – Ca and sulfate – are also shown (Fig. 3.15a).

It is also necessary to consider the dissolved speciation in these computations and thus there is not a simple relationship between the binding constant for adsorption and the pH at which adsorption occurs. As will be shown later and further in Chapter 4, the degree of adsorption is a function of the competition between the dissolved ligands and the "ligands" on the surface of the solid. Adsorption of anions would follow an analogous series, but with the maximum adsorption being at low pH, and little adsorption at high pH. Again, the strength of binding influences the pH of the adsorption edge with strongly bound anions such as As(V) (i.e., AsO_4^{3-}) (see Fig. 3.15c) being appreciably adsorbed over all typical pH values of natural waters [29].

In developing a quantitative model of adsorption to surfaces, it is necessary to consider the interactions involved. There is the transport of ions or complexes to the surface, the interaction with the surface binding sites, and the movement of any species formed/released by the interaction away from the surface into the bulk solution. These species may/may not be charged and given that the surfaces are charged, it is necessary to consider the interaction between the charged surface, the complexing substances and the bulk solution (water), which is a polar molecule. The following diagrammatic representation of the surface and the various interactions (see Fig. 3.16) [1, 2, 29] provides a theoretical picture of the interactions and of the charge distribution resulting from the interactions (Fig. 3.16). The *surface charge* (σ_P) is determined by the surface area, the number of sites and their form (i.e., uncharged, negatively or positively charged). The overall charge of the *diffuse layer* (σ_D) counterbalances that of the surface such that the net charge is zero (Fig. 3.16b). The *surface potential* (Ψo) decreases exponentially away from the surface (Fig. 3.16c). This relationship is only valid when the surface potential is low, which is typical for environmental situations. The layer where there is decreasing potential is termed the *diffuse double layer* and the layer is divided by the *shear plane*, the distance at which the potential has decreased to 1/e. This distance is used as a measure of the thickness of the double layer. Figures 3.16(d) and (e) provide representations of the charge distributions and the distribution of *co-ions* and *counter ions* in the double layer.

The electrostatic interactions can be determined by the application of the *Gouy–Chapman Theory*, within the framework of the *Poisson–Boltzman equation*, as detailed in Morel and Hering [2] and Stumm [29]. The assumption in solving the equations is that the charge on the surface is uniformly distributed. Solution of the equations leads to the definition of the extent of interaction, which is called the *double layer thickness*, which is related to the solution ionic strength, and the *dielectric constant* (of water). This defines the distance of electrostatic influence and the extent of influence the surface has over the ions in the immediate solution – increasing the concentration of counter ions, decreasing the concentration of similarly charged ions (Fig. 3.16). The higher the ionic strength, the smaller is the double layer. The Gouy–Chapman Theory also relates the surface charge (σ_P) and potential (Ψ_o), and it can be shown that, for small potentials, these two parameters are proportionally related [2]:

$$\sigma_P = 2.3 I^{1/2} \Psi_o \tag{3.84}$$

where I is ionic strength. As will be discussed further next, this formulation allows the inclusion of surface charge effects when considering the complexation of ions with charged surfaces.

The interaction of the dissolved species with the surface could involve a chemical bond (a so-called *inner-sphere complex*) or an ion pair interaction (an *outer-sphere complex*) or it can be present in the immediate solution at the surface. The interaction of a metal ion with the surface can occur by association of the free metal ion or a metal complex with the surface. To demonstrate this with an example, consider the interaction of $Pb(OH)_2$ with a neutral surface acid site. The following complexes are possible:

$\equiv X{-}O{-}Pb^+$ and $\equiv X{-}O{-}Pb{-}OH$
$\equiv X{-}OH{-}Pb^{2+}$ and $(\equiv X{-}O)_2{-}Pb$

The first three complexes are formed due to a 1 : 1 interaction of the metal complex and the surface while the final complex represents the binding of the Pb to two surface sites. Of course, these are all possible complexes that might form but there is no guarantee that all would form in any particular circumstance but this could be calculated if the binding constants are known. It is likely that the uncharged complexes are more likely when the surface is neutrally charged.

A similar range of interactions are possible when considering the complexation of an anion to the surface, which could

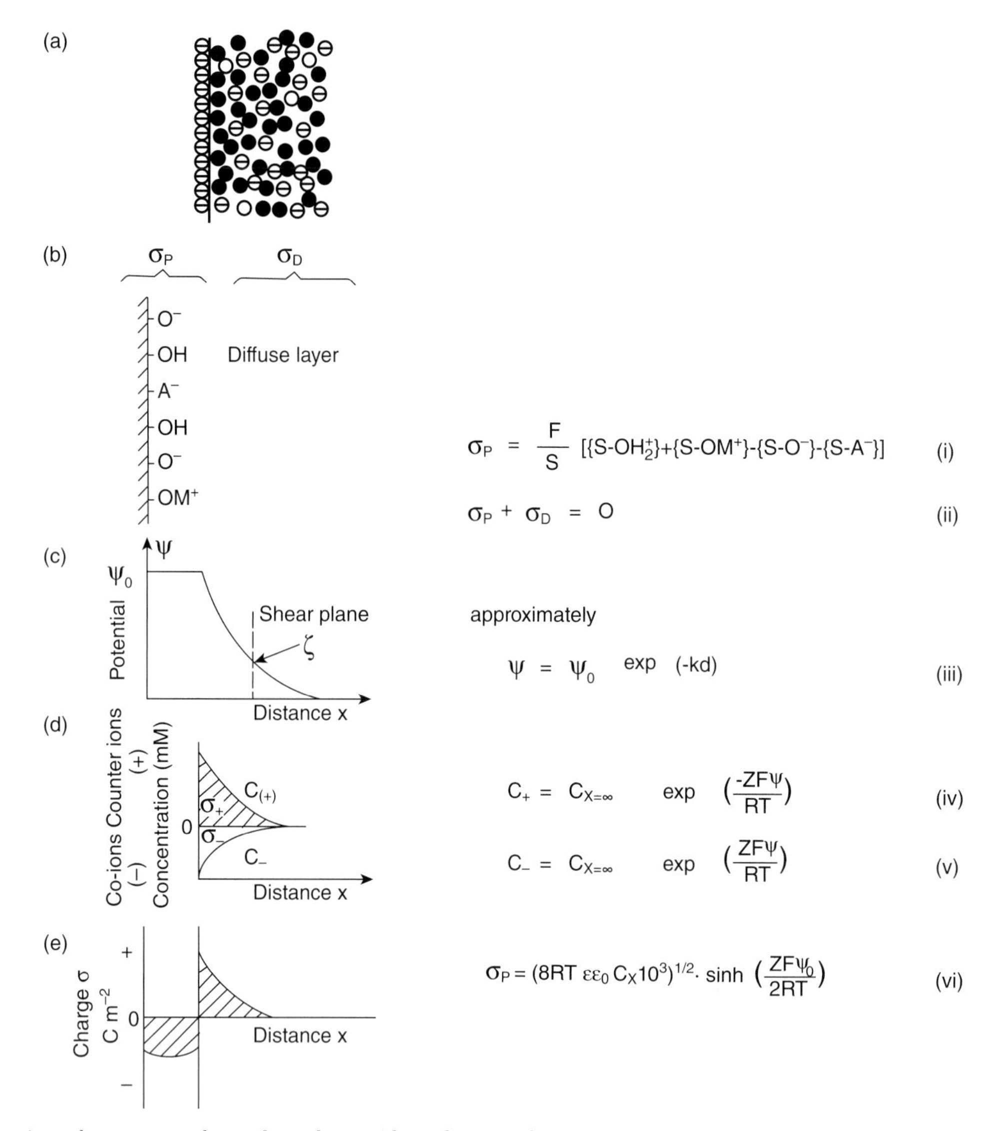

Fig. 3.16 Interactions that occur at the surface of an oxide surface in solution. Reprinted from Stumm (1992) *Chemistry of the Solid-Water Interface* [29] reproduced with permission of John Wiley & Sons, Inc.

occur, for example, if arsenic (arsonic acid, $AsO(OH)_3$) was to interact with the solid surface. Anions would have a stronger tendency to interact with the protonated sites and therefore the interaction of anions with the oxide surfaces is greatest at lower pH when the surfaces are charged. It should be noted that there is the possibility of the formation of *tertiary complexes*, that is, those having both a metal and a ligand attached to the surface, as suggested earlier for Pb. Generally, there are two types of tertiary complexes depending on whether the metal or the ligand is bound directly to the surface (M: metal ion, L: ligand):

$\equiv X{-}O{-}M{-}L$ or $\equiv X{-}L{-}M$

Finally, as the surface is charged, there is also the likelihood of ion pair interactions with the major ions in solution and the charged surfaces.

The approach to dealing with the interactions of metals with solid surfaces is to formulate relationships in a similar way as done with dissolved ligands. This allows for the overall incorporation of these interactions into a single framework which is advantageous when using computer programs or other approaches to solving complex problems. Also, as done with non-ideal effects, the effects and influence of surface charge are dealt with by including an additional term in the equations, rather than by using a completely different formulation. This approach is actually a generalization of the Langmuir formulation, except that a number of differing types of binding sites is allowed on the surface of the solid and further that such sites have different binding capacities for a particular metal. Furthermore, in examining the adsorption of a metal, there are a number of surface functional groups that have well-defined coordination properties that are specified and a total concentration of surface sites is defined, allowing for the formulation of a "typical" equilibrium equation. To account for electrostatic influences, the following free energy relationship is derived where there are two components, one relating to the interaction (complexation) of the metal cation or anion with the surface binding site (ΔG^o_{int}), which has an associated *intrinsic binding constant*, K_{int}, and another accounting for the effects of the overall surface charge (ΔG^o_{coul}) [1–3]:

$$\Delta G^o = \Delta G^o_{int} + \Delta G^o_{coul} \tag{3.85}$$

As suggested by the subscript, the formulation is accounting only for Coulombic interactions at the surface and therefore ΔG^o_{coul} is a function of the surface potential, Ψ_o. It is therefore representative of the electrostatic work involved in bringing an ion from the bulk solution to the reaction site, and for the transport of any charged products away from the site to the bulk solution. Given this, it is possible to formulate ΔG^o_{coul} in terms of the molar charge difference associated with the interaction ($F\Delta Z$), and the surface potential:

$$\Delta G^o_{coul} = F\Delta Z\Psi_o \tag{3.86}$$

The change in Z is for the overall interaction which includes the charge change due to the movement of the ion to the surface from the bulk solution, the subsequent binding reaction, and the charge change due movement of product species away from the surface. As discussed by Morel and Hering [2], it is more convenient to define an additional parameter, P, to simplify the calculations and formulations:

$$P = \exp(-F\Psi_o/RT) \tag{3.87}$$

Recall that there is a direct relationship between ΔG^o and K ($\Delta G^o = -RT\ln K$) and therefore it is possible to show that:

$$K = K_{int} \cdot K_{coul} \tag{3.88}$$

where K_{coul} represents the constant associated with the interaction of the ions and the solid charged surface. Combining Equations 3.86–3.88, the following is obtained:

$$K_{coul} = \exp[-\Delta G^o_{coul}/RT] = \exp[-\Delta Z \cdot F\Psi_o/RT] = P^{\Delta Z} \tag{3.89}$$

It is worthwhile reiterating the primary assumptions that are inherent in this formulation, that: (1) that reactions with the surface, forming inner-sphere (true) complexes and outer-sphere (ion pair) interactions, are analogous to those in solution, and can be defined by mass-action equations; (2) electrostatic effects can be accounted for by electric double-layer theory; (3) the surface charge and electrical potential of the surface are due only to interactions/chemical reactions involving surface functional groups; and (4) the binding constants are empirical (intrinsic) constants; that is, they are related to the thermodynamic equilibrium constants by the activity coefficients associated with the surface species and those in solution [3].

A few examples will demonstrate the method of dealing with surface interactions when determining complexation of ions in solution with solid surfaces. Again, the equilibrium constants are given in terms of the dissociation reactions, as typical for acid-base interactions. Consider first the acid-base reactions of the surface sites for HFO.

$$\equiv XOH \Leftrightarrow H^+ + \equiv X{-}O^- \quad K = K^{int}_{a2} \cdot K_{coul} = K_{int}/P$$

as the reaction results in an increase in the surface negative charge (i.e., $\Delta Z = -1$). Rearranging, the concentration of surface sites is given by:

$$[\equiv X{-}O^-] = K^{int}_{a2}[\equiv XOH]/[H^+]P \tag{3.90}$$

Similarly, the protonation of the surface results in:

$$[\equiv XOH_2^+] = [\equiv XOH][H^+]P/K^{int}_{a1} \tag{3.91}$$

It is noted that these formula are for the "molar concentration" of surface sites and therefore it is necessary to convert a site density for the surface sites into a concentration. This is achieved by multiplying the molar concentration of the solid (X) in solution (i.e., [X]) with the site density for the solid (i.e., mol sites/mol X). Therefore, it is required that the number of sites per mole of solid is known. This is the function of a number of variables, including the size of the particles, as the number of surface sites is a function of the surface area while the total number of moles of the solid is a function of the volume (mass) of solid. In many cases, the specific surface area of a particular solid is known or can be determined and the number of sites can also be assessed by titration and other means. Dzombak and Morel [30] examined the adsorption of both cations and anions to the surface of HFO, and provide a detailed examination of these interactions, and of the charge and impacts of ionic strength and other interactions on the surface charge. They determined that, for a "typical" method of preparation of HFO, the specific surface area is $600\,m^2\,g^{-1}$, which is much higher than that of other oxide phases. Values for a variety of different solids materials are given in Table 3.20 [3].

Table 3.20 Measured and estimated values for the specific surface area of a variety of common solid phases in aquatic systems and the related values for the surface site density. Adapted from Langmuir, D. (1997). *Aqueous Environmental Geochemistry*, with permission from Prentice Hall, Upper Saddle River [3].

Mineral/Solid	Surface Area (m^2/g)	Surface Site Density (sites/ nm^2)	Surface Site Density** (mol g^{-1} × 10^{-3})
Goethite (α-FeOOH)	45–170	2.6–16.8	1.35
Hematite (α-Fe_2O_3)	<5	5–22	–
Ferrihydrite ($Fe(OH)_3.nH_2O$	250–600	20	0.1–0.9
Birnessite (MnO_2)	140–290*	2–18	–
Quartz (SiO_2)	0.14	4.2–11.4	–
Gibbsite (α-$Al(OH)_3$)	120	2–12	–
Rutile (TiO_2)	5–20	6–12	–
Kaolinite	10–38	1.2–6.0	–
Illite	65–100	0.4–5.6	–
Soil humic material	26–1300	2.3	1–5 × 10^{-3}

Notes:
*Number of sites decreases with aging. This is also true for iron (hydr)oxides.
**Values from the literature. These values could be calculated from the data in the first two columns as mol g^{-1} = surface area (m^2/g) × site density (sites/m^2)/Avogadro's number. However, this will likely lead to an overestimate as the site density and the surface area are not likely to be simply related.

The distribution of sites on HFO over a pH gradient can be calculated using the equations above. Dzombak and Morel [30] concluded that there were two types of binding sites on the surface of HFO, or rather that the binding characteristics of the surface could be approximated in this fashion [31]. Most of the sites are weak (low affinity) binding sites (0.2 mol/mol Fe) but there is a smaller number of stronger (high affinity) sites (0.005 mol/mol Fe)(~2.5%). The acid-base characteristics are determined by the overall site density and the acid-base constants for these sites. However, the binding of trace metals and other cations is associated with both the high affinity and low sites, although given the differences in the binding strengths of these sites for most metals, it is adsorption to the high affinity sites that dominate, especially at the low concentrations typical of environmental solutions. Anions will bind mainly with the low affinity sites. These conclusions about the site characteristics and binding were determined by the overall examination of many published datasets that had studied the adsorption of cations and anions to the surface of HFO under a variety of conditions [31]. As noted previously, the specific surface area of $600\,m^2\,g^{-1}$ is assumed and given the formula for HFO of FeOOH, its molecular weight is $90\,g^{-1}$ mol.

Example 3.7

As a further example, consider the surface site characteristics for a 0.01 M ionic strength solution with $9\,mg\,l^{-1}$ of HFO. The number of weak binding sites is therefore:

$$9\times10^{-4}\text{ M Fe}\times0.2\text{ sites/mol}=2\times10^{-5}\text{ M sites.}$$

Similarly, the number of strong sites can be estimated at 5×10^{-7} M.

These will either be neutral, positively or negatively charged. Consider the situation at a pH of 7. The value of P from Morel and Hering [2] (see Appendix 3.1), for I = 0.01 M and pH = 7, is log P = −1.02. The acid-base constants for HFO [2, 31] are:

$$\text{Log }K_{a1}^{int}=-7.29\text{ and }\log K_{a1}^{int}=-8.93$$

The following equation needs to be solved:

$$\text{Total }[\equiv FeOH]=[\equiv FeOH]+[\equiv FeOH_2^+]+[\equiv FeO^-]=2\times10^{-5}\text{ M} \quad (3.92)$$

Now

$$[\equiv FeOH_2^+]=[\equiv FeOH][H^+]P/K_{a1}^{int} \quad (3.93)$$

or

$$[\equiv FeOH_2^+]=[\equiv FeOH]\cdot10^{-7}\cdot10^{-1.02}\cdot10^{7.29}=0.19[\equiv FeOH] \quad (3.94)$$

and similarly for $[\equiv FeO^-]$, so

$$\begin{aligned}\text{Total}[\equiv FeOH]&=[\equiv FeOH](1+[H^+]P/K_{a1}^{int}+K_{a2}^{int}/[H^+]P)\\ \text{Total}[\equiv FeOH]&=[\equiv FeOH](1+0.19+0.12)\\ &=1.31[\equiv FeOH]=2\times10^{-5}\text{ M}\end{aligned} \quad (3.95)$$

or

$$\begin{aligned}[\equiv FeOH]&=1.5\times10^{-5}\text{ M (76\%)}\\ [\equiv FeOH_2^+]&=0.29\times10^{-5}\text{ M (15\%)}\\ [\equiv FeO^-]&=0.18\times10^{-5}\text{ M (9\%)}\end{aligned} \quad (3.96)$$

Figure 3.15(a) shows the change in speciation across a pH gradient for the surface sites. As can be seen, the most dominant sites are the uncharged sites over the pH range except at low pH where the protonated sites are somewhat higher and high pH where the dissociated sites dominate (> pH 10). This is partly due to the effect of the surface charge interactions which can be seen to have a large impact on the relative number of charged sites. If these surface charge effects were not considered, a very different picture of the surface speciation of the solid phase would be estimated. Note that the effect of ionic strength is also important as the value of P decreases as I increases, and the charged sites are more dominant at higher I for a particular pH (Fig. 3.15d). For the range in pH values typical of natural waters it is the uncharged sites that are dominant and are the ones most likely to be involved in interactions with dissolved ions. For this reason, most equilibrium expressions are written in terms of the uncharged site interacting with the species of interest. The reaction constant for the same interaction with the other sites can be derived from these expressions if needed.

Example 3.8

It is useful to now consider the interaction of a metal with the surface of the HFO. For this example we shall consider Cd^{2+}, and to recall from the examples earlier in this chapter (Table 3.8), at a pH of 7 essentially all the Cd is present as the free ion in freshwater, and we shall therefore, for simplicity, assume that $[Cd^{2+}] \sim Cd_T = 10^{-8}$ M. The binding constant for Cd to the strong sites ($Fe_{st}OH$) is given as 0.47 [2] for:

$$Cd^{2+} + {\equiv}Fe_{st}OH = {\equiv}Fe_{st}OCd^{+} + H^{+}$$

As the concentration of Cd is much lower than that of the binding sites, and the overall surface charge is determined by the low affinity sites, and given that Cd is binding only to the high affinity sites that are a small part of the total, we will make the simplifying assumption that the Cd interactions will have no influence over the overall surface charge and those interactions can therefore be ignored in the equations for Cd.

So,

$$Cd_T = [Cd^{2+}] + [{\equiv}FeOCd^{+}] = 10^{-8}\ M$$

Or

$$Cd_T = [Cd^{2+}](1 + K_{cd}^{int} \cdot [{\equiv}Fe_{st}OH]/[H^{+}])$$

A list of the binding constants for cations and anions with HFO are given in Table 3.21 [2]. Recall from the calculation in Example 3.7 that 76% of the total sites were the uncharged site and so we can use this fact to solve for the speciation of Cd directly;

$$Cd_T = [Cd^{2+}](1 + 10^{0.47} \cdot (3.8 \times 10^{-7})/10^{-7}) = 11.2[Cd^{2+}]$$

and essentially all the Cd is adsorbed to the HFO, as shown in Fig. 3.15(b). Contrast this situation with that of the absence of solid phases, where most of the Cd is present as Cd^{2+}. In this example, in the presence of the oxide, $[Cd^{2+}]$ is about 9% of the total Cd. If the toxicity of Cd was determined by the free ion concentration, these calculations indicate that the toxicity could be lowered by an order of magnitude by such adsorption.

Table 3.21 The binding constants for cations and anions to the surface of hydrous ferrous oxide [2, 30]. For the cations, the K_1^{int} values refer to the binding to the strong sites while the K_2^{int} values refer to binding to the weaker sites. For the divalent anions, the complexes modeled are $\equiv FeOHA^{2-}$ and $\equiv FeA^{-}$. For the trivalent cations, the modeled complexes are $\equiv FeH_2A$, $\equiv FeHA^{-}$, $\equiv FeA^{2-}$ and $\equiv FeOHA^{3-}$. Lack of a constant indicates that the data could be modeled without invoking such an interaction.

Cation	Log K_1^{int}	Log K_2^{int}	Anion	Log K_1^{int}	Log K_2^{int}	Log K_3^{int}	Log K_4^{int}
Ca^{2+}	4.97	−5.95					
Ba^{2+}	5.46		H_3BO_3	0.62			
Sr^{2+}	5.01	−6.58	AsO_4^{3-}	29.31	23.51		10.58
Co^{2+}	−0.46	−3.01	AsO_3^{3-}	5.41			
Cu^{2+}	2.89		SeO_4^{2-}		7.73	0.80	
Cd^{2+}	0.47	−2.90	SeO_3^{2-}		12.69	5.17	
Ni^{2+}	0.37		CrO_4^{2-}		10.85		
Zn^{2+}	0.99	−1.99	PO_4^{3-}	31.29	25.39	17.72	
Pb^{2+}	4.65		SO_4^{2-}		7.78	0.79	
Hg^{2+}	7.76	6.45	$S_2O_3^{2-}$			0.49	
Ag^{+}	−1.72		VO_4^{3-}				13.57
Cr^{3+}	2.06						

These simple examples and calculations show how it is possible to incorporate the interaction of metal adsorption onto solid surfaces into equilibrium calculations. It is evident however that these calculations could be complex, especially if there is more than one metal in solution. It is worthwhile to return briefly to a further discussion of Fig. 3.15.

The calculations for this figure were solved by computer programs and these further considerations of the interactions of metals with surfaces will be dealt with in Chapter 4, which focuses on modeling approaches to solving thermodynamic equilibrium problems. However, before leaving the subject it is useful to present the results of various studies which show some of the interactions and complexities involved in these calculations and estimations.

One further example which illustrates the complexities of the systems is consideration of the interaction of metals with oxide surfaces in the presence of ligands in solution. One example is shown in Fig. 3.17(a) [2], which shows actual experimental data [32] rather than modeled results. These data provide a good illustration of the interaction of a number of anions with an Fe-oxide surface; amorphous iron oxide in this case. As can be seen, all anions have their peak adsorption at low pH, in contrast to cations, and there is very little adsorption at pHs >8. As the concentration of the anion increases, the fraction adsorbed decreases and this decrease results from the fact that the total number of binding sites is of the order of 10^{-4} M, based on the estimations detailed above, and so at the higher anion concentrations all sites on the solid surface are occupied at the lowest pH values. The different anions have a different binding capacity that is reflected in the relative value of their binding constants, as shown in Table 3.21. The acidity of the binding sites of the anions also needs to be considered; for example, sulfate is fully dissociated at all the pHs shown in the figure, while

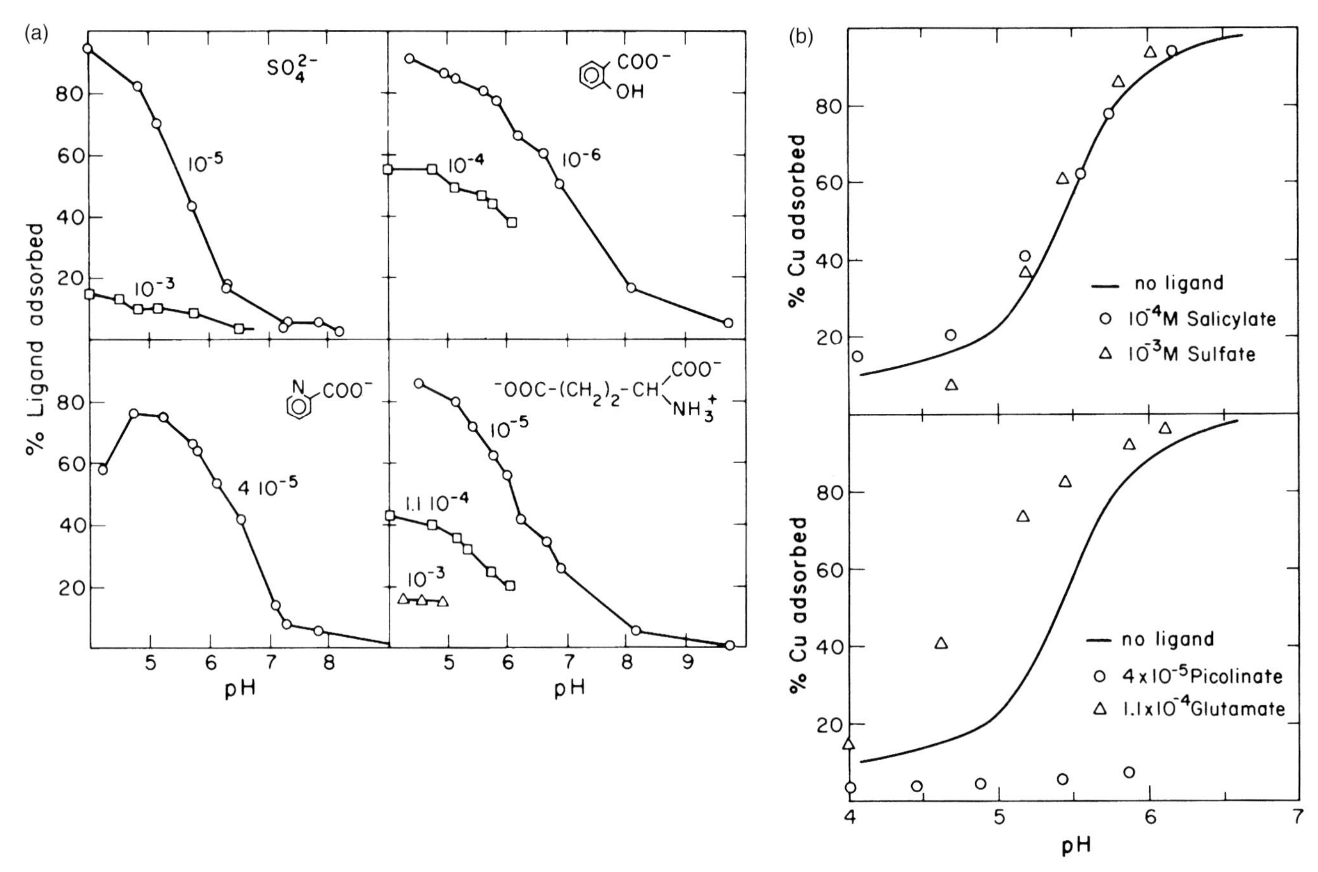

Fig. 3.17 (a) Adsorption of anions onto the surface of hydrous ferric oxide with concentrations as shown; total Fe = 10^{-3} M except for picolinate, 6.9×10^{-4} M; (b) Adsorption of copper to the surface of hydrous ferric oxide in the presence of different potential binding ligands, whose adsorption is show in Fig. 3.17(a). Figure reprinted with permission from Morel and Hering (1993) *Principles and Applications of Aquatic Chemistry* [2] but original data and figures from Davis and Leckie (1978) *Environmental Science and Technology*, **12**: 1309–15 [31]. Copyright (1978), American Chemical Society.

the acid dissociation constant (pK_{a1}) for picolinate is 13.7 (Appendix 3.1). Clearly, the extent of binding is a function of both the strength of the surface association constant and the acid dissociation constant of the acidic site on the anion.

As a contrast, the adsorption of copper (Cu) to the same oxide surface is shown in Figure 3.17(b) [2, 32]. Here the solid line represents the adsorption of Cu to the solid phase in the absence of any ligands except those present in water in equilibrium with the atmosphere (OH^- and carbonate species). Again, recall from the earlier discussions in this chapter that under the conditions of this experiment the carbonate and hydroxide complexes will dominate in solution at the higher pHs but that at lower pH, Cu is present mostly as the free metal ion. Assuming conditions similar to those in the example in Table 3.8, most of the Cu is present as a free ion at pH 6 (90.3% Cu^{2+}) while at a pH of 8, only 5.4% is as the free ion with the major complexation being with carbonate, and the hydroxide complexes being relatively minor. Using these two pH conditions, we can examine the complexation of Cu with an iron oxide surface and compare that with the experimental results.

Example 3.9

Consider the situation of 10^{-6} M Cu at a pH of 6 and 8 in the presence of HFO at a Fe total concentration of 10^{-3} M and the conditions used in previous examples (5×10^{-4} M total carbonate, 2×10^{-4} M Cl and 10^{-4} M sulfate). As sulfate complexation of the metals is not important even in the absence of HFO, its concentration is not critical to the solution to the problem. From consideration of the acid–base chemistry of HFO, using the approach above, we can conclude that at a pH of 6, 74% is as ≡FeOH, and similarly, the percentage at pH 8 is 77%. This is consistent with Fig. 3.15(a) which shows that there is little change in the overall speciation of the surface sites over this pH range. So, using the same approach as used for the Cd example previously:

$$Cu_T = [Cu^{2+}](1 + Cu_I + K_{Cu}^{int} \cdot [\equiv Fe_{st}OH]/[H^+])$$

where Cu_I reflects the relative fraction of the unbound Cu that is as inorganic complexes in solution. Given the speciation calculated previously, Cu_I = 0.11 for a pH of 6 and 17.5 for a pH of 8. Thus, for a pH of 6:

$$\begin{aligned} Cu_T &= [Cu^{2+}](1.11 + 10^{2.89} \cdot 0.74(5 \times 10^{-7})/10^{-6}) \\ &= [Cu^{2+}](1.11 + 287.2) = 287.3[Cu^{2+}] \end{aligned}$$

So, it can be concluded that all the Cu is bound to the oxide surface under these conditions and similarly, calculations will show this to be even more so at a pH of 8. This is consistent with the data in Fig. 3.17(b).

However, the data in Fig. 3.17(b) show the adsorption isotherm for Cu in the absence of ligands as well as in the presence of each of the ligands shown in Fig. 3.17(a). Clearly, there is minimal impact of sulfate and salicylate on the adsorption and for sulfate this is likely due to the fact that sulfate does bind strongly to Cu (Appendix 3.1; log K for CuL is 2.4) and so the presence of the anions in solution have no impact/do not compete with the oxide surface for Cu complexation. For salicylate, the log K for CuL is much larger (11.5) and so some impact of this ligand might have been expected. However, this constant refers to the binding of the anion with Cu, and the pK_a for the carboxylic acid group is large (13.7) and so within the pH range of this problem, most of the salicylate is present as the uncharged complex, and thus the ligand does not effectively compete with the oxide surface in terms of binding Cu.

For the other two ligands, glutamate and picolinate, it is clear that both have a large impact on the adsorption of Cu to the oxide surface. For glutamate, it appears that its presence enhances adsorption while for picolinate, its presence clearly decreases adsorption to the oxide surface. So, the question arises of how the presence of these organic ligands in solution impacts the complexation of Cu with the oxide surface? Note that the concentration of Cu is much lower than that of the anions and the metal binding is mostly to the strong sites on the HFO, and therefore there is the potential for binding of the anion to the weaker surface sites even if the cation is binding to the strong sites. For the picolinate, even though some fraction of the ligand is bound to the oxide surface at low pH, there is still sufficient ligand in solution to complex the Cu to prevent its adsorption. The pK_a for the carboxylic acid group for picolinate is 5.4 and so at the pH values examined for the Cu adsorption (4–7), all or some substantial fraction of the ligand is present as an uncharged complex in solution. In addition, the binding constant for CuL in this case is 8.4 and given the similar "binding site" concentrations of the solid surface and the picolinate, and the relatively much higher binding constant for Cu to picolinate, the Cu is completely bound by the organic ligand and thus does not associate with the oxide to any substantial degree even at the higher pH values.

For glutamate, the same situation could be expected to prevail (log K for CuL is 8.8) except that the carboxylic acids groups are more basic (pK_{a1} is 9.95 and pK_{a2} is 14.47), and so the situation is more similar to that of salicylate. However, there is also the potential for the formation of a tertiary complex which could impact the cation adsorption. This can occur as glutamate is a dicarboxylic acid that binds strongly to Cu. Such a tertiary interaction would likely manifest itself at lower pH values where the cation is not strongly adsorbed in the absence of organic ligands and the ligands themselves are strongly bound. Here the interaction would be that the ligand is both binding to the oxide surface and complexing the metal, and for this to occur, the ligand would need to have at least two relatively strong binding sites per molecule. This is the case with glutamate and therefore the interaction apparent in Fig. 3.17 is valid, that is, the presence of glutamate is increasing Cu adsorption at pH values where the ligand itself is strongly adsorbed but where the Cu cation does not strongly bind to the oxide surface [32]. The resultant complex is represented as:

$$\equiv Fe{-}OOC{-}R{-}COO{-}Cu^{+}$$

where the carboxylic acid groups are represented as $-COO^-$, and the remainder of the glutamate molecule as R. This formula is clearly generic for any dicarboxylic acid interacting with the oxide surface and binding a cation simultaneously. No clear interaction is apparent for the other ligands and this reflects the fact that these ligands do not have the necessary binding sites for such an interaction. At this stage, no further calculations of the interactions of metals with solid surfaces will be done but this topic will be further considered in Chapter 4 where modeling calculations will be used to investigate further these interactions of Cu adsorption in the presence of ligands.

A final example, which will also be further discussed in the modeling chapter (Chapter 4) is the interaction of metals with oxide surfaces in the presence of natural organic matter (NOM). In this case, there is clearly more than one ligand per organic molecule and so the type of tertiary interactions discussed previously with Cu and glutamate would more than likely occur with NOM. The results in Fig. 3.18 [2, 33], which present the interaction of Cu with the surface of alumina in the presence or absence of NOM, clearly show a typical adsorption curve for Cu in the absence of NOM, and a complex interaction in the presence of NOM. The resultant curve, which shows higher adsorption at lower pH and lower adsorption at higher pH, can be explained by the formation of tertiary complexes at the lower pH values where the NOM is binding strongly to the oxide surface. However, at the higher pHs, there is less adsorption of the NOM, and therefore more of it in solution and the increasing NOM in solution begins to outcompete the oxide surface for the Cu (i.e., it is a relatively stronger ligand) resulting in decreased adsorption and more of the Cu in solution. Again, modeling this interaction will be a topic of the next chapter of the book.

In summary, therefore, it is possible that the presence of inorganic solids, such as oxide surfaces, and other charged solids in solution will have an impact on the partitioning and behavior of metals. These interactions can be a

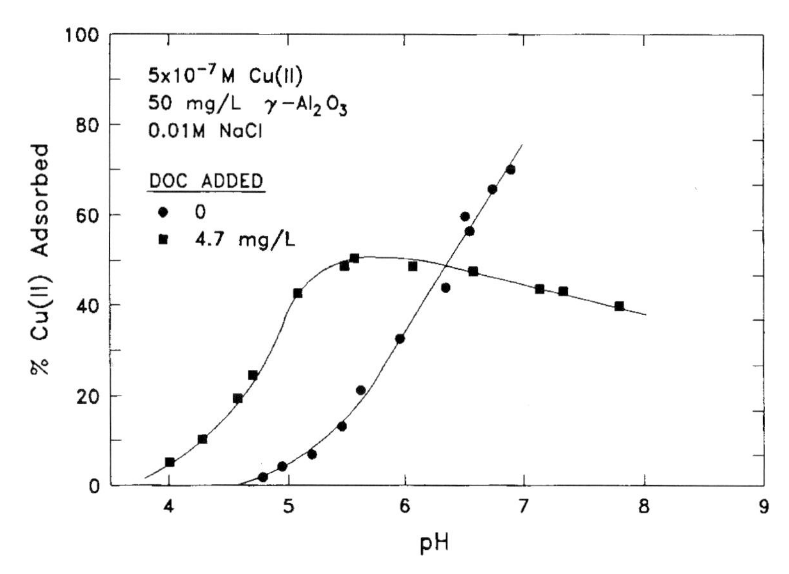

Fig. 3.18 The adsorption of copper to alumina both in the absence and presence of organic matter. Taken from Morel and Hering (1993) *Principles and Applications of Aquatic Chemistry* [2] but original data and figures from Davis (1984) *Geochimica et Cosmochimica Acta*, **48**: 679 [32]. Reprinted with permission from Elsevier. Copyright (1984), American Chemical Society.

straightforward interaction of the metal with the oxide surface or can involve much more complex interactions involving the metal ion, and dissolved ligands, as well as the ligands that may themselves be interacting with the solid phase. Such complexities will be discussed as appropriate in the later chapters of the book as they clearly have an important effect on the biogeochemistry of trace metals in solution.

3.5 Redox transformations and thermodynamic calculations

In the context of examining the speciation and equilibrium dynamics of trace metal(loid)s in solution there is one remaining consideration that needs to be discussed. For many of the important trace metals in natural waters, oxidation and reduction reactions form an important part of their aquatic chemistry, fate and transport. In addition, many of the processes that impact metal cycling are also clearly and closely linked to redox dynamics and microbial transformations that involve redox cycles and therefore it is necessary to incorporate an understanding of how to include redox transformations into speciation calculations and chemical modeling. The basic chemistry of redox transformations is a topic that is covered in undergraduate chemistry classes and therefore the principles and concepts will only be briefly discussed here. As with the other processes that we have considered, an equilibrium approach is used but it should be noted that these reactions are often far from equilibrium under environmental conditions [1–3]. Kinetic considerations should be considered as well in discussing and examining redox transformations and these will be dealt with in Chapter 4. The energy gained from redox transformations is used by many organisms, and thus in the environment, and especially at the microbial level, many redox reactions are microbially-mediated, for example, sulfate reduction. Thus, in considering the chemical reactions that occur it is crucial to remember that while these processes are often parameterized as chemical reactions occurring in solution it is likely that there are microbes that are mediating and potentially enhancing the processes above their abiotic rates in natural waters.

3.5.1 Electrochemistry and the equilibrium constant

While the transfer of electrons that is the basis of redox reactions is always in a coupled process where one species is being oxidized and one reduced in the exchange, it is useful in the formulation of the underlying chemical principles to consider a reaction to be the sum of two half reactions. Examples of such half reactions are:

$$O_2 + 4H^+ + 4e^- = 2H_2O$$

Oxygen is being reduced to water in this half reaction and the overall reduction involves the gain of electrons by the species being reduced. Of course, free electrons do not persist in the environment and therefore this reaction cannot occur in the absence of a counter oxidation reaction that is the source for the electrons being consumed by the reduction of oxygen. Reduced Fe is a reasonable environmental example of a species being oxidized, that is, losing the electrons:

$Fe^{2+} = Fe^{3+} + e^-$

Overall, if these two reactions were to occur simultaneously, then oxygen is the oxidant, and is reduced during the reaction – the oxidation state of molecular oxygen is 0, and that of the oxygen in water -II, so that there is a gain of two e^- for each O atom, or a gain of four e^- overall. In contrast, the reduced form of Fe, Fe^{2+}, is oxidized to Fe^{3+} that is, loses one e^-. The balanced reaction is the sum of these two reactions with the stoichiometry such that there is no net overall electron transfer in the net reaction. In this case, therefore, there is the equivalent of four Fe atoms oxidized for each oxygen molecule reduced:

$$O_2 + 4H^+ + 4Fe^{2+} = 4Fe^{3+} + 2H_2O$$

It is worthwhile to reiterate the "typical" oxidation states found in environmental systems: (1) O is usually in the –II oxidation state, except in H_2O_2 and other similar reactive species that are transient species in the environment; (2) H is +I in most cases; (3) all elements in their elemental state have an oxidation state of 0, for example, Fe, O_2; (4) most cations have a variety of oxidation states, with the oxidation state of the free metal ion given by its net charge, and the oxidation state of complexes being determined by the formula of the compound; for example: Cr(III) in Cr^{3+} and Cr(VI) in CrO_4^{2-} (O is –II and the overall charge is –2).

Example 3.10

An example of the range of oxidation states for a particular element is typified by sulfur. Consider, SO_4^{2-}, where S is +VI; H_2S where S is –II. For $S_4O_6^{2-}$, the overall oxidation state is not an integer suggesting that there is an equal mixture of oxidations states of II and III.

In the sulfate reduction reaction, SO_4^{2-} is converted to S^{2-} and requires an exchange of eight electrons for each atom of S reduced. Consider the degradation of organic matter during sulfate reduction:

$$SO_4^{2-} + CH_2O + H^+ = CO_2 + HS^- + H_2O \quad \text{(unbalanced equation)}$$

1. Balance electron exchange.
2. Balance S and C.
3. Balance O.
4. Finally balance H.

The overall exchange of electrons is eight per S atom and the organic matter (written as CH_2O for simplicity in this reaction) has C in a reduced state (0), while C in carbon dioxide is more oxidized (+IV). Photosynthesis involves reduction of C and respiration, the reaction being considered here, is the opposite reaction. Note that the most reduced form of C in the natural environment (–IV) is methane (CH_4). In the reaction as written there are four electrons transferred per C oxidized and so there needs to be two C atoms oxidized for each S atom reduced, and the remaining O and H atoms need to be balanced:

$$SO_4^{2-} + 2CH_2O + H^+ = 2CO_2 + HS^- + 2H_2O$$

Returning to the reduction of oxygen by the oxidation of Fe, this overall reaction can be written in terms of an equilibrium equation in the same fashion as done for metal complexation reactions. In addition, it is possible to write the individual half reactions in terms of an equilibrium equation, using the same conventions as used in defining the equilibrium constant and the relationships earlier in this chapter. So, therefore, for the reduction of oxygen:

$$K = (P_{O2} \cdot \{H^+\}^4 \cdot \{e\}^4)^{-1}$$

where {} refers to activities in this case, with water having an activity of 1, and writing the reactions involving gases in terms of their partial pressures. To make these types of relationships more useful, and to allow a direct link of the equilibrium equation to electrochemical measurements, this equation can be written, by taking the log of each side of the equation, as:

$$\log K = -\log(P_{O2}) - 4 \cdot \log\{H^+\} - 4 \cdot \log\{e\}$$

Now, a new concept is defined, the *electron activity*, which is a measure of the potential reducing power of a half reaction, as:

$$p\varepsilon = -\log\{e\} \quad (3.97)$$

This parameter provides a hypothetical estimate for the equilibrium situation for the half reaction. So, returning to the equation for the reduction of oxygen, that Equation 3.97 can now be written as:

$$p\varepsilon = (\log K - 4 \cdot pH - \log(P_{O2}))/4 \quad (3.98)$$

These equations can be generalized in the following fashion. Consider the reaction, where "Ox" refers to the oxidant and "Red", the reductant, and n is the number of electrons transferred:

$$Ox + ne^- = Red \quad (3.99)$$

Then

$$K = \{Red\}/\{Ox\}\{e^-\} \quad (3.100)$$

The overall electron activity can then be defined by rearranging Equation 3.100 as:

$$\{e^-\} = [K^{-1} \cdot \{Red\}/\{Ox\}]^{1/n} \quad (3.101)$$

and in terms of log values:

$$p\varepsilon = 1/n[\log K - \log\{Red\}/\{Ox\}] \quad (3.102)$$

The pε scale provides an indication of the reducing capacity of the reaction or, if used to refer to an environmental medium, the reducing capacity of the medium. The values of pε are determined by the principal redox reactions that control the overall reducing capacity of the environment.

Recall that there is a relationship between the equilibrium constant and the energy change, ΔG^0, and therefore if it is possible to define a value of ΔG^0 for individual half reactions, then the overall ΔG^0 for a reaction can be determined and the equilibrium constant can be calculated. As discussed earlier in Section 3.1, the definition of the equilibrium constants and the molar free energies was accomplished by defining a reference state and calculating these values relative to this reference state. The same approach is used here as this allows a free energy to be assigned to the individual half reactions. The reference state used in this instance is that for the conversion of water to hydrogen gas, when the reaction occurs under standard conditions and with a partial pressure of hydrogen of 1 atm.

It is therefore be useful and reasonable to define a relationship in a similar fashion for the electron activity in redox reactions. Using the relationship in Equation 3.102, we can then define a parameter related to K in a similar fashion. Using this approach, the standard electron affinity is defined as:

$$p\varepsilon^o = \log K \tag{3.103}$$

for a one electron transfer or

$$p\varepsilon^o = (1/n)\log K \tag{3.104}$$

For the oxidation of H_2 (g) to H^+, this reaction is assigned a value of $p\varepsilon^o = 0$ (for standard conditions) and therefore, the half reactions of all other redox processes can be assigned a value relative to this arbitrary reference value. This will be discussed further now. So, the reference reaction is:

$$2H^+ + 2e^- = H_2\,(g) \quad p\varepsilon^o = 0, \log K = 0 \tag{3.105}$$

Combining Equations 3.102 and 3.104 we therefore obtain the following:

$$p\varepsilon = p\varepsilon^o - (1/n)\log\{Red\}/\{Ox\} \tag{3.106}$$

This is the *Nernst Equation*, named after the person who derived this relationship, and it describes the overall electron affinity of a system in terms of a standard free energy parameter ($p\varepsilon^o$) and the ratio of the redox couple that is driving the system in each case. As with other standard entities the value is for 1 M solutions at standard temperature and pressure (Equation 3.106). Also, the formula as written is for one oxidant being converted into one product and unit stoichiometry of the reaction and so a more general formula for the following reaction would be:

$$p\varepsilon = p\varepsilon^o - (1/n)\log \Pi_i\{Red\}^{ni}/\Pi_i\{Ox\}^{ni} \tag{3.107}$$

In the laboratory, the *electrode potential* (E_H) rather than the electron affinity is often measured. This is an equivalent measure of the overall potential (in volts) for a particular redox half-reaction, which can be determined, relative to the reaction with the oxidation of H_2 (g). The subscript H indicates that the potential is referenced to the hydrogen reaction. The *standard electrode potential* (E^o_H) for a reaction can further be defined as the standard potential for that half reaction calculated when the reaction involves the simultaneous oxidation of H_2 (g) to protons in solution, under standard conditions. Thus, there is a relationship between E_H and $p\varepsilon$ and between E^o_H and $p\varepsilon^o$. The standard potential for a reaction, as a consequence of the relationships previously, is related to the standard electrode affinity by:

$$E^o_H = -\Delta G^o/nF = (2.3RT/F)p\varepsilon^o \tag{3.108}$$

where n is the number of electrons in the half reaction and F is the Faraday constant.

3.5.2 The range in electrode potential and the stability of water

The range of $p\varepsilon$ values that are found in the environment is limited by the overall stability of the primary controlling reactions [1–3]. In an aqueous system, the redox range is confined to that of the stability of water and the two pertinent reactions to be considered are the equilibrium between water and oxygen setting the upper limit, and the equilibrium between water and hydrogen at low $p\varepsilon$ values. The relevant reactions are:

$$O_2 + 4H^+ + 4e- = 2H_2O$$

Or, for a one electron transfer:

$$\frac{1}{4}O_2 + H^+ + e- = \frac{1}{2}H_2O \quad p\varepsilon^o = 20.75$$

So

$$p\varepsilon = p\varepsilon^o - \log((P_{O2})^{-1/4}\cdot[H^+]^{-1}) \tag{3.109}$$

The actual value is dependent on the pH and the partial pressure of oxygen (atmospheric $P_{O2} = 0.21$ atm) and ranges from about 12–14 for the pH of natural waters. For example, assuming 1 atm of O_2, the Equation 3.109 simplifies to:

$$p\varepsilon + pH = 20.75$$

So, at a pH of 1, $p\varepsilon = 19.75$, at pH 7, $p\varepsilon = 13.75$ for standard conditions. At a pH of 7 and atmospheric pressure, $p\varepsilon = 13.6$. This shows the relatively small impact of concentration on the overall $p\varepsilon$ and this is generally the case for redox reactions and is a function of the overall equation as the log of the concentrations is being computed.

The other reaction is the reduction of water to hydrogen:

$$H^+ + e- = \frac{1}{2}H_2\,(g) \quad p\varepsilon^o = 0$$

So

$$p\varepsilon = -\log((P_{H2})^{1/2}/[H^+])$$

Again, the actual value will therefore be a function of the partial pressure of hydrogen in the gas phase and the pH. For standard conditions (1 atm H_2), the equation simplifies to:

$$p\varepsilon + pH = 0$$

The atmospheric concentration of H_2 (g) is 6×10^{-7} atm [2], so in this case, the concentration will have an impact of the overall value. However, in deep environments where anoxic conditions exist, the hydrogen concentration is likely to be higher and so a higher value for the pε is likely. For the given atmospheric concentration and a pH of 7, the value is $p\varepsilon = -3.9$.

Often in diagrams of pH versus electron potential, the value is given in terms of E_H rather than pε. The relationship between the two variables is given previously and therefore the values can be easily estimated. Given the values of R, F and the standard temperature, $E_H = 0.059p\varepsilon$.

3.5.3 Equilibrium calculations involving redox reactions

When performing calculations, it is evident from Equation 3.107, as $p\varepsilon^o$ is a constant, that either the pε can be fixed, thus determining the concentrations of the various redox species for the specific reaction under the specific conditions, or if the ratio of the redox couple is fixed, and the pε is therefore prescribed. The values for the equilibrium constant, E_H and $p\varepsilon^o$ for a number of trace metals and their likely important counter reactions under environmental conditions are given in references [1–3]. With such information it is therefore possible to determine the speciation and redox status of any metal in solution if either the E_H or the pε of the system is known. Again, as was the case for the solubility product relationship, the values in the equations are for the specific forms that are oxidized and reduced and therefore complexation in solution may have an important impact on the overall distribution of a metal between its oxidized and reduced phase. Again, this is probably best illustrated with an example.

Example 3.11

Consider the redox couple for iron which was discussed briefly earlier. The equilibrium constant for this reaction is given as 13 for $Fe^{3+} = Fe^{2+} + e^-$ at 25°C. Using this $p\varepsilon^o$ value of 13, let us reconsider the problem in Example 3.6. We will specify a total concentration of 10^{-11} M total Fe and the conditions (seawater chemistry) given in the problem previously. Recall from the previous calculations that if there is precipitation, the chemistry of the dissolved phase is dominated by the formation of hydroxide species and that $[Fe^{3+}]$ was 2.36×10^{-11}. Fe_T (where this is the total Fe(III) concentration) under the conditions of the problem. Further, for the analytical calculations, we will consider the only important dissolved reduced species to be Fe^{2+}. So, therefore, for the Fe^{3+}/Fe^{2+} couple:

$$K = 10^{13} = [Fe^{2+}]/[Fe^{3+}]\{e^-\}$$

or

$$p\varepsilon = 13 - \log([Fe^{2+}]/[Fe^{3+}])$$

We will consider the following situations: (1) $p\varepsilon = 10$; (2) $p\varepsilon = 5$; (3) $p\varepsilon = 0$; (4) $p\varepsilon = -3$.

For these conditions, the $\log([Fe^{2+}]/[Fe^{3+}])$ is: (1) 3; (2) 8; (3) 13; (4) 16 and one could conclude from this that under all these conditions that the concentration of Fe^{II} is greater than Fe^{III}. This is the case for the free metal ions but not all the dissolved Fe is in a reduced state. Recall that if there is precipitation and the presence of the solid Fe^{III} oxide phase, then:

$$[Fe^{3+}] = 2.36 \times 10^{-11} \cdot Fe(III)_T$$

and, for example, for $p\varepsilon = 10$

$$[Fe^{3+}] = 10^{-3}[Fe^{2+}]$$

so that, solving these two simultaneous equations, we see that Fe^{III} is still the dominant form in solution. The other situations can be similarly calculated. To show the full results, the calculations were run using the MINEQL+ program and the results for these different electron affinities is shown in Table 3.22.

(Continued)

Table 3.22 Calculated speciation (log values) and redox status for iron in seawater for a number of different pε values. Calculations were done using MINEQL+.

Species	pε = 10	pε = 5	pε = 0	pε = −3
Fe(III) species				
Fe^{3+}	−23.7	−23.7	−23.7	−26.3
$Fe(OH)^{2+}$	−13.3	−13.3	−13.3	−15.8
$Fe(OH)_3$	−13.3	−13.3	−13.3	−15.9
$Fe(OH)_4^-$	−14.0	−14.0	−14.0	−16.6
Hematite (solid)	−11 (99.9%)	−11	−11.02 (96%)	None
Total Diss. Fe(III)	−12.9	−12.9	−12.9	−15.5
Fe(II) species				
Fe^{2+}	−21.5	−16.5	−11.5	−11.1
$Fe(OH)^+$	−23.3	−18.3	−13.3	−12.9
$FeCl^+$	−22.6	−17.6	−12.6	−12.2
$FeSO_4$	−22.2	−17.2	−12.2	−11.8
Total Diss Fe(II)	−21.4	−16.4	−11.4	−11
Ratios				
$[Fe^{2+}]/[Fe^{3+}]$	2.2	7.2	12.2	15.2
$Fe(II)_T/Fe(III)_T$	−8.5	−3.5	1.5	4.5

This is a somewhat artificial situation as the speciation is calculated for a closed system. However, it does illustrate very clearly the point that the ratio of the ions involved in the redox couple can be very different from the ratio of the total amount in each oxidation state. This is likely most pronounced for Fe because of the high solubility and lack of appreciable complexation of the reduced Fe(II) state and the very high insolubility and the high degree of solution complexation of the oxidized Fe(III). If one was to base the estimation of the dominant form of Fe purely on the ratio of the free ion concentrations, one would conclude that Fe(II) dominance would occur at a high pε. This, however, is not the case and the reduced form only dominates the total dissolved concentrations at a much lower value, around a pε of 1.5 for the conditions used in this problem. The problem was kept simple by not including the presence of sulfide complexes. At low pε values, the formation of S(−II) would lead to the precipitation of FeS and other sulfide phases and this would then lead to a shift in the speciation toward the reduced form of Fe, as discussed further next.

While the formulation of the Nernst equation and the calculation of redox equilibria appear simple based on the Equations 3.107 and 3.108 previously, the impact of speciation in the dissolved phase on the actual equilibrium situation is a very necessary and important consideration. There are many approaches to examining the redox status of trace metals in the environment. One approach is the construction of so-called *stability diagrams*. As shown with the discussion of the stability of water and the related equations, it is possible to convert a redox reaction into a formula and if the concentrations are taken to be at 1 atm for gases and at 1 M for dissolved constituents, that is, standard conditions, then the reactions simplify into reactions that involve only pε and pH. Alternatively, if the concentrations of the redox couple are set to equivalent values for any other concentration, then the same result is obtained, and this equation represents the situation that is the transition point between the two forms, be they a redox couple, and acid–base couple, or some other reaction. These reactions can therefore be plotted and if all the pertinent reactions for a particular metal can be calculated and plotted on a graph, a so-called *stability diagram* or *pε–pH diagram* can be drawn up which shows the regions where the various forms of the element are the dominant species. To reiterate the approach, consider the following reaction:

$$\begin{aligned} &Ox + ne^- + mH^+ = Red \\ &np\varepsilon - np\varepsilon^o + mpH = -(1/n)\log\{Red\}/\{Ox\} \end{aligned} \quad (3.110)$$

If {Red} = {Ox}, then this equation simplifies into an equation involving only pε, $p\varepsilon^o$ and pH and therefore can be plotted in two dimensions with the two unknowns as the axes. Clearly, this is a simple diagram and a simplification of the complexities in some situations, and the actual stability will depend to a large degree on the actual concentrations of the redox couples and other speciation, although, as noted previously, the pε value is often relatively insensitive to changes in the concentrations of the redox constituents.

An example of such a construction is shown for sulfur across the typical pε–pH range of the natural environment (see Fig. 3.19). Each of the lines are derived from equations that represent a relationship either between the two forms

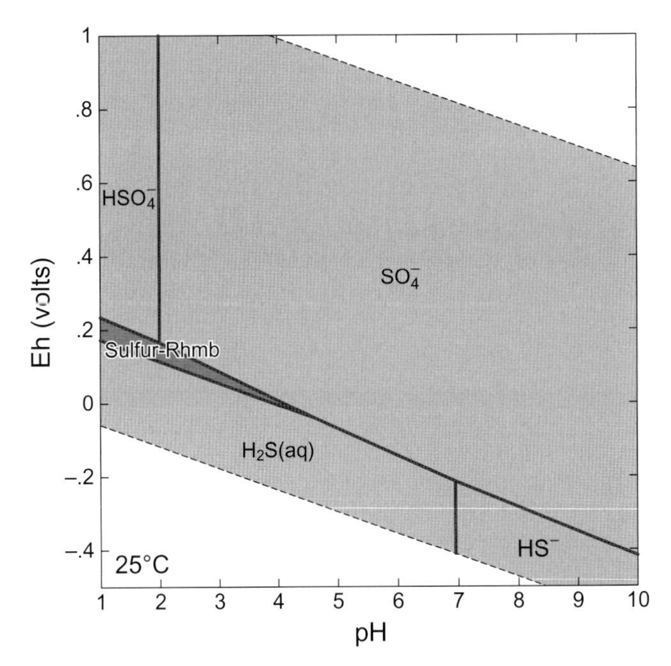

Fig. 3.19 The pε-pH diagram for the sulfur system with the total sulfur concentration at 10^{-3} M in the presence of atmospheric carbon dioxide and oxygen ($pCO_2 = 10^{-3.5}$ atm) at 25°C. Figure generated using the computer equilibrium program *The Geochemist's Workbench*, Release 9.

of a constituent forming a redox couple, or represent changes in the speciation of the dissolved constituents. For example, the vertical line between HSO_4^- and SO_4^{2-} represents the acid–base equilibrium of these two species and the value is equivalent to the pK_a for this reaction (1.9). The line separating SO_4^{2-} and HS^- represents the relationship for the redox transition between these two species:

$$SO_4^{2-} + 9H^+ + 8e^- = HS^- + 4H_2O \quad p\varepsilon^\circ = 4.23$$

or

$$8p\varepsilon + 9pH = 33.8$$

Similar lines can be constructed for the other redox couples involving the other forms of sulfide (H_2S and S^{2-}). Of course, in reality there are other potential forms of S with intermediate oxidation states, such as the formation of elemental S, and their inclusion would lead to a more complicated diagram. A similar situation can be developed for metals. An example will suffice to demonstrate the approach.

Example 3.12

Consider the stability of mercury in water with 10^{-3} M chloride over a pH and pε gradient. For simplicity, the formation of the +I oxidation state of Hg will not be considered. Also, the concentration of Hg is chosen such that precipitation of oxidized complexes is not likely but that the insoluble sulfide solid is the dominant phase in the reduced environment. There is some controversy in the literature about the value of this constant [15] but a consensus value results in the prediction of precipitation at low nM Hg concentrations in the presence of sulfide. We will therefore assume that this is the case for the reduced environment, and at nM concentrations there will not be precipitation of oxidized solids. It is necessary to determine the dominant complex in solution for the oxidized form of Hg at each pH, and this can be done by considering the various possible species in solution. Given that Hg forms strong complexes with Cl, it is apparent that at lower pH the dominant species will be $HgCl_2$ but at some pH value, the hydroxide complex will dominate. The overall speciation equation, using values for the constants from Appendix 3.1, is:

$$Hg_T = [Hg^{2+}](1 + 10^{7.2}[Cl] + 10^{14}[Cl]^2 + 10^{15.1}[Cl]^3 + 10^{15.4}[Cl]^4 + 10^{10.6} + 10^{21.8}[OH]^2)$$

At a Cl^- concentration of 10^{-3} M, the dichloride complex will dominate and the other Cl complexes can be ignored. As the pH increases, the hydroxide complex will become more important and the pH at which it becomes the dominant species can be estimated from the equation. As $10^{14}[Cl]^2 = 10^8$, the hydroxide will dominate when:

$$10^{21.8}[OH]^2 > 10^8 \text{ or } [OH] > 10^{-6.9}\text{ M (i.e., pH} > 7.1)$$

So, there will be a vertical line separating these two dominant species at a pH of 7.1. This is shown in Fig. 3.20. The main redox reaction is given by:

$$Hg^{2+} + 2e^- = Hg^0\,(aq);\ p\varepsilon^\circ = 11.15;\ \log K = 22.3$$

for the equation as written [1]. However, the reactions need to be written in terms of the dominant oxidized species as these are the relevant couples to be examined. So:

$$Hg^{2+} + 2e^- = Hg^0\,(aq) \quad \log K = 22.3$$
$$HgCl_2 = Hg^{2+} + 2Cl^- \quad \log K = -14$$

And therefore

$$HgCl_2 + 2e^- = Hg^0\,(aq) + 2Cl^- \quad \log K = 8.3$$

so

$$p\varepsilon = 4.15 + \frac{1}{2}\log([Hg^0][Cl^-]^2)/[HgCl_2]$$

Setting the quotient as equivalent to unity, the reduction of $HgCl_2$ to Hg^0 is represented by a horizontal line with a value of 4.15.

A similar relationship needs to be developed for the conversion of $Hg(OH)_2$ to Hg^0. This relationship will be pH dependent and is given by:

$$Hg(OH)_2 + 2H^+ + 2e^- = Hg^0\,(aq) + 2H_2O \quad \log K = 28.5$$

and

$$2p\varepsilon + 2pH = 28.5 \text{ or } p\varepsilon + pH = 14.25$$

So, at a pH of 14, pε = 0.25. At a pH of 7.1, where the speciation switches from the chloride complex to the hydroxide complex, the value of pε is 7.15 (Fig. 3.20).

Finally, the change between the reduced form and the sulfide form – where in fact the Hg is an oxidized form – occurs under the more reduced conditions. The following redox equation is pertinent [1]:

$$Hg^{2+} + SO_4^{2-} + 8e^- + 8H^+ = HgS\,(s) + 4H_2O \quad \log K = 70$$

When combined with the equation for the conversion of Hg^{2+} to Hg^0, the following overall equation is obtained:

$$Hg^0\,(aq) + SO_4^{2-} + 6e^- + 8H^+ = HgS\,(s) + 4H_2O \log K = 47.7$$

or

$$6p\varepsilon + 8pH = 47.7$$

The lines associated with this equation can be plotted and the overall results are shown in Fig. 3.20.

In the environment, where measurements have been made of the Hg speciation, there is little evidence for an accumulation or dominance of the reduced elemental form at intermediate pε values. However, it is found in most samples that some fraction of the total dissolved concentration throughout the redox gradient is Hg^0. This is because Hg(II) can be readily reduced photochemically or can be formed as the result of microbial biological reduction in the water column [34]. That there is not a region of Hg^0 dominance in the environment is likely a result of the kinetics of these interactions, and the fact that in many environments the redox transition occurs over a very small vertical gradient.

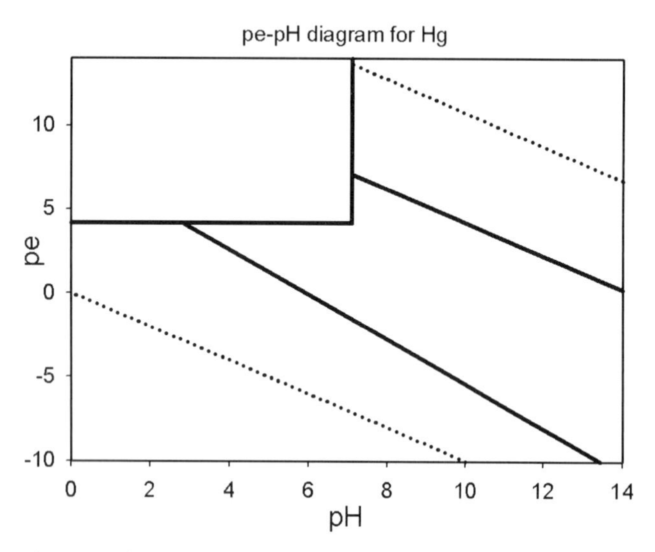

Fig. 3.20 The pε-pH diagram for mercury using the conditions and calculations discussed in Example 3.12 in the text. Figure created using data generated by MINEQL+, a computer program. Dotted lines show the stability of water; inside the box is $HgCl_2$, to the right $Hg(OH)_2$, while below is Hg^0 and finally HgS(s).

This is one aspect of the environmental interpretations of redox stability diagrams and related equilibrium equations that needs to be mentioned. For example, in most sediments, the transition from a low oxygen environment (dissolved oxygen <50 μM) to a sulfidic environment (measurable dissolved sulfide (>0.1 μM) occurs over 1–2 cm of depth and this represents almost the entire pε gradient. Thus, it is very difficult in an environmental situation to observe the intermediate speciation predicted by the stability diagrams as the system is not closed, and advective and diffusive process could easily be mixing the dissolved species on a similar timescale to the reactions that are predicted to be occurring, and therefore the assumption that the system is approaching an equilibrium environment is not valid. While this is a pertinent consideration for all the reactions discussed in this chapter, the situation is likely the farthest from equilibrium for many redox reactions.

3.5.4 Environmental considerations and controlling reactions

Therefore, before leaving the topic of redox transformations, it is worthwhile to discuss briefly the important environmental redox transitions that govern in many instances the fate and transport of trace metals, and discuss the issues associated with the use of an equilibrium approach to estimating the speciation of trace metals in environments where there are substantial gradients in electron affinity. In addition to the issue of the system being far from equilibrium, in most environments, many of these redox reactions are used by microbes as an energy source, or as a source or sink for electrons, and therefore many of the major reactions, which may not involve trace metals directly, are microbially mediated [1, 2]. For example, the reduction of Fe, Mn, and nitrate in low oxygen regions are likely mediated by organisms, and in the marine environment, sulfate reduction is a major process connected with organic matter degradation that is dominantly microbial. Such microbial processes can enhance the kinetics of the reactions to a degree that reactions that may not be favorable based purely on thermodynamic calculations occur.

Given the importance of sulfate reduction in the environment, the stability diagram of sulfur, introduced earlier (see Fig. 3.19), is worth further consideration. These reactions are important as many of the trace metals form strong complexes with dissolved sulfide and also many of the metals are highly insoluble in the presence of sulfide (Appendix 3.1). Thus, the speciation of many trace metals in the presence of sulfide is often dominated by these reactions. As shown in Fig. 3.19, over the pH range most often encountered in the environment (pH 4–9), the range of pε where the reduced forms of sulfur dominate is from just above zero at the lowest pHs to around −6 at pH of 9. Around a neutral pH, the value falls between −3 and −4 and the exact value would depend on the actual concentration of the species in solution. However, given that the reduction reaction involves the transfer of eight electrons, the resultant equation is very insensitive to the change in concentration. So, in the environment, if there is detectable sulfide and also measurable sulfate, it is reasonable to estimate the pε based on Fig. 3.19 if the pH is known. In examining the speciation of trace metals in environmental media it is typical that the concentration of sulfur is much greater than that of the trace metals, except perhaps for Fe, and therefore in many instances this is the controlling redox couple that determines the overall pε of the medium. In some instances, the redox couple of Fe will be the dominant reaction controlling the pε and this is the case in low sulfur environments. Thus, in ocean waters, where sulfate concentrations are high, the redox chemistry of sulfur is likely the main control over pε while in a low sulfur freshwater environment, it could be that Fe chemistry is the major reaction determining the overall redox state.

The speciation diagram for Fe under one set of conditions is shown in Fig. 3.21. Here, the system being modeled is relatively high in Fe (10^{-4} M) and therefore the solid phases of oxidized Fe are dominant except at the very low pHs. There is an intermediate region where the reduced form of Fe dominates in solution, again at the lower pH values. As there is the precipitation of Fe as a sulfide solid at higher sulfide levels, this is the dominant form of Fe(II) at lower pε values and higher pH. The pK_{a1} for H_2S/HS^- is 7.1 and the pK_{a2} for HS^-/S^{2-} is 17 (this value is uncertain and the value from Stumm and Morgan [1] is used here), and this controls to some degree the pH and pε of solid phase precipitation. The dominant stable form of the sulfide solid is however pyrite, which has the formula FeS_2, and is a more stable form than FeS (iron monosulfide) that often forms initially in environmental situations, and can be found in the environment. The presence of FeS is often inferred by measuring acid volatile sulfur (AVS), while chromium reducible sulfur (CRS) [35] is often taken as a measure of the pyrite present. These measurements have been widely used although their interpretation is somewhat controversial [35]. It is thought that AVS is a reasonable surrogate measurement for FeS while CRS is a measure of pyrite. This will be discussed further in later chapters.

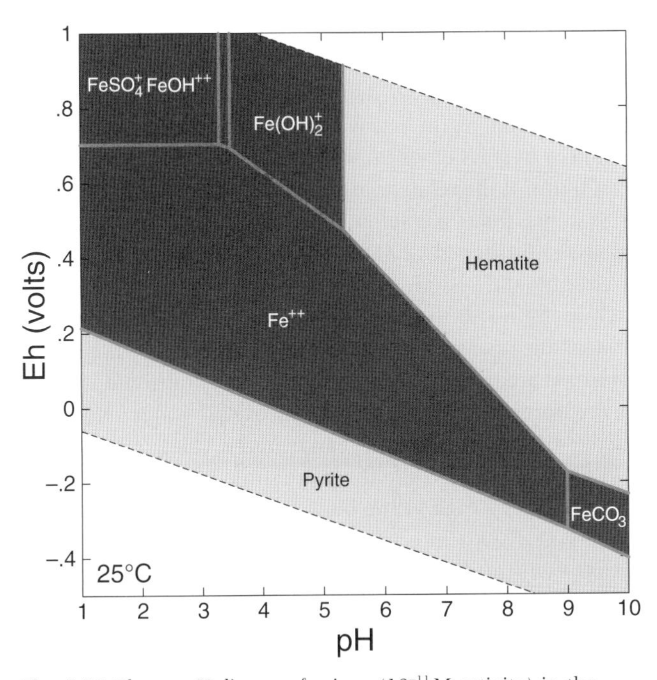

Fig. 3.21 The pε-pH diagram for iron (10^{-11} M activity) in the presence of carbonate (in equilibrium with $pCO_2 = 10^{-3.5}$ atm) and sulfur at 10^{-3} M in the presence of atmospheric oxygen at 25°C. Figure generated using the computer equilibrium program *The Geochemist's Workbench*, Release 9.

Given the speciation diagram, it can be seen that, under many conditions, the transition from an oxidized environment to a reduced environment leads to a dramatic change in the dissolved Fe concentration. Consider, for example, the situation at pH 6 in Fig. 3.21. Under oxic conditions, the Fe(III) is mostly precipitated, as discussed in the various examples earlier in the chapter. However, at lower pε values, the reduced form of Fe (Fe(II)) dominates and thus the dissolved concentration of Fe will increase dramatically. However, once the sulfide concentration builds up, the solid sulfide forms of Fe(II) precipitate and there is a resultant decrease in the dissolved concentration of Fe. This phenomenon is somewhat illustrated by the calculations in Example 3.11 where the Fe is completely dissolved at the lower pε values. Vertical profiles in sediments and in the water column of stratified systems often show this transition and the changes in concentration of Fe species through the redox transition zone.

This is also illustrated by the fact that, because of the high intermediate concentration of reduced dissolved Fe, there is diffusion of the Fe(II) into the oxic regions where it is subsequently oxidized and precipitated. The sinking of this solid material removes Fe from this region to the more reduced environment where it is reduced and solubilized. This recycling of Fe has been termed the "ferrous wheel" by analogy, and is illustrated diagrammatically and with actual data in Fig. 3.22 [1]. The data is from the Pettaquamscutt estuary in the USA (Fig. 3.22b) [36] is just one example of this. This is a permanently stratified system with highly sulfidic bottom waters (mM levels of dissolved sulfide) and restricted vertical advective mixing. As can be seen in Fig. 3.22(b) [36], the oxic-anoxic transition region is defined by the rapid increase in sulfide concentration and this is also the region where there is a sharp increase in dissolved total Fe and in Fe(II), as measured using a chemical assay. In the core of this region, it is clear that all the dissolved Fe is reduced, while in the surface waters above there is little reduced Fe. Reduced

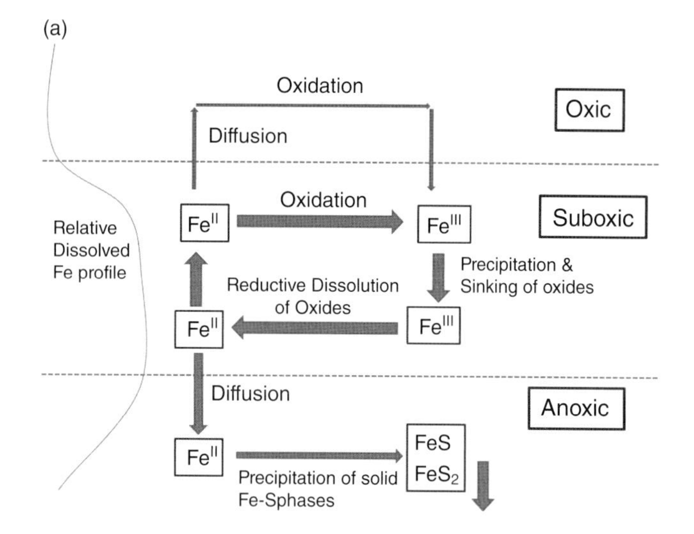

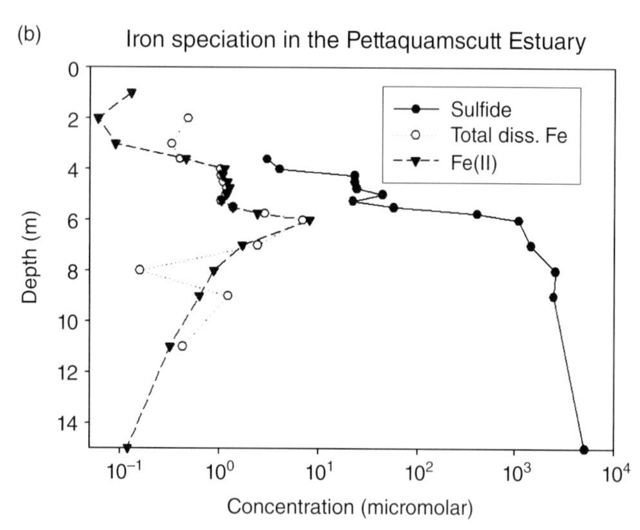

Fig. 3.22 (a) The hypothetical cycling of iron species across a redox interface in the water column where there is diffusion of dissolved species and settling of particles. Figure drawn using information in the literature. (b) The distribution of reduced and total iron and the concentration of sulfide through the water column of the Pettaquamscutt Estuary in Rhode Island, USA. Figure created using data from O'Sullivan *et al.* (1997) *Estuarine, Coastal and Shelf Science*, **45**: 769–88 [35].

Fe in the oxic waters is possible due to its formation via photochemical processes [36].

In the deeper waters of the Pettaquamscutt, there is a continual buildup of sulfide until a maximum concentration is reached. However, the same is not true for Fe and the concentration decreases to very low values in the deeper waters. The solubility product for the formation of FeS can be written in terms of the dominant species of sulfide at the pH of the estuary (pH ~8) as [1]:

$$FeS\,(s) + H^+ = Fe^{2+} + HS^- \quad \log K = -4.2$$

Or

$$[Fe^{2+}] = 10^{-4.2} \cdot 10^{-8}/[HS^-]$$

Note that there are a number of different forms of FeS in the environment and all have somewhat different K_{sp} values, and so the value used here is just one typical measured constant. Given the maximum dissolved Fe concentration in the estuary (~10^{-5} M), precipitation of FeS would begin when the sulfide concentration exceeded about 0.1 µM, and at 1 mM, the Fe(II) concentration is predicted to be at nM levels. These calculations, which assume that there are not dissolved Fe(II)-sulfide complexes forming, are consistent with the results. There are however other considerations that are resulting in a discrepancy between the measured values and the predicted data from the equilibrium calculations [3]. These include the fact that the rate of precipitation of FeS is relatively slow; the equilibrium constant used in the calculation may not reflect accurately the conditions in the estuary; and also that there is the potential for colloidal material being measured by the Fe(II) assay. Modeling of such processes will be dealt with further in Chapter 4, and discussed further later in the book. The data in Fig. 3.22(b) illustrate clearly the role of precipitation of metal sulfides in controlling the metal concentration in reduced environments. Many of the heavy metals also form very insoluble sulfide phases, although the formation of dissolved sulfide complexes can enhance their solubility. The Fe distribution found in the Pettaquamscutt estuary is also found in other environments that have been extensively examined, including the Black Sea, Framvaren Fjord in Norway, and many others. A similar situation is found in sediments but this is much more difficult to demonstrate experimentally given the large changes in redox occur over very small vertical distances in such environments.

In terms of the relative pH and pε of various environmental systems, the trend across systems covers a large range, as discussed further in Chapter 7 (Fig. 7.28) [3]. The range in environmental systems includes oxic, low pH environments such as found in locations where acid mine drainage is a controlling process, as was discussed briefly in Chapter 2. Intermediate pH oxic waters include rain and some streams in environments where buffering of the water chemistry changes is low. The pH of ocean waters is relatively constant and above neutral, while there are some higher pH environments, such as high saline lakes. Many groundwaters have low oxygen with intermediate pH. In such environments, nitrate reduction and trace metals (Fe, Mn) are often the controlling reactions resulting in relatively high trace metal concentrations.

As the source of oxygen to most aquatic systems is from the atmosphere, low oxygen conditions are found in any environment that is isolated from the atmosphere, such as deeper ocean waters and deeper reaches of sediments. This depletion is often driven by the decomposition of organic matter. The system redox status is often determined by the relative rate of input of oxygen, via diffusion, advection and mixing, compared to the input of organic matter, which is often delivered mostly due to particle settling [36]. In highly eutrophic coastal environments, NOM input greatly exceeds the rate of oxygen penetration into these sediments and oxygen depletion and anoxia is the resultant situation. The opposite occurs in offshore, deeper sediments where organic matter input is very low. Overall, environments where mixing is restricted, or organic matter production is high, or where organic matter levels are elevated, are often environments where very low pε values are found, and if sulfate is present, where sulfidic conditions dominate. These issues are further discussed in Chapter 8.

The processes of *photosynthesis* and organic matter degradation (*respiration*) often control the overall system redox. The photosynthetic reaction and the reverse reaction of organic matter degradation is:

$$CO_2 + H_2O = CH_2O + O_2$$

The energy required to drive the forward reaction is provided by sunlight – the reaction as written is not *spontaneous* (i.e., ΔG^0 is greater than zero) and the overall process is mediated by organisms. The overall reaction involves the exchange of four electrons per mole of organic C formed and could be written as two half reactions:

$$CO_2\,(g) + 4H^+ + 4e^- = CH_2O + H_2O \quad p\varepsilon^o = -0.2, \log K = -0.8$$

and

$$2H_2O = O_2\,(g) + 4H^+ + 4e^- \quad p\varepsilon^o = -20.75, \log K = -83$$

So, overall

$$CO_2 + H_2O = CH_2O + O_2 \quad \log K = -83.8$$

The overall reverse reaction is spontaneous as written and thus is an energy yielding reaction and many organisms use the degradation of organic matter in the presence of oxygen as an energy source. In the case of *oxic respiration*, oxygen is the *electron acceptor* for the reaction but in the absence of oxygen, other compounds can perform a similar role. The amount of energy gained from the reaction depends on the specific compounds involved, and the maximum energy is obtained from the reduction of oxygen. This is therefore the preferred reaction. Additionally, while water is the most common *electron donor* for photosynthesis, there are other compounds that are also used by microorganisms in carbon fixation, such as hydrogen sulfide (clearly an analog for water) and other reduced sulfide compounds, and some nitrogen compounds. Further, while the energy for carbon fixation is typically the sun, other energy sources, such as reduced inorganic compounds, can be used for carbon fixation (*chemotrophs*).

The relative importance of other pathways of organic matter degradation in the environment is determined by the relative amount of energy that can be obtained from the reactions and this can be presented as a sequence of preferred pathways for organic matter degradation [1–3]. Under typical environmental conditions, the following reactions are involved in the following preferential sequence. In each case, the $p\varepsilon^o$ for the reaction is given in Table 3.23 as well as the value under "typical" environmental conditions (pH = 7 and assuming the concentrations of the redox couples are equal) and denoted as $p\varepsilon^0_w$ (in a similar approach

Table 3.23 Values for the standard electron affinity ($p\varepsilon^o$) and the standard affinity under "typical" environmental conditions ($p\varepsilon^0_w$)(pH = 7, and other conditions as described in the text) for the reduction half-reaction associated with the degradation of organic matter, and the equivalent standard free energy associated with these processes (the combined reduction half reaction and the oxidation of organic matter) (ΔG^0_w). Data taken from [2].

Process	$p\varepsilon^o$	$p\varepsilon^0_w$ (pH = 7)	ΔG^0_w ($kJ^{-1}\,mol^{-1}$)
Aerobic respiration	20.75	13.75	–119
Denitrification (N_2 as product)	21.05	12.65	–113
Denitrification (NO_2^- as product)	14.15	7.15	–
Mn reduction	20.8	9.8*	–96.9
Fe reduction	16.0	1.0*	–46.7
Sulfate reduction (HS^- as product)	4.25	–3.6	–20.5
Sulfate reduction (elemental S as product)	5.9	–3.4	–
Methane formation	2.9	–4.1	–17.7
Hydrogen fermentation	0	–7.0	–1.1

to the calculations done when constructing the stability diagrams).

Aerobic respiration

$$CH_2O + O_2\,(g) = CO_2\,(g) + H_2O$$

Denitrification

$$5CH_2O + 4NO_3^- = 5CO_2\,(g) + 2N_2\,(g) + 7H_2O$$

Mn reduction

$$CH_2O + 2MnO_2\,(s) + 4H^+ = CO_2\,(g) + 2Mn^{2+} + 3H_2O$$

Fe reduction

$$CH_2O + 4Fe(OH)_3\,(s) + 8H^+ = CO_2\,(g) + 4Fe^{2+} + 11H_2O$$

Sulfate reduction

$$2CH_2O + SO_4^{2-} + H^+ = 2CO_2\,(g) + HS^- + 2H_2O$$

Methane formation

$$2CH_2O = CO_2\,(g) + CH_4\,(g)$$

Hydrogen fermentation

$$CH_2O + H_2O = CO_2\,(g) + 2H_2\,(g)$$

The calculated energy released for a one electron transfer under typical conditions is also given as ΔG_w^0 in the table. These values allow for an easy comparison of the relative potential energy released with the different reactions when coupled to the oxidation of organic matter, but the actual differences for any environment will depend on the system variables, and the final products. For example, during denitrification, the nitrate can be reduced to nitrite, nitrogen gas, and even ammonia, and each reaction has a different $p\varepsilon^o$ value under the same conditions.

In terms of trace metal chemistry, as noted earlier, the oxidation and reduction of Fe and Mn can have a large impact on the mobility of trace metals as the metals are often strongly associated with the solid phases of the oxidized metal. In addition, sulfate reduction is the main mechanism for the generation of reduced sulfide species in the environment and will also have a strong influence over the concentration and solubility of a number of metals. Thus, the metals and metalloids that bind strongly with solid oxide surfaces can be depleted in oxic waters due to such associations; their concentration will be enhanced by their release into solution as these oxide phases are reduced. An example is As, which is strongly bound to Fe-oxide in the oxidized state (As(V), see Table 3.21), and is therefore released into the environment in low oxygen environments; phosphate behaves similarly.

There are many situations where the thermodynamic estimations based on speciation, partitioning with the solid phase or redox transformations are not validated by field measurement. It is very necessary to consider the kinetics of the processes involved, and to understand the impact of processes that are driven by energy outside of the system under consideration. Therefore, these are systems that violate the closed-system assumption of the thermodynamic equilibrium approach. Photosynthesis is one example. Many reduction reactions in oxic environmental surface waters are also driven by solar energy (*photochemistry*). Many of these involve the reduction of metals, with Fe, Cu, and Hg being just a few examples. Thus, biological reduction and photochemical reduction are an important aspect of the chemistry of surface waters than needs to be considered in depth, and this will be dealt with in Chapter 4, as well as the later chapters. Other interactions, and the overall cycling of metal(loid)s in the various aquatic systems will be discussed further in the upcoming chapters.

It is clear that microbial processes have an important impact on trace metal biogeochemistry in terms of their redox chemistry and their mediation of the redox status of their environment. In Chapter 8 we will discuss further the impact of organisms on the concentration and distribution of metals in the environment, as many metals have a distinct biochemical role and therefore they are actively accumulated into organisms. This can have a significant impact on their aqueous concentration and form.

3.6 Chapter summary

1. Environmental systems are often analyzed based on equilibrium modeling although these systems are often only at steady state rather than true equilibrium. Both a kinetic and a thermodynamic approach are relevant for examining the chemistry of environmental systems.

2. Many cations in solution form complexes with dissolved anions and the extent to which such formation occurs depends on the element and its oxidation state. Many of the trace metals of environmental interest form strong complexes with inorganic and organic ligands. The extent of complex formation, and many other chemical attributes, can be explained by the outer electron configuration and distribution in the presence of the complexing ligands. This is highlighted in the text for the transition metals.

3. The approach taken to predicting the speciation of an element in environmental media depends on the complexity of the solution and the acid–base chemistry of the anions and the strength of complexation with the cations being considered. Numerous examples of this are provided in the chapter.

4. Many metals from strong complexes with NOM and while the exact nature of NOM in any environmental solution is not known there are approaches that can be taken to understand the extent of interaction of the metal with NOM.

5. Most elements and compounds interact with surfaces in the environment. Such interactions are most easily expressed as a partition coefficient. Another way to present the extent of the interaction is to calculate adsorption isotherms, which differ in formulation depending on the assumptions made.

6. Both cations and anions form complexes with the surfaces of many suspended solids or sediments and these interactions can be modeled using a similar approach to that used for other complexes. However, the influence of the surface charge on the complexation needs to be considered and details of the theory and method for doing such calculations was provided.

7. Oxidation-reduction reactions can also be assessed using equilibrium approaches even though in the environment many of these reactions are far from equilibrium. Many reactions are also microbially-mediated. Methods for examining the effect of system redox status (electron affinity) of the system on trace elements are discussed.

8. In many systems, the overall redox status is often determined by a major controlling reaction. For example, respiration of organic matter is often a determining reaction and this can occur through the use of a suite of reactions, each of which is less favorable energetically. Oxic respiration is the most favorable reaction.

Appendix 3.1 Formation constants for the most likely important complexes of various elements with inorganic and organic ligands and compounds. Note that the formation constants for the oxides are given for the reaction: $M^{n+} + mH_2O = M(OH)_m^{n-m} + mH^+$. Data is primarily taken from the MINTEQA2 database (http://www.epa.gov/ceampubl/mmedia/minteq/MinteqDocs/SUPPLE1.PDF), but supplemented and compared with databases in Morel and Hering [2], and Stumm and Morgan [1].

Complex	Zn(II)	Cd(II)	Hg(II)	Cu(II)	Cu(I)	Ag(I)	Ni(II)	Co(II)	Fe(III)	Fe(II)	Mn(II)	Sn(II)	Pb(II)
MOH	−9.00	−10.10	2.80	−7.50		−12.00	−9.90	−9.70	−2.19	−9.40	−10.60		
$M(OH)_2$	−17.79	−20.29		−16.19		−24.00	−18.99	−18.79	−4.59	−20.49			−17.09
$M(OH)_3$	−28.09	−32.51	−14.90	−26.88			−29.99	−31.49	−12.56	−28.99	−34.80		−28.09
$M(OH)_4$	−40.49	−47.29		−39.98				−46.29	−21.59		−48.23		−39.70
M_2OH		−9.40						−11.00					−6.40
$M_2(OH)_2$				−10.59					−2.85				
$M_3(OH)_4$									−6.29				−23.89
$M_4(OH)_4$								−30.53					−19.99
MF	1.30	1.20	7.76	1.80		0.40	1.40	1.50	6.04		1.60	11.58	1.85
MF_2		1.50							10.47			14.39	3.14
MF_3									13.62			17.21	3.42
MF_4													3.10
MF_5													
MF_6													
MCl	0.40	1.98	13.49	0.20	3.10	3.31	0.41	0.54	1.48		0.10	8.73	1.55
MCl_2	0.60	2.60	20.19	−0.26	5.42	5.25	−1.89	2.31	2.13		0.25	9.52	2.20
MCl_3	0.50	2.40	21.19	−2.29	4.75	5.20			1.13		−0.31	8.35	1.80
MCl_4	0.20		21.79	−4.59		5.51							1.46
MClOH	−7.48	−7.40	10.44										
MCl_2OH													
MBr	−0.07	2.15	15.80			4.60	0.50					8.25	1.70
MBr_2	−0.98	3.00	24.27			7.50						8.79	2.60
MBr_3			26.70			8.10						7.48	
MB_{r4}			27.93										
MBrCl			22.18										
MBrI			27.31										
MBrOH			12.43										
MI	−2.04	2.28	19.60			6.60							2.00
MI_2	−1.69	3.92	30.82			11.70							3.20
MI_3			34.60			12.60							
MI_4			36.53			14.23							
MSO_4	2.34	2.37	8.61	2.36		1.30	2.30	2.30	4.05	2.39	2.25		2.69
$M(SO_4)_2$	3.28	3.50					0.82		5.38				3.47
$MOH(SO_4)$													
$M2(OH)_2SO_4$													
$M_2(OH)_2(SO4)_2$													

Tl(I)	Tl+3	Cr+3	VO+2	VO2+	U+4	UO2+2	Mg	Ca	Sr	Na	K	Ba	Al
−13.21	1.897	5.9118			−0.597	−5.897	−11.397	−12.7	−13.18			−13.357	−4.997
2.69	−11.697				−2.27								−10.094
		−8.4222			−4.935								−16.791
		−17.819			−8.498								−22.688
						−5.574							
0.10		14.7688	3.778	3.244	9.3	5.14	2.05	1.038	0.548	−0.2			7
			6.352	5.804	16.4	8.6							12.6
			7.902	6.9	21.6	11							16.7
			8.508	6.592	23.64	11.9							19.4
					25.238								
					27.718								
0.51	11.01	9.6808	0.448		1.7	0.21							
0.28	16.77	8.658											
11.01	19.79												
	24.59												
	10.63												
		2.9627											
0.91	12.803	7.5519											
−0.38	20.711												
	27.024												
	31.153												
0.82													
2.19													
1.43		4.8289											
1.86													
	34.76												
1.37		12.9371	2.44	1.378	6.6	3.18	2.26	2.36	2.3	0.73	0.85		3.89
					10.5	4.3							4.92
		8.2871											
		16.155											
		17.9288											

Appendix 3.1 (Continued)

Complex	Zn	Cd	Hg	Cu	Cu+	Ag	Ni	Co+2	Fe+3	Fe+2	Mn	Sn	Pb
MCO_3	4.76	4.36	18.27	6.77			4.57	4.23		11.43	11.63		6.48
$M(CO_3)_2$		7.23	21.77	10.20									9.94
$M(CO_3)_3$													
$MHCO_3$	11.83	10.69	22.54	12.13			12.42	12.22					13.20
MSH		8.01				13.81							
$M(SH)_2$	12.82	15.21	44.52			17.91				8.95			15.27
$M(SH)_3$	16.10	17.11		25.90						10.99			16.57
$M(SH)_4$	14.64	19.31											
M_2SH													
$M_2(OH)SH$													
$M_2(OH)_2SH$													
MS(SH)	6.81												
$MS(SH)_2$	6.12												
MS_2			29.41										
MHS_2			38.12										
MSe											–5.39		
M_2Se						34.91							
$MSeO_3$						–5.59							
$M(SeO_3)_2$		–10.88				–13.04							
$MHSeO_3$									3.42				
$MSeO_4$	2.19	2.27					2.67	2.70			2.43		
$M(SeO_4)_2$	2.20												
MNH_3			5.75	–5.23		–5.93	–6.51	–7.16					
$M(NH_{3)2}$			5.51			–11.27	–13.60	–14.78					
$M(NH_3)_3$			–3.14					–22.92					
$M(NH_3)_4$			–11.48					–31.45					
$M(NH_3)_5$								–40.47					
$M(NH_3)_6$								–43.71					
MNO_3	0.40	0.50	5.76	0.50		–0.10	0.40	0.20	1.00		0.20	7.94	1.17
$M(NO_3)_2$	–0.30	0.20	5.38	–0.40				0.51			0.60		1.40
MCN		6.01	23.19					14.31					
$M(CN)_2$		11.12	38.94	21.91		20.48							
$M(CN_{)3}$		15.65	42.50	27.21		21.70							
$M(CN)_4$		17.92	45.16	28.71			30.20						
MCNOH						–0.78		23.00					
$M(CN)_6$									43.60	35.40			
MPO_4													
$MHPO_4$								15.41	22.29	15.98			
MH_2PO_4									23.85	22.27			
$M(HPO_4)_2$													
$M(HPO_4)_3$													
$M(HPO_4)_4$													
$M(H_2PO_4)_2$													
$M(H_2PO_4)_3$													
M9EDTA)	18.00	18.20	29.30	20.50		8.08	20.10	18.17	27.70	16.00	15.60	27.03	19.80
M(HEDTA)	21.40	21.50	32.90	24.00		15.21	23.60	21.59	29.20	19.06	19.10	29.93	23.00
$M(H_2EDTA)$				26.20				23.50				31.64	24.90
MOH(EDTA)	5.80			8.50			7.60		19.90	6.50			
$M(OH)_2(EDTA)$									9.85	–4.00			
M(NTA)	11.95	11.07	21.70	14.40		6.00	12.79	11.67	17.80	10.19	8.57		12.70
$M(NTA)_2$	14.88	15.03		18.10			16.96	14.97	25.90	12.62	11.58		
M(HNTA)				16.20						12.29			15.30
M(OH)(NTA)	1.46	–0.61		4.80			1.50	0.44	13.23	–1.06			
M(EN)	5.66	5.41	20.40	10.50		6.50	7.32	5.50	4.26		2.74		5.04
$M(EN)_2$	10.60	9.90	29.30	19.60		12.70	13.50	10.10	7.73		4.80		8.50

Tl+	Tl+3	Cr+3	VO+2	VO2+	U+4	UO2+2	Mg	Ca	Sr	Na	K	Ba	Al
						9.6	2.92	3.2	2.81	1.27		2.71	
						16.9							
						21.6							
							11.339	11.599	11.539	10.079		11.309	
2.47													
5.97													
1.00													
−11.07													
								−9.144	−9.344			−9.444	
								−18.79					
		−32.895											
0.33	7.0073	8.2094		−0.296		0.3		0.5	0.6			0.7	
						13.25	4.654	6.46					
					24.443	19.655	15.175	15.035	14.873	13.445	13.255		
		31.9068				22.833	21.256	20.923	20.402				
					46.833	42.988							
					67.564								
					88.483								
						44.7							
						66.245							
7.27		35.5					10.57	12.42	10.436	2.7	1.7	7.72	19.1
13.68		37.4					14.97	15.9	14.795				21.8
		27.7											12.8
5.39		21.2					6.5	7.608	6.2767			5.875	13.3
		29.5						8.81					
													15.2
													8
							0.37	0.11					
		22.57											

(Continued)

Appendix 3.1 (Continued)

Complex	Zn	Cd	Hg	Cu	Cu+	Ag	Ni	Co+2	Fe+3	Fe+2	Mn	Sn	Pb
$M(EN)_3$	13.90	11.60					17.60	13.20	10.17				
M(HEN)			34.70			11.99							
M(Form)	1.44	1.70	9.60	2.00			1.22	1.21					2.20
$M(Form)_2$								1.14					
M(Acet)	1.58	1.93	10.49	2.21		0.73	1.37	1.38	4.02	1.40	1.40	10.02	2.68
$M(Acet)_2$	2.64	2.86	13.83	3.40		0.64	2.10	0.76	7.57			12.32	4.08
$M(Acet)_3$				3.94					9.59			13.55	
MSal	7.71	6.20		11.30			8.20	7.43	17.60	7.20	6.50		
$M(Sal)_2$	15.50	16.00		19.30			12.64	11.80	29.30	11.60	10.10		
MHSal				14.80									
MTart	3.43	2.70	14.00	3.97			3.46	3.05	7.78	3.10	3.38	13.15	3.98
$M(Tart)_2$	5.50	4.10						4.00					
MHTart	5.90			6.70			5.89	5.75			6.00		
MGlut	6.20	4.70	19.80	9.17		4.22	6.47	5.42			4.95		6.43
$M(Glut)_2$	9.13	7.59	26.20	15.78		7.36	10.70	8.72			8.48		8.61
$M(Glut)_3$	9.80												
MHGlut				13.30									14.08
M(Phthal)	2.91	3.43		4.02			2.95	2.83			2.74		4.26
$M(Phthal)_2$	4.20	3.70		5.30									4.83
M(HPhthal)		6.30		7.10			6.60	7.23					6.98
$M(H_2Phthal)$													
MGly	5.38	4.69	17.00	8.57		3.51	6.15	5.07	9.38	4.31	3.19		5.47
$M(Gly)_2$	9.81	8.40	25.80	15.70	10.30	6.89	11.12	9.07		8.29	5.40		8.86
$M(Gly)_3$	12.30	10.70					14.63	11.60					
MHGly									11.55				
M(Pico)	1.40	1.59		5.65	2.88	2.03	2.11	1.56					
$M(Pico)_2$	2.11	2.40		8.20	5.16	4.39	3.59	2.51					
$M(Pico)_3$	2.85	3.18		8.80	6.77		4.34	2.94					
$M(Pico)_4$		4.00		9.20	8.08		4.70	3.17					
M(Prop)	1.44	1.60	10.59	2.22			0.91	0.67	4.01				2.64
$M(Prop)_2$	1.84	2.47		3.50				0.56					3.18
$M(Prop)_3$													
M(Buty)	1.43		10.35	2.14			0.69	0.59					2.10
$M(Buty)_2$								0.78					
M(Cit)	6.21	4.98	18.30	7.57			6.59	6.19	12.55	5.70	5.28		7.27
$M(Cit)_2$	7.40	5.90		8.90			8.77						6.53
M(HCit)	10.20	9.44		10.87			10.50	10.44	19.80	3.50	3.02		
$M(H_2Cit)$	12.84	12.90		13.23			13.30	12.79					
$M_2(Cit)_2$				16.90									
$M(HCit)_2$							14.90						
M(Benz)	1.70	1.80		2.19		0.91	1.86	1.05			2.06		2.40
$M(Benz)_2$		1.82											
MOH(Benz)													

Tl+	Tl+3	Cr+3	VO+2	VO2+	U+4	UO2+2	Mg	Ca	Sr	Na	K	Ba	Al
		29											
		1.07					1.43	1.43	1.39			1.38	
−0.11		15.0073					1.27	1.18	1.14			1.07	
		17.9963											
		20.7858											
							5.76	4.05					
							15.3	14.3				13.9	
1.40							2.3	2.8	2.55	0.9	0.8	2.54	
													9.37
4.80							5.75	5.86	5.8949	4.58		5.77	
		22.6					2.8	2.06	2.2278			2.14	
		30.7											
		25.2						11.13					13.07
		16.3					2.49	2.45		0.8	0.7	2.33	
		21.2											7.2
								6.43					
1.72		18.7					2.08	1.39	0.91			0.77	
		25.6											
		31.6											
								10.1					
		15.0773					0.9689	0.9289	0.8589			0.7689	
		17.9563										0.9834	
		20.8858											
							0.9589	0.9389	0.7889			0.7389	
												0.88	
1.48							14.4	6.1	5.534	1.03	1.1	4.1	9.97
										1.5			14.8
							4.28	10.2	9.442	6.45		8.74	12.85
							9.6	13.1	12.486			12.3	
							1.26	1.55					2.05
													−0.56

References

1. Stumm, W., and Morgan, J.J. (1996) *Aquatic Chemistry*, John Wiley & Sons, Inc., New York.
2. Morel, F.M.M., and Hering, J.G. (1993) *Principals and Applications of Aquatic Chemistry*, John Wiley & Sons, Inc., New York.
3. Langmuir, D. (1997) *Aqueous Environmental Geochemistry*, Prentice Hall, Upper Saddle River, NJ.
4. Klotz, I.M., and Rosenberg, R.M. (1986) *Chemical Thermodynamics: Basic Theory and Methods*. 4th edn. Benjamin/Cummings Publ. Co., Menlo Park, CA, USA.
5. Gibbs, J.W. (1906) *Scientific Papers of J. Willard Gibbs, Volume 1: Thermodynamics*, Longmans, Green and Co., London, http://books.google.com/books?id=-neYVEbAm4oC&oe=UTF-8 (accessed October 2, 2012).
6. Burgess, J. (1999) *Ions in Solution: Basic principles of Chemical Interactions*, Horwood Publishing Limited, Chichester.
7. Cotton, F.A., and Wilkinson, G. (1972) *Advanced Inorganic Chemistry*, 3rd edition. John Wiley & Sons, Inc., New York, 1143 pp.
8. Emsley, J. (1998) *The Elements*, 3rd edn. Oxford University Press, Oxford.
9. Basolo, F., and Pearson, R.G. (1967) *Mechanisms of Inorganic Reactions: A Study of Metal Complexes in Solution*, 2nd edition. John Wiley & Sons, Inc., New York, 701 pp.
10. Chester, R. (2003) *Marine Geochemistry*, Blackwell Science, Malden.
11. Broecker, W.S., and Peng, T.-H. (1982) *Tracers in the Sea*, Eldigio Press, Palisades, NY.
12. Morel, F., Milligan, A.J., and Saito, M.A. (2004) Marine Bioinorganic chemistry: The role of trace metals in the oceanic cycles of major nutrients. In: Elderfield, H.. (ed.) *The Oceans and Marine Geochemistry*. in Holland, H.D. and Turekian, K.K. (Exec. eds) *Treatise on Geochemistry*. Elsevier Pergamon, Amsterdam, pp. 113–143.
13. Lindh, U. (2004) Biological function of the elements. In: Selinus, E., Alloway, B., Centeno, J.A. et al. (eds) *Medical Geology: Impact of the Natural Environment on Public Health*. Elsevier, Amsterdam, pp. 115–160.
14. Mason, R.P. (2000) The bioaccumulation of mercury, methylmercury and other toxic trace metals into pelagic and benthic organisms. In: Newman, M.C. and Hale, R.C. (eds) *Coastal and Estuarine Risk Assessment*. CRC Press, Boca Raton, pp. 127–149.
15. Benoit, J.M., Gilmour, C.C., Mason, R., and Heyes, A. (1999) Sulfide controls on mercury speciation and bioavailability to methylating bacteria in sediment pore waters (vol. 33, p. 951, 1999). *Environmental Science & Technology*, **33**, 1780–1780.
16. Rashid, M.A. (1985) *Geochemistry of Marine Humic Compounds*, Springer-Verlag, New York, 300 pp.
17. MINEQL+ (2011) *Software*. http://www.mineql.com/ (accessed October 2, 2012).
18. Bruland, K., and Lohan, M.C. (2004) Controls on trace metals in seawater. In: Elderfield, H.. (ed.) *The Oceans and Marine Geochemistry*. in Holland, H.D. and K.K. Turekian (Exec. eds) *Treatise on Geochemistry*. Elsevier Pergamon, Amsterdam, pp. 23–47.
19. Rue, E.L., and Bruland, K.W. (1995) Complexation of iron(III) by natural organic-ligands in the central North Pacific as determined by a new competitive ligand equilibration adsorptive cathodic stripping voltammetric method. *Marine Chemistry*, **50** (1–4), 117–138.
20. Lalah, J.O., and Wandiga, S.O. (2007) Copper binding by dissolved organic matter in freshwaters in Kenya. *Bulletin of Environmental Contamination and Toxicology*, **79** (6), 633–638.
21. Lohan, M.C., Statham, P.J., and Crawford, D.W. (2002) Total dissolved zinc in the upper water column of the subarctic North East Pacific. *Deep-Sea Research Part II*, **49** (24–25), 5793–5808.
22. Sander, S., Ginon, L., Anderson, B., and Hunter, K.A. (2007) Comparative study of organic Cd and Zn complexation in lake waters – seasonality, depth and pH dependence. *Environmental Chemistry*, **4** (6), 410–423.
23. Saito, M.A., and Moffett, J.W. (2001) Complexation of cobalt by natural organic ligands in the Sargasso Sea as determined by a new high-sensitivity electrochemical cobalt speciation method suitable for open ocean work. *Marine Chemistry*, **75**, 49–68.
24. Lamborg, C.H., Fitzgerald, W.F., Skoog, A., and Visscher, P.T. (2004) The abundance and source of mercury-binding organic ligands in Long Island Sound. *Marine Chemistry*, **90**, 151–163.
25. Guentzel, J.L., Powell, R.T., Landing, W.M., and Mason, R. (1996) Mercury associated with colloidal material in an estuarine and an open-ocean environment. *Marine Chemistry*, **55**, 177–188.
26. Richard, D., and Luther, G.W., III (2006) Metal sulfide complexes and clusters. *Sulfide Mineralogy and Geochemistry, Reviews in Mineralogy & Geochemistry*, **61**, 421–504.
27. Stumm, W. (1977) Chemical interactions in particle separation, *Environmental Science and Technology*, **11**(12), 1066–1070.
28. Aiken, G.R., Hsu-Kim, H., and Ryan, J.N. (2011) Influence of dissolved organic matter on the environmental fate of metals, nanoparticles, and colloids. *Environmental Science and Technology*, **45**, 3196–3201.
29. Stumm, W. (1992) *Chemistry of the Solid-Water Interface*, John Wiley & Sons, Inc., New York.
30. Dzombak, D.A., and Morel, F.M.M. (1990) *Surface Complexation Modeling: Hydrous Ferric Oxide*, John Wiley & Sons, Inc., New York.
31. Davis, J.A., and Leckie, J.O. (1978) Effect of adsorbed complexing ligands on trace metal uptake by hydrous oxide. *Environmental Science & Technology*, **12**, 1309–1315.
32. Davis, J.A. (1984) Complexation of trace metals by adsorbed natural organic matter. *Geochimica et Cosmochimica Acta*, **48**, 679.
33. Fitzgerald, W.F., Lamborg, C.H., and Hammerschmidt, C.R. (2007) Marine biogeochemical cycling of mercury. *Chemical Reviews*, **107**, 641–662.
34. Rickard, D., and Morse, J.W. (2005) Acid-volatile sulfide (AVS). *Marine Chemistry*, **97**, 141–197.
35. O'Sullivan, D.W., Hanson, A.K., and Kester, D.R. (1997) The distribution and redox chemistry of iron in the Pettaquamscutt Estuary. *Estuarine, Coastal and Shelf Science*, **45**, 769–788.
36. Burdige, D.J. (2006) *Geochemistry of Marine Sediments*, Princeton University Press, Princeton.

Problems

Use the constants in Appendix 3.1 or other sources (indicate source) for these problems when required. Indicate sources of other information if not taken directly from the text.

3.1. Calculate the relative change in the value of the equilibrium (formation) constant in freshwater ($I = 10^{-3}$) compared to seawater ($I = 0.5\,M$) for the following compounds, when formed from the free ion and the associated anions:

$FeCl_3^0$; $HgCl_4^{2-}$; $AgCl$; $Co(OH)_2$

Indicate the method used to make the calculations.

3.2. A stream discharges into a $10\,m^3$ pond at a rate of $2\,m^3\,d^{-1}$. The stream contains $10^{-5}\,M\,Mn^{2+}$ which has an oxidation rate of $10^{-2}\,d^{-1}$ for the reaction $Mn^{2+} + O_2 = MnO_2$ (s). The microbial reduction of the solid is not known. What is the maximum value for the reduction rate if the concentration of Mn^{2+} in the lake is $10^{-6}\,M$?

3.3.

a. Discuss briefly the differences in the color of solution of the following cations: Mn^{2+}, Co^{2+}, Co^{3+} and Fe^{3+}.

b. Consider the complexes of Fe^{3+}, Mn^{2+}, Cu^{2+} and Zn^{2+} with chloride and ammonia. Which complexes form? Which ones are colored? If there is a difference between the color of the complexes, provide a reason for this.

3.4. Contrast the speciation in solution of the +2 cations of the metals of Group 12 (Zn, Cd and Hg) in $10^{-3}\,M$ chloride solution at pH = 7. Assume that the system is open to the atmosphere.

3.5. Estimate the inorganic speciation of Cu^{2+} in a rain droplet at equilibrium with the atmosphere. Assume that the atmospheric concentrations are: $p_{CO2} = 10^{-3.5}$ atm and $p_{NH3} = 5 \times 10^{-9}$ atm.

3.6. Estimate the speciation of Pb^{2+} and Ag^+ in estuarine waters with a salinity of 15. Calculate the ionic strength and make ionic strength corrections if necessary. Assume that the concentration of the major anions is obtained by conservative mixing and that the system is at equilibrium with the atmosphere.

3.7. Consider a seawater culture medium containing $EDTA_T = 5 \times 10^{-5}\,M$, $Fe(III)_T = 10^{-8}\,M$ and $Zn_T = 5 \times 10^{-9}\,M$. Ionic strength = 0.5 M, $Ca_T = 10^{-2}\,M$ and pH = 8.1.

a. Calculate the speciation and the free concentrations of EDTA, Cd and Fe.

b. A scientist wants to use another ligand besides EDTA and thinks about using picolinate. Is this feasible? How much picolinate would need to be used to obtain the same free ion concentration of Fe^{3+}?

3.8. A lake with the simple composition $Na_T = Ca_T = Mg_T = Cl_T = SO_{4T} = CO_{3T} = 10^{-3}\,M$ contains an organic complexing agent, X, that has the following interactions with metals:

$HX = H^+ + X^-$ $\quad pK_a = 6.0$

$CaX^+ = Ca^{2+} + X^-$ $\quad pK = 2.0$

$MgX^+ = Mg^{2+} + X^-$ $\quad pK = 1.0$

a. Consider all the relevant inorganic complexes but no solid phases. What is the speciation of the ligand at $X_T = 10^{-6}\,M$ and pH = 7?

b. Copper is added to the solution at $Cu_T = 10^{-9}\,M$. It is found using a specific ion electrode that only 1% of the Cu is present as a free ion under these conditions. Estimate the binding constant for the CuX^+ complex based on this information.

c. How will the Cu speciation change as a function of pH in the presence of this ligand?

d. Consider now that $Cu_T = 10^{-6}\,M$ and that both carbonate and/or hydroxide solid phases may form and precipitate with everything else the same as part (b). Would solids form in the presence of the organic ligand? If the ligand was not present, what would be the critical pH for precipitation of the Cu carbonate and hydroxide solids?

3.9. Precipitation of calcium carbonate can occur in lakes in late summer due to the increased pH of the system that results from enhanced primary productivity in association with increased temperature and water column stratification. For a lake with the following surface water characteristics:

$Ca = K = 10^{-3}$ M; $Mg = 5 \times 10^{-4}$; $Na = 10^{-4}$ M;
sulfate = 10^{-3} M and $Cl = 10^{-4}$ M.

a. What is the pH of calcium carbonate precipitation? What solid phase(s) forms?

b. Would the Mg concentration be an important factor? What would be the effect of changing the Mg_T concentration from 5×10^{-3} to $10^{-5}\,M$ on the Ca^{2+} concentration? Why is this?

3.10. The drinking water standard for dissolved lead (Pb) is $50\,\mu g\,l^{-1}$. If water is sitting in contact with Pb pipes, there is the tendency for Pb to solubilize and to increase the dissolved Pb concentration. Calculate the pε at which the total dissolved Pb concentration would be equivalent to the water quality standard under the following conditions: pH = 7 and the bicarbonate concentration is 3 mM. What is the concentration of Pb^{2+}?

The electrode potential, E^0 for $Pb^{2+} + 2e^- = Pb$ is −0.12 V.

3.11. Examine the changes in speciation of arsenic in freshwaters across a pH and pe gradient. Consider a system with 10^{-9} M As, 10^{-4} M total sulfur and 10^{-5} M Fe. Assume the system is open to the atmosphere, and that the major cations are present at "typical" freshwater concentrations. Assume a total ionic strength of 5×10^{-3} M and don't make ionic strength corrections. Show the differences in speciation for pH 5, 7 and 9 in oxic environments (pe = 10), low oxygen (pe = 5), suboxic environments (pe = 0) and anoxic environments (pe = −3) for As and Fe. This problem is likely best accomplished using a spreadsheet or computer program.

CHAPTER 4
Modeling approaches to estimating speciation and interactions in aqueous systems

4.1 Introduction

As discussed in Chapter 3, and detailed in the later sections of this book, there are many situations where is it necessary and useful to create an aquatic model of the system being investigated to allow for a more detailed assessment of the controlling reactions and processes and to test hypotheses about the most important mechanisms. In Chapter 3, the basic manipulation of the equilibrium equations was described and these were used to solve relatively simple systems. For complex systems, the following interactions could be occurring and would need to be simultaneously solved to understand the overall controlling processes: (1) complexation of cations and anions to dissolved species, both inorganic ligands and organic complexes, including the complex mixture of compounds in natural organic matter (NOM); (2) precipitation and dissolution reactions; (3) oxidation-reduction reactions when considering vertical distributions in the water column or in porewaters; (4) adsorption/desorption reactions with solid phases such as metal oxides and sulfides; and (5) potential interactions between dissolved species and organisms. Examples of these types of interactions and their importance in the biogeochemical cycling of trace elements in environmental waters are discussed specifically in Chapters 5–8 with examples for each water type (atmospheric waters; marine and estuarine waters, surface freshwater and groundwater), as well as examining the interactions at the chemical-biological interface. Here, we will focus on the modeling approaches used to estimate the interactions that occur, and not deal specifically with modeling the associated fate, transport and transformations that may be occurring in aquatic waters. These are discussed further in the subsequent chapters. As noted in Section 3.3, there are various approaches that can be used to examine the complexation of trace constituents with the major ions and in various media, and the details of how to perform such calculations was focused mostly on the consideration of cations in solution, and their interactions with solids and NOM, although there have been a number of studies examining the adsorption of metalloids to surfaces. While cation complexation has been the focus and thrust of much research, especially as many metal cations are potentially limiting nutrients given their presence in many important enzymes, the complexation of cations and their interactions with solid phases and biota needs also to be considered.

One approach taken computationally to solve such complex systems is based on the method of solving equations using the "Tableau" method espoused in detail in Morel and Hering [1] and discussed in many similar texts. According to an anecdote by Francois Morel, this approach was developed "on a napkin in a restaurant" with his student, John Westall, and the approach has been adopted in many of the models currently available. The approach is to set up a matrix of equations where all these equations are written in terms of a minimal number of principal components that can be used to describe the species [2]. This approach was discussed in Section 3.3.3 and an example that follows will illustrate the approach. The key to solving the equations using the Tableau approach is setting up the

Trace Metals in Aquatic Systems, First Edition. Robert P. Mason.

equations in terms of the principal components (supposed major species), as these are then eliminated from the resultant equations and the problem can be often more simply solved analytically by examining the minor components and their relative concentrations. For analytical solutions, if some of the minor components are present at very low concentration compared to others, they can be ignored for the initial calculation to obtain an estimate of one complex. By repeating this approach, through iteration if necessary, the final solution to the problem can be obtained.

As an example, consider the approach to estimating the composition of a carbonate-containing aquatic system. Consider for this example, the system created by the dissolution of 10^{-3} M sodium bicarbonate in distilled water. Sodium is a so-called "spectator ion" due to its unreactivity with other dissolved constituents and we will not consider it in the Tableau as its concentration is given by the added bicarbonate concentration. Similarly, any reactions with water will not change its overall concentration and so it is also not included in the Tableau [1]. Given this simple system, the species in solution will be H^+, OH^-, HCO_3^-, CO_3^{2-} and $H_2CO_3^*$ (this notation is used to include both dissolved CO_2 and H_2CO_3). Given the rules detailed in Section 3.3.3, the system can be described in terms of two principal components, and H^+ is always chosen as the principal component for water. Recall that OH– is equivalent to $H_2O - H^+$ in the Tableau. Given the addition of the bicarbonate, this is likely the major species even though the other acid-base pairs will form due to its interaction with water. The following Tableau is therefore generated (Table 4.1):

The two equations that can be generated from this Tableau are obtained by reading the expressions that are formed vertically in the Tableau. These are:

$$[H^+]-[OH^-]-[CO_3^{2-}]+[H_2CO_3^*]=0$$

$$[HCO_3^-]+[CO_3^{2-}]+[H_2CO_3^*]=10^{-4}\ M$$

The first equation contains the minor species and is the most useful to help solve the overall problem if this is to be done analytically. Given an initial pH of 7, it is likely that addition of the bicarbonate would yield the following reactions:

Table 4.1 The problem matrix for the situation where 10^{-4} M sodium carbonate is added to distilled water.

Species	$[H^+]$	$[HCO_3^-]$	Log K
$[H^+]$	1	–	–
$[OH^-]$	–1	–	14
$[HCO_3^-]$	–	1	–
$[CO_3^{2-}]$	–1	1	–10.3
$[H_2CO_3^*]$	1	1	+6.3
Total Concentration (M)	0	10^{-4}	–

$$HCO_3^- + H^+ = H_2CO_3^*$$

$$HCO_3^- = H^+ + CO_3^{2-}$$

Overall, the system would become more basic, but it cannot be assumed in all instances that $[OH^-] >> [H^+]$. However, given the concentration of bicarbonate added, it is likely that the acid-base dissociation reactions would lead to higher concentrations of all carbonate species compared to $[OH^-]$ and $[H^+]$, and so the likely solution to the problem is, using the first equation:

$$[H_2CO_3^*]=[CO_3^{2-}]$$

Then, as $[H_2CO_3^*]=[HCO_3^-][H^+]/K_{a1}$ and $[CO_3^{2-}]=K_{a2}[HCO_3^-]/[H^+]$, the following results from these equations:

$$[H^+]^2 = K_{a1}\cdot K_{a2} \text{ and pH} = 8.3$$

This problem is not simple to solve analytically if the total carbonate concentration is smaller as then the assumption that $[CO_3^{2-}]>[OH^-]$ would not be valid. A more rigorous solution would involve substituting for $H_2CO_3^*$ using the equilibrium equation and substituting for OH^- (= $K_w/[H^+]$), and solving the quadratic for $[H^+]$, assuming that the concentration of bicarbonate is similar to that of the added salt. This again illustrates the complexity of solving equations for even simple systems. However, these equations can be easily solved by a computer using an iteration technique, through using initial guesses, to estimate the concentrations of the various species.

This example is used to demonstrate the concepts that are used in the computer equilibrium programs to solve a set of complex equations using a mass balance approach for each chemical component. In the computer model, the components are pre-chosen and fixed and computing power is relied on to solve the equations. This does not always work if the initial chosen conditions are far from those at equilibrium, or if the solution has a major species that is close to saturation, as successive iterations may lead to the precipitation and then dissolution of this compound, which can have a marked impact on the overall solution chemistry and prevent the convergence of the answer to the preselected level of accuracy. For similar reasons, problems involving redox reactions often don't converge. There are a number of products on the market and in the literature that allow for such calculations to be performed. These include MINEQL [2], the program that will be discussed here and used in this book for calculating the solution to complex problems, as was done in Chapter 3. Use of this program is due to familiarity as the author has used it since being a post-doc at MIT, and in numerous classes taught since then, and does not indicate any value judgment on its superiority or lack thereof. A number of publications in the literature have compared the various programs [3–5] and these can be used as a better guide on suitability for a particular application.

Some programs are free while others require purchase of software and/or a license, which is the case for the Windows version of MINEQL (MINEQL+) [6].

Some of the other programs will be briefly described. As noted by Paquin et al. [5], the earlier versions of such models were RAND, REQEQL and WATEQ, and most of these used mass action expressions and were based on a thermodynamic database. MINETAQA1 is the latest version of a series of model codes produced by the US Environmental Protection Agency (USEPA) which can be obtained on-line (http://nepis.epa.gov/). The program manual and databases can be obtained from the web and the program is described in various publications [7, 8]. The approach and methodology of the program is very similar to that of MINEQL. The US Geological Survey (USGS) maintains a number of programs for chemical speciation determination (http://water.usgs.gov/software/lists/geochemical/). One program is PHREEQC version 2 which is designed to perform a wide variety of low-temperature aqueous geochemical calculations, in a similar manner to the other programs. Additionally, batch-reaction and one-dimensional (1D) transport calculations involving different scenarios are possible, as well as for inverse modeling (http://wwwbrr.cr.usgs.gov/projects/GWC_coupled/phreeqc/html/final-1.html).There are a number of publications describing this program [9, 10]. Another USGS program that allows speciation calculations is WATEQ4F [11]. The Geochemist's Workbench is a commercial product which allows the same complex speciation calculations and interactions and also has capabilities for reaction path modeling and other applications (http://www.rockware.com/product/overview.php?id=132) [12]. It is aimed more at the geochemistry community and is a useful tool for examining precipitation-dissolution reactions, creating pH-redox and activity diagrams and many other geochemical figures, as shown throughout this volume.

Other programs exist in and outside of the USA [3–5]. Finally, there has been more recent development of models that are more specifically designed to examine the complexation of metals in the presence of natural organic matter. One such model is the Windermere Humic Acid Model (WHAM) [13] which involves both monodentate and bidentate associations of metals with NOM, and includes electrostatic interactions in a similar manner to that of solid surfaces, as discussed in Chapter 3. The WHAM program has an extensive database of interactions and is used by many investigators if the focus is interactions with NOM. Another approach is termed the Non-Ideal Competitive Adsorption (NICA)-Donnan Model [14] which uses a different approach as detailed later in this chapter. One other model [5] is the Chemical Equilibrium in Soils and Solutions (CHESS) model [15]. This program is focused on examining interactions in the terrestrial environment, and on examining differences in the adsorption of metals to surfaces. Many of these models have also been incorporated into larger fate and transport and/or bioaccumulation models where the speciation is recalculated at each time step in the model to allow for changes in speciation to impact the fate and transport, and bioaccumulation. These approaches are discussed in more detail in the subsequent chapters in this book.

Paquin et al. [5] compared the various models available at that time (MINEQL, MINTEQ, MINEQL+, WHAM and CHESS) and noted that most use the FORTRAN programming language; WHAM uses BASIC and CHESS can use both. All make activity coefficient (ionic strength) corrections and most rely on the Davies equation; however, WHAM uses the extended Debye–Huckel equation and CHESS can use both. All make temperature corrections and all have some form of interaction with solid surfaces although there is variability amongst the models. All except WHAM and CHESS can incorporate redox equations and transformations. Incorporation of gas and solid solubility is possible with most formulations. As all models are being relatively rapidly updated, many of the specific differences may not continue to exist over time.

These programs are useful in examining a number of different processes, and answering geochemical questions such as calculation of the chemical equilibrium of a system that could include dissolved species, precipitation and dissolution, redox transformations, and adsorption to solid phases. Typical problems that are examined include the change in speciation with changes in pH and/or alkalinity, ionic strength (salinity and/or changes in the concentrations of major ions), simulation of titrations, calculation of charge balance, and examine the factors affecting precipitation of compounds. In addition, such a program could be used to estimate the binding constant of a chemical species, and can be used to probe the potential impact of various interactions, including the effects of temperature. All models rely on a database and there are differences in the list of species and the magnitude of equilibrium constants for some reactions – especially species that form very strong complexes, or are highly insoluble – and this can lead to differences in their predictions if different models are used to evaluate the same problem. Constants should always be checked to ascertain that they are consistent with the most recent and reliable literature. Such differences are particularly apparent when examining the binding to surfaces and the complexation with organic matter. This is discussed further next and also demonstrated through examples in the subsequent chapters in this book.

Temperature corrections typically rely on the Van't Hoff equation ($d\ln K/dT = \Delta H^0/(RT^2)$) and so the ΔH^0 value for the compound is needed for making these corrections. The databases can be undated with new chemical species in most models, and the existing information can be modified in light of new information. This is especially important given the increasing acknowledgment of the importance of metal(loid) interactions with NOM. As noted, these programs can include redox transformations and can also be used in systems open to the atmosphere as the concentration of a dissolved gas at equilibrium with the atmosphere can be determined from Henry's Law ($K_H = C_w/p_x$; C_w = molar water concentration; p_x = atmospheric concentration in atm) (Section 5.2.6).

The overall function and capabilities of such programs will be examined in more detail using the MINEQL+ program (http://www.mineql.com/) [6] as an example in the following section. After that, more complex modeling of adsorption reactions and interactions with NOM and other organic phases will be discussed. In all cases, the usefulness and application of these computer programs will be demonstrated with suitable examples. As always, caution is needed as a computer calculation is only as good as the information that is used to make the calculation and models should always be tested and verified with data. Many of the equilibrium constants in the database were calculated many years ago, and there could be more accurate and more detailed information available in the recent literature. Additionally, there may be reactions occurring that are not included in the database and the user should always verify and check the information in the database. However, even given these caveats, these models are useful for examining various scenarios, making future predictions, for hypothesis testing and sensitivity analysis.

4.2 The underlying basis and application of chemical equilibrium models

In constructing the overall thermodynamic database on which the programs rely, and on defining and maintaining the correct chemical formulations and calculations, the following types of species are separated and defined in the programs:

1. *Components*: These are the independent variables (principal components in the Tableau) in the mass balance equations. Components are used to describe all the chemical species in the system and should be such that any species is uniquely defined by a set of components; that is, a component cannot be described in terms of any other species. These components are pre-determined in the program and are used to scan the associated database for all the relevant species. Note that the MINEQL+ program relies on the thermodynamic database of MINETAQ. Additionally, it is possible to add components to the database, and their associated chemical species and the equilibrium equations.

2. *Species*: These are generated from the database after the components are chosen and are either dissolved or solid species. *Dissolved Species* exist in the aqueous phase and can be a component or a complex (of several components).

3. *Solids*: These are species that have a fixed activity, typically 1. In the model formulation, dissolved gases are also considered in this category as their concentration is fixed by the defined atmospheric concentration. In the mathematical framework, two types of solids are tracked in the calculations: so-called "dissolved solids" and precipitated solids.

4. *Dissolved Solids*: These exist in the mathematical framework as it is necessary to track the concentration of the potential precipitation reactions so that these species can form when their solubility criteria are exceeded. These solids do not contribute to the overall mass balance of the system.

5. *Precipitated Solids*: These form once the solubility criteria, as defined by their solubility product, are exceeded. Once they form, they are present at unit activity. If the system becomes unsaturated in a subsequent iteration, or while performing a "titration calculation" (i.e., one variable is changing), these solids will dissolve. Precipitation and dissolution are controlled by the saturation index (SI):

$$SI = \log Q/K_{sp} \tag{4.1}$$

where Q is the calculated ion product for the solid at that point and K_{sp} is the solubility product. When SI = 0, equilibrium exists; if SI > 0, precipitation occurs and vice versa.

6. *Fixed Entities*: These are parameters that are fixed as a constant in the calculations. It is possible to define the concentration of some species as constant, which is what occurs for solids and gases, and which could be done for example by defining a fixed pH or carbonate concentration for the problem, or by defining the pε of the system.

Precipitation is also controlled by Gibbs Phase Rule which states that:

$$F = C + 2 - P \tag{4.2}$$

where F is the numbers of degree of freedom of the system; C is the number of components and P is the number of phases present [1]. If too many species are set up as fixed entities then a phase rule violation will occur. In terms of precipitated solids, once one solid of a particular stoichiometry has precipitated it will then maintain the dissolved free ion concentration at a particular value and will not allow the precipitation of other solids of the same composition, so there will likely only to be a phase rule violation if the program is set up incorrectly with too many fixed entities. Note that water is always included as a fixed entity with an activity of 1. In many problems, the pH is also fixed but this is not necessary as the pH can be estimated from the charge balance (electrical neutrality) or from the alkalinity. The carbonate system is defined by three variables pH, carbonate alkalinity (Alk) and total carbonate concentration (CO_{3T}) which are not independent as if two are fixed, the other is predetermined [16]. The carbonate alkalinity equation is:

$$Alk = -[H^+] + [OH^-] + [HCO_3^-] + 2[CO_3^{2-}] \tag{4.3}$$

and

$$CO_{3T} = [H_2CO_3^*] + [HCO_3^-] + [CO_3^{2-}] \quad (4.4)$$

As noted before, in tracking these species in setting up the computer simulation, MINEQL+ has six different categories: Type-I species (components); Type-II species (aqueous complexes); Type-III species, which are entities of fixed activity; Type-IV species (precipitated solids) and Type-V species which are the "dissolved solids". In addition, MINEQL+ also allows the elimination of various species from the calculation (Typed-VI species). For a solid, defining it at unit activity is essentially decreeing that there is an infinite reservoir of this species, and the same is true for a gas (i.e., an infinite atmospheric reservoir);

In MINEQL, and in the other similar programs, the solution can be written mathematically as the following for each component:

$$C_i = K_i \prod_{j=1}^{n} X_j^{aij} \quad (4.5)$$

For $i = 1$ to m, and

$$Y_i = \sum_{i=1}^{m} a_{ij} C_i - T_i \quad (4.6)$$

where C_i is the concentration of species i and X_j is the concentration of component j; K_i is the equilibrium constant for the pertinent reaction for species i; T_i is the total concentration of component j; a_{ij} is the stoichiometric coefficient of component j in species i; Y_i is the mass balance equation for component j; n is the number of components and m is the number of species. The program solves for X_i such that Y_i is equal to zero, or within the limits set for the solution of the problem (convergence criteria are prescribed). The solution uses the Newton–Raphson method to solve the problem, by finding the roots to the set of non-linear equations, through iteration with a small adjustment in X_j being made after each iteration is complete, to minimize the value of Y_i [6].

Corrections to the equilibrium constant can be made for temperature using the Van't Hoff equation, as noted in Section 4.1, and corrections can also be made for ionic strength through the use of activity coefficients, as discussed in Section 3.3.4. MINEQL+ uses the Davies equation:

$$\log\gamma_j = -Az_j^2\left(\frac{\sqrt{I}}{1+\sqrt{I}} - bI\right) \quad (4.7)$$

where γ_j is the activity coefficient of species j, I is the ionic strength, z is the species charge and A and B are constants; A is $1.82 \times 10^6(\varepsilon T)^{-3/2}$ (ε is the dielectric constant, T is temperature in °C) and b is set at 0.24 in MINEQL+ [1, 6]. The dielectric constant can be corrected for temperature. In the program, I can be defined or calculated.

Given this basic description of the model framework and the approach to solving equations, it is useful to return to some of the examples in Chapter 3 and examine the calculations using MINEQL+. The basic principles and some of the uses of computer programming for understanding trace element cycling will be demonstrated using these examples.

Example 4.1

This is a re-examination of Example 3.1, which calculated the speciation of zinc (50 nM) in a solution of fixed pH (6), with carbonate (5×10^{-4} M), chloride (2×10^{-4} M) and sulfate (10^{-4} M) in the solution. The components chosen in MINEQL+ would be H_2O, H^+, HCO_3^-, Cl^-, SO_4^{2-}, and Cu^{2+}. The resultant matrix is shown in Table 4.2. The solution was generated for a system at 25°C and with an ionic strength of 10^{-3} M.

A number of comments can be made on the data in Table 4.2. Firstly, it is obvious that data on the enthalpies for various compounds is missing and this would impact calculations made at concentrations substantially different from 25°C. Secondly, there are some species, such as ZnOHCl, that were not considered in the calculations in Example 3.1. Also, the constants for some of the species are somewhat different, or reported to greater accuracy than those used in the calculations in Chapter 3, and presented in the related Appendix. None of these differences are large and change the overall conclusion that most of the Zn is present in solution as the free ion (Fig. 4.1). Such differences in the values for K in the databases may be important in some situations, and can be large based on the reported values for some constants in the literature. The calculation, which included the potential for precipitation of numerous Zn solids, determined that no solids precipitated and the free ion was the major species in solution. This is consistent with the results of the analytical calculations in Example 3.1. This example illustrates the approach to determining the speciation of a metal in solution using a computer program.

Table 4.2 The principal components and the species in the MINEQL+ database for the situation of zinc in a solution containing chloride, sulfate, and carbonate species. While not shown or discussed, the counter major ions are also present. Information extracted from the database.

Species	H_2O	H^+	Cl^-	CO_3^{2-}	SO_4^{2-}	Zn^{2+}	LogK	ΔH
OH^-	1	−1	0	0	0	0	−13.997	13.339
ZnOHCl	1	−1	1	0	0	1	−7.48	0
$Zn(OH)_4^{2-}$	4	−4	0	0	0	1	−40.488	0
$Zn(OH)_3^-$	3	−3	0	0	0	1	−28.091	0
$ZnOH^+$	1	−1	0	0	0	1	−8.997	13.339
$Zn(OH)_2$	2	−2	0	0	0	1	−17.794	0
H_2CO_3	0	2	0	1	0	0	16.681	−5.679
HCO_3^-	0	1	0	1	0	0	10.329	−3.490
$ZnHCO_3^+$	0	1	0	1	0	1	11.829	0
HSO_4^-	0	1	0	0	1	0	1.990	5.258
$ZnCl_3^-$	0	0	3	0	0	1	0.500	9.560
$ZnCl^+$	0	0	1	0	0	1	0.400	1.291
$ZnCl_4^{2-}$	0	0	4	0	0	1	0.199	10.960
$ZnCl_2$	0	0	2	0	0	1	0.600	8.843
$ZnCO_3$	0	0	0	1	0	1	4.160	0
$Zn(SO_4)_2^{2-}$	0	0	0	0	2	1	3.280	0
$ZnSO_4$	0	0	0	0	1	1	2.340	0
Total concentration	0	0	2E-4	5E-4	1E-4	5E-7	–	–

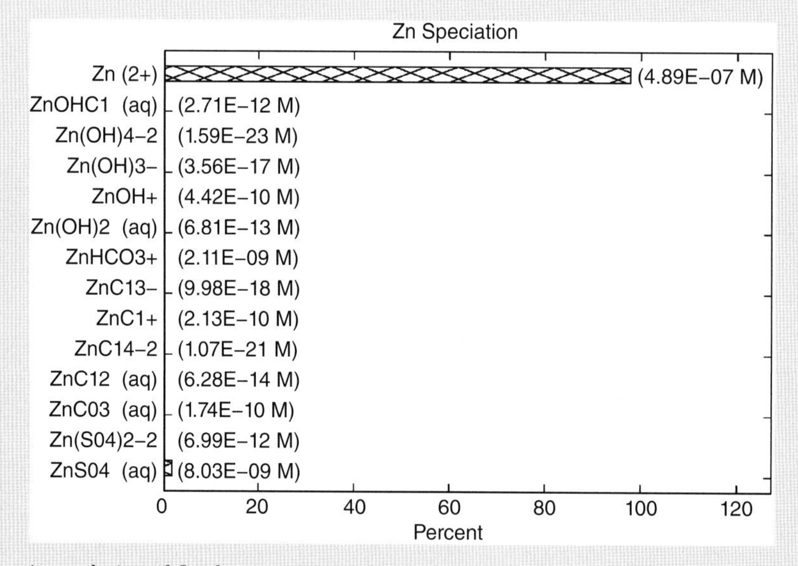

Fig. 4.1 Speciation of zinc in a solution of fixed concentration that is closed to the atmosphere. Details of the solution chemistry are given in the text. Results calculated using the MINEQL+ computer program.

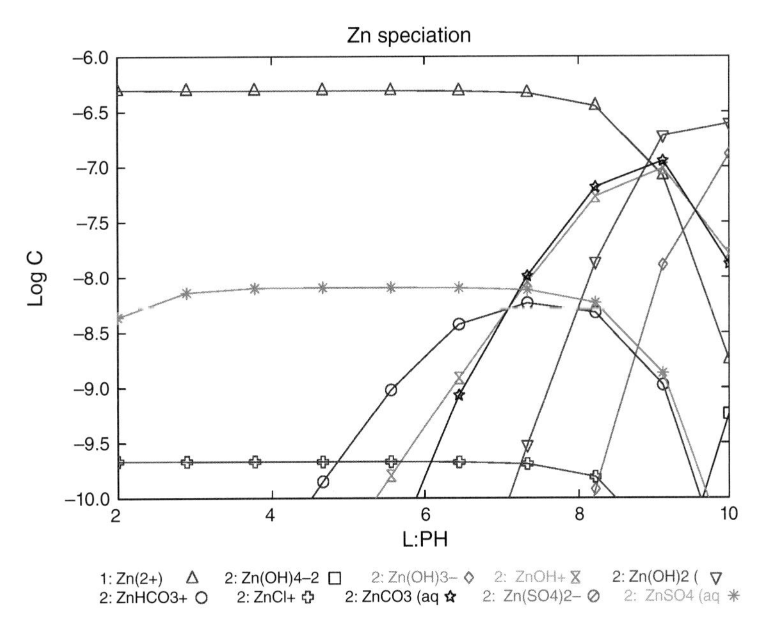

Fig. 4.2 Calculation of the changing speciation of zinc with pH under the solution conditions of Fig. 4.1 (Table 4.2). Results calculated using the MINEQL+ computer program.

It is relatively simple to do such calculations over a range of concentrations. For example, consider how the Zn speciation may change over a pH range. The same system shown in Table 4.1 is considered, but instead of a fixed pH, a pH range from 2–10 is modeled. The resultant output is shown in Fig. 4.2. The relative importance of the hydroxide and carbonate species increases with pH, above about pH 8, and that the complexes with chloride are never important – this is not surprising given that the changing pH has no impact on chloride ion content. The increased pH increases the relative complexation of Zn in solution, and the higher the pH, the more important the higher hydroxide species (e.g., $Zn(OH)_3^-$ and $Zn(OH)_4^{2-}$) become. Again, this result is similar to that shown in Table 3.8.

Example 4.2

Complexation with organic compounds is also possible. Many of the constants are already included in the databases, but can also be added if not. As an example, consider the complexation of Zn with citrate (Example 3.5), and over a range in pH. Such a calculation can be performed by further including citrate in the choice of components. The species in Table 4.2 are all still pertinent, but in addition, MINEQL+ added the species in Table 4.3. The results are shown in Fig. 4.3; the acid–base species of citrate and its complexes with Zn. The citrate complex (ZnCit) is important at intermediate pH values but decreases in importance both at low pH, as the concentration of Cit^{3-} decreases, and high pH, when the hydroxide species increase in importance.

Table 4.3 The additional species generated by the addition of citric acid as an additional component within the system. All the other conditions are as in Table 4.2.

Species	H_2O	H^+	Cl^-	CO_3^{2-}	SO_4^{2-}	Cit^{3-}	Zn^{2+}	LogK	ΔH
ZnHCit	0	1	0	0	0	1	1	10.200	0.800
ZnH_2Cit^+	0	2	0	0	0	1	1	12.840	0.0
H_2Cit^-	0	2	0	0	0	1	0	11.157	0.310
$HCit^{2-}$	0	1	0	0	0	1	0	6.396	0.800
H_3Cit	0	3	0	0	0	1	0	14.285	−0.660
$ZnCit^-$	0	0	0	0	0	1	1	6.210	2.000
$Zn(Cit)_2^{4-}$	0	0	0	0	0	2	1	7.400	6.000

(Continued)

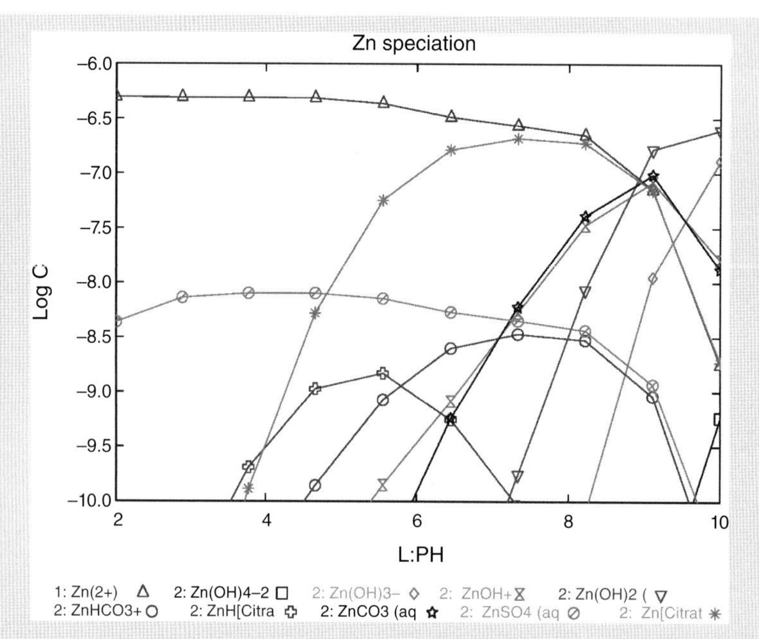

Fig. 4.3 Zinc complexation in a solution with a fixed total carbonate concentration in a closed system, with the addition of citrate to the solution. Results calculated using the MINEQL+ computer program.

Example 4.3

The considerations so far are for a system closed to the atmosphere. With the system open to the atmosphere the total carbonate concentration is fixed by the equilibrium dissolution of CO_2. For $p_{CO2} = 10^{-3.5}$ atm, with $\log K_H = -1.5$, the $[H_2CO_3^*]$, concentration is fixed at 10^{-5} M (Fig. 4.4). The total carbonate concentration is much higher at higher pH. As a result, the carbonate species of Zn are more important at higher pH given the higher concentrations of the carbonate and bicarbonate anions (Fig. 4.5). This example illustrates the potential effect of doing experiments with metals that bind strongly with carbonate or bicarbonate species. Clearly, the conditions of the experiment – an open or closed system – are crucial if one is to obtain the correct results.

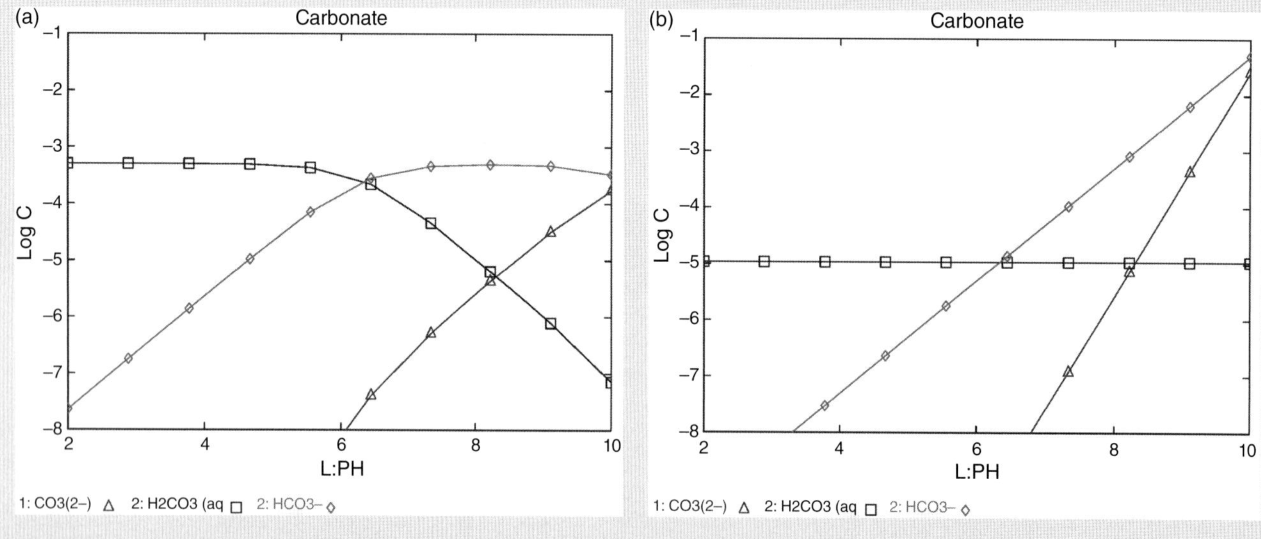

Fig. 4.4 (a) Speciation of dissolved carbonate in a system of fixed total carbonate concentration in a closed system; and (b) the speciation and concentration of the various acid–base pairs in an open container at equilibrium with the atmosphere. Results calculated using the MINEQL+ computer program.

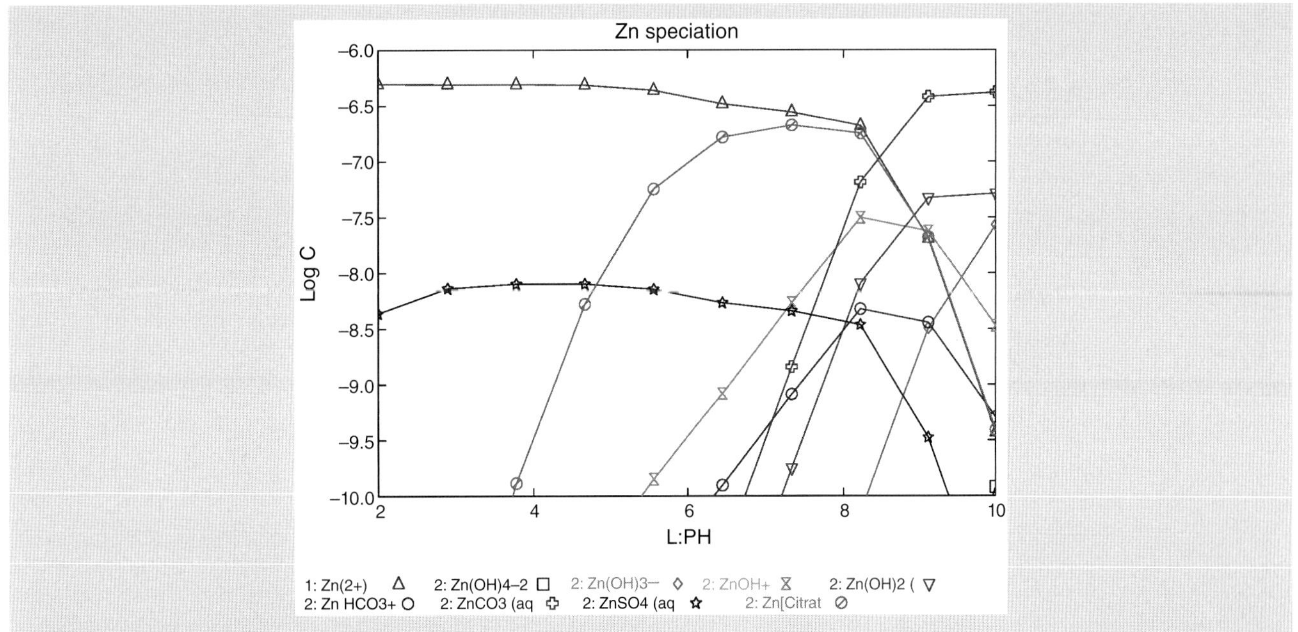

Fig. 4.5 Zinc complexation in a solution with the same conditions as in Fig. 4.2 except that the solution is at equilibrium with the atmosphere. Results calculated using the MINEQL+ computer program.

4.3 Adsorption modeling

The details of how the adsorption models are formulated were discussed in Section 3.4.5. The computer models can also typically be used to examine and calculate adsorption isotherms (Section 3.4.4) and partition coefficients (Section 3.4.3) but their particular usefulness is for modeling the complex interactions that can occur with dissolved and solid phases, especially if there is the potential for formation of tertiary species, typically involving a surface site, a cation and an anion [1, 5, 6]. An example is given in Fig. 3.17 and was discussed in Section 3.4.5. Additionally, potential competition for limited surface active sites is a common situation in low particulate environments, and there is the potential for competition between anions and cations at the intermediate pH ranges of many natural waters. Anions typically adsorb more strongly at low pH when the surface is overall more positively charged, and cations adsorb to the more negative high pH surfaces (e.g., Fig. 3.15). We will examine and model some of the interactions discussed in Chapter 3 in more detail next.

MINEQL+ and other computer programs have a number of different modeling approaches that can be used to examine dissolved-particulate interactions. The basic assumptions of these models are that the ions interact with specific functional groups on the surface of the solid forming surface complexes; true complexes (inner-sphere) or ion pairs (outer-sphere). Interactions alter the charge and potential of the surface and this is taken into account in the calculations. As discussed in Section 3.4.5, the theory of the double layer of interactions at the interface between the dissolved phase and solid surface (Figs. 3.14 and 3.16) is the most commonly used formulation for these interactions [1, 16–18]. In the Double Layer Model, the interactions with the solid phase occur in a single surface layer (termed the *0* layer) and there is a diffuse outer layer (termed the *d* layer) between this layer and the bulk medium. This model was discussed in detail in Chapter 3. MINEQL+ includes two formulations of this approach, the typical model in which the value of the surface potential decreases exponentially with distance outside the inner layer (Fig. 4.6a) or a more simple formulation where the decrease is linear from the solid surface (the Constant Capacitance Model) (Fig. 4.6b). The Constant Capacitance Model is representative of environments of high ionic strength but with low surface potential.

Additionally, MINEQL+ has a Triple Layer formulation that can be used (Fig. 4.6c). In this model, the adsorbing layer is divided into two surfaces, and inner *0* layer and a middle β layer, as well as a diffuse *d* layer. This formulation allows for a further distinction in the type and strength of the binding of the dissolved species to the surface, and allows for the inclusion of interactions of major ions and charged species with the surface, which can affect the overall charge balance of the system. Further, MINEQL+ has incorporated the formulation of Dzomback and Morel [17] for the adsorption of cations and anions to iron oxide surfaces, which uses the Double Layer Model. This approach was described in Section 3.4.5 and will be further discussed next.

Recall that when considering the interaction between the dissolved species and the solid phase surface, the

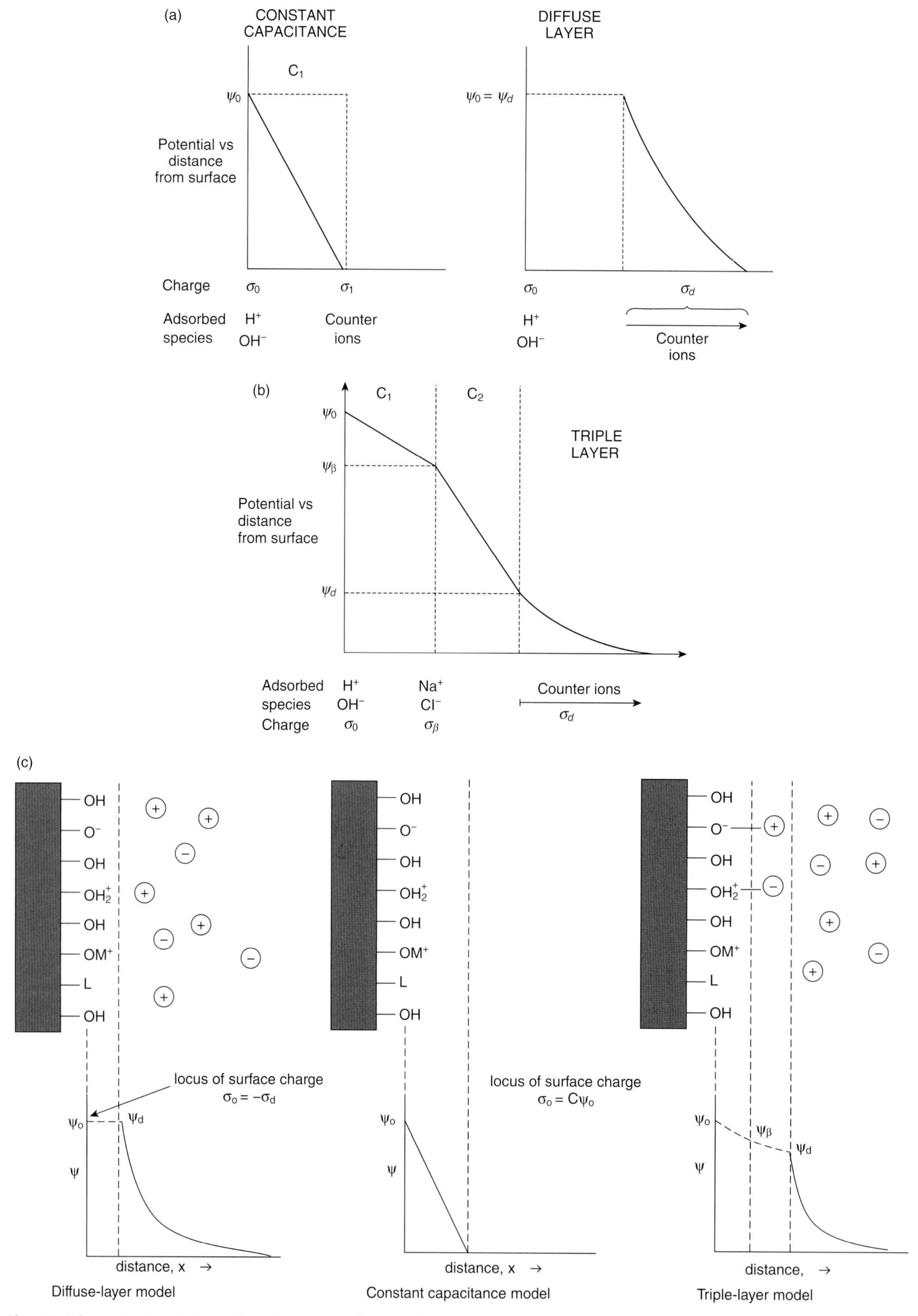

Fig. 4.6 Schematic plots of the surface charge (σ) and potential (ψ) with distance from the surface used in the various model formulations, and the location of the adsorbing species in (a) Constant Capacitance and Diffuse Double layer; (b) Triple Layer; and (c) Representations of the distribution of ions and interaction of dissolved species with a surface. Figures taken from Drever (2002) *The Geochemistry of Natural Waters* [81] and used with permission of the author.

electrostatic interaction is included, and the overall interaction is considered in the following manner [1, 16]:

$$\Delta G^o = \Delta G^o_{int} + \Delta G^o_{coul} \text{ and } \Delta G^o_{coul} = F\Delta Z\Psi_o \quad (4.8)$$

So, the activity of the adsorbing ion is influenced by the electrostatic potential at the surface and this relationship can be written as:

$$\{X^z_S\} = \{X^z\}\left[\exp\left(\frac{\Psi_0 F}{RT}\right)\right]^z \quad (4.9)$$

where $\{X^z_S\}$ is the activity of the ion near the surface and $\{X^z\}$ is the activity in the bulk solution, and z is the ion charge. Note that the designation of the subscript 0 in Ψ_o indicates the interaction of the ion at the solid surface forming an inner-sphere complex, as discussed further next.

The relationship between surface potential (Ψ_o) and surface charge (σ_0) is the difference between the Two-Layer and Constant Capacitance Models. The relationship can be written as:

$$\sigma_0 = f(\Psi_o)\cdot(A/F) \quad (4.10)$$

where A is the surface area. The function is a constant (C', a "fitting parameter") for the Constant Capacitance Model ($f(\Psi_o) = C'.\Psi_o$), while for the Two Layer Model, the value of $f(\Psi_o)$ is:

$$f(\Psi_o) = (8RT\varepsilon\varepsilon_0 10^3)^{0.5}\cdot\sinh(zF\Psi_o/2RT) \quad (4.11)$$

where ε is the dielectric constant and ε_0 the permittivity of free space (8.854×10^{-12} C/V.m) [16, 18].

Only the Triple Layer Model considers the impact of the major ions in solution which interact in the outer layer, while all models consider the impact of the inner-sphere complexes on the overall charge. The Hydrous Ferric Oxide (HFO) Model [17] is a specific case of the Double Layer Model and the parameters in this model were derived through a detailed and comprehensive analysis of the available data on the adsorption of cations and anions to oxide surfaces, and of the site density of interacting sites on the surface of the oxide. From the data analysis, these authors defined a set of parameters which were the "best fit" results of their modeling of the data, and they compiled and published extensive data on the complexation constants for cations and anions. The constants derived from this analysis are contained in Table 3.21 [1, 16].

The Three-Layer Model is more complex in that the strongly adsorbing ions, such as transition metal cations, some anions and protons, are adsorbed in the *0* layer while weakly adsorbing ions (that typically form ion pairs) are modeled as adsorbed in the β layer (major cations (alkali and alkali earth elements), sulfate, chloride), and this determines the overall potential and charge of this layer (σ_β and Ψ_β) (Fig. 4.6c). This formulation allows for inclusion of the impact of major ions on the overall potential of the layer around the oxide surface. The bonding for species in the outer β layer is due to weak Coulombic forces, and as $\Psi_\beta < \Psi_o$ the overall impact of the inclusion of the ion pairs in the calculation is greatest at high ionic strength. In most formulations, as shown in Fig. 4.6(b), the capacitance ($¢_1$) in the *0* layer and the capacitance ($¢_2$) in the β layer are considered to change at a constant rate with distance away from the surface, and with the change in the outer diffuse layer being exponential as in the Double Layer Model. Examples of adsorption using some of the Three Layer formulations in the literature will be discussed later in this section.

The HFO Model [17] includes models for both surface adsorption and at high ion concentrations, surface precipitation. Given the focus of the book, which is the consideration of trace species in environmental systems it is unlikely that surface precipitation will occur under such conditions. However, this approach is discussed in detail in Dzomback and Morel [17] and related texts [1, 16]. The overall details and the relationships that relate to the HFO model are shown in Table 4.4. After compiling the best fit data from

Table 4.4 Demonstration of the similarity and consistency of the two models (Gouy–Chapman (GC) Theory and the Debye–Huckel (DH) Theory) in terms of the relationships between activity coefficients and the interactions at the solid surface. Table created using information discussed in Dzomback and Morel [17].

Relevant Equations/Parameters	Solution and Equations (DH)	Solution and Equations (GC)
Balance of electrostatic and thermal forces: $RTLnC_i + Z_iF$ = constant Poisson's equation: $\nabla^2\Psi = p/\varepsilon\varepsilon_0$ Charge density: $p = F\Sigma Z_iC_i$ Ionic strength: $I = \frac{1}{2}\Sigma Z_i^2 Ci$ Double Layer thickness: $\kappa^2 = \frac{2F^2}{\varepsilon\varepsilon_0 RT}$ Poisson–Boltzmann equation: $\nabla^2\Psi = \kappa^2\Psi$	$\Psi_r = \frac{Z_ie}{\varepsilon\varepsilon_0}\left(\frac{1}{r}e^{-\kappa F} - \frac{1}{r}\right)$ where Z_ie is the charge on the ion and the $-1/r$ negates the contribution of the ion to the potential. To obtain the solution at the limit of r ($r\to0$) $\Psi_{DH} = \frac{-Z_ie}{\varepsilon\varepsilon_0}\kappa$ From the DH theory, the activity coefficient of an ionic solute considering an infinite dilution references state is: $ln\gamma_i = \frac{Z_iF}{2RT}\psi_{DH}$	$\Psi_x = \Psi e^{-kx}$ $\Psi_{GC} = \frac{\sigma}{\varepsilon\varepsilon_0}\kappa^{-1}$ From the GC theory, the activity coefficient of a surface species, considering a zero surface charge potential reference state is: $ln\gamma_i = \frac{Z_iF}{2RT}\psi_{GC}$

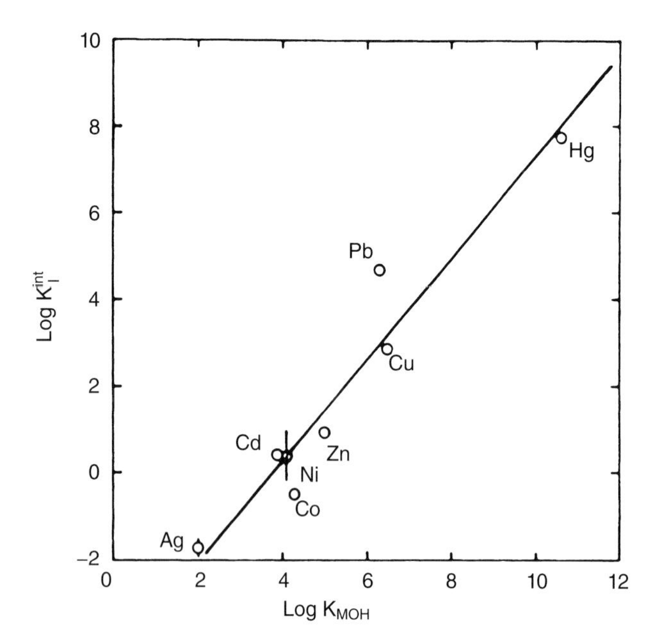

Fig. 4.7 Correlation between the first hydrolysis constant for metals and the predicted strong binding constant for the HFO Model. Figure reprinted from Dzomback and Morel (1990) *Surface Complexation Modeling: Hydrous Ferric Oxide* [17] and reproduced with permission from John Wiley & Sons, Inc.

the literature values, Dzomback and Morel [17] examined the data for its thermodynamic consistency. They were able to show a strong relationship between the derived constants for adsorption to the oxide surface and the first hydrolysis constant for the metal cations (Fig. 4.7) and similarly for anions, as shown in [17]. The importance of consideration of the overall HFO surface charge density because of the Coulombic effects is shown in Fig. 4.8 for both 1 : 1 and 2 : 2 electrolytes. The effect is more pronounced at high ionic strength.

Given this overview of the HFO Model it is worthwhile to return to some of the problems in Chapter 3 to examine the impact of metals, organic ligands and other factors that control the adsorption of metals to oxide surfaces (Example 4.4, Figure 4.9). In these calculations, the HFO model as included in MINEQL+ will be used. One potential problem with the way the HFO model is incorporated into MINEQL+ is that the HFO concentration is estimated based on the entered value for the total Fe^{III} concentration, not the calculated Fe^{3+} concentration and the related calculated concentration of insoluble (hydr)oxide. So, this can lead to errors if the model is run with conditions that may change the free ion concentration at a given total Fe^{III} concentration, such as organic complexation of the Fe^{III}, or a redox setting where most of the Fe is reduced, especially if the model is run across a range of pε values.

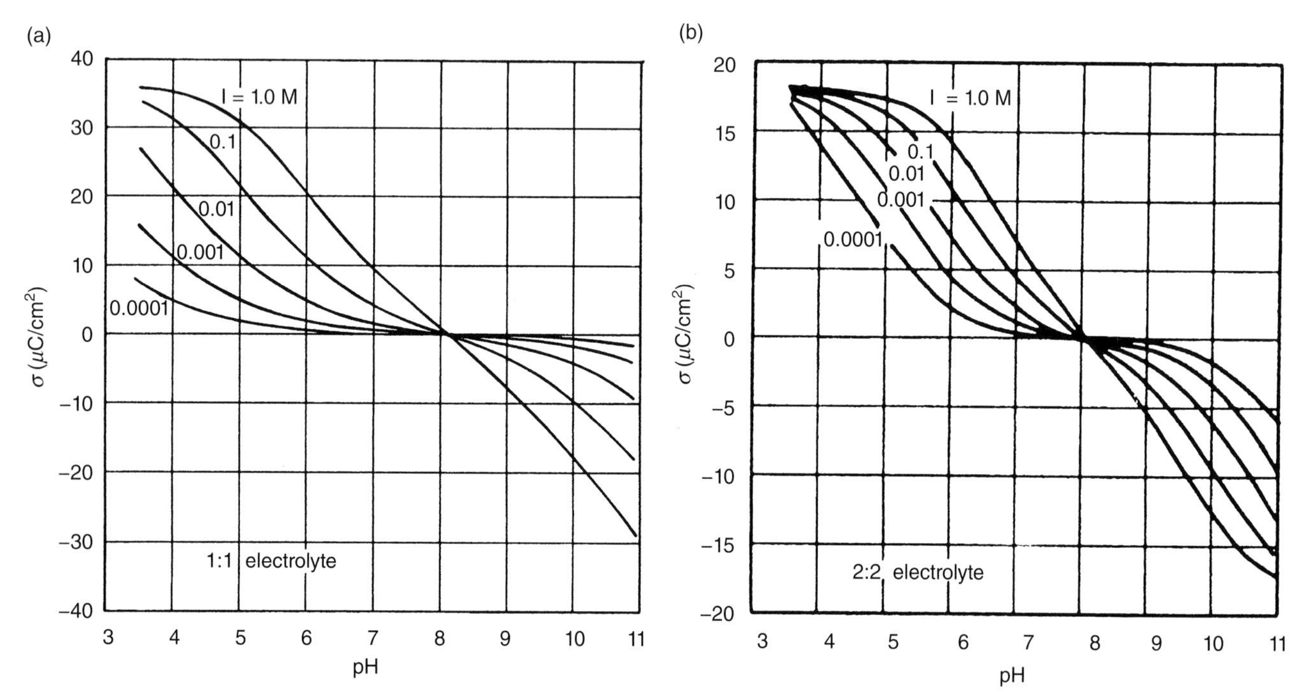

Fig. 4.8 Plots showing the change in surface charge density with pH as predicted by the HFO model for different ionic strength and in solution of (a) 1 : 1 electrolytes and (b) 2 : 2 electrolytes. Taken from Dzomback and Morel (1990) *Surface Complexation Modeling: Hydrous Ferric Oxide* [17] and reproduced with permission from John Wiley & Sons, Inc.

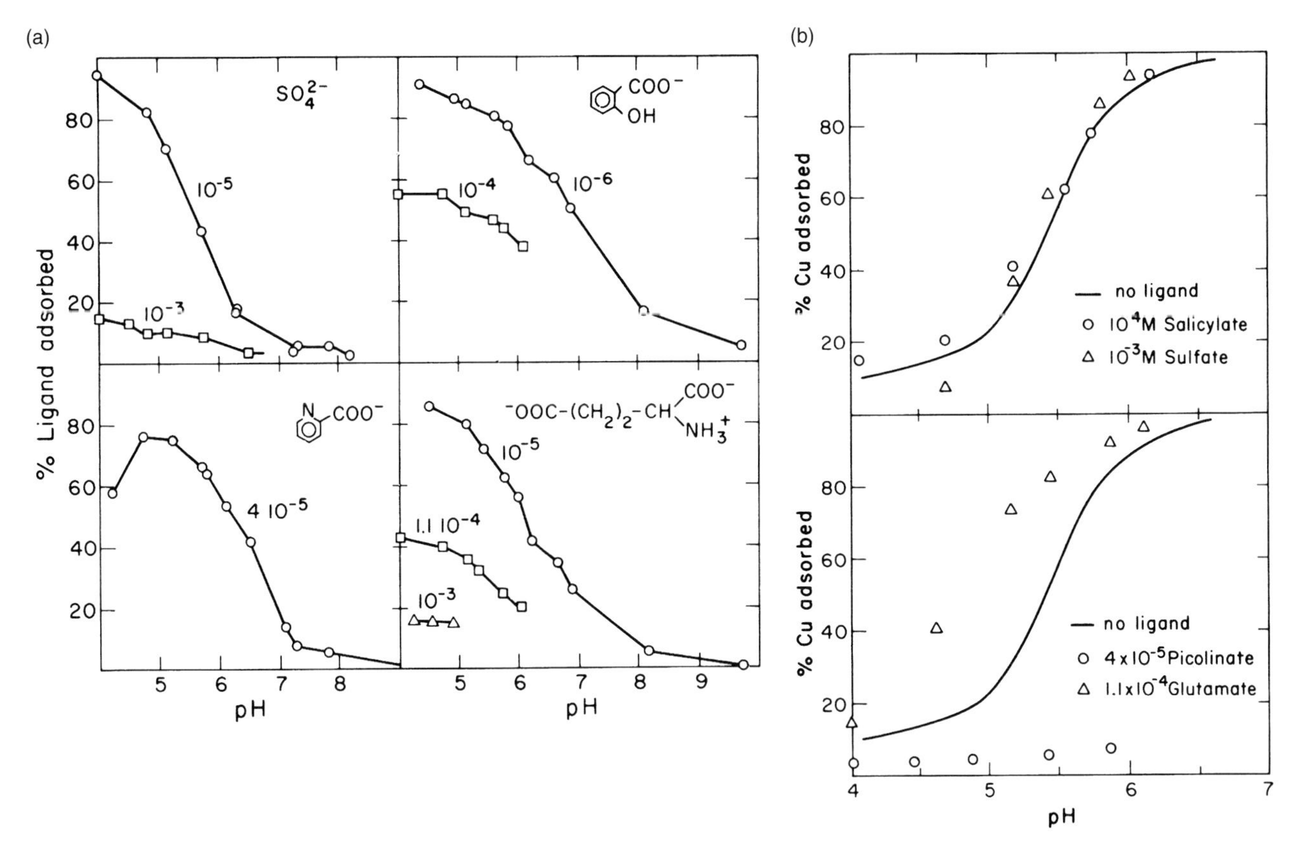

Fig. 4.9 (a) Adsorption of anions on to the surface of hydrous ferric oxide. (b) Adsorption of copper to the surface of hydrous ferric oxide in the presence of different potential binding ligands. Figures taken from Morel and Hering (1993) *Principles and Applications in Aquatic Chemistry*, reflecting data and figures from Davis and Leckie (1978) *Environmental Science and Technology* **12**: 1309–15 [19]. Reprinted with permission from the ACS. Copyright (1978), American Chemical Society.

Example 4.4

Let us consider the interactions discussed in Examples 3.7 and 3.9 (Fig. 3.17, presented again here as Fig. 4.10). Here the impact of a variety of organic ligands on the binding of Cu will be further discussed. The figure was taken from Ref. [1] and reports data published by Davis and Leckie [19]. The conditions of the experiment were: 10^{-3}M Fe, 2×10^{-4}M chloride, 5×10^{-4}M total carbonate, 10^{-3}M sulfate, 10^{-4}M salicylate and glutamate, and 4×10^{-5}M picolinate. The computer program will be used to model the data and reproduce the adsorption of Cu^{II} to HFO in the presence of the various ligands. We will focus on sulfate and glutamate. Dzomback and Morel [17] used a number of studies to define the constants that "best fit" the data and so we will first examine the specific instance of how the model reproduces the data of Davis and Leckie [19]. MINEQL+ does not have the binding constants for all the organic acids to HFO in the database but has Cu constants for all the acids except picolinate; constants are given for Cu and 2-picoline (C_6H_7N) but not the carboxylic acid, picolinate. After modeling the adsorption of sulfate we will then confirm that the presence of sulfate has no impact on the complexation of Cu to the HFO surface. Then, we will examine how to use the model to estimate the complexation of glutamate to HFO as these constants are not in the database, as an example. Furthermore, the interactions demonstrated in Fig. 4.9 – that the presence of glutamate enhances the adsorption of Cu to the HFO surface – will be examined. Clearly, the same approach could be used to examine the adsorption of picolinate and salicylate to HFO and to estimate the binding constants between Cu and salicylate that would reproduce the data – this is set as a problem at the end of the chapter.

The properties of the HFO surface are fixed in MINEQL+ as: surface area 600 $m^2 g^{-1}$; molecular weight of HFO (FeOOH) 90 g mol^{-1}; the number of strong sites is 0.005 sites/mol Fe and the number of weak sites is 0.2 sites/mol [17]. In the HFO formulation, anions bind to the weak sites only while the cations bind to both the strong and weak sites although the binding constants for the strong sites are much larger and so the binding to the weak sites is only important at high cation : surface site ratios. This does not typically occur in environmental waters. While there is likely no direct competition between the anions and Cu for the adsorption sites on the oxide surface under the conditions of this example, it is clear that some of the ligands form strong complexes in solution with Cu and therefore their impact on Cu binding to HFO and speciation will depend on their degree of adsorption.

(Continued)

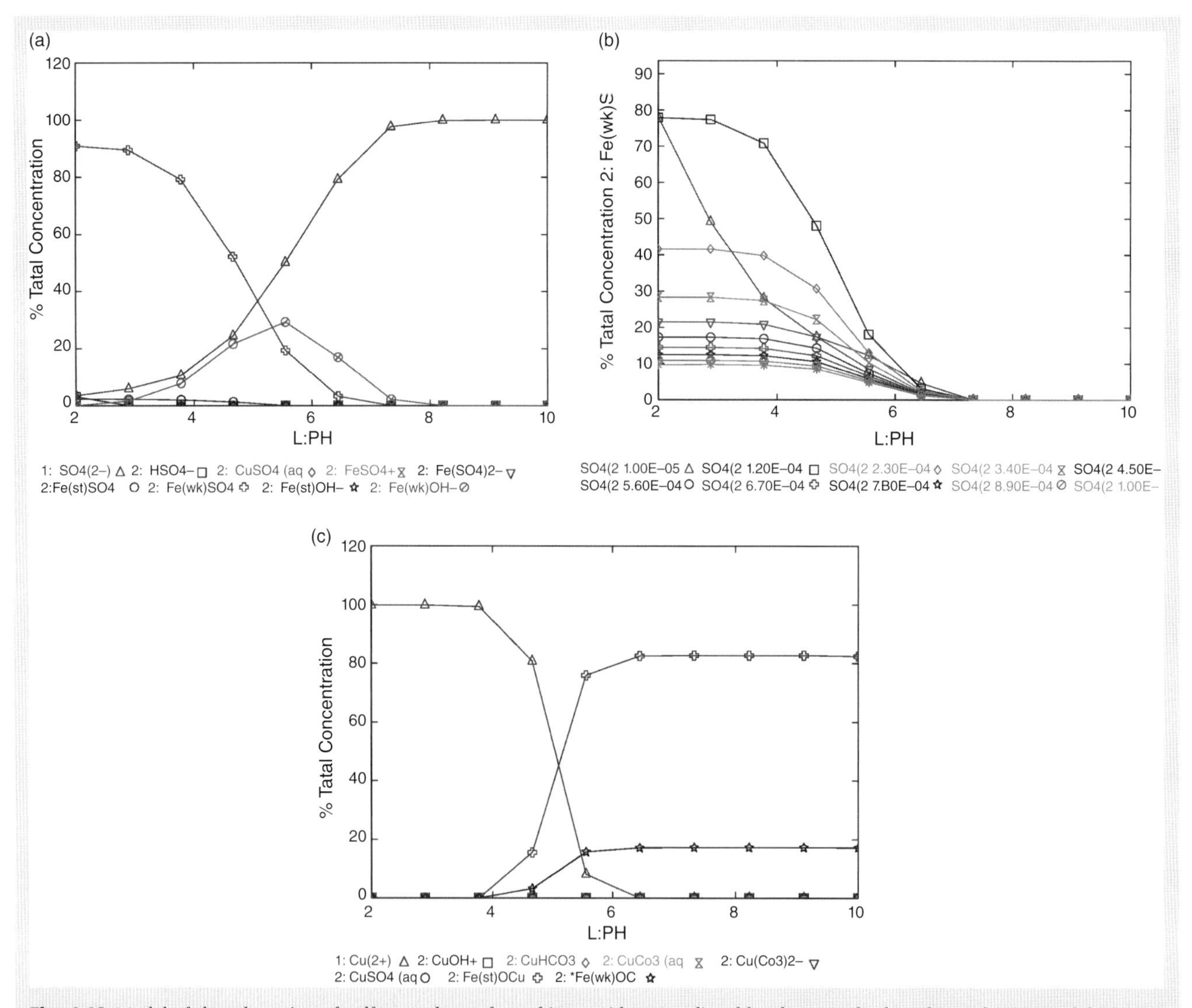

Fig. 4.10 Model of the adsorption of sulfate to the surface of iron oxide as predicted by the Dzomback and Morel HFO Model. Data produced using the MINEQL+ computer program: (a) The changing speciation and adsorption at a concentration of 10^{-5} M sulfate; (b) The effect of concentration on the relative amount of sulfate adsorbed and (c) Adsorption of copper to the HFO surface in the presence of 10^{-4} M sulfate.

The acid formation constants for the surface sites are the same for the strong and weak sites, as noted in Section 3.4.5. In the model, the surface site components (≡Fe(st)OH and ≡Fe(wk)OH) need to be added to the problem matrix, as well as the Coulombic term to take into account Coulombic interactions for any species formed with the surface sites. Scanning the MINEQL+ database with sulfate and Cu as components, results in the following metal-ligand (carbonate, bicarbonate and chloride complexes not shown), surface-ligand, and surface-cation associations (Table 4.5). Sulfate forms two associations with the oxide surface that can be written as:

$$\equiv FeOH + SO_4^{2-} = \equiv FeOH\text{-}SO_4^{2-}$$

and

$$\equiv FeOH + SO_4^{2-} + H^+ = \equiv Fe\text{-}SO_4^- + H_2O$$

The adsorption of 10^{-5} M sulfate in the presence of 10^{-3} M total Fe, forming HFO, is shown in Fig. 4.10(a). The adsorption is high at low pH, due to the formation of $\equiv Fe\text{-}SO_4^-$, and begins decreasing between pH 3 and 4. For this species, there is no adsorption above a pH of 8 according to the model. Note that the model also predicts the formation of the other sulfate-HFO complex at intermediate pH values and this is relatively important around pH 6. The effect of concentration on the fraction of the sulfate adsorbed to the surface, as shown in Fig. 4.9, can also be simulated with the model. MINEQL+ allows for so-called "two-way analysis" where two variables can be changed simultaneously during the calculation and in this case both the pH and the sulfate concentration are varied. The results of this model are shown in Fig. 4.10(b). As expected, given the dominance of adsorption at low pH, increasing the total concentration of sulfate leads to a decrease in the fraction adsorbed. This is because the limitation to binding is the number of surface adsorption sites. At a total Fe con-

Table 4.5 Formation constants from the MINEQL+ database for the complexes formed with copper and iron and the three ligands (glutamate (Glu), salicylate (Sal), and sulfate), and for the association of these ligands and copper with the hydrous ferric oxide surface.

Component	HX	H_2X	H_3X	FeX	FeX_2	
Glu^{2-}	9.96	14.26	16.42	–	–	
Sal^{2-}	13.70	16.8	–	17.6	29.3	
SO_4^{2-}	–	–	–	4.05	5.38	
	CuX	CuX_2	CuHX	$SO_4^{2-} + H^+$	SO_4^{2-}	$yH^+ + Glu^{2-}$
Glu^{2-}	9.17	15.78	13.3	–	–	–
Sal^{2-}	11.3	19.3	14.8	–		–
SO_4^{2-}	2.36	–	–	–	–	–
≡Fe(st)OH	2.89	–	–	7.78	0.79	ND
≡Fe(wk)OH	0.60	–	–	7.78	0.79	ND

ND: constants not in the database.

centration of 10^{-3} M, the maximum number of adsorption sites is 2×10^{-4} M, so at the higher concentrations of sulfate, the decreasing relative adsorption is due to this fact, and the assumption of monolayer adsorption.

The model for the adsorption of 10^{-7} M Cu in the presence of 10^{-4} M sulfate and 10^{-3} M total Fe is shown in Fig. 4.10(c). Adsorption of Cu begins around pH 4 and increases to a maximum around pH 6 where all the Cu is adsorbed. Sulfate has no impact on the Cu adsorption, which fits with the data [19]. At the concentrations used in this simulation, none of the inorganic ligands bind sufficiently strongly to outcompete the adsorption of Cu to HFO at high pH. Recall from Table 3.8 that most of the Cu is present in solution as the free ion at pH 6, but most is complexed as carbonate and hydroxide species at pH 8 in the absence of the oxide surface. Thus, the presence of the HFO is sufficient to outcompete the other inorganic complexes. While 5.4% of the Cu is present as a free ion in the simple solution considered for Table 3.8, in the simulation with HFO, the Cu^{2+} concentration is ~10^{-10} M.

The Example 4.4 results indicate there is a strong potential for adsorption of Cu to oxide surfaces to reduce the bioavailability and potential toxicity of Cu in solution, as toxicity is due to the presence of the free ion. Environmental regulations have taken into account the impact of "hard waters" – those with high total carbonate – on Cu toxicity due to the formation of the inorganic complexes which would reduce bioavailability. The model discussed here suggests that adsorption to surfaces could also alleviate toxicity, especially if the oxides were present as colloids that would pass through filters as dissolved concentrations are typically used to estimate the bioavailability and toxicity of environmental waters.

Example 4.5

Let us now consider the adsorption of glutamate to the surface of HFO under the conditions used in the experiments [19] (Fig. 4.9). As there are no data in MINEQL+ for these associations, they need to be entered and two formulations will be considered, in line with those for sulfate outlined earlier. These could be generalized for a diprotic acid, H_2A as:

$$\equiv FeOH + A^{2-} = \equiv FeOH\text{-}A^{2-}$$

and

$$\equiv FeOH + A^{2-} + H^+ = \equiv Fe\text{-}A\text{-} + H_2O$$

Other possibilities could be:

$$\equiv FeOH + HA^- + H^+ = \equiv Fe\text{-}AH + H_2O$$

which is equivalent to:

$$\equiv FeOH + A^{2-} + 2H^+ = \equiv Fe\text{-}AH + H_2O$$

and would be entered in the MINEQL+ database in this way, due to the components in the model.

To investigate which adsorption is the most dominant for glutamate (Glu^{2-}), the model was run at 10^{-5} M, with the same conditions for HFO, and species in solution as in Fig. 4.9, and assuming the

(Continued)

formed species is ≡Fe-Glu-. The acid-base constants for Glu^{2-} are shown in Table 4.5. As done with other anion associations, the same formation constant is used for the strong and weak sites. The results of this simulation are shown in Fig. 4.11(a). As the equilibrium constant is not known, a range of values were used in a two-way simulation with changing pH and varying the equilibrium constant. The results show the expected impact of the changing binding constant of the amount adsorbed, but in all cases the adsorption peaks at a pH of 6–8, which is inconsistent with the data (Fig. 4.9).

As the direct interaction of Glu^{2-} with the oxide surface forming ≡FeOH-Glu^{2-} would have an even higher pH maximum in adsorption, as the free glutamate concentration increases with pH, it seemed apparent that the third potential interaction suggested earlier is the most important, through the reaction of the protonated glutamate ion with the surface, forming an uncharged species. The Coulombic effects would be less for the uncharged complex. In MINEQL+, this is represented as:

$$\equiv FeOH + Glu^{2-} + 2H^+ = \equiv Fe\text{-}GluH + H_2O$$

Again, a range in the values of the equilibrium constant is used to provide a simulation of the data in Fig. 4.9, as shown in Fig. 4.11(b). The shape of this curve with pH is more in line with that of the data, suggesting that this likely best represents the dominant complex in this case. It is probable that more than one species forms, but for the example here, only the major species will be considered, as this example is presented mainly to illustrate the approach. Based on the comparison of the model and data, the equilibrium constant for the reaction is about $10^{20.6}$. The model was then run with changing pH

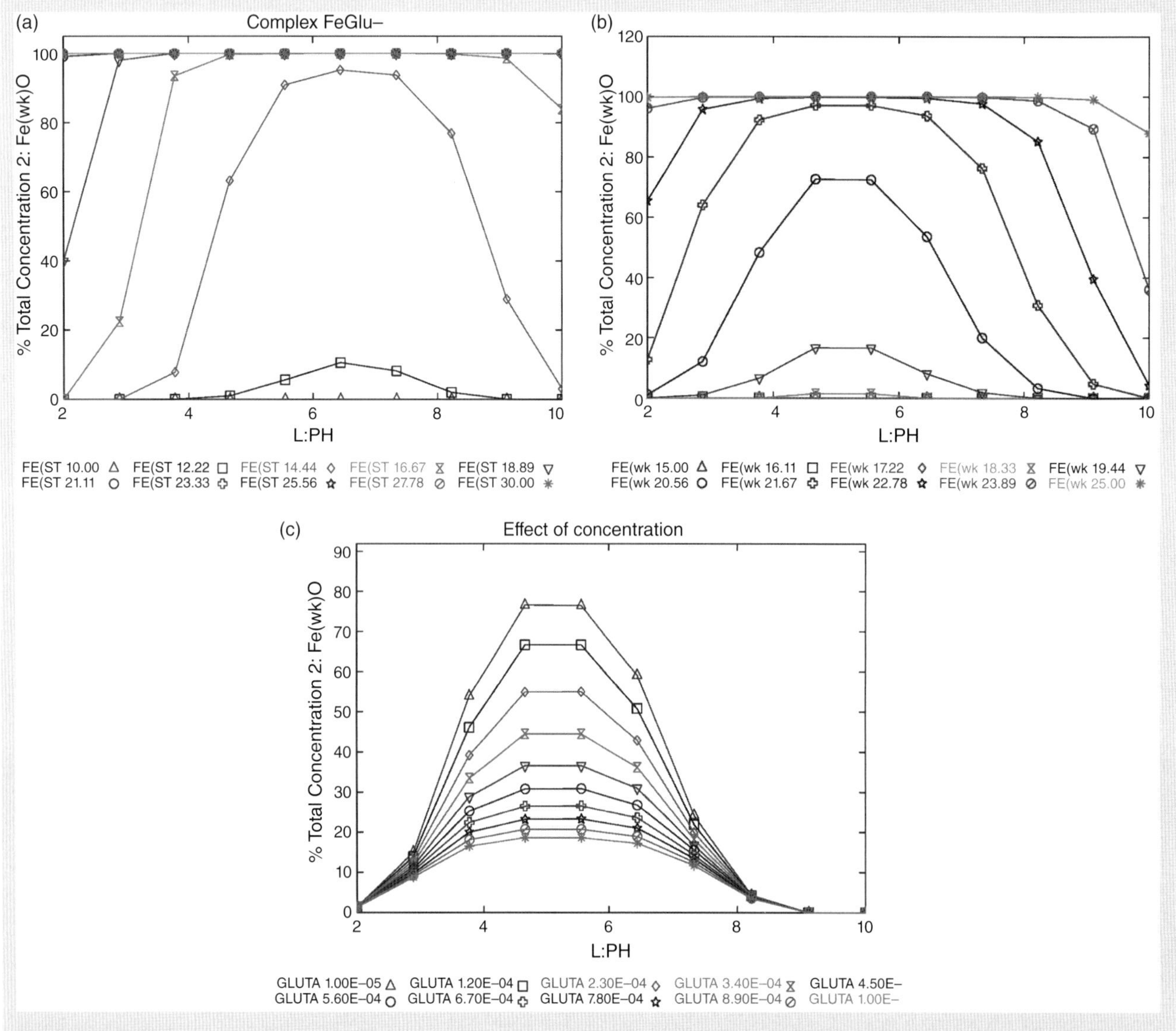

Fig. 4.11 Model of the adsorption of glutamate to the surface of iron oxide using the Dzomback and Morel HFO Model. Data produced using the MINEQL+ computer program: (a) Effect of changing values for the equilibrium constant for one formulation of the reaction of glutamate with the oxide surface – see text for details; (b) Another formulation of this association; and (c) The impact of glutamate concentration on the relative amount adsorbed.

and changing glutamate concentrations to further simulate the data [19]. The model results are again relatively consistent with the data (Fig. 4.11c) suggesting this is a reasonable first order approximation of the interaction of glutamate with the oxide surface. The overall extent of adsorption is controlled by the limited number of binding sites on the surface of the oxide.

Finally, the data of Cu adsorption in the presence of glutamate and HFO is simulated. The experimental data indicates an increase in adsorption and the assumed reason for this is the formation of a complex between HFO, glutamate, which has multiple acid groups, and Cu, which can be represented as:

$$\equiv FeOH + Glu^{2-} + H^+ + Cu^{2+} = \equiv Fe\text{-}Glu\text{-}Cu^+ + H_2O$$

The potential value of the equilibrium constant for this interaction was changed across the pH range to get the range in increased adsorption indicated in Fig. 4.9(b) – overall, an increase in adsorption between pH 3 and 6. The results in Fig. 4.12(a) show that the best match to the data is obtained with an equilibrium constant around $10^{21.8}$–$10^{22.3}$. MINEQL+ also allows the total adsorbed cation or anion to be plotted and this confirms the range in values for the equilibrium constant as the expected increase in adsorption is demonstrated (Fig. 4.12b).

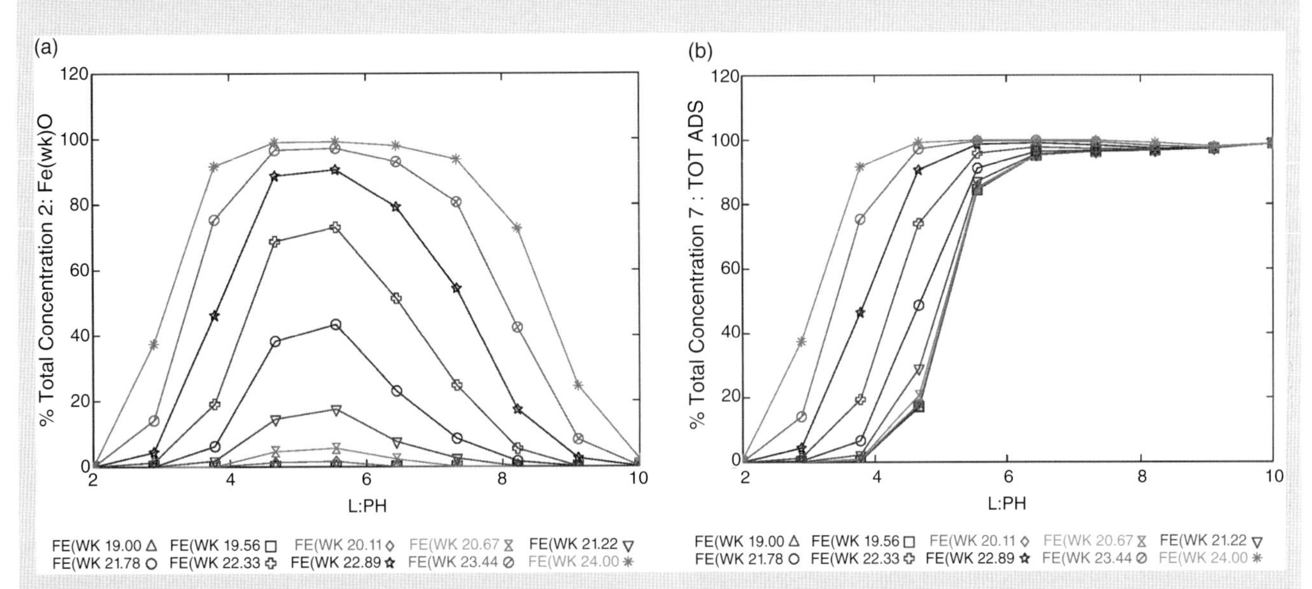

Fig. 4.12 The adsorption of copper to the HFO surface in the presence of glutamate using the Dzomback and Morel HFO Model. Data produced using the MINEQL+ computer program: (a) Impact on the value of the equilibrium constant for the tertiary complex (HFO-glutamate-copper) on the amount of this species adsorbed; (b) Demonstration of how changing the binding constant for the tertiary complex impacts the total copper adsorption.

There have been many studies and models developed to examine the adsorption of cations and anions to oxide surfaces and the above examples demonstrate the overall approach. A number of books [1, 16, 18, 20] have many worked and discussed examples. Papers in the literature have also studied the adsorption of non-transition metal cations [21], and anions such as As and Se to HFO, and have shown the effects of solution concentration and the presence of other anions on adsorption [22]. These, and other studies have investigated how the presence of anions affect the adsorption of metal cations [23, 24] and the transformation of HFO into more crystalline phases over time [25, 26]. It appears that some transition metals can be incorporated into the oxide matrix, impacting their release during reductive dissolution [27, 28]. Additionally, other oxides have been studied [29, 30]. Further examples of the impact of oxide sorption on metal(loid) fate and transport will be discussed in the latter chapters of this book.

Returning to the examples discussed earlier, it is difficult to confirm the model predictions of the HFO approach for glutamate adsorption based on the original study [19]. Other investigators have also examined and modeled this data (e.g., Sverjensky et al. [31–33]. These authors developed a detailed model to examine the adsorption of glutamate to titanium dioxide and HFO. The modeling used an extended Triple Layer Model, which was developed in the late 1990s [31, 32, 34] in an effort to relate the various parameters required for the Triple Layer Model to physical and chemical properties rather than being empirically determined.

For the Triple Layer Model, the following parameters are required: the two surface protonation constants ($K_{s,1}$ and $K_{s,2}$), the electrolyte binding constant ($K_{s,M+}$ and $K_{s,L-}$ for a 1 : 1 electrolyte), the site density (N_s) and the capacitance of the two layers ($¢_1$ and $¢_2$). The model developed by Sverjensky and co-workers [31–33] assumed a single surface site in contrast to the two site HFO Model [17]. The surface protonation constants are related to the properties of the bulk mineral and can be predicted using a function that includes the dielectric constant and the short range interactions between the surface and the absorbing proton. Additionally, it is assumed that electrolyte adsorption constants are determined in a similar manner to the protonation and surface binding inner-sphere constants.

Details of how the parameters were developed [34] will not be discussed in detail here. For example, the surface protonation constants were derived from the pristine point of zero charge (pH_{ppzc}) for each surface, which is related to the difference in the value for the protonation constants (ΔpK), and can be calculated using the dielectric constant, and the Pauling bond-strength per unit bond-length based on experimental data. The electrolyte adsorption constants were derived from an equation involving constants related to each mineral surface and the inverse of the dielectric constant (ε_1). The inner capacitance was estimated as: $¢_1 = \varepsilon_0\varepsilon_1/d$ where ε_0 is the absolute permittivity of a vacuum and d is the distance of the layer, and is related to the aqueous effective radius of the adsorbed electrolyte. A summary of some of the derived constants for electrolyte adsorption to some surfaces likely found in environmental systems is shown in Table 4.6, which compares the experimental and predicted constants based on the model, for a NaCl electrolyte.

With the model, the $\log K_{s,M+}$ and $\log K_{s,L-}$ adsorption constants for other monovalent cations and anions were predicted [34]. For example, for solid FeOOH, values ranged from 2.70 for Li^+ to 2.19 for Cs^+ for Group I cations, while the values for anions ranged from 2.48 for F^- to 2.12 for I^- (Group 7 anions). Values for other monovalent anions ranged from 1.8–2.22. For other surfaces, the values for K^+, for example, ranged from 1.11 for amorphous silica to 2.89 for Fe_3O_4. While the authors make these predictions of the binding constants, they acknowledged that differences in solid sample preparation and other experimental protocols may influence surface properties, and therefore the accuracies of these predictions are related to the accuracy of the original data. This is clearly true for this model and for other formulations based on experimental data, such as HFO Model [17].

Sverjensky et al. [33] used this model to examine the adsorption of glutamate to HFO and TiO_2. The 1 molar standard state protonation constants used were 3.7 for $\log K_1^0$ and 12.1 for $\log K_2^0$ and the electrolyte constants ($\log K_{s,M+}^0$ and $\log K_{s,L-}^0$ for the electrolyte ($NaNO_3$)) were respectively −7.8 and 8.2. In addition to modeling, these authors were able to determine, based on spectroscopic investigations using attenuated total reflectance Fourier transform infrared spectroscopy (ATR-FTIR), the binding of glutamate to the oxide surfaces [33]. They found three potential interactions in examining the glutamate association with titanium oxide, which they termed: (1) *chelating monodentate* (this involved three points of interaction with two inner-sphere bonds, and a hydrogen bond); (2) *bridging bidentate* (four points of contact with two inner-sphere bonds, and two hydrogen bonds); and (3) *chelating* (two inner-sphere complexes to the same surface site). These interactions can be written, respectively, as the following reactions:

$$2{\equiv}FeOH + HGlu^- + 2H^+ = {\equiv}Fe_2\text{-}HGlu + 2H_2O$$

$$4{\equiv}FeOH + HGlu^- + H^+ = ({\equiv}FeOH)_2{\equiv}Fe_2\text{-}Glu + 2H_2O$$

$${\equiv}FeOH + HGlu^- + H^+ = {\equiv}Fe\text{-}HGlu + H_2O$$

These authors note that in their surface adsorption model that both the major ions interacting with the surface and the water dipole molecules contribute to the surface potential. The results of this modeling exercise are shown in Fig. 4.13 and can be compared with the results generated using the more simple model discussed previously (Fig. 4.11). The *Sverjensky* et al. model [33] demonstrated that the complex that likely forms in the interaction of glutamate with HFO at low glutamate concentrations is represented by $\equiv Fe_2$-DHGlu and is due to the association of the glutamate with two sites on the surface of the solid. This is similar in overall formulation to the complex suggested by the HFO model, except in this instance the glutamate is forming two inner-sphere bounds with the surface. Moreover, the more complex modeling done in this instance predicts that there is a shift in the overall relative importance of the surface species with changes in the total concentration of glutamate. This is probably due to the limitation in the number of inner-sphere binding sites, as noted earlier, at the higher concentration levels. Overall, the interaction of anions with surfaces

Table 4.6 A comparison of modeled and measured values for the adsorption constants for Na^+ and Cl^- for a number of common solid phases, using the Triple Layer Model discussed in the text. Taken from Sahai and Sverjensky [34]]. Reprinted with permission of Elsevier.

Mineral	Exp. $\log K_{s,M+}$	Predicted $\log K_{s,M+}$	Exp. $\log K_{s,L-}$	Predicted $\log K_{s,L-}$	Exp. $¢_1$	Predicted $¢_1$
Anatase (TiO_2)	2.9	2.81	2.9	2.72	1.30	0.97
Hematite (Fe_2O_3)	1.6	2.87	2.0	2.81	0.90	0.97
Goethite (FeOOH)	2.4	2.49	2.4	2.24	0.60	0.97
Am. Silica	0.8	1.0	–	0.00	1.00	0.97
γ-Al_2O_3	2.3–2.5	2.40	1.9–2.5	2.10	0.90–1.10	0.97

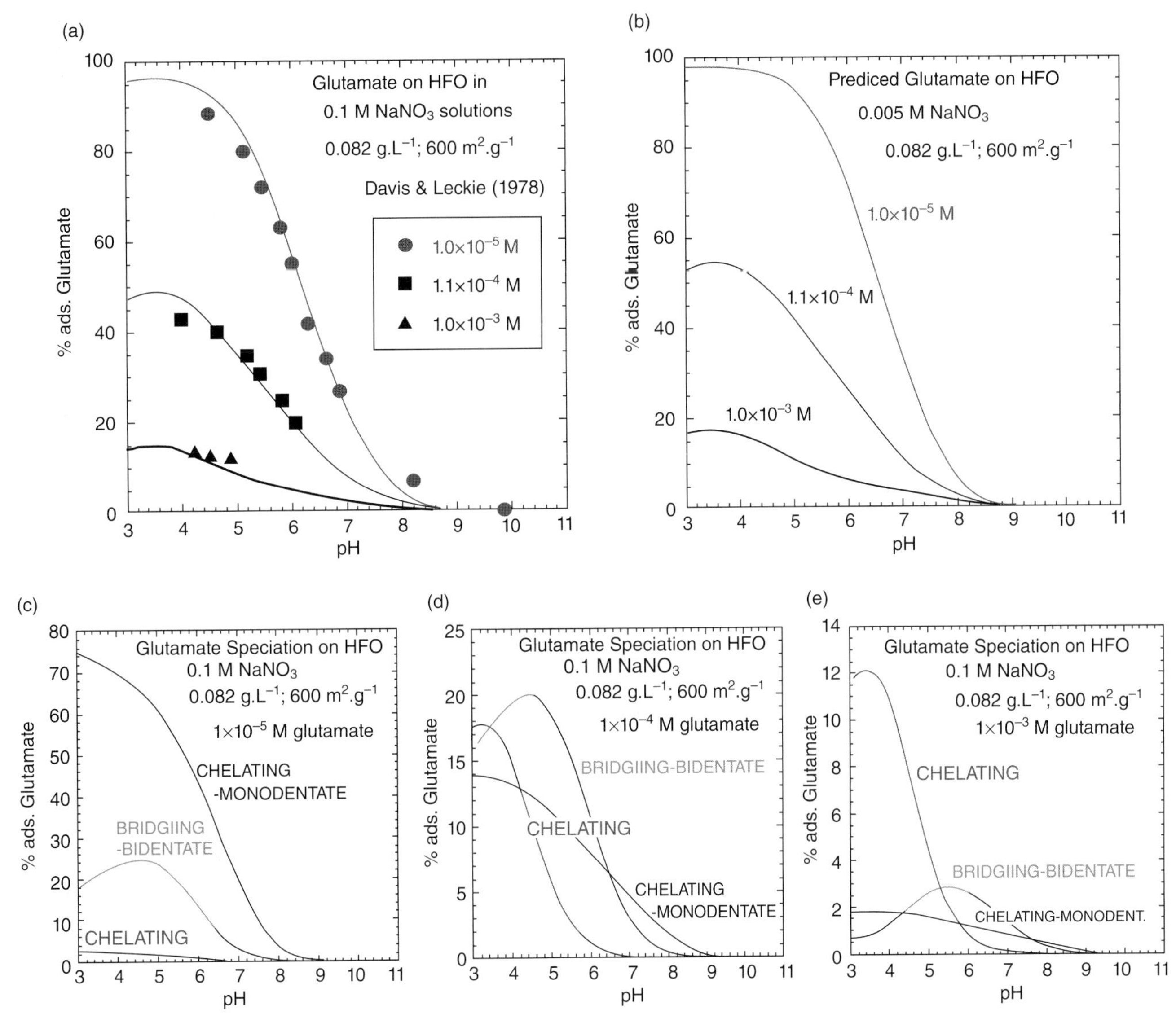

Fig. 4.13 Surface complexation calculations for the adsorption of glutamate onto HFO with comparison to the data of Davis and Leckie (1978) [19]. Curves show the predicted complexation under different types of association of the glutamate with the solid surface, as described in the text: (a) comparison of model and data at different glutamate concentrations in a 0.1 M $NaNO_3$ solution; (b) in a 0.05 M $NaNO_3$ solution; and (c–e) the depiction of the relative amount of the various types of complexes at each concentration. Model results reprinted with permission from Sverjensky et al. (2008) *Environmental Science and Technology* **42**: 6034–9 [33]. Copyright (2008), American Chemical Society.

is not simple and can include a number of different associations, especially for polyprotic acids such as glutamate.

The modeling in the previous examples demonstrates the usefulness of the computer models to predict the potential interactions that are occurring and confirm our understanding of the results of a particular dataset. The model allows for complex calculations and simulations to be performed and for the testing of various hypotheses. The examples show that it is possible to conclude which species are forming during an interaction and allows the examination of complex interactions such as the formation of tertiary complexes between solid surfaces, cations and ligands.

More complex models are being developed in an effort to provide a more representative model of the interactions of ions with the oxide surface. However, there is always a compromise in developing a more detailed model as it may not be possible to obtain reliable parameters for all the required variables, or these may need to be measured prior to the application of the model. The Double Layer Model described earlier, and used in the calculations in Examples

4.4 and 4.5, is the simplest approach to examine the interaction of ions with the solid surface. Even when focusing on the Double Layer Model for the interactions, more detailed models have been developed for ion interactions with surfaces [35]. For example, such models have a variety of sites, rather than being more limited, such as in the HFO Model, which has only two types of sites: strong and weak, with the majority of the sites being weak sites. One approach is the Charge Distribution Multisite Ion Complexation (CD-MUSIC) Model [36] which still invokes the double layer formulation but includes a larger number of sites [37]. This model includes the potential for interactions at a number of faces on the solid surface, and for the formation of both inner-sphere and outer-sphere complexes.

The CD-MUSIC approach for modeling metals on oxides is based on the premise derived from crystallographic studies that there are many different sites on the mineral surface which is consistent with the results of numerous studies that show that different cations interact with the surface in different ways [37]. Thus, the surface is considered heterogeneous with several groups of different reactivity. For example, goethite, which consists of double chains of Fe-octahedrals cross-linked by corner linkages, has two crystal planes: (1) the [110] surface which is parallel to the long axis of the mineral, and which are low affinity sites; and (2) the [021] end termination planes, which are high affinity sites. Because of the crystal structure, the oxygen atoms can be either unprotonated or protonated. Overall, the specific surface area of goethite is 30–100 $m^2 g^{-1}$. In contrast, the HFO surface, having shorter chains, has a high specific surface area (200–800 $m^2 g^{-1}$) and has a higher number of high affinity sites.

In the CD-MUSIC Model both planes are treated as having their own electrostatic potential and double layer. The overall proton affinity constants are calculated based on the surface characteristics and whether the oxygen atoms on the surface are protonated or not, and whether there is an adsorbed water molecule. For goethite, for each plane, four types of species are described, each with its own set of proton association constants (K_{i1} and K_{i2}) and site density, that is, a total of eight different surface sites. From most experimental data, it is found that the [110] plane is >90% of the surface area for goethite. Ponthieu et al. [37] summarize the data from numerous studies and provide binding constants for various metals for each plane and for different complexes (Appendix 4.1). For each plane, different surface groups are designated FeO_IH, $FeO_{II}H$ (or as $FeOH_m$ which is the mean of these two sites) and Fe_2O_IH depending on whether the oxygen atom is protonated (O_I) or unprotonated (O_{II}), or single, doubly or triply coordinated to Fe. It can be seen that for most metals there are numerous potential binding sites and the importance of each will depend on a number of factors such as the cation concentrations, the mineral concentration and other factors. As noted, the main differences between geothite and HFO are the relative densities of the high affinity and low affinity sites and so the constants in Appendix 4.1 can be used for either surface.

A comparison of the model predictions at low mineral concentrations in solution and relatively low cation concentrations are shown in Fig. 4.14 for Pb and for Cu. For Pb at 500 nM, at HFO concentrations of 89 (~10^{-3} M) and 8.9 $mg\,l^{-1}$ (~10^{-4} M), the adsorption appears to be mostly controlled by edge sharing in the [021] plane. Overall, the model is able to predict the adsorption isotherms at both HFO concentrations; at the lower HFO concentration the adsorption begins at a higher pH given the interaction of the cations with the acidic surface sites. In contrast, Cu adsorption occurs through multiple sites, with edge sharing in both planes. Comparing the data when the metal concentration and HFO concentrations are the same, it is evident that Pb adsorption begins at a lower pH than Cu. Other experimental data is also reasonably modeled with the CD-MUSIC approach and the model is also extended to predict adsorption for other metals for which data is not available [37]. Overall, the model demonstrates that cation adsorption to either goethite or HFO can be adequately modeled using the same set of parameters as the main difference between the two forms of iron oxide is the relatively amount of the high affinity and low affinity sites. As noted, this is mainly due to the differences in the structure of these two minerals and the composition of their surfaces.

Other applications of the CD-MUSIC Model are in the literature as it has been used to investigate the adsorption of a variety of metals to iron oxide surfaces [38–41]. The model has also been used to examine the adsorption of anions, such as arsenate and selenate, to surfaces [42–44] and that of other organic ligands [45], in a similar manner to that discussed previously for glutamate. Additionally, it has been used to examine the adsorption to carbonates [46] and other surfaces such as TiO_2 [47].

Overall, therefore there are a number of approaches with different levels of complexity, which have a corresponding difference in the need for information or parameters for the modeling effort. In most situations, the decision on which model to use will be based on the required level of certainty in the prediction, and whether the modeling effort is to confirm a suspected interaction between metals, ligands and surfaces, such as examined here with the interactions between Cu, glutamate, other anions and HFO. The Dzomback and Morel [17] approach was sufficient to demonstrate the possibility of a tertiary interaction with the glutamate acting as a bridging ligand between the Cu and HFO at the intermediate pHs and that the adsorption at higher pH was primarily due to the direct interaction of Cu with the surface. These complex interactions likely occur in many instances, especially in the presence of NOM. For this reason, we will discuss in more detail modeling efforts to examine the complexities of the interaction of NOM with solid surfaces, and the interactions of NOM, solid surfaces and cations.

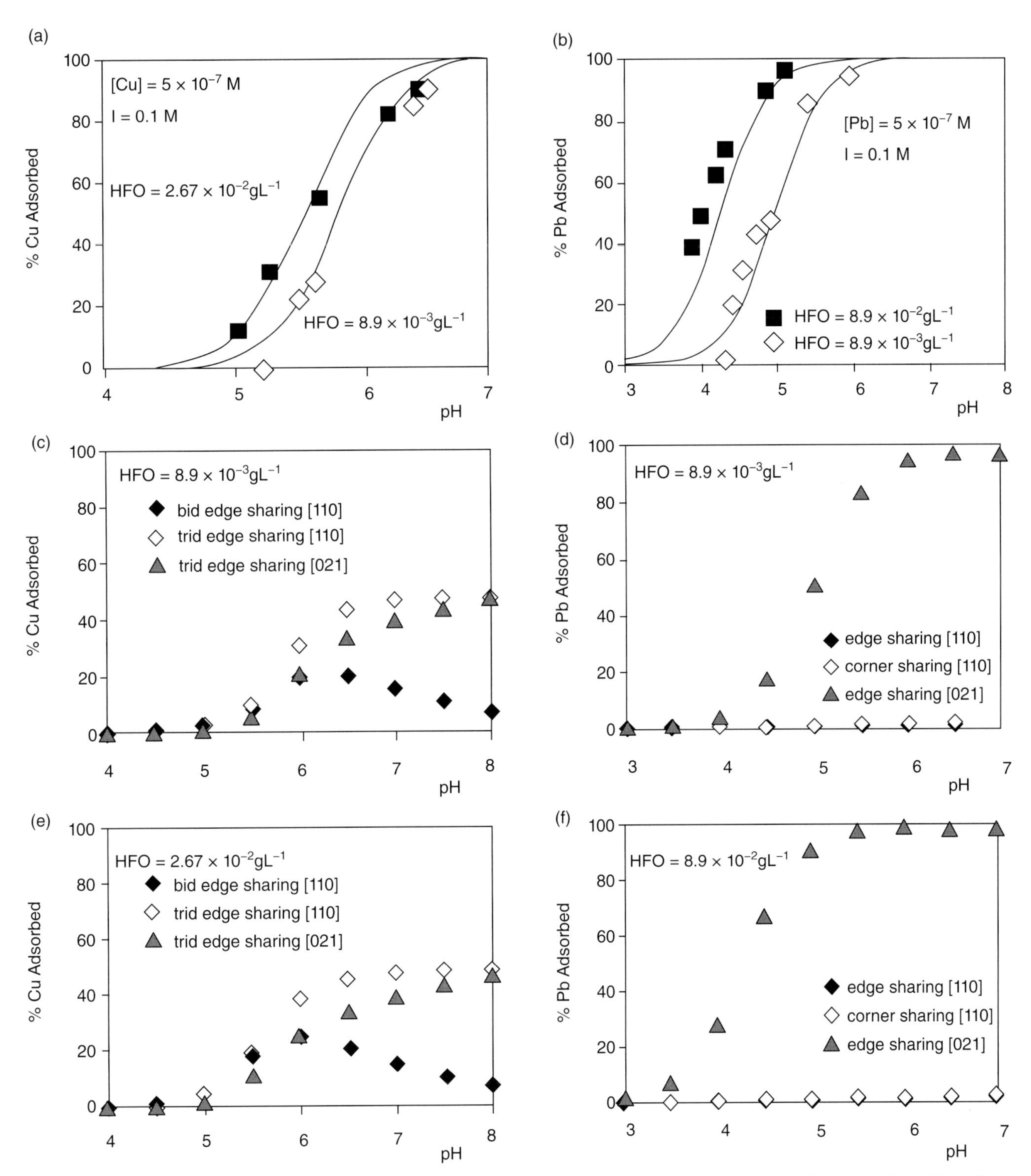

Fig. 4.14 Modeled adsorption of (a) copper (Cu) (three *left-hand* figures) and (b) lead (Pb) (three *right-hand* figures) to the surface of HFO under different concentrations of the solid; and (c–f) the remaining figures show the various associations that occur with the metals and the solid surface at each concentration across a pH range. Reprinted from Ponthieu et al. (2006) *Geochimica et Cosmochimica Acta* **70**: 2679–98 [37] with permission from Elsevier.

4.4 Modeling interactions between cations and organic matter, and inorganic surfaces

A number and variety of approaches have been used to examine the complexation of cations, primarily, with natural organic matter. For specific low molecular weight organic compounds, as modeled previously, the interaction with cations or with anions can be determined if the equilibrium constants for the various interactions are known. However, in most environmental systems, most of the organic matter "in solution" (NOM) is not well-characterized and while there may be some important measurements of the overall characteristics, such as the number of carboxylic acid groups per mole, and/or the number of other acidic or basic sites (e.g., thiol, amine, phenol groups), as discussed in Section 3.3.5, these acidic groups likely have a range of pK_as that makes it difficult to devise a reasonable model for such a complex mixture. Additionally, the number of binding sites varies with humic acid source, their distributions are heterogeneous and there is the potential for lateral interactions and steric hindrance. The number of proton binding sites, for example, range from 4–14 mol kg^{-1}, with humic acids tending to have a higher number of reactive sites [48]. Such variability must be included in any model formulation. Many of the studies that have developed models and approaches to examine the interactions with NOM have also included the potential for interactions of NOM, such as humic and fulvic acids, with oxide surfaces [49]. We will consider the complexities of these interactions which are very important in controlling the bioavailable fraction and dissolved concentration of many cations and anions in environmental waters.

While it is often stated that NOM is "dissolved", given the overall range in molecular weight, it is possible that some of the NOM is not truly dissolved and may be present in the colloidal fraction, either as a pure organic colloid or as an organic coating of an inorganic entity. Some models consider NOM as permeable micelles. If it was possible to know the fraction colloidal and its size characteristics, it could then be possible to use a similar approach to that discussed for adsorption to oxide surfaces, for NOM colloids, and take into account the surface charge of the colloid and its likely impact on cation and anion binding. Some models do this. There are a number of approaches to examining the complexation of trace elements to NOM, and these will be discussed and compared to illustrate their similarity and differences, and their usefulness in examining environmental waters. The specific situation where the "surface" is a biological membrane can also be examined in a similar manner by considering the complexation of dissolved species to sites on the membrane surface. This modeling approach, which has been termed the Biotic Ligand Model, is discussed in Chapter 8.

4.4.1 The WHAM modeling approach

Of the models developed to examine the interactions of ions with NOM, the models that have been most widely used are WHAM [13], which has appeared in a number of versions, and the NICA-Donnan Model [49]. WHAM Version V, for example, considers humic acids to be size homogeneous molecules which have surface acid sites and considers the interaction in terms of an intrinsic binding constant and an electrostatic term, in a similar fashion to solid surface binding models. Two types of groups are considered: those that are more acidic (Type A), and which are reflective of carboxylic acid groups, and those that are less acidic (Type B), such as phenolic groups. Both types of sites are considered to heterogeneous with a range of pK_a values. Metals can bind through a single (monodentate) interaction or via formation of a bidentate association. Metals can bind to the NOM as a free ion and as a single hydrolyzed species (i.e., MOH^{n-1} for metal, M^{n+}).

WHAM Version VI [13] is similar to Model V, as it considers humics to be rigid spheres of uniform size, with ion-binding groups positioned on the surface. In Model VI, there are eight different types of binding sites; four strong (Type A, more acidic) and four weak (Type B, more basic) sites and these are parameterized in the following way in terms of four constants (pK_A, pK_B, ΔpK_A and ΔpK_B):

$$for\ i = 1-4,\ pK_i = pK_A + \left[\frac{2i-5}{6}\right]\Delta pK_A \tag{4.12a}$$

$$for\ i = 5-8,\ pK_i = pK_B + \left[\frac{2i-13}{6}\right]\Delta pK_B \tag{4.12b}$$

Within a group, the number of sites is constant, but there are twice as many Type A sites than Type B sites. Monodentate metal binding to each site is parameterized in a similar manner, for example:

$$for\ i = 1-4,\ pK_i = pK_{MA} + \left[\frac{2i-5}{6}\right]\Delta LK_1 \tag{4.12c}$$

where ΔLK_l is a constant that is estimated from data fitting, and K_{MA} corresponds to binding of the free ion to the unprotonated ligand site. If there is sufficient accurate information, it is possible to define a constant for each acid/base site. Bidentate and tridentate binding is also possible with the overall binding constant being the sum of the log values for the constants for the monodentate associations, plus a fitting parameter (x.ΔLK_2 for bidentate; y.ΔLK_2 for tridentate). Again, ΔLK_2 is determined from fitting the model to data, with x set as 0 for 90.1% of the sites, as 1 for 9% and as 2 for 0.9% of the sites; similarly, y is set as 0, 1.5 and 3% for the same distribution across sites. This makes it possible for the cations to bind to multiple sites of different strength, and by various associations, but most of the sites are monodentate sites. The bidentate and tridentate sites are fixed entities with regard to metal binding and their distribution is set by the following parameters: For fulvic acids (FA), the fraction of sites that are bidentate (f_{prB}) is set at 0.42, that for tridentate (f_{prT}) at 0.03; for humic acids (HA), f_{prB} is 0.5, f_{prT} is 0.065. To simplify the overall model, a subset of possible sites are typically allowed leading to, including the eight monodentate sites, a total of 80, with different sites types and strengths in the standard model. Electrostatic interactions are taken into account in a similar manner to the surface binding models discussed previously. However, while the NOM size is considered homogeneous (FA radius 0.8 nm; Molecular weight 1500; HA radius 1.72 nm; molecular weight 15,000), the differences in site type and density leads to an overall uneven charge density on the surface, and therefore, to calculate electrostatic effects, the average surface charge is used. Finally, rather than using a Boltzmann equation to calculate the balancing counter ions in the double layer surrounding the NOM surfaces, this is taken into account using a Donnan Model formulation, which is discussed further later and which calculates the size of the Donnan volume based on the molecular weight and molecular radius of the NOM [13].

In using the WHAM VI Model, the majority of the required parameters are fixed and in the literature, and there are up to four fitting parameters that can be used to adjust the model. Through extensive examination of data in the literature, many of the parameters are relatively well-constrained and can be listed and used for most modeling applications (see Appendix 4.1), and these include the binding constants for many of the metals to HA and FA. Thus, if the default parameters are used, the model is relatively straightforward to implement using the values in the database, which are based on an extensive assessment of the available data [13], in a similar fashion to detailed analysis behind the Dzomback and Morel HFO Model. The standard model provides a reasonable simulation of the experimental data in most situations. A brief examination of some examples will illustrate this point.

Tipping [13] used the WHAM VI Model to simulate the binding of Cu and Cd to humic and fulvic acids, based on published data [50]. As shown in Fig. 4.15, there is a good correlation between the model and the data for this system for all the pHs. In this instance the experiments was at relatively high cation concentrations. The system was modeled for the four high affinity and four low affinity sites but the value of ΔLK_1 was set at zero for some of the sites to obtain the fit shown in the figure. It was concluded from further examination of a range of data that the value of ΔLK_1 was relatively insensitive to the metal or to the NOM type (fulvic versus humic acid) and therefore proposed that a fixed value could be used in many situations ($\Delta LK_1 = 2.8$) [13]. Additionally, it was found that the value of ΔLK_2 was linearly related to the binding constant of the metal to NH_3 and therefore it is possible to define the value of ΔLK_2 by the following equation: $\Delta LK_2 = 0.55.logK_{NH3}$. Similarly, it was found that $logK_{MA}$ and $logK_{MB}$ were linearly related: $logK_{MB} = 3.39.logK_{MA} - 1.15$, again further reducing the number of parameters that need to be specified or derived from data. The various relationships between these values are shown in Fig. 4.16. These correlations, especially with lactic acid, a simple carboxylic acid, suggest that the WHAM VI model is providing a good representation of the relative binding strength of cations to NOM.

It is also possible to investigate potential cation interactions using the model, and Tipping [13] discussed various examples. One set of data that was modeled is shown in

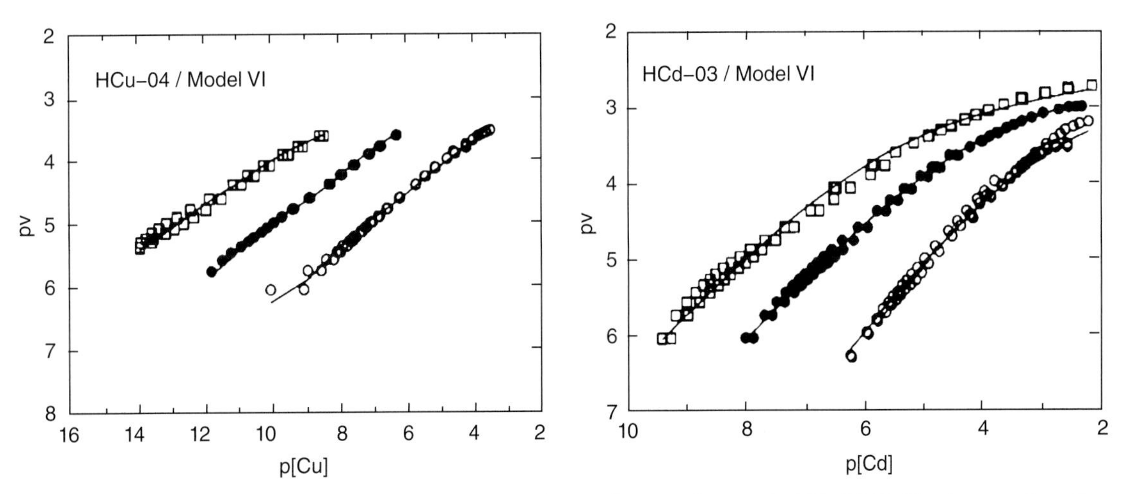

Fig. 4.15 Copper and cadmium binding by humic acid (HA) and fulvic acid (FA) using WHAM Model VI. The data are from Benedetti et al. (1995) *Geochemical Exploration* **88**(1–3): 81–5 [41]. In the figure, v is the moles of metal bound per g FA or HA, and p signifies $-\log_{10}$. Key: (open circle) pH 4, (closed circle) pH 6, (square) pH 8. Taken from Tipping (1998) *Aquatic Geochemistry* **4**: 3–48 [13] and reprinted with kind permission of Springer Science+Business Media.

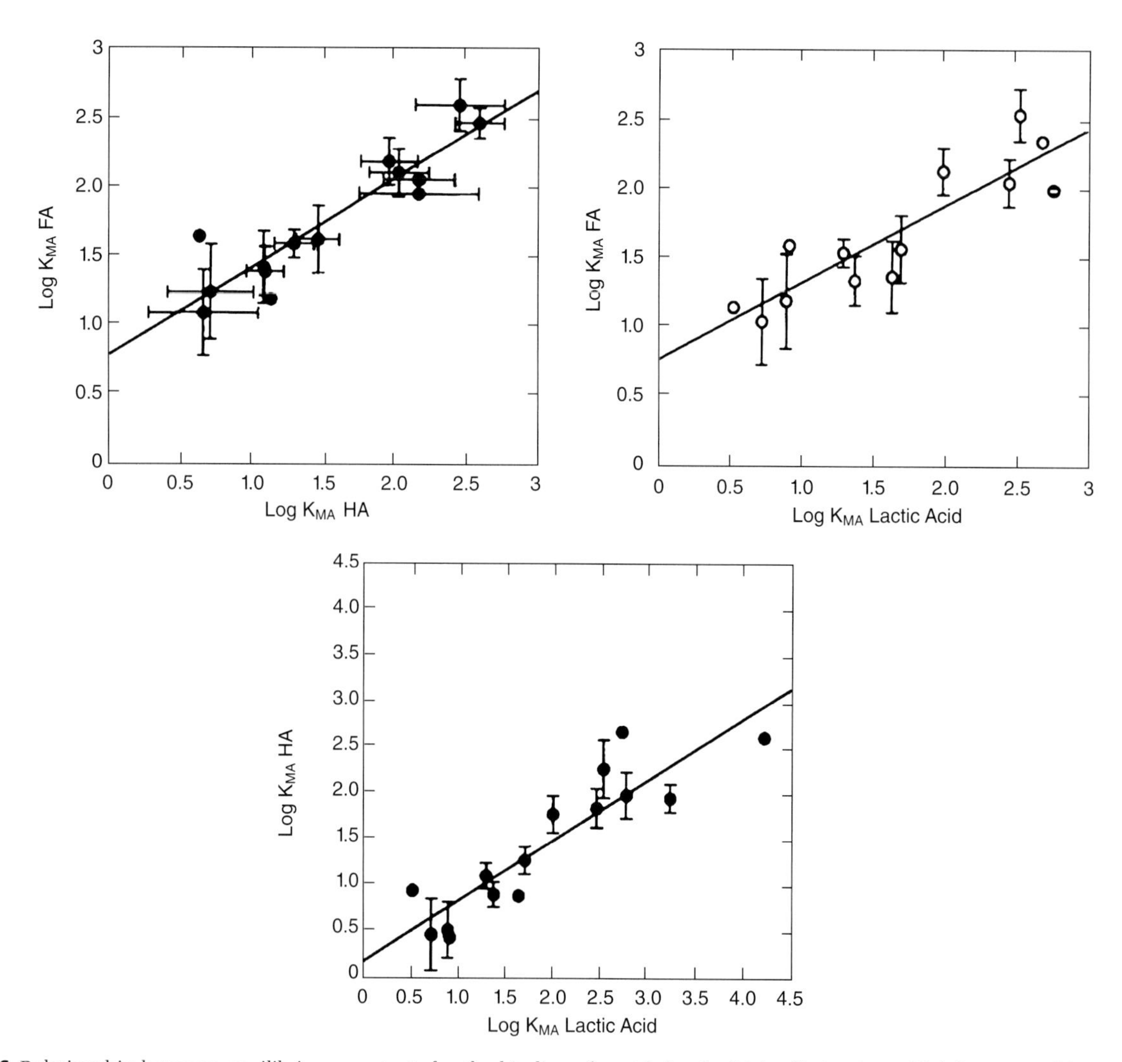

Fig. 4.16 Relationship between equilibrium constants for the binding of metals by the high affinity sites of fulvic (FA) and humic (HA) acids and with lactic acid. The regression lines are shown on the graph. Taken from Tipping (1998) *Aquatic Geochemistry* **4**: 3–48 [13] and reprinted with kind permission of Springer Science+Business Media.

Fig. 4.17 in which a number of metals were investigated to examine the simultaneous binding of the metals to a humic acid. The values for the constants were from the derived database (Appendix 4.1b) although there was not data for Hg and the value for the binding constant for Hg was determined by fitting the model to the data. As the cation concentrations exceeded the solubility of Fe, the result for Fe likely reflects precipitation rather than adsorption. The results show the preferential binding of Hg to the humic acid, followed by Pb and Cu, which are adsorbed to a similar degree, then the other metals. Overall, the model is able to predict the data to a relatively high degree of accuracy suggesting that the formulation is robust over a wide range of NOM concentrations. The relative binding of the metals fits with expectations based on their relative propensity to bind to NOM, as discussed in Chapter 3 and in the later chapters of this book. However, it should be noted that it's likely that Hg, and perhaps Cu, are not binding to carboxylic acid sites but to thiol sites, as discussed in Section 7.2.2. Modifying the WHAM model to compensate for such differences in binding to different types of acid sites in NOM has been relatively successful and has mostly confirmed the empirical constants derived from data fitting [51].

The application of the WHAM Model to Hg and methylmercury (CH_3Hg) [51] confirmed one aspect of the binding of metals to NOM that has received little attention in the

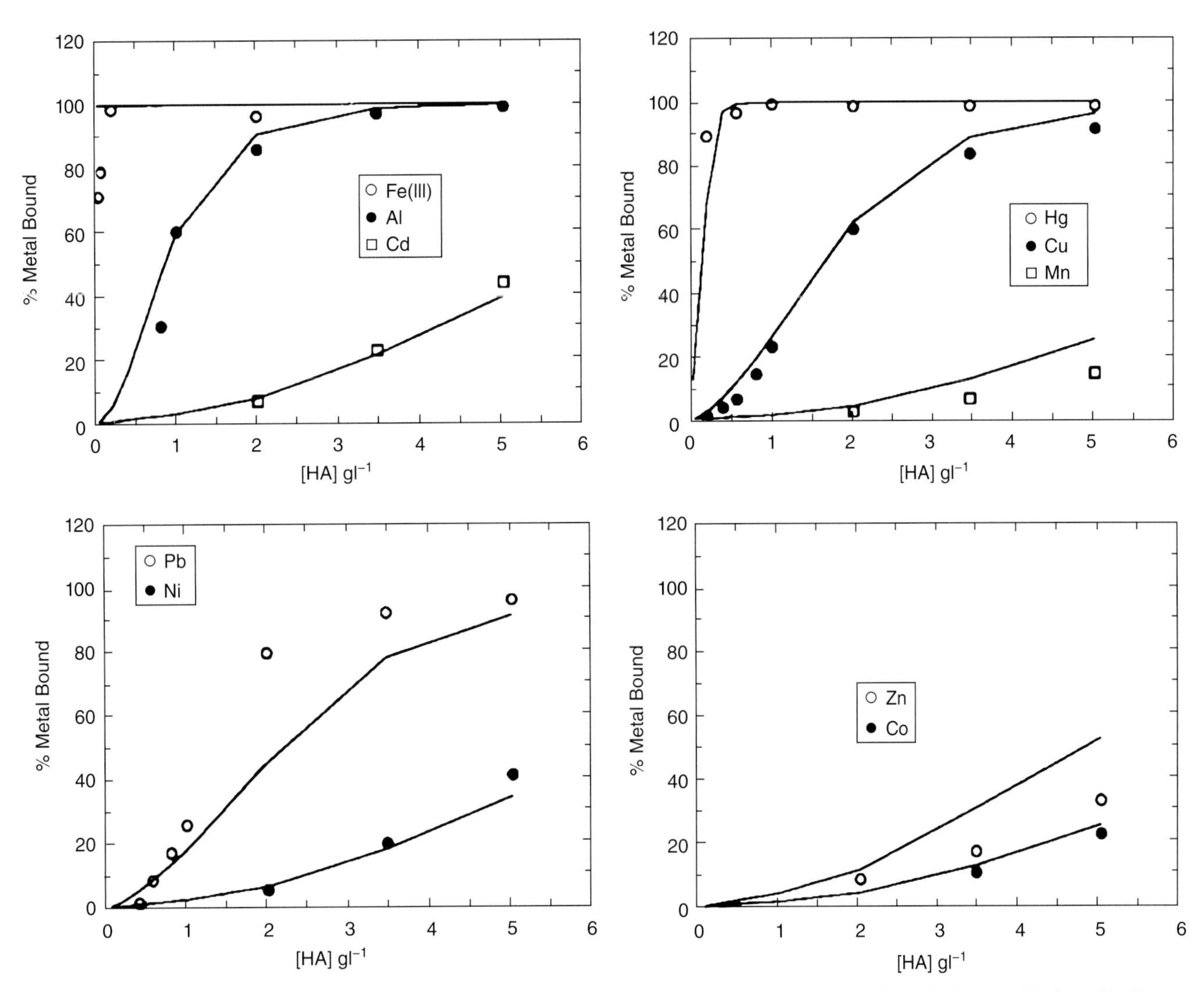

Fig. 4.17 Data of Kerndorff and Schnitzer (1980) for the simultaneous binding of metals by HA modelled using the WHAM model. Except for Hg, the lines are Model VI simulations using default values of $\log K_{MA}$ in the program. The logKMA value for Hg (3.5) was optimized using the data. Precipitation of $Fe(OH)_3$ was assumed to occur, and to be indistinguishable from binding to HA for the experimental data. Taken from Tipping (1998) *Aquatic Geochemistry* **4**: 3–48 [13] and reprinted with kind permission of Springer Science+Business Media.

modeling studies discussed in this section which have interpreted data collected at relatively elevated metal and NOM concentrations. Most metal concentrations in natural waters are nM or lower, while many modeling studies are interpreting data at micromolar or higher concentrations. Overall, it is likely the ratio of metal/ligand in the experiments that really matters. For example, various studies of Hg complexation have shown that the value obtained for the equilibrium constant is a strong function of the relative ratio of Hg to NOM, with much lower values being obtained at higher ratios [52, 53]. This confirms the expectation that there are a small number of high affinity sites in NOM, such as thiol groups, and a larger number of weaker binding sites, which likely reflect the more dominant carboxylic acid sites [54]. Additionally, the study design is important as it is likely that metals will initially bind to the relatively abundant weak sites initially, and will be transferred to the less abundant sites over time. For example, studies of Hg binding to NOM suggest that it could take up to 24 h before the steady state system is reached [55].

The WHAM model has been used to model environmental systems by a number of investigators and two examples will

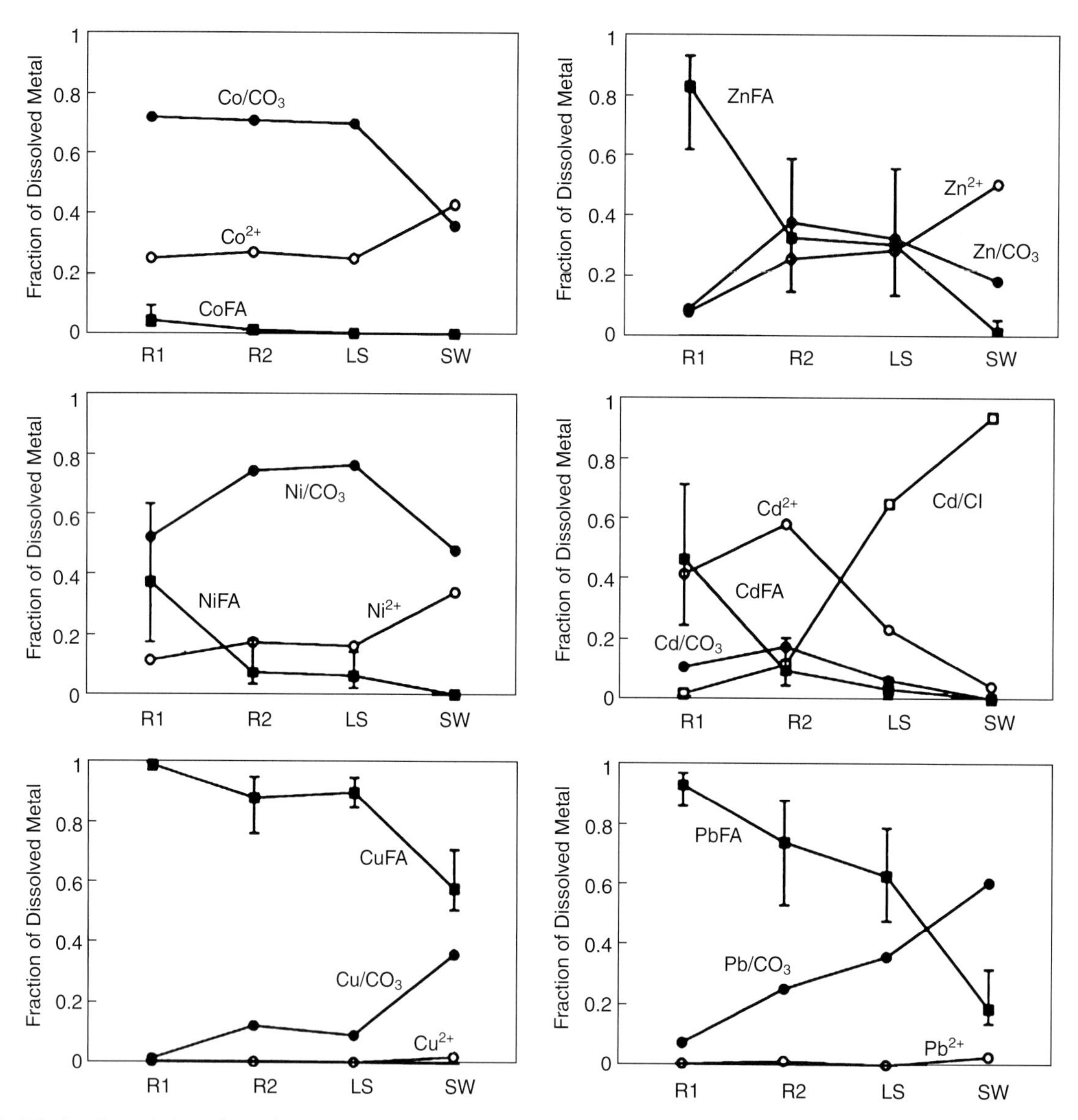

Fig. 4.18 Calculated speciation of metals in waters of the Humber River, denoted as RA, R2, LS and SW in the figures. In seawater (SW) the speciation is different than the riverine waters for most of the metals. Minor complexes are not shown on the graphs to enhance clarity. Error bars reflect the uncertainty in the magnitude of the binding constants for the fulvic acid (FA) fraction. Figure reprinted from Tipping et al. (1998) *Science of the Total Environment* **210**: 63–77 [57], with permission of Elsevier.

be discussed here [56, 57]. Tipping et al. [57] modeled the speciation of various metals (Co, Ni, Cu, Zn, Cd, Pb) in the Humber river/estuary system, and was able to simulate the change in speciation from a dominance of the free ion concentration of Co, Ni, Zn and Cd in the freshwaters reaches to that of increased complexation for these metals in the higher salinity water. In this modeling exercise, adsorption of metals to suspended particulate was done using a modified partition coefficient, that was modeled in terms of metal adsorption using a cation/H^+ exchange reaction and assuming low metal loading. Complexation to organic matter (modeled as FA) dominated for Cu and Pb in the low salinity waters (Fig. 4.18). Modeling suggested that these two metals are fully complexed to FA at 10–15 $mg\,l^{-1}$ FA. For the other metals, the importance of complexation to NOM increased with NOM concentration: Zn was ~80% complexed to FA at 15 $mg\,l^{-1}$ FA, while 40–60% of the Cd and Ni, and <20% of the Co were complexed at 15 $mg\,l^{-1}$ FA.

The surface adsorption modeling using a simple adsorption isotherm approach provided a reasonable estimate of the measured K_Ds for the various metals in both the fresh and saline waters of this system. Additionally, the modeling demonstrated the expected result that the K_D decreases as the NOM concentration increases as this leads to a higher fraction of the metal being in the dissolved fraction relative to adsorbed on the surface of the suspended solids. Some dependence of K_D on pH was also demonstrated and this is consistent with the modeling approach where adsorption of a cation leads to the release of protons into solution. This simple adsorption modeling is clearly less complex and detailed than that discussed in the previous section of this chapter. Most computer models in the literature do not allow for the modeling of metal adsorption to both surfaces and to NOM in the same framework (i.e., with the NOM treated as a "particle" with surface sites), although some examples will be discussed later. This, for example, is a limitation of the MINEQL+ Model. Modeling the partitioning and the dissolved phase relative speciation separately is one approach but clearly this would require numerous iterations to obtain an accurate assessment of a complex system.

The WHAM Model was also used to study metal (Co and Cu) binding in freshwater and saline waters in the presence of NOM, and the results of the model were compared to data obtained using equilibrium dialysis experiments, which enable the distinction between inorganic and NOM complexation of the metals [56]. One aspect of these studies was to examine the interactions between the trace metals studied (added at ~5×10^{-7} M) and the major ions present in seawater, especially Ca and Mg which can bind with NOM to some degree, and which are present in seawater at relatively elevated concentrations (10^{-2} M range for Ca and Mg). Thus, even though these metals do not bind strongly, they can outcompete other trace metals present at relatively low concentrations for the NOM binding sites. As discussed earlier, and in the later chapters, Co does not bind strongly to NOM and very little was complexed under the conditions of these experiments. In contrast, Cu is strongly bound and mostly associated with the NOM even at low NOM concentrations. However, for both metals the extent of complexation to NOM was reduced at the higher major ion concentrations at low pH (4.6). While this is interesting, it is not exactly environmentally relevant as the pH of seawater is around 8.1. Predictions under conditions more relevant to the environment, and with the presence of other trace metals that could also bind to NOM confirm that Co is mostly present as the free ion or as inorganic complexes while Cu is mostly complexed to NOM, and that the major ions have little impact on Cu binding under these conditions.

The above examples show potential applications of this model. The WHAM Model has been used in a number of environments and applications [58]. It has been modified to be used with a Biotic Ligand Model which includes interactions with organisms, as discussed in Chapter 8. The model has been used both for surface waters, such as lakes [59, 60] and soil solutions [61–63], and has been used to examine speciation in acid mine drainage and mining impacted waters [64–66]. In addition to transition and heavy metals, the model has been applied to rare earth elements [67]. In a number of these studies, as also discussed earlier, the model output has been compared to other methods of measuring the metals speciation in solution, such as dialysis membranes, diffusive gradient in thin film (DGT) gels and other approaches. Some of these applications are discussed in the remaining chapters of this book.

4.4.2 The NICA-Donnan modeling framework

The NICA–Donnan Model is described by Kinniburgh et al. [14]. In this model, the NOM (humic and fulvic acids) are considered as a gel-like phase of specific volume, the Donnan volume (V_D), which depends on the ionic strength (I) of the solution:

$$\log V_D = b(\log I - 1) - 1 \tag{4.13}$$

where b is an empirical parameter. The overall net negative charge in the Donnan phase is compensated for by the cations, which are increased in concentration in this phase (c_{Di}) relative to the bulk solution (c_i). The NICA formulation represents the binding sites as a distribution with both low and high affinity. The overall relationship for the amount of a specific component i (Q_i) that is bound to these sites is [14, 49]:

$$\begin{aligned} Q_i = {} & \frac{n_{i1}}{n_{H1}} Q_{\max 1,H} \frac{(\tilde{K}_{i1} c_{Di})^{n_{i1}}}{\sum_i (\tilde{K}_{i1} c_{Di})^{n_{i1}}} \frac{\left[\sum_i (\tilde{K}_{i1} c_{Di})^{n_{i1}}\right]^{p_1}}{1 + \left[\sum_i (\tilde{K}_{i1} c_{Di})^{n}_{i1}\right]^{p_1}} \\ & + \frac{n_{i2}}{n_{H2}} Q_{\max 2,H} \frac{(\tilde{K}_{i2} c_{Di})^{n_{i2}}}{\sum_i (\tilde{K}_{i2} c_{Di})^{n_{i2}}} \frac{\left[\sum_i (\tilde{K}_{i2} c_{Di})^{n_{i2}}\right]^{p_2}}{1 + \left[\sum_i (\tilde{K}_{i2} c_{Di})^{n_{i2}}\right]^{p_2}} \end{aligned} \tag{4.14}$$

where $Q_{\max\ 1,H}$ and $Q_{\max\ 2,H}$ are the maximum proton site densities of each distribution of affinities (units of mol kg^{-1}), p_1 and p_2 reflect with width of the distributions, and for each component i there are four parameters: the median affinities for each distribution (K_{i1} and K_{i2}) and their non-idealities (n_{i1} and n_{i2}). The total amount of component i associated with the organic phase is the combination of that from the Donnan interaction ($V_D(c_{Di} - c_i)$) and Q_i.

The various constants optimized in a study of Cu and Pb binding to humic acids in 0.1 M $NaNO_3$ were compared with estimates of binding constants using electrochemical methods [48], that are typically used to examine organic complexation in natural waters (see Chapter 6) [68]. The humics were obtained using standard extraction procedures from a soil sample. These authors were able to demonstrate, through a comparison of the measured complexation using a variety of electrochemical techniques and the model

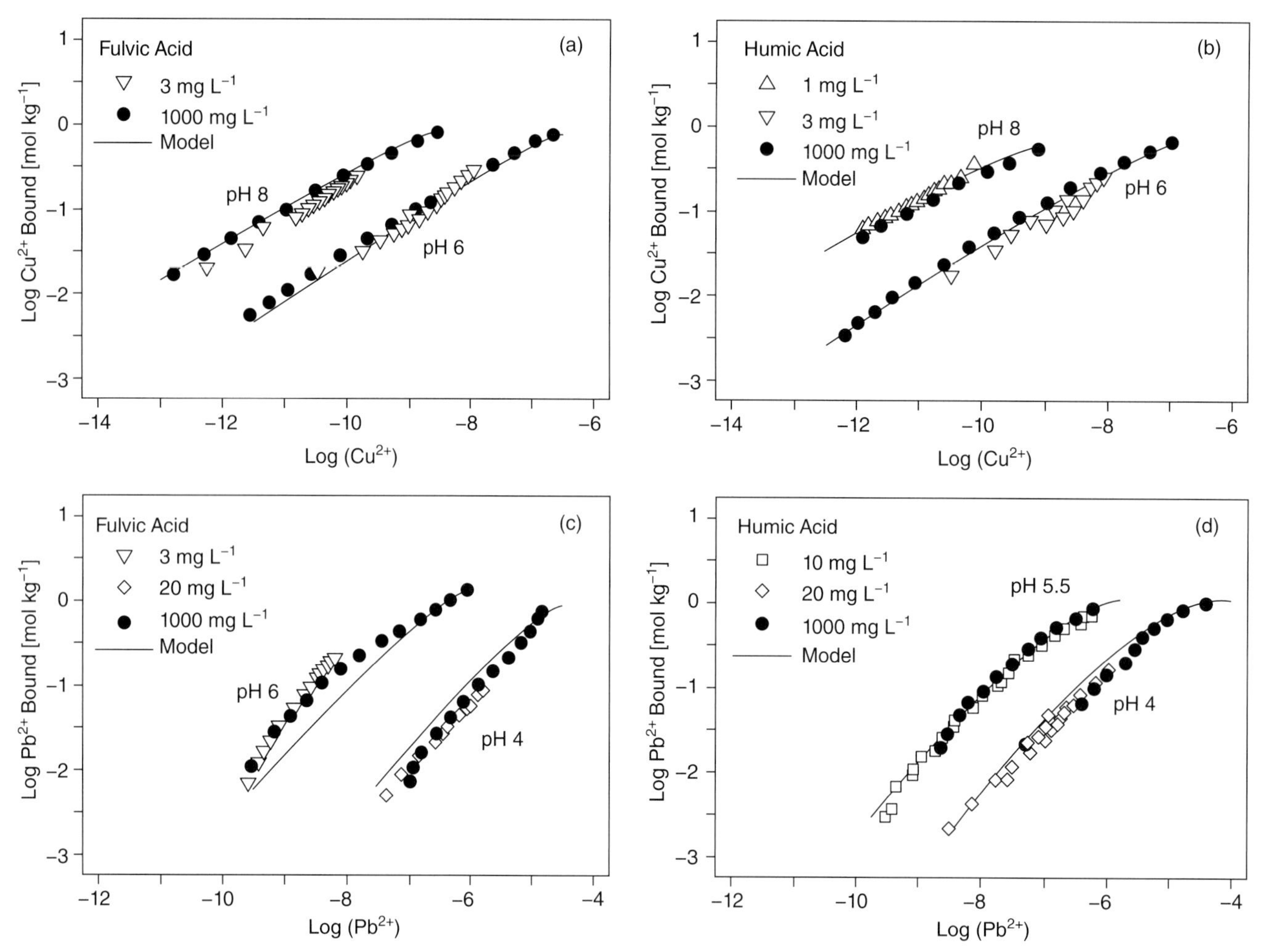

Fig. 4.19 Comparison of data obtained for the binding of copper (Cu) and lead (Pb) to humic (HA) and fulvic acids (FA) at different concentrations, over a pH gradient, as estimated by electrochemical techniques (data points) and estimated using the NICA-Donnan model (lines). See text and Christl et al. (2005) *Environmental Science and Technology* **39**: 5319–26 [48] for details. Reprinted with permission; Copyright (2005), American Chemical Society.

predictions that the model could simulate the binding of Cu and Pb over a range of total concentrations and at two different pH values, for different amounts of NOM [48] (Fig. 4.19). The overall agreement between the model and the data suggest that the NICA-Donnan model provides a reasonable approximation of the data and can therefore likely be adjusted to provide a model for the interaction of other metals with NOM recovered from a variety of environments. As discussed in Chapter 3 and elsewhere in this chapter, the amount of metal complexed increases with pH. The results in Fig. 4.19 also show that at low ligand and metal concentrations and higher pH, Cu and Pb complexation to HA is greater than to FA, but this effect is less at higher metal concentrations. Additionally, the model recreates the data in that the relative binding per unit concentration of NOM is similar at the different exposure concentrations. There are a number of studies in the literature that have further examined this model and its predictions for environmental solutions.

Merdy et al. [35] discuss the advantages and disadvantages of having a continuous distribution of site types rather than a discrete set of sites. For discrete sites, a large number of parameters have to be assigned and these can make the system cumbersome but there is the advantage that with a discrete formulation it is possible to add specific reactions. These authors discuss and compare the various potential formulations for the local binding function, how to deal with heterogeneity and how to relate the electrostatic potential to the charge on the humic acid or NOM entities as these cannot be assumed to be hard spheres. The Donnan formulation takes this into account by assuming that there is charge balance both on the surface and due to penetration of ions into the organic matrix. The NICA-Donnan approach is designed to minimize the number of specific parameters that are required.

Tipping compared the WHAM VI and NICA-Donnan Models using various datasets and concluded at that time that both provided a similar accuracy in predicting the data when four adjustable parameters were used in each model [13]. Each model has its advantages and limitations and both models need to be more extensively tested on field data to demonstrate their application under real-world situations.

4.4.3 Modeling the adsorption of humic acids to surfaces and the interaction with metal(loid)s

There have been models developed to characterize the adsorption of large molecular weight organic matter to oxide and other surfaces. One approach, the Ligand and Charge Distribution Model (LCD) [69] is an integration of the CD-MUSIC modeling framework and the NICA model for binding to oxide surfaces. This combined model has four modules: the NICA-Donnan, the NICA-LD, the CD-MUSIC and the ADAPT (Adsorption and Adaptation) module. The NICA–Donnan module, which calculates the state of the NOM in solution, and the CD-MUSIC approach, which calculates the interactions at the oxide surface, will not be discussed in detail as the approach is similar to that outlined earlier. Recall that the Donnan approach represents the FA or HA as Donnan "particles" and these would interact with the oxide surface. Additionally, it is possible for interactions to occur between other ions and these NOM particles, and the LD-MUSIC module calculates the interaction of both small ions and surface sites for the reactive parts of the adsorbed molecules (i.e., fulvic or humic acids), and compiles the overall average state of the adsorbed HA or FA. It is assumed that the adsorbed layer is the thickness of the HA or FA particles and laboratory studies suggest it is only carboxylic acid groups that interact with the oxide surface, and so the model considers only these inner-sphere complexes, and a monodentate association [69].

The ADAPT module calculates the distribution of the HA or FA between the dissolved and surface phases. The module estimates this using the free energy changes associated with changes in the surface potential due to adsorption or complexation of the NOM entities, and ensures that the charge balance of the overall system is maintained with the electrochemical potential of the NOM particles in solution and those adsorbed being equal. The details of this formulation are contained in the reference and the approach has been included into the computer code called ORCHESTRA. For the ADAPT module, the parameters needed are the total particle (HA or FA) concentration, the average chemical state of the particles and the Boltzmann factors for the Donnan phase and at the d-plane of the oxide surface. These values are computed by the NICA-Donnan and CD-MUSIC modules. The ADAPT module will calculate the volume fractions of the NOM particles in solution and in the adsorption phase.

The LCD Model was used to examine the adsorption of a fulvic acid to goethite which had been experimentally determined using a series of batch experiments over a range of FA concentrations (75–450 $mg\,l^{-1}$) and with a 5–6 $g\,l^{-1}$ goethite suspension. The model provides a good prediction of the data when both electrostatic adsorption and binding of the carboxylic acid groups are included in the model framework (Fig. 4.20) [69]. The model predictions agree at both high and low FA concentrations and at both low and high pH, suggesting that the combined model is a suitable estimation of the various interactions that are occurring. The average chemical state of the carboxylic acids groups of the FA in these simulations is shown in Fig. 4.21 for the situation where 50% (volume fraction) of the FA is adsorbed. As the pH increases, the fraction of the FA's carboxylic acid groups in the solution phase increases, and, as found for any acid, the degree to which the solution phase FA's carboxylic acid groups is protonated is a function of pH (Fig. 4.21a) [69]. As with simple anions, the amount of the carboxylic acids groups that is complexed to the surface decreases with pH. The fraction of the FA that is adsorbed to the goethite surface through an electrostatic interaction of the deprotonated form increases with pH as the surface becomes more positive. As more of the FA is adsorbed, there is a change in the relative ratio of inner-sphere complexes compared to electrostatic adsorption, primarily as the overall surface becomes more negative as more FA is adsorbed. This is shown in Fig. 4.21(b) for a pH of 5.5. These results demonstrate the complexity of the different interactions that can occur for the interaction of NOM with oxide surfaces.

Comparable studies also examined the adsorption of either or both HA and FA to oxide surfaces and have demonstrated the overall nature of the interaction. For example, a comparison of adsorption of humic and fulvic acids to goethite was measured so that the overall factors involved could be determined [69, 82]. The experimental results showed that the adsorption was greater for HA, but this was also more dependent on ionic strength. The data was modeled using the LCD approach and it was shown that it predicted the adsorption of both HA and FA.

The results suggest the importance of proper characterization of the interactions between surfaces, NOM and charged ions in solution. The extension of the LCD Model to the adsorption of trace metals and metalloids to oxide surfaces is needed, and there are likely to be other approaches. Overall, the impact of these interactions on the degree of binding of cations and anions to oxide surfaces is important and complex, and as discussed in the examples with low molecular weight ligands and metals with HFO, there is the potential for such interactions to both increase or decrease adsorption. Increased adsorption of cations is likely in many instances given the multiple available sites on NOM and therefore the potential for the formation of tertiary complexes. Conversely, the presence of NOM likely reduces anion adsorption through competition for the binding sites on the oxide surface. Such interactions between humics,

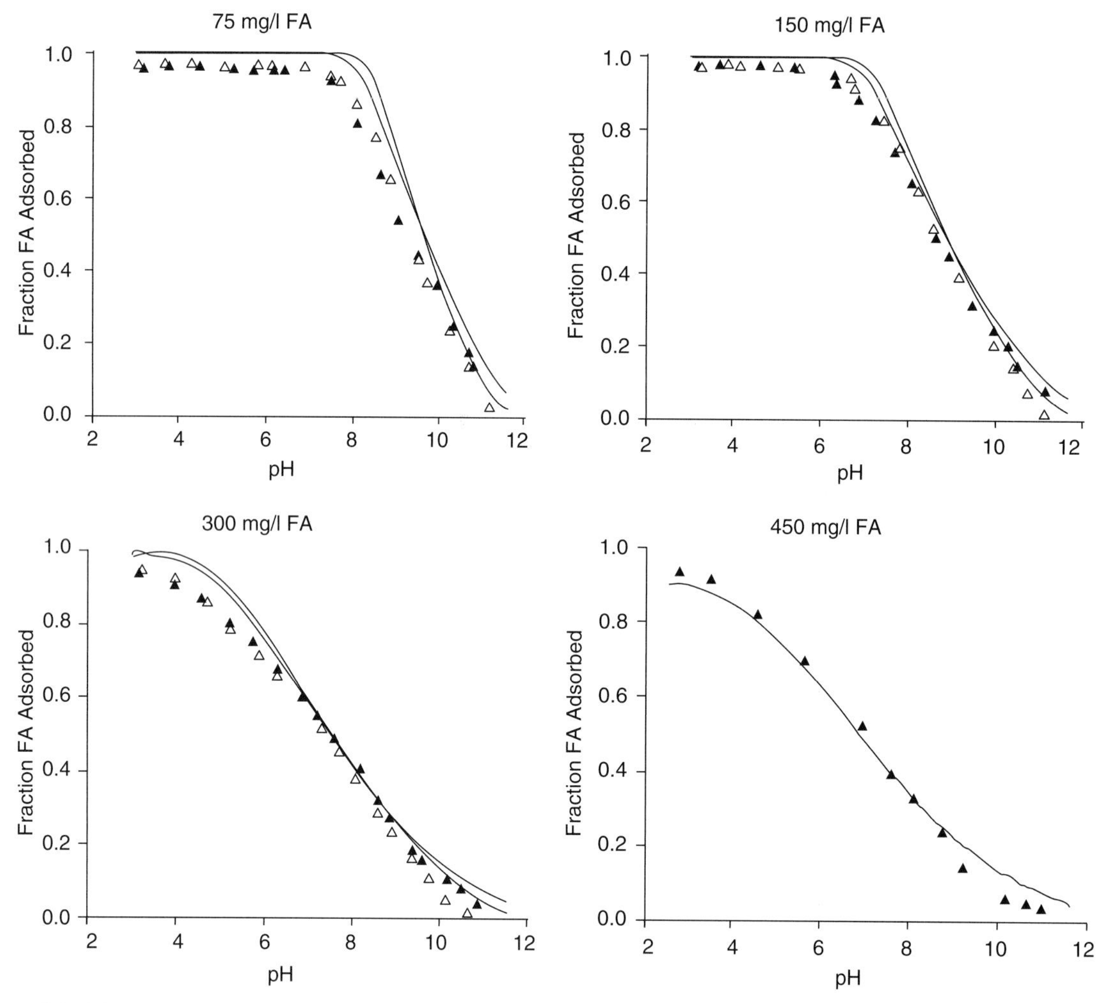

Fig. 4.20 Model of the adsorption of fulvic acid (FA) at different concentrations to the surface of goethite in different ionic strength solutions: 0.1 M (open symbols) and 0.015 M (closed symbols) $NaNO_3$ solutions. Data modeled taken from various sources as indicated in Weng et al. (2006) *J. Colloidal and Interface Science* **302**: 442–57 [69]. Reprinted with permission of Elsevier.

ions and mineral surfaces has been acknowledged and discussed in the literature [70].

While the focus of these sections has been the impact of FA and HA on metal interaction with surfaces, it is worth reiterating that NOM found in many natural waters is a combination of both high molecular weight compounds and low molecular weight species that are more directly derived from microbial production. This is especially true for the marine environment. There has been an improvement in the methods for the extraction of NOM from natural waters that collect a larger range and fraction of the NOM, and not just the humic and fulvic acids. These studies have shown that in many environments the concentrations of small molecular weight compounds are sufficient that these species are important in the complexation of metals in environmental waters. Modeling studies examining metal complexation need to consider such potential interactions more closely.

Furthermore, there are few studies attempting to model the interaction of metals with other surfaces besides oxides. As noted earlier, there have been models developed to examine the association of metals with carbonates but there are other important phases, especially in reduced environments. In many environments, amorphous and crystalline forms of FeS, as well as pyrite (FeS_2), are important phases for metal adsorption. Future studies should examine how well the current models predict such associations and develop models that are suitable to describe these associations in detail.

4.5 Modeling redox transformations

As noted in Section 3.5, the equations governing redox reactions can also be written in terms of equilibrium equations for the balanced reactions. However, in order to accurately

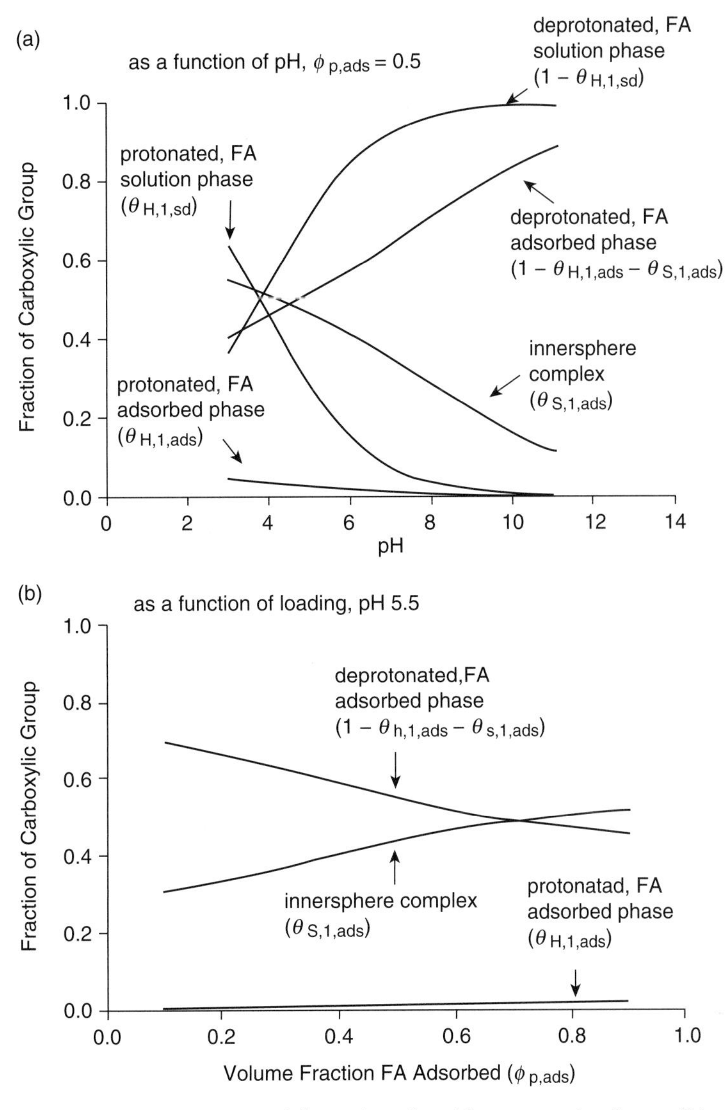

Fig. 4.21 Demonstration of the changes in the speciation of the carboxylic acid groups under the conditions of the model depicted in Fig. 4.20: (a) across a pH gradient at a fixed concentration and (b) at a fixed pH as a function of concentration. Reprinted from Weng et al. (2006) *J. Colloidal and Interface Science* **302**: 442–57 [69] with permission of Elsevier.

model the redox reactions, it is necessary to derive a formulation for the individual half reactions as it may not be possible in all instances for complex media to know the exact redox couple that is occurring and this should not always be pre-defined. The solution to this is to assign the "concentration" of the electron $\{e^-\}$ as a principal component as this removes it from the governing equations. The two forms of the compounds involved in the redox couple are also chosen as components. For example, to describe the reduction of Fe^{III} to Fe^{II} would require the following principal components in MINEQL+: Fe^{3+}, Fe^{2+}, e^-, H^+ and H_2O. The system would be fully described as the default would be for oxygen to act as the electron source in this instance, and would be a species of fixed composition. The problem would be solved by either defining the pε of the system, or the ratio of $[Fe^{3+}]/[Fe^{2+}]$. Recall, from Equation 3.106 that:

$$p\varepsilon = p\varepsilon^{o} - (1/n)\log\{Red\}/\{Ox\} \quad (4.15)$$

so if the pε is set, the ratio of the redox species is set, and vice versa. Such problems can be included into thermodynamic models and it is easy to investigate the changes in speciation across a pε gradient. As the equilibrium constants, which are related to $p\varepsilon^0$, are typically large for redox processes, there is often the potential for greater instability in the model calculations when crossing a large redox gradient. A further problem should help demonstrate the procedures.

Example 4.6

Consider the oxidation-reduction reactions of 10^{-8} M As in the presence of 10^{-4} M sulfate/sulfide in 0.5 M NaCl solution. Arsenic can exist as As^{V} and As^{III} in natural waters, and both exist as oxyanions in solution: H_3AsO_4 ($pK_{a1} = 2.2$; $pK_{a2} = 7.0$; $pK_{a3} = 11.5$) and H_3AsO_3 ($pK_{a1} = 9.3$). Consider the change in redox state across a pH gradient of 2–10 and pε values for 10 (near oxygen saturation) to –4 (highly anoxic). Both As and S will go through redox transitions across these ranges. The overall pH–pε diagram of sulfur is shown in Fig. 3.19 and that for As in Fig. 7.15 later. Additionally, the speciation of the As species with pH are also shown in Fig. 7.15. Over a pH range of 2–10, As^{III} is primarily as H_3AsO_3 while As^{V} will be mostly as $HAsO_4^{2-}$ and as $H_2AsO_4^{-}$. The pertinent As redox couples is:

$$AsO_4^{3-} + 2H^+ + 2e^- = AsO_3^{3-} + H_2O \quad \log K = 9.7$$

The concentrations as noted were entered into MINEQL+ after the selection of the relevant components (e^-, H^+, H_2O, AsO_3^{3-}, AsO_4^{3-}, SO_4^{2-}, HS^-, Na^+, Cl^-) and scanning the database. The model was then run across the range of pH (2–10) and pε (10—4), with a fixed ionic strength of 0.5 M. The following output was obtained (Fig. 4.22).

There is a decrease in the concentration of As^{V}, shown as $H_2AsO_4^{-}$ in Fig. 4.22(a), which begins at about a pε of around 3 at pH 5, where this is the dominant species. As the pH is changing, the relative importance of the various acid species changes for As^{V} and so the plots in Fig. 4.22(a) reflect both changes due to redox reactions as well as acid dissociation. The figure shows that the transformation begins at a higher pH for a lower pε, which is consistent with Fig. 7.15, which shows the data in terms of E_H rather than pε. In contrast, the concentration of As^{III} increases as the concentration of As^{V} decreases (Fig. 4.22b). Note that at the lower pε values, the concentration of H_3AsO_3 decreases. This is due to the precipitation of solid phases in the presence of sulfide and the model predicts the major species is orpiment (As_2S_3) (Fig. 4.22c). The following reaction describes what is occurring:

$$2AsO_3^{3-} + 3HS^- + 9H^+ = As_2S_3\,(s) + 6H_2O \quad \log K_{sp} = 130.6$$

The very large equilibrium constant reflects the fact that the undissociated As anion is present at a very small concentration under typical environmental conditions. Rewriting the equation in terms of the dominant species, which is the unprotonated acid, the reaction is:

$$2H_3AsO_3 + 3HS^- + 3H^+ = As_2S_3\,(s) + 6H_2O \quad \log K_{sp} = 61.2$$

The equilibrium constant is still large and will cause precipitation, if all the As is reduced at a pH of 7, at a sulfide concentration of around 10^{-7} M. So this accounts for the rapid decrease in soluble As^{III} at the lower pεs, as shown in Fig. 4.22(c). The increase in total dissolved sulfide with changing pH and pε is shown in Fig. 4.22(d). The sulfide levels increase when the pε reaches relatively low values, and this is again dependent on the pH of the system. The relevant equation is:

$$SO_4^{2-} + 8e^- + 9H^+ = HS^- + 4H_2O \quad \log K = 34$$

At the lowest pε (–4), this redox reaction occurs at a relatively high pH (~8) while at the lower pHs, this reaction can occur at a pε near zero or slightly positive. The MINEQL+ model output is consistent with the figures in the literature and demonstrates the impact of pH and pε on the redox speciation of As and S. Further examples will be explored in the examples at the end of the chapter and in the discussions in the remainder of the book.

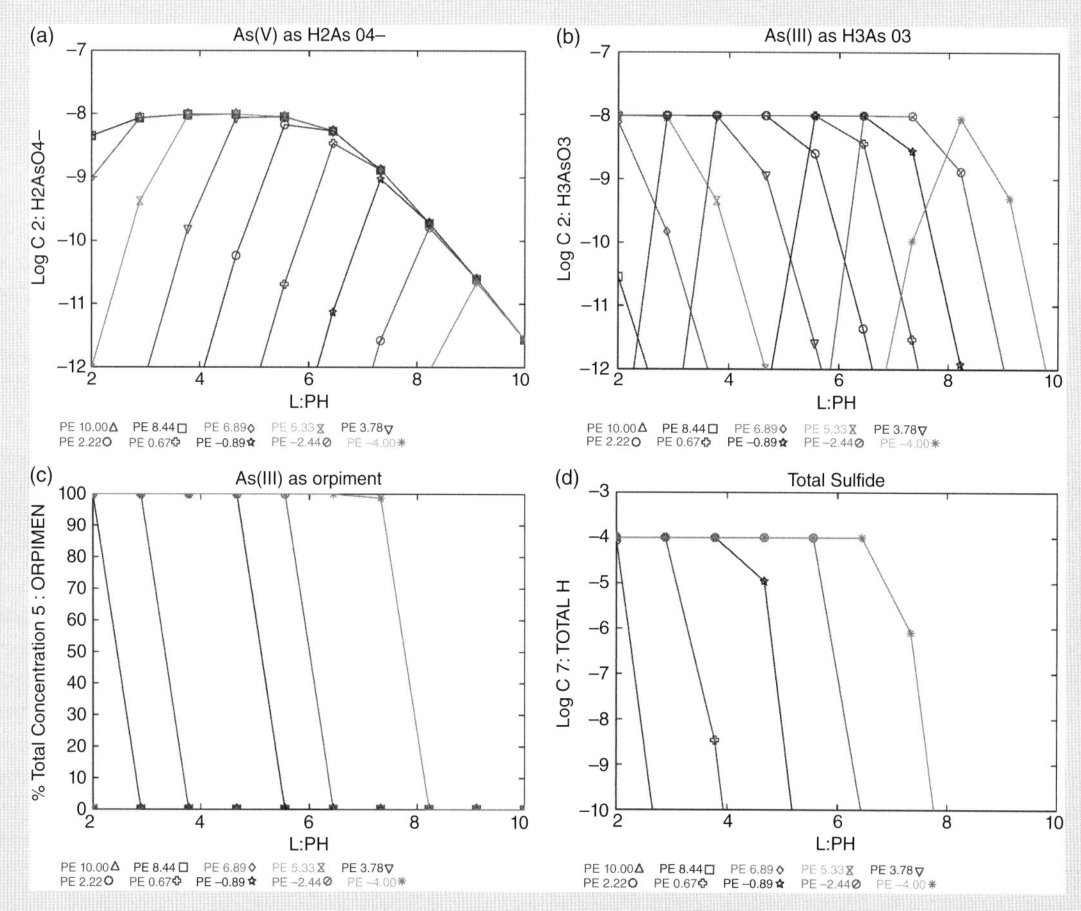

Fig. 4.22 A demonstration of the effect of pH and pe on the speciation and oxidation state of arsenic (As) in the presence of sulfur (S). Figures (*clockwise*) show the changing concentration of the major form of As(V) and As(III) as well as the precipitation of orpiment () that occurs at high sulfide concentration. Data generated using MINEQL+.

4.6 Modeling the kinetics of reactions

The focus on the modeling in this chapter and in Chapter 3 has been on the system at equilibrium as it is considered that in many environmental situations that even though the system may not be at true equilibrium, the steady state that exists is close to the equilibrium condition as the rate of reactions are much faster than the rate of exchange of chemicals between the water body of interest and the surroundings, or that the amount of exchange occurring in the "flow-through reactor" is small compared to the mass of reactants and products within the reactor (Section 3.1.3). Under these conditions, the equilibrium calculations provide a realistic prediction of the steady state system. However, this is not always the case and the kinetics of the reactions can be important, especially when laboratory studies are being used to mimic the field situation. Reactions involving some metals, and reactions involving adsorption/desorption reactions with surfaces or interactions with NOM are often relatively slow compared to the reactions that occur with inorganic ligands, and therefore these can prevent the situation from achieving steady state conditions on short timescales. These kinetic effects need to be considered but are often ignored. A few examples will illustrate the importance of assessing if equilibrium has been achieved.

For example, various studies investigating the interaction of metals such as Hg and Pb with NOM have shown that the reactions occur slowly, over hours rather than minutes or seconds [1, 55, 71–73]. Additionally, such experiments have shown that that the rate of reaction is rapid for metals bound to inorganic ligands or to small molecular weight organic ligands such as EDTA. For example, it was shown that Hg^{II} complexed to Cl and EDTA were "easily reducible" by tin chloride ($SnCl_2$) while Hg bound to NOM was not [74]. Additionally, they found, as others have [53, 55, 72], that equilibrium was achieved within 4–24 h after the addition of Hg to natural waters containing NOM. Similar results have been found for Cu and Pb, as reported in Morel and Hering [1]. Also, in making media for laboratory experiments with phytoplankton [75], the metals must be added in order of their relative rate of reactivity to EDTA (Fe^{III} is added first), the "typical" ligand used to maintain a constant free ion concentration, otherwise the solution will not achieve rapid equilibrium. This further illustrates the differences in the reaction timescale of different metals as Hg bound to EDTA is considered "reactive" (i.e., able to be reduced or participate rapidly in ligand exchange reactions) while Fe^{III} does not react rapidly with EDTA, as discussed further later.

The results of Miller et al. [55] are instructive of what has been observed in the environment. Studies downstream from a large known Hg source (mostly inorganically complexed: easily reducible) discharging into a flowing stream showed that the relative concentration of this "reactive" fraction decreased with distance, while the fraction that was strongly complexed Hg increased (Fig. 4.23a). These results where mimicked in the laboratory by adding inorganically complexed Hg to waters from different sources as well as a solution containing different amounts of Suwannee River fulvic acid (Fig. 4.23b). These results show that the equilibrium condition is not reached until many hours have elapsed. Other studies have demonstrated similar results. Similarly, Jones et al. [76] compared the kinetics of the dissociation of NOM complexed to either Fe^{III} and Al^{III} and found that the dissociation rate was lower for Fe, with half-times for dissociation of Fe bound to the strong sites being 6–7 h, compared to around 1.5 h for Al. This suggests that the complexes with NOM are kinetically more stable with Fe.

Often, the slow achievement of equilibrium is also because the added metal will react with the most abundant ligands first, which may not be the strongest and therefore most thermodynamically stable, especially in natural waters, and repartitioning of the metal to the stronger ligands will occur over time, resulting in a slower approach to the steady state or equilibrium condition. The reasons for this can be examined using the examples outlined next for solutions only, and for systems that also have solid phases present and/or microorganisms. Each situation will be dealt with in the following sections.

4.6.1 Reactions in solution

In much of the discussion in this book, and especially in the following chapters, the actual state of uncomplexed metal in solution is ignored. In Section 3.2.3 it was noted that cations do not exist in solution as uncomplexed "naked" ions, but are complexed to water even in the absence of other ligands and in the absence of hydrolysis, that is, as $M(H_2O)_n^{x+}$, where n = 4, 5 or 6 typically and octahedral complexes are the most common for the transition metals. Thus, reactions of free ions with ligands actually involve an exchange reaction where a water molecule is replaced by a stronger binding ligand. This is implicit but mostly ignored in most depictions of these reactions.

Let us consider one scenario for the reaction of a divalent free metal with a dissolved ligand, assuming an octahedral complex with six water molecules. The ligand will initially form an intermediate outer-sphere (ion association) complex that is at equilibrium with the free ion and ligand (K_{OS}), but which then forms an essentially "irreversible" product through the loss of a water molecule with rate, k_{-w}; that is, the final dissociation reaction is the rate limiting step [1]:

$$M(H_2O)_6^{2+} + L^{n-} = M(H_2O)_6L^{2-n}$$

$$M(H_2O)_6L^{2-n} \rightarrow M(H_2O)_5L^{2-n} + H_2O$$

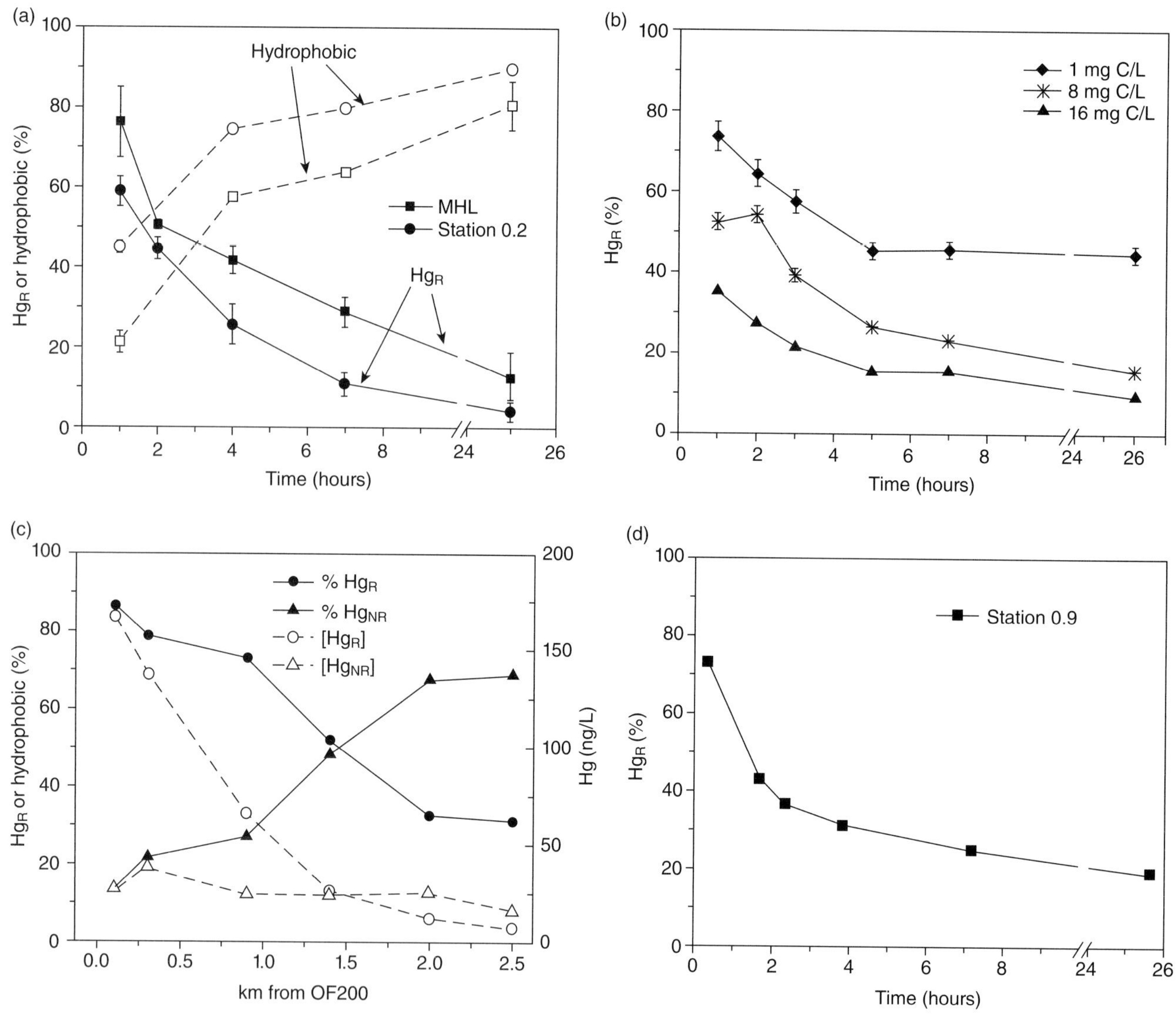

Fig. 4.23 (a) Changes in the speciation of mercury in water over time after its addition to a stream from two locations showing the kinetics of binding of mercury to the hydrophobic fraction (Hg_{NR}); (b) changes in the labile mercury fraction (Hg_R) in spiked solutions with different concentrations of Suwannee River organic matter; (c) changes in speciation with distance downstream from the pollution source and (d) in the labile fraction, both illustrating the time delay in complexation of mercury with organic ligands. Reprinted with permission from Miller et al. (2009) *Environmental Science & Technology* **43**(22): 8548–53 [55]. Copyright (2009), American Chemical Society.

The rate of formation of the product is given by:

$$\frac{d[ML^{2-n}]}{dt} = k_a[M^{2+}][L^{n-}] \tag{4.16}$$

where $k_a = K_{OS}k_{-w}$. The value of K_{OS} depends on the solution ionic strength and the charge of the cation while the rate of water loss (k_{-w}) is characteristic of a metal and can vary over many orders of magnitude. This is shown in Fig. 4.24(a) and clearly illustrates why the reactivity of Hg^{II} is much higher than that of Fe^{III} as the values of k_{-w} differ by many orders of magnitude, depending on the form of Fe^{III} in solution. Hydrolysis of the metal ion can dramatically increase the water exchange rate for Fe(III) by 3 or more orders of magnitude [1]. However, complexation to other ligands does not have a similar effect as found with hydrolyzation. In the case of Hg, it is a relatively "soft" metal and therefore does not strongly bind the water molecules, and thus their rate of release is high. The metals that tend to form complexes with OH groups are the opposite extreme and have relatively slow water exchange constants. The rates are also related to the ratio of charge to size, as shown in Fig. 4.24(a). Finally, the rate of association and release of water, or other ligands, is related to the overall reaction equilibrium constant and the rate of dissociation (k_b): $k_b = k_a/K$.

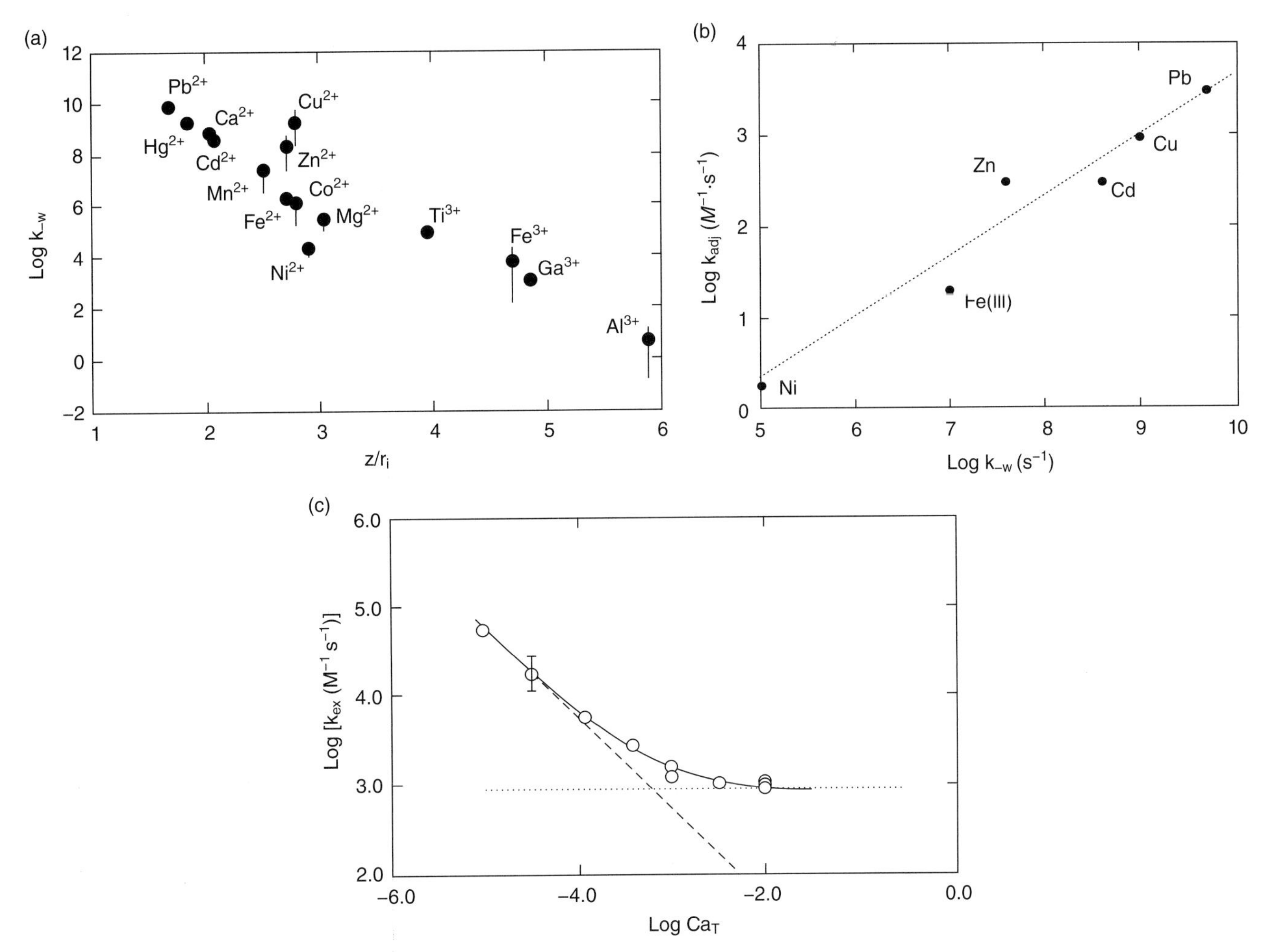

Fig. 4.24 (a) Rate of water loss (log k_{-w}) against the charge to radius ratio (Z/r_i) for a variety of metals; (b) the correlation between the measured adjunctive rate constant (k_{adj}) and the water loss rate constant for reactions involving CaEDTA complexes interacting with other metals in solution; and (c) the rate constant for the metal exchange reaction for copper (Cu) with CaEDTA as a function of the calcium concentration. Parts (a) and (b) taken from Morel and Hering (1993) *Principles and Applications of Aquatic Chemistry* and reprinted with permission of John Wiley & Sons, Inc. Part (c) reprinted from Hering and Morel, (1990) Kinetics of trace metal complexation: Ligand-exchange reactions. *Environmental Science & Technology*, **24**, 242–52 [71] with permission. Copyright (1993) American Chemical Society.

Consider now a specific ligand exchange reaction. For a ligand exchange reaction (where charges are ignored for simplicity), the metal–ligand complex, ML interacts with another ligand X, with the formation of the subsequent complex, MX. This can occur through two pathways, termed a *disjunctive* or *adjunctive* mechanism [1]. The adjunctive mechanism is what is described above when there is formation of an intermediate complex, while in a disjunctive mechanism, the complex first dissociates prior to the addition of the new ligand (X). Similar mechanisms can occur for metal exchange reactions: $M_1L + M_2 = M_2L + M_1$. Morel and Hering [1] show that there is a relationship between the rate of the adjunctive process for metal exchange reactions, as exemplified by the reaction of calcium–EDTA complexes with other cations, and the value of k_{-w} for the exchanged metal, and suggest this is because the formation of the intermediate, rather than the final dissociation reaction, is rate limiting. Various other possible interactions [1] include soluble exchange reactions when two different metal–ligand complexes exchange. The overall result depends on the specifics of the various interactions.

The importance of such processes is exemplified by the data shown in Fig. 4.24(c). Here the reaction of Cu^{II} with a mixture of EDTA and humic acid is examined. In the presence of Ca, there is initial formation of the Ca-EDTA complex and therefore the formation of the Cu-EDTA associations on the addition of Cu requires the metal exchange processes described earlier. The Cu initially complexes with the humic acid resulting in a slow equilibrium due to the necessity of a double-exchange reaction occurring. This example demonstrates the complexity of the interactions that may occur in natural waters in the presence of high concentrations of major ions such as Ca and Mg which form relatively strong complexes with NOM and other organic ligands, and the fact that there is often an inverse relationship between the strength of a binding site and its relative abundance, as discussed previously in characterizing the various models for NOM complexation with metals.

More recent studies have taken these ideas further with the derivation of a model to fit a variety of ligand exchange reactions [77]. The relationship outlined earlier is a special case and a more general equation can be derived for the rate of formation of the intermediate outer-sphere complex (k_a^{OS}), and its dissociation (k_d^{OS}), as exemplified by the reaction of a cation M^{2+} with a ligand L^-. The rate of interaction depends on the electron pair interaction energy (U^{OS}) and the mobilities of the hydrated ion and the ligand, which are related to their diffusion coefficients, respectively D_M and D_L as [77]:

$$k_a^{OS} = 4\pi N_{AV} a(D_M + D_L)(U^{OS}/(\exp(U^{OS}) - 1)) \qquad (4.17a)$$

$$k_d^{OS} = (3(D_M + D_L)/a^2)(U^{OS}\exp(U^{OS})/(\exp(U^{OS}) - 1)) \qquad (4.17b)$$

where a is the distance of closest approach and U^{OS} for a single ion pair is the product of the primary Coulombic energy and a term that corrects this value for electrostatic screening effects:

$$U^{OS} = (z_M z_L e^2/4\pi\varepsilon\varepsilon_0 akT)(1 - \kappa a/(1 + \kappa a)) \qquad (4.18)$$

where z is the numerical charge on the metal and ligand and κ is the reciprocal Debye length, and the other parameters are defined earlier. As U^{OS} approaches zero, the expressions simplify to those representing the maximum diffusive rate. The overall equilibrium constant for the formation of the intermediate is the ratio of the rate equations. For small ligands, the values of D are around $10^{-9}\,m^2s^{-1}$ and $a < 1\,nm$, so the overall interaction rate is about $10^{10}\,s^{-1}$, which is similar to the k_{-w} for Hg and Pb. For these metals, the special case outlined earlier is valid [77] that is, $k_a = K^{OS}k_{-w}$.

This model is amenable to the formation of complexes with multidentate ligands as the overall equation can be generalized as:

$$U^{OS} = (z_M ze^2/4\pi\varepsilon\varepsilon_0 akT)\sum_i^n z_i/a_i[1 - \kappa a_i/(1 + \kappa a_i)] \qquad (4.19)$$

where n is the total number of charged groups in the ligand. The situation is complicated for ligands that can be present in both protonated and unprotonated forms. If it is assumed that the protonation/deprotonation reactions of the ligand are at equilibrium and the rate is independent on whether the ligand is protonated or not, the overall rate is the sum of the rates for the various potential forms of the ligand. A similar approach can be used to examine the dissociation of multidentate ligand–metal complexes [77].

It is possible to extend this approach to the rate of interaction with colloids, specifically to model the reaction with NOM, which can be considered colloidal, as discussed earlier (based on the Donnan theory) [77, 78]. In these cases, the appropriate approach depends on the separation distance between sites on the surface and in situations where the charge density is low; the approach used for the simple ions is valid. This could occur in many instances of environmental interest. High and intermediate charge density regimes also can be found and need to be treated differently, as discussed in detail in [77].

As noted above, the complexation of ligands with major ions may impact the rate of their reaction with trace metals, and this was investigated for Fe^{III} [79]. The impact of Ca and Mg on the reaction of Fe with a variety of ligands (citrate, EDTA and fulvic acid) under conditions representative of seawater showed that there is indeed a dramatic impact as the rate constants determined ranged from 3.3×10^4 to $3.2 \times 10^6\,M^{-1}s^{-1}$. Under comparable conditions at pH 8 in carbonate buffered 0.5 M NaCl solution, the formation rate constant was highest with EDTA, then DFA (a fulvic acid extract), with the slowest rate with citrate. This is partially due to the fact that citrate is mostly as a sodium complex under these conditions (62%) while EDTA is mostly uncomplexed by Na (11%). Under these conditions, Fe is mostly hydrolyzed, with a small fraction as a carbonate complex (5%), and the rates of the reaction are higher than expected for the free ion, as discussed previously. Additionally, the rate is dependent on the type of complex formed as it is likely that the more stable $Fe(OH)_2(Cit)_2^-$ is formed compared to the 1 : 1 complexes with EDTA ($Fe(OH)_x(EDTA)$, $x = 1,2$).

To further examine the impact of major ion binding, Fujii et al. [79] investigated the impact if the ligand was first bound to a major +2 ion when Fe is added to the solution, and if the rate of reaction is different than that of a simple solution. The impact of the major ions was greatest for EDTA, as both Ca and Mg form relatively strong complexes with EDTA compared to other organic ligands. The type of reaction pathway – disjunctive or adjunctive – was found to be dependent on the relative ratio of the various cation concentrations. At low major ion concentrations (less than those typically found in natural waters), and where the ligand concentration exceeded that of the major ion, a disjunctive mechanism dominated, while at higher major ion concentrations the reaction followed an adjunctive pathway (Fig. 4.25; data for Ca). These results suggest that the adjunctive mechanisms are the most likely in natural waters ($Ca_T \sim 10^{-2}$–$10^{-4}\,M$) given the typical concentrations of ligands in solution. Modeling results suggest that the rate of reaction also decreases as the ligand concentration decreases for both

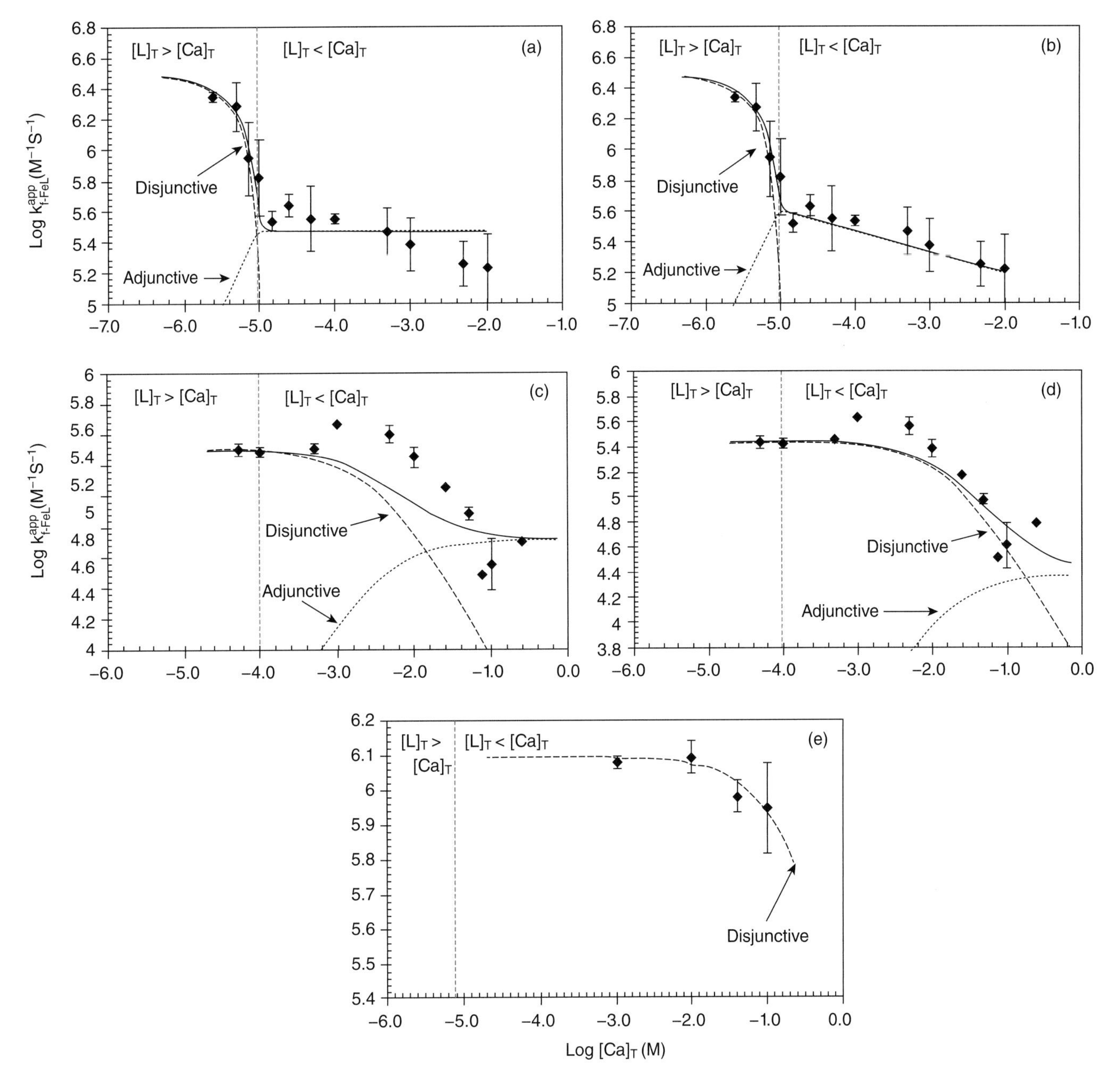

Fig. 4.25 Graphs showing the changes in the pathways of the exchange reactions of iron (Fe) and the rate constants for the reaction with changing calcium concentration. Figures show experiments in the presence of different ligands: (a) and (b) EDTA; (c) and (d) citrate; and (e) DFA. Experimental data are shown with symbols and are the same for (a) and (b), and for (c) and (d). Difference for the figures for the same ligand represent different model assumptions as described in Fujii et al. (2008) *Geochimica et Cosmochimica Acta* **72**(5): 1335–49 [79]. Reprinted with permission of Elsevier.

citrate and EDTA. At higher ligand concentrations, the disjunctive pathway becomes more important for both EDTA and citrate.

For the fulvic acid (DFA), the disjunctive pathway is dominant even at the lower experimental concentration [79]. This suggests that there is hindrance of the formation of the outer sphere complex which is consistent with the lack of sensitivity of the rate to the concentration of the major ion. The results for Fe complexation with fulvic acid in this study were comparable to previous rates measured by other investigators and these relatively high rates suggest that complexation to organic matter is likely preventing the precipitation of Fe in ocean waters, where measured concentrations are higher than those predicted based on solubility of Fe-hydroxides. This is discussed further in Chapter 6.

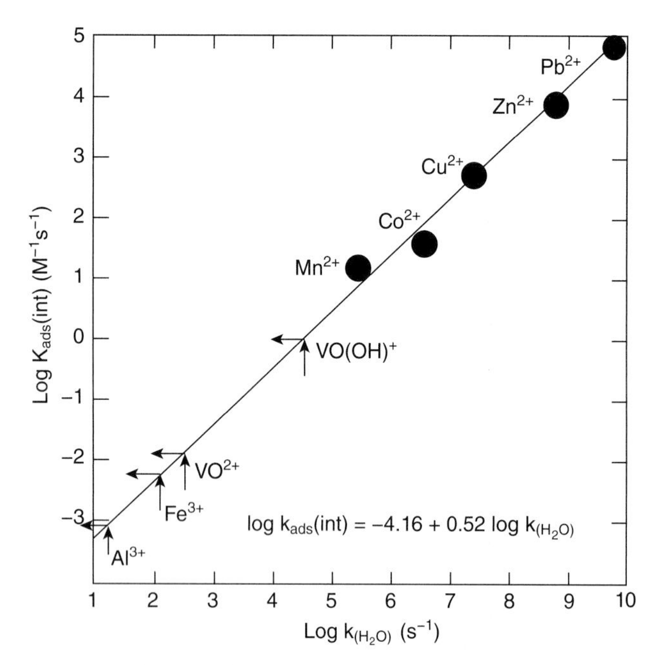

Fig. 4.26 Relationship between the water loss rate constant (log $k_{(H2O)}$ in the figure) and the rate constant for the binding of the metal to the surface (log k_{ads}) for an uncharged surface. Figure shows both data and extrapolations for ions with lower water exchange rate constants. Figure reproduced from Stumm (1992) *The Chemistry of the Solid-Water Interface* with permission of John Wiley & Sons, Inc.

4.6.2 Adsorption/desorption kinetics

The kinetics of adsorption to surfaces can be developed in a similar manner [20]:

$$\frac{d[XM]}{dt} = K_{OS}k_a[M][X] \tag{4.20}$$

where X represents the surface site for adsorption of metal, M. Here K_{OS} is equal to exp(-zψ/RT), which is approximately the same as exp(−zF^{2}Q/¢sRT) where ¢ is the capacitance of the double layer, s is the specific surface area, and Q is the accessible surface charge. The term F^{2}Q/CsRT is that used for the Constant Capacitance Model discussed previously. The value of k_a for surface adsorption is related to the value of k_{-w}, as shown in Fig. 4.26. This demonstrates that adsorption to surfaces is similar overall to that of solution reactions in that the rate of loss of water has a strong impact over the kinetics of the reaction. From the data in Fig. 4.26 (circles), a relationship was derived for logk_a:

$$\log k_a = -4.16 + 0.92 \log k_{-w} \tag{4.21}$$

Equation 4.21 can be converted to units of $M^{-1}s^{-1}$ by dividing through by the concentration of water (55.6 M). The resultant intercept value, −2.42, demonstrates that water loss reactions with surfaces occur about 250 times slower than the comparable reactions in solution. So, for the ions that have low values for k_{-w}, adsorption is predicted to be slow. However, many of these ions (Fe, Cr, Co, V, Al) are hydrolyzed in solution and the rates of their adsorption will be higher than predicted here for the free ion.

Stumm [20] worked through an example of the adsorption of Co^{2+} to 10 mg l^{-1} Al_2O_3 (s = 10^4 m^{2}kg^{-1}; Q = 0.1 mol sites kg^{-1}). For this reaction, the value of K_{OS} is 1.25×10^{-3} (ψ = 0.086 V in this example), and k_a is 3.2 M^{-1}s^{-1} from Figure 4.26. The rate of change in concentration is equivalent to 4.03×10^{-6} [Co^{2+}] s^{-1}, and therefore is slow for typical concentrations of Co in natural waters. The half-time for reaction is on the order of days rather than hours for this interaction.

Others have also discussed how the approach can be extended to the interaction of metals with surfaces [77], where the formation of the outer-sphere initial complex is related to U^{OS}. Again, in low site density cases, the approach outlined (see Equation 4.19) earlier can provide a reasonable estimate. In the other situations and in cases where the particles may consist of both a porous outer shell and an impermeable inner core the modeling approach outlined previously can be modified to consider such situations. Suffice to note that the kinetics of such interactions are complex but are readily examined using existing theory [77].

Gelabert et al. [80] examined the interaction of Cd and Pb ions with the surface of various phytoplankton to determine the mechanism of interaction at the cell surface. They concluded that the loss of a water molecule from the inner-sphere complex of the metal with the cell surface ligand was the rate determining step. They used a modified Constant Capacitance Model, in a similar manner as described above, for adsorption onto inorganic surfaces to model their results although there was not the need to invoke different site reactivity on the cell surface.

4.6.3 Uptake kinetics for microorganisms

The basis for the assimilation of most trace elements by microorganisms is the reaction of the molecule (cation or anion) with a surface site (transporter) followed by the internalization of the molecule or ion of interest as a complex, which then dissociates within the cell releasing the required ion or molecule for use by the organism. Such assimilation is essentially equivalent to the process whereby enzymes catalyze reactions, as the formation of the intermediate complex at the cell surface is often the rate limiting step. Complexation between the constituents in solution (the cellular boundary layer) and the surface sites is rapid and at equilibrium (a reversible reaction), while the internalization process is the slower reaction and irreversible. Thus, the surface complexed form reaches a steady state concentration and the reaction is controlled by the rate of the internalization process. Additionally, after internalization, the surface site is regenerated to further interact with dissolved species. The overall process can be expressed by the following reactions, where E represents the surface site, and S is the species of interest in solution, and P is the internalized species or product of the process. The two stage reaction can be expressed as:

$$E + S = ES \rightarrow P + E \quad (4.22)$$

where k_1 is the rate of formation of the surface complex and k_{-1} is the rate of dissociation, and k_2 is the rate of internalization. Therefore:

$$d[P]/dt = k_2[ES] \quad (4.23a)$$

$$d[ES]/dt = k_1[E][S] - (k_{-1} + k_2)[ES] \quad (4.23b)$$

After an initial reaction period, the concentration of [ES] becomes constant (i.e., d[ES]/dt = 0) as its concentration is controlled by the rate of dissociation. As the total number of surface sites is fixed (E_T), then [E] = E_T − [ES], and by substitution, Equation 4.23(b) can be solved to show that, at steady state:

$$[ES]_{SS} = (k_1E_T[S]/\beta) \text{ where } \beta = (k_{-1} + k_2 + k_1[S]) \quad (4.24)$$

and

$$d[P]/dt = k_2k_1E_T[S]/\beta = k_2E_T[S]/([S] + (k_{-1} + k_2)/k_1) \quad (4.25)$$

This is the *Michaelis-Menten Expression* which also applies to reactions involving enzymes, and is typically written in terms of the reaction velocity, V:

$$V = V_{max}[S]/K_M + [S] \quad (4.26)$$

where $V_{max} = k_2E_T$ and $K_M = (k_{-1} + k_2)/k_1$. If $S << K_M$, then the reaction becomes essentially first order, that is, $V = (V_{max}/K_M)[S]$. This situation is typical for the accumulation of metals by microorganisms in environmental waters as the metal concentrations are typically low (nM or lower).

Morel and Hering [1] discuss the potential limiting cases in the accumulation of cations by microorganisms. They consider the uptake of metal M in the presence of a solution ligand Y, forming a surface complex MY with the cell surface transporter site (rate constants: k_L for association; k_{-L} for dissociation; total sites, L_T). If the uptake (internalization) rate (k_{in}) is slow relative to the rate of the complexation reactions, the metal reaches pseudo-equilibrium with the both the surface site and the ligands in solution, and the internalization is the rate limiting step. Here the uptake rate depends on:

$$\text{Uptake rate} = k_{in}(K_L)L_T[M]_{SS} \quad (4.27a)$$

where $[M]_{SS}$ is the steady state metal free ion concentration, and $K_L = k_L/k_{-L}$. Overall, the uptake rate is controlled by the system thermodynamics. Alternatively, if the reaction with the surface site is essentially irreversible, then the surface site concentration is determined by the relative rate of binding compared to internalization:

$$\text{Uptake rate} = k_LL_T[M]_{SS} \quad (4.27b)$$

In this case, the uptake rate depends on the relative kinetics of the various reactions. If the dissociation of the MY complexes in solution is slow and on similar order to the irreversible uptake, then the uptake rate depends on the complexed ligand concentration rather than the free ion concentration:

$$\text{Uptake rate} = k_{-y}[MY] \quad (4.27c)$$

The first two scenarios are most relevant to environmental conditions. The uptake of metal(loid)s by microorganisms is discussed in more detail in Chapter 8.

4.7 Incorporating kinetics and thermodynamics into fate and transport modeling

The equilibrium and kinetics reactions discussed previously can be included into more comprehensive models to examine the biogeochemistry and changes in an aquatic system over time. The generalized version of the one dimensional reaction transport model [4, 81] that incorporates physical (advection and diffusion), chemical and biological processes, and how these affect the chemical composition is:

$$\frac{\partial C_i}{\partial t} = D_i\frac{\partial^2 C_i}{\partial x} + \omega\frac{\partial C_i}{\partial x} + \sum_{k=1}^{n} R_i \quad (4.28)$$

where C_i is the chemical of interest, D_i is the *diffusion coefficient* in the medium (units: $m^2 s^{-1}$), ω is the advection velocity for the system under study (units: m s^{-1}) and R_i represents the reactions that are occurring in the medium, as discussed in this chapter. The diffusion term is based on *Fick's Law* which states that the flux, $J = -D\partial C/\partial x$. the diffusion coefficient is the *molecular diffusion coefficient* in environments where the fluid movement is slow or stationary, while in faster moving systems, the value represents the hydrodynamic dispersion and is therefore considered a *dispersion coefficient*. If the flow or mixing is turbulent, then the coefficient is the *eddy diffusivity*. Molecular diffusion coefficients of simple ions in solution are ~$10^{-9} m^2 s^{-1}$.

In complex media, such as in sediment porewater and in groundwater flow, the diffusion and advection coefficients are modified by the complexity of the system, defined by the *porosity* as the flow will be constrained by the paths around the particles, and will be circuitous, and this is defined as the system's *tortuosity*, which is related to the porosity. These impacts are discussed in detail in Chapter 6. In such systems, where flow is typically in one direction, the dispersion coefficient accounts for both lateral and longitudinal dispersion. The overall dispersion can be estimated through the use of a *retardation factor*, which is related to the porosity of the system, and reflects interactions such as adsorption with the solid phase [81]. The reactions that occur need to maintain a charge balance within the system and this is an important consideration in any model.

The model is typically solved using a numerical integration typically though a finite difference approach, with a set timestep. After each timestep where the impact of transport is calculated, the effects of the reactions and equilibrium processes are then reevaluated using the equations and models described earlier. These new concentrations are then used to estimate the changes due to transport during the next timestep. The details of reaction transport modeling are beyond the scope of this book but in the subsequent chapters there will be further discussion of various applications of the general model to specific systems, such as transport and diagenetic reactions in sediments, and the transport of chemicals in groundwater. Examples will also be given for surface waters. Finally, in Chapter 8, the application of such models to chemical accumulation and trophic transport will be discussed.

4.8 Chapter summary

1. A number of computer models are available that allow the estimation of the speciation of elements in solution. The precipitation and dissolution of solids is also included in the models as is gas exchange and redox chemistry. Ionic strength effects on the equilibrium constant can be included in the calculations. These models often also include formulations for the interaction of ions with solid phases. These models are based on thermodynamic equilibrium equations and the interactions with solids are represented as a surface reaction with acidic surface sites. The influence of the surface charge on the reaction is also taken into account in these calculations.
2. Typically, calculations are made using the double layer model for the interaction with surfaces although the more complex triple layer model can also be used. There are even more detailed models that consider ion pair interactions as well as surface complexation. Often these complex models are not well constrained as the various parameters needed for the model are not available. However, a number of studies have shown their validity and applicability.
3. Models have also been developed to examine the binding of ions to large molecular weight organic matter that is not well characterized. These models again vary in complexity from taking an approach similar to that for interactions with surfaces where the organic molecule is considered to be solid. In an alternative approach the molecule is considered porous and interactions occur both on the surface and within the molecular matrix. Both models have been successfully used to examine the interaction of ions with NOM. Finally, the three-way interactions between dissolved ions, solid surfaces and NOM have been modeled using a compilation of the various approaches.
4. While most modeling relies on a thermodynamic approach, it is necessary in many situations to consider the kinetics of the processes involved and the approaches to doing these evaluations were discussed in detail. The importance of kinetics in uptake of ions into microbes was one focus of discussion.

Appendix 4.1

Appendix 4.1a

Median metal binding constants for the various surface complexes and the charge distribution for the charged complexes for metal adsorption on ferric oxides (goethite and HFO) using the CD-MUSIC approach. Standard deviations are given when appropriate. The complexes formed in the [110] plane are the first four listed while the last two are for the [021] plane. The subscripts M-1and OH-1 refer to monodentate complexes, and similarly the other subscripts refer to bidentate and tridentate associations. See the text for more details of the model. Data extracted from Ponthieu et al. [37].

Constant	Cd	Pb	Cu	Zn-Goeth	Zn-HFO	Co	Ni
$\log K_{M\text{-}1}$	7 ± 0.9	8.6 ± 0.8	8.5± 0.8	7 ± 0.6	7.2 ± 0.4	4.3	4.1
z_0/z_1	1.05/0.95	1/1	1.1/0.9	0.7/1.3	0.9/1.1	–	–
$\log M_{OH\text{-}1}$	12 ± 1	15.5 ± 1	15.5 ± 0.8	13.8 ± 0.8	13.9 ± 0.4	–	–
$\log K_{M\text{-}2}$	–	11.5 ± 0.6	12 ± 0.4	–	–	7	6.9
z_0/z_1	–	1.1/0.9	1.2/0.8	–	–	–	–
$\log M_{OH\text{-}2}$	–	17.8 ± 0.6	18.8 ± 1.2	–	–	12.7	12.4
$\log K_{M\text{-}3}$	11.5 ± 0.6	15.5 ± 1	13.2	12.5 ± 0.5	9.5 ± 0.4	11.9	11.7
z_0/z_1	1.05/0.95	1.1/0.9	1.2/0.8	0.9/1.1	0.9/1.1	–	–
$\log M_{OH\text{-}3}$	14.5 ± 0.9	19.5 ± 0.7	19.2	18 ± 0.8	15.8 ± 0.4	15.8	15.4

Appendix 4.1b

Fitting parameters for the WHAM Model VI for the various cations modeled. Values are given for the intrinsic constants for the association of the metals with the dominant (strong) metal binding sites (K_{MA}) of fulvic (FA) and humic (HA) acids, and the fitting parameter (ΔLK_2), which defines the spread of values around the medians. The weaker sites are related to the stronger sites by the following relationship: $\log K_{MB} = 3.39$: $\log K_{MA} - 1.15$. A standard deviation is given if the value is based on two or more datasets. Values are compiled and summarized from Tipping [13, 51].

Parameter	Mg	Ca	Sr	Ba	Al	Mn	Fe(II)	Fe(III)	Co	Ni
ΔLK_2	0.12	0	0	0	0.46	0.58	0.81	2.20	1.22	1.57
Log K_{MA} FA	1.1	1.3	1.2	0.6	2.5	1.7	1.6	2.4	1.4	1.4
std dev	0.3	0.3	–	–	0.1	–	–	–	0.2	0.3
Log K_{MA} HA	0.7	0.7	1.11	-0.2	2.6	0.6	1.3	2.5	1.1	1.1
std dev	0.4	0.3	–	–	0.2	–	–	–	0.1	–

Parameter	Cu	Zn	Cd	Pb	Hg	CH_3Hg	Th	UO_2^{+}	Cr(III)	VO^{2+}
ΔLK_2	2.34	1.28	1.48	0.93	5.1	3.6	0.23	1.16	1.97	1.74
Log K_{MA} FA	2.1	1.6	1.6	2.2	3.1	0.3	2.7	2.1	2.2	2.4
std dev	0.2	0.2	0.1	0.2	0.4	–	0.3	0	–	–
Log K_{MA} HA	2.0	1.5	1.3	2.0	3.6	0.3	2.8	2.2	2.2	2.5
std dev	0.2	0.2	0.1	0.2	0.6	–	0	0.3	–	–

References

1. Morel, F.M.M. and Hering, J.G. (1993) *Principles and Applications of Aquatic Chemistry*. John Wiley & Sons, Inc., New York.
2. Westall, J.C., Zachary, J.L., and Morel, F.M. (1976) MINEQL, a computer program for the calculation of chemical equilibrium composition of aqueous system. Cambridge, MA., Dept. Civil Eng., MIT.
3. Bruno, J., Duro, L., and Grive, M. et al. (2002) The applicability and limitations of thermodynamic geochemical models to simulate trace element behaviour in natural waters: Lessons learned from natural analogue studies. *Chemical Geology*, **190**, 371–393.
4. Zhu, C. and Anderson, G. (2002) *Environmental Applications of Geochemical Modeling*. Cambridge University Press, Cambridge.
5. Paquin, P.R., Farley, K., Santore, C.D. et al. (2003) *Metals in Aquatic Systems: A Review of Exposure, Bioaccumulation and Toxicity Models*. SETAC Press, Pensacola, FL, USA.
6. MINEQL+ (2007) •Chemical Equilibium Modeling System. Available at http://mineql.com/ (accessed October 22, 2012).
7. Felmy, A.R., Girvin, D.C., Girvin, D.C., and Jenne, E.A. et al. (1984) *MINTEQ-A Computer Program for Calculating Aqueous Geochemical Equilibria*. U.S. Environmental Protection Agency, Athens, GA.
8. Peterson, S.R., Hostetler, C.J., Deutsch, W.J., and Cowan, C.E. et al. (1987) *MINTEQ User's Manual*. Pacific Northwest Laboratory, Richland, USA.
9. Parkhurst, D.L. (1995) User's guide to PHREEQC – a computer model for speciation, reaction-path, advective-transport and inverse geochemical modeling. USGS Water Resources Investigations USGS: 95-4227.
10. Parkhurst, D.L. (1997) Geochemical mole-balance modeling with uncertain data. *Water Resources Research*, **33**, 1957–1970.
11. Ball, J.W. and Nordstrom, D.K. (1991) User's manual for WATEQ4F, with revised thermodynamic data base and test cases for calculating speciation of major, trace and redox elements in natural waters. Open-file report, USGS.
12. Bethke, C.M. (1996) *Geochemical Reaction Modeling*. Oxford University Press, New York.
13. Tipping, E. (1998) Humic ion-binding model VI: An improved description of the interactions of protons and metal ions with humic substances. *Aquatic Geochemistry*, **4**(1), 3–48.
14. Kinniburgh, D.G., van Riemsdijk, W.H., Koopal, L.K., et al. (1999) Ion binding to natural organic matter: Competition, heterogeneity, stoichiometry and thermodynamic consistency. *Colloids and Surfaces A – Physicochemical and Engineering Aspects*, **151**(1-2), 147–166.
15. Van Der Lee, J. (1998) *Thermodynamic and mathematical concepts of CHESS*. CIG Ecole des Mines de Paris, Fontainebleu, France.
16. Stumm, W. and Morgan, J.J. (1996) *Aquatic Chemistry*. John Wiley & Sons, Inc., New York.
17. Dzomback, D.A. and Morel, F.M.M. (1990) *Surface Complexation Modeling: Hydrous Ferric Oxide*. John Wiley & Sons, Inc., New York.
18. Langmuir, D. (1997) *Aqueous Environmental Geochemistry*. Prentice Hall, Upper Saddle River.
19. Davis, J.A. and Leckie, J.O. (1978) Effect of adsorbed complexing ligands on trace metal uptake by hydrous oxide. *Environmental Science & Technology*, **12**, 1309–1315.
20. Stumm, W. (1992) *The Chemistry of the Solid-Water Interface*. John Wiley & Sons, Inc., New York.
21. Jang, J.H., Dempsey, B.A., and Burgos, W.D. et al. (2007) A model-based evaluation of sorptive reactivities of hydrous ferric oxide and hematite for U(VI). *Environmental Science & Technology*, **41**(12), 4305–4310.
22. Wilkie, J.A. and Hering, J.G. (1996) Adsorption of arsenic onto hydrous ferric oxide: Effects of adsorbate/adsorbent ratios and co-occurring solutes. *Colloids and Surfaces A-Physicochemical and Engineering Aspects*, **107**, 97–110.
23. Bryce, A.L., Kornicker, W.A., Elzerman, A.W. and Clark, S.B. et al. (1994) Nickel adsorption to hydrous ferric-oxide in the presence of EDTA – effects of component addition sequence. *Environmental Science & Technology*, **28**(13), 2353–2359.
24. Romero-Gonzalez, M.R., Cheng, T., Barnett, M.O. and Roden, E.E. et al. (2007) Surface complexation modeling of the effects of phosphate on uranium(VI) adsorption. *Radiochimica Acta*, **95**(5), 251–259.
25. Khalil, L.B., Alaya, M.N., Petro, N.S., and Abo Elenein, R.M.M. et al. (2002) Changes in the porous texture of hydrous ferric

oxide on adsorption of transition metal ions: Adsorption mechanism. *Adsorption Science & Technology*, **20**(5), 501–509.
26. Jang, J.H., Dempsey, B.A., Catchen, G.L., and Burgos, W.D. et al. (2003) Effects of Zn(II), Cu(II), Mn(II), Fe(II), NO3-, or SO42- at pH 6.5 and 8.5 on transformations of hydrous ferric oxide (HFO) as evidenced by Mossbauer spectroscopy. *Colloids and Surfaces A – Physicochemical and Engineering Aspects*, **221**(1–3), 55–68.
27. Ainsworth, C.C., Pilon, J.L., Gassman, P.L., and Van der Sluys, W.G. et al. (1994) Cobalt, cadmium, and lead sorption to hydrous iron-oxide – residence time effect. *Soil Science Society of America Journal*, **58**(6), 1615–1623.
28. Davranche, M. and Bollinger, J.C. (2000) Release of metals from iron oxyhydroxides under reductive conditions: Effect of metal/solid interactions. *Journal of Colloid and Interface Science*, **232**(1), 165–173.
29. Karthikeyan, K.G. and Elliott, H.A. (1999) Surface complexation modeling of copper sorption by hydrous oxides of iron and aluminum. *Journal of Colloid and Interface Science*, **220**(1), 88–95.
30. Landry, C.J., Koretsky, C.M., Lund, T.J., and Das, S. et al. (2009) Surface complexation modeling of Co(II) adsorption on mixtures of hydrous ferric oxide, quartz and kaolinite. *Geochimica et Cosmochimica Acta*, **73**(13), 3723–3737.
31. Sverjensky, D.A. (2003) Standard states for the activities of mineral surface sites and species. *Geochimica et Cosmochimica Acta*, **67**(1), 17–28.
32. Sverjensky, D.A. (2005) Prediction of surface charge on oxides in salt solutions: Revisions for 1 : 1 (M+L) electrolytes. *Geochimica et Cosmochimica Acta*, **69**(2), 225–257.
33. Sverjensky, D.A., Jonsson, C.M., Jonsson, C.L., et al. (2008) Glutamate surface speciation on amorphous titanium dioxide and hydrous ferric oxide. *Environmental Science & Technology*, **42**(16), 6034–6039.
34. Sahai, N. and Sverjensky, D.A. (1997) Evaluation of internally consistent parameters for the triple-layer model by the systematic analysis of oxide surface titration data. *Geochimica et Cosmochimica Acta*, **61**(14), 2801–2826.
35. Merdy, P., Huclier, S., and Koopal, L.K. et al. (2006) Modeling metal-particle interactions with an emphasis on natural organic matter. *Environmental Science & Technology*, **40**(24), 7459–7466.
36. Hiemstra, T. and Van Riemsdijk, W.H. (1996) A surface structural approach to ion adsorption: The charge distribution (CD) model. *Journal of Colloid and Interface Science*, **179**(2), 488–508.
37. Ponthieu, M., Juillot, F., Hiemstra, T., et al. (2006) Metal ion binding to iron oxides. *Geochimica et Cosmochimica Acta*, **70**(11), 2679–2698.
38. Venema, P., Hiemstra, T., and Van Riemsdijk, W.H. et al. (1996) Multisite adsorption of cadmium on goethite. *Journal of Colloid and Interface Science*, **183**(2), 515–527.
39. Weerasooriya, R., Aluthpatabendi, D., and Tobschall, H.J., et al. (2001) Charge distribution multi-site complexation (CD-MUSIC) modeling of Pb(II) adsorption on gibbsite. *Colloids and Surfaces A – Physicochemical and Engineering Aspects*, **189**(1–3), 131–144.
40. Weng, L.P., Koopal, L.K., Hiemstra, T., et al. (2005) Interactions of calcium and fulvic acid at the goethite-water interface. *Geochimica et Cosmochimica Acta*, **69**(2), 325–339.
41. Benedetti, M.F. (2006) Metal ion binding to colloids from database to field systems. *Journal of Geochemical Exploration*, **88**(1–3), 81–85.
42. Rietra, R., Hiemstra, T., and Van Riemsdijk, W.H. et al. (2001) Comparison of selenate and sulfate adsorption on goethite. *Journal of Colloid and Interface Science*, **240**(2), 384–390.
43. Weerasooriya, R., Tobschall, H.J., Wijesekara, H.K.D.K., et al. (2003) On the mechanistic modeling of As(III) adsorption on gibbsite. *Chemosphere*, **51**(9), 1001–1013.
44. Antelo, J., Avena, M., Fiol, S., et al. (2005) Effects of pH and ionic strength on the adsorption of phosphate and arsenate at the goethite-water interface. *Journal of Colloid and Interface Science*, **285**(2), 476–486.
45. Filius, J.D., Hiemstra, T., and Van Riemsdijk, W.H. et al. (1997) Adsorption of small weak organic acids on goethite: Modeling of mechanisms. *Journal of Colloid and Interface Science*, **195**(2), 368–380.
46. Wolthers, M., Charlet, L., and Van Cappellen, P. et al. (2008) The surface chemistry of divalent metal carbonate minerals; a critical assessment of surface charge and potential data using the charge distribution multi-site ion complexation model. *American Journal of Science*, **308**(8), 905–941.
47. Jing, C.Y., Meng, X.G., Liu, S., et al. (2005) Surface complexation of organic arsenic on nanocrystalline titanium oxide. *Journal of Colloid and Interface Science*, **290**(1), 14–21.
48. Christl, I., Metzger, A., Heidmann, I., and Kretzschmar, R. et al. (2005) Effect of humic and fulvic acid concentrations and ionic strength on copper and lead binding. *Environmental Science & Technology*, **39**(14), 5319–5326.
49. Koopal, L.K., van Riemsdijk, W.H., and Kinniburgh, D. et al. (2001) Humic matter and contaminants. General aspects and modeling metal ion binding. *Pure and Applied Chemistry*, **73**(12), 2005–2016.
50. Benedetti, M.F., Milne, C.J., Kinniburgh, D., et al. (1995) Metal-ion binding to humic substances – application of the nonideal competitive adsorption model. *Environmental Science & Technology*, **29**(2), 446–457.
51. Tipping, E. (2007) Modelling the interactions of Hg(II) and methylmercury with humic substances using WHAM/Model VI. *Applied Geochemistry*, **22**(8), 1624–1635.
52. Haitzer, M., Aiken, G.R., and Ryan, J.N., et al. (2002) Binding of mercury(II) to dissolved organic matter: The role of the mercury-to-DOM concentration ratio. *Environmental Science & Technology*, **36**(16), 3564–3570.
53. Ravichandran, M. (2004) Interactions between mercury and dissolved organic matter – a review. *Chemosphere*, **55**(3), 319–331.
54. Smith, D.S., Bell, R.A., and Kramer, J.R., et al. (2002) Metal speciation in natural waters with emphasis on reduced sulfur groups as strong metal binding sites. *Comparative Biochemistry and Physiology C-Toxicology & Pharmacology*, **133**(1–2), 65–74.
55. Miller, C.L., Southworth, G., Brooks, S., et al. (2009) Kinetic controls on the complexation between mercury and dissolved organic matter in a contaminated environment. *Environmental Science & Technology*, **43**(22), 8548–8553.
56. Hamilton-Taylor, J., Postill, A.S., Tipping, E., and Harper, M.P., et al. (2002) Laboratory measurements and modeling of metal-

humic interactions under estuarine conditions. *Geochimica et Cosmochimica Acta*, **66**(3), 403–415.

57. Tipping, E., Lofts, S., and Lawlor, A.J., et al. (1998) Modelling the chemical speciation of trace metals in the surface waters of the Humber system. *Science of the Total Environment*, **210**(1–6), 63–77.
58. Atalay, Y.B., Carbonaro, R.F., and DiToro, D.M. et al. (2009) Distribution of proton dissociation constants for model humic and fulvic acid molecules. *Environmental Science & Technology*, **43**(10), 3626–3631.
59. Meylan, S., Odzak, N., Behra, R., and Sigg, L. et al. (2004) Speciation of copper and zinc in natural freshwater: Comparison of voltammetric measurements, diffusive gradients in thin films (DGT) and chemical equilibrium models. *Analytica Chimica Acta*, **510**(1), 91–100.
60. Guthrie, J.W., Hassan, N.M., Salam, M.S.A., et al. (2005) Complexation of Ni, Cu, Zn, and Cd by DOC in some metal-impacted freshwater lakes: A comparison of approaches using electrochemical determination of free-metal-ion and labile complexes and a computer speciation model, WHAM V and VI. *Analytica Chimica Acta*, **528**(2), 205–218.
61. Nolan, A.L., McLaughlin, M.J., and Mason, S.D., et al. (2003) Chemical speciation of Zn, Cd, Cu, and Pb in pore waters of agricultural and contaminated soils using Donnan dialysis. *Environmental Science & Technology*, **37**(1), 90–98.
62. Tipping, E., Rieuwerts, J., Pan, G., et al. (2003) The solid-solution partitioning of heavy metals (Cu, Zn, Cd, Pb) in upland soils of England and Wales. *Environmental Pollution*, **125**(2), 213–225.
63. Almas, A.R., Lofts, S., Mulder, J., and Tipping, E., et al. (2007) Solubility of major cations and Cu, Zn and Cd in soil extracts of some contaminated agricultural soils near a zinc smelter in Norway: Modelling with a multisurface extension of WHAM. *European Journal of Soil Science*, **58**(5), 1074–1086.
64. Balistrieri, L.S. and Blank, R.G. (2008) Dissolved and labile concentrations of Cd, Cu, Pb, and Zn in the South Fork Coeur d'Alene River, Idaho: Comparisons among chemical equilibrium models and implications for biotic ligand models. *Applied Geochemistry*, **23**(12), 3355–3371.
65. Gopalapillai, Y., Chakrabarti, C.L., and Lean, D.R.S., et al. (2008) Assessing toxicity of mining effluents: Equilibrium- and kinetics-based metal speciation and algal bioassay. *Environmental Chemistry*, **5**(4), 307–315.
66. Sondergaard, J., Elberling, B., and Asmund, G., et al. (2008) Metal speciation and bioavailability in acid mine drainage from a high Arctic coal mine waste rock pile: Temporal variations assessed through high-resolution water sampling, geochemical modelling and DGT. *Cold Regions Science and Technology*, **54**(2), 89–96.
67. Pourret, O., Davranche, M., Gruau, G., and Dia A., et al. (2007) Competition between humic acid and carbonates for rare earth elements complexation. *Journal of Colloid and Interface Science*, **305**(1), 25–31.
68. Bruland, K.W., Rue, E.L., Donat, J.R., et al. (2000) Intercomparison of voltammetric techniques to determine the chemical speciation of dissolved copper in a coastal seawater sample. *Analytica Chimica Acta*, **405**(1–2), 99–113.
69. Weng, L.P., Van Riemsdijk, W.H., Koopal, L.K., et al. (2006b) Ligand and Charge Distribution (LCD) model for the description of fulvic acid adsorption to goethite. *Journal of Colloid and Interface Science*, **302**(2), 442–457.
70. Van Riemsdijk, W.H., Koopal, L.K., Kinniburgh, D.G., et al. (2006) Modeling the interactions between humics, ions, and mineral surfaces. *Environmental Science & Technology*, **40**(24), 7473–7480.
71. Hering, J.G. and Morel, F.M.M. (1990) Kinetics of trace metal complexation: Ligand-exchange reactions. *Environmental Science & Technology*, **24**, 242–252.
72. Louis, Y., Garnier, C., Lenoble, V., et al. (2009) Kinetic and equilibrium studies of copper-dissolved organic matter complexation in water column of the stratified Krka River estuary (Croatia). *Marine Chemistry*, **114**(3–4), 110–119.
73. Bligh, M.W. and Waite, T.D. (2010) Formation, aggregation and reactivity of amorphous ferric oxyhydroxides on dissociation of Fe(III)-organic complexes in dilute aqueous suspensions. *Geochimica et Cosmochimica Acta*, **74**(20), 5746–5762.
74. Lamborg, C.H., Tseng, C.M., Fitzgerald, W.F., et al. (2003) Determination of the mercury complexation characteristics of dissolved organic matter in natural waters with "reducible Hg" titrations. *Environmental Science & Technology*, **37**(15), 3316–3322.
75. Morel, F.M.M., Reuter, J.G.M., Anderson, D.M., and Guillard, R.R.L., et al. (1978) AQUIL: A chemically defined phytoplankton culture medium for trace metal studies. *Journal of Phycology*, **15**, 135–141.
76. Jones, A.M., Pham, A.N., Collins, R.N., et al. (2009) Dissociation kinetics of Fe(III)- and Al(III)-natural organic matter complexes at pH 6.0 and 8.0 and 25 degrees C. *Geochimica et Cosmochimica Acta*, **73**(10), 2875–2887.
77. van Leeuwen, H. and Buffle, J. (2009) Chemodynamics of aquatic metal complexes: From small ligands to colloids. *Environmental Science & Technology*, **43**(19), 7175–7183.
78. van Leeuwen, H., Town, R.M., and Buffle, J., et al. (2007) Impact of ligand protonation on eigen-type metal complexation kinetics in aqueous systems. *Journal of Physical Chemistry A*, **111**(11), 2115–2121.
79. Fujii, M., Rose, A.L., Waite, T.D., and Omura, T., et al. (2008) Effect of divalent cations on the kinetics of Fe(III) complexation by organic ligands in natural waters. *Geochimica et Cosmochimica Acta*, **72**(5), 1335–1349.
80. Gelabert, A., Pokrovsky, O.S., Schott, J., et al. (2007) Cadmium and lead interaction with diatom surfaces: A combined thermodynamic and kinetic approach. *Geochimica et Cosmochimica Acta*, **71**(15), 3698–3716.
81. Drever, J.I. (1997) *The Geochemistry of Natural Waters*, 3rd edn. Prentice Hall, New Jersey, 436 pp.
82. Weng, L.P., Van Riemsdijk, W.H., Koopal, L.K., et al. (2006) Adsorption of humic substances on goethite: Comparison between humic acids and fulvic acids. *Environmental Science & Technology*, **40**(24), 7494–7500.

Problems

4.1. Examine the changing speciation of mercury and cadmium across a salinity gradient from zero to 35 in the presence of inorganic ions. How would the speciation change if there was 10^{-8} M cysteine in the waters?

4.2. You want to do some phytoplankton culture experiments with Cu and Cd and you talk to your professor who gives you this formula for a seawater culture medium: total carbonate = 2.4×10^{-3} M, total SO_4 = 3×10^{-2} M, total Cl = 0.5 M, total Na = 0.5 M, total K = 1×10^{-2} M, total Ca = 1×10^{-2} M and total Mg = 5×10^{-2} M, pH = 8.2 containing $EDTA_T = 1 \times 10^{-5}$ M. The EDTA of course is added to buffer the concentration of the metals in solution. You want to add Fe as an essential nutrient and your professor suggests adding $Fe(III)_T = 1 \times 10^{-7}$ M. For your experiments you wish to have around 10 pM of Cu and Cd as the free metal ions. Your professor does some calculations in his head, mumbling incoherently in a fashion you cannot follow, and then says you should add $Cu_T = 5 \times 10^{-10}$ and $Cd_T = 10^{-10}$ M. You want to check the medium for consistency and also the suggested metal concentrations.

a. Firstly you determine the speciation and free ion concentrations of EDTA, Fe, Cu and Cd. What are they? Was the professor right in terms of Cd and Cu? What is the dissolved inorganically-complexed Fe concentration? Is EDTA effective in its role as buffering the metals in solution?

b. What would be the impact of changing the EDTA concentration over a range of 10^{-3} to 5×10^{-5} M? Discuss the effect on the speciation of Fe, Cd, Cu, Ca, and Mg.

4.3. Precipitation of calcium carbonate can occur in lakes in late summer due to the increased pH of the system that results from enhanced primary productivity in association with increased temperature and water column stratification. For a lake with the following surface water characteristics: Ca = K = 10^{-3} M; Mg = 5×10^{-4}; Na = 10^{-4} M; sulfate = 10^{-3} M and Cl = 10^{-4} M.

a. What is the pH of calcite precipitation? What solid phase(s) forms?

b. Would the Mg concentration be an important factor? What would be the effect of changing the Mg_T concentration from 5×10^{-3} to 10^{-5} M on the Ca^{2+} concentration? Why is this?

4.4. Fit the data which shows the impact of Pic on the adsorption of copper to iron oxide by adjusting the binding constant for the following reaction:

$$Cu^{2+} + Pic^{-} = CuPic^{+}$$

The chapter discussed the impact of ligands on the adsorption of Cu to amorphous iron oxide. Using the experimental data discussed, estimate:

a. the required binding constant for picolinate with Fe oxide necessary to reproduce the data. The pK_a for this acid is 4.5. Use the experimental conditions of $Fe_T = 6.9 \times 10^{-4}$ M, $Pic_T = 4 \times 10^{-5}$ M. Assume that it binds to the surface of the Fe oxide in the following manner:

$$\equiv FeOH + HPic = \equiv FePic + H_2O \quad K = ?$$

Use the experimental conditions – $Fe_T = 6.9 \times 10^{-4}$ M, $Pic_T = 4 \times 10^{-5}$ M, $Cu_T = 10^{-6}$ M.

b. Fit the data which shows the impact of Pic on the adsorption of copper to iron oxide by adjusting the binding constant for the following reaction:

$$Cu^{2+} + Pic^{-} = CuPic^{+}$$

4.5. The data shown in Fig. 3.10 indicate that there are large differences in the speciation of copper between surface and subsurface waters. Estimate what the binding constant for an organic ligand complex would need to be to account for the speciation in the surface waters. Assume that the ligand is a carboxylic acid molecule of the formula RCOOH with an acid dissociation constant, pK_a = 5.8 and a concentration of 2 nM. For the deep waters, what would the binding constant need to be to account for the speciation at 400 m, assuming that the ligand is present at 1 nM in concentration?

4.6. Estimate the solubility of silver and mercury in water at pH = 7 and with the typical concentrations of the major ions found in freshwater

4.7. Determine the effective binding constant for Cu and NOM to alumina that would result in the differences shown in Fig. 3.18.

CHAPTER 5
Metal(loid)s in the atmosphere and their inputs to surface waters

5.1 Introduction

Atmospheric transport and dispersion of metals and metalloids is an important mechanism for their redistribution at the Earth's surface, especially for those elements that are stable as volatile species, such as methylated compounds or hydrides. Furthermore, for many metal(loid)s, atmospheric input is the dominant source to aquatic systems, especially for large lakes and the ocean. In addition, much of the input from riverine sources is derived from atmospheric inputs that have been processed through the watershed. The global mass balances discussed in Chapter 2 provide a clear example of the importance of atmospheric dispersion of metal(loid)s from sources. The release of metal(loid)s into the atmosphere as a result of human activity has dramatically increased the atmospheric burden and the concentration of many elements in remote locations, as outlined earlier (Table 2.4). Many metal(loid)s are dispersed through the atmosphere mostly in the particulate phase. For most of the elements discussed in this book, natural sources, such as dust and volcanic eruptions, do not dominate over anthropogenic sources. Additionally, close to sources, *dry deposition* (i.e., the flux of particle and gas phase species to the surface) of metal(loid)s likely dominates over *wet deposition* (rain, snow etc.), while in remote locations, wet deposition is probably the more important flux [1].

In this section the main processes involved with the exchange of metal(loid)s between the Earth's surface and the atmosphere, primarily through wet and dry deposition and gas exchange, will be discussed (Fig. 5.1). The aqueous chemistry of some of the important metal(loid)s in the atmosphere, in clouds, fog, and in wet deposition, will be discussed as well as the inputs of atmospheric metal(loid)s and their photochemistry in surface waters. The focus of the discussion will be the *troposphere*, which is the lower region of the atmosphere (<10 km), and therefore the region that mostly interacts with the remainder of the biosphere. In considering the exchange of material between the atmosphere and the Earth's surface, many investigators often separate the processes of dry deposition and gas exchange, but the theory behind the estimation of the rates of these processes is similar. Overall, the sections will deal with the processes of deposition with discussion of the factors that influence these processes, as well as a presentation of representative data to indicate the magnitude of the concentration and deposition fluxes, spatial and temporal variability, and to further illustrate the impact of anthropogenic activities on these processes.

5.2 Atmospheric transport and deposition

5.2.1 Dry deposition

Dry deposition refers to the exchange of material and elements between the atmosphere and the surface (land or water) and covers a number of processes. Particulate material in the atmosphere can be deposited and the rate of deposition is a function of parameters such as the wind speed and turbulence of the atmospheric boundary layer, and the particulate diameter. Larger particles are generally deposited more rapidly that small particles, although the relationship is not linear for ultrafine particles, and besides size, density and shape are also important parameters [1, 2].

Trace Metals in Aquatic Systems, First Edition. Robert P. Mason.

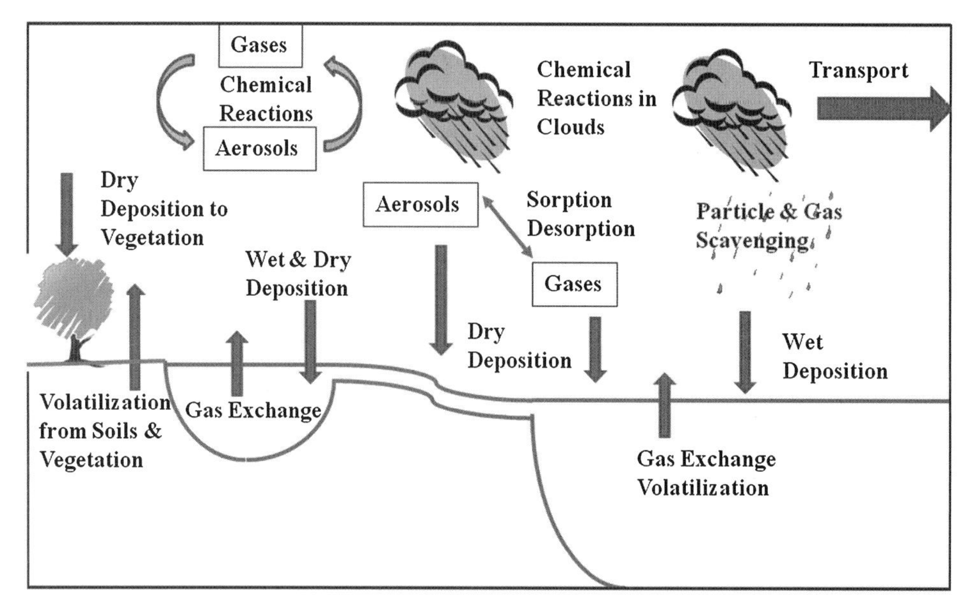

Fig. 5.1 Conceptual diagram showing the main processes involved in the exchange of materials between the atmosphere and the surface of the biosphere (water and terrestrial environments) through wet and dry deposition and as a result of gas exchange processes.

Natural particles, such as dust and sea spray, tend to be larger than those derived from anthropogenic sources, such as high temperature combustion (coal, oil burning, waste incineration). Any species in the gas phase can also be deposited to surfaces and the rate of deposition is a function of the chemistry and reactivity of the gaseous species, as well as the atmospheric physical conditions. For deposition to a water (wet) surface, the solubility of the gas is an important parameter. Overall, for all surfaces, the surface characteristics, such as roughness, or the presence of vegetation, have an important impact on the rate of deposition [1]. The overall mechanisms involved in atmospheric deposition to aquatic systems are depicted in Fig. 5.1 [3].

Dry deposition can also refer to the uptake of gaseous species by vegetation, and this has been shown to be an important process for Hg uptake into plants [4]. In addition, gas exchange of volatile species across the air-water interface can occur with the direction of exchange being a function of the relative concentration difference between the volatile species in water compared to the air. Finally, it is also possible for species present as a gas in the pore spaces of soils to be exchanged with the atmosphere, and again, the direction of exchange is a function of the relative differences in concentration. Gas evasion from a surface can be thought of as negative deposition.

While a detailed discussion of the parameters and methods of estimating dry deposition of metals and metalloids is beyond the scope of this book, a brief account of the background theory will be given. The reader is referred to Seinfeld and Pandis [1] (*Atmospheric Chemistry and Physics*) and other similar texts [2], and to review articles, such as [5] for details. The flux due to dry deposition is typically calculated based on the concentration of the species of interest in the gas and/or particulate phase, C, at some height above the surface, and the appropriate *deposition velocity*, υ_d [1]:

$$\text{Flux, } F = -\upsilon_d \cdot C \tag{5.1}$$

The flux is given in units of concentration per time per surface area, and thus the units of υ_d are length per time, as a velocity. The deposition velocity term incorporates all the complexities and mechanisms involved in the transfer of a gas or particle between the atmosphere and the surface of interest. These can be broken down into three categories: (1) the *aerodynamic transport* to the atmospheric boundary; (2) the transport of the species through the *stagnant layer* to the surface itself, by diffusion for gaseous species, by Brownian motion for particles; and (3) the uptake by the surface (irreversible or reversible absorption for gases; adhesion for particles) [1]. These considerations arise from the basics of fluid dynamics and the assumption that there is not air movement at the air-solid surface and therefore that there is a boundary layer in which transport is via diffusive processes. Turbulent diffusion controls the rate of transport from the bulk air to the stagnant sublayer, whose thickness is millimeters, and where the air is temporarily stationary (no-slip surface condition assumed). The stationary layer exists at the surface, which in most cases is not flat over a large scale due to the presence of ocean waves, and for terrestrial waters, the presence of buildings, trees and other obstacles on the shoreline and/or which extend over the water surface. The surface roughness length scale defines the extent of interaction and the importance of the surface conditions in deposition. Moisture content of the surface can also have a large effect on absorption and adhesion, especially for gases. The surface roughness can play a role in dry deposition as particles in the stagnant layer can be deposited due to sedimentation, interception, impaction, in addition to the diffusive processes.

Based on this, a *resistance model* concept to describing dry deposition has been developed [1], which draws analogy from the conceptualization of electrical processes (velocity (current) and resistance). In this conceptualization, the three processes noted above contribute to the overall resistance to deposition. For gases:

$$1/\upsilon_d = \text{total resistance, } r_t = r_a + r_b + r_c \tag{5.2}$$

where r_a is the *aerodynamic resistance*; r_b is the *quasi-laminar layer resistance* and r_c is the *canopy resistance*. For particles, it is assumed that a particle that contacts the surface is removed and thus r_c is zero. In addition, however, the *settling rate* of the particle (υ_s) needs to be included in the formulation and the deposition velocity for particles is given by [1]:

$$\upsilon_d = 1/r_t = (r_a + r_b + r_a r_b \upsilon_s)^{-1} + \upsilon_s \tag{5.3}$$

The turbulent intensity determines the rate of transfer from the bulk solution to the surface and this is the same for gases and particles, and is independent of the chemistry of the constituent. It depends on the atmospheric stability, which is influenced by wind speed, temperature, and the surface roughness. The atmosphere is often well-mixed during the day so the resistance is low and there is an ample reservoir of material to be deposited and dry deposition rates are relatively high. At night, when the atmospheric boundary layer is more stable, the dry deposition is less and the boundary layer can become depleted due to deposition to the surface, and the lack of mixing of the air. For quasi-laminar resistance, it can be shown for gases that the deposition velocity is related to the *Schmidt number* (Sc) for the species of interest, the concentration gradient across the layer, and the *friction velocity* (u_*). The Schmidt number is the ratio of the diffusivity (diffusion coefficient) to the kinematic viscosity of the fluid phase.

The surface and canopy resistance for gases depends to a large degree on the surface moisture content (especially for soluble gases) and other chemical characteristics that may enhance reaction and uptake (e.g., pH), and is typically low for water surfaces. For most metal(loid)s, gaseous dry deposition is not important but this is not the case for Hg, which can exist in the atmosphere as both elemental Hg (Hg^0) and as gaseous ionic species, such as $HgCl_2$ [6, 7]. It is estimated that the dry deposition velocity of these gaseous ionic Hg species is as high as that found for other reactive gases, such as nitric acid (0.5–$2\,cm\,s^{-1}$). In contrast, the dry deposition velocity of Hg^0 is typically much lower ($<0.1\,cm\,s^{-1}$). Other metal(loid)s found in the gas phase are Se, primarily as dimethylselenide [$(CH_3)_2Se$], and methylated compounds of As, Pb, Sn and Hg [8–10].

For particles, transfer through this layer depends on particle size and in fact, due to inertial effects, the resistance to transport through this layer is higher for larger particles so that there is a minimum in the particle size range of 0.1–1 μm where the resistance is the smallest. The gravitational settling rate, υ_s, is determined by Stokes Law and increases with increasing particle size, and is the dominant term in Equation 5.3 for large particles, above a few μm in diameter. This is shown by the relationship between deposition velocity and particle diameter in Fig. 5.2 [1]. While small particles (<0.05 μm) are transported through the quazi-laminar layer by Brownian diffusion, the larger particles possess enough inertia that their transport is not that of random diffusion, and this may enhance or even retard the deposition of the particles. A minimum in the value of υ_d is therefore found

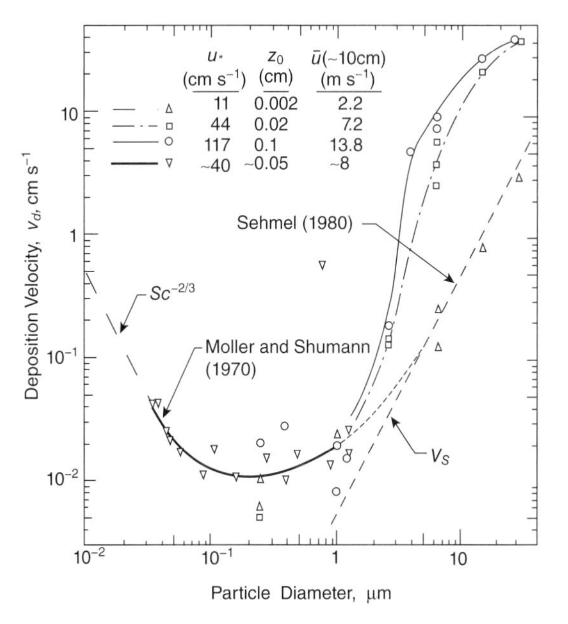

Fig. 5.2 Relationship between dry deposition velocity for particles and the particle diameter, derived from a number of studies (as shown by the symbols) and a number of parameterizations. The dotted line labeled v_s reflects the deposition velocity due to Stokes Law. Reprinted from Seinfeld and Pandis (1998), *Atmospheric Chemistry and Physics* [1] with permission from John Wiley & Sons, Inc.

in the particle size range of 0.1–0.5 μm, depending on the exact conditions that exist at a particular time [1, 11] (Fig. 5.2). This range of particle sizes has been referred to the *accumulation mode* as these particles are typically formed by the aggregation of smaller particles [2] and their relatively low deposition velocity results in their relative net accumulation within the atmosphere. Overall, it can be stated that the very small particles behave similarly to gases while the large particles settle according to Stokes Law. Deposition to the water surface can be enhanced if the wind speed is high enough to cause sea spray and breaking waves as the sea spray ejected into the atmosphere can scavenge particles and gases [11]. The estimated and measured deposition velocities to water surfaces from a number of studies have been compiled in Fig. 5.2. A large range in deposition velocities are found, from relatively low values for sub-micron particles ($<0.1\,cm\,s^{-1}$) to high values for larger particles ($>1\,cm\,s^{-1}$). It is possible that the deposition velocity of larger mineral particles could be similar to that of pollutant aerosols even though they are in different particle size classes.

Most elements are not evenly distributed across the particle spectrum given that different size particles have different origins (e.g., crustal, anthropogenic, sea salt). Therefore, it is likely that the average deposition for a particular metal(loid) associated with the particulate phase will be a function of the particle size, which would be dependent on the source, and would likely be different for different locations, as the size distribution of urban aerosols is different from that of the remote atmosphere. As an example of the differences in deposition velocity, it is worthwhile to examine the result of Ondov et al. [12] who measured the distribution of metal(loid)s within the particulate phase of the atmosphere in coastal Maryland (MD), USA (Solomons, on the shores of the Chesapeake Bay), a site that is within 100 km of both Washington, DC and Baltimore, MD (Fig. 5.3). These investigators estimated from their measurements that the effective deposition velocity at low relative humidity of the crustal elements Al and Cr was $0.1–0.15\,cm\,s^{-1}$; that of Fe and Mn $0.15–0.2\,cm\,s^{-1}$; while the deposition velocity of some of the heavy metals and metalloids was either

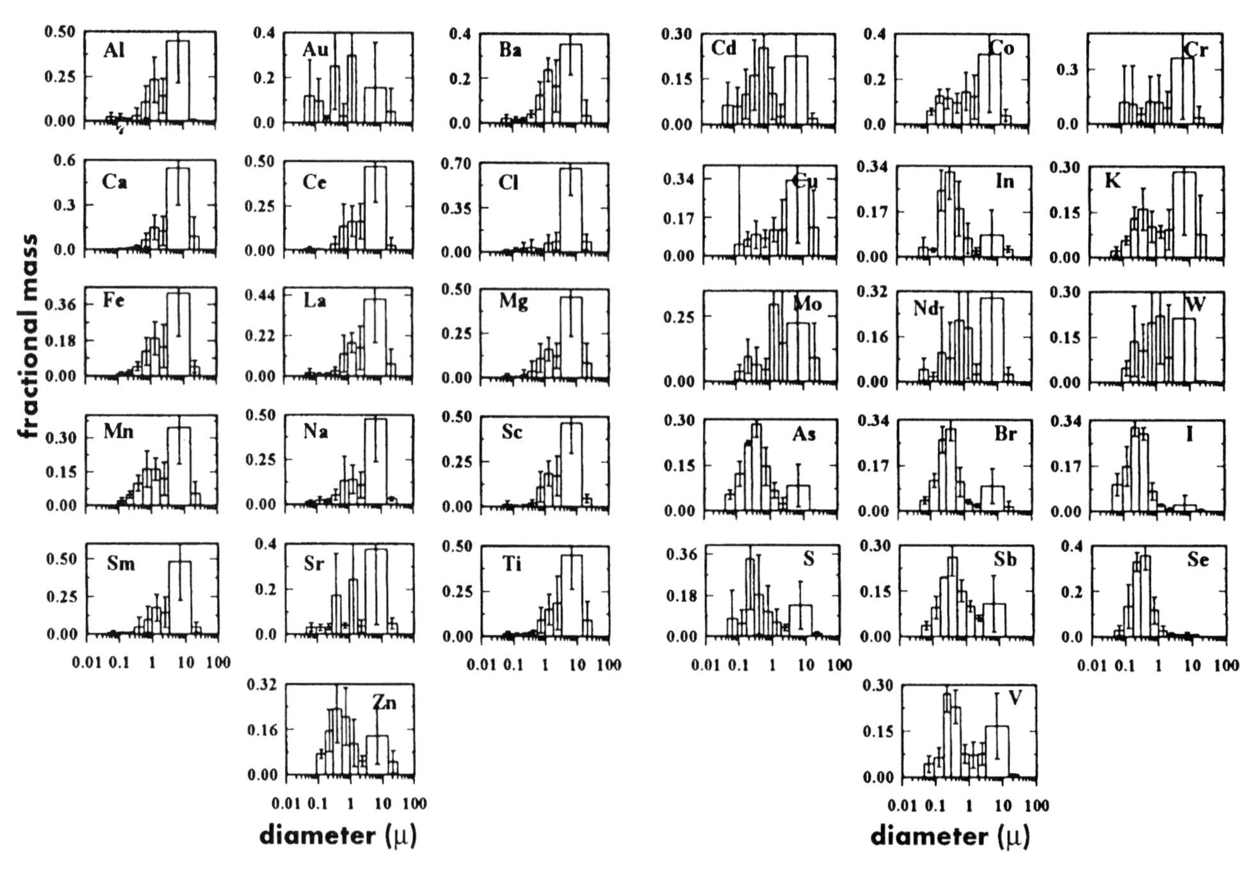

Fig. 5.3 The distribution of metals on particles collected using a cascade impactor at the Chesapeake Biological Laboratory (CBL) over water site. The bars reflect the size fractions of the impactor, as indicated on the x axis (in μm), and the y axis is the fraction of the total mass in each size range. Reprinted from Ondov [12] with permission from Atmospheric Deposition of Contaminants to the Great Lakes and Coastal Waters, copyright 1997, Society of Environmental Toxicology and Chemistry (SETAC), Pensacola, FL, USA, ISBN 978-1-880611-10-4.

higher (Cd 0.1–0.25; Cu 0.25–0.3 $cm\,s^{-1}$); or lower (As 0.04–0.05; Sb and Zn 0.075–0.125 $cm\,s^{-1}$). These data are consistent with Fig. 5.2 in that the larger particles and the ultra-fine particles have similar deposition velocities and that the elements that have intermediate deposition velocities are found in both the coarse and fine fraction [12], as shown for example in Fig. 5.3 for the Chesapeake Bay data, and discussed in Section 5.2.2.

It is worthwhile to discuss briefly approaches to estimating dry deposition and dry deposition velocities in the environment. Clearly, if the value of υ_d can be accurately determined, then it is sufficient to measure the atmospheric concentration and compute the flux. However, it is also possible, and potentially more accurate, to measure the flux directly. Thus, the measurement of dry deposition falls broadly into two approaches – the direct measurement of the deposition, and the indirect method. For the direct method, fluxes are measured directly using the collection of the material on a surface – typically a *surrogate surface* that mimics the actual surface – or by measurement of the flux in the air close to the surface. A surrogate surface is a device that is used to collect either gases or particles from the atmosphere through deposition under natural flow conditions. In addition, measurement of the direct deposition to a surface is possible in some circumstances. For example, for deposition to the canopy, leaf washing and analysis can be used to determine the change in concentration over time, and with the surface area known, such techniques can be used to estimate deposition. Additionally, the use of surrogate surfaces, initially developed for reactive gases and organic contaminants, have provided further approaches to estimate metal(loid)s fluxes. Typical surfaces include recirculating water in a container of known surface area (to mimic the water surface) and, for particles, aerodynamic surfaces that are coated to capture particles effectively. Recently, windows have been tested as deposition collectors in urban environments with encouraging results [13]. Surrogate surfaces are most effective in situations where the aerodynamic resistance is the dominating factor in deposition. For gases, uptake onto a surface contained in a chamber has been used to estimate dry deposition as well as the absorption of the chemicals of interest onto treated filter surfaces. For metal(loid)s, there are examples in the literature of surrogate water surfaces, with circulating water to renew the surface layer [14, 15], and aerodynamically designed surfaces [16–18] being used for particle collection and dry deposition estimates. These have been used for most trace elements. For Hg, devices have been tested that can estimate the dry deposition of both gaseous and particulate ionic Hg [19–21]. All approaches appear to provide a reasonable estimate of dry deposition when compared with the alternative approach of measuring atmospheric concentrations and estimating fluxes based on an estimation of υ_d.

Additionally, to estimate dry deposition, many investigators have compared the concentration in wet deposition collected in an open area to that of precipitation collected beneath the forest canopy (called *throughfall deposition*) and have inferred dry deposition as the difference between these two values. The results from a number of studies using such approaches, as noted in Chapter 2, suggest that dry deposition can be important relative to wet deposition for most elements. Most trace metal(loid)s, such as Hg, Pb, As and Se, are enriched by a factor of 1–2 in throughfall deposition relative to wet deposition, while Fe and Mn are substantially enriched (ratio >5) [22–26]. Of the metals studied, Cd appears to be enriched to the least extent.

So-called *micrometeorological* approaches have become more widely used and are based on two approaches – *eddy correlation* and *eddy accumulation*. Eddy correlation measurements rely on the simultaneous measurement and correlation of the concentration and wind field to directly obtain flux estimates [1]. High speed measurements of the vertical velocity and concentrations are used to obtain the time series of the fluctuations of both. The instrumentation needs to be able to resolve the turbulent fluctuations that contribute to the vertical flux and thus requires resolution on the sub-second scale. This time series can be used to estimate the time averaged vertical flux. The approach is based on the assumption that the vertical flux is the major factor controlling changes in concentration (i.e., no reactions are occurring). The approach works even if the sampling period for the concentration is not instantaneous as long as the long-term average value is consistent over time scales longer than that of the measurement.

An alternative approach is termed the *eddy accumulation* technique. Here, the same assumptions and conditions apply but the air is sampled through two lines, which are set up to sample either the positive vertical air movement and flux, or the negative flux. The system samples each according to the direction of movement of the air mass. The net flux over a period is computed as the difference between the two values measured.

For most indirect methods, atmospheric concentrations are measured and the deposition velocity is estimated and the overall flux determined using Equation 5.1. As noted, most studies mentioned earlier have used this approach to compare with surrogate surfaces. Typically the concentration of constituents in the atmosphere is measured using particulate air samplers or gaseous trapping devices such as denuders and thus it is necessary to estimate the flux based on the concentration and a measured or estimated deposition velocity. For gases, various empirical estimates have been made but one reasonable method that has been derived for acid gases (SO_2 and NO_x) [17] could be extended to other small molecular weight gaseous species. The formulation relates the deposition velocity to the diffusion coefficient in air (D_A) and the wind speed at 10 m (u_{10}):

$$\upsilon_d = D_A^{0.95}[(0.98 \pm 0.1) \cdot u_{10} + (1.26 \pm 0.3)] \quad (5.4)$$

The D_A can be estimated from either the molar volume or molar mass [27]. Other similar formulations exist in the literature.

An alternative approach is to measure the concentration at one or more heights and then estimate the flux based on the concentration difference. While this approach appears to be reasonable, the concentration differences are often small and therefore the accuracy of this approach is dependent on the accuracy of the measurement technique. Also, the approach is not valid for situations where the surface is rough as the assumption is that the measurements are being made in the surface constant flux layer.

5.2.2 Aerosol distributions and metal(loid) concentrations

The distribution of aerosol size is highly variable and forms a continuous spectrum for most environments. Therefore, atmospheric scientists often divide the continuous distributions into size fractions, either for ease of comparison between datasets or as a consequence of the methods of collecting aerosol samples. Firstly, there is the need to determine the effective diameter of a particle given that they are not spherical, and can be formed from the agglomeration of smaller particles [2]. The most-used parameterization, the *aerodynamic diameter* (D_a) is defined as the diameter of a sphere of unit density (ρ_0; $1\,g\,cm^{-3}$) which has the same terminal velocity in air as the particle being considered. This parameterization can be represented as:

$$D_a = D_s k(\rho_p/\rho_0)^{1/2} \quad (5.5)$$

where D_s is the geometric diameter, ρ_p is the density of the particle (neglecting the buoyancy effects of air); and k is a shape factor which is 1 for a sphere. Heavier particles will therefore have a higher D_a compared to their D_s, but the effect is typically small as the relationship is related to the square root of the difference. Overall, for atmospheric particles, the value of D (for simplicity the subscript is dropped in the further discussions) can range over several orders of magnitude from low values (~0.002 µm) to >10 µm. The particle distribution is also often divided into *modes* which are termed the *coarse* (>2.5 µm) and the *fine* mode, which is further broken down into the *accumulation mode* (0.08–2.5 µm) and the *ultra-fine* (Aiken nuclei or transient) *mode* (<0.08 µm) [1, 2]. The sources and processes important for each mode are shown in Fig. 5.4 [2].

A brief mention of common particulate sampling devices is given at this juncture. The sampling devices fall into four broad categories: *filters, sedimentation collectors, impactors,* and *precipitators.* Filters can be used to collect particles, and if

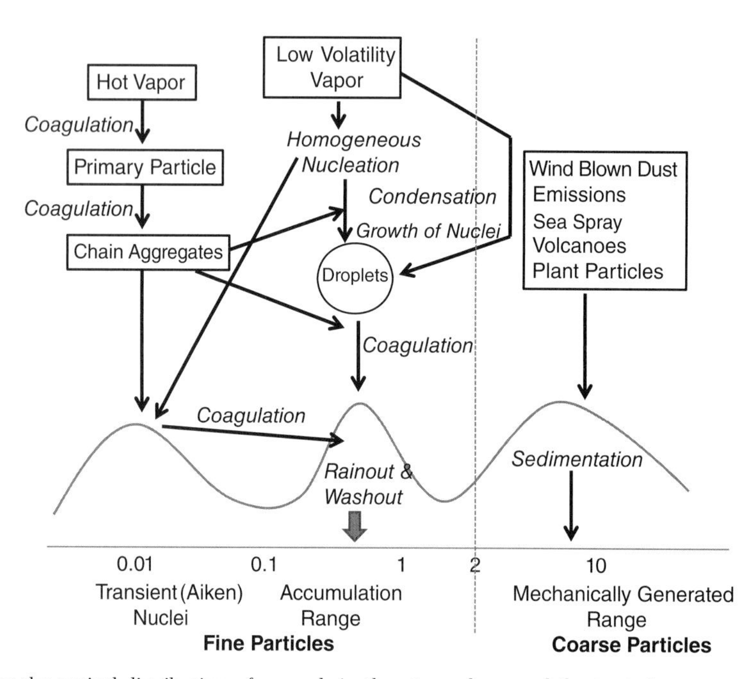

Fig. 5.4 A schematic showing the typical distribution of aerosols in the atmosphere and the typical processes that are occurring, with an indication the likely removal mechanisms for the different size fractions. Figure created based on information from Finlayson-Pitts (1986), *Atmospheric Chemistry: Fundamentals and Experimental Techniques* [2]. Reprinted with permission from B.J. Finlayson-Pitts and J.N. Pitts.

impregnated with chemicals, gases as well. Particle collection relies mostly of contact of the particle with the surface (impaction, interception, and electrostatic attraction) and thus the pore size of the filter is not the main criteria for determining usage. Impaction is most important at high flow and for larger particles. Fibrous mat filters (e.g., glass fiber or cellulose) allow higher flow rates and are often used with so-called high volume (*high-vol*) samplers (>1 $m^3 min^{-1}$) [2]. Membrane filters have the advantage that they can be more rigorously cleaned (e.g., polycarbonate, Teflon, polysulfone) and some materials are hydrophobic (e.g., Teflon), which may be advantageous in some instances. Some filter types have a large pressure drop across them, which is less desirable as the collection efficiency, especially for the smaller particles, decreases with increasing pressure drop [2].

Sedimentation collectors are used mainly for large (coarse) particles and rely on gravitational settling or centrifugation. A so-called *cyclone collector* uses the principle that air brought into a stationary cylinder will form a vortex spiral and the particles associated with the air will be deposited on the walls in a location relative to their size [2]. Most often cyclones are used as "pre-filters" to remove large particles if the fine fraction only is to be sampled. *Impactors* rely on the fact that particles in the airstream will continue in a straight line due to their inertia if the air stream is forced in a circuitous path and can therefore be collected on an impaction plate placed in their path. When placed in series, a number of impactors (a *cascade*) can be designed to remove particles in relation to their effective diameter. The collection surface can have an impact on retention and bounce-off and re-entrainment are both potential artifacts associated with the use of impactors. Finally, *precipitators* use either electrostatic or thermal devices to collect the particles from the air stream.

Given the sampling approaches discussed, another approach to dealing with the distribution in particle sizes is to define the distribution based on certain cutoff fractions. This approach is the result of various experimental methods of particle collection, as specific size fractions are often collected. These are often defined using the maximum particle size as a reference, so-called *PM_x terminology*, where the value of x represents the upper cutoff size (in microns). The most common fractionations used in pollutant sampling are PM_{10} (<10 μm) and $PM_{2.5}$ (<2.5 μm) (i.e., the fine particle fraction). Often these fractions are collected simultaneously using a separation device and in this instance the PM_{10} refers to the fraction between 2.5–10 μm. The chosen size ranges are both a reflection of the typical cutoff of the sampling devices and, more importantly, are chosen because smaller particles are more of a human health (respiratory) concern. The range in the number of particles and their distribution for various atmospheric settings is shown in Table 5.1 [1, 2]. The ratio of the larger particle fraction to the finer particles gives an indication of the differences for the different environments. As may be expected, the continental aerosol is relatively enriched in larger particles (likely of crustal origin) compared to the urban environment where there is more evidence of enrichment in smaller particles, likely from anthropogenic sources. The marine aerosol is enriched in large particles (sea spray) and deficient in the smaller particles derived from human activity. Overall, the choice of how to represent and present the information gathered from field collections of ambient aerosol is determined by the potential use of the information and the particular need in a specific instance.

Table 5.1 The distribution of aerosols for different air masses in different locations at the Earth's surface. Adapted from [1]. Reprinted with permission of John Wiley & Sons, Inc.

Location	Number, N (cm^{-3})	PM_1 ($\mu g\ m^{-3}$)	PM_{10} ($\mu g\ m^{-3}$)	Ratio PM_{10}/PM_1
Polluted, urban	10^5–4×10^6	30–150	100–300	2–3
Rural	2–10 × 10^3	2.5–8	10–40	~5
Remote continental	50–10^4	0.5–2.5	2–10	~4
Marine	1–4 × 10^2	1–4	~10	2.5–10

Because of the range in sizes, and due to methods of particle collection, such as the use of cascade impactors, often the data are "*binned*" into different size classes and the parameter, ΔD, is used to refer to the side of the chosen or arbitrary size classes. For example, if a cascade impactor was used for collection these size ranges would be defined by the instrument. As the distribution of particles by number is skewed toward the smaller particles, plots often are presented with the x-axis parameter being logΔD rather than ΔD. Representative plots showing the distribution by number, volume and surface area for an urban and remote continental air is shown in Fig. 5.5 [1]. In this figure the various parameters describing the number, size and volume distribution of aerosols are expressed using the following parameters, which are derived by assuming that the aerosol size distribution can be expressed as a power relationship, an approach which is typically used only for size ranges greater than 0.1 μm:

$$n_N^o(\log D) = C/(D)^\alpha \tag{5.6}$$

where n_N^o is the number of particles; C and α are constants. The parameter ΔD represents the size range chosen for a particular population. Values of α from 2–5 have been suggested for ambient aerosols. The volume distribution and size distribution are calculated from the number distribution defined earlier [1].

The plot in Fig. 5.5(a) shows that a typical urban aerosol population is dominated in terms of particle number by the

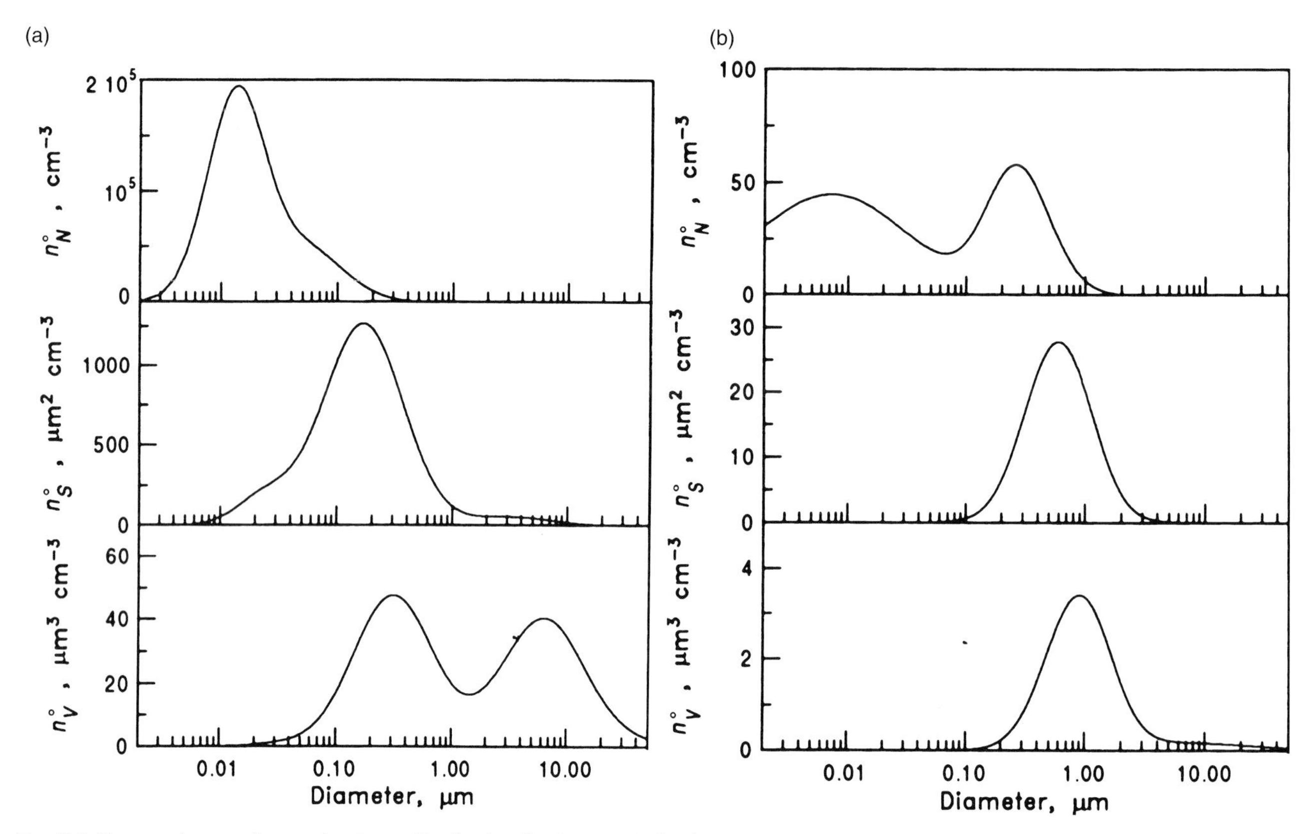

Fig. 5.5 The number, surface and volume distribution for (a) a typical urban air mass (*left*) and (b) for that of a remote continental aerosol (*right*). The values are plotted against the size distribution of the aerosols. Reprinted from Seinfeld and Pandis (1998), *Atmospheric Chemistry and Physics* [1] with permission from John Wiley & Sons, Inc.

small size fraction, with the *mode number* (peak) being around 0.01 μm. In contrast, the remote continental atmosphere (Fig. 5.5b) has a bimodal distribution with overall larger particles – modes of the number distribution being around 0.02 and 0.1 μm [1]. Note, however, that the number of particles is significantly smaller for the remote air mass. For the marine atmosphere (not shown), the distribution is also bimodal with peaks around 0.05 and 0.2 μm. There is of course a gradation over the range of terrestrial environments with the more rural and remote environments containing relatively more larger particles (crustal material) and less fine particles if the environments are remote from anthropogenic sources. Examples of a variety of aerosol distributions are discussed in [1].

The volume distributions (Fig. 5.5) show the differences between the urban and remote aerosols noted previously. The remote environments are more dominated by larger particles and this contributes most to the volume distribution. Again, given the relatively high contribution of large particles to the overall volume, those environments with dominance of the larger particles will have that fraction dominate the volume and surface area. Note that the relative difference in volume is much less than that of particle number given the shift in particle size to larger diameters for the remote atmosphere.

In addition to a distribution of particles across the aerosol size range, different elements are also enriched in the particles to a different degree. The crustal elements are enriched in the larger continental aerosols, the constituents of salt are elevated in the large marine aerosols and the heavy metals and other contaminant elements are enriched in urban fine particulate matter. The differences in concentration for Fe, Mn and the "pollutant" derived metals are shown in Table 5.2 [12, 18]. The data in the table also reinforces the discussion previously of the large differences in the concentrations of particles (and therefore the elements associated with the particles) between remote, rural and urban environments.

It can be seen that a large fraction of the particulate in the remote environment is derived from crustal material (as indicated by the relative Fe/mass concentration) by the data in Fig. 5.6, which shows the relative mass distribution of

Table 5.2 Range in concentrations of metals in aerosols collected in various locations. Data compiled from [1, 2, 18]. All concentrations in nmol m^{-3} except Hg (pmol m^{-3}).

Element	Mode**	Remote	Rural	Urban	Urban/Rural*
Fe	F & C	<20–70	0.9–270	1.8–250	1–2
Mn	F & C	0.2–0.27	0.07–1.8	0.07–9	1–5
Cd	F	<0.01	0.004–9	0.01–60	1–7
Cu	F & C	0.01–0.25	0.05–5	0.05–80	1–15
Ni	F & C	0.003–1	0.02–1.3	0.003–5	1–4
Zn	F	0.01–7	0.15–6	0.4–125	1–20
Hg	F	0.01–0.1	<0.5	0.25–5	1–10
Pb	F	10^{-3}–0.3	10^{-3}–9	0.15–440	15–70
As	F	<0.01–0.03	0.01–0.3	0.025–33	2–85
Se	F & C	<0.01–0.025	<0.01–0.38	0.025–0.38	1–20

Notes: *Calculation based on the concentration ranges in each case.
**F: Fine, C: Course Fraction.

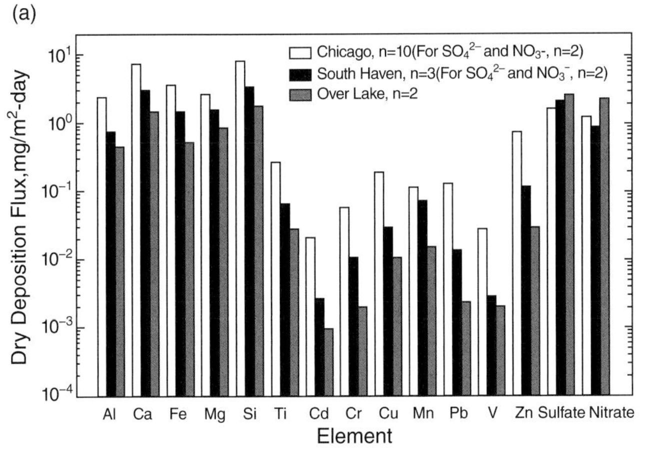

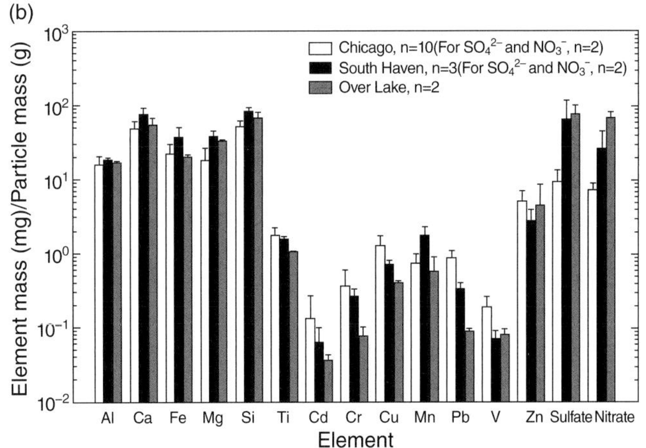

Fig. 5.6 The concentration of elements (mass/mass) on aerosols collected at three locations in Michigan, USA: in the city of Chicago, at an urban site in the city of South Haven and for samples collected from a ship on Lake Michigan. Figure reprinted from Holsen et al. (1997), in *Atmospheric Deposition of Contaminants to the Great Lakes and Coastal Waters* [28]. Reprinted with permission from Atmospheric Deposition of Contaminants to the Great Lakes and Coastal Waters, copyright 1997, Society of Environmental Toxicology and Chemistry (SETAC), Pensacola, FL, USA, ISBN 978-1-880611-10-4.

elements in particulate collected in downtown Chicago, IL, USA, and at a more rural location on Lake Michigan (South Haven) and over the lake [28]. Elements, such as Al, Ca, Fe, Si, and Ti, and sulfate and nitrate, are present at between 1–10% of the overall mass and there is little difference between sites in these crustal elements. A much more significant difference in the concentrations is seen for those elements with a strong pollution source, such as Cd, Cu, and Pb, and to a lesser extent, Cr and V. This is expected given the likely urban source of these elements (e.g., coal burning, metal processing, and waste incineration). For the crustal metals, the urban/remote ratio is lowest. The higher value for Mn likely indicates that there are some relatively important urban sources for Mn relative to the crustal input. For the heavy metals and metalloids, the ratios are much higher (>100). Interestingly, the ratio for Hg and Se is relatively low and this may not be expected given that they do not have a substantial crustal source. More likely the low ratio likely reflects that these elements are mostly associated with small particles that are not rapidly removed from the atmosphere, as discussed further later. Alternatively, the differences may reflect the manner in which the ratio was calculated, using average data.

As another example of the association of various elements with different particles, it is worthwhile to re-examine the data of [12], discussed peviously (Fig. 5.3), for collections made in coastal MD, USA. The overall size distributions are for elements measured during the sampling campaign using a cascade impactor. The difference in the size distribution of the various elements is clearly evident. The crustal elements (e.g., Al, Fe, Mn, Cr) are obviously dominant in the larger size fractions while the heavy metals and pollutant-derived elements (e.g., As, Sb, Se, Cd, Zn) are mostly associated with the finer material, although some of these elements also have a bimodal distribution. Interestingly, the heavy metals Zn, Cd and the metalloid As all have bi-modal distributions that likely reflect a number of different sources of these elements to the air masses sampled at CBL.

The concentrations on a mass basis for various types of particles has been collated by Deguillaume et al. [29] (Table 5.3). These results also highlight the differences for the different metal(loid)s. Concentrations of Fe and Al vary by less than a factor of five while for some elements, the variability is greater than an order of magnitude, with some metals being more enriched in dust (e.g., Mn, Cr) and others more enriched in the urban aerosol (e.g., Co, Zn). These data reinforce the results presented in Figs. 5.3 and 5.5. In addition, these results show the usefulness of some metal(loid)s as tracers of specific sources. For example, oil fly ash is strongly enriched in V and Ni, and to a lesser extent Co, compared to the other particles and these metals have been, or could be, used as tracers of such sources. While not shown in Table 5.3, the volatile metals such as Se are good tracers of coal burning. Other metal(loid)s are potential tracers of other sources.

The overall deposition velocity for a particular element will be an average value that depends on the overall particulate size distribution for that element. The overall distribution patterns in particle size in Fig. 5.5 account for the differences in the estimated deposition velocities. These differences also impact the long term fate and transport of the aerosols as the large particles tend to deposit more rapidly, and are also potentially removed from the atmosphere more efficiently by wet deposition.

Another way of characterizing the enrichment of an element in the particulate phase due to anthropogenic activity is to calculate the *enrichment factor*, EF. Such a concept was discussed in Chapter 2 in terms of sediments and the same approach can be used for atmospheric aerosol, in normalizing the concentration to that of crustal aerosols (dust), or crustal material. For the atmosphere, it is assumed that Al is derived almost exclusively from crustal material and so this is the element most often used for the normalization. Thus, the EF is defined for element, E, as:

$$EF = (E/Al)_{air}/(E/Al)_{crust} \quad (5.7)$$

The EF for a number of metal(loid)s for a variety of particle types is given in Table 5.4. These values illustrate that for even some of the more abundant metals such as Mn and Cr, there is a strong anthropogenic component to aerosols collected in polluted locations, or downwind of such a location (e.g., North Atlantic Westerlies). Also, it is worth noting that the EF for Pb is probably dependent on the date of the sample collection given the large changes in atmospheric Pb concentration that has accompanied the phasing out of Pb in gasoline [30]. Similar changes are also likely for other metals when reviewing the historical record. A comparison

Table 5.3 The average trace metal composition in weight percent of different types of particulate matter. Information collated from a number of sources [1, 18, 29, 30].

Element	Saharan Dust	Urban particles	Coal fly ash	Oil fly ash
Units	*wt %*	*wt %*	*wt %*	*wt %*
Al	8.4	3.4	15.8	1.4
Fe	7.7	3.9	8.8	3.5
Mn	0.13	0.08	0.22	0.04
V	0.05	0.013	0.013	4.7
Units	*wt % × 10^{-3}*	*wt % × 10^{-3}*	*wt % × 10^{-3}*	*wt % × 10^{-3}*
Cr	30	3.2	21	30
Co	0.085	1.8	1.5	23
Ni	10	8.2	4.3	1280
Cu	5	61	7	21
Zn	10	476	21	120
Cd	5	9.7	0.25	3

Table 5.4 Enrichment Factors (EF) for a number of metals based on the collection of aerosols over various ocean regions [30] and for Los Angeles (LA), USA [18] as an example of a polluted environment.

Element	EF LA	EF NAtl Westerlies	EF NAtl Trades	N. Pacific (Eniwetok)
Mn	–	3.9	1.0	–
Cr	~1	13	1.5	–
Cu	35–50	30	1.2	3.2
Pb	20–45	1490	9.3	40
Ni	2–4	16	1.3	–
Zn	45–65	290	3.8	14
Cd	–	1260	9.5	130

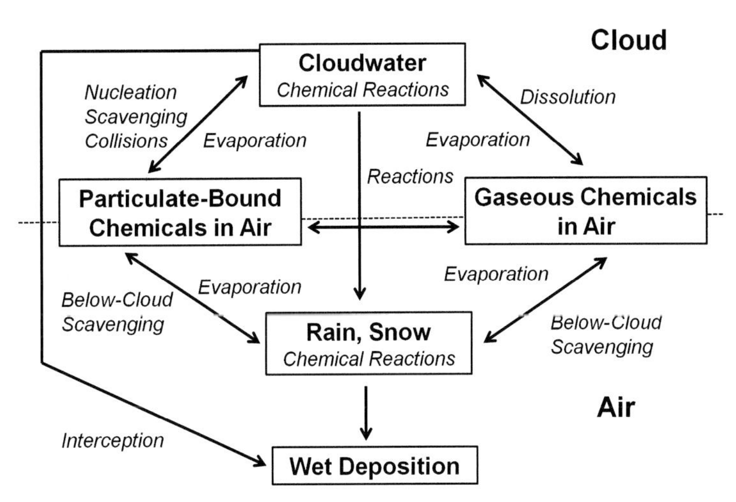

Fig. 5.7 A diagram illustrating the major processes involved in wet deposition. Figure created using information in Seinfeld and Pandis (1998), *Atmospheric Chemistry and Physics* [1], Finlayson-Pitts (1986), *Atmospheric Chemistry: Fundamentals and Experimental Techniques* [2] and other sources.

between the North Atlantic Westerlies and the NE Trade Winds (dominant aerosol source being Saharan dust) illustrates the extent of the anthropogenic signal as only the pollutant metals are enriched in the trade winds (e.g., Pb and Cd) (Table 5.4).

Also, it is apparent that there are differences between the North Pacific Ocean aerosol and that of the North Atlantic, again a reflection of different pollutant sources to these two remote regions. Generally, the EF values for the crustal elements are <10 while those of the elements with a strong anthropogenic signal are generally greater than 100, with some having values >1000 (e.g., Pb, Cd, Se, Sb). Marine scientists have also used similar approaches to examine the relative enrichment of elements in marine aerosols relative to sea salt. In this case, the normalization is to Na, the major cation in sea salt, rather than to a crustal element [30]. When discussing aerosols over the ocean or in coastal regions, there is also often discussion of the *non-sea salt sulfate* concentration as sulfate has a strong anthropogenic signal, mainly from combustion sources (e.g., coal burning), but is also abundant in seawater. Thus, by normalizing to Na, the contribution of sulfate from seawater can be determined and subtracted from the total, allowing estimation of the anthropogenic signal. Such approaches can be used for the metal(loid)s, and this is another way to estimate and refine the source signal of elements found in aerosols. Finally, these EF values are estimated based on the bulk aerosol and there are undoubtedly larger differences if the size distribution and different modes are compared as the fine fraction would be substantially enriched in most metal(loid)s.

5.2.3 Wet deposition

Wet deposition refers to the removal of species from the atmosphere in precipitation, both rainfall and snowfall, and all their various intermediate forms (e.g., sleet, ice, freezing rain). The sources of trace metal(loid)s to wet deposition include the scavenging of gaseous and particulate material from the atmosphere by raindrops, and within clouds, and also by the interconversion of some species due to reactions that occur primarily during cloud formation, but also in droplets. One depiction of the various processes that contribute to the removal of species from the atmosphere by wet deposition is shown in Fig. 5.7 [1]. Clearly, the incorporation of species into wet deposition involves the complex interaction of the dissolved phase with both the atmospheric gaseous and particulate phases. Most processes that occur are reversible – for example, dissolution and evaporation – and therefore it is possible for an aerosol particle to go through various wetting and drying cycles before it is finally removed by wet deposition. This is important for species whose chemistry can be changed by reactions occurring in the wet phase as this may prolong the extent to which such reactions may occur, and enhance their impact. In clouds, and similarly in fog and dew, many chemical reactions occur that can impact the concentration or form of the species in wet deposition.

While most metal(loid)s are clearly associated with the particulate phase in the atmosphere, chemical transformations can lead to their enhanced solubility. This is true for redox sensitive metals such as Fe (insoluble as Fe^{III}, soluble as Fe^{II}), copper (both Cu^{II} and Cu^{I} oxidation states can exist in the atmosphere) and Hg (Hg^0 is a dissolved gas, Hg^{II} is much more soluble). In addition, complexation with organic acids, such as oxalate, may enhance the solubility of metals in wet deposition. The factors that influence metal solubilization, and the degree of solubilization of various elements will be discussed in Section 5.4. The transformations of these species in wet deposition have an important impact on their

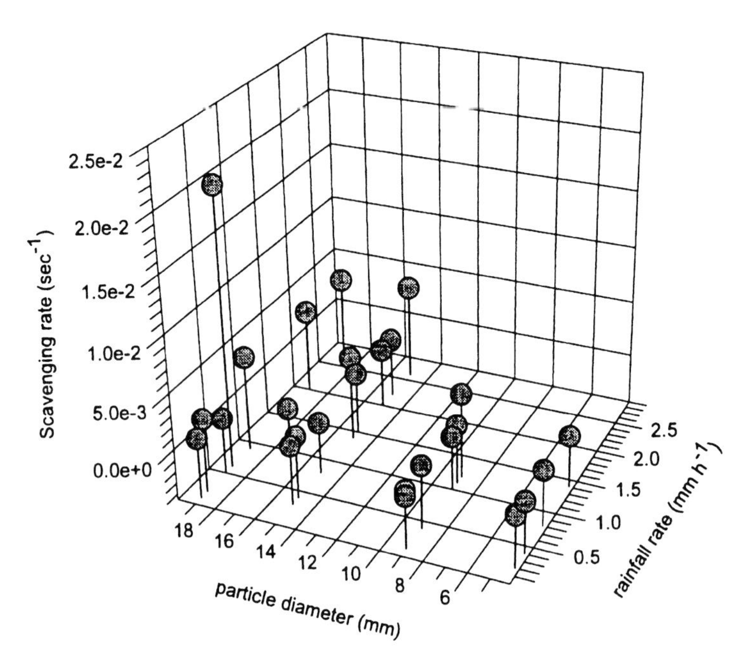

Fig. 5.8 Influence of particle size and rain rate on the scavenging rates of atmospheric particles as determined from field measurements of below-cloud scavenging of dye-labeled particles during precipitation events. Figure from Poster and Baker (1997) [3] but original figure published by Nicholson et al. (1991), *Atmospheric Environment* **25**, 771–7 [31], reprinted with permission from Elsevier.

fate in the atmosphere, their bioavailability upon deposition, especially to surface waters, and these transformations can also impact the fate of other constituents.

The scavenging of particles from the atmosphere by rainfall depends on both the size of the raindrops and the size of the particulate material. While such interactions can be examined in detail, and discussed in a theoretical framework, a more simplistic approach will be taken here. As shown in Fig. 5.8 [3, 31], field measurements show that there is a relationship between particle size and scavenging rate with the larger particles being scavenged more efficiently at any rainfall rate. Clearly, the particles used in this experiment are larger than those found in the atmosphere but the overall trend is valid. Additionally, the particle size is influenced by the humidity of the air with their overall size increasing as the humidity increases. This appears reasonable as the aerosols take on moisture, especially in the oceanic atmosphere where the salt content of the particulates is high [12]. Somewhat surprisingly, the data in Fig. 5.8 do not suggest that there is a strong relationship between scavenging rate and rainfall rate for a particular size fraction, and this appears somewhat counterintuitive.

While there is obvious complexity in determining all the processes involved in the removal of species from the atmosphere by wet deposition, there are some approaches that yield a useful parameterization of this process. One approach is to define a *scavenging ratio, SR*, which can be represented by the following relationship [30]:

$$SR = C_{precip}/C_{air} \tag{5.8}$$

where C_{precip} is the concentration of the element of interest in wet deposition and C_{air} is the concentration of the element in the air entering the storm, or in the atmosphere that is interacting with the precipitation event, and for most metal(loid)s this would be the concentration of the metal(loid) within the aerosol fraction (mol m^{-3}). Clearly, such a simple approach would not provide a reasonable parameterization for an element that is present both in the particulate phase and in the gas phase, as these two phases would interact differently with the precipitation and be scavenged to a different extent. In reality, it is often difficult, or indeed not possible, to measure the concentration of the element in the air mass at the location and elevation of the cloud or raindrop, and thus most investigators have used the more easily calculated *washout ratio*, WR, which is defined based on the ground level measurement of air concentration and the concentration in the precipitation:

$$WR = C^0_{precip}/C^0_{air} \tag{5.9}$$

where both values are measured at the surface [30]. Clearly, it is being assumed with this calculation that the concentrations measured at the ground are representative of those in the free troposphere and this is likely never to be a completely valid assumption.

However, the WR does give some indication of the relative efficiency of removal of elements and compounds from the atmosphere by deposition. A collection of values reported in the literature is given in Table 5.5 [3, 30]. From first principles, for elements that have a similar size distribution in the aerosol, their SR or WR should be similar. This is indeed the case overall, given that these values are empirically derived from field data. Even more apparent for most of the metal(loid)s is that the range in values covers the same order suggesting that the particle size differences between the larger mineral aerosols and the finer pollutant aerosols do not lead to major differences in the scavenging rates. This is consistent with the data in Fig. 5.8. The high WR value for Hg, much higher than most other heavy metals, suggests two possibilities: (1) that gas phase species are being scavenged by wet deposition or (2) oxidation of Hg^0 is occurring in the clouds. Both processes are likely occurring, as discussed in detail in Section 5.5.2. Overall, these processes account for the higher scavenging ratio of Hg compared to other trace metals and the metalloids. Indeed, various modeling exercises suggest that the scavenging of gas phase Hg species is much more important that particulate scavenging, and that in-cloud reactions are a substantial source as well [7, 23, 32, 33]. Also, the similarity of the WR values for the metalloids to that of the other metals suggests that these compounds do not have significant gas phase species that are highly soluble under typical atmospheric conditions.

Table 5.5 Washout ratios for metals based on measured particulate concentrations and the concentration of the total metal in wet deposition. Compiled from [3] and references therein, as well as [30].

Element	WR	Element	WR	Element	WR
Al	10–400	V	110	As	110
Fe	10–250	Cu	30–400	Pb	80–150
Mn	50–400	Ni	125	Hg	500–1000
Cr	150	Zn	30–800	Cd	75–450

The relative concentrations of metal(loid)s in wet deposition, and wet deposition fluxes, reflect the various sources – natural and anthropogenic – contributing to their accumulation in the atmosphere in a similar fashion to that of the particulate phase. Thus, in a similar fashion, there is enrichment of the "pollutant" metals in wet deposition relative to their concentration in crustal material. Again, many investigators have found correlations between crustal metal(loid)s in rain, and between metal(loid)s that have similar sources and these ratios can be useful in identifying source contributions. A compilation of data is shown in Table 5.6 for the shores of the Chesapeake Bay and for the Great Lakes region [34, 35]. This comparison highlights differences that should be apparent in deposition collected in rural versus urban locations – none of these sites can be considered to be uncontaminated however. The comparison of metal(loid) concentrations in precipitation collected in downtown Baltimore, MD, USA (SC in Table 5.6) to that of Solomons, MD, about 60 miles away on the coast (CBL), shows that there is little enrichment in the urban wet deposition in terms of concentration for the crustal elements but the heavy metals (Hg, Pb and Zn) are indeed enriched. Similar relationships have been found for other locations; see for example, Refs [36–38].

One additional difficulty in examining and compiling data is that much of the data has been collected to either examine a regional problem, or there has not been consistent collection of data over sufficient time to allow for trends to emerge. For wet deposition, changes in precipitation amount on an annual basis can be an important variable controlling the magnitude of the concentration and annual flux and

Table 5.6 The annual wet deposition fluxes for metals for various locations around the Chesapeake Bay. Elms and CBL (Solomons) are on the western shore and Wye is on the eastern shore of the Chesapeake Bay. The Baltimore site is SC (the Science Center). The western Maryland site was near Frostburg, MD. Samples were collected at different times by different investigators, as indicated by the dates. All data are in mg m^{-2} yr^{-1} and taken from [34].

Element	Great Lakes[1] 1993–1994	Elms[2] 1990–93	Wye[2] 1990–1993	West MD[3] 1996–1998	CBL[4] 1998	SC[4] 1998	Ratio SC/CBL
Al	–	6.96	12.7	31.9	18.5	18.0	0.97
Fe	–	5.49	11.4	27.2	14.9	12.3	0.83
Mn	1.9–2.4	1.15	1.33	–	3.04	3.67	1.21
Cr	0.06–0.08	0.035	0.15	0.78	0.24	0.25	1.04
Ni	0.23–0.29	0.20	0.33	0.80	0.59	0.52	1.13
Cu	0.57–0.85	0.38	0.27	0.64	1.49	0.71	0.48
Zn	3.5–5.5	1.56	1.56	4.41	3.06	4.11	1.34
Cd	0.07–0.09	0.035	0.038	0.13, 0.12	0.063	0.033	0.52
Pb	0.55–1	0.42	0.47	0.68, 0.64	0.89	2.52	2.83
Hg	–	–	–	0.015	0.014	0.030	2.14

Notes: [1]Taken from [35]; [2]Taken from [36];: [3]Taken from [24];: [4]Taken from [34].

therefore many investigators normalize the data to the total precipitation, or calculate a volume weighted mean concentration (total metal(loid) deposited/total precipitation). These normalizations can account to some degree for differences in the rainfall amount, and the intensity and distribution of deposition seasonally. As noted, rain and snow have different abilities to scavenge particles from the atmosphere, and the particle size and precipitation rate may also influence the overall removal (Fig. 5.8). Also, for metal(loid)s that are mostly associated with particulate, there is likely to be a *washout effect* in that the initial precipitation will remove disproportionally more particles than the latter stages of the precipitation event if there is no source of additional particles to the air mass during the event. This is shown for the radionuclides ^{137}Cs and ^{7}Be emitted as a result of the Chernobyl incident [39] (Fig. 5.9a). Both are particle reactive and Cs is very soluble and there is a clear indication of their rapid removal; and a resultant decrease in concentra-

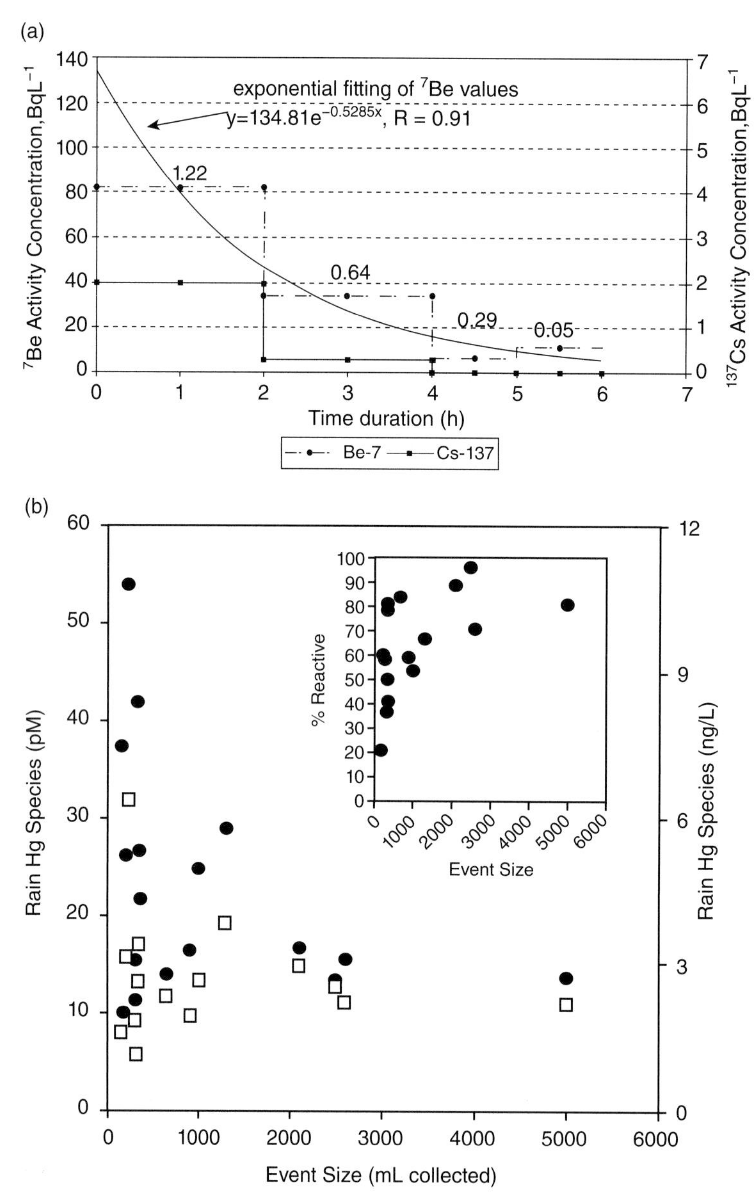

Fig. 5.9 Washout curves for particulate-associated metals in the atmosphere. (a) Change in rainfall concentration with time for ^{137}Cs and ^{7}Be during one rainfall event; from Loannidou and Papasterfanou (2006), *J. Environmental Radioactivity* **85**: 11–136 [39], reprinted with permission from Elsevier; (b) the relationship between mercury concentration (total: solid circle and reactive: square) and rainfall amount for a number of different precipitation events over the ocean; from Lamborg et al. (1999), *Deep-Sea Research II* **46**: 957–77 [40], reprinted with permission from Elsevier. In part (a), the variability of ^{7}Be and ^{137}Cs activity concentrations is for sequential samples of a rainfall event with the numbers indicating the precipitation rate, in mm h^{-1}.

Table 5.7 Range in the flux of metals from the atmosphere to the terrestrial surface. All values are given in μmol m^{-2} yr^{-1}. Data taken from the literature and units converted as necessary [36, 37, 41–43, 46]. Note that there is overlap as data for coastal regions are appropriate for both this table and Table 5.8 later.

Element	Rural	Rural	Rural	Urban	Urban	Urban
	Wet	Dry	Total	Wet	Dry	Total
Al (× 10^3)	0.26–1.2	0.01–1.2	1.4–5.6	0.67–9.3	0.03–0.10	–
Cr	0.7–15	1.5–4.0	2.9–8.7	1.3–4.8	0.2–1.3	–
Mn	20–44	13–22	22–51	70	–	–
Fe (x10^3)	0.1–0.27	0.90–1.0	0.97–1.4	220–630	27–63	–
Ni	3.4–14	8.5–12	2.2–26	8.9	–	–
Cu	3.1–24	4.7–7.9	2.5–17.3	0.3–11.2	0.3	–
Zn	18.3–53.5	1.0–33	–	12–63	2.3–12	–
Cd	0.3–1.2	0.2–2.6	0.1–1.6	0.2–0.3	0.2	0.4–1.3
Pb	1.0–4.8	2.9–4.3	1.0–16	12.2	–	3–67
Hg	0.03–0.10	<0.01	0.03–0.10	0.15	<0.01	0.1–0.5
As	0.5–0.8	1.3	1.3–2.1	–	–	–
Se	0.9–1.5	–	4.1–5.3	–	–	–

tion during a single rain event. The same effect is shown by comparing a number of rain events over the ocean where the Hg concentration is a strong function of the size of each rain event, suggesting again rapid removal in the early stages of the precipitation event (Fig. 5.9b) [40]. The dynamics of mercury are further discussed in Section 5.5.2.

The impact of anthropogenic emission on the concentration of metal(loid)s in deposition has been mentioned already and this is clearly shown in the comparison of the concentrations over time. As noted, there is likely to be strong variability within years in concentrations and fluxes and therefore a long record is typically necessary to shown any trends. Overall, a dramatic decrease in concentration with time is apparent for both metal(loid)s. There are many other examples in the literature, and some of are discussed elsewhere in the book, and in Section 5.2.4, which focuses on atmospheric deposition fluxes to the Earth's surface.

5.2.4 Atmospheric deposition fluxes

The concentrations of metal(loid)s in deposition and the fluxes associated with these concentrations have been measured for a number of locations [41–52] and more data has accumulated for the ocean environment than for the terrestrial realm. Studies of deposition over land have been driven by more regional issues, such as impacts in Europe, or have even had a country-specific focus. In Europe, many of the studies have been focused on the Scandinavian and northern regions, around the marginal seas, and many of these efforts have had a regional component as well. Furthermore, there have been studies focused on examining impacts of deposition on the Mediterranean Sea. In North America, there has been more focus on the eastern portion of the country and the Midwest, and this was driven to some degree by a program funded through legislation and the US EPA to look at deposition and inputs to the "Great Waters", including the Great Lakes and the Chesapeake Bay, and other major lakes and estuaries, which resulted in many publications and lead to the publication of *Atmospheric Deposition of Contaminants to the Great Lakes and Coastal Waters* (edited by J.E. Baker). The resultant estimated fluxes from some of these studies have been compiled in Table 5.7 [36, 37, 41–43], which gives the average concentrations of metal(loid)s in wet and dry deposition, and the total flux. This tabulation of data is meant to be illustrative rather than complete, and is collated to show the variability in the deposition for different locations, in the manner of Table 5.6. A similar table is constructed for ocean deposition (Table 5.8) [7, 41–45].

It is difficult to summarize and collate the information for two reasons. One, sampling has occurred over a span of nearly 30 years and during these times there have been large changes in the deposition of metal(loid)s to the ocean due to changes in anthropogenic inputs and loadings. For example, data from Bermuda suggest that the concentration of Pb in deposition has decreased by about 90% between 1981/82 and 1996/97 [46]. This decrease is even larger than the change in ocean surface water concentrations over the same period (see Fig. 2.19; ~60% decrease). Similarly, Cd has decreased by 80%, Zn by about 55%, Cu and Ni by about 60%, and the decreases for the coastal east coast USA are of the same order [46]. Also, the data suggest that the rate and timing of the changes are not all similar, reflecting the different anthropogenic sources for the metal(loid)s. Similarly, data of Pb and Cd in rain in Europe have shown

Table 5.8 Range in the flux of metals from the atmosphere to the ocean surface. All values are given in μmol m^{-2} yr^{-1}. Data taken from the literature and units converted as necessary [7, 30, 36, 37, 41–46, 50].

Element	Open Ocean	Open Ocean	Open Ocean	Coastal/Seas	Asian Coastal
	Wet	Dry	Total	Total*	
Al	40–555	7–110	100–2000	590–10^4	–
V	0.28	–	1.0–3.3	95	–
Cr	1.2	–	1.7–3.3	1.7–40	–
Mn	0.55–1.6	0.11	0.73–10.4	10–170	730–1370
Fe	5.0–100	3.0	132–600	220–4570	–
Co	0.003–0.017	0.001	0.042–0.46	0.60–6.6	–
Ni	–	–	1.4–5.1	6.3–66	–
Cu	0.32–1.1	0.55	0.63–16	5.0–205	–
Zn	0.31–10	0.30–0.15	0.37–21	50–1370	310–920
Cd	0.036	–	0.21–0.80	0.20–38	–
Pb	0.97–4.1	0.005	0.068–7.0	1.0–130	–
Hg	0.001–0.048	<0.001	0.001–0.048	0.05–0.15	–
Ag	0.28	0.009	0.18–0.32	0.28–0.83	–
As	0.13–0.4	0.09	0.36–0.4	1.3–38	–
Se	0.051–0.51	0.13	0.13–1.8	2.8–6.1	–
Sb	0.003–0.27	0.02	0.008–0.29	3.9–4.8	0.8–33

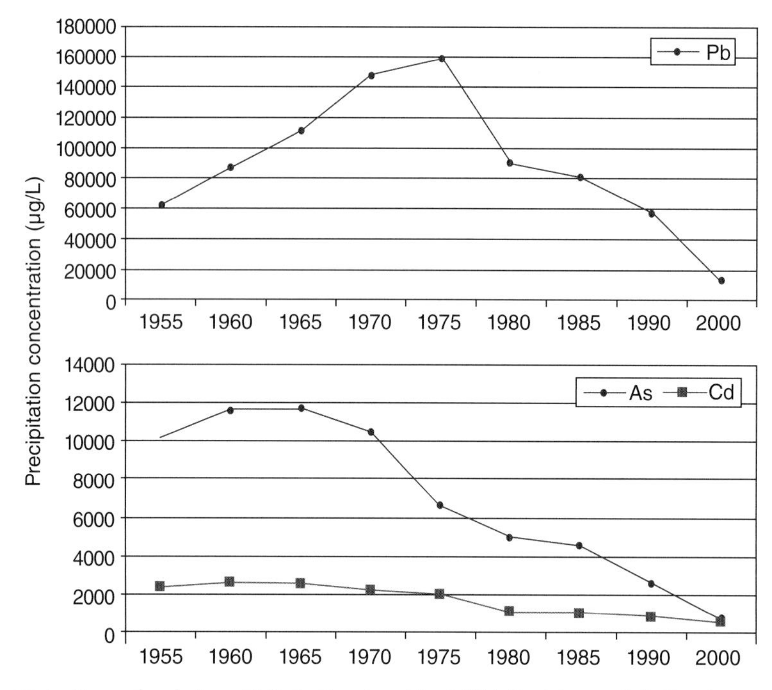

Fig. 5.10 The measured concentrations of cadmium (Cd) (bottom) and lead (Pb) (top) in precipitation for a number of sites in Europe. For conversion of units, 1 μg l^{-1} Cd = 8 nM; for Pb, 1 μg l^{-1} = 4.8 nM. Figure from Pacyna et al. (2007), *Atmospheric Environment* **41**: 8557–66 [47], reprinted with permission from Elsevier.

a decrease of similar order – a factor of 5–10 – between the late 1970s and the early part of this century [47], with a somewhat larger decrease for Pb compared to Cd (Fig. 5.10), in agreement with the Bermuda and East Coast, USA data. Studies in Sweden have documented a similar decline in Hg levels in deposition, of more than 50% in some locations between the mid-1980s and mid-1990s [48, 49]. The data show higher variability in the 1980s compared to more recent measurements but this may be the combination of a lack of data for these earlier times and a general improvement in the accuracy and precision of environmental measurements over time.

Secondly, compiling data is difficult as it is often reported in different units and not all the same metal(loid)s have been the focus of the various investigations. Thus, for some metal(loid)s there is a relatively large number of datasets to work from, while for other metal(loid)s, data limitation is a problem. Also, for the crustal elements, the concentrations vary depending on the relative importance of dust input and therefore there are large ranges in the reported concentrations. For example, studies on Bermuda have shown that there are large seasonal differences in the concentration of metal(loid)s in rain and in aerosols [46] and this reflects the dominance of inputs from the east, with a relatively high dust component, in summer, and inputs from the west, off the North American continent, being more dominant in fall-winter. Additionally, seasonal differences in rainfall amounts are also important.

The compilation of data (Table 5.7) shows that there are similar levels of the crustal elements in rural and urban atmospheres and this is a reflection of the sources and the fact that metals such as Fe, Al and Mn do not have a substantial anthropogenic component for most samples. This is less the case for the other "crustal" metals (Cr, Co, Ni) but there is little evidence of strong differences between urban and rural locations in the same region, as also illustrated in Table 5.6 for wet deposition. In concert with the data in Table 5.6, the metals which show the largest differences between urban and rural locations are the heavy metals (Pb, Hg, Cu, Zn) which have the largest anthropogenic component, and the metalloids (As, Se). Overall, the results from the wet and dry deposition flux estimates are consistent with the differences in enrichment of metal(loid)s in particles, discussed earlier in this chapter.

For the ocean, estimation of the deposition flux has also received substantial attention given the relative importance of atmospheric inputs to ocean metal(loid) content, as discussed in Chapter 2, and especially for the metal(loid)s with a large anthropogenic component. Two efforts have led to much of the data in the literature. For the ocean environment, the Sea Air Exchange (SEAREX) program provided substantial information on the concentrations and fluxes of trace metals and other constituents to the ocean surface and also quantified ocean surface emissions (sea spray, gaseous emissions). Use was made of island locations so that clean open ocean air could be sampled. Studies were completed at Eniwetok Atoll (11°20′N, 162°20′E) in the North Pacific and American Samoa (14°15′S, 170°34′W) in the South Pacific, and on the northern tip of the north island of New Zealand (34°33′S, 172°45′E) in the Southern Hemisphere [37, 50–52]. Sampling in the North Atlantic Ocean occurred under the Atmosphere-Ocean Chemistry Experiment Program (AEROCE) that included sampling on Bermuda (33°22′N, 64°41′W) and at Barbados (13°10′N, 59°32′W). Both these research programs have led to a number of papers and to a large amount of information that resulted in relatively well constrained fluxes for metals and metalloids to the open ocean realm [30, 50]. Other studies, and more recent studies occurring on Bermuda are part of a recent effort to characterize output of contaminants and other constituents from Asia due to both natural (dust) input and due to anthropogenic sources resulting from rapid industrialization in the 1990s in this region. Sites used in these studies include Okinawa, an island off mainland Japan, and other coastal and island locations, through a number of monitoring programs and studies. Activities have been coordinated and reported through various programs such as the Hemispheric Transport of Air Pollutants (HTAP) (http://www.htap.org/) initiative and others. The World Meteorological Organization's (WMO) Global Atmospheric Watch (GAW) is a global initiative to integrate atmospheric monitoring programs (http://www.wmo.ch/pages/prog/arep/gaw/gaw_home_en.html). A more limited dataset has been generated from collections made on open ocean research cruises. Overall, these initiatives and investigations have had a strong Northern Hemisphere focus and there is little data available on open ocean regions in the Southern Hemisphere. Again, this data is compiled in Table 5.8.

The data in Table 5.8, while compiled from a number of papers, as indicated in the table caption, also took advantage of two compilations in the literature [30, 50]. An examination of the data indicates firstly that for the open ocean, wet deposition is the more important source, especially for the "anthropogenic metals" (Pb, Zn, As, Hg) and this reflects the typically low aerosol concentrations found over the open ocean (Table 5.2). Thus, the dry deposition component is relatively small and scavenging of the particles from the atmosphere by precipitation is the main mechanism of their removal. For the crustal metals (Al, Fe, Mn), wet deposition is of somewhat lesser importance but still the dominant atmospheric flux, while for the other metals, wet is typically greater than 70% of the total deposition. The wet flux is somewhat less for the metalloids and this may reflect the importance of gaseous sources and their adsorption to aerosols in the boundary layer (see Section 5.2.5). Total deposition fluxes range by more than an order of magnitude for the main crustal elements, while the variability appears to be lower for metals such as Cr, V, Ni, and Co, which have both a crustal source and an anthropogenic source (Chapter 2). Metals such as Pb, Cu, and Cd show a large range in their values and this reflects the changes in their inputs over time, as noted earlier. For the crustal metals, changes have been relatively small in the last 30 years.

The contrast between the fluxes for the open ocean and those for the coastal areas, and marginal seas, is readily apparent as the fluxes are an order of magnitude or more higher (range 5–50 times higher) (Table 5.8). This is true both for the crustal metals and for the anthropogenically

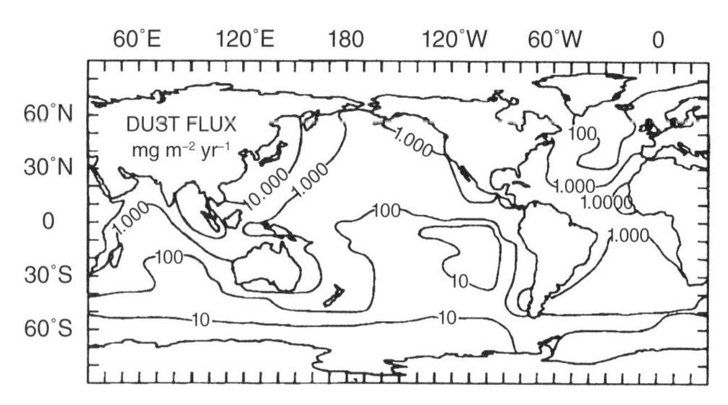

Fig. 5.11 Estimated dust deposition to the ocean as proposed by Duce et al. (1991). Figure reprinted from Duce et al. (1991), *Global Biogeochemical Cycles* **5**, 193–259 [37], with permission of the American Geophysical Union, copyright (1991).

derived metal(loid)s. The increased flux for the crustal metals is primarily due to the higher aerosol concentrations in the coastal environments, while the higher fluxes for the other metal(loid)s are mostly a consequence of their proximity to anthropogenic sources (Fig. 5.1 and 5.2). Estimates of the dust deposition to the open ocean readily demonstrate the heterogeneous nature of the inputs (Fig. 5.11) [37]. Dust fluxes decrease exponentially offshore and most of the open ocean has inputs that are two orders of magnitude lower that the surrounding coastal environments. Dust inputs are highest to the North Pacific from Asia, and the tropical Atlantic due to inputs from Saharan Africa. Overall, the remote Atlantic, being a smaller ocean, has higher inputs, while there is clearly a strong gradient between the Northern and Southern Hemispheres.

Additionally, many cities are located in coastal areas or close to the coast, and therefore contribute substantially to the coastal fluxes, which are highest within a local radius of 50–100 km. This is highlighted with the wet deposition data in Table 5.6. As noted, it is primarily for Hg, Pb and Zn that the fluxes measured in the urban environment are substantially greater than those representing the regional average value. In addition, continents with high urbanization and which typically have offshore air mass flow for all or part of the year (Asia and North America in the Northern Hemisphere, in particular) result in a substantial input of metal(loid)s to the coastal waters. The much higher fluxes for the coastal regions of Asia is a result of the current rapid urbanization in the region, as well as the substantial input of dust, particularly during the spring "dust period" (March–May) when a large fraction of the annual dust input occurs. In the case of Europe, it has been shown that due to meteorological conditions, there are periods where there is substantial transport from the continent to the Arctic region, and also clear evidence of contamination of the North Atlantic coastal waters and the Mediterranean, due to anthropogenic inputs [53, 54].

Overall, there is much uncertainty in the flux estimates as a result of the sporadic nature of the measurements and the lack of consistency in the metal(loid)s that are quantified in specific studies. Also, many studies are of relatively short duration. For Hg, there is more data and evidence of the distribution of deposition fluxes at a continental level. For example, there is a well-developed network of sampling sites in North America, the Mercury Deposition Network (MDN), under the auspices of the National Atmospheric Deposition Network (http://nadp.sws.uiuc.edu/mdn/) which monitors the concentration, and calculates deposition fluxes for many sites. Results for 2007 are shown in Fig. 5.12. The data clearly illustrates the large differences that can occur in deposition on a regional basis. For Hg, sources in North America are concentrated in the Midwest and southwest, and given the typical air flow from west to east, the two factors result in the dominance of deposition in the eastern section of the continent. The somewhat higher fluxes in the southeast portion is partially a reflection of higher rainfall amounts, and the lack of snow, and may also reflect the importance of atmospheric chemistry that results in the production of gaseous ionic Hg species. In Tables 5.7 and 5.8, the dry deposition estimate is for Hg aerosols only, and this is a small fraction of the total flux. The dry deposition of gaseous ionic Hg species is thought to be very important, and can rival that of wet deposition in some locations, as discussed in Section 5.5.2. Additionally, there is the potential for uptake of gaseous Hg species by vegetation (Section 5.5.2).

Modeling of deposition in Europe and North America has also provided similar indications of the spatial variability in deposition. While there are a number of models for Hg [55], the results for Europe for Pb and Cd (Fig. 5.13) are discussed briefly to illustrate that the comments earlier for Hg are generally applicable for all pollutant-derived metals [53, 56]. In addition to the temporal changes discussed above, deposition is highly variable spatially and this reflects the heterogeneous nature of the anthropogenic inputs. Maximum deposition occurs in central Europe and much of the deposition occurring in Scandinavia is the result of inputs elsewhere and long range transport. The model results show that emissions in Europe have a substantial impact on deposition to the Mediterranean Sea and to the North Sea, other marginal seas and the coastal North Atlantic. Some differences in the relative deposition between metal(loid)s are due to the mode of deposition and the primary aerosol size fraction for the metal(loid), but the model reflected in Fig 5.13 is clearly illustrative of the local component to deposition, reflective of the sources, and the more diffuse and relatively consistent regional signal [53].

For many metal(loid)s, there has been little attempt to construct global atmospheric budgets that are capable of demonstrating the impact of changes over time. The relative magnitude of change can be examined using lake sediment cores, ice cores and other historical archives, as discussed in Chapter 2. Overall, there has been little recent coordinated study of the global concentration and deposition of metal(loid)s and while there is a strong focus on inputs to

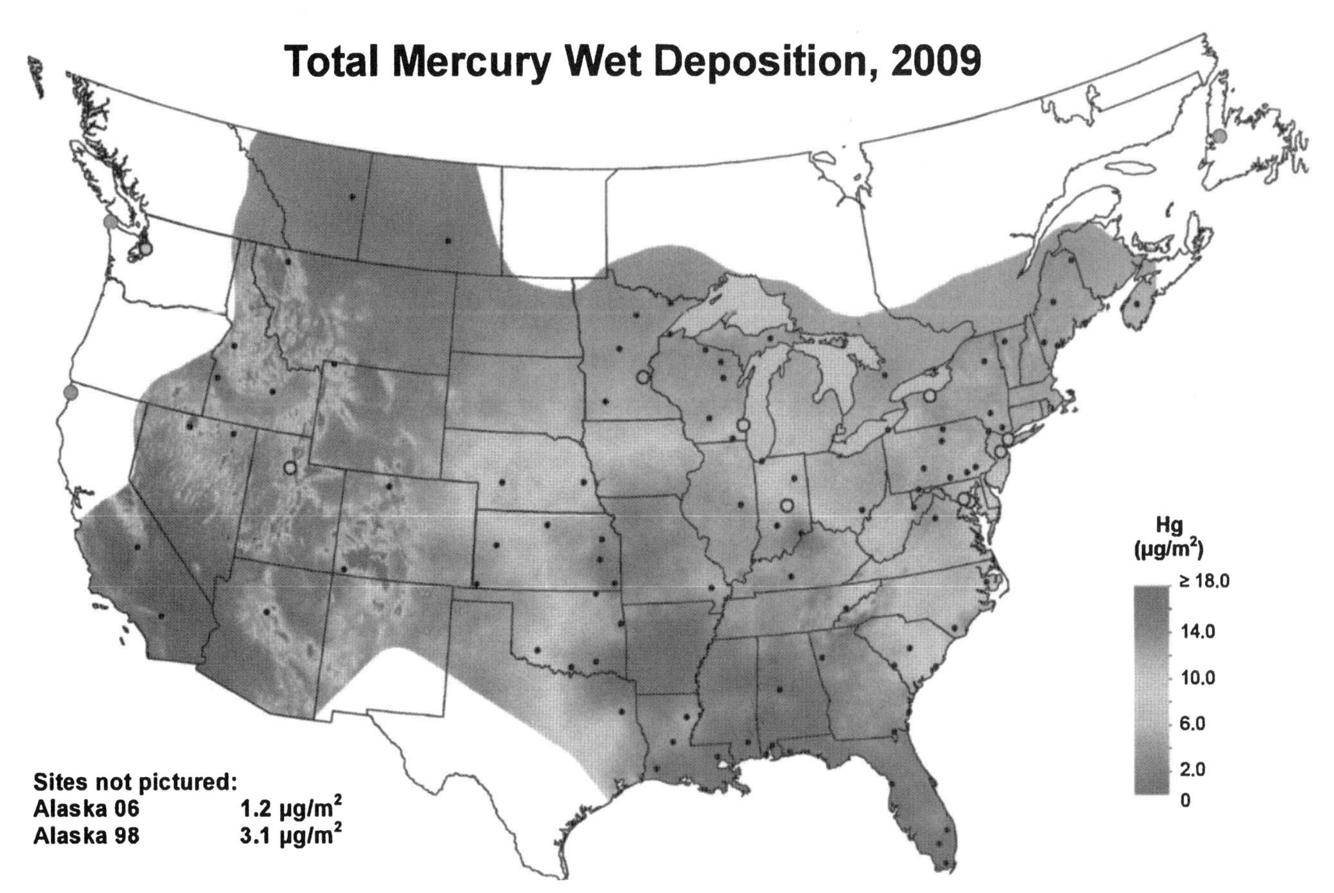

Fig. 5.12 Estimated wet deposition of total mercury for 2007 from the Mercury Deposition Network, part of National Atmospheric Deposition Program (Program Office, Illinois State water Survey, Champaign, IL, USA). Dots indicate sampling sites and contours represent extrapolation of the data. Data are given in $\mu g\,m^{-2}\,yr^{-1}$ which can be converted to $nmol\,m^{-2}\,yr^{-1}$ by multiplying by 5. Values therefore range from <20–>90 $nmol\,m^{-2}\,yr^{-1}$. Taken from the NADP/MDN website (http://nadp.sws.uiuc.edu/mdn/) and used with permission (National Atmospheric Deposition Program, (NRSP-3), 2007. NADP Program Office, Illinois State Water Survey, 2204 Griffith Dr., Champaign, IL, 61820).

the ocean, this is driven more by concerns of limitations of supply of essential metals, in contrast to examination of the metals that are likely most enriched by human activity. Two exceptions are Hg and Pb, whose anthropogenic input and change over time has been a focus of much research. The impact of changes in the amount of Pb deposition on concentrations of Pb in the surface Atlantic Ocean was discussed in Chapter 2. The input of Fe to the ocean is discussed in detail in Section 5.5.1. These discussions are illustrative of the overall mechanisms and cycling that is typical for most metal(loid)s that are strongly associated with particulate material in the atmosphere, and whose inputs to the ocean and other large water bodies is mostly dominated by direct and indirect atmospheric inputs.

5.2.5 Source apportionment of atmospheric metal(loid)s

One of the more difficult aspects in understanding sources and sinks is synthesizing and interpreting the information gained from studies of atmospheric metal(loid) chemistry, and of metal(loid) concentrations and fluxes to water bodies of interest. In determining and understanding the sources of particulate metal(loid)s and the metal(loid)s in wet deposition, it is necessary to have information on the sources, their relative contribution, and the change that occurs during transport and their rate of removal from the atmosphere. While there is substantial evidence for higher concentrations of metal(loid)s in urban environments, as discussed in the initial sections of this chapter, and such elevated concentrations are a human health concern, it is difficult to formulate regulations if the sources and fate of the metal(loid)s cannot be clearly demonstrated and understood. Additionally, there is the need to understand the impact of metal(loid) deposition on nearby waters. Overall, it has often been difficult to accurately assess the dominant sources and attribute contamination to a particular source. The overall complexity of the problem is illustrated in Fig. 5.14 [1]. Short-term measurements provide a snapshot of a complex system, and even longer-term datasets cannot provide sufficient information with measurement alone. As noted in Section 5.2, information about changes over time require a long-term dataset of collections.

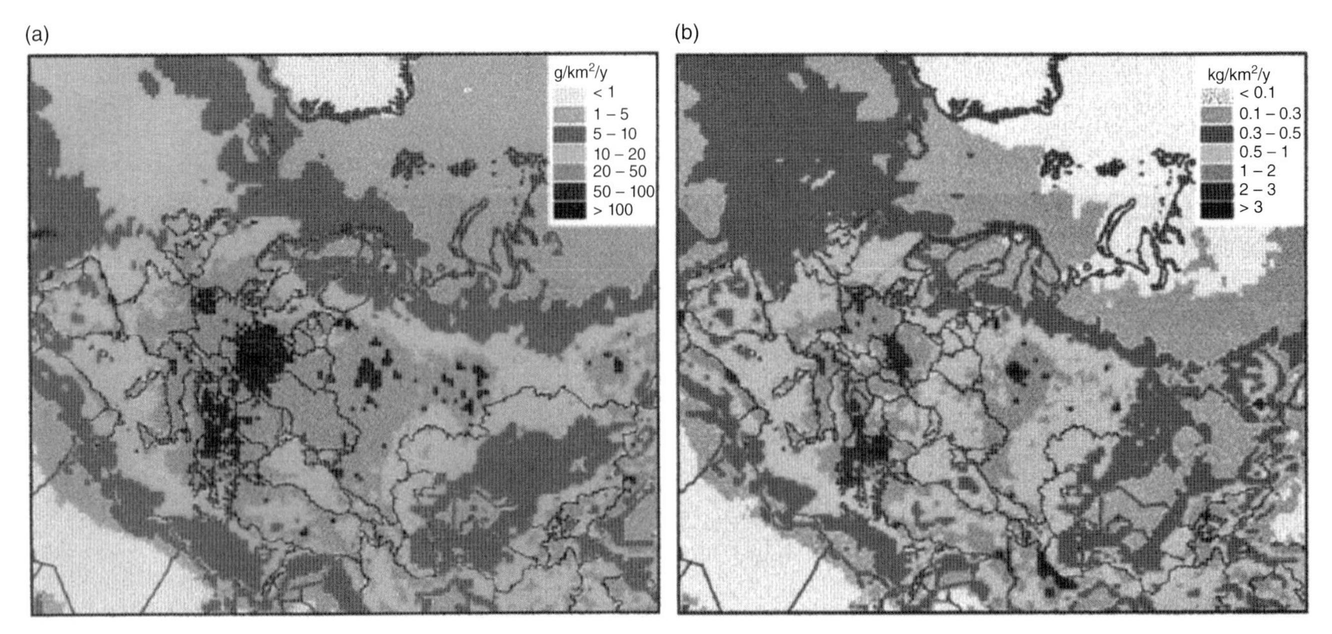

Fig. 5.13 Modeled total deposition of lead (Pb) (a) and cadmium (Cd) (b) for Europe for 1990 (within the so-called EMEP region). Data for Pb is given as $kg\,km^{-2}\,yr^{-1}$ and can be converted to $\mu mol\ m^{-2}\,yr^{-1}$ by multiplying by 4.83. Levels therefore range from <0.5–>50 $\mu mol\,m^{-2}\,yr^{-1}$. Similarly, the conversion for Cd is to multiply by 9×10^{-3}. So the concentrations range from <0.2–>4.4 $\mu mol\,m^{-2}\,yr^{-1}$. Figure reprinted from Ryaboshapko et al. [53, 56], and used with permission from the European Monitoring and Evaluation Programme.

In a particular instance, when a high concentration is measured at a receptor site, investigators use atmospheric models, such as the Hybrid Single Particle Lagrangian Integrated Trajectory (HYSPLIT) Model (http://www.arl.noaa.gov/ready.php) to estimate the source region of the air mass through back trajectory analysis. The HYSPLIT model is a *Lagrangian* model, which means that the model framework moves with the particular air mass and calculates and predicts its trajectory through space over time. Atmospheric models vary in complexity and have three main components – a model framework for emissions into the air mass, for transport of the air mass, and for physicochemical reactions and changes during transport [1]. The HYSPLIT model provides a simulation of the projected source and pathway for the air mass impacting the receptor but does not provide confirmation that a particular source or location is the major input of metal(loid)s measured at a downwind site. A complete model with emissions, reactions, transport and transformations is needed to determine the source and impact of any particular emission. A discussion of detailed atmospheric chemistry models is not within the scope of this book but their usefulness derives from their ability to integrate information, provide insight into the limitations of understanding of atmospheric chemistry and process, and to provide a predictive tool of the impact of changes in emissions or other factors.

Models broadly fall into two categories: (1) models that attempt to accurately reflect the physical and chemical processes of the atmosphere (numerical models); and (2) statistical models. Models are either Lagrangian, or *Eulerian*, (which means they are fixed in space, and the model follows change over time). Models are either *zero order* (a well-mixed box with no vertical or horizontal stratification; and changes are a function of time only); *first order* (when there is vertical changes with height as well), *second* or *third order* (three dimensional space and time) [1]. Three dimensional models were used to generate the output in Fig. 5.13 [56], and global models have been developed to examine cycling of contaminant metals such as Pb, Hg, and Cd. Many global models exist for other atmospheric species and examples are the Harvard GEOS-Chem model (http://www.as.harvard.edu/chemistry/trop/geos/), the Canadian GRAHM model, the Community Multi-scale Air Quality (CMAQ) model, and others. Specific mercury models have been developed [57–59].

Mathematical models, which can be used for projection and prediction of the impact of changes, are the models most used by environmental managers and regulators as they are an important policy tool. In contrast, statistical models have been developed that use correlation between variables to assess sources and their importance in contributing to contaminant loading at a specific location [1, 60]. These models have the advantage as they can be used in the absence of detailed information on emissions inventories and of meteorological data. Given their usefulness in examining sources to impacted environments, some discussion of statistical models and their application is warranted. The approach taken is to back calculate from the measured concentration at the receptor site to the combination of sources that could provide the chemical signature measured. Such *receptor models* require knowledge of the *chemical fingerprint* of

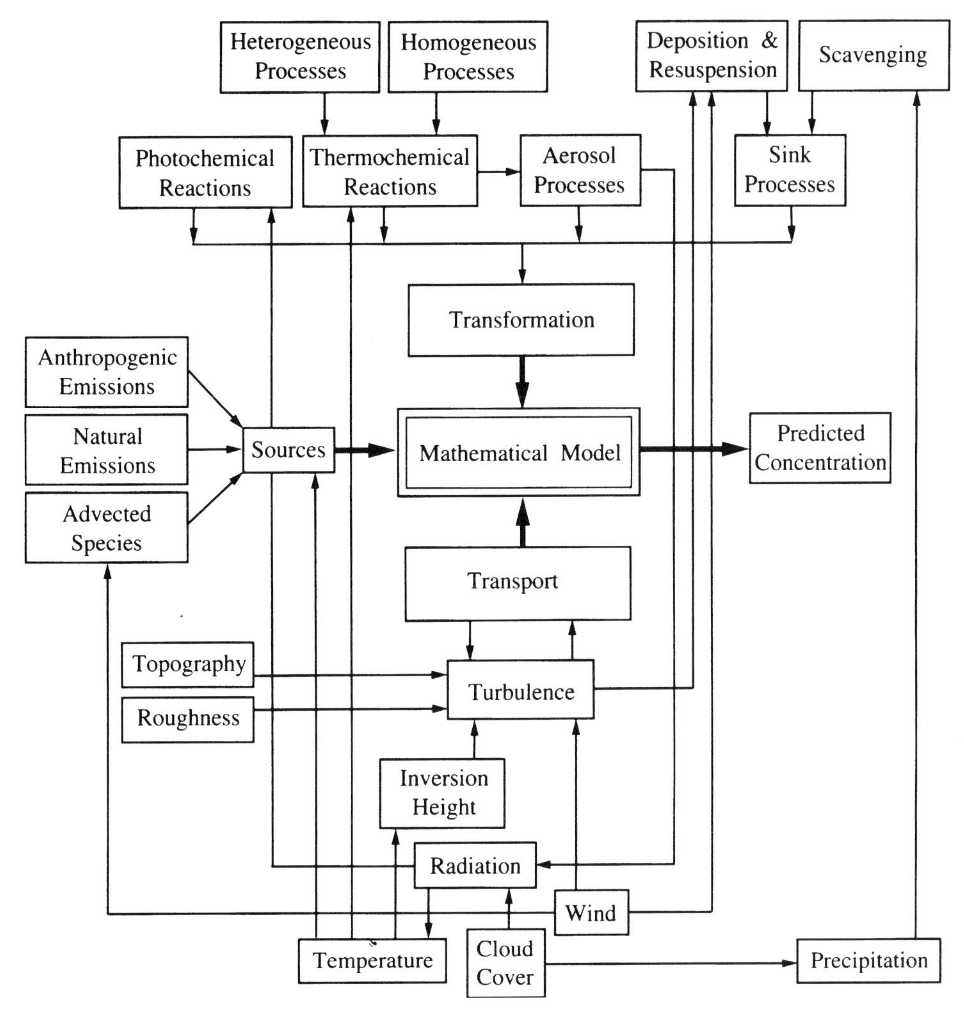

Fig. 5.14 Diagram illustrating the major components for a numerical computer model for modeling the atmospheric chemistry, fate and transport of chemicals in the atmosphere. Figure created using information from Seinfeld and Pandis (1998), *Atmospheric Chemistry and Physics* [1], and other sources. Reprinted with permission of John Wiley & Sons, Inc.

the sources but do not need to know the details of the dispersion processes that could have altered the actual, but not the relative concentrations from a particular source. This approach is powerful if there are a large number of different chemicals measured, or if some sources have a specific identifying signature, such as V being mostly emitted from oil combustion, and Se being emitted from coal burning [12]. The idea is to apportion the various contributions to the measured concentration to various sources, and the approach has been termed *source apportionment* receptor modeling.

The approach is formulated by Equation 5.10 [1]. For element i, with concentration c_i (moles m^{-3}) at a specific location, derived from j sources:

$$c_i = \sum f_{ij} \cdot a_{ij} \cdot s_j \quad j = 1 \ldots m, i = 1 \ldots n \tag{5.10}$$

where f_{ij} is the fraction representing any modification of the source composition aij during transit from source to receptor due to atmospheric processes, so the product is the fraction of species i from a specific source to the receptor, while sij is the total contribution (in concentration units) of the particles from source j. In many cases $f_{ij} = 1$, or is assumed to be 1. Thus, the concentration of any element at the receptor is a linear combination of the various sources and their fractional contribution. If c_{ij} and a_{ij} is known for each source, it is possible to minimize the error in the equation, and this provides the best estimate of the relative source contribution to the receptor site. Furthermore, if a number of samples have been collected, then the analysis can be performed for each case and the overall average value obtained from all the information available. Alternatively, the information can be used to examine differences due to changes in wind direction, season, or other variables. The various methods based on this approach are: chemical mass balance (CMB), which is used for source apportionment; principal component analysis (PCA) which is used for source identification, and an empirical approach that is a combination of these two.

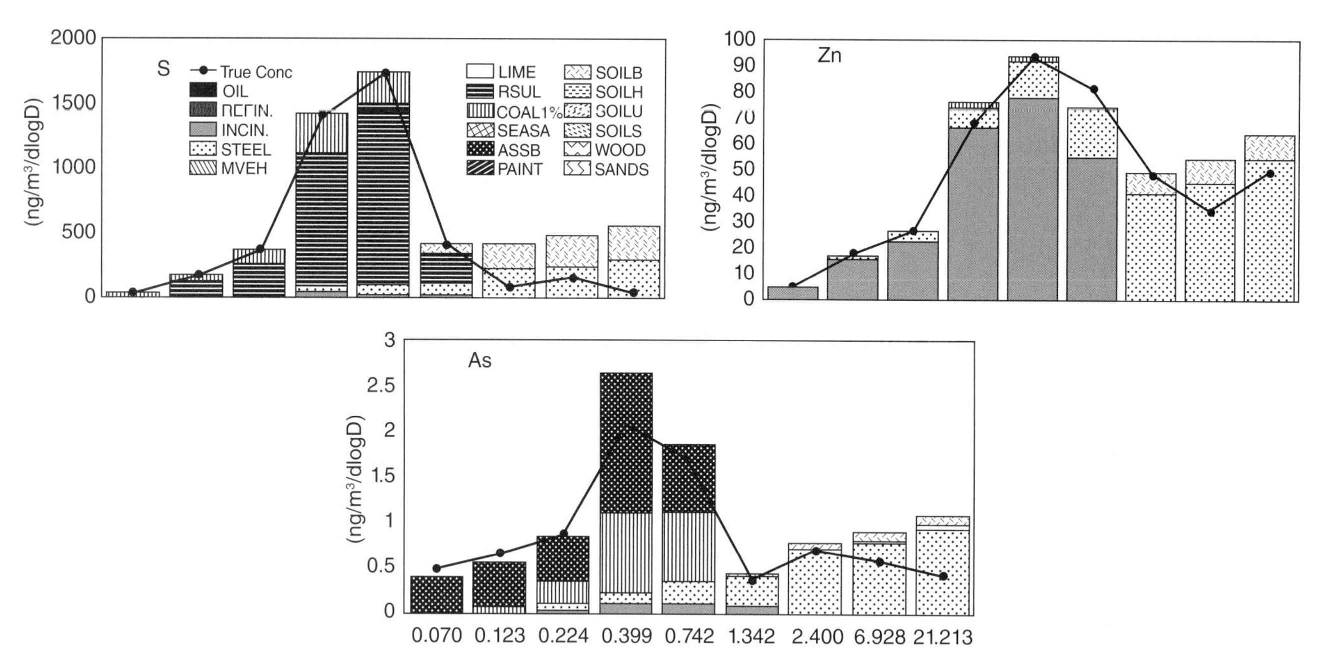

Fig. 5.15 The size distribution and attributed sources for aerosols collected in Baltimore, MD, USA for the metal(loids) zinc and arsenic and sulfur. Concentrations can be converted to molar units by dividing by the atomic mass. Reprinted with permission from Suarez and Ondov (2002) *Energy & Fuel* **16**: 562–8 [62]. Copyright (2002) American Chemical Society.

The CMB approach assumes that all possible sources have been identified and their profiles are known, and so the concentration c_i is the sum of its actual concentration plus an error term, and that the error terms are normally distributed and therefore have a zero average value, and can be statistically characterized by the standard deviation. The major assumptions of this approach are that sources are constant in composition, species are non-reactive during transport, all sources are known and uncertainties are random, uncorrelated and normally distributed. The number of sources must be less than the number of variables (elements i). These approaches were implemented by Gordon and co-workers [61] and used to examine sources in cities in the mid-Atlantic states of the USA. The CMB approach is more powerful if size segregated aerosols are sampled using cascade impactors or other devices as this gives further power to the discrimination between sources. For example, the data in Fig. 5.3 were collected and used in such an analysis [12]. A careful choice of the elements for the analysis can increase the resolving power enormously. For example, the soil signature is enriched in the crustal elements (e.g., Al, Sc, Cr, Mn, Fe, Co), sea salt is enriched in Na and Cl, oil-fired power plants are enriched in V, incinerators with Zn, and perhaps Cd, power plants with Se.

As an example, Suarez and Ondov [62] examined their data in terms of its human health impact. They analyzed size-segregated aerosol samples, collected with an impactor in Baltimore, MD, for up to 32 elements for CMB analysis with a 15-source model (Fig. 5.15). Agreement between measured and predicted concentrations was within 40% for all elements except Co, Cr, Cs, Ti, and W. The results suggest that coal-fired power plants are minor sources of the fractions of several metals, including Cr (19%), V (5%), and Zn (<1%). Coal combustion contributions to ambient levels were <30% for As, Fe, Co, Mn, and Sb. Major fractions of airborne V and Co (each about 55%), and Zn (70%) were attributed to aerosol emitted from fuel oil combustion and incinerators, respectively. Steel production was the major source for airborne Cr (45%), Fe (52%), and Mn 52%). Large (70 and 57%) fractions of airborne As and Sb were attributed to an unidentified source, which was resolved in the size spectra. Metal(loid) concentrations attributed to long-range transport, owing to their association with secondary sulfate were typically negligible, except for Mn (15%), Se (28%), and Zn (5%) [62].

In an application to data collected on the southern tip of the Korean peninsula, a remote site downwind of Asia, the techniques of positive matrix factorization (PMF) was used to analyze data collected for 19 elements in 8 size ranges, for samples collected over time [63] (Fig. 5.16a). The PMF approach uses the uncertainty in the data and in the measurements to constrain the limits of the values and this approach is better suited if there is missing data or values are below the detection limit. The approach pro-

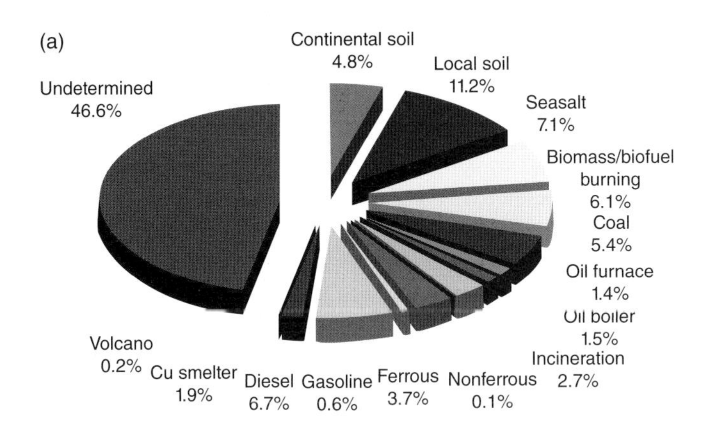

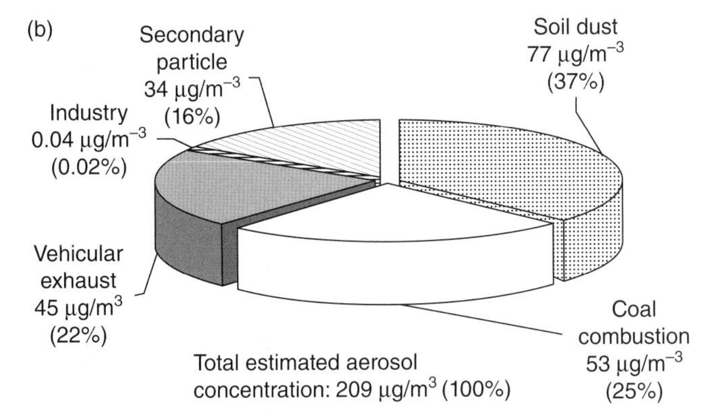

Fig. 5.16 (a) Attributed sources of aerosols collected at the southern tip of the Korean peninsula. Reprinted from Han et al. (2006) *Atmospheric Chemistry and Physics* **6**: 211–23 [63] with permission of the European Geosciences Union and the author; (b) a plot of the attributed sources of aerosols collected in Beijing, China. From Okuda et al. (2004) *Science of the Total Environment* **330**: 145–58, [64] reprinted with permission from Elsevier.

vides an optimal weighting of the errors inherent in the data. In the coarse factions, soil and sea salt sources accounted for 79%, while the fine fraction (0.56–2.5 μm) was dominated (60%) by coal combustion and biomass/biofuel burning aerosols. The ultra-fine fraction was attributed mostly to diesel sources (52%). Thus, there was both a long-range and local contribution to the aerosols collected at this remote site.

The Korean results are overall in agreement with a study done in Beijing, China [64], which could be considered in the upwind source region (Fig. 5.16b). This study used CMB approaches and focused on the PM_{10} fraction and found that there was a dominant soil component (37%), which likely accounted for most of the larger particles, as found in the Korean study. Coal combustion (25%), and vehicular sources (22%) were also major components (Fig. 5.16b), with secondary aerosols also contributing (16%). The total particulate concentration was high and variable ($170 \pm 120\,\mu g\,m^{-3}$). This is a high aerosol concentration compared to what is found in other cities, but overall the relative amount of natural versus anthropogenic sources is similar to other studies done in this region of the world. Relatively similar results were found in a study done in Seattle where $PM_{2.5}$ samples only were collected [65]. However, about 380 samples and 36 variables, including a number of carbon fractions, were used in the analysis and 11 sources were identified. Secondary S and N-containing sources accounted for 31% of the aerosol, diesel for 22%, natural soils and sea salt 19%, wood burning 16%, gasoline vehicles 10%, with all other sources contributing about 6%. The data differed by season. These examples illustrate the power of these modeling approaches for examining and identifying the major sources of particulate material in the downwind local field of anthropogenic and other emission sources.

5.2.6 Gaseous volatilization and gas exchange of metal(loid) compounds

As mentioned in Section 5.2.1, many compounds of metals and metalloids are volatile at room temperature and thus if these are formed in the atmosphere, or in surface waters, or terrestrial environments, there is the potential for these compounds to be exchanged between the surface and the atmosphere. The term *volatilization* is usually used to refer to the emission of gases from the terrestrial environment to the atmosphere while the term *gas exchange* is usually used to refer to exchange of gases across the air-water interface. Such processes are bi-directional and the direction is related to the relative concentration of the compound in the surface water relative to that of the atmosphere. To deal with this bi-directional exchange, Equation 5.1 needs to be modified to account for the processes involved. However, for many compounds, and in many instances, the flux will be only in one direction if there is supersaturation of the concentration of the compound in water relative to the air, and vice versa.

The expression governing the exchange depends on the depiction of the processes occurring at the water-air surface and while various model approaches have been used in the literature, here we will focus exclusively on the most widely used model, the *stagnant two-film model* (see Fig. 5.17) [27, 66]. In this model, the rate of gas exchange is assumed to be limited by the rate of molecular diffusion of the compound through the unstirred layer (air or water) that exists at the interface between the two media. Above these unstirred layers, the reservoirs are considered well-mixed with a constant bulk concentration. Thus, the coefficient for exchange in the unstirred layer (v_a for air or v_w for water) is given by $v = D/z$, where D is the relevant diffusion coefficient for that medium (typical units are $cm^2 s^{-1}$) and z is the thickness of the boundary layer (Fig. 5.17). The flux, F, according to Fick's Law, is related to these parameters and the concentration gradient between the interface (thin film)

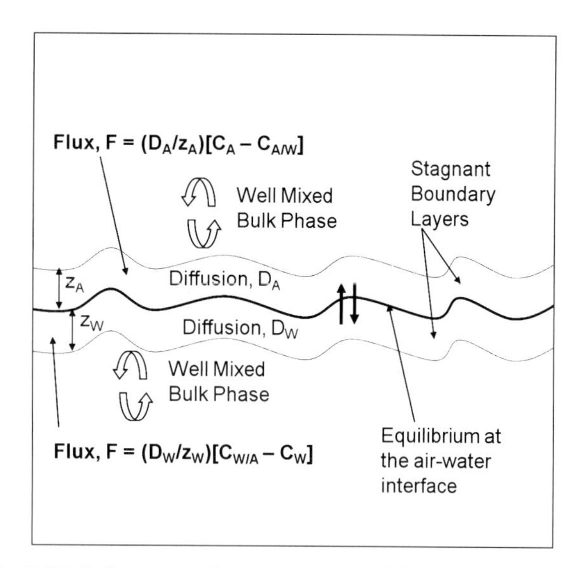

Fig. 5.17 A diagrammatic representation of the air-water exchange process as parameterized in the thin film model.

concentration and the bulk solution. At steady state, the fluxes on either side of the interface are equal and thus:

$$F = -(D_A/z_A)[C_A - C_{A/W}] = -(D_W/z_W)[C_{W/A} - C_W] \quad (5.11)$$

where C_A is the concentration in the bulk air, C_W is the concentration in the bulk water, $C_{A/W}$ is the concentration in the stagnant film on the air side of the interface and $C_{W/A}$ is the concentration in the stagnant film on the water side of the interface. The units of flux are mol $m^{-2}s^{-1}$. It is assumed that no reactions are occurring in the stagnant layers, which would obviously impact their concentration, and that the concentration in the stagnant layer is at steady state. The concentrations $C_{A/W}$ and $C_{W/A}$ are related to each other by the Henry's Law Constant (in the form of a dimensionless equilibrium constant (molar units), K_H) for the compound, which is given by:

$$K'_H = C_{A/W}/C_{W/A} \quad (5.12)$$

It is possible to combine the Equations 5.10 and 5.11 to derive an overall expression for the exchange at the interface:

$$F = [(z_W/D_W) + (z_A/D_A \cdot K'_H)]^{-1} \cdot [C_W - C_A/K'_H] \quad (5.13)$$

The direction of the flux will be determined by the concentration of the volatile compound in water relative to its equilibrium concentration determined by the bulk air concentration of the compound. The flux is positive and to the atmosphere is this is so, and negative and into the water if the reverse is true. As can be seen from Equation 5.13, the transfer coefficient is related to the transfer through both stagnant phases. Often, however, it is the transfer through one phase that dominates the overall exchange and therefore depending on the compound, a simpler formulation can be used. Also, the transfer coefficients are often referred to as *transfer velocities* or *piston velocities*, respectively, v_W and v_A, as their units are length $time^{-1}$.

The parameters z or v can be estimated based on a variety of approaches. Much work has been done on determining the transfer velocity of CO_2 and other highly volatile gases from water to air and this has been related to wind speed by a number of investigators. There are different formulations and approaches that have been used (Fig. 5.18) [66, 67]. For the ocean and large water bodies, the impact of breaking waves and bubble entrainment needs to be also considered and it is found that the relationship between wind speed and v_W is not linear, but increases relatively quickly at higher wind speeds (above about $10\,m\,s^{-1}$) [67]. A trace metal chemist, studying the gas exchange of a volatile compound, can use the information for these gases to estimate a value of v_W for the compound of interest if D_W is known (or alternatively the *Schmidt number*, Sc ($Sc = D_W/k$ where k is the kinematic viscosity of the solution)). This allows the calculation of the transfer velocity by using the information in Fig. 5.18 and the ratio of the diffusion coefficient to CO_2, for example [66]. This is the approach that has been taken in many of the studies of water-air exchange of mercury (elemental Hg (Hg^0)) being a dissolved gas in water. However, the various parameters are often not well-known and some of the estimation methods used for calculating either D or Sc are not well-constrained [68]. For example, the following relationships have been derived for organic compounds and simple gases [27]:

$$D_W\ (cm^2\ s^{-1}) = 2.3 \times 10^{-4}/V^{0.71} \text{ and } D_W = 2.7 \times 10^{-4}/m^{0.71} \quad (5.14)$$

where V is the liquid molar volume of the compound ($cm^3 mol^{-1}$) and m is the molecular mass (g mol^{-1}). For Hg^0, these formulations result in, respectively, values for D_W of $3.3 \times 10^{-5} cm^2 s^{-1}$ and $0.62 \times 10^{-5} cm^2 s^{-1}$. It is noted that the second formulation does not work for liquids of high density and therefore this likely accounts for the differences for Hg^0. Alternatively, Loux [68] in his review of the literature notes the wide range in values of D_W used by various authors and calculated, based on a new formulation, a value of $0.72 \times 10^{-5} cm^2 s^{-1}$. Clearly, the choice of the D_W could have a large impact on the estimation of gas exchange. Similar formulations have been derived for D_A [27]:

$$D_A\ (cm^2\ s^{-1}) = 2.35/V^{0.73} \text{ and } D_A = 1.55/m^{0.65} \quad (5.15)$$

The typical values for D_W are on the order of 0.5–$5 \times 10^{-5} cm^2 s^{-1}$ while those for D_A are around 0.05–0.5 $cm^2 s^{-1}$, or a ratio (D_A/D_W) of about 10^{-4}. Typical values for z_W are 0.2–2×10^{-2} cm and for z_A, 0.1–1 cm. Thus, the typical ratio of v_W/v_A is around 0.02. Recall, from Equation 5.11, that the

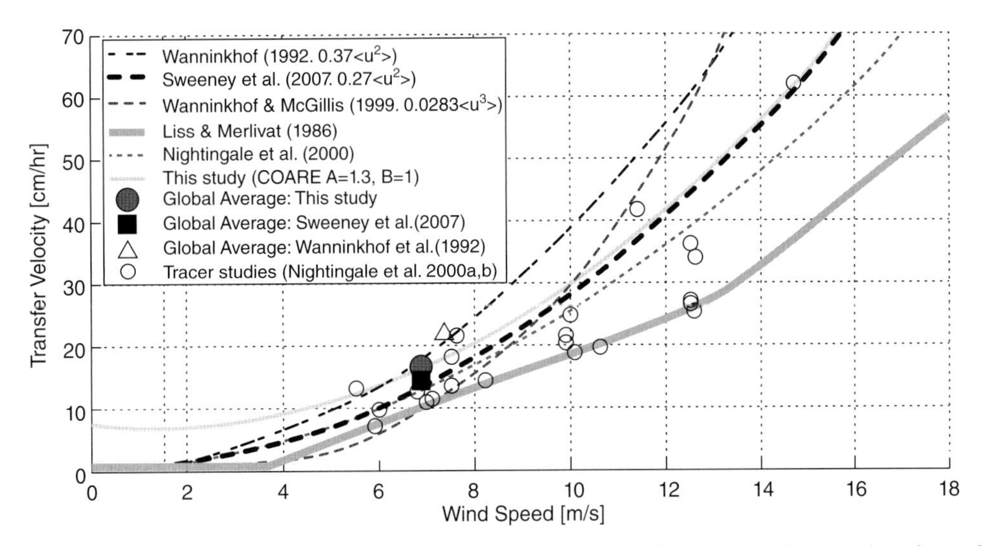

Fig. 5.18 The relationship between carbon dioxide (CO_2) gas exchange coefficient and wind speed showing the relationship for various formulations published in the literature. From Jeffrey et al. (2010) *Ocean Modelling* **31**: 28–35, reprinted with permission from Elsevier.

overall transfer coefficient is determined by the transfer between both stagnant layers. If we use the same resistance approach and the formulation discussed previously for dry deposition, of which the current discussion is a special case, then:

$$1/v_{tot} = 1/v_W + 1/v_A K'_H \quad (5.16)$$

Thus, if v_W is much smaller than $v_A K'_H$, then transfer will be dominated by the water boundary layer, and vice versa. Given the ratio of the values of v, it is clear that v_W will only dominate under conditions where K'_H is small ($<10^{-2}$). Note that this value of K'_H is in terms of molar concentrations and values in the literature are normally given with the air concentration in atmospheres (K_H). Thus, there is a need to convert the value to the appropriate units ($K'_H = K_H/RT$). Values of K'_H for gases such as CO_2 and even H_2S are $<10^{-2}$ and so their gas exchange is dominated by water side transfer. This would be the same for Hg^0 and most neutrally-charged methylated or other volatile metal complexes.

For many of the heavy metals/metalloids that form volatile species, there is more evidence for their existence and release from terrestrial environments than there is for their presence in aquatic systems. The discussion here is focused only on the volatile compounds and there is a more thorough and general discussion of organometallic compounds in aquatic systems in Chapter 8, which covers the formation and cycling of methylated compounds in more detail. A compilation of most common identified volatile metal(loid) species in the environment and the conditions of their detection are given in Table 5.9 [8, 10, 69]. Mostly the species are methylated compounds, hydrides or related species.

Table 5.9 Listing of volatile compounds of metals and metalloids that have been detected in environmental media, and especially in aquatic environments. Data taken from [8, 10, 69] and references therein.

Element	Species	Locations
Group 14 (Ge, Sn, Pb)	$(CH_3)_2SnH_2$, $(CH_3)_4Sn$, $(C_4H_9)_xSn(CH_3)_{4-x}$ (x = 1–3), $(C_2H_5)_xSn(CH_3)_{4-x}$ (x = 1–3), $(CH_3)_4Pb$, $(CH_3)_xPb(C_2H_5)_{4-x}$ (x = 1–3), $Pb(C_2H_5)_4$	Anaerobic, field and lab data, landfill gas (LG), sewer gas (SG), sediment (tributyltin is anthropogenically introduced), estuarine waters
Group 15 (As, Sb, Bi)	AsH_3, $(CH_3)AsH_2$, $(CH_3)_2AsH$, $(CH_3)_3As$, SbH_3, $(CH_3)_3Sb$, $(CH_3)_3Bi$	Anaerobic, field and lab data, landfill gas (LG), sewer gas (SG), hydrothermal locations (HT). marine sediments
Group 16 (Se, Te, Po)	$(CH_3)_2Se$, $(CH_3)_2Se_2$, $(CH_3)_2SSe$, $(CH_3)_2Te$	Anaerobic, field and lab data, LG, SG, wetlands (WT), estuarine waters, marine sediments
Hg	$(CH_3)_2Hg$, Hg^0	Lab and field, LG, SG, HT, WT, fresh and marine waters
Transition Metals	$Ni(CO)_4$, $Mo(CO)_6$, $W(CO)_6$	Field data, LG, SG

Besides Hg, which has been extensively studied, there is little detailed information on the release of these compounds from environmental waters, except for information on the concentration of some species in surface waters [8, 69, 70]. Given the detailed study and interest in volatilization of Hg, it is dealt with as a special focus topic in Section 5.5.2.

The sources of these compounds in the aquatic environment are two-fold. Some of the compounds, such as $(CH_3)_4Sn$ and $(C_4H_9)_3SnX$ (x = halide or other anion; TBT), were introduced into the environment by human activity. Alkyl Pb additives in gasoline and tributyltin and other alkylated Sn compounds have been used as antifouling paints in aquatic environments, as agricultural products and wood preservatives, for example. Other compounds are produced as a result of chemical reactions and microbial transformations. For example, abiotic alkyl group transfer reactions are possible, but it is more likely that the formation of alkylmetal compounds in the environment is microbially mediated. For example, Meyer et al. [69] discuss in detail the formation of volatile organometallic compounds of As, Se and Te by Archea, while other studies have clearly shown the role of bacteria, and even phytoplankton in the production of such compounds. It is not difficult to conceive that pathways that result in the formation of dimethylsulfide (DMS, $(CH_3)_2S$), which is known to be produced as a result of phytoplankton activity and photochemical degradation in ocean surface waters [71]. Such pathways could also produce $(CH_3)_2Se$, $(CH_3)_2SSe$, and $(CH_3)_2Te$ [72], given that Se and Te are found in the same group of the periodic table as S. These authors showed that photodecomposition of selenoamino acids produced significant amounts of volatile selenium species in both light and dark conditions in the laboratory, with $(CH_3)_2SSe$ and $(CH_3)_2Se_2$ being the major products, with small amounts of $(CH_3)_2Se$ being formed. Inorganic selenium oxyanions did not produce any volatile products. It was hypothesized that formation of H_2O_2 under the laboratory conditions used reacted with the organic Se compounds producing the products.

In ocean surface waters, during a phytoplankton bloom, volatile Se compounds were also found [70]. The relative production of the methylated Se compounds compared to DMS was constant further suggesting a link between the production pathways of volatile S and Se compounds. In contrast to the laboratory experiments, $(CH_3)_2SSe$ and $(CH_3)_2Se$ were the major products, with small amounts of $(CH_3)_2Se_2$ being formed. These authors estimated the total volatile Se flux to be $6.4\,nmol\,m^{-2}\,d^{-1}$, which are of the same order as the estimated Se deposition to this ocean region ($1.5–2.5\,nmol\,m^{-2}\,d^{-1}$). The presence of gaseous Se compounds has been determined in the atmosphere and measurements showed that about 20–30%, on average, of the total Se in the surface atmosphere was in the vapor phase. Locations studied include open ocean sites, in the Atlantic and Pacific, as well as coastal locations. Vapor Se concentrations ranged from $<0.1–9.5\,pmol\,m^{-3}$, with much higher concentrations in coastal locations.

Therefore, it is likely that there is production of volatile organometallic compounds in many regions of the aquatic environment, and that their production is mediated by microbes, and that the lack of information on their fate and transport in aquatic systems is due to a lack of study, rather than a lack of their presence at measurable concentrations. As noted in Table 5.9, and by Feldman [10] and others, these compounds are easily detected in environments that have received an enhanced input of the precursors, or which have anoxic conditions and elevated microbial activity, such as landfills and sewage treatment facilities. For example, it appears that methane-producing organisms are also the dominant organisms producing alkylated metalloids compounds [10, 69].

Cima et al. [73] discuss the presence of alkylated Sn compounds in the environment. While TBT is degraded in the environment to the di and mono-butyl compounds, it can also be reduced, forming volatile hydrides, or even methylated to form the mixed alkyl species shown in Table 5.9. Water to air fluxes of Sn compounds have been estimated to range from $20–510\,nmol\,m^{-2}\,yr^{-1}$, and these compounds are generated by the various pathways mentioned. The estimated sediment to water flux is estimated to be of comparable magnitude so the atmosphere appears to be an important sink for organotin compounds released into the water column. Water concentrations have decreased from 200–400 nM since the introduction of legislation banning TBT use, and concentrations in coastal waters are now in the range of <1–10 nM, even in harbors, and levels of the di and mono compounds are of the same order [73]. River concentrations are somewhat lower, typically <5 nM. Given the uses of TBT, much of the data collected is for coastal waters and freshwaters, and there is little information about concentrations and fluxes for the open ocean.

Tessier et al. [8] developed methods for the simultaneous measurement of a number of volatile metal(loid) species (cryogenic trapping, GC separation, ICP-MS) and used these techniques to investigate concentrations of these compounds in various estuaries in France. The results of their investigations are shown in Fig. 5.19, and these results build on a number of earlier studies. They were able to detect the methylated Se species, a variety of Sn compounds, as well as Hg^0. Many of the compounds were found at low concentrations (e.g., $(CH_3)_2Se_2$, $(CH_3)_2SSe$, $(C_4H_9)_xSn(CH_3)_{4-x}$ ($x = 1 - 3$), $(CH_3)_4Sn$ and $(CH_3)_2Hg$) and the dominant species for each element is shown in Fig. 5.19. As discussed further in Section 5.5.2, Hg^0 concentrations varied seasonally, being low in winter. Conversely, the concentration of $(CH_3)_2Se$ was higher in spring and may reflect the likely production in conjunction with phytoplankton activity. For Sn and Hg, concentrations of volatile compounds were in the sub-pM range while concentrations were in the pM range for $(CH_3)_2Se$. In each case, there was some evidence of a gradient across the estuary that is likely related to water mixing and changes in ecosystem productivity. Concentrations of alkylated Sn compounds in the estuarine sediments were elevated, especially in the Gironde, suggesting that the source of the volatile compounds in the water is flux from

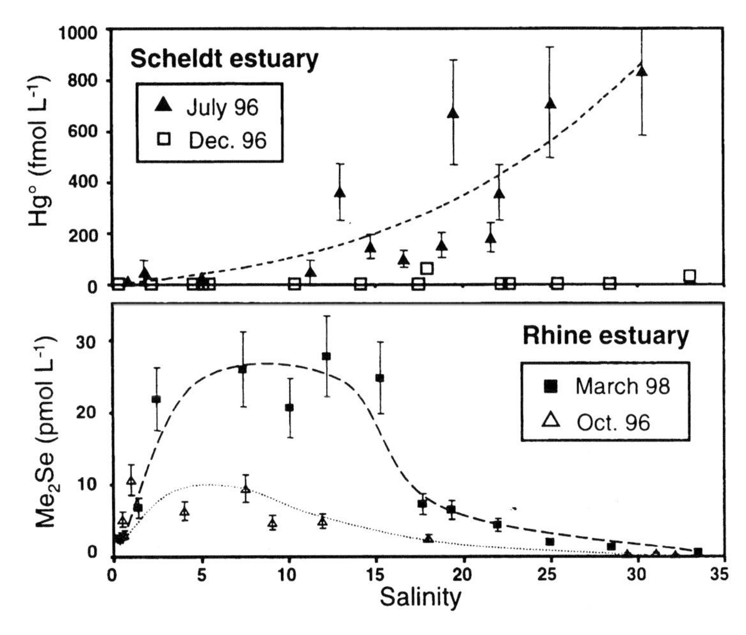

Fig. 5.19 Measurements of various dissolved volatile species in European Rivers. Reprinted with permission from Tessier et al. (2003) in *Biogeochemistry of Environmentally Important Trace Elements*, ACS Publications [8]. Copyright (2003), American Chemical Society.

sediments, and probably methylation of TBT and its breakdown products within the sediment realm.

These investigators [8] estimated fluxes to the atmosphere for various European estuaries, of 0.5–0.7 $nmol\,m^{-2}d^{-1}$ for Hg^0, 3–30 $nmol\,m^{-2}d^{-1}$ for volatile Se compounds, and 0.04–1 $nmol\,m^{-2}d^{-1}$ for Sn compounds. Thus, the estuarine fluxes of volatile Se compounds was higher than that of the North Atlantic [70]. The authors estimate that the emissions from European estuaries to the atmosphere are comparable to a few percent of the emissions from fossil fuel combustion for Hg and Se, but not for Sn. Clearly, however, the volatilization of Sn from these locations to the atmosphere and the likely subsequent decomposition and deposition of Sn back to the Earth's surface provides a mechanism and pathway for the transfer of Sn trapped in estuarine sediments to other environmental locations.

A study in the hydrothermal environment at Yellowstone National Park, USA found that there were volatile As compounds emitted in locations with high aqueous As (up to 50 μM levels) [74]. Concentrations of volatile As in surface waters were from <1–2.5 μM. Species identified were $(CH_3)_2AsCl$, $(CH_3)_3As$, $(CH_3)_2AsCCH_3$ and $(CH_3)_2AsCl_2$. There are also some reports of volatile As compounds in some marine environments [75]. It has also been shown that volatile Se compounds are emitted from freshwaters, as a study in Great Salt Lake, USA, found evidence for volatile Se compounds, although the exact species present was not determined [76].

Studies of volatile (gaseous) metal(loid) compounds in the atmosphere, besides Hg and its methylated compounds, are limited, although there were many early studies looking into volatile Pb compounds during its use as a gasoline additive, and some studies of Se, where it was shown that about 20% of the Se was in the gas phase [75]. A recent study of the atmosphere in an urban setting [77] still found the major species to be alkyl lead compounds in outdoor air. The presence of such compounds is consistent with their presence in Arctic ice cores [78], and their historical use. The major components were $(Alk)_4Pb$ (Alk = CH_3 or C_2H_5; respectively 31% and 58% of the total) and there was evidence for all the mixed alkyl species [$(C_2H_5)_xSn(CH_3)_{4-x}$, (x = 1–3)], although the relative ratios were not exactly consistent with that expected from gasoline. Total alkyl lead (TAL) concentrations ranged widely, depending on location, from around 1 $ng\,m^{-3}$ to values 50 times greater for open air sites. The authors discuss these concentrations (average TAL 15.5 $ng\,m^{-3}$) in terms of historical levels found in other cities during times when Pb was added to gasoline and show that these levels are lower on average by about an order of magnitude than values reported in the 1960s. Mercury was found in all samples and there was also evidence for $(CH_3)_2Se$ in some indoor samples. A volatile Cr species, likely a carbonyl, was suggested based on the data collected, and this is consistent with the information in Table 5.9.

While the information in the literature on volatile metal(loid) species in aquatic systems, and their gas exchange with the atmosphere, is limited, there is clear evidence for their production within the sediment, and likely within other anoxic regions, such as seasonally low oxygen waters. It does appear that microbial production of these compounds is the main route of their formation, although more studies are needed to confirm this notion. However, extrapolation of data from other environments, such as landfills and sewage systems, and from laboratory culture experiments, suggests that this is a relatively common occurrence in the aquatic environment. More research should be aimed at examining the sources and cycling of volatile metal(loid) compounds in aquatic systems.

5.3 Atmospheric chemistry and surface water photochemistry of metals

The principal chemistry of the atmosphere is driven by atmospheric reactions and processes that are little impacted by trace metals [1, 79], except perhaps the chemistry of Fe, and is mostly driven by other photochemical reactions. The most important photochemical reaction for life is photosynthesis but there are many abiotic photochemical processes that have a substantial impact on the chemistry of the atmosphere. While water is a minor component of the atmosphere ($<10^{-6}$ %), it does provide the medium for many homogeneous and heterogeneous photochemical reactions and these interactions, which are mostly redox reactions, do often involve trace metals and other reactive metal-containing constituents. In the gas phase, there is little chemistry that is important in terms of trace metals, although such atmospheric reactions lead to the generation of acidic species, due to the oxidation of trace gases (SO_2, NO_x; x = 0.5, 1, 2 etc.), and the formation of acidic aerosol. These reactions impact the overall aerosol composition, and therefore alter factors such as the degree of dissolution of trace metals in precipitation and other natural waters, as discussed in Section 5.4 later. Mercury can exist in the gas phase in either the elemental form, or as gaseous ionic species, or as methylated Hg, and this is due to the low, but measurable vapor pressure of Hg compounds, such as $HgCl_2$ and CH_3HgCl; $(CH_3)_2Hg$ is highly volatile, but relatively unstable to photochemical decomposition (see Section 5.5.2). While there has been little study of other trace metal(loid)s in the gas phase in the atmosphere, there is also likely to be methylated compounds of other metals present, and some of these compounds are soluble in atmospheric waters (see Section 5.2.5). Overall, many atmospheric reactions, and anthropogenically-impacted situations, such as smog and ozone generation, are due to processes which are photochemical. Of all the metals, Fe is the most important in terms of its impact and role in these photochemical transformations [79], but other redox-sensitive metals that exist in the atmosphere in sufficient concentration, especially Cu but also Mn, can influence the overall chemical reactions that are occurring.

Initiation of photochemical reactions occurs when a substance absorbs solar radiation and is transformed into an *excited* (more reactive) state [79]. Such excited entities can decompose directly into products, or could produce further reactive species which would be involved in further chemical reactions. Thus, the overall process is either a direct (*primary*) *photolytic* reaction or the reaction can occur as a result of *secondary* reactions. In natural surface waters and in atmospheric waters, the primary adsorbing species (*chromophores*) are: (1) organic chemicals, which are either small molecular weight compounds, such as organic acids (e.g., formic, acetic and oxalic acid) or larger, complex NOM, both dissolved and particulate; (2) oxides and other reactive surfaces (aerosols); and (3) dissolved constituents, such as nitrate, which can be photolyzed by the shorter wavelength (UV) radiation. The absorption of radiation occurs at a very high rate and therefore the reactions involved, and the species produced, play a fundamental role in the chemistry of atmospheric waters, as well as the chemistry of surface waters.

The absorption (A) of light is expressed in terms of *Beer's Law*:

$$A = \log(I_0/I) = \varepsilon l C \tag{5.17}$$

where I_0 and I refer to the incident and transmitted radiation, ε is the molar extinction coefficient for the molecule of interest ($l\,mol^{-1}\,cm^{-1}$), and l is the transmission length (cm). In photochemical reactions, the energy of the absorbed radiation determines the extent of reaction. The *quantum yield* (φ_λ) for a particular molecule absorbing at a particular wavelength (λ) is the ratio of the number of moles reacting relative to the number (moles) of photons absorbed. The direct *photolysis rate* (ν_λ) depends on the incident light intensity and φ_λ, and is related to the concentration of the chromophore (assuming low concentrations in solution) [79].

The primary photochemically produced species in natural waters include singlet oxygen, the superoxide ion ($O_2^{-\bullet}$), hydrogen peroxide (H_2O_2), the hydroperoxyl radical ($HO_2^\bullet$), ozone (O_3), the hydroxyl radical ($OH^\bullet$), the hydrated electron (e^-_{aq}) and organic peroxy radicals ($ROO^{-\bullet}$, where R is the remainder of the organic molecule). The main reactions forming these compounds are the interaction of light with NOM, with Fe(III) complexes, and the decomposition of dissolved species such as H_2O_2, NO_2^- and NO_3^- [79]. Most of these photochemically-derived species are highly transient and reactive in natural waters and therefore their concentrations are typically low, but at a steady state, given their high rates of formation and decomposition. Their atmospheric concentration varies over a diurnal cycle and with location in concert with the factors influencing their rate of formation and destruction. In most cases, the steady state concentration can be estimated if the important controlling reactions that are occurring in a particular solution can be determined.

For Fe, it has been shown that $Fe(OH)^{2+}$ is the principal photoactive species with the following reaction occurring with the adsorption of light [79, 80]:

$$Fe(OH)^{2+} \rightarrow Fe^{2+} + OH^{\bullet}$$

This reaction can be thought of as a ligand to metal electron transfer reaction and the $OH^{\bullet}$ produced is highly reactive, reacting with organic compounds, such as benzoic acid [81]. Products were H_2O_2 and Fe^{II}. The ligand is initially oxidized and then the charge transfer reaction leads to the reduction of the Fe. The values of φ_λ for the complex decrease from 0.14 at ~300 nm to 0.017 at 360 nM [66], and thus this reaction requires higher energy radiation (UV) to proceed. Similar reactions can occur with other organo-Fe complexes, as shown here for Fe-oxalate, and mostly occur at low wavelengths. The reactive product in this case can further react with oxygen, forming a series of reactive species [79]:

$$Fe(O_2C_2O_2)^+ \rightarrow Fe^{2+} + C_2O_4^{-\bullet}$$

$$C_2O_4^{-\bullet} + O_2 \rightarrow O_2^{-\bullet} + 2CO_2$$

$$2O_2^{-\bullet} + 2H^+ \rightarrow H_2O_2 + O_2$$

Such reactions can also occur with Fe complexation to more complex organic molecules. Besides organic acids, EDTA and other similar complexing ligands and large organic molecules (NOM), if complexed to metals, can all mediate the charge transfer reactions. Therefore, while NOM can be an important source of reactive species through homogeneous photochemical production and NOM degradation, it has been shown that Fe and other dissolved metal-organic matter complexes can also be important chromophores, because of the potential internal charge transfer reactions occurring in these complexes, leading to the reduction of the metal and the production of a reactive organic species. In addition to Fe, Cr, Mn, Cu, Co, and Hg are all metals that could be involved in similar reactions as the metals are reducible under typical light conditions encountered in natural waters, and these reactions are enhanced by their complexation to organic molecules. The major role of the complex is to shift the wavelength of absorption to longer wavelengths (i.e., increase the φ_λ at longer wavelengths), so that, given the relatively low penetration of the shorter UV wavelengths into natural waters, this extends the spectrum over which absorption can take place and enhances the overall extent of photochemical reaction.

In atmospheric waters, concentrations of low molecular weight organic acids, such as oxalate, are relatively high, as is the Fe concentration, and therefore these processes are important in the overall aqueous chemistry of the atmosphere (see Fig. 5.20) [79, 82, 83], especially as the

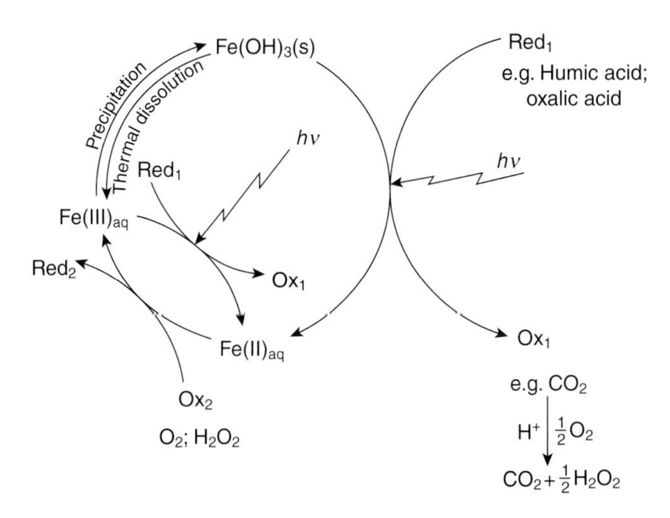

Fig. 5.20 The photochemical redox cycle of iron (Fe) and the principal mechanisms whereby the oxidation and reduction reactions occur in surface waters and in atmospheric waters. Figure published in Stumm and Morgan [79] but from Siffert and Sulzberger (1991), *Langmuir* **7**: 1627–34 [83]. Reprinted with kind permission of Springer Science+Business Media.

concentration of the organic acids can be exacerbated by anthropogenic inputs, and the amount of Fe in aerosols in the atmosphere is a function of natural and human-induced processes. Also, it is known that Cu^{II} can be complexed and photoreduced; the most cited examples being that the Cu^{II}-cysteine complex is unstable to Cu reduction [66] and the complex with EDTA and other aminopolycarboxylic acids are photoactive complexes [84]. A study of the speciation of Cu in rainwater [85] showed that the Cu^{II}/Cu^{I} ratio was higher in summer than in winter, and there was evidence of complexing ligands for both oxidation states of Cu, and this appeared to account for the stability of Cu^{I}, stabilizing it and hindering its oxidation. Also, it appears that H_2O_2 is the major oxidant for Cu^{I} in natural waters [84]. The type of complex formed is important to the rate of reaction, as shown by the differences in the quantum yields for the Cu complexes (malonate > succinate > glutarate > adipate~pimelate) [86] (Fig. 5.21). These results were explained in terms of the differences in the types of complexes formed (increasing degree of outer-sphere coordination of the complexes). Overall, the mechanism is similar to that outlined above for the reaction of Fe with oxalate, but in this instance, CO_2 is released prior to the decomposition of the Cu-organic radical complex. In complex mixtures containing both Cu and Fe, there is also a potential interaction if both Fe and Cu are present at sufficient concentration through the reduction of $FeOH^{2+}$ by Cu^{I}.

In addition to direct photochemical reduction, Fe^{III} compounds can also react with $HO_2/O_2^{-\bullet}$. In cloud droplets and

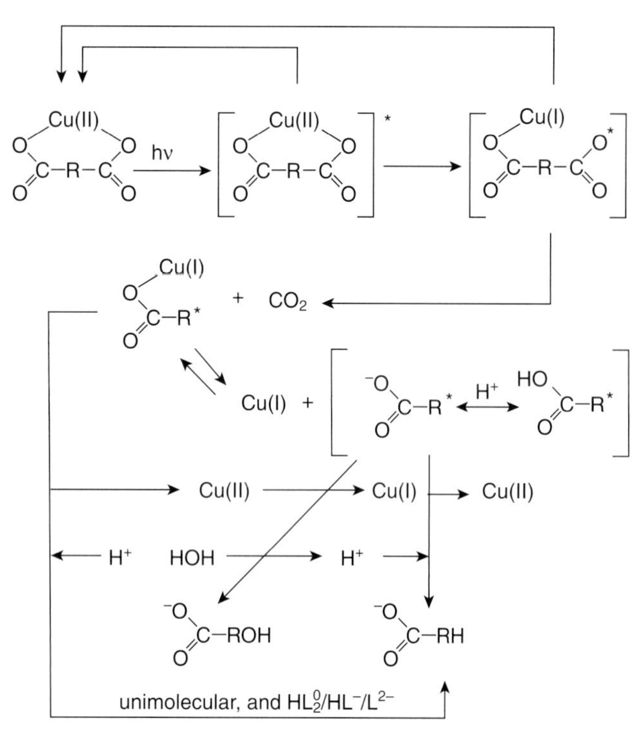

Fig. 5.21 Reaction mechanism for the photochemical reduction of copper in the presence of dicarboxylic acids. Reprinted with permission from Wu et al. (2000), *J. Physical Chemistry A* **104**: 4989–96 [86]. Copyright (2000), American Chemical Society.

surface waters, the sources of $HO_2/O_2^{-\bullet}$ include the transfer of HO_2 from the gas phase and the photolysis of Fe^{III}-polycarboxylate complexes, and their consumption is dominantly by Fe^{II} if present at sufficient concentration. However, when the concentration of Cu in the cloudwater or surface water is elevated, this reaction dominates as the reaction rate with Cu^{I} is five times faster [87]. In all cases, H_2O_2 is produced. These reactions are also major mechanisms for the oxidation and solubilization of acid gases, such as SO_2 [79]. Further discussion of the atmospheric and photochemistry of Fe is given in Section 5.5.1.

Contrasting this, in a study of cloudwater [88], and from the model results, it was concluded that Cu^{I} and Cu^{II} were predominantly as free metal ions (Cu^{+} and Cu^{2+}), with less than 10% as Cu^{II} organic complexes. The model results also indicated that Fe^{II} was the dominant oxidation state of Fe during daylight, in agreement with the field measurements, and that Fe^{III} dominated at night, again in agreement with observations. For other metals, model predictions were that Cu^{II} and Mn^{II} were dominant oxidation states but that the Cu^{I} and Mn^{III} fraction increased during daylight. It was predicted that Cr^{VI} in cloudwater would be reduced to Cr^{III} if free S^{IV} is present.

Studies of Cu in streams have also shown the importance of photochemistry on Cu-NOM complexation [89]. In most cases, photochemical degradation of NOM resulted in less complexation of Cu. This was also true in estuarine waters where photochemical reactions appeared to destroy the so-called "stronger" Cu NOM ligands (those that have higher conditional stability constants) and thiols but did not impact the "weaker" ligand class [90, 91]. Rates of photolysis were highest with UV radiation but still occurred at wavelengths above 400 nm. Studies have shown a different outcome for the photolysis of Fe-binding ligands in estuarine waters [92]. In this case, it was shown that the ligand was not destroyed and there was no change in the estimated binding constant after irradiation. Therefore, the impact of radiation likely depends on the specifics of the individual sites in terms of organic matter content, light intensity and wavelength, and the mix of redox sensitive metals present. However, even if there is no change in ligand binding capacity, these studies showed evidence of Fe^{II} production although the amount produced was small compared to the amount of organically-complexed Fe^{III}. There is also evidence that biologically-produced Fe ligands (siderophores) are photoactive and are degraded releasing reduced Fe [93]. It is not clear, however, whether these reactions are more important than the reactions discussed earlier, and for those with reactive oxygen species in solution. Overall, in terms of organic matter degradation, it appears that complexation with Cu, and probably other metals, tends to slow its rate of degradation. The role of heterogeneous reactions with solid surfaces in the presence of adsorbed NOM can also enhance these reactions, and these need to be considered in both atmospheric waters and in surface waters.

Recently, the speculation that similar reactions were important in Hg^{II} reduction [94] was confirmed by laboratory photochemical studies of the Hg-oxalate complex, and a similar mechanism of Hg reduction has been suggested [95]. These reactions are discussed in more detail in Section 5.5.2, but the outcome is that Hg^{0}, as a dissolved gas, can be lost from the water to the atmosphere and thus the reduction can result in its loss from the system. Oxidation reactions do also occur and result in the cycling of Hg between the two oxidation states in surface and atmospheric waters [96]. Similar reactions also occur with Cr^{III} as the oxalate complex is decomposed to Cr^{II} and $C_2O_4^{-\bullet}$ [97]. The Cr^{II} is oxidized by oxygen back to the +3 oxidation state, or can be further oxidized to Cr^{VI}, especially at higher pH. It should be noted that direct photoreduction of Cr^{VI} does not occur but is possible in the presence of an oxide solid or other catalytic phase, as discussed later in this section. Overall, a number of potential interactions can occur in the environment as all reactions do not take place in isolation in the complex mixture of natural waters. These interactions are either due to the competitive binding of metals to the photoactive ligands, as was shown by the impact of Cr^{III} on Fe-carboxylate complex degradation [98] as the Cr-carboxylate complexes are not photoactive, or the complexation of the reduced form results in its enhanced stability relative to oxidation.

In liquid droplets in the atmosphere, the role of the aerosols is important and the heterogeneous photochemistry that occurs often involves metal oxide phases. The absorption of radiation in these instances can lead to the reductive dissolution of the aerosol for metals, such as Fe and Mn, which are more soluble in their reduced state. In addition to these phases, many species absorbed onto or coating the surfaces of the solids (e.g., organic matter) can also take part in photochemical reactions, being primarily involved in the initial energy absorption with the subsequent energy transfer (or electron transfer) to the underlying oxide phase. Cwiertny et al. [99, 100] conclude that the necessary chromophores, such as organic matter, nitrate and nitrite, are plentiful in the atmosphere, as are the necessary mineral dust aerosol constituents (e.g., Fe-oxides). These therefore may represent important pathways for the production of reactive species in the atmosphere and in the formation of the $OH^{\bullet}$ radical, for example [99, 100].

Stumm and Morgan [79] discuss the process of photoreductive dissolution of Fe oxide (hematite, α-Fe_2O_3) in the presence of oxalate, which occurs in a number of steps (Fig. 5.22) [83]. Initiation occurs through the formation of a surface complex between the oxalate ion and the oxide. Excitation of the complex occurs, followed by a charge transfer reaction that results in the decomposition of the oxalate to CO_2 and the $C_2O_4^{-\bullet}$ radical. The reduced Fe atom is released from the surface into solution and this is thought to be the rate-limiting step in the overall reaction. As this restores the original surface configuration of the oxide, the process can therefore be repeated. In addition, the $C_2O_4^{-\bullet}$ radical is a strong reductant and this could further react with the oxide surface, causing further reduction, or could react with other species in solution. A number of absorbed species can initiate such reactions and one recent study links the Fe and S cycle in this way [101]. Photoreductive dissolution of solid phase Fe, in the presence of methane sulfinic acid, which was oxidized, occurred through a ligand-to-metal charge transfer reaction, leading to the formation of Fe^{II}, methane sulfonic acid and H_2O_2. This reaction may be important given the role of alkylated S species in the formation of atmospheric aerosols, especially over the ocean.

The reaction pathways discussed above all involve metal reduction, but there is also the potential for oxidation reactions that are photochemically mediated. For Fe^{II}, as shown in Fig. 5.20 [83], the oxidation can proceed directly through reaction with oxygen, and with oxidants such as H_2O_2, which is likely the more important reaction pathway. Overall, the stability of the Fe^{II} formed in rainwater appears

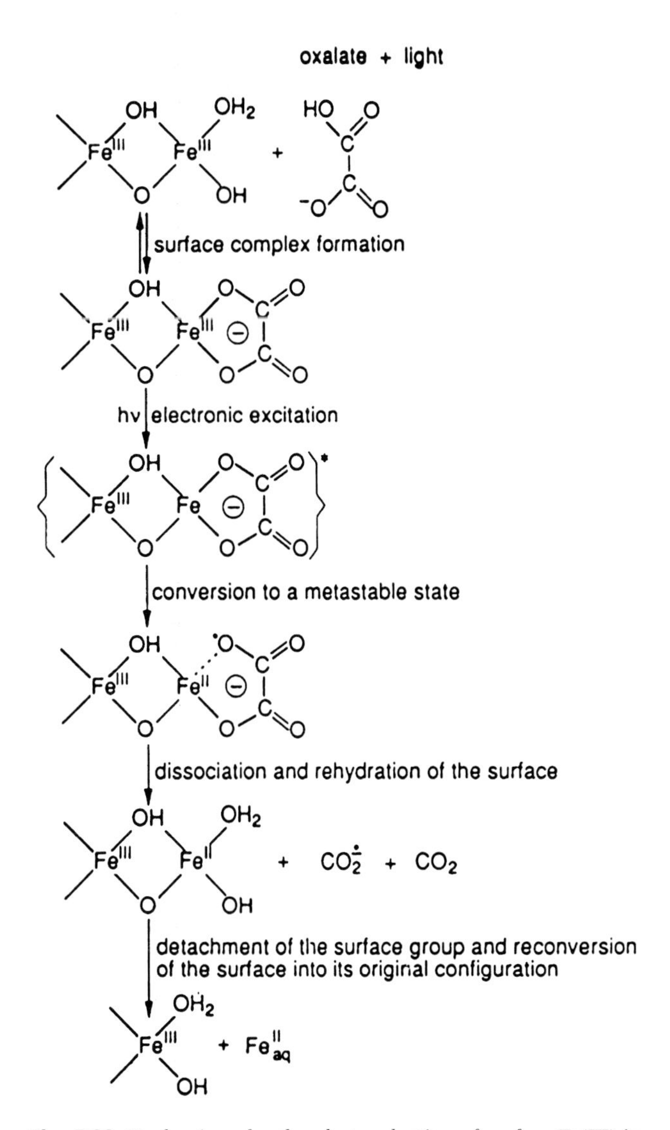

Fig. 5.22 Mechanisms for the photoreduction of surface Fe(III) in the presence of oxalate, which results in the dissolution of the solid phase, the release of Fe(II) and the decomposition of the oxalate. Figure published in Stumm and Morgan [79] but reprinted with permission from Siffert and Sulzberger (1991), *Langmuir* **7**: 1627–34 [83]. Copyright (1991), American Chemical Society.

to be a function of a number of variables [102] with the rate of oxidation linked to the H_2O_2 concentration. Rates of oxidation were slower in rainwater than in pure water solutions and it is suggested that complexation of the Fe^{II} to organic matter stabilizes it against oxidation. Thus, there may be a dual role of NOM in atmospheric waters as both an enhancer of Fe^{III} reduction and a hindrance to Fe^{II} oxidation, and the same is true for other redox sensitive metals. While complexation may stabilize Fe^{II} against oxidation, this is not always the case as some ligands can enhance oxidation [80].

Multiple oxidation pathways are also probable for the other reduced metal species that are thermodynamically and kinetically unstable in oxic waters, such are Mn^{II} and Cu^{I}. In contrast, some species are relatively stable in their reduced form, such as Cr^{III} and Hg^{0} and therefore their oxidation pathway in the atmosphere and in surface waters is more likely to be photochemically mediated. The oxidation of As^{III} by H_2O_2 was shown to be impacted by the presence of other metal ions in solution [103]. It was found that some metals, such as Fe, Pb and Cu influenced the rate of oxidation but most other metals did not. It was proposed that the formation of metal-arsenic complexes ($MeAsO(OH)_2^{n-1}$; Me = metal, n = metal oxidation state) enhanced the rate of oxidation as H_2O_2 does not react with the uncharged $As(OH)_3$ complex. The complexation therefore results in an increase in the fraction of the As that is reactive. Fe^{III} also enhances the oxidation if present as colloidal or solid Fe-oxides, as will Fe^{II}. Another study showed that As^{III} is oxidized in the presence of fulvic acid and light [104] and the rate was proportional to NOM concentration and increased with pH. Oxidation was enhanced by the presence of Fe^{III} which suggests that the formation of reactive intermediates from NOM and the Fe-NOM complexes was the main mechanism for the enhancement of oxidation.

In summary, the reduction of Fe^{III}, either present in solution as dissolved organic or hydroxide complexes, and the reductive dissolution of oxide phases, often mediated by the presence of absorbed species, are major photochemical pathways that are important in the atmospheric chemistry of metal(loid)s, and in the overall production and destruction of reactive species in the atmosphere. Similarly, the reactivity and complexation of Cu^{II} leads to similar pathways and reactions and in some instances the reactions of Cu may be more important than those of Fe. Other metals that have multiple oxidation states and whose complexes are photoactive include Mn, Cr and Hg, and all these species are likely reduced from the more oxidative states (e.g., Mn^{IV}, Cr^{VI}, Hg^{II}) by one or more of the pathways discussed previously. In terms of oxidation, unless the reduced form of the metal is stabilized, the oxidation occurs rapidly through reaction with species such as H_2O_2, or even with oxygen. Complexation has been shown to enhance the stability of Fe^{II} and Cu^{I}. For Hg^{0}, a dissolved gas, stability is due to the overall redox potential and the two step (2 electron transfer) nature of the oxidation compared to Cu and Fe where the reactions only involve one electron transfer. This is likely the reason for the relative stability of Hg^{0} and for the stability of other reduced species in the presence of light and an oxic environment.

As the chemistry of Fe is important in the overall atmospheric cycle, and given the limitation of ocean productivity by Fe, and that the atmospheric pathway is its dominant source, the chemistry of Fe in the atmosphere and surface waters is discussed further in Section 5.5.1. Of all the trace metals, Hg is the only element to exist dominantly in the gas phase in the atmosphere, as Hg^{0}. Also, given the importance of air–sea exchange in the global Hg cycle, the reactions that occur with regard to Hg are highlighted in Section 5.5.2.

5.4 Solubilization of aerosol metal(loid)s in natural waters

There has been much research investigating the dissolution of aerosols in water, and specifically in the release of metal(loid)s into solution upon deposition. This research was driven initially by the study of the solubilization and bioavailability of Fe and other metals in seawater as it was realized that the main source of Fe to ocean waters was the atmosphere, and that most Fe in atmospheric aerosol was from crustal sources. Furthermore, the pioneering work of John Martin and others showed that Fe, as a required nutrient for photosynthetic organisms, was often present in limiting concentrations in ocean waters even given its abundance in the aerosol and in wet deposition. Thus, it was evident that it was not the total supply of Fe to the ocean that mattered, but the availability of this Fe to these microorganisms. The dissolution of Fe in seawater was a primary controlling factor. Moreover, there are locations in the ocean where atmospheric Fe supply is relatively high (e.g., the North Pacific and North Atlantic) but productivity is relatively low, which could be due to limitation in Fe and/or other nutrients. These initial studies in seawater were later extended to the examination of the release of other metal(loid)s as a result of particle dissolution. The wealth of literature on the subject is discussed briefly here and some of the studies highlighted, but the interested reader should consult the broad literature on the topic.

An examination of the dissolved-particulate partitioning of metal(loid)s in rainwater provides a hint at the degree of solubility of aerosol metal(loid)s. For example, a compilation of data for precipitation around the Mediterranean Sea showed that the dissolved fraction for Al and Fe was <20%, while that of Mn, Ni, Co, Cu, Zn, Pb, and Cd was >50 %, and that the solubility of metal(loid)s in rain depends on the source region [30]. The solubility of Pb in rain appears to be a function of the rainwater pH, which reflects both the impact of pH on the dissolution of aerosol metal(loid)s and also the source of the aerosols. Overall, the anthropogenic metal(loid)s are found mostly in the dissolved phase while Fe is mostly in the particulate fraction, as defined by 0.4 μm filtration. The results of numerous studies of aerosol dissolution, summarized in Table 5.10, show similar trends overall [105–110]. Firstly, the solubilization of metal(loid)s associated with crustal material is much lower than that for metal(loid)s typified as being from anthropogenic sources. Furthermore, it was found that there was a greater solubi-

Table 5.10 Solubility in seawater of elements attached to aerosols sampled over the remote ocean. From [50] and incorporating references therein. Also from [29, 105–110].

Element	Oceanic aerosol solubility* %	Crust/dust solubility* %	Urban aerosol Solubility* %
Al	<1–10	–	<10
Fe	<1–50	<2	<1–40
Mn	25–50	<4	>10
V	30–85	<5	>30
Zn	25–75	10–20	>20
Cd	80–85	10	>40
Cu	15–85	–	20–50
Ni	30–50	2	10–80
Co	–	–	>20
Pb	15–90	–	>20
Hg	–	–	<10
As	–	–	>50
Se	–	–	>20

lization of metal(loid)s from fine particles compared to the coarser particles [105], again suggesting that metal(loid)s associated with anthropogenically-derived particles may be more soluble/bioavailable. There are a number of potential reasons for these differences. It is probable that metal(loid)s on aerosols derived from high temperature combustion sources have become associated through condensation processes and are therefore most likely bound to the surface of the "particle" rather than incorporated into its matrix. Also, it is possible that other constituents (organic acids, anions) associated with the anthropogenic particles assist in the dissolution, and also that the lower pH of the solutions derived from such particles enhances the dissolution. The metal(loid)s are likely not only present as insoluble oxide or other mineral phases, which is the dominant form for the crustal elements.

For redox sensitive metal(loid)s, and especially for Fe, atmospheric chemical "processing" can lead to an enhancement of the reduced Fe^{II} form, which would lead to enhanced solubility, as shown by a number of studies in the early 1990s [50, 111] and in numerous more recent studies. Complexation of metals in deliquescent particles likely also enhances solubility. Finally, while not mentioned in many papers, it is also possible, given that solubility is typically defined as 0.4 μm filtration that some fraction of the "solubilized metal" is associated with sub-micron particles or colloids generated by the disintegration of the aerosol, which are often derived from the coagulation of smaller particulate phases during atmospheric transport.

It can be seen from Table 5.10 that the release into solution of the crustal metals Fe, Al, and Mn, and to a lesser extent the other "crustal" transition metals (e.g., V, Co, Cr, and Ni), is relatively low for all aerosol types while the other elements are relatively soluble even for crustal material, and highly soluble from urban aerosol. However, the literature also indicates that there is high variability in the extent of dissolution and this is due, to some degree, to the differences in the methodologies used in the solubility experiments (water type, pH, dissolution time, degree of agitation). It also likely reflects the differences in the sources of the aerosols studied in each instance. While it may not be readily apparent from the table, a review of the literature shows that elements such as Zn, As, Pb, and Cd can have solubilities of >80% in some cases, and thus these elements must be associated with particles that are highly soluble, or be complexed to compounds that enhance their dissolution. The results of the dissolution studies are clearly consistent with the observation that most of these metals are predominantly in the dissolved fraction in precipitation [50] even though their principal source is the scavenging of aerosols by clouds and droplets/snow. Hsu et al. [106] recently confirmed these results by demonstrating that there was a strong correlation between the degree of solubilization for some metals and the enrichment factor (EF) of the metal in the aerosol, reflecting again the notion that the anthropogenically-derived metals had higher solubility than crustal-bound metals.

The degree of solubilization depends on the source location as well as physical and chemical factors. Some of the differences in the various studies could be the result of experimental artifacts. The relative amount of aerosol to solution used in the experiment impacts the degree of solubilization as does the extent and severity of the mixing of the solutions under experimental conditions. The degree of solubilization has been shown in laboratory experiments to decrease with an increase in the particle to solution ratio. Wu et al. [112] concluded from their studies that aerosol Fe dissolution in seawater is a time-dependent process that can be underestimated using some experimental approaches, such as when a short leaching time is used. Also, the amount of Fe added could be sufficient to overwhelm the complexation capacity of natural Fe-binding organic ligands in the leaching seawater solution. This would occur if experiments are conducted with high aerosol : water ratios. Absorption of Fe onto the leaching chamber wall is also a potential artifact, and appears to be more problematic with prolonged leaching times. To minimize these artifacts, investigators have developed different methods relying on batch exposures or continuous flow through reactor designs [112, 113]. These approaches appeared to have reduced artifacts and have led to some consensus in the extent of leaching of Fe from aerosols, and the controlling factors.

Thus, there are a number of primary factors that influence the extent of dissolution of metal(loid)s from aerosols upon mixing with natural waters. The source and type (mineralogy, origin) of the aerosol appear to be the most important factors. Secondary factors are the particle size, and the overall metal(loid) content, and the mixing ratio, which impact the degree of solubilization. Finally, for Fe, and likely for some other metal(loid)s, the solubility is related to the extent of weathering of the particles due to atmospheric reactions (oxidation/reduction and complexation) and processing due to the wetting and drying of aerosols. The fraction of the metal(loid) derived from anthropogenic relative to natural sources has a large impact on the solubility. Overall, most of the non-crustal derived aerosol metal(loid)s are relatively soluble in rain and surface waters and are therefore found in the filtered fraction in rainwater, or are found in solution after laboratory dissolution experiments with aerosols. This is not true for the crustal metals which are generally relatively insoluble in wet deposition and in surface waters.

5.5 Focus topics

5.5.1 Focus topic: Atmospheric inputs and atmospheric chemistry of iron

The atmosphere is considered to be one of the major, if not the major, input of Fe to the open ocean environment [50, 114]. The largest global source of mineral dust aerosol, the dominant Fe source, is windblown soils from arid desert regions. Asia and Africa contribute most of the aerosol mass, which is estimated to be on the order of 800–2000 Tg emitted annually (see Fig. 5.11) [114]. Mineral dust particles link the terrestrial realm, the atmosphere and the ocean in a different way compared to other types of tropospheric aerosol, such as anthropogenically-derived particles, which are emitted mostly from combustion sources and are enriched in the fine particulate fraction. The dust transported in the atmosphere is mostly derived from mid-latitude desert regions such as North Africa and the Arabian Peninsula, and China and other regions in Asia, and is often generated during large, episodic events. For the long-range transport of dust and other aerosols, it is necessary for the particulate to be transported to the upper troposphere (1–8 km) and then its deposition will occur once these higher altitude air masses are returned to the boundary layer atmosphere. This allows for long-range transport, and for example, for the measurement of dust aerosol from Asia at high latitude sampling sites on the west coast of North America. It is typically only the finer fraction of the dust aerosol (<10 μm) that travels large distances from its source. Human disturbance has likely enhanced the remobilization of dust and other crustal material, with some estimates suggesting that the enhancement is as much as 50% [114]. Natural variability in dust input is also related to drought, storm frequency, wind strength and other climate factors.

The dust aerosols, and the speciation of the Fe, is modified as these particles encounter reactive gas phase species, such as ozone, nitrogen oxides, sulfur oxides, and organics during transport, and undergo chemical reactions with these trace gases. Because reactions can occur on the surface and in the bulk of these particles as they are transported through the atmosphere, and there is repeated wetting and drying cycles prior to deposition, changes in the gas phase concentration of important atmospheric constituents can occur, as can changes in the physicochemical properties of the particles. Changes in aerosol particle properties alter both the direct and indirect climate impact of the mineral dust aerosol. Mostly, Fe in the atmosphere is associated with crustal material but there is evidence that Fe, derived from pollution sources, also forms a small component of the remote aerosol as there is more evidence for its enrichment in urban aerosol. As discussed in Section 5.4, the pollution derived fraction of the atmospheric aerosol appears to be more soluble in natural waters, and therefore more bioavailable than Fe associated with mineral dust [114].

It is estimated that the input of Fe to the world's oceans is within the region of 0.25–0.63 × 10^{12} mol yr^{-1}, with about 30% of the input due to wet deposition [114]. Clearly the overall distribution is related somewhat to ocean surface area, but not closely due to the heterogeneous distribution of the major source regions in the Northern Hemisphere (Fig. 5.11) [50]. Therefore, about 40–50% of the Fe deposition is to the North Pacific Ocean, about a third to the North Atlantic Ocean, and only around 10% is deposited to the Southern Oceans (Atlantic, Pacific and Indian) even though these regions constitute a large fraction of the total ocean surface area. The overall estimates of inputs are constrained by measurements at a number of island locations but overall there is still some uncertainty in the magnitude of the input. Records preserved in sediments and ice cores suggest that Fe inputs to the ocean were likely higher during the last glacial period, perhaps as much as 10-fold or more, but with a similar distribution to that which is currently occurring.

The atmospheric fate, transport and cycling of Fe is strongly related to the degree of reactivity during transport. Photoreduction of Fe can occur, as detailed in Section 5.3 (Figs. 5.20 and 5.22) [83], especially in low pH waters, such as atmospheric cloud water and rain, as the low pH enhances the solubility of the Fe, as discussed in Chapter 3, and also results in the dominance of the more photoactive species. Furthermore, low pH results in a slower rate of oxidation [66]. Both inorganic and organic complexes of Fe can participate in the redox reactions. It appears that the particular Fe^{III} species, $FeOH^{2+}$, which is present at high relative concentrations at low pH, is the dominant chromophore, and therefore can lead to enhanced reduction (Section 5.3). It is worth recalling that these reactions produce the $OH^{\bullet}$ radical, and through subsequent reaction, H_2O_2 and other reactive organic and inorganic species, which all can be important overall in the chemistry of the atmosphere. Morel and

Hering [66] estimate the half-life of $FeOH^{2+}$ to be about 8 h in sunlit low pH waters. The ratio of oxidized to reduced forms of Fe have been measured in atmospheric waters and in surface waters and their ratio is indicative of the extent of the photochemical reactions that are occurring, and these concentrations represent the steady state situation given the rapidity of the reduction reactions generally. For example, studies in cloudwater [82] suggest that the overall ratio depends on the simultaneous presence of complexing and reducing substances in the atmosphere and the degree and rate of solubilization of Fe from aerosol particles, which is pH dependent. A correlation between Fe^{II} in the liquid phase and the intensity of the solar irradiation was observed, and this is likely the result of both photochemical reduction of the Fe^{III}-complexes and the photochemical reductive dissolution of Fe (hydr)oxides, via mechanisms discussed in Section 5.3. Oxidation of Fe^{II} was mostly by H_2O_2, and during the night. Dissolved Fe correlated with the oxalate concentration and the extent of H_2O_2 production was related to the Fe^{III} content of the cloudwater.

The processing and cycling of Fe and dust particles through the atmosphere can lead to an enhancement of the solubility of Fe (Section 5.4). The reaction of the crustal material with acids derived from anthropogenic and other sources (e.g., H_2SO_4, HNO_3) can lead to the formation of hydration layers of very low pH (<3) and high ionic strength, and the more wet-dry cycles that occur, the higher the solubility of the Fe (though enhancement of low pH, and Fe reduction). Estimates suggest that some aerosols, especially those associated with anthropogenic sources, can have a pH as low as 1. These conditions enhance the solubilization of Fe from the solid phase [111, 114]. Some estimates suggest that aerosols that have been highly "weathered" during atmospheric transport can have the majority of the Fe in the reduced oxidation state [111]. As reported by Jickells and Spokes [114], various studies have determined the Fe^{II} content during aerosol leaching studies and have found this to range between <10 to >70% in the light, with 2–4 times lower values found in the dark. Therefore, the enhancement of solubility is due to the photochemical reduction of particulate and colloidal Fe^{III} to Fe^{II} which is much more soluble. The rate and extent of this process appears to be pH dependent and is therefore enhanced by the presence of mineral or other acids within the particle phase. Overall, given the higher fraction of soluble Fe in wet deposition compared to aerosols, the dominant source of soluble Fe to the ocean appears to be through wet deposition [114].

Recent research has focused on further evaluating the solubility of Fe in aerosols in seawater [106, 112–114]. Overall, the solubility for Fe ranges from <0.1–5% for crustal aerosols exposed to seawater solutions at pH ~8 [114]. For urban or coastal aerosols, leaching at pH ~8 does not appear to exceed 8% solubility, although higher values, up to 20% can be found for leaching at lower pH values. Wu et al. [112] found that the aerosol Fe solubility during the high dust season, when the atmospheric particulate is dominated by crustal material, was 3.5 ± 1.5% for North Atlantic Ocean aerosol and somewhat higher, at 5.7 ± 2.0% for North Pacific Ocean particulate. Results from Buck et al. [113] are similar for the North Pacific (6 ± 5%). These investigators also found that the solubility was higher for deionized water than for seawater, and that the degree of solubilization correlated with the soluble Fe(II) fraction in the aerosol and with the overall aerosol acidity. Furthermore, the degree of dissolution of Fe was correlated with that of Al, suggesting that the overall impact of Fe redox state was small for these samples. Clearly, while there may be initial enhanced dissolution of Fe due to the presence of Fe(II) in the aerosol, this would be a transient signal if the reduced Fe is rapidly oxidized and then complexed and/or precipitated if solubility is exceeded. Such an outcome was suggested by the results of Hsu et al. [106] who found that the dissolution reached a steady state after about an hour of contact and that for some metals there was actually a small decrease in concentration subsequently, reflecting repartitioning and other processes occurring. In the experiments of Wu et al. [112], where multiple extractions were performed, about 40–50% of the total Fe was released in the initial extractions of the aerosol and subsequent extractions lead to decreasing amounts of Fe being released. Thus, there is a finite amount of soluble Fe in the aerosol, and the degree of solubilization is not determined primarily by equilibrium partitioning.

Another study [115] reached the same conclusions for the North Atlantic. Under conditions dominated by Saharan dust, the aerosol was enriched in Fe and the solubility of the aerosol was low (<2%). While the Fe content was about 50 times lower (per m^3 of air) when the aerosol originated from North America, the solubility was around 20%. These more soluble particles also had higher concentrations of "pollutant-derived" metals, indicative of the source of the particulate material. Interestingly, Sedwick et al. [115] also examined the solubility of Fe in open ocean rain and found it similar to that found in their leaching experiments, suggesting that the ability of rainwater to leach Fe from aerosol is not substantially different from that of seawater, even given its lower pH.

Once deposited through wet and dry deposition, it would be expected that the reduced Fe and even the soluble Fe in rainwater would rapidly precipitate at the high pH of seawater where Fe^{III} is highly insoluble. As noted before, however, the concentration of Fe in seawater is much higher than that predicted based on solubility of the hydroxide phases, and this is due to the strong complexation of Fe in seawater by organic ligands [116]. The overall cycling of Fe in the surface waters of the ocean and the role of photochemistry is illustrated in Fig. 5.23 [80, 117]. The figure shows the role of photochemistry, and the role of organic

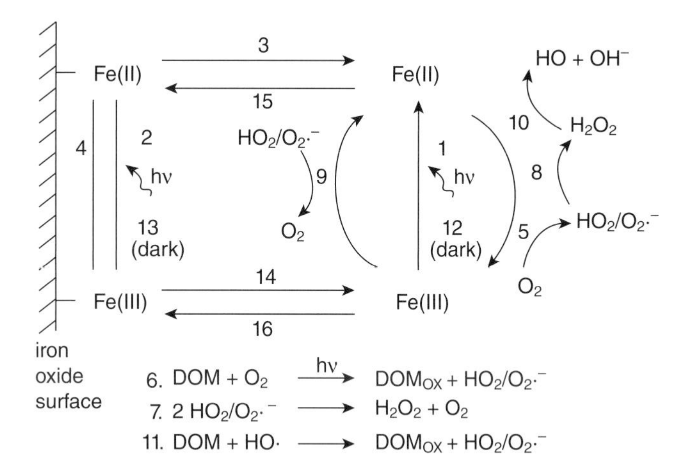

Fig. 5.23 Overall scheme of major processes occurring with Fe in the surface waters of the ocean and illustrating the role of photochemical processes and other interactions. Reprinted with permission from Voelker et al. (1997) *Environmental Science and Technology* **31**: 1004–11 [117]. Copyright (1997), American Chemical Society.

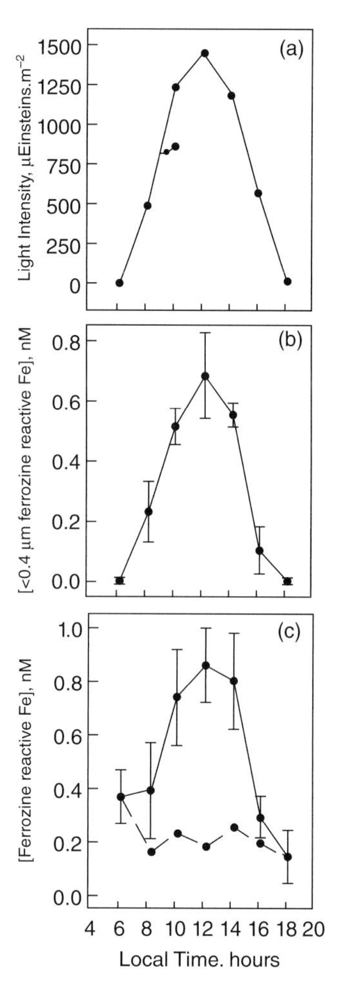

Fig. 5.24 Diel change in light intensity and surface water redox speciation of Fe(ferrozine reactive Fe) in filtered and unfiltered water, showing changes in concert with changes in light intensity. From Waite et al. (1995) *Marine Chemistry* **50**: 79–91 [118]. Reprinted with permission from Elsevier.

complexation. The overall complexity and reversibility of all the major pathways suggests that there is a dynamic steady state and the relative ratio of Fe^{II}/Fe^{III} will be a strong function of the light intensity. This is shown in Fig. 5.24 [118], which shows the diurnal change in the Fe^{II} concentrations, as measured by the ferrozine method, in surface waters and clearly shows the dependence on photochemistry. These data also illustrate that most of the Fe^{II} is in the dissolved (<0.4 μm fraction) [80]. For further discussion, many of these interactions are discussed in section of the book dealing with ocean biogeochemistry (Chapter 6) and the interaction of Fe and organisms (Chapter 8). These Fe-complexing ligands are either abiotically or biotically produced and estimates suggest that greater than 95% of the Fe dissolved in seawater is organically complexed [116]. The fate and transport and accumulation of Fe into marine microbes and its removal to the deep ocean is a topic for later discussion (Chapter 8).

5.5.2 Atmospheric chemistry and air–water exchange of mercury

In contrast to Fe, whose atmospheric chemistry is dominated by the particulate phase and its dissolution, Hg is dominantly in the atmosphere as a gas (Hg^0). This is also the dominant volatile Hg species in the surface waters of aquatic systems [42, 119–121]. It can be formed by biotic and abiotic reduction of ionic Hg (Hg^{II}) and most aquatic ecosystems appear to be saturated related to the atmospheric concentration. Given the relative surface area of the ocean compared to freshwater ecosystems it can be concluded that the gaseous evasion flux from the ocean will dominate over freshwater systems unless there is a dramatic difference in concentration between these ecosystems. This is not the case for Hg [7]. While open ocean total Hg concentrations are about a factor of 5 lower than those of the terrestrial waters, this difference is less for large water bodies such as the Great Lakes, which constitute the bulk of the fresh water surface area. A number of box model and numerical modeling papers have recently focused on the estimation of the evasion of Hg^0 from the ocean to the atmosphere [42, 119–121] and as these papers have reviewed and examined the available literature, they provide a reasonable summary and consensus of the current understanding. It is worth discussing briefly the processes whereby Hg^0 is produced. An examination of the redox chemistry of Hg, and the thermodynamic stability of Hg^0 in oxygenated environments, as presented in Chapter 3, would indicate that Hg^0 is not stable in oxic waters and should be degraded. However, a number of studies have shown that both photochemical oxidation and reduction of Hg can occur in aquatic waters [7, 122–125]. Reduction in surface waters is also biotically-mediated. Overall, there is net reduction and formation of Hg^0 in most surface waters and this is driven by the removal of the product (Hg^0) from the water via loss to the atmosphere as concentrations are mostly saturated relative to the atmosphere. For most aquatic systems, direct atmospheric deposition is the main source of the Hg^{II} that is being reduced. The overall cycling of Hg at the water–air surface of the ocean is depicted in Fig. 5.25(a) [42].

Evasion of Hg^0 from the ocean is thought to be an important part of the global Hg cycle, and the magnitude of the oceanic evasion is of similar magnitude to the input from anthropogenic point sources [42, 119–121]. While $(CH_3)_2Hg$ has been quantified in ocean waters [40, 126–129], essentially all the dissolved gaseous Hg is Hg^0 in most freshwaters examined, and in the surface ocean, as $(CH_3)_2Hg$ is photochemically unstable and relatively rapidly degraded [130–132]. For the ocean, many recent studies have focused on air–sea exchange, especially in Atlantic waters [40, 126–129, 133, 134], the Pacific [129, 135], the coastal regions [136, 137] and the Mediterranean [133, 138–140].

Besides Hg^0, a dissolved gas, oxidized Hg (principally Hg^{II}) exists in both the dissolved phase and attached to particulate matter, both living (phytoplankton, zooplankton) and detritus. Photochemical processes dominate the reduction of Hg^{II} in most instances, and studies in freshwater and shallow coastal waters have demonstrated short-term changes in Hg^0 concentration and a diurnal cycle linked to photosynthetically active radiation [122, 141, 142]. The reduction of Hg^{II} appears to be enhanced in the presence of organic matter [125], as discussed in Section 5.3, and recent studies suggest the mechanism involves charge transfer reactions involving Hg complexes with photoactive ligands. Therefore dissolved Hg speciation plays an important role in determining the rate and extent of this transformation [94, 95, 122]. This is similar to what occurs with Fe reduction [117]. Other ligands, such as Cl, can impact reduction rates through complexation competition while the presence of oxygen appears to enhance oxidation and therefore limit the next reduction [94]. The mechanisms of oxidation are less well understood but the reaction is clearly enhanced in seawater by the removal of reaction products through complexation with Cl^-, forming relatively stable complexes such as $HgCl_4^{2-}$, and by other ligands. However, organic matter can also lead to a decrease in Hg^{II} reduction [143], possibly through the inhibition of UV penetration, or through the formation of Hg complexes that are not photoactive. Additionally, a number of studies have also shown that Hg^{II} reduction occurs in the presence of algae and bacteria [144, 145]. Overall, higher DOC concentration is one of many factors that controls the overall lower concentrations and rates of Hg^0 evasion from coastal and estuarine environments compared to the open ocean.

A compilation of data in a number of recent publications [119, 130–132] show that there is variability across ocean basins in both the concentration of Hg^{II} and Hg^0, with the Atlantic Ocean and Mediterranean Sea having overall higher concentrations than the Pacific and Indian Oceans. The average variability in concentration is a factor of 3, with total Hg concentrations being 2–3 pM in the more impacted waters, and <1 pM for Antarctic waters. However, there is much spatial and temporal variability in the surface water concentrations. Such differences make sense in terms of the historical inputs of Hg to the atmosphere, which were previously highest for North America and Europe, and which therefore impacted the North Atlantic and Mediterranean more than the other ocean basins [146–148]. There is also a clear difference in the concentration of Hg in the North versus the South Atlantic, which is also likely a reflection of historic Hg inputs to the atmosphere. More recently, inputs to the North Pacific due to increased industrialization in Asia have been occurring but given the basin size, and the timing of these inputs, the ocean water concentrations have not yet changed to reflect these increased inputs. Again, the modeling [119–121] suggests that the response time of the surface North Pacific Ocean waters to these changes in atmospheric inputs will take many decades rather than years.

Published values of the fluxes of Hg^0 to the atmosphere vary over a wide range and some of the estimates cannot be truly representative given the constraints on the overall flux provided by the global budget [119]. Overall, average fluxes in the region of $0.05–0.5\,nmol\,m^{-2}\,d^{-1}$ appear reasonable across the ocean basins, and the globally averaged value is about $0.1\,nmol\,m^{-2}\,d^{-1}$ ($13\,mmol\,yr^{-1}$). To date, most measurements have been made in warmer regions and seasons and the higher flux may reflect high evasion due to higher photochemical and biological reduction rates. A recent global model, constrained by measurement to the extent possible, predicts values up to $0.3\,nmol\,m^{-2}\,d^{-1}$, and also suggests that there are instances of net deposition of Hg^0 at high latitudes in winter [148].

Fluxes for coastal areas are often somewhat lower [42, 149] and this is likely the impact of complexation of the Hg to large molecular weight NOM that is not photoactive, and/or the binding on the Hg to the solid phase reducing its reactivity. Also, the presence of elevated NOM and particles in the water column hinders light penetration and therefore the extent of photochemical processes. In large freshwater systems such as the Great Lakes, the fluxes can be important in terms of the overall system budget in that atmospheric inputs are the dominant sources and evasion is the major sink. Fluxes for lakes are on average lower than for the ocean [149] and one reason for this is the overall lower gas

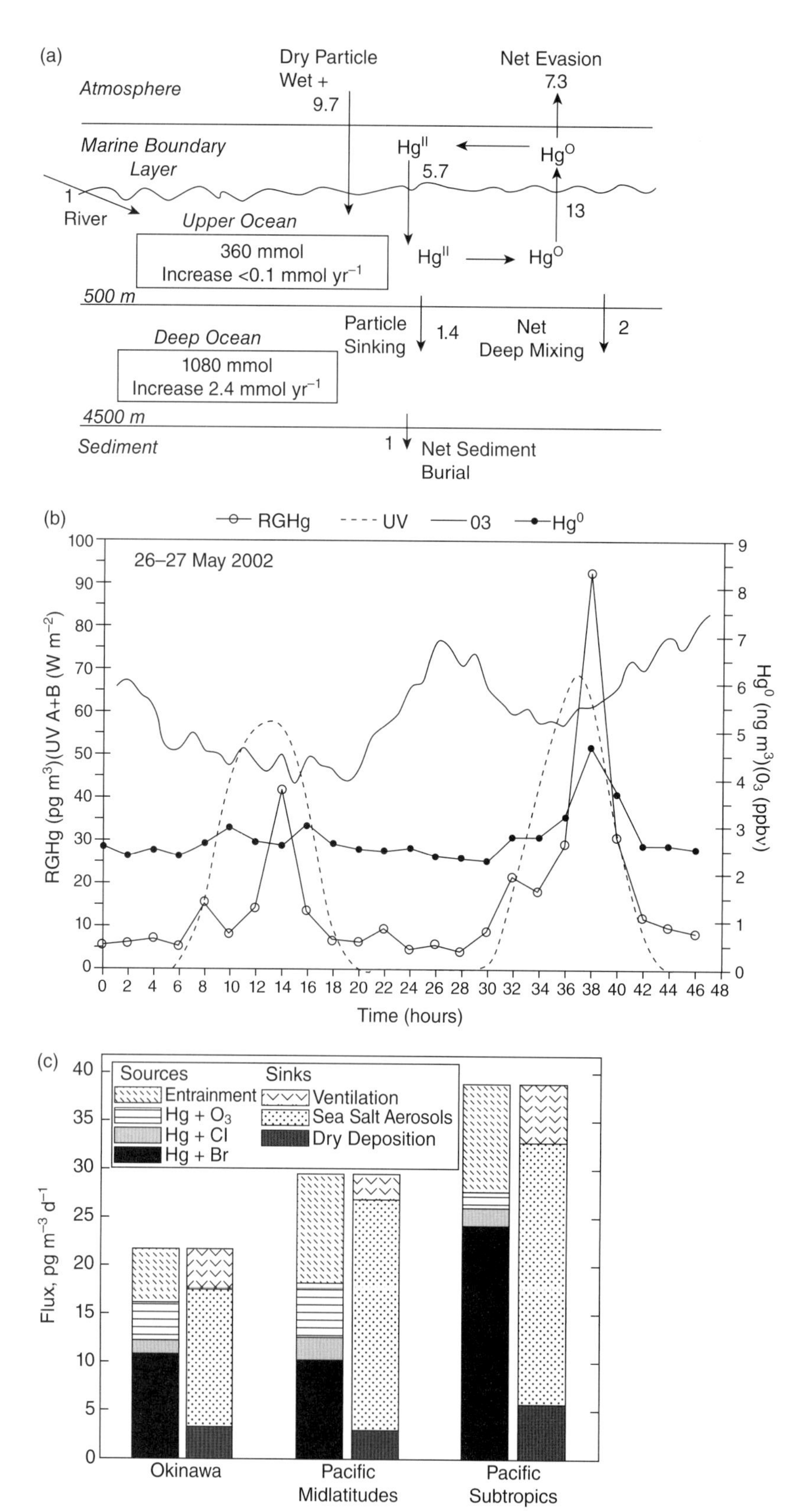

Fig. 5.25 (a) The cycling of mercury across the air-sea interface showing the primary reactions that are occurring in the surface waters and in the atmosphere, and the dominant fluxes. Modified from Mason and Sheu, Role of the ocean in the global mercury cycle. *Global Biogeochemical Cycles* 2002, **16**(4) [42] and reprinted with permission of the American Geophysical Union (2002); (b) Data over a 48 h period in the marine boundary layer of the North Pacific Ocean showing the increase in oxidized mercury (RGHg in the figure) that coincides with the highest UV radiation, and a depletion in ozone concentration. Reprinted from Laurier et al. (2003) *JGR – Atmospheres* **112**(D6) [209] with permission of the American Geophysical Union (2003); (c) Modeled fluxes within the marine boundary layer, including exchange fluxes with the free troposphere. Reprinted from Holmes et al. (2009) *Atmospheric Environment* **43**: 2278–85, with permission from Elsevier. Note: 40 $pg\,m^{-3}\,d^{-1}$ is equivalent to 0.2 $pmol\,m^{-3}\,d^{-1}$.

exchange coefficient due to lower winds speeds (Section 5.2.6). Estimates of emissions from lakes vary and are dependent on factors such as light intensity, lake depth and stratification, lake size, NOM levels and total Hg concentrations (range <0.01–0.1 $nmol\,m^{-2}\,yr^{-1}$) [149]; and references therein).

The chemistry of Hg in the atmospheric boundary layer is important in the overall cycling and net loss of Hg from surface waters to the atmosphere. A number of studies have shown that Hg^0 is readily oxidized in the atmosphere in the presence of UV radiation and a variety of oxidants although in many cases it is probable that the reaction pathways examined in these laboratory studies are not truly homogeneous, with reactions being enhanced by the presence of moisture, and potentially being catalyzed by the reaction chamber walls. Early studies examined the oxidation of Hg by ozone and the hydroxyl radical, and other potential oxidants [7] but there has been recent theoretical work that questions the likelihood of these reactions occurring homogeneously. From a thermodynamic point of view, it appears that the only homogeneous reactions that are likely in the atmosphere involve Br and other reactive halogen species (RXS) [150].

Gas phase oxidation of Hg^0 by halogen atoms and molecules (RXS) (e.g., Cl, Br, Br_2, Cl_2, BrCl, BrO) [33, 151–156] has been demonstrated, but it appears that the reaction with atomic Br is the most likely given measured rate constants and the presence of elevated oxidized Hg in regions with elevated Br atmospheric concentrations [157–161]. There are many studies and publications on both Hg, halogen, and ozone chemistry in the polar atmosphere and a representative but incomplete number of these studies is discussed here. The first demonstration of enhanced Hg^0 oxidation was through measurement of its depletion in surface-level Arctic air during the three month period following polar sunrise [162] (see Fig. 5.26a). The fluctuation of total atmospheric Hg^0 strongly resemble the depletion of ambient ozone [163–165], suggesting that both species were being removed by similar mechanisms. Other studies [166–169] have confirmed such rapid Hg depletion events in both the Arctic and Antarctic, and the depletion of Hg^0 coincides with the increase in gaseous and particulate Hg^{II} concentrations. Note that in many papers the gaseous oxidized Hg species are collectively referred to as *reactive gaseous mercury* (RGM or RGHg). It is currently thought that ozone destruction is initiated by Br atoms, and to a lesser extent, by Cl atoms and other RXS [170–185] (Fig. 5.27), with the sea salt particles that deposit to and accumulate in the snow pack during winter being a source of the precursors of Br and Cl atoms in polar regions, for example, Br_2 and BrCl. The reaction of Br with ozone is the primary sink for this species [1, 2] while Cl is also removed by other mechanisms. Instead of reacting with ozone, both Br and Cl could oxidize Hg^0.

The RXS precursors are also liberated from sea salt particles in the marine boundary layer (MBL), and ozone destruction has also been measured and described for the MBL and observed in laboratory studies [186–203]. Large diurnal variations in ozone concentration in the MBL have been found, and modeling studies suggest that ozone destruction is through reaction with RXS. Other sources of RXS include the destruction of organic halides, such as CH_3Cl and CH_3Br [1], which can be released to the atmosphere from the ocean surface [204–206]. The *in situ* oxidation of Hg^0 in the MBL in remote locations is predicted from the observed diurnal variation in its concentration [207–209] (Fig. 5.25b), which suggests that Hg^{II} is being produced photochemically. Laboratory studies confirm the potential importance of such reactions [210], and these observations are supported by modeling studies [96, 211, 212].

A recent modeling study examined the importance of the various processes involved in Hg^{II} formation in the MBL (Fig. 5.25c) [96]. The model was able to predict the daily increase observed in a number of studies [207–209] for the MBL gaseous Hg^{II} concentration through considering the primary reaction to be with Br. However, the model was unable to predict the removal of Hg^{II} from the boundary layer in the later afternoon/evening without invoking the uptake of Hg^{II} into sea salt particles. While such a process has not been examined experimentally, deliquescent salt particles could be a substantial sink for gaseous Hg^{II} given the likely formation of Cl complexes within the salt solution. Additionally, the model suggested that while much of the Hg^{II} was formed in situ, there was an important exchange of Hg^{II} between the boundary layer and the free troposphere (Fig. 5.25b). The evasion of Hg^0 from the ocean surface more than compensates for the depletion of its concentration in the boundary layer in most circumstances and therefore, in contrast to the polar region, no strong Hg^0 depletion events have been observed for the MBL.

The oxidation of Hg^0 in the MBL, coupled with the net reduction in surface ocean waters and the evasion of Hg^0 to the atmosphere results in a rapid recycling of Hg at this interface [42]. Most estimates of ocean evasion are based on measurements of Hg^0 in surface waters and gas exchange modeling and therefore the true net loss from the MBL to the free troposphere will be less. The extent to which ocean evasion contributes to the global tropospheric Hg pool needs to be further constrained through measurement and modeling. It is worth noting that while most measurements of atmospheric Hg are made in the boundary layer close to the water or terrestrial surfaces, the majority of the Hg is in the free troposphere and there may be a disconnect between free tropospheric concentrations and those of the boundary layer in regions where evasion from the surface is substantial. While the focus in this section has been evasion from the water surface and reactions in the

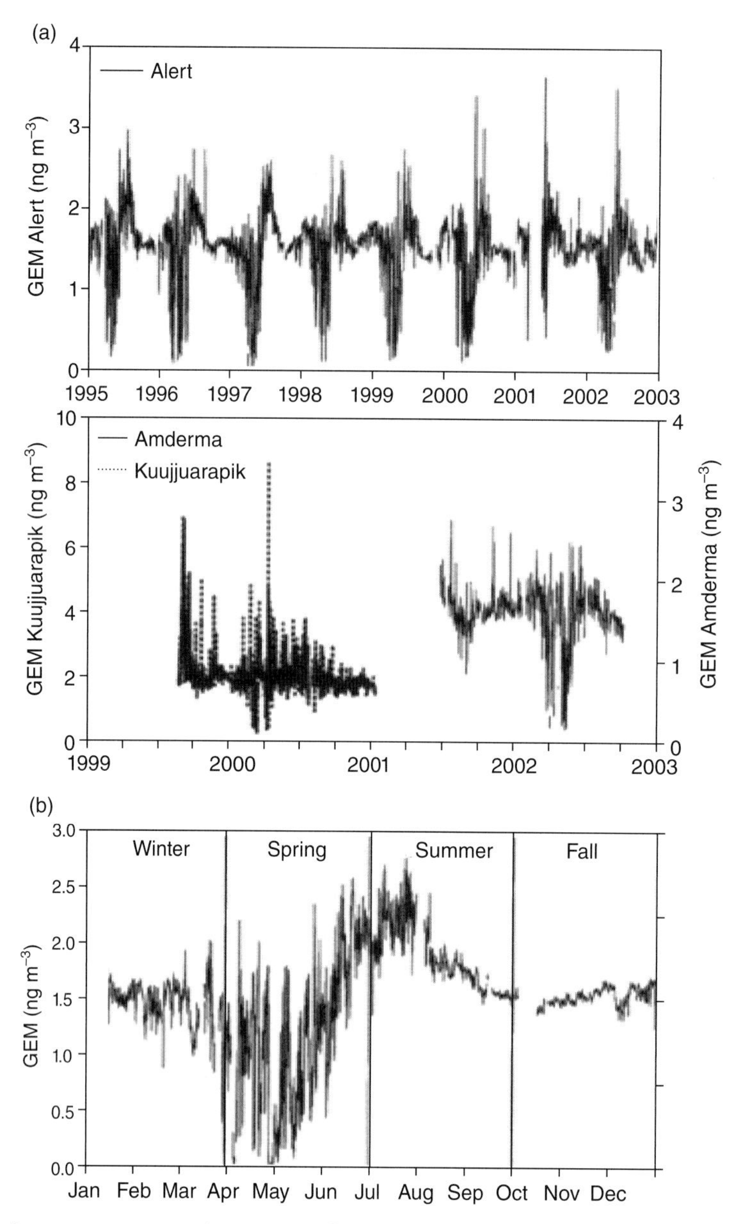

Fig. 5.26 Changes in elemental mercury concentration (GEM in figures) at (a) Alert (1995–2002) and Kuujjuarapik (1999–2000) in the Canadian Arctic and Amderma, Russia (2001–2003) showing the periods of low mercury concentrations (mercury depletion events) that occur during the period of polar sunrise, reprinted with permission of Elsevier; and (b) details of one year of data at Alert for 1997. Figure reprinted from Stefen et al. (2002) *Science of the Total Environment* **342**: 185–98 [164]. Reprinted with permission of Elsevier.

associated atmosphere, evasion from terrestrial surfaces is also important in the global Hg cycle, and there is strong regional variability in the extent of evasion, and this also varies with vegetation type [4, 149]. As noted earlier, there is also the potential for uptake of Hg^0 into plants, most likely through stomatal gas exchange. The exchange of Hg at the land-air interface is a topic of much research and the interested reader should examine a number of recent reviews and modeling studies in the literature [4, 149, 213–215].

5.6 Inputs of atmospheric metals to the biosphere

It is probably apparent from the discussions in this chapter so far and from the changes in the rate of deposition over time for many metal(loid)s, and the somewhat patchy distribution of sampling sites on a global basis, that it is a difficult task to estimate the total deposition of metal(loid)s to the ocean and to the terrestrial environment directly from

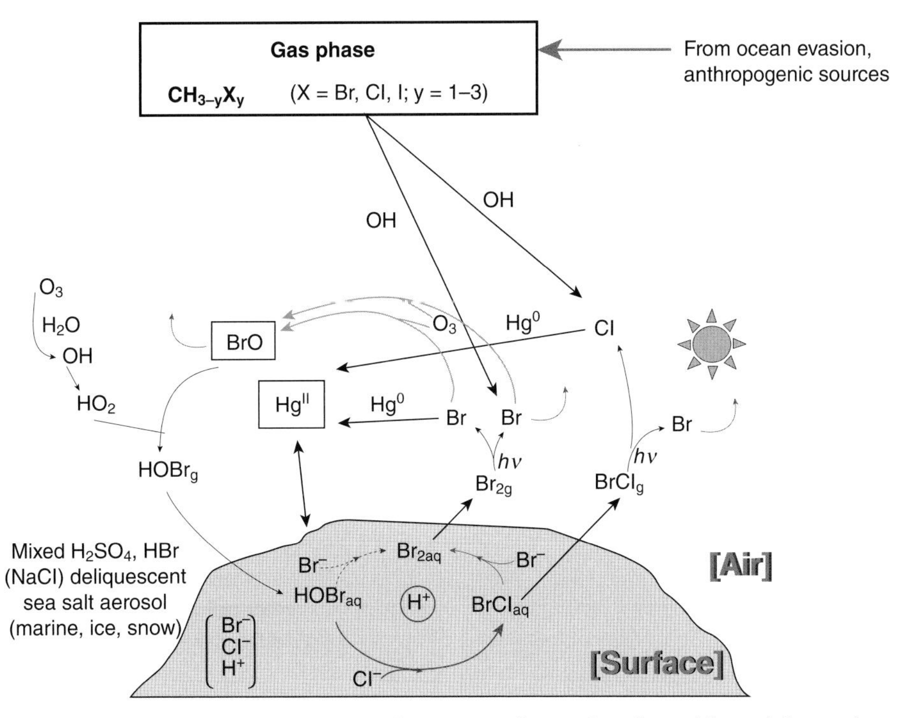

Fig. 5.27 The proposed reaction scheme whereby bromine oxide (BrO) is taken up by salt particles and the resultant catalytic formation of Br atoms, and the resultant ozone destruction that occurs. Modified from Vogt et al. (1996) to include the potential reactions for mercury. Reproduced from Sheu and Mason (2004) [210] with permission from Springer Science+Business Media.

the measurements themselves. These measurements provide useful scaling data and give a strong indication of the variability across space and time but another approach is needed to obtain a reasonable estimate of the atmosphere–land and air–sea fluxes of the metal(loid)s of interest. Indeed, the approach described here was used to constrain the flux values for the global budgets discussed in Chapter 2. The following approach was based on recent re-evaluations of important factors in the overall flux estimates, and in particular the scrutiny that has been given to estimates of the crustal (dust) deposition given its importance to the fate and transport of Fe, and to ocean productivity [80, 114], and the related issues of atmospheric CO_2 and global climate change.

The following model approach was therefore used to estimate the fluxes of metal(loid)s from the atmosphere to the Earth's surface. The natural inputs to the atmosphere for the crustal metals were estimated from the dust flux and used the reported crustal abundance for the metals Al, Fe, Mn, Ni, and Co. For these elements, the dust flux estimates of Jickells and Spokes [114] (a total flux of 800–2000 $Tg\,yr^{-1}$) were considered, but for the values reported in Table 5.11 a best estimate value of 1500 $Tg\,yr^{-1}$ was used. For the other metal(loid)s, their natural fluxes, which are typically a minor component, were taken from the literature, as reported in Table 2.4. Ocean input of natural dust sources was also based on the relative dust deposition to the ocean of Jickells and Spokes [114] (range 400–1000 $Tg\,yr^{-1}$); with a best estimate 600 $Tg\,yr^{-1}$. Anthropogenic inputs for all metal(loid)s were taken from the earlier data compilation, discussed in Chapter 2 (Table 2.4). Thus, these estimates define the inputs – natural and anthropogenic – in Table 5.11, and these are almost exclusively dominated by input from the terrestrial environment. This is not so for Hg given the importance of evasion from the ocean and the same is potentially true for Se. The discussion of the Hg cycle in Section 5.5.2 is used to derive the values reported in Table 5.11.

The deposition of metal(loid)s to either the land or the ocean is obviously a key parameter in determining the overall distribution of the outputs from the atmosphere. The relative deposition will not precisely reflect the ratio of surface area (land/ocean area = 0.4) for a number of reasons. If the deposition was evenly distributed, then the relative fluxes would match the relative area. However, as noted earlier, arid areas and the dust flux is heterogeneously distributed and concentrated in the Northern Hemisphere, and more of the ocean is in the Southern Hemisphere. Additionally, not all desert areas are close to the ocean, nor are the air flows always favorable for dust transport offshore. Overall, it is likely that the deposition, given that much of the dust input are relatively large aerosols that are locally

Table 5.11 Model estimates of the fluxes of metal(loid)s to the surface ocean and the terrestrial realm, determined as described in the text, and the relative fraction of the deposition that is to the ocean. Comparative values from the literature are also presented. All fluxes are in Gmol yr^{-1}.

Metal	Natural Input	Anthro. Input	Terrestrial Deposition	Ocean Deposition	% Dep. to Ocean	Ocean Literature
Al	4320	776	2940	2156	42	250
As	0.16	0.067	0.14	0.083	37	0.1
Cd	0.011	0.067	0.038	0.040	51	0.031
Co	0.51	0.051	0.35	0.23	41	0.04
Cr	3.0	0.59	2.1	1.5	42	1.0
Cu	0.44	0.95	0.74	0.65	47	0.6
Fe	1020	109	1090	469	41	250–630
Hg	0.011	0.022	0.013	0.017	56	0.017
Mn	25.9	0.70	26.3	10.7	40	1.4
Ni	1.50	1.7	1.8	1.5	48	0.68
Pb	0.06	2.3	1.1	1.28	54	1.3
Se	0.12	0.049	0.11	0.062	36	–
V	0.88	1.56	1.3	1.1	45	0.70
Zn	0.69	2.0	1.4	1.3	48	1.1

deposited, is larger to the land, even given its smaller area. The relative flux can be constrained if there is sufficient and detailed information for one of the major crustal elements – Al or Fe. The literature values for ocean Al deposition appear low based on its ratio so the relative deposition was estimated using Fe data, of which there is reasonable data and the flux values appear to be well-constrained. This lead to a relative deposition of 60% of the crustal-derived material to the land and 40% to the ocean based on the best estimate values above. For the other trace metal(loid)s, given that their natural sources are more evenly distributed, but not less prone to regional deposition, a value of 70% deposition to land was used.

For the anthropogenic component, it is likely that the metal(loid) inputs will have a similar land : ocean deposition ratio, except those with a gaseous evasion component, assuming relatively similar global source functions. As the anthropogenic metal(loid)s are mostly associated with a finer fraction of the aerosol, they are likely to be transported longer distances, and given that many of the larger cities and industrialization are in coastal locations, it seems that the relative ocean input for the anthropogenically-dominated metal(loid)s should be a higher fraction than the crustal metals. However, as the inputs are entirely in the terrestrial realm, the ratio will not likely be that of the surface area ratio, that is, by assuming that the fine aerosol is globally relatively well mixed in the atmosphere. Examining the available data, and especially for those metals that have been relatively well studied (e.g., Hg, Pb, and Cd) lead to the conclusion that a reasonable ratio for the anthropogenic deposition was 45% to land, and 65% to the ocean.

These estimates therefore lead to the data gathered in Table 5.11. The final column in the table is the average reported value for ocean deposition in the literature, as reported in Tables 2.8 and 5.8. Firstly, it is noted that due to the formulation of the model, the fraction of a metal(loid) deposited to the ocean increases with the fraction of the input that is anthropogenic and is highest for Pb and Cd, and lowest for the crustal metals (excluding the metalloids). The low values for the metalloids reflect the fact that while their inputs are dominated by anthropogenic sources, the fluxes are not estimated using the formulation for the crustal material but rather using the distribution for the anthropogenically-dominated elements.

The model values agree relatively well with the ocean literature data for most of the metals but there are some notable exceptions. The literature values for Al, Mn, and Co are low relative to the model, and this suggests that the measurement locations that these estimates are based on have not adequately sampled the crustal inputs, and this likely reflects the lack of data reflecting inputs from the Asian continent. Alternatively, as suggested by one study, some of the difference may reflect that the crustal value used may not be truly representative of the aerosol. For example, the value used for Al (7.8% by mass) may be higher than the actual aerosol content, which was found to be 2.6% for crustal material sampled over the Arabian Sea [216]. Further refinement of the model could be made if more accurate values for the crustal ratios of the major metals are derived. However, overall, the model appears to provide a reasonable estimate of the flux of the heavy metals and metalloids and gives reasonable estimates that can be used to develop and constrain global models and global budgets for these metals.

An examination of the information in Table 5.11 suggests that the highest fractional deposition to the ocean is for Hg, Cd, and Pb (>50%) while the lowest is for the metalloids (<40% for As and Se). The percent deposition to the ocean for the major crustal metals is in the low 40%. Clearly, these estimates suggest that direct anthropogenic inputs and human activity, which has likely exacerbated the dust input, have had a large impact on the inputs of metals to the ocean. This is further discussed in Chapter 6.

5.7 Chapter summary

1. For most of the metal(loid)s, wet and dry deposition contribute to the input to aquatic systems with wet deposition being more important in remote locations. The various approaches to modeling these deposition processes was discussed and how this varies with location and aerosol composition and size. Methods of measurement were briefly discussed.

2. Gas exchange was also discussed as this can be an important mechanism for metal(loid) cycling, especially for Hg. In addition, a number of metal(loid)s form alkylated compounds through natural processes (e.g., Hg, As, Se), or these compounds have been added to the atmosphere from anthropogenic sources (e.g., Pb and Sn compounds). These compounds are typically volatile and cycle through the atmosphere before being deposited, often after being decomposed. However, for most of the metals, particulate transport and deposition are the major processes for their cycling in the atmosphere.

3. Wet and dry deposition measurements, and associated flux estimates, show the impact of anthropogenic sources on the environment and allow for an examination of the contrasts between the crustal elements and the metal(loid)s that have a strong anthropogenic component. For many metal(loid)s there is evidence for the recent decrease in deposition in remote locations such as the North Atlantic, with increases for the North Pacific. These differences track differences in the distribution of anthropogenic inputs over time. Methods have been developed to use the relative concentration of metal(loid)s to determine the sources for these elements to a particular location.

4. Metals, especially Fe and Cu, play an important role in photochemical processes in natural waters, and additionally, many metals (e.g., Cr, As, Hg, Mn) are redox sensitive and can be transformed through photochemical reactions. Additionally, oxide surfaces and the presence of NOM and organic acids can enhance these reactions that involve metal(loid) transformations.

5. Aerosols deposited to aquatic systems have variable solubility. The crustal elements, especially when associated with crustal material, have low solubility while some of the heavy metals, primarily added to the atmosphere from anthropogenic sources are relatively soluble in natural waters. Differences in aerosol size and the location of the element within the particle, and other factors likely contribute to these differences in solubility.

6. The cycling of Fe across the air–water interface is complex as photochemistry and other processes can enhance the dissolution and bioavailability of Fe to microbes. For the ocean, atmospheric dust is a major source but the Fe associated with dust is not very soluble. Photochemical reduction of Fe in aerosols during transport, and in surface waters after deposition, can enhance its bioavailability.

7. The cycling of Hg across the air–water interface is complex because of the potential for both photochemical oxidation and reduction of Hg in surface waters, and the oxidation of Hg^0 in the atmosphere. The reactions in the atmosphere appear to be linked with the reactions of halides and it is thought that globally Br is the major oxidant of Hg^0 in the atmosphere.

8. A simple box model evaluation of the cycling of various metal(loid)s globally provides insights into the similarity and differences between elements and highlights the major factors that control the fate and transport of metal(loid)s, and the important role of the atmosphere in transporting these metals to remote locations.

References

1. Seinfeld, J. and Pandis, S. (1998) *Atmospheric Chemistry and Physics*. John Wiley & Sons, Inc., New York.
2. Finlayson-Pitts, B. and Pitts, J.N. (1986) *Atmospheric Chemistry: Fundamentals and Experimental Techniques*. John Wiley & Sons, Inc., New York.
3. Poster, D. and Baker, J.E. (1997) Mechanisms of atmospheric wet deposition of chemical contaminants. In: Baker, J. (ed.) *Atmospheric Deposition of Contaminants to the Great Lakes and Coastal Waters*. SETAC Press, Pensacola, pp. 51–72.
4. Gustin, M. and Lindberg, S.E. (2005) Terrestrial mercury fluxes: Is the net exchange up, down or neither? Pirrone, N. and Mahaffey, K.R. (eds) *Dynamics of Mercury Pollution on Regional and Global Scales*. Springer, New York, pp. 241–259.
5. Berg, T. and Steinnes, E. (2005) Atmospheric transport of metals. In: Sigel, A., Sigel, H. and Sigel, R.K.O. (eds) *Biogeochemsitry, Availability and Transport of Metals in the Environment*. Vol. 44. Metal Ions in Biological Systems. Taylor & Francis, Boca Raton, pp. 1–19.
6. Landis, M.S., Stevens, R.K., Schaedlich, F. and Prestbo, E.M. (2002) Development and characterization of an annular denuder methodology for the measurement of divalent inorganic reactive gaseous mercury in ambient air. *Environmental Science & Technology*, **36**(13), 3000–3009.
7. Mason, R.P. (2005) Air-sea exchange and marine boundary layer atmospheric transformations of mercury and their

importance in the global mercury cycle. In: Pirrone, N. and Mahaffey, K.R. (eds) *Dynamics of Mercury Pollution on Regional and Global Scales*. Springer, New York, pp. 213–239.

8. Tessier, E., Amouroux, D. and Donard, O.F.X. (2003) Biogenic volatilization of trace elements from European estuaries. In: Cai, Y. and Braids, O.C. (eds) *Biogeochemistry of Environmentally Important Trace Elements*. Vol. 835. ACS Symposium Series. ACS, Washington, DC, pp. 151–165.
9. Hirner, A. (2003) Volatile metal(loid) species associated with waste materials: Chemical and toxicological aspects. In: Cai, Y. and Braids, O.C. (eds) *Biogeochemistry of Environmentally Important Trace Elements*. Vol. 835. ACS Symposium. ACS, Washington, DC, pp. 141–150.
10. Feldman, J. (2003) Volatilization of metals from a landfill site. In: Cai, Y. and Braids, O.C. (eds) *Biogeochemistry of Environmentally Important Trace Elements*. Vol. 835. ACS Symposium Series. ACS, Washington, DC, pp. 128–140.
11. Zufall, M. and Davidson, C.I. (1997) Dry deposition of particles to water surfaces. In: Baker, J. (ed.) *Atmospheric Deposition of Contaminants to the Great Lakes and Coastal Waters*. SETAC Press, Pensacola, pp. 1–16.
12. Ondov, J., Quinn, T.L. and Han, M. (1996) *Size-distribution, growth and deposition modeling of trace element bearing aerosol in the Chesapeake Bay airshed*. Maryland Department of Natural Resources.
13. Liu, Q.-T., Diamond, M.L., Gingrich, S.E., Ondov, J.M., Maciejczyk, P. and Stern, G.A. (2003) Accumulation of metals, trace elements and semi-volatile organic compounds on exterior window surfaces in Baltimore. *Environmental Pollution*, **122**, 51–61.
14. Qi, J.H., Li, P.L., Li, X.G., Feng, L.J. and Zhang, M.P. (2005) Estimation of dry deposition fluxes of particulate species to the water surface in the Qingdao area, using a model and surrogate surfaces. *Atmospheric Environment*, **39**(11), 2081–2088.
15. Sakata, M. and Marumoto, K. (2004) Dry deposition fluxes and deposition velocities of trace metals in the Tokyo metropolitan area measured with a water surface sampler. *Environmental Science & Technology*, **38**(7), 2190–2197.
16. Tasdemir, Y. and Kural, C. (2005) Atmospheric dry deposition fluxes of trace elements measured in Bursa, Turkey. *Environmental Pollution*, **138**(3), 462–472.
17. Shahin, U.M., Holsen, T.M. and Odabasi, M. (2002) Dry deposition measured with a water surface sampler: A comparison to modeled results. *Atmospheric Environment*, **36**(20), 3267–3276.
18. Lim, J.H., Sabin, L.D., Schiff, K.C. and Stolzenbach, K.D. (2006) Concentration, size distribution, and dry deposition rate of particle-associated metals in the Los Angeles region. *Atmospheric Environment*, **40**(40), 7810–7823.
19. Malcolm, E.G., Keeler, G.J., Lawson, S.T. and Sherbatskoy, T.D. (2003) Mercury and trace elements in cloud water and precipitation collected on Mt. Mansfield, Vermont. *Journal of Environmental Monitoring*, **5**(4), 584–590.
20. Marsik, F.J., Keeler, G.J. and Landis, M.S. (2007) The dry-deposition of speciated mercury to the Florida Everglades: Measurements and modeling. *Atmospheric Environment*, **41**(1), 136–149.
21. Caldwell, C.A., Swartzendruber, P. and Prestbo, E. (2006) Concentration and dry deposition of mercury species in arid south central New Mexico (2001–2002). *Environmental Science & Technology*, **40**(24), 7535–7540.
22. Avila, A. and Rodrigo, A. (2004) Trace metal fluxes in bulk deposition, throughfall and stemflow at two evergreen oak stands in NE Spain subject to different exposure to the industrial environment. *Atmospheric Environment*, **38**(2), 171–180.
23. Hou, H., Takamatsu, T., Koshikawa, M.K. and Hosomi, M. (2005) Trace metals in bulk precipitation and throughfall in a suburban area of Japan. *Atmospheric Environment*, **39**, 3583–3595.
24. Lawson, N.M. and Mason, R.P. (2001) Concentration of mercury, methylmercury, cadmium, lead, arsenic, and selenium in the rain and stream water of two contrasting watersheds in Western Maryland. *Water Research*, **35**(17), 4039–4052.
25. Scudlark, J., Rice, K.C., Conko, K.M., Bricker, O.P. and Church, T.M. (2005) Transmission of atmospherically derived trace elements through an undeveloped, forested Maryland watershed. *Water, Air, and Soil Pollution*, **163**, 53–79.
26. Itoh, Y., Miura, S. and Yoshinaga, S. (2006) Atmospheric lead and cadmium deposition within forests in the Kanto district, Japan. *Journal of Forest Research*, **11**(2), 137–142.
27. Schwarzenbach, R., Gschwend, P.M. and Imboden, D.M. (1993) *Environmental Organic Chemistry*. John Wiley & Sons, Inc., New York.
28. Holsen, T.M., Zhu, X., Khalili, N.R., Lin, J.J., Lestari, P., Lu, C.-S. and Noll, K.E. (1997) Atmospheric particles size distributions and dry deposition measured around Lake Michigan. In: Baker, J.E. (ed.) *Atmospheric Deposition of Contaminants to the Great Lakes and Coastal Waters*. SETAC Press, Pensacola, FL, pp. 35–50.
29. Deguillaume, L., Leriche, M., Desboeufs, K., Mailhot, G., George, C. and Chaumerliac, N. (2005) Transition metals in atmospheric liquid phases: Sources, reactivity, and sensitive parameters. *Chemical Reviews*, **105**(9), 3388–3431.
30. Chester, R. (2003) *Marine Geochemistry*. Blackwell Science, Malden.
31. Nicholson, K.W., Branson, J.R. and Giess, P. (1991) Field-measurements of the below-cloud scavenging of particulate material. *Atmospheric Environment Part A-General Topics*, **25**(3–4), 771–777.
32. Lawson, S.T., Scherbatskoy, T.D., Malcolm, E.G. and Keeler, G.J. (2003) Cloud water and throughfall deposition of mercury and trace elements in a high elevation spruce-fir forest at Mt. Mansfield, Vermont. *Journal of Environmental Monitoring*, **5**(4), 578–583.
33. Lin, C.J. and Pehkonen, S.O. (1999) The chemistry of atmospheric mercury: A review. *Atmospheric Environment*, **33**(13), 2067–2079.
34. Mason, R., Lawson, N.M. and Sheu, G.-R. (2002) The urban atmosphere: An important source of trace metals to nearby waters? Lipnick, R., Mason, R.P., Phillips, M.L. and Pittman, C.U., Jr. (eds) *Chemicals in the Environment: Fate, Impacts and Remediation*. Vol. 806. ACS Symposium Series. ACS, Washington, DC, pp. 203–223.
35. Sweet, C., Weiss, A. and Vermette, S.J. (1998) Atmospheric deposition of trace metals at three sites near the Great Lakes. *Water, Air, and Soil Pollution*, **103**, 423–439.

36. Baker, J., Poster, D.L., Clark, C.A., Church, T.M., Scudlark, J.R., Ondov, J.M., Dickhut, R.M. and Cutter, G. (1997) Loadings of atmospheric trace elements and organic contaminants to the Chesapeake Bay. In: Baker, J. (ed.) *Atmospheric Deposition of Contaminants to the Great Lakes and Coastal Waters*. SETAC Press, Pensacola, FL, pp. 171–194.
37. Arimoto, R., Gao, Y., Zhou, M.-Y., Soo, D., Chen, L., Gu, D., Wang, Z. and Zhang, X. (1997) Atmospheric deposition of trace elements to the western Pacific Basin. In: Baker, J. (ed.) *Atmospheric Deposition of Contaminants to the Great Lakes and to Coastal Waters*. SETAC Press, Pensacola, FL, pp. 195–208.
38. Hillery, B.R., Hoff, R.A. and Hites, R.A. (1997) Atmospheric contaminant deposition to the Great Lakes determined from the Integrated Atmospheric Deposition Network. In: Baker, J.E. (ed.) *Atmospheric Deposition of Contaminants to the Great Lakes and Coastal Waters*. SETAC Press, Pensacola, FL, pp. 277–292.
39. Loannidou, A. and Papastefanou, C. (2006) Precipitation scavenging of Be-7 and (CS)-C-137 radionuclides in air. *Journal of Environmental Radioactivity*, **85**(1), 121–136.
40. Lamborg, C.H., Rolfhus, K.R. and Fitzgerald, W.F. (1999) The atmospheric cycling and air-sea exchange of mercury species in the south and equatorial Atlantic Ocean. *Deep-Sea Research Part II*, **46**, 957–977.
41. Scudlark, J. and Church, T.M. (1997) Atmospheric deposition of trace metals to the mid-Atlantic Bight. In: Baker, J. (ed.) *Atmospheric Deposition of Contaminants to the Great Lakes and Coastal Waters*. SETAC press, Pensacola, pp. 195–208.
42. Mason, R.P. and Sheu, G.R. (2002) Role of the ocean in the global mercury cycle. *Global Biogeochemical Cycles*, **16**(4), Article # 1093.
43. Pohl, C., Loffler, A., Schmidt, M. and Seifert, T. (2006) A trace metal (Pb, Cd, Zn, Cu) balance for surface waters in the eastern Gotland Basin, Baltic Sea. *Journal of Marine Systems*, **60**(3–4), 381–395.
44. Arimoto, R., Duce, R.A., Ray, B.J., Ellis, W.G., Cullen, J.D. and Merrill, J.T. (1995) Trace-elements in the atmosphere over the North Atlantic. *Journal of Geophysical Research-Atmospheres*, **100**(D1), 1199–1213.
45. Arimoto, R., Duce, R.A., Ray, B.J. and Tomza, U. (2003) Dry deposition of trace elements to the western North Atlantic. *Global Biogeochemical Cycles*, **17**(1), Article # 1010.
46. Kim, G., Alleman, L.Y. and Church, T.M. (1999) Atmospheric depositional fluxes of trace elements, Pb-210, and Be-7 to the Sargasso Sea. *Global Biogeochemical Cycles*, **13**(4), 1183–1192.
47. Pacyna, E.G., Pacyna, J.M., Fudala, J., Strzelecka-Jastrzab, E., Hlawiczka, S., Panasiuk, D., Nitter, S., Pregger, T., Pfeiffer, H. and Friedrich, R. (2007) Current and future emissions of selected heavy metals to the atmosphere from anthropogenic sources in Europe. *Atmospheric Environment*, **41**(38), 8557–8566.
48. Johansson, K., Bergback, B. and Tyler, G. (2001) Impact of atmospheric long range transport of lead, mercury and cadmium on the Swedish forest environment. *Water, Air, and Soil Pollution: Focus*, **1**, 279–297.
49. Munthe, J., Bodaly, R.A., Branfireun, B.A., Driscoll, C.T., Gilmour, C.C., Harris, R., Horvat, M., Lucotte, M. and Malm, O. (2007) Recovery of mercury-contaminated fisheries. *Ambio*, **36**(1), 33–44.
50. Duce, R., Liss, P.S., Merrill, J.T., et al (1991) The atmospheric input of trace species to the World Ocean. *Global Biogeochemical Cycles*, **5**, 193–259.
51. Duce, R.A. (1989) SEAREX: The sea-air exchange program. In: Riley, J.P. and Chester, R. (eds) *SEAREX: The Sea-Air Exchange Program*. Vol. 10. Academic Press, London, pp. 1–14.
52. Arimoto, R., Duce, R.A. and Ray, B.J. (1989) Concentrations, sources and air-sea exchange of trace metals in the atmosphere over the Pacific Ocean. In: Riley, J.P. and Chester, R. (eds) *SEAREX: The Sea-Air Exchange Program*. Vol. 10. Academic Press, London, pp. 107–151.
53. Ryaboshapko, I.I., Gusev, A., Afinogenova, O., Berg, T. and Hjellbrekke, A.-G. (1999) *Monitoring and modeling of lead, cadmium and mercury transboundary transport in the atmosphere of Europe*; EMEP July 1999, 125 pp.
54. Akeredolu, F.A., Barrie, L.A., Olson, M.P., Oikawa, K.K., Pacyna, J.M. and Keeler, G.J. (1994) The flux of anthropogenic trace-metals into the arctic from the midlatitudes in 1979/80. *Atmospheric Environment*, **28**(8), 1557–1572.
55. Ryaboshapko, A., Bullock, O.R., Christensen, J., Cohen, M., Dastoor, A., Ilyin, I., Petersen, G., Syrakov, D., Travnikov, O., Artz, R.S., Davignon, D., Draxler, R.R., Munthe, J. and Pacyna, J. (2007) Intercomparison study of atmospheric mercury models: 2. Modelling results vs. long-term observations and comparison of country deposition budgets. *The Science of the Total Environment*, **377**(2–3), 319–333.
56. Ilyin, I., Ryaboshapko, A., Ravnikov, O., Berg, T., Hjellbrekke, A.-G. and Larsen, R. (2000) *Heavy Metal Transboundary Air Pollution in Europe: Monitoring and Modelling Results for 1997 and 1998*; Co-operative Programme for Monitoring and Evaluation of the Long-Range Transmission of Air Pollutants in Europe.
57. Pirrone, N. and Mason, R.P. (2009) *Mercury Fate and Transport in the Global Atmosphere*. Springer, Dordrecht, The Netherlands.
58. Ryaboshapko, A., Bullock, O.R., Christensen, J., Cohen, M., Dastoor, A., Ilyin, I., Petersen, G., Syrakov, D., Artz, R.S., Davignon, D., Draxler, R.R. and Munthe, J. (2007) Intercomparison study of atmospheric mercury models: 1. Comparison of models with short-term measurements. *The Science of the Total Environment*, **376**(1–3), 228–240.
59. Ryaboshapko, A., Bullock, R., Ebinghaus, R., Ilyin, I., Lohman, K., Munthe, J., Petersen, G., Seigneur, C. and Wangberg, I. (2002) Comparison of mercury chemistry models. *Atmospheric Environment*, **36**(24), 3881–3898.
60. Dias, G.M. and Edwards, G.C. (2003) Differentiating natural and anthropogenic sources of metals to the environment. *Human and Ecological Risk Assessment*, **9**(4), 699–721.
61. Gordon, G.E. (1988) Receptor models. *Environmental Science and Technology*, **22**, 1132–1142.
62. Suarez, A.E. and Ondov, J.M. (2002) Ambient aerosol concentrations of elements resolved by size and by source: Contributions of some cytokine-active metals from coal- and oil-fired power plants. *Energy & Fuels*, **16**(3), 562–568.
63. Han, J.S., Moon, K.J., Lee, S.J., Kim, Y.J., Ryu, S.Y., Cliff, S.S. and Yi, S.M. (2006) Size-resolved source apportionment of ambient particles by positive matrix factorization at Gosan

background site in East Asia. *Atmospheric Chemistry and Physics*, **6**, 211–223.

64. Okuda, T., Kato, J., Mori, J., Tenmoku, M., Suda, Y., Tanaka, S., He, K.B., Ma, Y.L., Yang, F., Yu, X.C., Duan, F.K. and Lei, Y. (2004) Daily concentrations of trace metals in aerosols in Beijing, China, determined by using inductively coupled plasma mass spectrometry equipped with laser ablation analysis, and source identification of aerosols. *The Science of the Total Environment*, **330**(1–3), 145–158.
65. Kim, E., Hopke, P.K., Larson, T.V., Maykut, N.N. and Lewtas, J. (2004) Factor analysis of Seattle fine particles. *Aerosol Science and Technology*, **38**(7), 724–738.
66. Morel, F.M.M. and Hering, J.G. (1993) *Principals and Applications of Aquatic Chemistry*. John Wiley & Sons, Inc., New York.
67. Jeffery, C.D., Robinson, I.S. and Woolf, D.K. (2010) Tuning a physically-based model of the air-sea gas transfer velocity. *Ocean Modeling*, **31**, 28–35.
68. Loux, N.T. (2004) A critical assessment of elemental mercury air/water exchange parameters. *Chemical Speciation and Bioavailability*, **16**(4), 127–138.
69. Meyer, J., Michalke, K., Kouril, T. and Hensel, R. (2008) Volatilisation of metals and metalloids: An inherent feature of methanoarchaea? *Systematic and Applied Microbiology*, **31**(2), 81–87.
70. Amouroux, D., Liss, P.S., Tessier, E., Hamren-Larsson, M. and Donard, O.F.X. (2001) Role of oceans as biogenic sources of selenium. *Earth and Planetary Science Letters*, **189**(3–4), 277–283.
71. Yang, G.P., Li, C.X., Qi, J.L., Hu, L.G. and Hou, H.J. (2007) Photochemical oxidation of dimethylsulfide in seawater. *Acta Oceanologica Sinica*, **26**, 34–42.
72. Amouroux, D., Pecheyran, C. and Donard, O.F.X. (2000) Formation of volatile selenium species in synthetic seawater under light and dark experimental conditions. *Applied Organometallic Chemistry*, **14**(5), 236–244.
73. Cima, F., Craig, P.J. and Harrington, C. (2003) Organotin compounds in the environment. In: Craig, P.J. (ed.) *Organometallic Compounds in the Environment*. John Wiley & Sons, Ltd, Chichester, pp. 101–150.
74. Planer-Friedrich, B., Lehr, C., Matschullat, J., Merkel, B.J., Nordstrom, D.K. and Sandstrom, M.W. (2006) Speciation of volatile arsenic at geothermal features in Yellowstone National Park. *Geochimica et Cosmochimica Acta*, **70**(10), 2480–2491.
75. Weber, J.H. (1999) Volatile hydride and methyl compounds of selected elements formed in the marine environment. *Marine Chemistry*, **65**(1–2), 67–75.
76. Diaz, X., Johnson, W.P., Oliver, W.A. and Naftz, D.L. (2009) Volatile Selenium Flux from the Great Salt Lake, Utah. *Environmental Science & Technology*, **43**(1), 53–59.
77. Pecheyran, C., Lalere, B. and Donard, O.F.X. (2000) Volatile metal and metalloid species (Pb, Hg, Se) in a European urban atmosphere (Bordeaux, France). *Environmental Science & Technology*, **34**(1), 27–32.
78. Boutron, C., Rosman, K., Barbante, C., Bolshov, M., Adams, F., Hong, S.M. and Ferrari, C. (2004) Anthropogenic lead in polar snow and ice archives. *Comptes Rendus Geoscience*, **336**(10), 847–867.
79. Stumm, W. and Morgan, J.J. (1996) *Aquatic Chemistry*. John Wiley & Sons, Inc., New York.
80. Moffett, J.W. (2001) Transformations amongst different forms of iron in the ocean. In: Turner, D.R. and Hunter, K.A. (eds) *The Biogeochemistry of Iron in Seawater*. John Wiley & Sons, Ltd, Chichester, pp. 343–372.
81. Deng, Y.W., Zhang, K., Chen, H., Wu, T.X., Krzyaniak, M., Wellons, A., Bolla, D., Douglas, K. and Zuo, Y.G. (2006) Iron catalyzed photochemical transformation of benzoic acid in atmospheric liquids: Product identification and reaction mechanisms. *Atmospheric Environment*, **40**(20), 3665–3676.
82. Deutsch, F., Hoffmann, P. and Ortner, H.M. (2001) Field experimental investigations on the Fe(II)- and Fe(III)-content in cloudwater samples. *Journal of Atmospheric Chemistry*, **40**(1), 87–105.
83. Siffert, C. and Sulzberger, B. (1991) Light-induced dissolution of hematite in the presence of oxalate – A case-study. *Langmuir*, **7**(8), 1627–1634.
84. Sykora, J. (1997) Photochemistry of copper complexes and their environmental aspects. *Coordination Chemistry Reviews*, **159**, 95–108.
85. Kieber, R.J., Skrabal, S.A., Smith, C. and Willey, J.D. (2004) Redox speciation of copper in rainwater: Temporal variability and atmospheric deposition. *Environmental Science & Technology*, **38**(13), 3587–3594.
86. Wu, C.H., Sun, L.Z. and Faust, B.C. (2000) Photochemical formation of copper(I) from copper(II)-dicarboxylate complexes: Effects of outer-sphere versus inner-sphere coordination and of quenching by malonate. *Journal of Physical Chemistry A*, **104**(21), 4989–4996.
87. Fan, S.-M. (2008) Photochemical and biochemical controls on reactive oxygen and iron speciation in the pelagic surface ocean. *Marine Chemistry*, **109**, 152–164.
88. Siefert, R.L., Johansen, A.M., Hoffmann, M.R. and Pehkonen, S.O. (1998) Measurements of trace metal (Fe, Cu, Mn, Cr) oxidation states in fog and stratus clouds. *Journal of the Air & Waste Management Association*, **48**(2), 128–143.
89. Brooks, M.L., McKnight, D.M. and Clements, W.H. (2007) Photochemical control of copper complexation by dissolved organic matter in Rocky Mountain streams, Colorado. *Limnology and Oceanography*, **52**(2), 766–779.
90. Laglera, L.M. and van den Berg, C.M.G. (2006) Photochemical oxidation of thiols and copper complexing ligands in estuarine waters. *Marine Chemistry*, **101**(1–2), 130–140.
91. Witt, M.L.I., Skrabal, S., Kieber, R. and Willey, J. (2007) Photochemistry of Cu complexed with chromophoric dissolved organic matter: Implications for Cu speciation in rainwater. *Journal of Atmospheric Chemistry*, **58**(2), 89–109.
92. Rijkenberg, M.J.A., Gerringa, L.J.A., Velzeboer, I., Timmermans, K.R., Buma, A.G.J. and de Baar, H.J.W. (2006) Iron-binding ligands in Dutch estuaries are not affected by UV induced photochemical degradation. *Marine Chemistry*, **100**(1–2), 11–23.
93. Barbeau, K., Rue, E.L., Bruland, K.W. and Butler, A. (2001) Photochemical cycling of iron in the surface ocean mediated by microbial iron(III)-binding ligands. *Nature*, **413**, 409–413.

94. Gardfeldt, K. and Jonsson, M. (2003) Is bimolecular reduction of Hg(II) complexes possible in aqueous systems of environmental importance. *Journal of Physical Chemistry A*, **107**(22), 4478–4482.
95. Si, L. and Ariya, P.A. (2008) Reduction of oxidized mercury species by dicarboxylic acids (C-2-C-4): Kinetic and product studies. *Environmental Science & Technology*, **42**(14), 5150–5155.
96. Holmes, C.D., Jacob, D.J., Mason, R.P. and Jaffe, D.A. (2009) Sources and deposition of reactive gaseous mercury in the marine atmosphere. *Atmospheric Environment*, **43**(14), 2278–2285.
97. Mytych, P., Ciesla, P. and Stasicka, Z. (2005) Photoredox processes in the Cr(VI)-Cr(III)-oxalate system and their environmental relevance. *Applied Catalysis B-Environmental*, **59**(3–4), 161–170.
98. Wang, Z.H., Ma, W.H., Chen, C.C. and Zhao, J.C. (2008) Photochemical coupling reactions between Fe(III)/Fe(II), Cr(VI)/Cr(III), and polycarboxylates: Inhibitory effect of Cr species. *Environmental Science & Technology*, **42**(19), 7260–7266.
99. Cwiertny, D.M., Young, M.A. and Grassian, V.H. (2008) Chemistry and photochemistry of mineral dust aerosol. *Annual Review of Physical Chemistry*, **59**, 27–51.
100. Cwiertny, D.M., Baltrusaitis, J., Hunter, G.J., Laskin, A., Scherer, M.M. and Grassian, V.H. (2008) Characterization and acid-mobilization study of iron-containing mineral dust source materials. *Journal of Geophysical Research-Atmospheres*, **113**(D5), Article # D05202.
101. Key, J.M., Paulk, N. and Johansen, A.M. (2008) Photochemistry of iron in simulated crustal aerosols with dimethyl sulfide oxidation products. *Environmental Science & Technology*, **42**(1), 133–139.
102. Willey, J.D., Whitehead, R.F., Kieber, R.J. and Hardison, D.R. (2005) Oxidation of Fe(II) in rainwater. *Environmental Science & Technology*, **39**(8), 2579–2585.
103. Pettine, M. and Millero, F.J. (2000) Effect of metals on the oxidation of As(III) with H2O2. *Marine Chemistry*, **70**(1–3), 223–234.
104. Buschmann, J., Canonica, S., Lindauer, U., Hug, S.J. and Sigg, L. (2005) Photoirradiation of dissolved humic acid induces arsenic(III) oxidation. *Environmental Science & Technology*, **39**(24), 9541–9546.
105. Heal, M.R., Hibbs, L.R., Agius, R.M. and Beverland, L.J. (2005) Total and water-soluble trace metal content of urban background PM10, PM2.5 and black smoke in Edinburgh, UK. *Atmospheric Environment*, **39**(8), 1417–1430.
106. Hsu, S.C., Lin, F.J. and Jeng, W.L. (2005) Seawater solubility of natural and anthropogenic metals within ambient aerosols collected from Taiwan coastal sites. *Atmospheric Environment*, **39**(22), 3989–4001.
107. Qureshi, S., Dutkiewicz, V.A., Khan, A.R., Swami, K., Yang, K.X., Husain, L., Schwab, J.J. and Demerjian, K.L. (2006) Elemental composition of PM2.5 aerosols in Queens, New York: Solubility and temporal trends. *Atmospheric Environment*, **40**, S238–S251.
108. Guerzoni, S., Molinaroli, E., Rossini, P., Rampazzo, G., Quarantotto, G., De Falco, G. and Cristini, S. (1999) Role of desert aerosol in metal fluxes in the Mediterranean area. *Chemosphere*, **39**(2), 229–246.
109. Desboeufs, K.V., Sofikitis, A., Losno, R., Colin, J.L. and Ausset, P. (2005) Dissolution and solubility of trace metals from natural and anthropogenic aerosol particulate matter. *Chemosphere*, **58**(2), 195–203.
110. Birmili, W., Allen, A.G., Bary, F. and Harrison, R.M. (2006) Trace metal concentrations and water solubility in size-fractionated atmospheric particles and influence of road traffic. *Environmental Science & Technology*, **40**(4), 1144–1153.
111. Zhu, X. (1993) Photoreduction of iron(III) in marine mineral aerosol solutions. *Journal of Geophysical Research-Atmospheres*, **98**, 9039.
112. Wu, J., Rember, R. and Cahill, C. (2007) Dissolution of aerosol iron in the surface waters of the North Pacific and North Atlantic oceans as determined by a semicontinuous flow-through reactor method. *Global Biogeochemical Cycles*, **21**(4), Article # GB4010.
113. Buck, C.S., Landing, W.M., Resing, J.A. and Lebon, G.T. (2006) Aerosol iron and aluminum solubility in the northwest Pacific Ocean: Results from the 2002 IOC cruise. *Geochemistry Geophysics Geosystems*, **7**, Article # Q04M07.
114. Jickells, T.D. and Spokes, L.J. (2001) Atmopsheric iron inputs to the oceans. In: Turner, D.R. and Hunter, K.A. (eds) *The Biogeochemistry of Iron in Seawater*. John Wiley & Sons, Ltd, Chichester, pp. 85–121.
115. Sedwick, P.N., Sholkovitz, E.R. and Church, T.M. (2007) Impact of anthropogenic combustion emissions on the fractional solubility of aerosol iron: Evidence from the Sargasso Sea. *Geochemistry Geophysics Geosystems*, **8**, Article # Q10Q06.
116. Bruland, K. and Lohan, M.C. (2004) Controls of trace metals in seawater. In: Elderfield, H. (ed.) *The Oceans and Marine Geochemistry*. Vol. 6. Elsevier, Amsterdam, pp. 23–47.
117. Voelker, B.M., Morel, F.M.M. and Sulzberger, B. (1997) Iron redox cycling in surface waters: Effects of humic substances and light. *Environmental Science & Technology*, **31**(4), 1004–1011.
118. Waite, T.D., Szymczak, R., Espey, Q.I. and Furnas, M.J. (1995) Diel variations in iron speciation in northern Australian shelf waters. *Marine Chemistry*, **50**(1–4), 79–91.
119. Sunderland, E.M. and Mason, R.P. (2007) Human impacts on open ocean mercury concentrations. *Global Biogeochemical Cycles*, **21**, GB4022.
120. Selin, N.E., Jacob, D.J., Park, R.J., Yantosca, R.M., Strode, S., Jaegle, L. and Jaffe, D. (2007) Chemical cycling and deposition of atmospheric mercury: Global constraints from observations. *Journal of Geophysical Research – Atmospheric*, **112**, D02308.
121. Strode, S.A., Jaegle, L., Selin, N.E., Jacob, D.J., Park, R.J., Yantosca, R.M., Mason, R.P. and Slemr, F. (2007) Air-sea exchange in the global mercury cycle. *Global Biogeochemical Cycles*, **21**, Article # GB002766. doi: 10.1029/2006GB002766
122. Amyot, M., Mierle, G. and McQueen, D.J. (1997) Effects of solar radiation on the formation of dissolved gaseous mercury in temperate lakes. *Geochimica et Cosmochimica Acta*, **61**, 975.
123. Lalonde, J.D., Amyot, M., Kraepiel, A.M.L. and Morel, F.M.M. (2001) Photooxidation of Hg(0) in artificial and natural waters. *Environmental Science & Technology*, **35**(7), 1367–1372.

124. Whalin, L., Kim, E.-H. and Mason, R.P. (2007) Factors influencing the oxidation, reduction, methylation and demethylation of mercury in coastal waters. *Marine Chemistry*, **107**, 278–294.
125. Costa, M. and Liss, P. (2000) Photoreduction and evolution of mercury from seawater. *The Science of the Total Environment*, **261**(1–3), 125–135.
126. Mason, R.P., Rolfhus, K.R. and Fitzgerald, W.F. (1998) Mercury in the North Atlantic. *Marine Chemistry*, **61**, 37–53.
127. Mason, R.P. and Sullivan, K.A. (1999) The distribution and speciation of mercury in the South and Equatorial Atlantic. *Deep-Sea Research Part II-Topical Studies in Oceanography*, **46**(5), 937–956.
128. Cossa, D., Martin, J.-M., Takayanagi, K. and Sanjuan, J. (1997) The distribution and cycling of mercury species in the western Mediterranean. *Deep-Sea Research*, **44**, 721–740.
129. Kim, J.P. and Fitzgerald, W.F. (1988) Gaseous mercury profiles in the tropical Pacific Ocean. *Geophysical Research Letters*, **15**, 40–43.
130. Fitzgerald, W.F., Lamborg, C.H. and Hammerschmidt, C.R. (2007) Marine biogeochemical cycling of mercury. *Chemical Reviews*, **107**(2), 641–662.
131. Fitzgerald, W. and Lamborg, C.H. (2005) Geochemsitry of mercury in the environment. In: Lollar, B. (ed.) *Environmental Geochemistry*. Vol. 9. Elsevier, Amsterdam, pp. 107–147.
132. Mason, R.P., Lawson, N.M. and Sheu, G.R. (2001) Mercury in the Atlantic Ocean: Factors controlling air-sea exchange of mercury and its distribution in the upper waters. *Deep-Sea Research Part II-Topical Studies in Oceanography*, **48**(13), 2829–2853.
133. Gardfeldt, K., Sommar, J., Ferrara, R., Ceccarini, C., Lanzillotta, E., Munthe, J., Wangberg, I., Lindqvist, O., Pirrone, N., Sprovieri, F., Pesenti, E. and Stromberg, D. (2003) Evasion of mercury from coastal and open waters of the Atlantic Ocean and the Mediterranean Sea. *Atmospheric Environment*, **37**(Suppl. 1), S73–S84.
134. Andersson, M.E., Sommar, J., Gardfeldt, K. and Lindqvist, O. (2008) Enhanced concentrations of dissolved gaseous mercury in the surface waters of the Arctic Ocean. *Marine Chemistry*, **110**(3–4), 190–194.
135. Mason, R.P. and Fitzgerald, W.F. (1993) The distribution and biogeochemical cycling of mercury in the Equatorial Pacific-Ocean. *Deep-Sea Research Part I-Oceanographic Research Papers*, **40**(9), 1897 1924.
136. Mason, R.P., Lawson, N.M., Lawrence, A.L., Leaner, J.J., Lee, J.G. and Sheu, G.-R. (1999) Mercury in the Chesapeake Bay. *Marine Chemistry*, **65**(1–2), 77–96.
137. Baeyens, W. and Leermakers, M. (1998) Elemental mercury concentrations and formation rates in the Scheldt estuary and the North Sea. *Marine Chemistry*, **60**, 257–266.
138. Ferrara, R., Mazzolai, B., Lanzillotta, E., Nucaro, E. and Pirrone, N. (2000) Temporal trends in gaseous mercury evasion in the Mediterranean seawaters. *The Science of the Total Environment*, **259**, 183–190.
139. Ferrara, R., Ceccarini, C., Lanzillotta, E., Gardfeldt, K., Sommar, J., Horvat, M., Logar, M., Fajon, V. and Kotnik, J. (2003) Profiles of dissolved gaseous mercury concentration in the Mediterranean seawater. *Atmospheric Environment*, **37** (Suppl 1), S85–S92.
140. Andersson, M.E., Gardfeldt, K., Wangberg, I., Sprovieri, F., Pirrone, N. and Lindqvist, O. (2007) Seasonal and daily variation of mercury evasion at coastal and off-shore sites at the Mediterranean Sea. *Marine Chemistry*, **107**(1), 103–116.
141. Amyot, M., Mierle, G., Lean, D.R.S. and McQueen, D.J. (1994) Sunlight-induced formation of dissolved gaseous mercury in lake water. *Environmental Science and Technology*, **28**, 2366–2371.
142. Lanzillotta, E., Ceccarini, C. and Ferrara, R. (2002) Photo induced formation of dissolved gaseous mercury in coastal and offshore seawater of the Mediterranean basin. *The Science of the Total Environment*, **300**(1–3), 179–187.
143. Rolfhus, K.R. and Fitzgerald, W.F. (2001) The evasion and spatial/temporal distribution of mercury species in Long Island Sound, CT-NY. *Geochimica et Cosmochimica Acta*, **65**(3), 407–418.
144. Mason, R.P., Morel, F.M.M. and Hemond, H.F. (1995) The role of microorganisms in elemental mercury formation in natural waters. *Water, Air, and Soil Pollution*, **80**, 775–787.
145. Lanzillotta, E., Ceccarini, C., Ferrara, R., Dini, F., Frontini, E. and Banchetti, R. (2004) Importance of the biogenic organic matter in photo-formation of dissolved gaseous mercury in a culture of the marine diatom Chaetoceros sp. *The Science of the Total Environment*, **318**(1–3), 211–221.
146. Pacyna, E.G., Pacyna, J.M., Steenhuisen, F. and Wilson, S. (2006) Global anthropogenic mercury emission inventory for 2000. *Atmospheric Environment*, **40**, 4048–4063.
147. Pirrone, N., Keeler, G.J. and Nriagu, J.O. (1996) Regional differences in worldwide emissions of mercury to the atmosphere. *Atmospheric Environment*, **30**(19), 3379.
148. Soerensen, A.L., Sunderland, E.M., Holmes, C.D., et al. (2010) An improved global model for air-sea exchange of mercury: High concentrations over the North Atlantic. *Environmental Science and Technology*, **44**(22), 8574–8580.
149. Mason, R.P. (2009) Mercury emissions from natural processes and their importance in the global mercury cycle. In: Pirrone, N. and Mason, R. (eds) *Mercury Fate and Transport in the Global Atmosphere: Emissions, Measurements and models*. Springer, Norwell, MA, USA, pp. 173–191.
150. Hynes, A.J., Donohoue, D.L., Goodsite, M.E. and Hedgecock, I.M. (2009) Our current understanding of major chemical and physical processes affecting mercury dynamics in the atmosphere and at the air-water/terrestrial interfaces. In: Pirrone, N. and Mason, R.P. (eds) *Mercury Fate and Transport in the Global Atmosphere*. Springer, Dordrecht, The Netherlands, pp. 427–457.
151. Raofie, F., Snider, G. and Ariya, P.A. (2008) Reaction of gaseous mercury with molecular iodine, atomic iodine, and iodine oxide radicals – Kinetics, product studies, and atmospheric implications. *Canadian Journal of Chemistry-Revue Canadienne De Chimie*, **86**(8), 811–820.
152. Donohoue, D.L., Bauer, D., Cossairt, B. and Hynes, A.J. (2006) Temperature and pressure dependent rate coefficients for the reaction of Hg with Br and the reaction of Br with Br: A pulsed laser photolysis-pulsed laser induced fluorescence study. *Journal of Physical Chemistry A*, **110**(21), 6623–6632.
153. Pal, B. and Ariya, P.A. (2004) Gas-phase HO center dot-Initiated reactions of elemental mercury: Kinetics, product

studies, and atmospheric implications. *Environmental Science & Technology*, **38**(21), 5555–5566.

154. Pal, B. and Ariya, P.A. (2004) Studies of ozone initiated reactions of gaseous mercury: Kinetics, product studies, and atmospheric implications. *Physical Chemistry Chemical Physics*, **6**(3), 572–579.
155. Khalizov, A.F., Viswanathan, B., Larregaray, P. and Ariya, P.A. (2003) A theoretical study on the reactions of Hg with halogens: Atmospheric implications. *Journal of Physical Chemistry A*, **107**(33), 6360–6365.
156. Ariya, P.A., Khalizov, A. and Gidas, A. (2002) Reactions of gaseous mercury with atomic and molecular halogens: Kinetics, product studies, and atmospheric implications. *Journal of Physical Chemistry A*, **106**(32), 7310–7320.
157. Dastoor, A.P., Davignon, D., Theys, N., Van Roozendael, M., Steffen, A. and Ariya, P.A. (2008) Modeling dynamic exchange of gaseous elemental mercury at polar sunrise. *Environmental Science & Technology*, **42**(14), 5183–5188.
158. Seigneur, C. and Lohman, K. (2008) Effect of bromine chemistry on the atmospheric mercury cycle. *Journal of Geophysical Research-Atmospheres*, **113**(D23), Article # D23309.
159. Hedgecock, I.M., Pirrone, N. and Sprovieri, F. (2008) Chasing quicksilver northward: Mercury chemistry in the Arctic troposphere. *Environmental Chemistry*, **5**(2), 131–134.
160. Brooks, S.B., Saiz-Lopez, A., Skov, H., Lindberg, S.E., Plane, J.M.C. and Goodsite, M.E. (2006) The mass balance of mercury in the springtime arctic environment. *Geophysical Research Letters*, **33**(13), Article # L13812.
161. Calvert, J.G. and Lindberg, S.E. (2003) A modeling study of the mechanism of the halogen-ozone-mercury homogeneous reactions in the troposphere during the polar spring. *Atmospheric Environment*, **37**(32), 4467–4481.
162. Schroeder, W., Anlauf, K., Barrie, L., Lu, J., Steffen, A., Schneeberger, D. and Berg, T. (1998) Arctic springtime depletion of mercury. *Nature*, **394**, 331–332.
163. Steffen, A., Douglas, T., Amyot, M., et al. (2008) A synthesis of atmospheric mercury depletion event chemistry in the atmosphere and snow. *Atmospheric Chemistry and Physics*, **8**(6), 1445–1482.
164. Steffen, A., Schroeder, W., Macdonald, R., Poissant, L. and Konoplev, A. (2005) Mercury in the Arctic atmosphere: An analysis of eight years of measurements of GEM at Alert (Canada) and a comparison with observations at Amderma (Russia) and Kuujjuarapik (Canada). *The Science of the Total Environment*, **342**(1–3), 185–198.
165. Steffen, A., Schroeder, W., Bottenheim, J., Narayan, J. and Fuentes, J.D. (2002) Atmospheric mercury concentrations: Measurements and profiles near snow and ice surfaces in the Canadian Arctic during Alert 2000. *Atmospheric Environment*, **36**(15–16), 2653–2661.
166. Sprovieri, F., Pirrone, N., Landis, M.S. and Stevens, R.K. (2005) Oxidation of gaseous elemental mercury to gaseous divalent mercury during 2003 polar sunrise at Ny-Alesund. *Environmental Science & Technology*, **39**(23), 9156–9165.
167. Temme, C., Einax, J.W., Ebinghaus, R. and Schroeder, W.H. (2003) Measurements of atmospheric mercury species at a coastal site in the Antarctic and over the south Atlantic Ocean during polar summer. *Environmental Science & Technology*, **37**(1), 22–31.
168. Ebinghaus, R., Kock, H.H., Temme, C., Einax, J.W., Lowe, A.G., Richter, A., Burrows, J.P. and Schroeder, W.H. (2002) Antarctic springtime depletion of atmospheric mercury. *Environmental Science & Technology*, **36**(6), 1238–1244.
169. Lindberg, S.E., Brooks, S., Lin, C.J., Scott, K.J., Landis, M.S., Stevens, R.K., Goodsite, M. and Richter, A. (2002) Dynamic oxidation of gaseous mercury in the Arctic troposphere at polar sunrise. *Environmental Science & Technology*, **36**(6), 1245–1256.
170. Zhao, T.L., Gong, S.L., Bottenheim, J.W., McConnell, J.C., Sander, R., Kaleschke, L., Richter, A., Kerkweg, A., Toyota, K. and Barrie, L.A. (2008) A three-dimensional model study on the production of BrO and Arctic boundary layer ozone depletion. *Journal of Geophysical Research-Atmospheres*, **113**, Article # D24304.
171. Simpson, W.R., von Glasow, R., Riedel, K., et al. (2007) Halogens and their role in polar boundary-layer ozone depletion. *Atmospheric Chemistry and Physics*, **7**(16), 4375–4418.
172. Helmig, D., Oltmans, S.J., Carlson, D., Lamarque, J.F., Jones, A., Labuschagne, C., Anlauf, K. and Hayden, K. (2007) A review of surface ozone in the polar regions. *Atmospheric Environment*, **41**(24), 5138–5161.
173. Saiz-Lopez, A., Mahajan, A.S., Salmon, R.A., Bauguitte, S.J.B., Jones, A.E., Roscoe, H.K. and Plane, J.M.C. (2007) Boundary layer halogens in coastal Antarctica. *Science*, **317**(5836), 348–351.
174. Tackett, P.J., Cavender, A.E., Keil, A.D., Shepson, P.B., Bottenheim, J.W., Morin, S., Deary, J., Steffen, A. and Doerge, C. (2007) A study of the vertical scale of halogen chemistry in the Arctic troposphere during Polar Sunrise at Barrow, Alaska. *Journal of Geophysical Research-Atmospheres*, **112**(D7), Article # D07306.
175. Carpenter, L.J., Hopkins, J.R., Jones, C.E., Lewis, A.C., Parthipan, R., Wevill, D.J., Poissant, L., Pilote, M. and Constant, P. (2005) Abiotic source of reactive organic halogens in the sub-Arctic atmosphere? *Environmental Science & Technology*, **39**(22), 8812–8816.
176. Simpson, W.R., Alvarez-Aviles, L., Douglas, T.A., Sturm, M. and Domine, F. (2005) Halogens in the coastal snow pack near Barrow, Alaska: Evidence for active bromine air-snow chemistry during springtime. *Geophysical Research Letters*, **32**(4), Article # L04811.
177. Hara, K., Osada, K., Kido, M., Hayashi, M., Matsunaga, K., Iwasaka, Y., Yamanouchi, T., Hashida, G. and Fukatsu, T. (2004) Chemistry of sea-salt particles and inorganic halogen species in Antarctic regions: Compositional differences between coastal and inland stations. *Journal of Geophysical Research-Atmospheres*, **109**(D20), Article # D20208.
178. Hara, K., Osada, K., Matsunaga, K., Iwasaka, Y., Shibata, T. and Furuya, K. (2002) Atmospheric inorganic chlorine and bromine species in Arctic boundary layer of the winter/spring. *Journal of Geophysical Research-Atmospheres*, **107**(D18), Article # 4361.
179. Honninger, G. and Platt, U. (2002) Observations of BrO and its vertical distribution during surface ozone depletion at Alert. *Atmospheric Environment*, **36**(15–16), 2481–2489.
180. Bottenheim, J.W., Fuentes, J.D., Tarasick, D.W. and Anlauf, K.G. (2002) Ozone in the Arctic lower troposphere during

winter and spring 2000 (ALERT2000). *Atmospheric Environment*, **36**(15–16), 2535–2544.

181. Spicer, C.W., Plastridge, R.A., Foster, K.L., Finlayson-Pitts, B.J., Bottenheim, J.W., Grannas, A.M. and Shepson, P.B. (2002) Molecular halogens before and during ozone depletion events in the Arctic at polar sunrise: Concentrations and sources. *Atmospheric Environment*, **36**(15–16), 2721–2731.
182. Boudries, H. and Bottenheim, J.W. (2000) Cl and Br atom concentrations during a surface boundary layer ozone depletion event in the Canadian High Arctic. *Geophysical Research Letters*, **27**(4), 517–520.
183. McElroy, C.T., McLinden, C.A. and McConnell, J.C. (1999) Evidence for bromine monoxide in the free troposphere during the Arctic polar sunrise. *Nature*, **397**(6717), 338–341.
184. Sander, R., Vogt, R., Harris, G.W. and Crutzen, P.J. (1997) Modeling the chemistry ozone, halogen compounds, and hydrocarbons in the arctic troposphere during spring. *Tellus, Series B, Chemical and Physical Meteorology*, **49**(5), 522–532.
185. Ge, M.F. and Ma, C.P. (2009) Reactive halogen chemistry. *Progress in Chemistry*, **21**(2–3), 307–334.
186. von Glasow, R. (2008) Atmospheric chemistry – Sun, sea and ozone destruction. *Nature*, **453**(7199), 1195–1196.
187. Finley, B.D. and Saltzman, E.S. (2008) Observations of Cl-2, Br-2, and I-2 in coastal marine air. *Journal of Geophysical Research-Atmospheres*, **113**(D21), Article # D21301.
188. Read, K.A., Mahajan, A.S., Carpenter, L.J., et al. (2008) Extensive halogen-mediated ozone destruction over the tropical Atlantic Ocean. *Nature*, **453**(7199), 1232–1235.
189. Keene, W.C., Stutz, J., Pszenny, A.A.P., Maben, J.R., Fischer, E.V., Smith, A.M., von Glasow, R., Pechtl, S., Sive, B.C. and Varner, R.K. (2007) Inorganic chlorine and bromine in coastal New England air during summer. *Journal of Geophysical Research-Atmospheres*, **112**(D10), Article # D10512.
190. von Glasow, R. (2006) Importance of the surface reaction OH+Cl- on sea salt aerosol for the chemistry of the marine boundary layer – A model study. *Atmospheric Chemistry and Physics*, **6**, 3571–3581.
191. Saiz-Lopez, A., Shillito, J.A., Coe, H. and Plane, J.M.C. (2006) Measurements and modelling of I-2, IO, OIO, BrO and NO3 in the mid-latitude marine boundary layer. *Atmospheric Chemistry and Physics*, **6**, 1513–1528.
192. Thomas, J.L., Jimenez-Aranda, A., Finlayson-Pitts, B.J. and Dabdub, D. (2006) Gas-phase molecular halogen formation from NaCl and NaBr aerosols: When are interface reactions important? *Journal of Physical Chemistry A*, **110**(5), 1859–1867.
193. von Glasow, R., von Kuhlmann, R., Lawrence, M.G., Platt, U. and Crutzen, P.J. (2004) Impact of reactive bromine chemistry in the troposphere. *Atmospheric Chemistry and Physics*, **4**, 2481–2497.
194. Sander, R., Keene, W.C., Pszenny, A.A.P., et al. (2003) Inorganic bromine in the marine boundary layer: A critical review. *Atmospheric Chemistry and Physics*, **3**, 1301–1336.
195. Platt, U. and Honninger, G. (2003) The role of halogen species in the troposphere. *Chemosphere*, **52**(2), 325–338.
196. Finlayson-Pitts, B.J. and Hemminger, J.C. (2000) Physical chemistry of airborne sea salt particles and their components. *Journal of Physical Chemistry A*, **104**(49), 11463–11477.
197. Galbally, I.E., Bentley, S.T. and Meyer, C.P. (2000) Mid-latitude marine boundary-layer ozone destruction at visible sunrise observed at Cape Grim, Tasmania, 41 degrees 5. *Geophysical Research Letters*, **27**(23), 3841–3844.
198. Jacob, D.J. (2000) Heterogeneous chemistry and tropospheric ozone. *Atmospheric Environment*, **34**(12–14), 2131–2159.
199. Stutz, J., Hebestreit, K., Alicke, B. and Platt, U. (1999) Chemistry of halogen oxides in the troposphere: Comparison of model calculations with recent field data. *Journal of Atmospheric Chemistry*, **34**(1), 65–85.
200. Vogt, R., Sander, R., Von Glasow, R. and Crutzen, P.J. (1999) Iodine chemistry and its role in halogen activation and ozone loss in the marine boundary layer: A model study. *Journal of Atmospheric Chemistry*, **32**(3), 375–395.
201. Spicer, C.W., Chapman, E.G., Finlayson-Pitts, B.J., Plastridge, R.A., Hubbe, J.M., Fast, J.D. and Berkowitz, C.M. (1998) Unexpectedly high concentrations of molecular chlorine in coastal air. *Nature*, **394**(6691), 353–356.
202. Oum, K.W., Lakin, M.J., DeHaan, D.O., Brauers, T. and Finlayson-Pitts, B.J. (1998) Formation of molecular chlorine from the photolysis of ozone and aqueous sea-salt particles. *Science*, **279**(5347), 74–77.
203. Vogt, R., Crutzen, P.J. and Sander, R. (1996) A mechanism for halogen release from sea-salt aerosol in the remote marine boundary layer. *Nature*, **383**(6598), 327–330.
204. Warwick, N.J., Pyle, J.A. and Shallcross, D.E. (2006) Global modelling of the atmospheric methyl bromide budget. *Journal of Atmospheric Chemistry*, **54**(2), 133–159.
205. Kerkweg, A., Jöckel, P., Pozzer, A., Tost, H., Sander, R., Schulz, M., Stier, P., Vignati, E., Wilson, J. and Lelieveld, J. (2008) Consistent simulation of bromine chemistry from the marine boundary layer to the stratosphere – Part 1: Model description, sea salt aerosols and pH. *Atmospheric Chemistry and Physics*, **8**, 5899–5917.
206. Butler, J.H., King, D.B., Lobert, J.M., Montzka, S.A., Yvon-Lewis, S.A., Hall, B.D., Warwick, N.J., Mondeel, D.J., Aydin, M. and Elkins, J.W. (2007) Oceanic distributions and emissions of short-lived halocarbons. *Global Biogeochemical Cycles*, **21**(1), Article # GB1023.
207. Laurier, F. and Mason, R. (2007) Mercury concentration and speciation in the coastal and open ocean boundary layer. *Journal of Geophysical Research-Atmospheres*, **112**(D6), Article # D06302.
208. Sigler, J.M., Mao, H. and Talbot, R. (2009) Gaseous elemental and reactive mercury in Southern New Hampshire. *Atmospheric Chemistry and Physics*, **9**(6), 1929–1942.
209. Laurier, F.J.G., Mason, R.P., Whalin, L. and Kato, S. (2003) Reactive gaseous mercury formation in the North Pacific Ocean's marine boundary layer: A potential role of halogen chemistry. *Journal of Geophysical Research-Atmospheres*, **108** (D17), Article # 4529.
210. Sheu, G.R. and Mason, R.P. (2004) An examination of the oxidation of elemental mercury in the presence of halide surfaces. *Journal of Atmospheric Chemistry*, **48**(2), 107–130.
211. Hedgecock, I.M. and Pirrone, N. (2004) Chasing quicksilver: Modeling the atmospheric lifetime of Hg-(g)(0) in the marine boundary layer at various latitudes. *Environmental Science & Technology*, **38**(1), 69–76.

212. Hedgecock, I.M., Pirrone, N., Sprovieri, F. and Pesenti, E. (2003) Reactive gaseous mercury in the marine boundary layer: Modelling and experimental evidence of its formation in the Mediterranean region. *Atmospheric Environment*, **37**, S41–S49.
213. Smith-Downey, N.V., Sunderland, E.M. and Jacob, D.J. (2010) Anthropogenic impacts on global storage and emissions of mercury from terrestrial soils: Insights from a new global model. *Journal of Geophysical Research-Biogeosciences*, **115**, Article # G03008.
214. Millhollen, A.G., Gustin, M.S. and Obrist, D. (2006) Foliar mercury accumulation and exchange for three tree species. *Environmental Science & Technology*, **40**(19), 6001–6006.
215. Xin, M. and Gustin, M.S. (2007) Gaseous elemental mercury exchange with low mercury containing soils: Investigation of controlling factors. *Applied Geochemistry*, **22**(7), 1451–1466.
216. Schussler, U., Balzer, W. and Deeken, A. (2005) Dissolved Al distribution, particulate Al fluxes and coupling to atmospheric Al and dust deposition in the Arabian Sea. *Deep-Sea Research Part II-Topical Studies in Oceanography*, **52**(14–15), 1862–1878.

Problems

5.1. For a particular element, it could be proposed that the relative removal via wet versus dry deposition is a function of the amount of rainfall. Using the formula for fluxes due to wet and dry deposition, and the washout ratio, and assuming that there is no gas phase for the element in the atmosphere, only aerosols, show how the ratio of wet/dry deposition changes with rainfall amount if the dry deposition velocity remains constant. Discuss how your answer may differ for a crustal metal (mostly associated with large particles) compared to an element released primarily from high temperature combustion. For example, use the distributions in Fig. 5.3 and compare Fe and Se.

5.2. The evasion of elemental Hg (Hg^0) to the atmosphere is dominated by the ocean. How much larger would the concentration of Hg^0 need to be in surface freshwaters for the flux from these systems to be 10% of the total flux? Use a value of $3.6 \times 10^{14}\,m^2$ for the ocean surface, $2 \times 10^{12}\,m^2$ for the total freshwater surface area. Assume a wind speed of $3\,m\,s^{-1}$ at the freshwater surface and $10\,m\,s^{-1}$ for the ocean, and use the average value from the models in Fig. 5.18 to determine the relevant gas exchange coefficient. Assume the air concentration is negligible compared to that of the surface waters.

5.3. Use Equation 5.14 in terms of molar mass to compute the D_W for $(CH_3)_2S$, $(CH_3)_2S_2$, $(CH_3)_2Se$, $(CH_3)_2Se_2$ and $(CH_3)_2SSe$. Estimate the flux to the atmosphere for a "typical ocean surface" (wind speed = $10\,m\,s^{-1}$) based on the following surface water concentrations, and assuming that the atmospheric concentration is negligible:

Compound	Conc.	Compound	Conc.
$(CH_3)_2S$	7.8 nM	$(CH_3)_2Se$	1.1 pM
$(CH_3)_2SSe$	1.2 pM	$CH_3)_2Se_2$	0.1 pM

How does the total flux of Se compounds to the atmosphere compare to open ocean deposition (Table 5.11)?

5.4. Measurements have found that the solubility of Fe is different for crustal aerosols versus anthropogenic aerosols. Calculate the dry deposition flux of soluble Fe into the surface ocean during the winter and summer using the following information: (a) the aerosol Fe concentration is $10\,nmol\,m^{-3}$ in summer and $2\,nmol\,m^{-3}$ in winter; (b) the summer aerosol is dominated by crustal matter (80% of the total) while the winter is dominantly anthropogenic (90%); (c) the dry deposition velocity is $0.05\,cm\,s^{-1}$ in summer and $0.01\,cm\,s^{-1}$ in winter because of differences in the overall particle sizes; and (d) the solubility is 1% for crustal aerosol and 20% for anthropogenic aerosol. How important are anthropogenic aerosols as a source of Fe to the ocean under these conditions?

5.5. Cloudwater samples were collected and found to contain Fe^{III}, Fe^{II}, Cu^{II}, and Cu^{I}, as well as other metal redox species. The total concentration of Fe was 1 μM, while the total concentration of Cu was 100 nM.

a. Calculate the pε assuming that the Fe^{III}/Fe^{II} couple controls the redox status of the cloudwater for the condition where Fe^{II} is 30% of the total dissolved Fe. Other constituents are: pH = 4.5, acetate = 30 μM, oxalate = 20 μM, Cl = 200 μM, sulfate = 200 μM. The carbonate concentration is that of equilibrium with the atmosphere. Assume that none of the Fe^{II} is complexed to dissolved ligands and that the only solid phase Fe^{III} species that could form is ferrihydrite ($\log K_{sp}$ = −3.19). The log values for the acid dissociation constants for oxalate (H_2Ox) are: K_{a1} = −4.19, K_{a2} = −5.42. The log formation constants for the Fe^{III}-oxalate complexes are, respectively: 9.4 for $FeOx^+$, 16.2 for $Fe(Ox)_2^-$ and 20.4 for $Fe(Ox)_3^{2-}$. The total ionic strength is 10^{-3} M.

b. Estimate the concentration and the speciation of Cu^{II} in the cloudwater under the conditions of the above problem. Assume for the calculation that Cu^{+} is not complexed by any ligands under these conditions. The log formation constant for Cu(Ox) is 5.1. How much of the Cu is in the reduced state?

5.6. Estimate the speciation of the metals in rainwater given the following conditions (Cl = 100 μM, sulfate = 100 μM, carbonate species at equilibrium with the atmosphere, pH 5.5). Are any of the metals complexed to the organic acids under these conditions?

Table P.5.1

Metal	C_T Metal (nM)	Acetic	Oxalate	Succinate	Malonate
		Log K_1, K_2	Log K_1, K_2	Log K_1, K_2	Log K_1, K_2
Cd^{II}	0.2	1.9, 3.2	4.0, 5.77	2.62	3.45
Zn^{II}	100	1.6, 1.8	4.85, 7.55	2.67	3.82, 5.95
Hg^{II}	0.05	6.1, 10.1	9.66, 17.4	–	–
Pb^{II}	15	2.7, 4.1	5.49, 7.46	4.04, 6.11	3.08
Ni^{II}	20	1.4	5.16, 6.50	2.34	4.10
Co^{II}	4	1.5	4.79, 7.16	2.6	3.76
Mn^{II}	100	1.4	3.97, 5.25	2.26	3.29
	C_T ligand (μM)	15	10	5.0	1.0
	Log K_{a1}	–4.76	–4.19	–4.21	–5.70
	Log K_{a2}		–5.42	–5.64	–8.55

CHAPTER 6
Trace metal(loid)s in marine waters

6.1 Introduction

The transport of metal(loid)s from the terrestrial environment to the open ocean occurs through atmospheric deposition of metal(loid)s, discussed in Chapter 5, through transport in riverine discharge (Chapter 7), as well as from inputs of metal(loid)s from hydrothermal sources (Section 6.3.7). Groundwater inputs are also likely important in some instances (Section 7.3.2). These sources and the overall material cycle were discussed in Chapter 2. For some metal(loid)s (e.g., Hg, Pb, Cd, and Se), the atmosphere is the dominant pathway for their addition to the ocean, while for the others, terrestrial inputs dominate [1]. The relative importance of each pathway is summarized in Table 2.8, and discussed in Section 2.3, and reiterated here in Table 6.1 [2, 3]. For most metal(loid)s, there is substantial removal in the coastal zone in combination with particulate material removal that occurs during the mixing of river water and seawater in the estuary, and due to changes in the physical regime [4, 5]. The estuarine zone is therefore an important location for the removal and trapping of metal(loid)s in the sediment. Coastal and deep ocean sedimentation and burial is the long-term sink for most metal(loid)s. In pre-industrial times, the extent of this deep ocean removal would have balanced the net surface erosion and volcanic inputs.

The extent that the trace metals and metalloids cycle through the various pathways in the biosphere to the ocean depends on their chemistry, their abundance and their usefulness to humans. Metals are purposefully extracted from the Earth's interior and can be transported to the ocean after release in combination with human activity via rivers and the atmosphere. Once entering the ocean, it is the strength of metal complexation and the tendency to form complexes with dissolved natural organic matter (NOM), the solubility of the hydroxide, carbonate and other phases, the propensity of the metal(loid)s to adsorb onto inorganic and organic solids, or be taken up by organisms, that determine the overall concentration and distribution of the metal in coastal and offshore ocean waters. These interactions and dynamics are reiterated in Fig. 6.1 which illustrates the important processes impacting metal concentration and form in coastal waters.

The export of material from rivers to the coastal zone is not evenly distributed globally because of the dominance of export from large rivers and because of spatial differences in terrain (Fig. 2.10). The export of particulate material from rivers to the ocean was discussed in Chapter 2. Major inputs are the Amazon River region (~14% of the total input), southern Asia (34%) and eastern Asia (36%) [4]. These regions all have large rivers and periods of high rainfall that drive the high particulate discharge [5]. Additionally, in these regions, inputs have been exacerbated in recent times due to enhanced erosion of the landscape as a result of biomass burning, forest removal and intensive agricultural development. In other regions, such activities are also leading to changes in the extent of particulate input to the coastal zone. This increased riverine particulate transport however is countered in many locations by the presence of reservoirs that act as particle traps and prevent the input of material to the estuarine and coastal waters. Such trapping is thought to have resulted in a decrease in silica inputs in some locations, with resultant impacts on the abundance of diatoms, which are important phytoplankton in coastal waters [6, 7]. There has also been some suggestion of coastal Fe limitation which could also be partially a result of decreased terrestrial material flux to the coastal zone.

While it is most likely that many particle-reactive elements will be removed from the water column during

Trace Metals in Aquatic Systems, First Edition. Robert P. Mason.

Table 6.1 Estimated relative input of metals and metalloids to the ocean from the atmosphere, compared to other sources. Also listed are the range and average concentrations for open ocean waters, the estimated residence time of each metal in the global ocean, and the major dissolved inorganic species in solution. Data taken from the literature [1–3].

Element	%Pluvial Input to the Ocean*	Ocean Conc Range (nM)	Average Conc (nM)	Residence time $\times 10^3$ yrs	Major dissolved inorganic species in seawater
Al	28	0.3–40	20	0.6	$Al(OH)_x^{3-x}$, x = 3, 4
V	16	30–40	30	45	HVO_4^{2-}
Cr	32	3–5	4	8.2	CrO_4^{2-}
Mn	16	0.08–5	0.3	1.3	Mn^{2+}
Fe	25	0.01–2	0.5	0.05	$Fe(OH)_x^{3-x}$, x = 2, 3
Co	24	<0.01–0.3	0.02	0.34	Co^{2+}
Ni	34	2–12	8	8.2	Ni^{2+}
Cu	14	0.5–4.5	4	0.97	$CuCO_3^0$
Zn	13	0.05–9	5	0.51	Zn^{2+}
As	2	20–25	23	39	$HAsO_4^{2-}$
Mo	6	–	105	820	MoO_4^{2-}
Ag	8	<0.01–0.04	0.02	0.35	$AgCl_x^{1-x}$, x = 1–4
Cd	45	<0.01–1	0.5	–	$CdCl_x^{2-x}$, x = 1–4
Hg	81	<0.01	0.002	0.56	$HgCl_x^{2-x}$, x = 1–4
Pb	62	<0.01–0.15	0.1	0.81	$PbCO_3^0$
Ga	–	0.01–0.03	0.3	9	$Ga(OH)_4$
Se	–	0.5–2.3	1.7	26	SeO_4^{2-}
Sb	–	–	1.2	5.7	SbO_6^-
W	–	–	0.06	~500	WO_4^{2-}
U	–	–	14	500	UO_4^{2-}

**Estimates of percentage are the fraction of the total flux.*

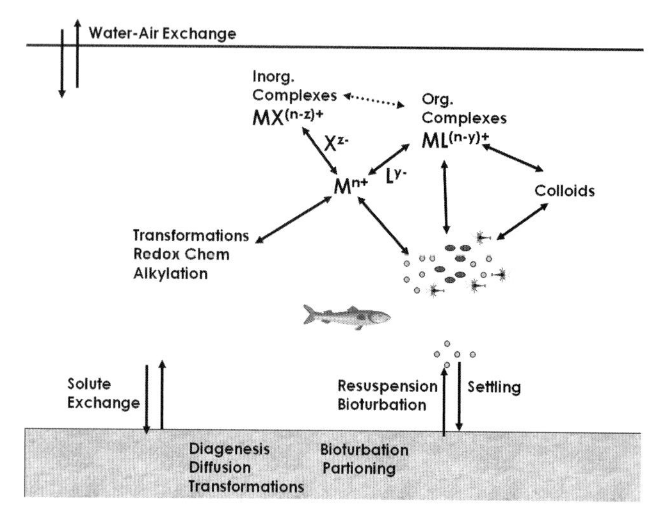

Fig. 6.1 Conceptual diagram representing the main factors controlling metal(loid) fate and transport in coastal waters focusing on the interactions that occur between different phases.

estuarine mixing, it is possible for them to pass through without their concentration changing dramatically (so-called *conservative* elements) or they may even be added to the water column as a result of estuarine processing. These different scenarios are illustrated in Fig. 2.11 and discussed in Chapter 2. The details of metal(loid) input to the coastal zone from rivers will not be discussed further here and the focus of the chapter will be an examination of the major chemical reactions and pathways that influence their concentration and speciation in both coastal and offshore waters. While some mention will be made of organometallics, these are discussed in detail in Chapter 8. Additionally, the mechanisms of accumulation of metal(loid)s by microbes are also discussed in that chapter and will not be discussed in detail here.

6.2 Metal(loid) partitioning in coastal and open ocean waters

Overall, the factors that have the most impact on the fate of metal(loid)s in the estuarine zone are their propensity to attach to and/or be taken up by particulate material (living and abiotic) (Fig. 6.1). In many large estuaries and rivers, the composition of the particulate material changes in the estuarine mixing zone, where the inter-mixing of saline and freshwaters results in rapid changes in water chemistry, flocculation of material due to changes in ionic strength, and where physical changes in water flow and circulation lead to particle settling. These dynamic interactions often leads to a region of high particulate load and settling, termed the *turbidity maximum* zone, and this is where removal of inorganic (terrestrial) material is highest. Due to the lack of light penetration, this is often not the most productive region of the estuary even though rivers are important nutrient sources, and the highest productivity therefore occurs somewhat downstream (in the *chlorophyll maximum* zone). The exact details vary depending on estuarine type, circulation, freshwater flow and a number of other factors, but overall, these two regions account for most of the particulate settling and metal(loid) removal in the estuarine mixing zone.

6.2.1 The mechanisms of partitioning

As noted in Chapter 2, the degree of removal of a metal(loid) from the dissolved phase through particle scavenging varies depending on the metal and its affinity for the solid phase, which can be represented as a partition coefficient, K_D. Recall that while this coefficient should be independent of particulate load there is often a strong inverse relationship with TSS [8–10]. The accepted explanation for this trend is that the experimentally separated dissolved fraction (0.4 μm filtered typically) contains both truly dissolved and "colloidal" material and that the relative amount of colloidal material is related to the TSS, and that there is a dynamic steady state between the truly dissolved, colloidal and particulate phases and exchange between the dissolved and particulate involves the colloidal fraction; so-called *colloidal pumping*, illustrated in Figs. 2.3(b) and 6.1. Overall, many particles or colloids are not discreet entities but aggregates and can be formed by the combination and coagulation of smaller particles and colloids, especially in coastal waters, in addition to the microbial (biotic) particles. In the open ocean, it is generally accepted that in addition to colloids, there are two distinct groups of particles, the suspended smaller particles, which consist of phytoplankton and other microbes, small detrital material, and "marine snow" (suspended large flocs) and larger sinking particles, which comprise zooplankton fecal pellets, skeletal material and other biogenic debris (Fig. 6.2) [11–13].

As discussed earlier in the book, in many instances the absorption of metal(loid)s to particulate and colloidal material is often considered to be through surface complexation with active acidic sites (Section 3.4) (reversible and essentially at instantaneous equilibrium) but this is not always the case as discussed in Section 2.2.2 [14]. A number of studies of both dissolved-particulate interactions and interactions in sediments between the solid phase and porewater also suggest that while adsorption may be rapid, desorption is often a much slower process [9, 15]. If metal(loid)s become incorporated into the inner matrix of the particulate, then desorption could be slow and complex. Additionally, metal(loid)s likely first bind to the more abundant but relatively weaker sites and this is rapid compared to desorption where the metal(loid)s are being released from the less abundant, but stronger binding sites where they have migrated over time since adsorption.

Therefore, as concluded in Section 2.2.2, the partitioning between the dissolved and particulate phase is not a simple reversible equilibrium situation. Such an idea makes sense in terms of complexation as it is known that there are many more weakly binding sites on the surfaces of aquatic solids than strongly binding sites, as discussed in Chapters 3 and 4, and therefore, as the ion binds initially to the first site encountered, this is more likely to be a weaker site. However, over time, the metal will migrate from weaker sites to stronger sites. Additionally, there could be penetration of the metal(loid) into the particle, or re-association of particles into flocs and other conglomerates that would result in the adsorbed element being less available for desorption. In envisioning adsorption and desorption, the particle is often thought of, or parameterized as, a sphere of uniform shape and density when in reality, especially in coastal waters, the particles are a conglomerate of colloids and particles derived from both inorganic and biological sources. Electron microscope studies confirm the heterogeneous nature of marine particles but scientists studying processes do not often acknowledge this. Thus, given an understanding of the likely fate of an ion after adsorption,

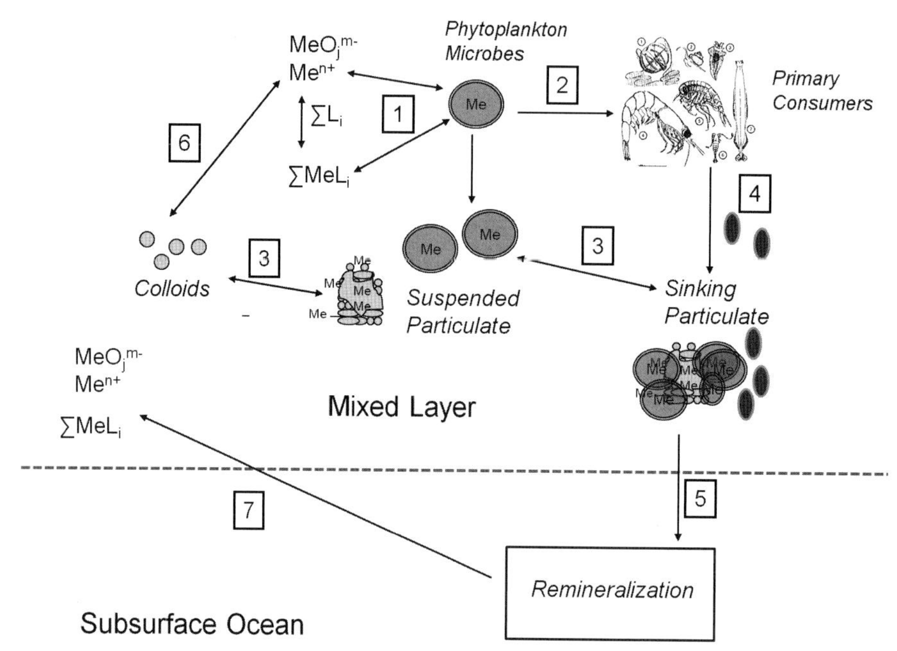

Fig. 6.2 Conceptual diagram representing the interactions that can occur between microorganisms, particles, colloids and dissolved species and how this impacts metal(loid) cycling in open ocean waters. Me refers to metalloids. The major processes are indicated by numbers as: (1) uptake of free metal ions and complexes by phytoplankton and microorganisms; (2) grazing on microbes by primary consumers; (3) coagulation and disintergration of particulate; (4) metal removal via sinking fecal pellets and other large debris; (5) removal of metals from the mixed layer by sinking; (6) uptake of metals into colloids from the dissolved phase and their release; (7) return of metals to the mixed layer through physical transport processes.

it is not difficult to understand that desorption is slower, and that desorption of recently adsorbed metal(loid) is much faster than that of metal that has been associated with the solid phase for an extended period.

If one considers that adsorption is due to chemical binding to sites within the solid material, then it is reasonable to conclude that the amount of adsorption will also depend on the amount of the metal(loid) already present in the solid phase. Additionally, it is possible that exchange reactions may occur and one metal(loid) could out-compete another for a particular site and therefore the results of adsorption studies with multiple metal(loid)s in solution could be different depending on the actual conditions used. Therefore it is difficult to extrapolate from studies done at high metal(loid) solution concentrations to the natural environment and studies should endeavor to examine changes in partitioning at levels as close to natural levels as possible.

A more detailed hypothesis for the differences in the rate of desorption and adsorption, and for the changing value of K_D with time after addition of a metal(loid) to solution, is based on the colloidal pumping mechanism described earlier, and in Chapter 2. As the colloidal fraction is measured with the filtered fraction, and given that colloidal material can coagulate and form particles, and that particles themselves can disintegrate to form colloids, such processes can explain changing K_D values over time, especially in environments where there is substantial organic matter in the filtered fraction, or colloidal inorganic oxides and other material. The colloidal pumping model (Fig. 2.3) postulates that there are rapid adsorption and desorption reactions occurring between the truly dissolved species and the colloidal material in solution, and that the particulate phase is formed by the relatively slow coagulation of this colloidal material, and that there is also disintegration of particles to produces colloids. These processes are continually occurring but reach a steady state so that particles are being continuously created and destroyed, and adsorption/desorption which involves surface complexation and other interactions is continually occurring. Clearly, this is a purely physical interpretation of the environment in the absence of living microbes as these "particles" take up metal(loid)s directly from the dissolved phase and not from colloidal material. Additionally, it is possible for direct adsorption and desorption to occur from the dissolved phase to the particulate material and therefore the model shown in Figs. 6.1 and 6.2 is likely a simplification of a more complex set of interactions in ocean waters. However, it does provide a clear conceptual framework of the processes involved in the interactions of metal(loid)s with particles.

Wen et al. [16] tested the framework of the colloidal pumping model with samples, collected using ultrafiltration and other separation techniques, from Galveston Bay and surrounding waters by spiking with metal radioisotopes (Fig. 6.3). In one experiment, metal spiked colloidal material was added to an unfiltered seawater solution and the fate of the radioisotopes monitored over time. Filtration was used to

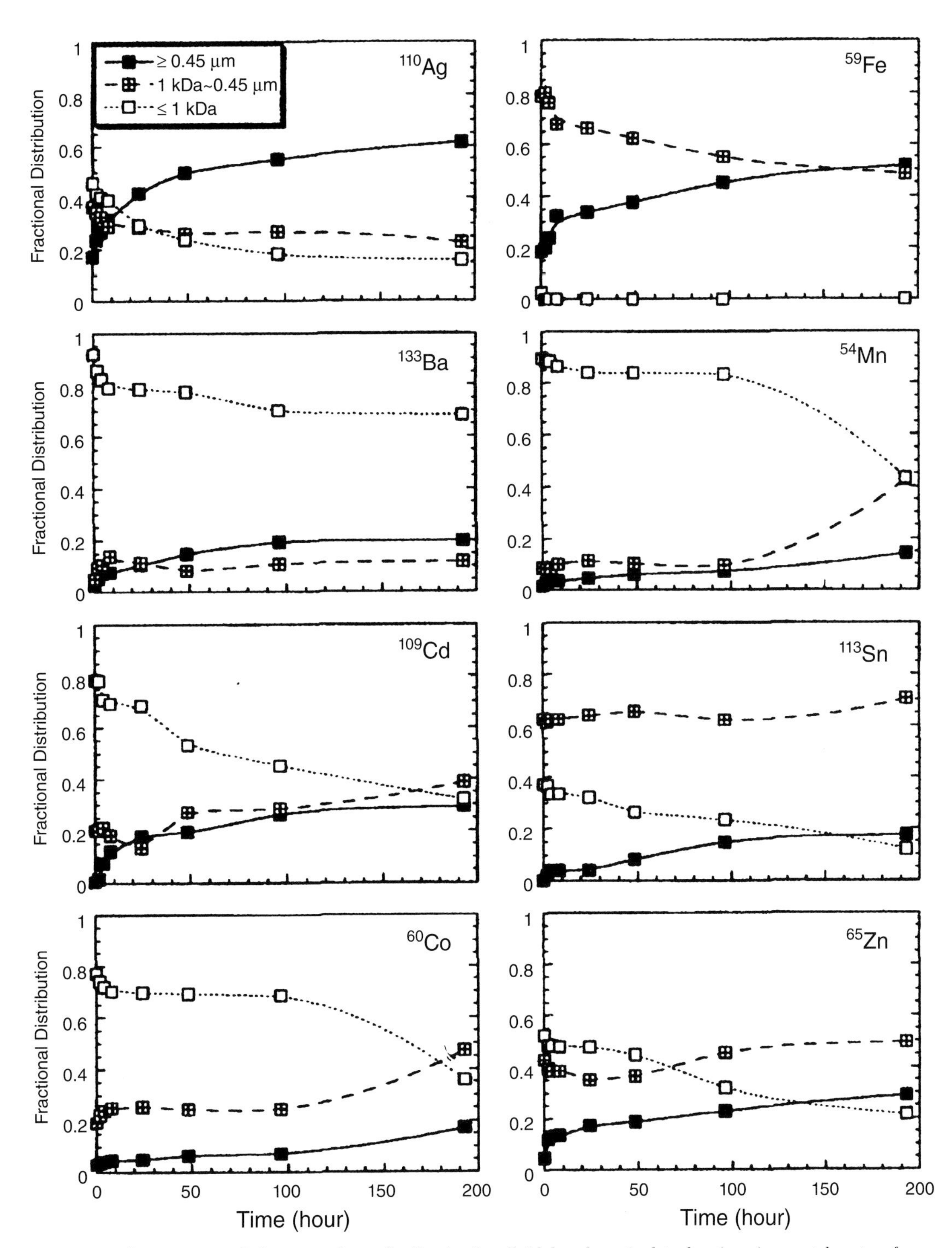

Fig. 6.3 Fractionation of various metals between the truly dissolved, colloidal and particulate fractions in coastal waters for experiments done using metal isotopes. Reprinted from Wen et al. (1997) *Geochimica et Cosmoschimica Acta* **61**: 2867–78 [16], with permission of Elsevier.

separate the metals into the various fractions. In this instance, the net rate of uptake of the metals into particulate is a proxy for the net exchange rate between colloids and particles. The isotope transfer to particulate occurred to a different degree for different metals, being initially rapid in all cases and decreasing over time. After 10 days, essentially all the Fe isotope spike (>99%) and Mn (~80%) was in the particulate fraction while the fractions for the other metals were lower, ranging from ~30% for Zn to 40–60% for Ba, Co, Cd, and Ag, and ~75% for Sn [16]. For some of these metals it was clear that a steady state had not been attained even after 10 days. These results confirm the notion that there is a two way exchange interaction between the colloidal phase and the particulate phase, but that the dynamics are metal dependent, depending on metal speciation in solution, ion size and reactivity, and the relative affinity of different metals for different phases in the colloids and particulate (e.g., oxides, NOM).

Mixing of filtered and spiked riverine and seawater was carried out to simulate the processes that occur within an estuary and to study the resultant coagulation processes. The results of this experiment are shown in Fig. 6.3. There was formation of particulate material upon mixing, as evidenced by the presence of metals in this fraction, and after an initial rapid formation of particulate-bound metals, there was a slow further accumulation over time. The removal of the Fe isotope from solution occurred rapidly and after about eight days, the Fe was evenly distributed between the particulate and the colloidal fractions. For the other metals, their partitioning reflected their propensity to be adsorbed to particulate (<30% of the Ba, Co, Zn, Cd, Sn, and Mn was in particulate, compared to 60% of the Ag) (Fig. 6.3), and their distribution between dissolved and colloidal fractions was similar overall to that found in the experiment described earlier. An examination of the partitioning within different colloidal fractions showed that the metals were predominantly in the smaller colloidal fractions (<10 kDa). There were no NOM measurements so the exact associations of the metals with different colloidal material could not be directly ascertained. From this study, rates of adsorption to colloidal material for the different metals were determined, assuming first order kinetics and two different uptake processes, as well as the rates of formation of particulate. These data were used to estimate values for K_D for each metal. Overall, the estimated values were comparable to values obtained from field measurements, suggesting for this ecosystem that the colloidal pumping model is a reasonable description of the overall system [16]. The differences for the different metals is likely due to differences in their affinities for different phases in both colloidal and particulate material, as discussed further in Sections 6.2.2 and 6.2.3.

The ocean contrasts the estuarine and coastal waters which have much of the particulate material mostly derived from riverine or sediment sources. The fraction of the total that is recently formed biogenic particles is small, compared to the open ocean environment, especially in surface waters, as these are dominated by microbial particles and other biogenic material, such as fecal pellets. Here the colloidal pumping model describing the interactions of metals has less validity as there is direct and facilitated uptake of metal(loids) by living organisms (Chapter 8). Therefore, exchange between dissolved, colloidal and living organisms (small particles) involves a number of different pathways. A modified model is subsequently needed given that the export of particulate material from the surface ocean is a crucial part of most global models and budgets, and because it is important to understand the removal of material from the surface ocean to deep waters. The models that have been developed are focused primarily on explaining the export of particles from the euphotic zone and the passage of particles through the deep ocean.

In open ocean environments, it has been found that there are two types of particles, in addition to colloids, with different attributes. Most of the small particles (microbes and typically small phytoplankton) do not rapidly sink out the mixed layer, while the particles contributing most to particulate, carbon and metal export from the surface waters are large, heavier particulate (e.g., flocculated materials, fecal pellets, and the skeletal and other materials of organisms) that sink rapidly [17] (Fig. 6.2). So, there is a not a direct connection between the uptake of metals into microbial particulate and the export of metals from the ocean mixed layer. A three compartment model has been proposed to account for these interactions consisting of the dissolved phase, the suspended particulate and the sinking fraction (Fig. 6.4) [18], although even more complex models with four or multiple compartments have been suggested [17]. This model approach was first described in the early 1990s [11, 19]. The model figure shows the interactions for Th, which has been used extensively as a tracer for these processes in ocean studies, as it is particle reactive. Thorium is rapidly removed to the particulate phase after being formed by radioactive decay of its parent, U, which is highly soluble and has a conservative distribution in ocean waters. All radioactive isotopes of U have long half-lives. The removal of Th causes a disequilibrium that can be estimated based on the known U concentration (see Fig. 6.5). The model premise is that in surface waters there is direct uptake of Th, and other metals, from solution by the small particulate fraction, due to both active and passive processes, and that there is net formation of large particles from these smaller non-sinking particles. The simplest model assumption is that these processes are essentially irreversible, while in reality it is likely that the rate constants that are determined from the Th distribution reflect the net rate of these processes. The more complex models assume that all adsorption (uptake) processes are reversible; that is, there is desorption and/or release due to remineralization.

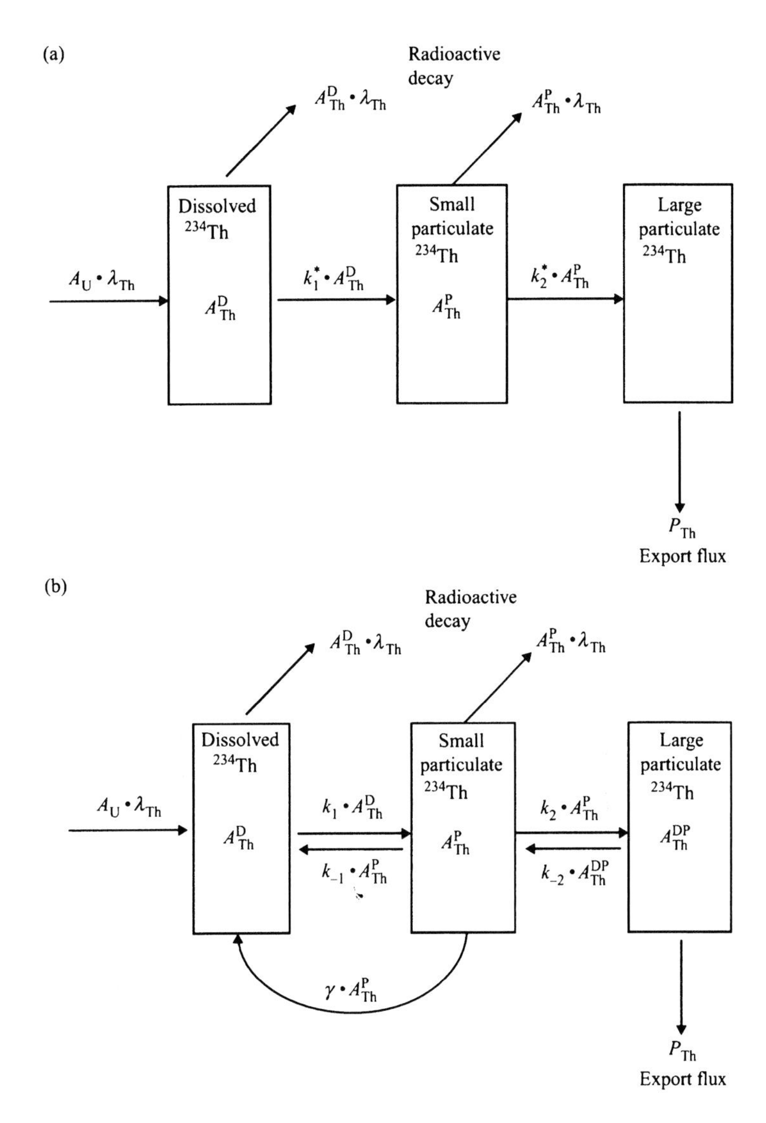

Fig. 6.4 Representation of the model for assessing ocean vertical fluxes based on the partitioning of thorium between the dissolved, small particle and sinking particle phases under two scenarios: (a) irreversible and (b) reversible conditions. Reprinted from Anderson (2004), in The Oceans and Marine Geochemistry, vol. 6, *Treatise of Geochemistry* [17], with kind permission from Springer Science+Business Media.

The most used radioisotopes and their potential uses are gathered in Table 6.2 [17]. Other than Th, radioisotopes that are used in ocean geochemistry are those of Ra, formed from Th, and highly soluble, and ^{222}Rn, a dissolved gas formed from decay of ^{226}Ra, used to estimate gas exchange in the surface ocean. Radon decays in the atmosphere forming ^{210}Pb, which attaches to particles and can be used to measure sedimentation rates, and sediment accumulation in shallow ecosystems. The most widely used isotope for studies of particulate removal is ^{234}Th, which has a half-life of 24.1 days, and is therefore suitable for the examination of processes that occur over the timescale of months (~5 half-lives), such as occur in the upper ocean. As ^{234}Th is in secular equilibrium with its parent, ^{238}U, given its short half-life compared to that of its parent, any disequilibrium in the Th concentration reflects its removal from processes other than its radioactive disintegration. This fact allows for the calculation of its rate of removal from the water column due to uptake and settling of particles (overall scavenging) and can therefore be used as a proxy for the removal of particulate material from the water column. Values for this net scavenging from surface waters have been estimated to range from as low as $0.003\,d^{-1}$ (mean mixed layer lifetime for ^{234}Th of up to 300 days) for oligotrophic waters to values as high as $0.125\,d^{-1}$ (mean lifetime for ^{234}Th of 6 days) [20]. It has been argued that there is a relationship between this constant and

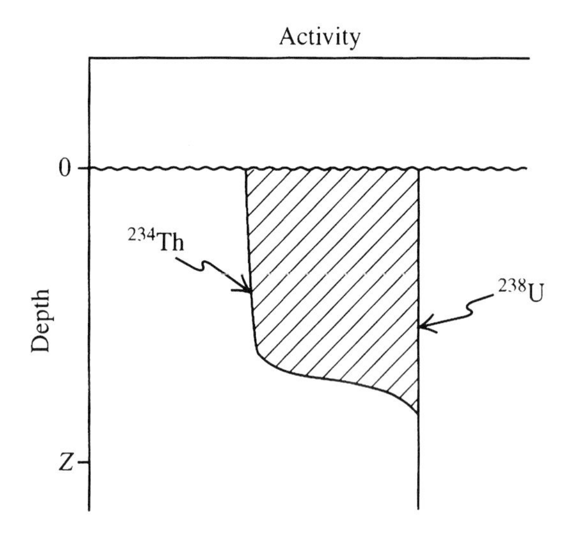

Fig. 6.5 A representation of the thorium deficit that results from the particulate removal of thorium produced *in situ* by uranium decay. Taken from Bruland and Coale (1986) in *Dynamic Processes in the Chemistry of the Upper Ocean*, Plenum Press [18]. Reprinted with permission of Elsevier.

Table 6.2 Radioisotopes and their uses. Unless otherwise noted the isotopes are formed through the decay of the various uranium isotopes and their products. Data taken from various sources [3, 16, 20].

Isotope	Half-life	Comment
^{234}Th	4.1 d	Widely used to estimate vertical particulate transport: export production
^{230}Th	75,400 yr	Used for long timescales: deep ocean sedimentation rates and distribution
^{228}Th	1.91 yr	Used as a particulate tracer: colloidal aggregation
^{231}Pa	32,500 yr	Used for long timescales: deep ocean sedimentation rates and scavenging
^{228}Ra	5.75 yr	Diffuses from sediments, useful tracer of advection, vertical mixing
^{226}Ra	1599 yr	Used to estimate advection of water masses: groundwater inputs
^{210}Pb	22.3 yr	Used to estimate sedimentation rates
^{222}Rn	3.8 d	Used to estimate upper ocean mixing: gas exchange
^{14}C	5730 yr	Cosmogenic and bomb-produced: used over long timescales
^{7}Be	53 d	Cosmogenic: used to estimate bioturbation, mixing and recent sedimentation
^{137}Cs	30.3 yr	Bomb-produced: used as a specific age marker (~1963) for sediments

new primary productivity and it is therefore highest for highly productive waters.

The usefulness of many of these tracers stems from the differences in the particle reactivity of the parent and daughter; either mobile parent/particle reactive daughter (U/Th, U/Pa, Ra/Th) or mobile daughter/particle reactive parent (Th/Ra) [21]. The partition coefficient (K_D) for U is 500 and so it is mostly dissolved in low particulate waters and it has a long residence time (Section 6.4.8). In contrast, Pa, Th, Pb and Po have K_Ds of $>10^6$ and therefore fall within the range of the most highly reactive metals. Radium is intermediate ($K_D \sim 10^4$). Therefore, in examining the fate of other metals in open ocean waters using the information from these tracers, it is important to consider the similarity and differences in their behavior relative to that of Th. For example, Al and Ga have similar chemistry and therefore the behavior of Th is likely a good surrogate for the behavior of these elements which are highly scavenged in surface and deeper waters [20]. In contrast, some of the metalloids and other oxyanions will behave very differently to Th in terms of their rate of uptake and removal. Also, Fe, while having a similar inorganic chemistry to Th, is taken up actively in the euphotic zone and therefore its removal may be poorly simulated by Th in highly productive waters. Additionally, the chemistry and strength with which various metals form inorganic and organic complexes will determine their uptake into the particulate phase relative to Th.

In the deep ocean, there is a dynamic adsorption-desorption cycle that occurs as timescales are longer and steady state is achieved, as illustrated in Fig. 6.4, and thus metals likely are taken up into particulate and released numerous times before they are removed to the sediment surface. The net scavenging that occurs in the deeper waters is a function of the relative binding characteristics of the individual metals. Strongly scavenged metals, such as Al, will be depleted in the deeper ocean waters relative to their mid-depth concentrations while more weakly bound metals, such as Cd and Zn, could have a higher deep water concentration, especially in the Pacific Ocean, as these are the oldest waters given ocean circulation patterns.

Thus, the most important processes governing the uptake of metal(loid)s by particles in ocean waters are different for different ocean regions. In productive open ocean waters, the majority of the particulate is of biogenic origin and the uptake of many elements involves active or passive uptake into microbes, by need or accident. The formation of small suspended particulate is primarily due to *in situ* biological formation, and not due to the coagulation of colloidal material. The opposite is the norm in the estuarine environment, especially those connected to rivers that have a high particulate load. In such systems, much of the suspended particulate is derived from river input and from coagulation of colloidal organic and inorganic material, and the dynamics of the system do not lead to the formation of large particulate material that rapidly sinks, which is the primary removal mechanism for material and metals in open ocean waters. Besides the source of the particulate, its composition will have a large impact on its propensity to accumulate metals from solution. This aspect of the interaction of the dissolved and the particulate phase is discussed in the next sections of this chapter.

6.2.2 Examination of metal speciation in the particulate phase

A better understanding of the dynamics of metal partitioning requires knowledge of the binding and form of the metal in the solid phase. The theoretical framework of metal binding was discussed in Chapters 3 and 4. Chemical extraction techniques have been used to examine the partitioning of metals within the solid phase, and fractions are often defined as exchangeable metal, leachable metal, and the residual fraction with extractions relying on ever harsher extraction solutions (low pH, more oxidizing acids, etc.) for each subsequent step. A number of approaches have been used [22–26]. Data from Galveston Bay are illustrative [26]. Here, two fractions were measured: a methane sulfonic acid (MSA) extractable fraction, which represents weakly bound or labile metal, and a HCl/HNO_3 leach solution. Most of the metals were predominantly in the MSA leach except for Fe which was primarily in the acid leach fraction. Some of the metals studied (Cd, Cu and Ni) were almost entirely in the dissolved phase while Zn was evenly divided between the dissolved and particulate phases, with Fe and Mn being dominantly in the particulate phase.

The binding of metals to particulate and colloidal material is determined by a number of factors such as NOM content, the presence of oxide or sulfide phases, and the biogenic component. A study in Galveston Bay found that the partition coefficient between colloidal and dissolved species (K_C) was larger than K_P (the ratio between particulate and the truly dissolved fraction), by almost an order of magnitude, for most of the metals studied [27]. This seems reasonable given the much larger surface area : volume ratio of colloidal material compared to larger particulate, and assuming that partitioning is controlled by the relative abundance of binding sites on the surface of the colloidal and particulate material. Also, the colloidal fraction is likely to be of higher organic content than particulate and therefore may have a higher affinity for some metals as a result. The data from Galveston Bay indicate that overall, for this system, the colloidal material had a higher organic content (OC) compared to particulate (based on the Fe/OC ratio) (Fig. 6.6), and both were much higher than the crustal ratio [25, 27]. Also, through the estuary, the relative ratio of colloidal to particulate organic matter (COC : POC) changed from high values in the high salinity shelf waters (COC : POC = 5–10 for >25 ppt) to intermediate values in the mesohaline region (3–5 for 10–25 ppt) and low values in the low salinity waters

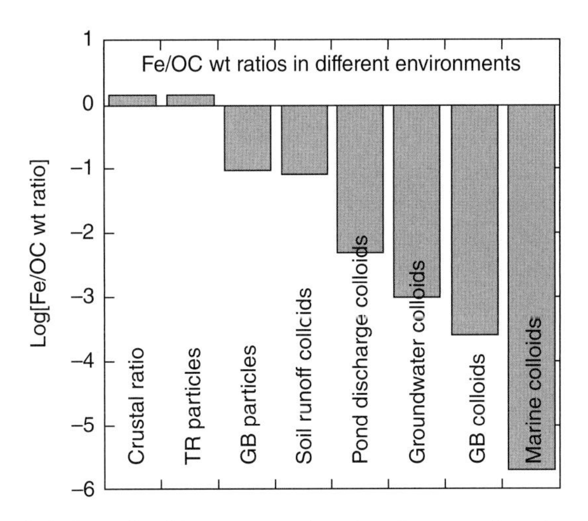

Fig. 6.6 The ratio of iron to organic carbon for a number of fractions collected from Galveston Bay (GB) and the Trinity River (TR) relative to that of other materials (average crust; various colloidal material). Reprinted from Wen et al. (1999) *Marine Chemistry* **63**: 185–212 [27], with permission of Elsevier.

(0.5–1 for <10 ppt). Interestingly, in this system, the particulate at low salinity had a much higher %POC (4–6%) than the offshore higher salinity waters (<2%), which is likely not truly representative of all systems, and certainly not all coastal waters, where phytoplankton biomass is an important fraction. In this system, coastal waters were low in phytoplankton (chlorophyll a $<1\,\mu g l^{-1}$) compared to the inland waters where concentrations were as high as $10\,\mu g l^{-1}$ [27].

Many studies have used ultrafiltration techniques to separate out colloidal fractions of different size ranges. A study of the colloidal partitioning of metals in Narragansett Bay [28, 29] provides one example that reflects many larger estuarine/coastal systems (Fig. 6.7). The colloidal fraction made up the majority of the filtered (0.2 μm filtration) fraction and the concentrations decreased with salinity for the metals studied (Fe, Mn, Ni, Cu, and Zn). The majority of the colloidal Fe, Ni, and Zn were in the large colloidal fraction while most of the colloidal Cu was in the smaller fraction suggesting that different metals interact with different types of colloidal material, as noted in Section 6.2.1. The changes with salinity likely reflect the decreasing organic content of the filtered fraction, as shown in a study of the Mersey estuary [22].

Another approach is to use resins of different hydrophobicity or binding characteristics to separate out the different types of metal-containing colloidal material (Fig. 6.8). Jiann et al. [29] used this approach in their study of the Danshuei River in Taiwan and compared the results for the column separations with the results of ultrafiltration. The Danshuei River flows through Taipei and its watershed is heavily

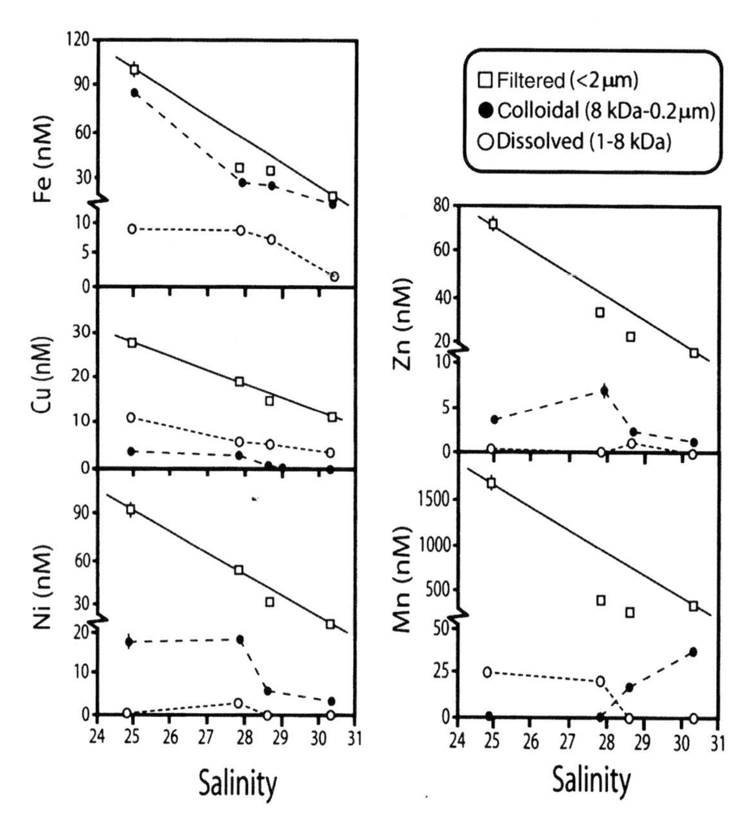

Fig. 6.7 Distribution of metals between the dissolved, and various colloidal size fractions against salinity for Narragansett Bay. The original data was taken from Wells et al. (1998) *Marine Chemistry* **62**: 203–17 [29]. Figure reprinted from Bianchi (2007) *The Biogeochemistry of Estuaries*, Oxford University Press [13]. Reprinted with permission of Elsevier and Oxford University Press.

urbanized [30]. The low salinity region of the estuary can become anoxic and the study contrasted an anoxic and oxygenated period in the estuary. While a number of metals were studied, only Cd and Cu will be discussed here as they provide contrasting behavior for metals. For Cd, which typically has a low K_D relative to other metals, the dissolved fraction (<0.45 μm) dominated and Cd was mostly in the <0.1 μm fraction (Fig. 6.8). This correlates with the fact that most of the Cd was retained on a Chelex column, which is interpreted to be a measure of Cd bound in labile complexes. Very little of the Cd was bound to the anionic column that was used to estimate the organically-bound fraction, or in the inert fraction, that was not retained. This changed for the anoxic reach where Cd, although still primarily in a labile fraction, was associated with larger material (0.1–0.45 μm). There was more evidence however that a fraction of the Cd was associated with the inert fraction at low salinity, and this may represent association of Cd with sulfide-containing colloids.

For Cu, there was a higher relative fraction associated with the particulate phase, mostly around 50%, and the particulate Cu was much higher in the anoxic period (Fig. 6.8). Dissolved Cu was similar over the two sampling times and showed an exponential decrease with salinity [30]. In the absence of anoxia, filtered Cu was concentrated in the <0.1 μm fraction and was associated with the labile fraction and, to a lesser extent, with the organic fraction. During anoxia, most of the Cu was in the inert pool and was of higher molecular weight in the low salinity region. Again, this likely reflects association of Cu with larger colloidal S-containing entities. The formation of, and association of metals with inorganic colloidal material is discussed further below.

As a comparative example, in Galveston Bay, which is also a highly urbanized system, ultrafiltration was used to collect three different fractions: dissolved (≤1 kDa), small colloidal material (1–10 kDa) and larger colloids (10 kDa – 0.45 μm) [27]. A large fraction of the Cd was in the truly dissolved fraction (≤1 kDa), especially at the higher salinities (Fig. 6.9), with most of the remainder equally divided between the large colloidal fraction (>10 kDa – <0.45 μm) and the smaller colloidal fraction. For Cu, there was little found in the larger colloidal fraction with the Cu being evenly partitioned between the truly dissolved and smaller colloidal fractions (Fig. 6.9). The distribution of other metals ranged from Co being mostly dissolved (>60% in the ≤1 kDa fraction) to Fe being essentially colloidal (<40% in the ≤1 kDa fraction) and in the larger colloidal fraction. While Zn was primarily in the colloidal fraction, it was associated with the smaller colloidal material. About 50% of the "filtered" OC was in the ≤1 kDa fraction (i.e., truly DOC) with the remainder being split relatively evenly between the two colloidal fractions.

In summary, the differences in the partitioning of the metals are determined by the propensity of the metals to associate with ligands in organic matter compared to their ability to bind to the oxide surfaces. For the transition metals, both colloidal phases are important, but to differing degrees, as shown by the data. Similarly, most of the metals are associated with the surface absorbed fraction while the

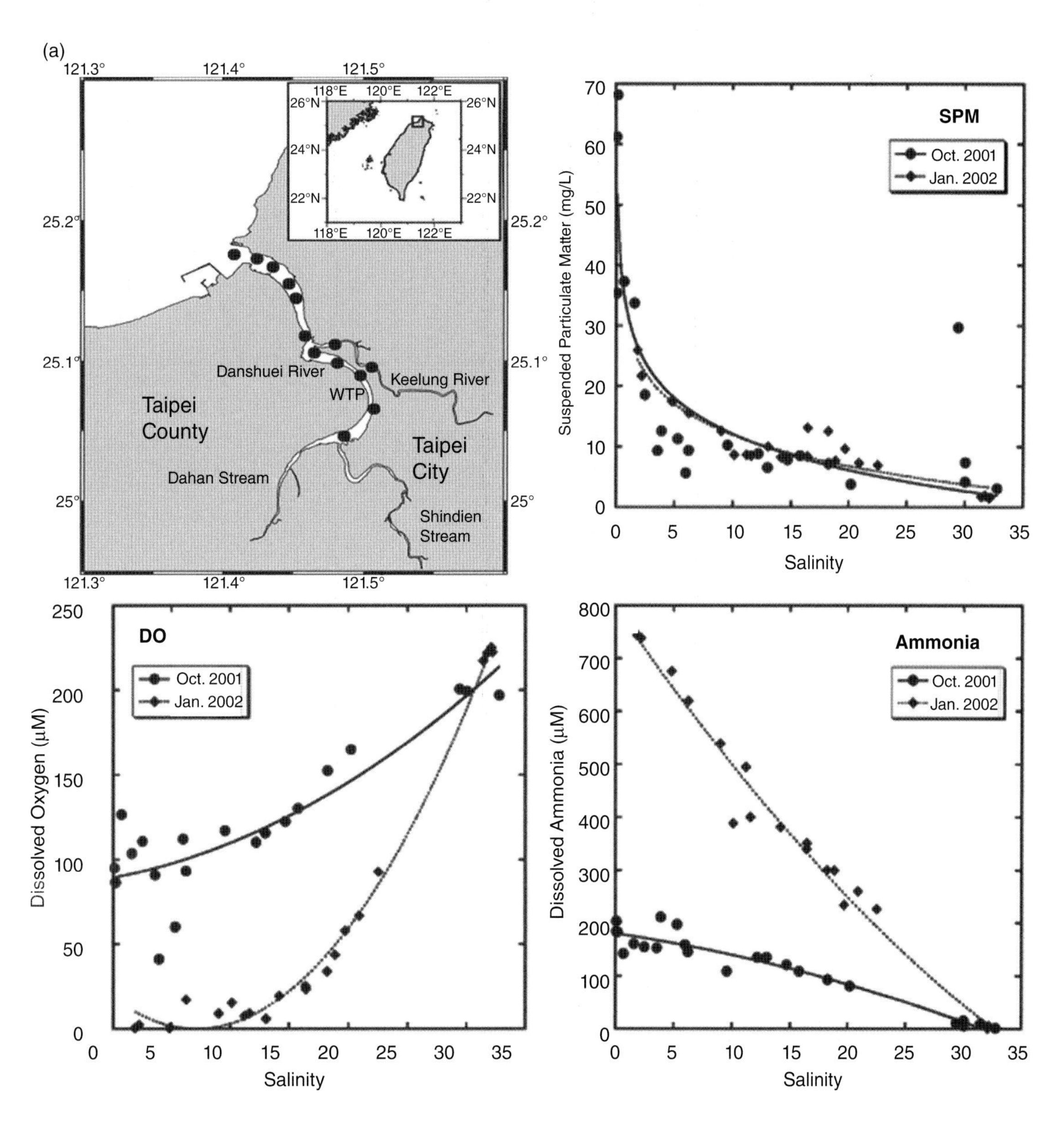

Fig. 6.8 (a) Ancillary data from a study of the Dansheui River, Taiwan; (b) Partitioning data for copper and cadmium from the Dansheui River, Taiwan. Figures extracted from Jiann et al. (2005) *Marine Chemistry* **96**: 293–313 [30], and reprinted with permission of Elsevier.

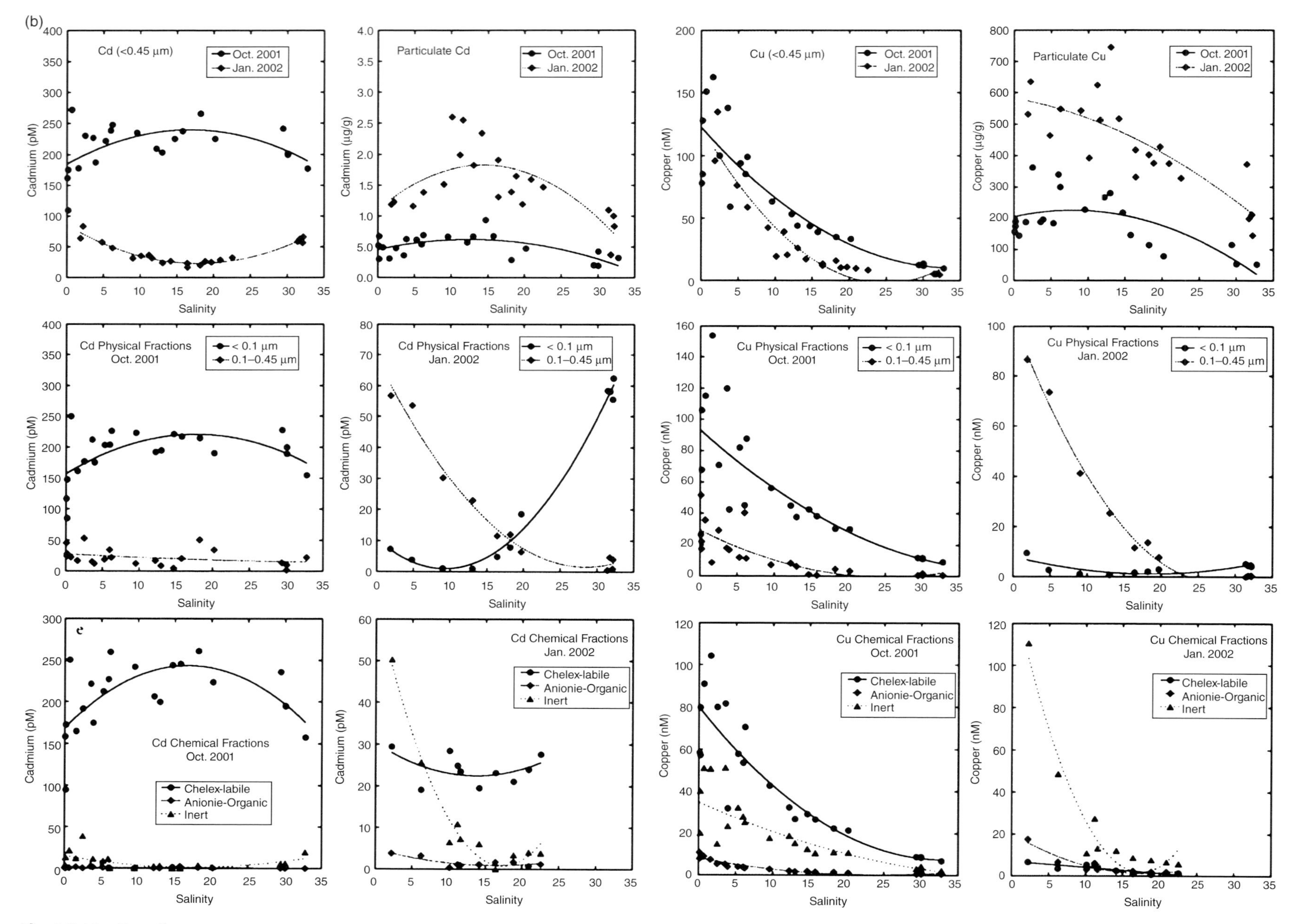

Fig. 6.8 (Continued)

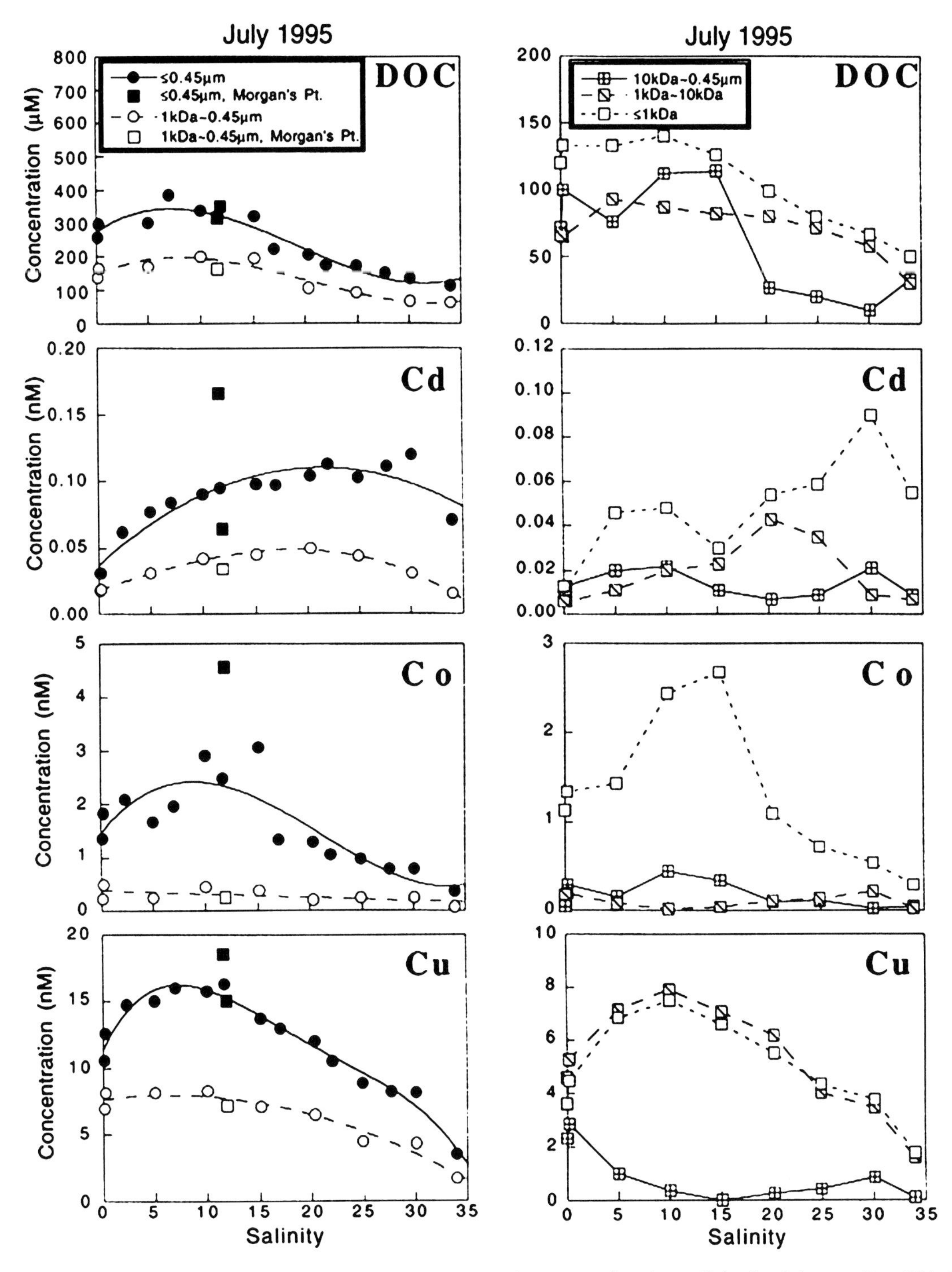

Fig. 6.9 The distribution of organic matter and various metals between fractions and against salinity for Galveston Bay, USA. Reprinted from Wen et al. (1999) *Marine Chemistry* **63**: 185–212 [27], with permission of Elsevier.

Fe is mostly in the acid-leachable fraction. This further suggests that the metals are not directly associated with the Fe fraction. More likely, the metals are associated with organic matter, which may be a covering over inorganic particulate material. While the sequential extraction approaches provide some information about partitioning in the colloidal and particulate phase, this is not the complete picture given the complex nature of the estuarine particles, which are likely conglomerations of smaller particles of both biological and abiotic material. Higher resolution approaches, using techniques such as scanning electron microscopes or microprobes, are required to understand more clearly what controls the degree of partitioning of metals in the estuarine environment.

6.2.3 Examination of the complexation of metals with natural ligands in the filtered fraction

The previous discussions have focused on methods of characterize the partitioning and speciation of metals to colloids and particulate in estuarine and ocean waters in terms of physical separations based on size (e.g., ultrafiltration) or methods relying on chemical reactivity, such as the sequential leaching procedures. The approach for the filtered fraction focuses on determining more specifically the binding of the metals to the ligands in solution. In these studies it cannot be ascertained whether the metal-ligand entity is truly dissolved or colloidal but these applications can give an estimate of the potential lability and bioavailability of the metal under consideration. Most of these studies have focused on examining filtered waters. One approach to examine metal-ligand associations involves electrochemical detection methods. A number of measurement approaches are used and the principal behind the method is the titration of the binding sites of the ligand with the metal of interest, or *vice versa,* while measuring the amount of metal that is present as labile complexes, which is considered to be the metal present as a free ion or associated with inorganic or small labile organic complexes [31–35]. From such titrations, it is possible to determine the average strength of the binding constant for the metal and the nominal concentration of the ligand (actually, the number of binding sites). The equations used in these approaches are derived from the speciation approach detailed in Chapter 3, and are summarized here, considering a +2 cation forming a di-complex with a monoprotic ligand, in the following equations:

$$M^{2+} + 2L^{-} = ML_2 \quad K_L = [ML_2]/[M^{2+}][L']$$

$$M_T = [M^{2+}] + \sum \beta_i[X_i] + [ML_2] \quad L_T = [HL] + [ML_2] + [L^{-}]$$

where $\Sigma\beta_i[X_i]$ represents the inorganic complexation of the metal, and $([M^{2+}] + \Sigma\beta i[Xi])$ is often referred to as the labile fraction, M′; similarly for L′. Depending on the relative concentrations, various simplifying assumptions are possible, but given these equations and the titration of the ligand with added metal, or through the use of competitive ligand extraction approaches, the amount of metal bound to the unknown ligand, L, can be determined.

In marine chemistry studies two approaches have been often used: anodic stripping voltammetry (ASV) and adsorptive cathodic stripping voltammetry (AdCSV). Both approaches have been used for Zn and Cu, while ASV has also been used for Cd and Pb complexation studies and AdCSV for studies of Fe and Co. It must be emphasized that these approaches do not give any information about the structure or nature of the binding ligands and while most investigators assume these to be organic molecules, this does not need to be so. With ASV, the concentration of the labile metal is determined directly and therefore stepwise addition of metal to the sample then allows determination of the changing labile concentration, and this can then be used to estimate the concentration and binding strength of the unknown ligand(s). In reality, there are likely many ligands present of different binding strength and this is often therefore parameterized as a two ligand system, containing "strong" and "weak" ligand classes.

For AdCSV, a ligand of known binding strength, which additionally is electrochemically active, is added to the sample and the concentration of this ligand is then determined. Here instead of a titration with addition of the metal, the titration involves addition of the known ligand, and with the information gained from this titration, the concentration and binding strength of the unknown ligands can be determined.

In a study by Bruland et al. [33], estimations of Cu speciation in Narragansett Bay waters were compared using the two approaches with different ligands (salicylaldoxime, benzoylacetone and 8-hydroxyquinoline) for AdCSV, and with the measurements being made by three different research groups. The study showed that the results were determined to a degree by the complexation capacity of the added competing ligand, as the value determined for the natural Cu-binding ligand concentration decreased with increasing strength of the added ligand, while the estimated conditional stability constant increased. The results were interpreted to illustrate that there is indeed not distinct ligands with a specific complexation capacity but a continuum of Cu-binding ligands, as suggested above, and that the natural ligands with the highest binding strength are present in lower concentration [33]. Overall, the results from the different approaches were similar when similar competitive ligands were used. All of the approaches determined that >99.7% of the dissolved Cu was complexed to the strong Cu-binding ligands. Further examples of these approaches are given later in this chapter, where the approach has been used to examine the complexation of metals in open ocean waters.

In addition to the electrochemical approaches, the method of using a competitive ligand can be used with detection or measurement of the complexed or labile or free ion concentration by other means. Examples are the measurement of the fluorescent signal of the added ligand-metal complex, measurements of the metal concentration after separation of the inorganic and organic complexes (such as using liquid-liquid extraction) or direct quantification of the labile pool, which can be done by chemical reduction for Hg, or using diffusion gradient in thin film (DGT) devices [35–38].

Recently, there has been more study of the potential for metals in natural waters to be associated with reduced sulfur ligands or entities in solution. A number of potential biochemicals could be present in natural waters due to their release during phytoplankton decay or via their excretion from microbes, such as cysteine, glutathione and phytochelatins, which are metal-binding compounds found in plants and phytoplankton (see Chapter 8). Tang et al. [25] measured across the estuarine gradient in Galveston Bay the total amount of reduced sulfur (TRS) in solution, which includes small molecules such as glutathione, labile sulfide complexes, and higher molecular weight compounds containing thiol groups. Potentially all of the TRS could complex to dissolved metals, and there was a strong correlation between the filtered concentrations of Cd, Pb, and Cu and either the glutathione or TRS concentration, reinforcing this notion. Another study in the Elizabeth River [39] correlated the concentration of Cu-binding ligands determined using the competitive ligand-AdCSV approach with the total measured concentration, using HPLC, of 9 individual thiols. Similar relationships were found for the major thiols, such as 2-mercaptoethanol and mercaptosuccinic acid (Fig. 6.10). The total thiol content (20–~350 nM) was greater than the estimated concentration of Cu-binding ligands (20–60 nM), but the determined binding constants for the Cu-binding ligands were similar to the literature values for Cu binding to the measured thiols. Thiol concentrations were lowest in winter, suggesting their biochemical formation in the river.

The role of thiol groups in complexation of Hg in solution has been clearly demonstrated [40, 41] and the idea that other metals are also strongly associated with thiol groups needs more study. The importance of thiols in the complexation of other metals such as Cu and Ag in natural waters has been suggested [42–44]. However, in many models and simulations of metal binding, it is the carboxylic acid groups, which are more abundant in NOM, that are thought to be the major binding sites for metals with NOM and are considered the dominant metal complexing ligands (see Chapters 3 and 4).

Discussing the interactions in the filtered phase in terms of colloidal material and organic metals-binding ligands is reasonable, especially in the coastal region, as it is known that many metals bind strongly with NOM and this accounts for their higher accumulation in colloidal material. However, not all colloidal material is primarily organic and there has been much recent research examining the binding of metals to sulfide-containing solids and smaller entities such as *nanoparticles* and *metal-sulfide clusters*, which would appear in the colloidal fraction, or the filtered fraction, of most environmental separation techniques.

Metal-sulfide clusters are *polynuclear* metal complexes (i.e., more than one metal atom per molecule) that have been isolated and identified from natural waters [45]. These clusters are likely metastable intermediates that form during

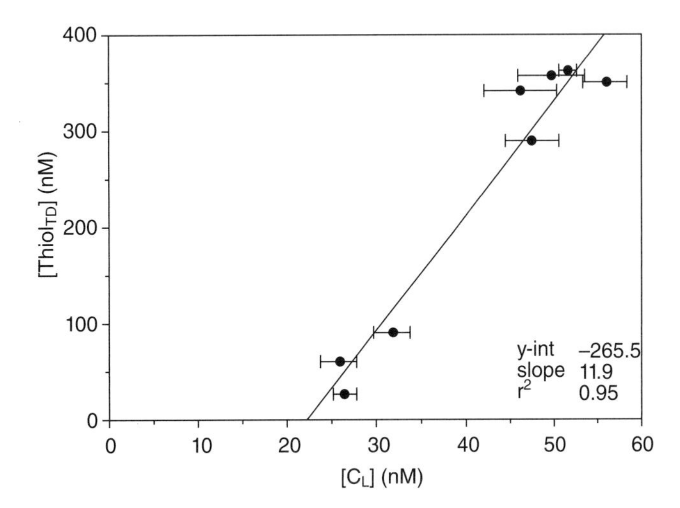

Fig. 6.10 Relationship between the measured total thiol content and the estimated concentration of copper binding ligands for waters of the Elizabeth River, VA, USA. Taken from Dryden et al. (2007) *Marine Chemistry* **103**: 276–88 [39] with permission of Elsevier.

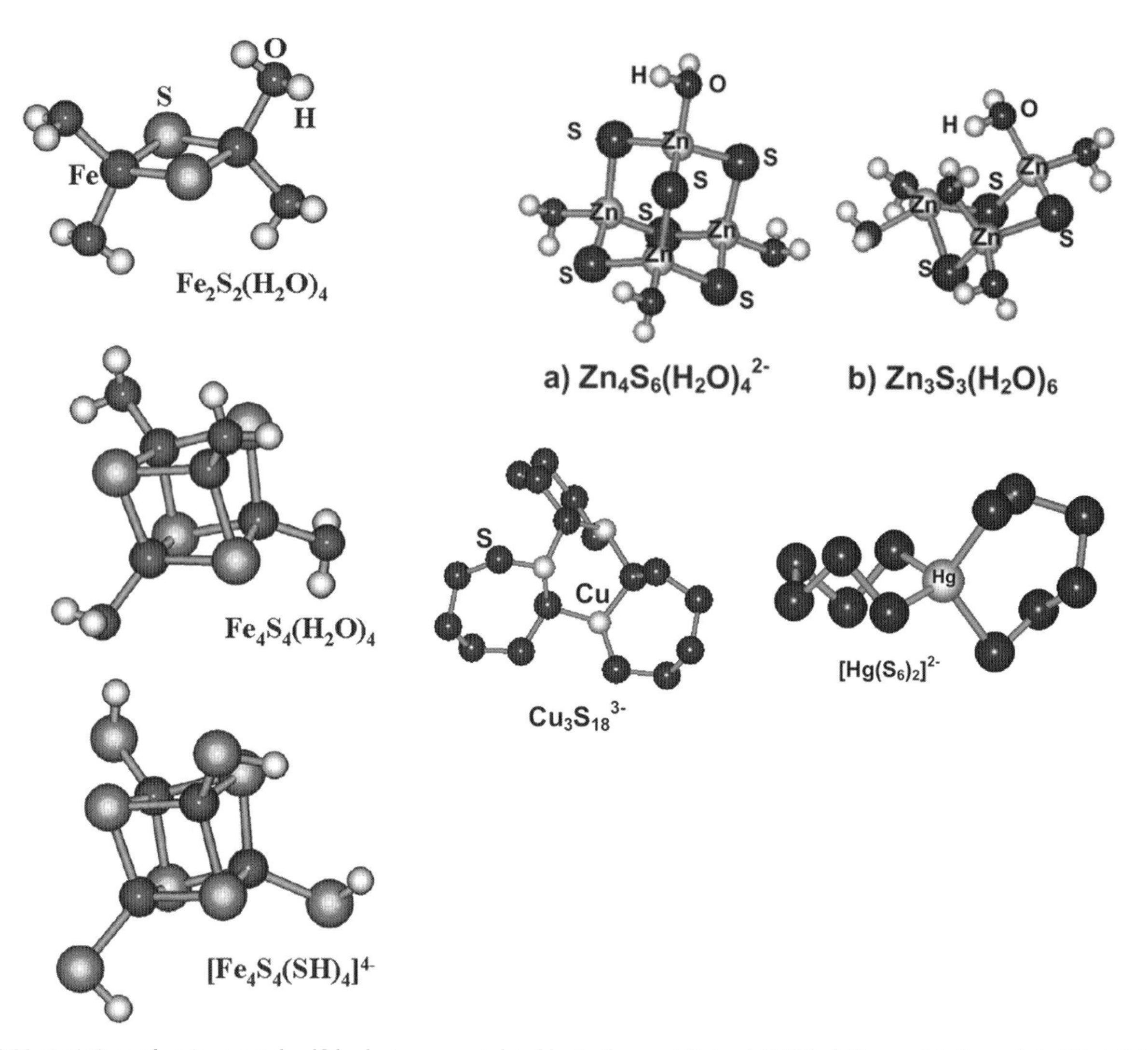

Fig. 6.11 Depictions of various metal-sulfide clusters as postulated by Luther and Richard (2005) *J. Nanoparticle Research* **7**: 389–407 [47] and reprinted with kind permission of Springer Science+Business Media.

the precipitation of the metal-sulfide solid. It is known that such metal-sulfide clusters can be stabilized in the presence of organic matter [46]. Their general formula is M_xS_y, although there may be water molecules also associated within the cluster. Some examples are shown in Fig. 6.11 [47], and clusters have been detected for Fe, Cu, Zn, Ag, Pb, and Hg, and likely are formed with other metals. Most of the known metals that form clusters are those that from strong complexes with S^{2-} (Class B metals), and are also metals which form relatively insoluble sulfides. Both charged and neutral complexes are possible and for the clusters, the molecular weight is typically <1000. Formation constants for some metal-sulfide clusters are given in Table 6.3. It can be seen that the K values for the Ag complexes are greatest, then Cu, followed by Pb and Zn, and this order fits with the overall relative strength of complexation of these metals with ligands.

It has been shown that metal-S clusters are kinetically stable toward oxidation in surface waters, especially in the dark, and it is suspected that they can persist for extended period in the environment. These complexes are potentially less toxic to organisms than the free metal ion although neutral complexes could partition passively across membranes and likely have relative high hydrophobicity, as discussed in Chapter 8. In addition to metal-sulfide clusters which are of relatively low molecular weight (typically <1 kDa) and involve S^{-II} completion, there are also other metal-sulfide entities that exist in the environment, and that can play an important role in metal speciation and chemistry. One such class of compounds involves the interaction of

Table 6.3 Equilibrium (binding) constants for metal-sulfide clusters of a variety of metals. Taken from [47]. Reprinted with kind permission from Springer Science+Business Media.

Metal complex	Ag	Metal complex	Cu	Pb	Zn
M_6S_3	78.3	M_3S_3	54.7	62.9	48.5
M_8S_4	106.2	M_4S_4	–	–	–
–	–	M_4S_6	96.4	–	84.4

metals with polysulfides, which are ligands in the form S_x^{2-} where x = 3–6, and where S exists in mixed oxidation states. Polysulfides can form readily in sulfidic waters either by reaction of sulfide with elemental sulfur or directly as intermediates in the biological oxidation or reduction of sulfur. Although there is some uncertainty in the literature regarding the existence of S_6^{2-} and other polysulfides, as well as the values for the equilibrium constants for particular polysulfide species, and their metal complexes, their impact on metal chemistry in salt marshes and other high sulfidic marine environments has been documented [45, 48–51].

In addition to the presence of dissolved polynuclear metal-S compounds, there is the possibility of the formation of metal-S nanoparticles in the environment. Additionally, there has been an explosion in the industrial production of metal-sulfide and metal-selenium nanomaterials (primarily with Zn, Cd, Hg, and Ag) for a wide variety of industrial applications [52–55] and their presence in the environment as a result of their release is likely.

The accumulating evidence for the formation of metal-sulfide nanoparticles in the environment suggests that such materials form in the initial stages of metal sulfide precipitation, and this makes sense given the understanding of the process of precipitation [56]. These nanoparticles can be stabilized by the presence of organic matter and organic compounds [45, 57]. For example, the hindrance of HgS precipitation in the presence of sulfide suggested a role for organic matter in stabilizing metal-sulfide nanoparticles [58], and recent studies of the formation of ZnS and HgS nanoparticles confirm these ideas and show that in particular it is the thiol groups in simple organic molecules, and by inference on NOM, that are involved in the stabilization of these nanomaterials (Fig. 6.12) [59, 60]. Supersaturated solutions of ZnS or HgS form nanoparticles, in the 50–150 nm range, that persist for extended periods (>1 h) in solutions containing either cysteine, thioglycolate or NOM. However, without added ligands, or in the presence of serine and glycolate, the solid precipitates rapidly form. It is not known whether these nanoparticles contain a metal-sulfide core surrounded by an organic "film" or whether they are aggregates of smaller "clusters" which are coagulated through their interaction with organic matter. Recent evidence suggests the latter [61, 62]. Whatever their form, such nano-

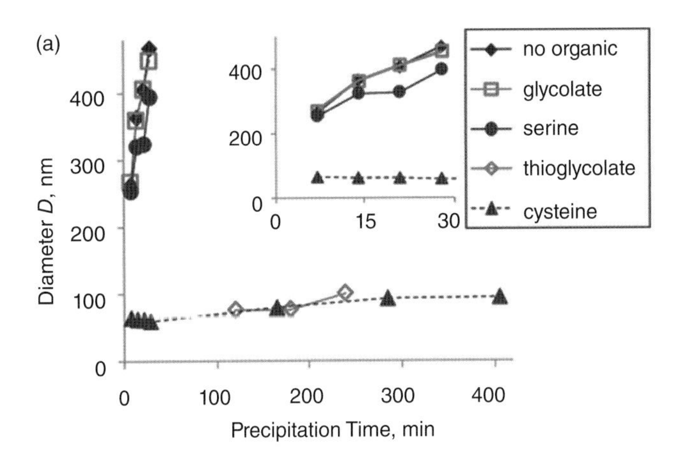

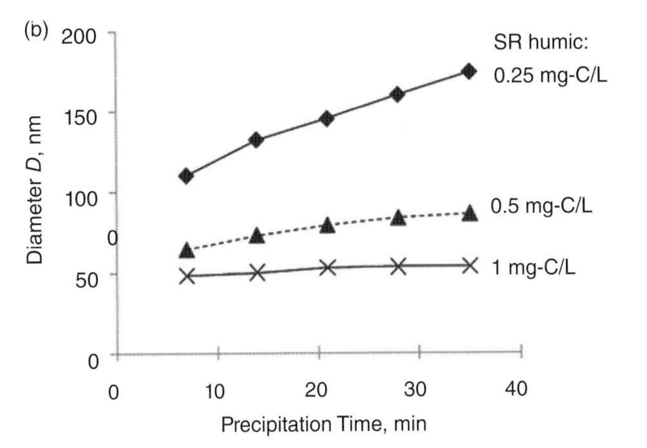

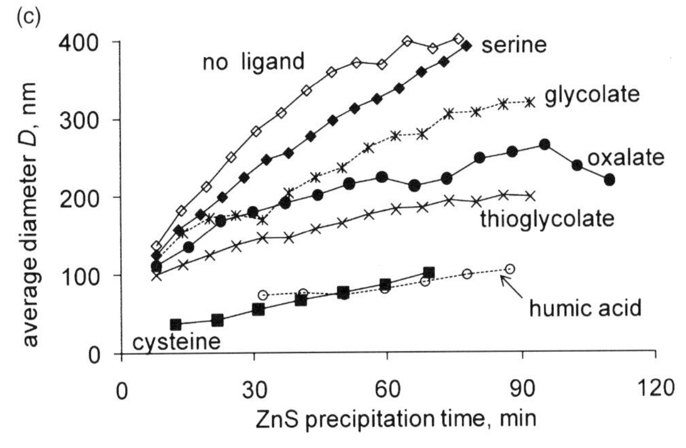

Fig. 6.12 The relationship between the rate of precipitation of (a) and (b) mercury and (c) zinc in the presence of excess sulfide and in the presence and absence of various organic compounds (carboxylic acids and thiols) and natural organic matter. Reprinted with permission from Lau and Hsu-Kim (2008) *Environmental Science and Technology* **42**: 7236–41 [60]; and Deonarine and Hsu-Kim (2009) *Environmental Science and Technology* **43**: 2368–73 [59]. Copyright (2008), (2009) American Chemical Society.

particles would pass through filters nominally used to separate the filtered and particulate fraction and would be considered "dissolved" in many of the assay approaches used and discussed throughout this book.

Recent studies by Slowey [61] have demonstrated that the Hg-S nanoparticles formed in the presence of NOM are actually agglomerations of smaller <5 nM HgS crystals surrounded by organic matter and held together by the NOM. Additionally, there is a dynamic exchange between these materials and the solution with a large fraction of the HgS being "reactive" in the sense that it is continually transferred between the dissolved and nanoparticulate fractions [61]. Additionally, Slowey states that similar interactions occur with solid HgS, and this is consistent with the earlier research of others [58, 63]. Given the dynamic nature of the interactions between "dissolved" constituents and the solid phase, Slowey concludes that Hg bioavailability to microorganisms in the presence of HgS(s) is controlled by kinetic (intermediate) rather than equilibrium partitioning of Hg species, and that there could be a relation between the rate of these kinetic interactions and the rate of Hg methylation in soils and sediments.

The effect of thiols on sulfide nanoparticle stabilization is strongest for HgS, and the resultant nanoparticles are smaller, and this likely reflects the much stronger binding of Hg to reduced sulfide and thiols compared to Zn. The presence of such stabilized colloidal metal-sulfide-organic entities in the environment is likely, and the presence of such compounds can be inferred from the results of some studies that have found the presence of metal-sulfide "complexes" in natural waters.

The idea of the formation of such colloidal metal-sulfide clusters is not new, and the example discussed below provides a useful exercise in the various potential relationships that exist, and how the various interactions that occur in solution can be estimated, and the relationship between measurement and theory. Dyrssen and Wedborg [64] used a number of thermodynamic approaches to estimate the binding constants for Hg complexes and found a large discrepancy in the values depending on the approach taken in the estimation of the constant for dissolved HgS^0, actually HOHgSH, the hydrated form [64, 65]. The constant for this mixed complex could be derived using a previously-demonstrated equation which relates the formation constant of the mixed complex to those of the individual di-ligand complexes:

$$\log\beta_{11} = \log 2 + 0.5(\log^A \beta_2 + \log^B \beta_2) \quad (6.1)$$

where in this case A is OH and B is SH. As the constants for the formation of $Hg(OH)_2$ and $Hg(SH)_2$ are known (respectively, $\log\beta_2 = 22.$ and 37.7), this allows the estimation of the value for the mixed constant for HOHgSH. A value of −22.3 was derived for log β_{OHSH} for the following reaction:

$$HgS_{(S)} + H_2O = HOHgSH \quad \log\beta_{OHSH} = -22.3$$

As this appeared to be a low value relative to data in the literature for Zn and Cd, Dyrssen and Wedborg [64] also estimated the constant using a thermodynamic relationship developed based on estimated and measured constants for the solubility of the other Group 12 metals (Zn and Cd). For Zn, the estimated constant for the formation of HOZnSH was log $\beta_{OHSH} = -12.95$ while the measured value in the literature was −5.87. Similarly, for Cd, the estimated value for log β_{OHSH} was −16.04 while the measured value was −6.85. Based on these discrepancies, Dyrssen and Wedborg [64] concluded that if a similar relative difference between the estimated and measured constant existed for Hg, then the value for HOHgSH should be −10:

$$HgS_{(S)} + H_2O = HOHgSH^0 \quad \log\beta_{OHSH} = -10$$

Such a value is consistent with the measurements of Hg concentration in equilibrium with the solid phase. However, as the exact speciation in solution cannot be determined, there is no conclusive proof that the complex exists. Other potential complexes of Hg in solution in the presence of sulfide include $Hg(SH)_2$ and its dissociated forms:

$$Hg^{2+} + 2HS^- = Hg(SH)_2^0 \quad \log K = 37.7$$

$$Hg(SH)_2 = HgS(SH)^- + H^+ \quad \log K_{a1} = -6.2$$

$$HgS(SH)^- = HgS_2^{2-} + H^+ \quad \log K_{a2} = -8.3$$

and the single sulfide form:

$$HOHgSH + H^+ = HgSH^+ + H_2O \quad \log K = -10.3$$

Given that the exact form of the complexes in solution cannot be determined, the formation constant values for each complex are derived from modeling the results of experiments across a range of pH and sulfide levels. The apparent contradiction between estimated values and measured values noted previously could be explained in terms of the formation of colloidal or nanoparticles in solution. Such a suggestion was previously made [64] and this appears to be consistent with the experimental results discussed here.

Additionally, much of the experimental work examining the impact of speciation on the bioavailability of Hg to sulfate-reducing bacteria have used the larger value for the constant in the estimations and correlations, and have shown that this provides a suitable explanation for the trends in methylation rate across a sulfide gradient [66]. Also, the modeled speciation is consistent with experimental determinations of the hydrophobic Hg fraction at different sulfide levels, assumed to be only neutral Hg complexes, using octanol-water partition experiments [67]. Furthermore, as discussed later in this chapter, the more recent field data for sediments are consistent with the higher value for the binding constant, suggesting that such colloidal (or nanoparticulate) Hg-S entities are present in the environment.

Finally, while the focus of the discussion has been the formation of nanomaterials and multinuclear metal entities in the environment, especially in the presence of sulfide, such compounds have been found to have numerous applications in industry and commerce and these compounds are now manufactured, and likely released to the environment. Many of these compounds contain metals and metalloids such as Ag, Se, Cd, and Zn. The explosion in the manufacture of nanomaterials for industrial applications and the likely formation of such compounds under environmental conditions has resulted in concerns about their toxicity and environmental impact. Bacteria and living cells can take up nanosized particles, providing the basis for potential bioaccumulation in the food chain [68]. The bioavailability of specific nanomaterials in the environment will depend in part on the particle characteristics and the chemical speciation of metallic nanomaterials may have important interactive effects on their biological availability and photochemical reactivity [68, 69]. Environmental fate processes (coagulation, settling) are too slow for effective removal of persistent nanomaterials before they can be taken up by an organism. Studies using fluorescent nanomaterials have shown the accumulation of these materials observed into zooplankton (*Daphnia*) [69, 70]. Besides such examples, there is currently little information on the impact of metal nanomaterials on organisms [71, 72], but given the discussion earlier about nanoparticle and colloidal aggregation, there is the potential for their uptake directly and also after their potential accumulation in larger flocs and other material that may be consumed by filter feeding and benthic organisms [73]. Overall, the bioavailability of metal clusters and other metal nanoparticles is an important area for future study.

6.2.4 Metal concentrations in coastal waters

The focus of this chapter is on mechanisms and cycling of metal(loid)s in marine systems but the discussion would be incomplete without some mention of the range of concentrations found in coastal waters. Overall, as many metal(loid)s are strongly associated with particles, the total concentration in the water is a strong function of the TSS loading. Thus, it is not reasonable to compare directly the total concentrations but rather to compare the dissolved concentrations. For many elements even these values are highly variable between estuaries and coastal waters depending on the extent of anthropogenic input. Numerous studies in the literature have quantified these values and it is beyond the scope of this book to examine these in detail. Some indication in the variability can be gained by an examination of the concentrations in the figures discussed in the previous sections, especially Figs. 6.7–6.9 and Fig. 6.13. For example, concentrations of Cu vary from <10 nM to over 150 nM, compared to an open ocean average value of 4 nM (Table 6.1). Similarly, Cd and Zn also vary over a substantial range (Cd <0.06–0.25 nM, Zn 20–200 nM). The crustal elements, such as Mn and Fe, may be expected to show lower variability but overall there concentrations are also highly variable. This is partly due to the factors discussed earlier in that a large fraction of these metals may be present in the filtered fraction as colloidal phases and not as truly dissolved species. This is true for both the crustal elements that are relatively insoluble (e.g., Fe, Mn) and for the metals that form strong complexes with organic matter that can enhance their concentration above that predicted based on equilibrium with the solid phase (e.g., Fe) or result in a correlation between dissolved concentration and NOM concentration (e.g., Cu, Hg).

There is little information on the levels of metals in background waters as most of the studies are completed in

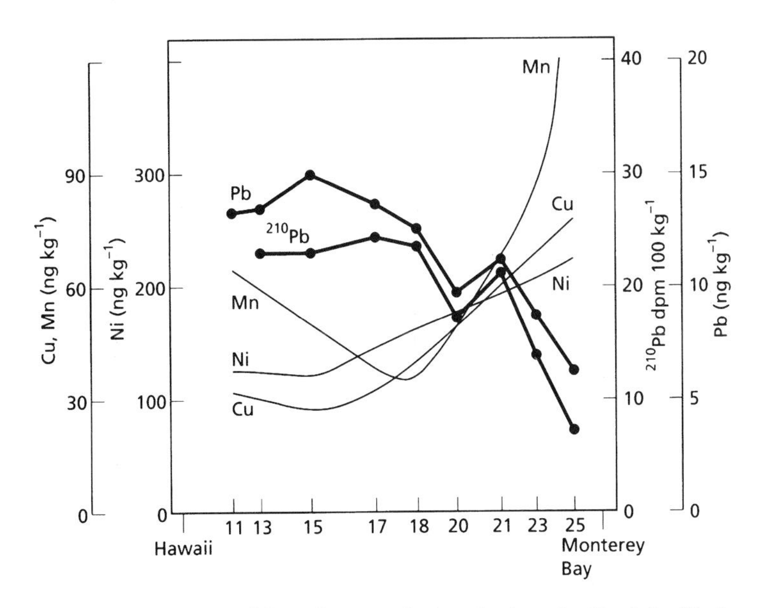

Fig. 6.13 Distribution of trace metal concentrations offshore from North America into the North Pacific Ocean. Figure taken from Chester (2003) *Marine Geochemistry* [1] which was drawn using data from Schaule and Patterson (1981) [75]. Used with permission of John Wiley & Sons, Inc.

response to some concern or mitigation study, and therefore the published data likely provides a "worst case scenario". In an attempt to account for this, Tueros et al. [74] examine data from a wide variety of systems and developed a methodology to estimate the background concentrations of metals in coastal waters. They derived values for a limited number of metals: As 10–20 nM; Cu 15–30 nM; Mn 20–110 nM; Ni 15–35 nM; Pb 5–7 nM and Zn 180–340 nM [74]. These concentrations are typically at the lower range of the values reported in the studies earlier which were mostly done on impacted ecosystems (e.g., Galveston Bay and the Dansheui River). Interestingly, the As concentration range is lower than the open ocean concentration (Table 6.1) while, for most of the other metals, the concentrations are higher (3–10 times for Cu and Ni; 20 times for Mn and >30 times for Zn and Pb). This demonstrates the impact of the coastal environment on metal concentrations, even when accounting for the anthropogenic signal. Higher concentrations are likely for those metals that bind strongly to organic matter, are significant components of the colloidal phase, or have a substantial anthropogenic component that has permeated even remote locations, probably as a result of significant atmospheric input, such as Pb.

The interaction between sources, sinks and the role of internal cycling leads to some unexpected results in terms of horizontal offshore concentrations of metals in surface waters. As may be expected, metals that have a strong terrestrial input via rivers compared to the atmosphere are likely to be present in higher concentrations in coastal waters than offshore. Additionally, metals that have relatively strong anthropogenic sources, which are primarily terrestrial in origin, should have higher coastal water concentrations due to local sources and enhanced atmospheric input (i.e., a "pollutant signal"). However, as illustrated in Fig. 6.13, this is not exactly the case when comparing concentrations across different water masses that have substantial differences in their ability to remove metals from surface waters. The increase in the levels of Mn, Cu and Ni in coastal waters reflects both their relatively high riverine input (Table 6.1) and their relative solubility, and for Cu the distribution reflects the anthropogenic component [1]. For Pb, it's strong scavenging from surface waters, and its relatively small riverine input, results in its decrease in coastal waters even given local anthropogenic inputs [75].

Important aspects of the horizontal concentration gradients shown in Fig. 6.13 are the decrease in Pb in the region of the shelf break – where there is higher productivity driven by upwelling of nutrients from depth – and an increase in the low productivity waters near Hawaii [1]. Additionally, the higher concentrations in the mid-Pacific Ocean may reflect enhanced inputs due to anthropogenic inputs from Asia and long-range transport. The similarity in the profiles of total Pb and ^{210}Pb, which is added from the atmosphere, confirm the importance of atmospheric sources combined with differences in rates of removal from the surface waters as the two main variables driving the Pb distribution. Mn also shows an increase near Hawaii which also reflects the lower scavenging of these oligotrophic waters, but the lack of a strong atmospheric signal results in an overall decreasing concentration trend offshore [1]. Such a trend would typically be found for most metals, and would be more pronounced for oceans where coastal waters are "downwind" of regional anthropogenic sources and inputs, and also where there are large riverine sources.

6.3 Metals in coastal and offshore sediments

6.3.1 Metals in the bulk phase

The distribution of metals in estuarine and coastal sediments is related to two main factors: the strength of the sources versus the strength of the binding phases in sediments (the retention factor). The inputs of metals from the atmosphere is discussed in Chapter 5 and this leads to a gradient of input from the coastal zone offshore but the major differences in concentration are also likely driven by more local sources, either point source inputs from current or historical industrial discharges, sewage effluent and other anthropogenic point sources, and from areal sources, such as urban runoff. Often, in coastal waters, elevated inputs of metals occur in conjunction with either higher inputs of nutrients or carbon (NOM), or both, and therefore it is often the case that coastal waters and sediments receiving enhanced metal inputs are also environments where there is likely to be elevated binding capacity for metals in sediments due to the higher organic loading, and the resultant sediment oxygen depletion close to the surface and the presence of reduced sulfide phases in the sediments. All these factors enhance binding in sediments. Therefore, there is often a correlation between sediment metal concentration and sediment organic content, and sediment OM and reduced sulfide content are often also correlated. For example, in Baltimore Harbor, sediment %C correlated with %S, %N, %Fe, acid volatile fraction (AVS), and many of the metals, and therefore there is likely not one unique binding phase present for a particular metal [76]. The %Fe was often greater than the %S, suggesting that not all the Fe was bound as sulfide phases. In low carbon environments, Fe and Mn oxides are likely the most important binding phases, while in high carbon environments, both OM and reduced sulfide are the dominant metal binding phases.

In the presence of sulfides, metals are either adsorbed to the surface or co-precipitated in the sulfide matrix. The relevant phases for these interactions are the AVS and the pyrite phase [77], often estimated as the chromium reducible fraction (CRS). The AVS fraction consists principally of amorphous FeS, mackinawite (FeS) and greigite (Fe_3S_4).

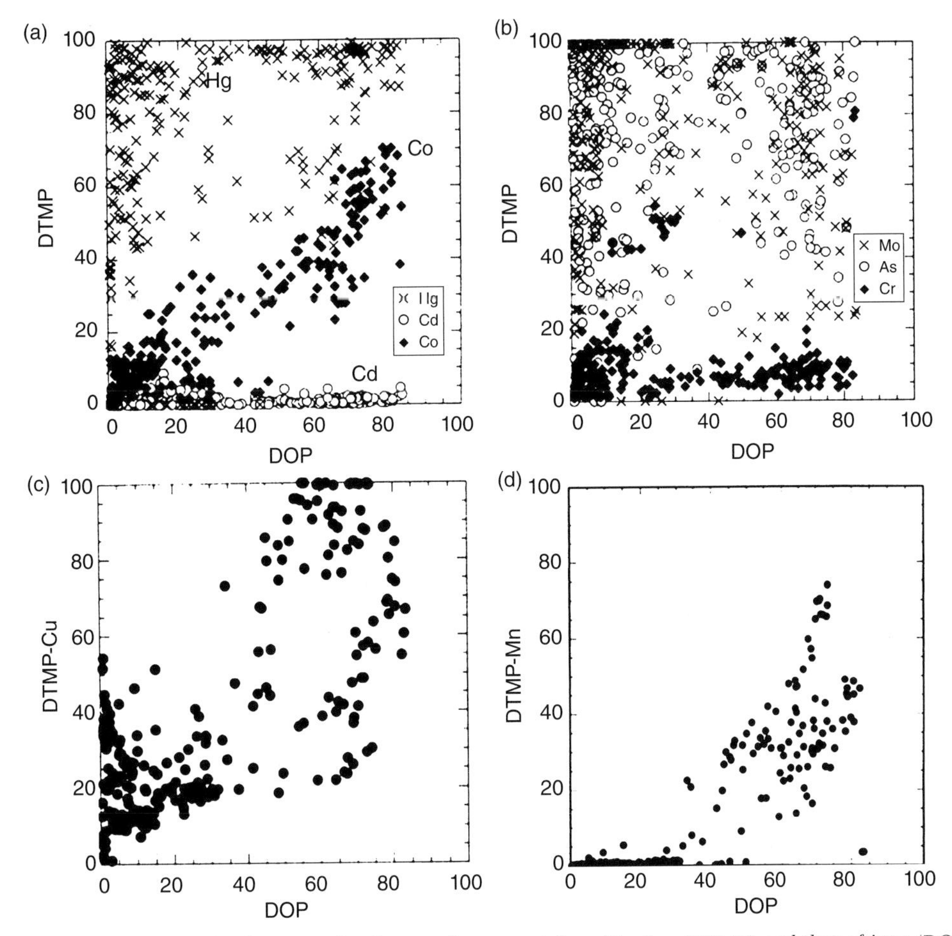

Fig. 6.14 Graphs showing the relationship between the degree of trace metal pyritization (DTMP) and that of iron (DOP) for various estuarine and coastal sediments. Data shown for (a) mercury, cobalt, cadmium; (b) arsenic, molybdenum and chromium; (c) copper; and (d) manganese. Taken from Morse and Luther (1999) *Geochimica et Cosmochimica Acta* **63**: 3373–8 and reprinted with permission from Elsevier.

Metals associated with AVS are more labile, and likely adsorbed to the surface of the mineral phase while the incorporation of metals during the formation of pyrite (FeS_2) often is through co-precipitation and incorporation into the sulfide solid matrix, as well as adsorption to the mineral surface. Pyrite is a thermodynamically more stable form than the AVS phases and is formed in the presence of elemental sulfur:

$$AVS + S^0 \rightarrow FeS_2$$

Given the incorporation of Fe into pyrite, Berner defined the degree of pyritization (DOP) [77] as:

$$DOP = \text{pyrite Fe/Total Fe-S phases} \qquad (6.2)$$

and similarly, the incorporation of other metals into pyrite can be defined as the degree of trace metal pyritization (DTMP) as:

$$DTMP = \text{pyrite metal/Total metal in S phases} \qquad (6.3)$$

Based on extensive analyses of sediments, it has been shown that As and Hg are enriched in pyrite, even relative to Fe (Fig. 6.14). Overall, the Class A metals are not strongly incorporated, the transition metals are incorporated to an intermediate extent, while the Class B metals form the strongest associations. These generalities are consistent with the relative binding strengths of metals to sulfide, as discussed in Chapter 3.

As there are multiple phases that can complex metals in estuarine sediments, several approaches have been developed to assess metal bioavailability and toxicity in sediments. Ankley et al. [78] proposed that metal toxicity in sediments (Cd, Ni, Pb, Zn and Cu) was related to the presence or absence of AVS [78], as discussed further in Chapter 8. Since AVS is an important phase for metal binding, metal speciation in sediment and therefore toxicity was proposed to be related to the relative amounts of metal and AVS in the sediments. The metal recovered during the AVS extraction step has been termed the simultaneously extractable

Table 6.4 Typical concentration ranges for metals in nearshore and offshore sediments. All values are given in $nmol\,g^{-1}$ dry weight of sediment. Data obtained from a variety of sources.

Element	Highly Contaminated	Urban Influenced	Remote Estuarine/Coastal	Deep Ocean
Cr	>2.0	0.1–0.2	0.01–0.07	0.3–2
Co	>1.0	0.1–0.5	0.2–1.0	0.1–2
Ni	>2.0	0.01–1.0	0.6–1.0	0.5–5
Cu	>2.0	0.2–2.0	0.2–0.9	0.5–9
Zn	>20	0.05–10	0.05–1.0	0.05–1
Cd	>1.0	0.1–1.0	<0.01	0.01–0.02
Hg	>0.025	0.01–0.05	<0.01	0.001–0.01
Pb	>2.5	0.1–1.0	0.01–0.1	0.02–0.5
Ag	>0.02	0.001–0.01	<0.01	<0.01
Sn	>0.1	0.01–0.1	<0.01	<0.03
As	>2.0	0.5–1.0	<0.2	0.3
Se	>0.025	0.005–0.01	<0.01	<0.01

metal (SEM) fraction and was proposed as a measurement of potential bioavailability, lability or toxicity of the metal in sediment. The hypothesis was that if AVS/SEM was >1, the sediments were not toxic because the metals were not bioavailable as they were incorporated into AVS. This initial approach has been refined because AVS is not the only important binding agent found in the solid phase, including sulfidic sediments [79]. For example, in Baltimore Harbor, where the AVS concentration of the sediments was greater than that of the metals being examined (Hg, CH_3Hg, Ag, Cu, Pb, and Cd), sedimentary metals accumulated in bivalves [80]. Others also concluded that SEM-AVS approaches do not provide a good measure of the bioavailability of Cu to benthic invertebrates [81]. These issues are discussed in more detail in Section 8.5.

Overall, for most metal(loid)s, the lability and bioavailability is dependent on the distribution and interaction with the two most important binding phases in environmental solutions – organic matter and sulfide, or for the solid phase, NOM and AVS [82, 83] – and the complexation strength of each phase with the specific element. Metals such as Hg and Cu bind strongly to both the sulfide phases and organic matter, while the sulfide binding to Cd, Ni and possibly Zn is probably stronger than their interaction with organic matter. Since it is unlikely any one "binding phase" dictates behavior in all situations, quantification of the available fraction in a specific situation remains complex. Computer thermodynamic equilibrium programs, such as MINEQL or WHAM, can be used to predict the speciation but such an approach often requires knowledge of detailed sediment chemistry, such as metal redox states and the concentrations of the important phases (oxides, AVS and pyrite, and POC). Often, the ability to model and predict the solid phase partitioning is limited by the lack of fundamental data such as binding constants and adsorption constants for the metals to the solid phases. In addition, the role of POM in modifying adsorption to oxide and AVS or other sulfide phases needs further study. Organic matter could either increase adsorption through formation of tertiary complexes (oxide-POM-metal), or to decrease adsorption because of complexation of the metal to NOM compounds in porewater. Finally, POM coatings on the surface of inorganic phases could alter the interaction with the oxide phases. Such interactions were discussed in detail in Chapters 3 and 4.

The concentration ranges for metals in impacted, remote and offshore sediments are given in Table 6.4 [1, 2, 84]. Mostly, metal concentrations in nearshore environments are elevated above crustal abundances (see Chapter 2) due to the impact of anthropogenic inputs and the tendency of coastal sediments to be enriched in organic matter and sulfide phases. For Fe, and some of the other first row transition metals, concentrations are controlled by the type of terrestrial material. Therefore, their concentration is not included in Table 6.4. As might be expected, there is a wide range in concentration that reflects the local inputs from anthropogenic sources. Also, there is high variability and this is primarily due to the differences in the capacity of sediments of different type and characteristics (grain size, oxide and organic content, reduced sulfur content) to retain metals. Overall, considering surface area and particle makeup, it is the smaller grain size particles that contain the highest fraction of most trace metals. Therefore, in many instances, investigators have limited their analysis to the smaller size fractions, and the fraction <63 μm is the size fractionation that is normally considered. In contaminated environments, there are often strong gradients in concentration over small spatial scales and this reflects the high capacity of sediments for metals.

The legacy of previous anthropogenic inputs is particularly acute for estuaries and the coastal zone where current inputs are now often lower than those of the past century, or even earlier, and there have been numerous studies that have demonstrated that inputs typically peaked in the last 30 years or so for the more industrialized nations, but may not have yet peaked in developing countries. Such changes are recorded in sediments in estuarine and coastal environments, especially locations where sediments have accumulated undisturbed, such as salt marshes, and provide a record of the current and past environmental insult. Radioisotopes, such as ^{210}Pb (Table 6.2), added to surface waters from atmospheric sources, or ^{137}Cs, a tracer added in high concentration during the period of heightened above-ground nuclear bomb testing prior to the Test Ban Treaty (~1963), are often used to determine the age of these sediments as they provide appropriate markers (the half-life of ^{210}Pb is 22 yrs, and its usefulness as a geochronometer is about 110 yrs) of period of industrialization in the last century. For example, sediment cores collected in the vicinity of New York and New Jersey, USA (the Hudson River and Long Island Sound) (Fig. 6.15a) show that the concentrations of metals were highly elevated in many locations in the past, but that the concentrations have been steadily decreasing in recent years [85].

In addition, measurements of dissolved metals in the water column of the Hudson River in the vicinity of New York City provide a similar story – concentrations were higher in the 1970s compared with more recent data [86, 87] (Fig. 15b). Overall, there have been notable reductions (typically >80%) in the concentration of filtered and total metals in the Lower Harbor and NY Bight since the mid-1970s, with the majority of current inputs being from rivers (38 to 88% of totals) [88]. Total inputs are significantly decreased for Cd (85–95%), Ni (50–80%), Cu (30–70%), and Zn (0–75%) since the mid-1970s (25–30 years) along with the decreased relative contribution of metals from sewage, industrial discharge and other local and regional sources, and decreases in metal inputs to the watershed of the Hudson River.

While concentrations are often highly elevated in estuaries, they decrease rapidly with distance from shore. However, interestingly, the concentrations of some metals are comparable or higher in deep sea sediments than in nearshore uncontaminated environments (Table 6.4). For Mn, the reasons for this include the fact that sediment accumulation rates are much slower in deep ocean sediments (~mm kyr^{-1} rather than cm yr^{-1}) so precipitation of metals from the water column can continue to occur. Additionally, increased concentrations in upper sediment could result from redox cycling and solid Mn oxide precipitation. The presence of solid Mn precipitates and other oxide phases also tends to scavenge and concentrate other transition metals. Close to hydrothermal inputs, sediments are often enriched due to precipitation of metals released in the reduced fluid emitted from such systems. For other metals, the low sediment accumulation rate and the overall destruction of organic matter during early diagenesis is the mechanism for their concentration in deep ocean sediments. During degradation of accumulating organic matter in deep sediments, many metals will be released back to the water column, but some metals that bind strongly to the inorganic phases (oxides and carbonates) will be retained in deep ocean sediments and therefore can exist in high concentration. This is discussed further later in terms of metal accumulation in one particular phase – as manganese nodules (Section 6.3.6).

The exchange of material at the sediment-water interface occurs for both solid material and dissolved constituents. The exchange of dissolved constituents will be discussed further in Section 6.3.2. In many locations, the deposition of solid particulates from the overlying water is the principal input but in shallow ecosystems such as estuaries, and in environments where there is strong currents or physical mixing, sediment resuspension can reintroduce sedimentary particles to the water column. Sedimentation rates vary over many orders of magnitude from rates as high as centimeter year for estuarine locations to less than a millimeter per thousand years for open ocean deep sediments. In estuaries, sedimentation is often the primarily removal mechanism for metals introduced via riverine transport, and from local point source inputs, as discussed in Chapter 2. For the open ocean, even though sedimentation is much lower, it is still the primary long-term sink for most metals.

Measurements of metals in sediments tend to fall into different categories, depending on the approach to analysis. For the determination of the total metal concentration it is often necessary to use strong acid digestion using hydrofluoric or perchloric acids. However, in many studies it is more relevant to determine the fraction of the metal in sediments that is not associated with the granular matrix of the sediment, and under these conditions, less stringent, but still strong acid digestion techniques are used, such as sediment dissolution using hot nitric or sulfuric acids, or acid mixtures. Additionally, there have been a number of studies that have attempted to characterize the sediments further using a series of sequential extraction approaches that use consistently more harsh conditions to extract the metals associated with different fractions. Typical fractions that are assayed are the labile (reactive), typically oxide fraction, the organic matter fraction, the AVS and/or pyrite fractions, and finally the metal in residual material. These approaches were discussed in Section 6.2.2.

It is often advantageous to normalize the concentration of the metal in sediment to a particular size fraction, or to a specific element, or to compare the concentration to the crustal abundance. Such approaches were discussed in Chapter 2 and often a metal enrichment factor can be calculated (Equation 2.4) by normalizing to the Al concentration in the sample versus the relative crustal abundance. Alternatively, correlations can be determined with other sedimentary or watershed parameters. Often, for example, there is a correlation between metal content and

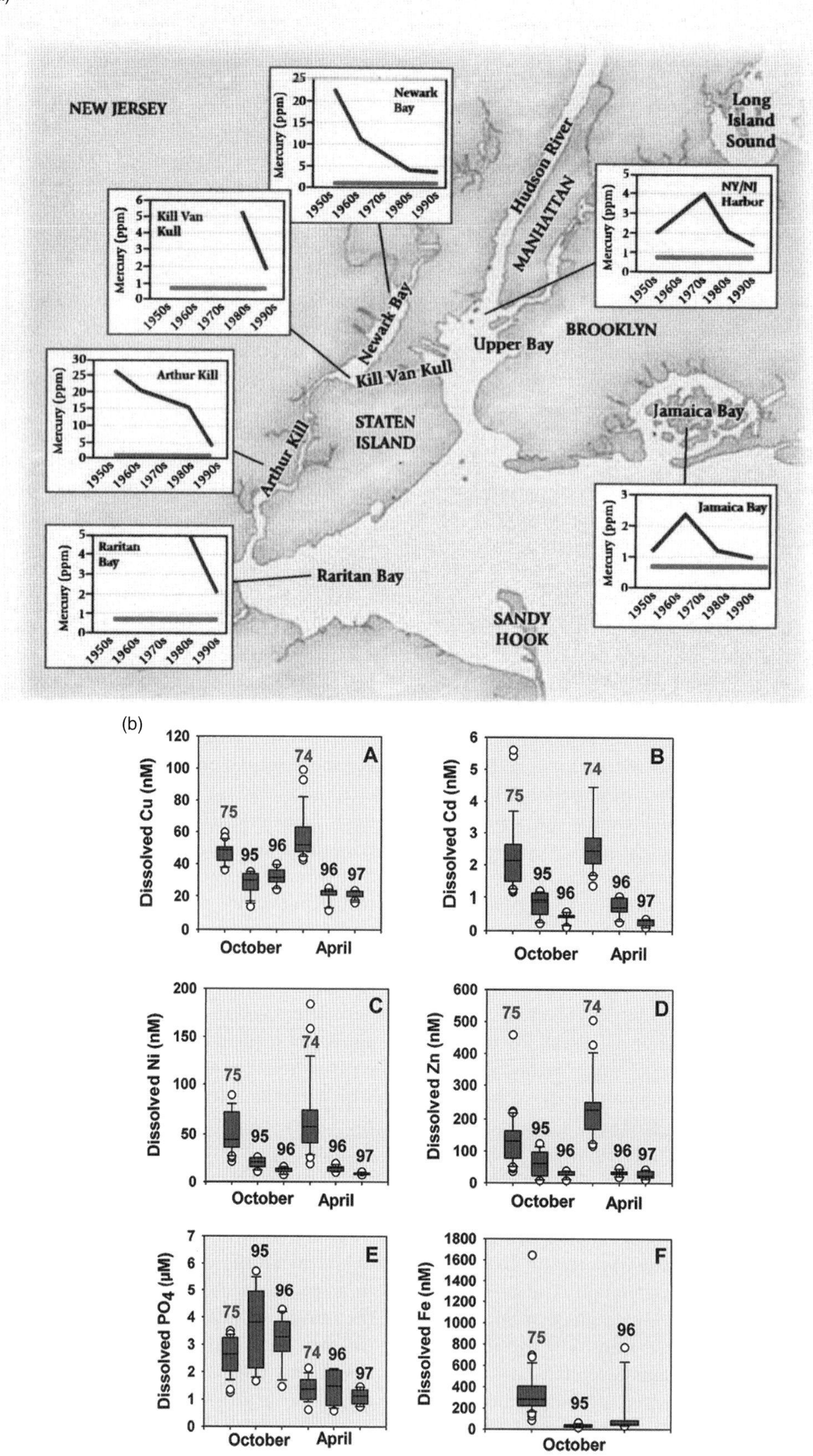

Fig. 6.15 (a) Sediment concentration data for mercury for various locations in New York/New Jersey Harbor, USA. Taken from The Health of the Harbor Report [85], reprinted with permission from Hudson River Foundation; and (b) concentrations of metals in the dissolved fraction for harbor waters in different seasons and over time, reprinted with permission from Hudson River Foundation and also reprinted with permission from Sanudo-Wilhelmy and Gill (1999) *Environmental Science and Technology* **33**: 3477–81 [86]. Copyright (1999) American Chemical Society.

sediment organic content. A recent publication examining the relationship between metal sediment content and a variety of factors for estuaries throughout the eastern US (Maine to South Carolina) concluded that there was actually a small number of variables that could be used to estimate relatively accurately the surface sediment concentration for a variety of metals [89]. The variables included the bulk characteristics of the sediment (OC, grain size, silt/clay content) and the characteristics of the estuary and watershed (e.g., estuary and watershed surface area, fraction urbanized, as agriculture, forests, etc.; tidal volume and range, river flow, and point source inputs). The final best fit overall model used functions that were transformed to normalize the metal data, and included the following variables: %urban, %agricultural, point source loading, estuary area, river flow, tidal range and sediment %silt+clay. The results in Fig. 6.16 confirm the importance of both local sources and sediment binding characteristics in determining metal levels in sediments. In the model, %silt+clay overall provided a better correlation than sediment OC. Many of these variables are easily estimated or the information is readily available. If this modeling approach can be demonstrated to be widely applicable, it will be a useful tool for providing a first order estimate of the sediment concentration in an estuarine location where no measurements have been made.

Understanding and modeling the partitioning of metals to sediments is an active area of research that has been discussed more generally in Chapter 4. The main phases that need to be considered for marine and estuarine sediments are those of oxides, organic matter and sulfides. The approaches discussed previously can be used to estimate the bulk partitioning in terms of organic matter or solid sulfide phases, but it is also possible to examine the speciation in terms of binding constants and interactions between dissolved species and the bulk phase. All the phases can be represented as a particle with surface absorption sites, which are acidic, as discussed in detail in Section 3.4.5. For example, considering a 2+ cation (M^{2+}) and a 1 : 1 association with a solid particle, represented as ≡XH:

For oxides (e.g., Fe(OOH)): $M^{2+} + {=}FeOH = {=}FeO\text{-}M^{+} + H^{+}$

For sulfides (e.g., AVS or FeS):

$$M^{2+} + {\equiv}FeSH = {\equiv}FeS\text{-}M^{+} + H^{+}$$

For organic matter (e.g., RCOOH)

$$M^{2+} + {\equiv}RCOOH = {\equiv}RCOO\text{-}M^{+} + H^{+}$$

For organic thiols (e.g., RSH) $M^{2+} + {\equiv}RSH = {\equiv}RS\text{-}M^{+} + H^{+}$

Thus, if it is possible to determine the relative concentration of each phase, the surface reactive site density or the number of adsorption sites per mole of material, or some other normalizing factor, then it is possible to represent the surface sites in a similar way to calculating the speciation in the dissolved phase, as discussed in detail in Chapter 3. There are examples of such calculations in the literature and mostly these take a somewhat empirical approach as in most instances, an effective binding constant is calculated for the specific conditions (pH, particulate material characteristics) that include the impact of ionic strength and surface potential effects. Alternatively, the modeling can be formulated in a rigorous manner as discussed in Chapter 4. Additionally, it is also be possible to include co-precipitation into the model formulation. Examples of modeling will be discussed further later in this chapter.

6.3.2 Metals in sediment porewater

There are many reasons why scientists are interested in the concentration of metals in the waters in the pore spaces of sediments (sediment porewater) and thus there has been substantial investigation of their concentration and distribution, and of the factors, such as redox status, organic content and sulfide levels, that may affect the overall concentration and form of the metal in the porewater. The collection of porewater samples for studying the concentration and distribution of metals is an exacting task both because sample contamination is always a potential issue, especially for remote or deep water sediments, and because there is also the potential for redox changes during sampling and handling altering the concentration of metals in porewater, as a result of the oxidation of reduced sediments. It is critical that the redox status of the sediments is maintained during sampling and handling [90, 91], and therefore sampling should be carried out in a glove box or under an inert atmosphere. Methods of extracting porewater include: (1) sediment core sectioning followed by centrifugation or filtration; (2) pressure filtration of the whole core using an inert gas and collection of porewater using sampling ports on the core perimeter; and (3) use of *in situ* dialysis membrane samplers (peepers) or diffusive gradient in thin film (DGT) gels [90–92]. In some instances, concentrations can be measured in situ in intact cores or in the field using microelectrodes. The advantage of microelectrodes is that this is a nondestructive technique. Microelectrodes have been used to measure oxygen, Fe^{II} and Fe^{III} and speciation, Mn^{II}, S species and metal complexes [93, 94]. The use of peepers, DGT devices or microelectrodes allows for much higher resolution measurements to be made than can be obtained using devices where the sediment must first be sub-sectioned. However, the need for sufficient volume for determination of the concentration sometimes precludes the use of these devices for the metal(loid)s that are present in porewaters at nM or sub-nM concentrations [91, 92].

The concentration and speciation of metals in porewater are controlled by their oxidation state, their complexation capacity by dissolved inorganic and organic ligands and their propensity to be associated with the solid phase

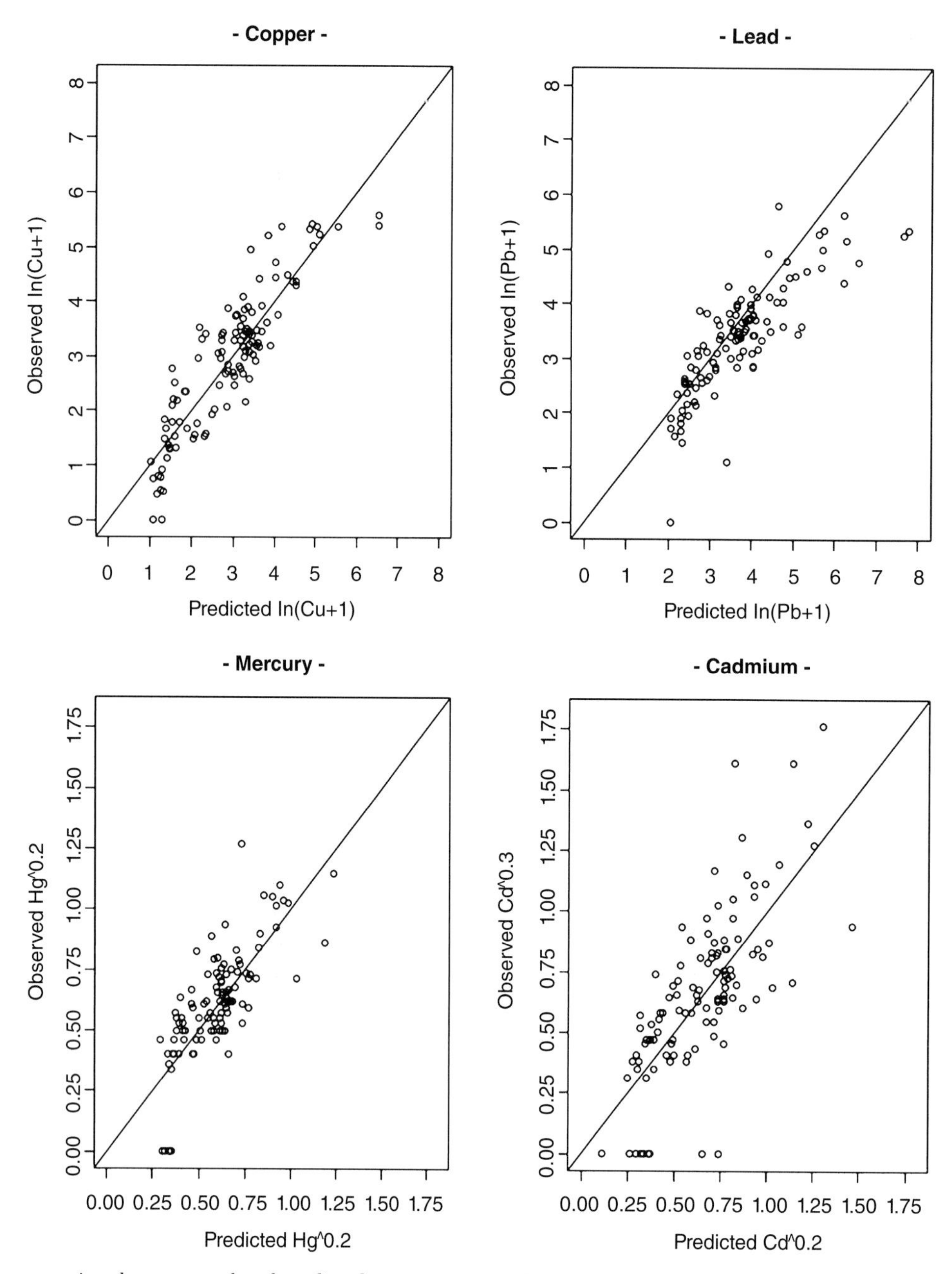

Fig. 6.16 Graphs comparing the measured and predicted concentrations of various metals in sediments on the east coast of the USA. Data derived using the model of Hollister et al. (2005) *J. Environmental Quality* **37**: 234–44 [89] and used with permission.

Table 6.5 The range and average concentration for the fraction of the metal in the truly dissolved phase for filtered water from San Francisco Bay. Data from [92].

Metal	Fe (μM)	Mn (μM)	V (nM)	Cr (nM)	Co (nM)	Cu (nM)	Zn (nM)	Cd (nM)	Pb (nM)	Ag (pM)
Truly Dissolved Minimum	<0.09	<0.02	6.3	3.0	2.4	1.3	119	<0.1	<0.01	<1
Maximum	40	0.28	91	27	29	20	602	1.2	0.28	15
Average	5.1	46	25	11	11	4.2	432	0.26	0.09	5.1

material (oxide and sulfide phases and particulate organic matter). For some transition metals, such as Fe, Mn, and Cr, there are large differences in solubility between the reduced forms (Fe and Mn are substantially more soluble in the reduced (+II) oxidation state; Cr is less soluble when reduced). Many metals persist in solution (in the filtered fraction) at concentrations above their solubility limit based on oxide or sulfide precipitation due to their formation of inorganic and organic complexes, and their presence in colloidal fractions. Most of the Fe in porewater extracts is complexed in comparison to only a small fraction of the Mn [90]. For the metals that form strong associations with sulfide, or with NOM, their concentrations in sediment porewaters can be elevated compared to the overlying waters.

Measurement of porewater concentrations are typically made for three reasons: (1) to calculate the partitioning and K_D, as discussed above for the water column; (2) to study the speciation and impact of redox status and other changes in bulk parameters on metal distributions and fate; and (3) to estimate the input of metals from the sediments to the water column, or the potential for uptake of metals from the dissolved phase. In some instances, the fluxes of dissolved constituents across the sediment-water interface can be an important source or sink, especially in shallow water environments, or in locations where sediment concentrations are elevated.

Overall, for most metals, the K_D value for the porewater is typically lower than that for the water column and this is likely related to the presence of colloids in the sediment porewaters due to the generally higher concentrations of NOM in porewater compared to overlying water. For metals that form strong complexes with sulfides, the value of the K_D will likely change with redox status as the presence of dissolved and solid phase sulfides will lead to a difference in the controlling parameters for the partitioning. However, while the strength of the binding to the solid phase for Hg and other metals that form strong associations with inorganic reduced S increases as anoxia increases, the concentration of dissolved sulfide also increases thereby providing stronger binding ligands in solution to compete with the solid phase for the metal in question.

Huerta-Diaz et al. [92] measured the concentrations of a variety of trace metals in sediment porewaters in San Francisco Bay using different approaches and used these data to investigate the distribution of metals between truly dissolved and colloidal fractions. The measured range and average truly dissolved concentrations (<3 kDa) are shown in Table 6.5. The elements could be divided into two groups – those that were mostly in the colloidal phase – up to >99% on occasion (Ag, Cd, Cu, Fe, Pb, and Zn) – and those that were mostly in the truly dissolved phase (<50% colloidal) (Co, Mn, V, and Cr). The oxyanions were found less in the colloidal phases. For most of the metals in the colloidal phases, they were dominantly in the smaller colloidal fraction (3 kDa to 0.1 μm). These authors compared the fraction in colloidal matter in porewater to seawater and showed that in almost all cases the average colloidal fraction was higher for porewater, although due to the high variability the error bars overlapped in most instances. For Mn and Zn, the differences were the smallest. Finally, the concentration of these metals in porewater was lower than average seawater [92].

In comparing these concentrations to the average range in values for ocean seawater (Table 6.1) it can be seen that for some metals these concentrations are not particularly elevated relative to the ocean waters, but for Fe, Mn, Zn, and Cd the porewater concentrations appear elevated. This is probably due to the impact of redox status and its impact on Fe and Mn concentration, and for Cd and Zn, the higher concentrations likely reflect the relatively low K_D values for these metals. In sediment porewater, the concentration is determined to a large degree by the overall redox status as dissolution of oxide phases could release trace elements into solution, and this would be more important in low organic matter or offshore environments. In locations where sediment organic content is high, and reduced environments persist, metals that form strong associations with organic matter or sulfide phases are likely to have concentrations that are not largely different from the overlying water. Of course, the extent to which these comments are true is determined by the method of separation as the presence of colloidal material in sediment porewaters could increase the filtered concentration of many trace elements by 1–2 orders of magnitude above the truly dissolved concentration.

As noted earlier, many studies have examined the potential for metals to be released from the sediment to the overlying waters. The diffusive sediment flux can be estimated using values for the porewater and overlying water concentrations and Fick's Law, or the exchange can be modeled using a more complex computer formulation, as discussed in the next section. These estimations rely on the measurement of the concentrations and the calculation of the flux based on the difference in concentration and a diffusion coefficient for the system under examination. Measurement of the porewater and overlying water concentrations can be done using the methods discussed previously. Often, such diffusive calculations tend to underestimate the flux compared to its measurement by other means and this reflects the fact that there are many advective processes and mechanisms that can enhance the flux (physical mixing and enhanced transport due to related pressure gradients and biological activity (bioturbation and bioirrigation). Other approaches that attempt to take into account such processes include the use of a benthic flux chamber device or sediment core incubations. Benthic chambers are enclosed containers that are deployed in situ over the sediment surface and the change in concentration over time is then used to estimate the flux into or out of the sediment. Core chamber incubations use sediment cores that are maintained under the sediment conditions with overlying water, and again the change in concentration in the water over time is monitored.

All of the methods suffer from deficiencies in that there are potential artifacts associated with each approach. The benthic chambers isolate the sediment environment and therefore potentially reduce the flux that may be enhanced by physical advective processes, and current flow. Also, the chamber material may impact light penetration and therefore photosynthesis at the sediment surface [95, 96]. Additionally, oxygen depletion in the chambers often occurs and this may enhance or decrease the flux compared to that of the unenclosed sediment. Flux chamber construction and deployment can be expensive, and there are also potential artifacts associated with their use due to the need to stir the chamber contents to ensure these are well-mixed. Changes in oxygen content can have a dramatic impact on the sediment flux for redox-sensitive constituents and for metals bound to solid oxide phases [97]. Also, benthic chambers are best deployed in muddy sediments and flat environments as there is the need to have a seal between the chamber and the sediment surface. Core incubation studies can lead to artifacts due to the small sediment surface being studied given the known heterogeneity of most sediments, the need to withdraw a substantial amount of overlying water during sampling, with potential changes in redox status and concentration with its replacement, walls effects (uptake of solutes and microbial growth effects) and the potential for sediment disturbance during sampling. Therefore, it is often useful to use more than one approach as this may allow a better appreciation of the most important processes controlling the flux.

Tengberg et al. [96] compared the impact of various benthic flux chamber design parameters (size, shape, stirring speed etc.) of 14 different designs and concluded that variations in chamber design and hydrodynamic settings did not have a measurable impact on the fluxes of oxygen and silicate. However, they recommend that the circulation patterns in any benthic chamber design should be examined to determine the mixing rate at which flow becomes turbulent, and to calculate the chamber mixing time and other parameters. Detailed methods for making such estimates are included in the paper. The advantages of using a pump for mechanical stirring/mixing and other aspects of the chamber design are also discussed. Overall, it was concluded that a design with a centrally placed stirrer was preferable. In these studies, estimates of Si flux based on porewater gradients collected at high resolution (4 mm sections) were within a factor of two of the benthic chamber estimates. However, overall the system studied here was not highly dynamic and these results may not be applicable to all ecosystems.

Recently, eddy correlation techniques, as discussed in the Atmospheric Chapter 5, have been applied to measuring benthic boundary fluxes of oxygen [95, 98] and could be used for estimates of metal flux if a suitable high resolution, in situ measurement device exists. As measurements are made above the sediment surface, this is a non-invasive approach. The technique relies on measuring two parameters simultaneously at the same point in the water above the sediment (10–50 cm and above). One measurement is the chemical of interest and the other, the fluctuating vertical velocity, which can be measured using an acoustic Doppler velocimeter. Studies with oxygen used a measurement period of 10–20 min and at a frequency of 15–25 Hz to provide data for the flux estimates [95]. The technique was compared with chamber measurements. Overall, this approach has wide applicability for a variety of sediments and to locations where deployment of benthic flux chambers is not possible. The main drawback of such an approach for metal measurements is the need for a quick response sensor but electrochemical techniques are potentially available for such measurements of redox-sensitive metals such as Fe.

While the diffusive calculations do not estimate the total flux, they do provide a relatively simple method to estimate the magnitude of inputs from the sediments to the overlying water. Additionally, a number of studies have compared diffusive estimates with other methods, and there is the possibility that the overall flux can be estimated based on the diffusive flux by extrapolation from these studies, or through modeling of the advective component using advective eddy coefficients, formulated based on sediment type, physical environment and biological mixing of the sediments. Such estimations are documented in the literature for a number of solutes and show that the fluxes derived from benthic chambers were greater than the estimated diffusional fluxes by a factor of 30 or less, but were elevated in all cases. For example, the total measured DOC flux was ~10 times higher than the calculated diffusive flux from the permeable sediments at the mouth of Chesapeake Bay [99], while the measured $Si(OH)_2$ and NH_4^+ fluxes were, on average, ~30 times higher for offshore sediments on the southern Atlantic Bight [100]. However, for similar sediments, the measured DIC flux was only ~two times higher than the calculated diffusive flux [101]. Studies of the Hg and CH_3Hg flux from sediments have similarly have shown that diffusional calculations underestimate the total flux by up to an order of magnitude in some estuarine locations, when comparing diffusive calculations to flux chamber studies [97, 102–104]. For example, Benoit et al. [105] showed that fluxes of CH_3Hg, estimated from sediment core incubations, were up to 10 times higher than the diffusive flux estimates for sediments in Boston Harbor. Physical pumping and biological activity are often quoted as the reasons for these differences and it was shown in Boston Harbor that fluxes were strongly correlated with the burrow density in the sediment and bioirrigation [105]. Overall, it is apparent that the diffusion calculation provides a minimum estimate of the flux from sediments to the water column.

Diffusive fluxes (F) have been calculated for a number of metals for estuarine environments [97, 106–108]. Fick's first law of diffusion, applicable for estuarine sediments is:

$$F = -\frac{\phi D_W}{\theta^2}\frac{\Delta C}{\Delta x} \tag{6.4}$$

where ϕ is the porosity (dimensionless), θ is the tortuosity (dimensionless), D_W is the sediment diffusion coefficient (e.g., in $cm^2 s^{-1}$), ΔC is the difference in concentration between the filtered overlying water and the surficial sediment pore water ([M]), and Δx is the average depth of the pore water sample (cm). Tortuosity can be calculated using porosity, as suggested by Boudreau [109]:

$$\theta^2 = 1 - ln(\phi)^2 \tag{6.5}$$

To determine the diffusive fluxes of a metal it is necessary to know the speciation in the porewater so that the correct D_W can be used. The value of D_W varies by more an order of magnitude, being lowest for metals bound to NOM and highest for small inorganic metal species or for the free ion. Diffusion coefficients also differ to some degree depending on the charge of the metal ion or complex [97, 110]. The highest values are for the neutral complexes with slightly lower values for the charged inorganic species, and values also vary with molecular size. This is why complexes with NOM have lower D_w than the inorganic complexes.

The diffusion coefficient at 25°C (D_w) can be estimated using an inverse linear relationship between D_w and the molar volume (V) for neutrally charged molecules [111], as discussed in Chapter 5. For metals complexed to macromolecular organic matter (NOM) a value of $2 \times 10^{-6}\,cm^2 sec^{-1}$ for D_w is often used [97]. Methods for estimating the D_w for charged species exist [112–114]. Temperature corrections to D_w can be made using the Stokes–Einstein equation [115]. The net flux calculation should be made using the sum of the concentrations of the individual species:

$$F = (\phi/\theta^2)\sum D_{wi}(C_{i,pw} - C_{i,bw})/\Delta x \tag{6.6}$$

where $C_{i,pw}$ is the porewater concentration of each species and $C_{i,bw}$ is the bottom water concentration of the same species. In a number of investigations, flux estimates have been made without taking into account the detailed speciation in the water column and sediment and this has lead to an incorrect estimation of the flux. For example, considering an oxic water column but low oxygen sediment, it may be that the overlying water column speciation is dominated by complexes of the metal with NOM, while the surface porewater speciation may be dominated by sulfide speciation. In such a scenario, there will be a flux of organic complexes into the sediment and a flux of sulfide species into the water column. The flux of sulfide species will further be maximized as there could be a sink in the overlying water due to the recomplexation of the metal to NOM upon efflux. Such a situation may not always occur and in many instances it is likely that both the water column and sediment porewater speciation is similar, and likely dominated by NOM complexes.

Again, a specific example will further demonstrate the approach. Using Hg and CH_3Hg data collected in the Chesapeake Bay and Mid-Atlantic Bight, Hollweg et al. [116] developed a model including the following CH_3Hg species (CH_3HgSH^0, CH_3HgS^-, CH_3HgSR and CH_3HgCl^0) and Hg^{II} species ($HOHgSH^0$, HgS_2H^-, HgS_2^{2-}, $Hg(RS)_2$), where –SR refers to NOM binding. For these waters, Cl complexation to Hg^{II} is not important given the strong Hg-NOM interaction in oxic waters, and the strong binding to sulfide when present in the porewater. The diffusion coefficients at 25°C (D_w) of $HOHgSH^0$, CH_3HgSH^0 and CH_3HgCl^0 were estimated using the inverse linear relationship between D_w and the molar volume (V) for neutrally charged molecules (Equation 6.7) [111, 117] and the estimated molar volumes [118, 119]:

$$D_w = \left(\frac{2.3\times10^{-4}}{V^{0.71}}\right) \tag{6.7}$$

Values were estimated respectively as: 1.7, 1.2 and $1.3 \times 10^{-5}\,cm^2 sec^{-1}$ [97, 120]. The D_w of the organic complexes was assumed to be $2 \times 10^{-6} cm^2 sec^{-1}$, based on the diffusion coefficient of organic matter (5000 Da) [97]. The D_w of HgS_2^{-2}, HgS_2H^- and CH_3HgS^- were estimated to be 9.5, 8.0 and $8.0 \times 10^{-6} cm^2 sec^{-1}$, based on the average diffusion coefficient of other singly or doubly charged anionic species of similar mass [112–114]. Temperature corrections to the D_w were applied by the Stokes-Einstein equation [115].

The results of the estimations are shown in Table 6.5 where the major species at each station in the porewater and overlying water are shown, as well as the net flux in each case. The calculated diffusive flux of CH_3Hg ranged from 0.06 to 2.2 pmol $m^{-2} day^{-1}$, with an average flux of 0.8 pmol $m^{-2} day^{-1}$ [110, 116]. The variation in the flux across sites was strongly related to the modeled speciation of CH_3Hg in the pore water. Dissolved sulfide concentrations were clearly important in controlling the flux. For example, in sediments where these CH_3Hg-S complexes (CH_3HgSH^0 and CH_3HgS^-) dominated, the overall diffusive flux was significantly higher than for sediments where the organically bound complexes dominated, due essentially to the differences in the D_w for different complexes. For Hg^{II}, the calculated diffusive flux ranged from 3.4–60 pmol $m^{-2} day^{-1}$, with an average flux of 26 pmol $m^{-2} day^{-1}$. The variations in flux between sites were primarily controlled by differences in concentration, since modeled Hg complexation did not vary substantially across sites.

Others studies have also examined the factors that influence the flux of metals from a variety of sediments [1, 121, 122]. For example, a five-month study, using mesocosms with intact sediment from Baltimore Harbor, MD, USA, examined the fluxes of Mn, As, Cu, and Cd from sediments [107] and showed that during hypoxic conditions the flux of Mn and As was into the water column but overall fluxes of Cd and Cu were into the sediment. The As flux remained elevated during the hypoxic period while the Mn flux decreased over time. Conversely, with oxic water column conditions, there was little or no Mn or As flux, but there was measurably fluxes of Cd and Cu into the water column. Based on estimated oxic flux rates, the authors concluded that the benthic fluxes of Cu and Cd were comparable to that of point sources and storm-water inputs. Comparable results were found in another study in Baltimore Harbor [123] which included a wide range of metals, including Hg, and nutrients and NOM. It did not appear that the flux of most of the metals and metalloids was related strongly to the NOM flux. The flux of CH_3Hg appeared to increase when the surface sediment and water column became hypoxic.

Studies of the flux of Co, Ni, Cu, Zn, Cd, and Pb from sediments from New Bedford Harbor and Buzzards Bay, MA, USA [124] found that generally the net flux of metals was into the water column. Regression analysis of the fluxes, normalized to sediment metal content, showed a relationship with benthic oxygen demand. The magnitude of the fluxes generally followed the relationship: Cd > Zn > Co, Ni, Cu > Pb. As found by Mason et al. [13], the estimated annual fluxes were a small fraction (<2%) of the metal content of the sediment (top 2 cm). Fluxes were sufficient to replace the water column inventory within weeks to months [124].

A study by Point et al. [104] using benthic chambers in various locations in the Thau lagoon in France found that the sediment redox status, as indicated by the Mn flux, was a good predictor of the flux of the other metals (Fig. 6.17). A strong positive correlation was found between the Mn flux and that of Co, Pb, and Cu. For Cd, U, and Hg, and for CH_3Hg, the relationship was more complex, with positive correlations in some instances but with the opposite trend also apparent. These results suggest that conditions exist where uptake of these metals can occur during anoxic conditions. This study showed that, in their shallow coastal system, the presence and activity of photosynthetic organisms on the sediment surface and the activity of invertebrates had a significant impact on the overall flux and its direction, especially for the redox sensitive metals. For one site, which was coated with a macroalgal mat, negative fluxes were found in the dark for Cd and U in contrast to positive fluxes during the day, while the opposite occurred for Mn and Pb. For the other metals, fluxes were reduced in the dark but the flux direction was the same. Importantly, this study showed that the estimated fluxes based on the benthic chambers were almost two orders of magnitude higher than those estimated based on Fick's Law for Mn, Cd, Hg, and CH_3Hg; again clearly illustrating the importance of physical and biological processes in enhancing the sediment flux, as found by others [97, 123], and as discussed previously.

The importance of organic complexes of the metals in controlling the flux is not evident in some of the studies, and this is likely the result of the relatively low rate of diffusion of larger molecules and therefore the NOM-related metal flux is likely less important for systems where there is not substantial biological activity or physical forcing. However, a number of studies have looked at the flux of Cu and Zn, and of ligands that bind these metals, measured using electrochemical approaches, from estuarine sediments and have shown that there is a correlation between the ligand flux and the NOM flux, and between the flux of these metals and that of the metal-binding ligands [125, 126]. The metal-binding ligand fluxes did not appear to be important sources to the studied ecosystems, and it appeared that riverine inputs were a larger source.

In summary, the research to date suggests that fluxes of metals from sediments can be important in some situations and more studies are needed to understand the controls over these fluxes. While estimation of fluxes using diffusion models and measured concentrations has advantages, these approaches clearly underestimate the overall magnitude of

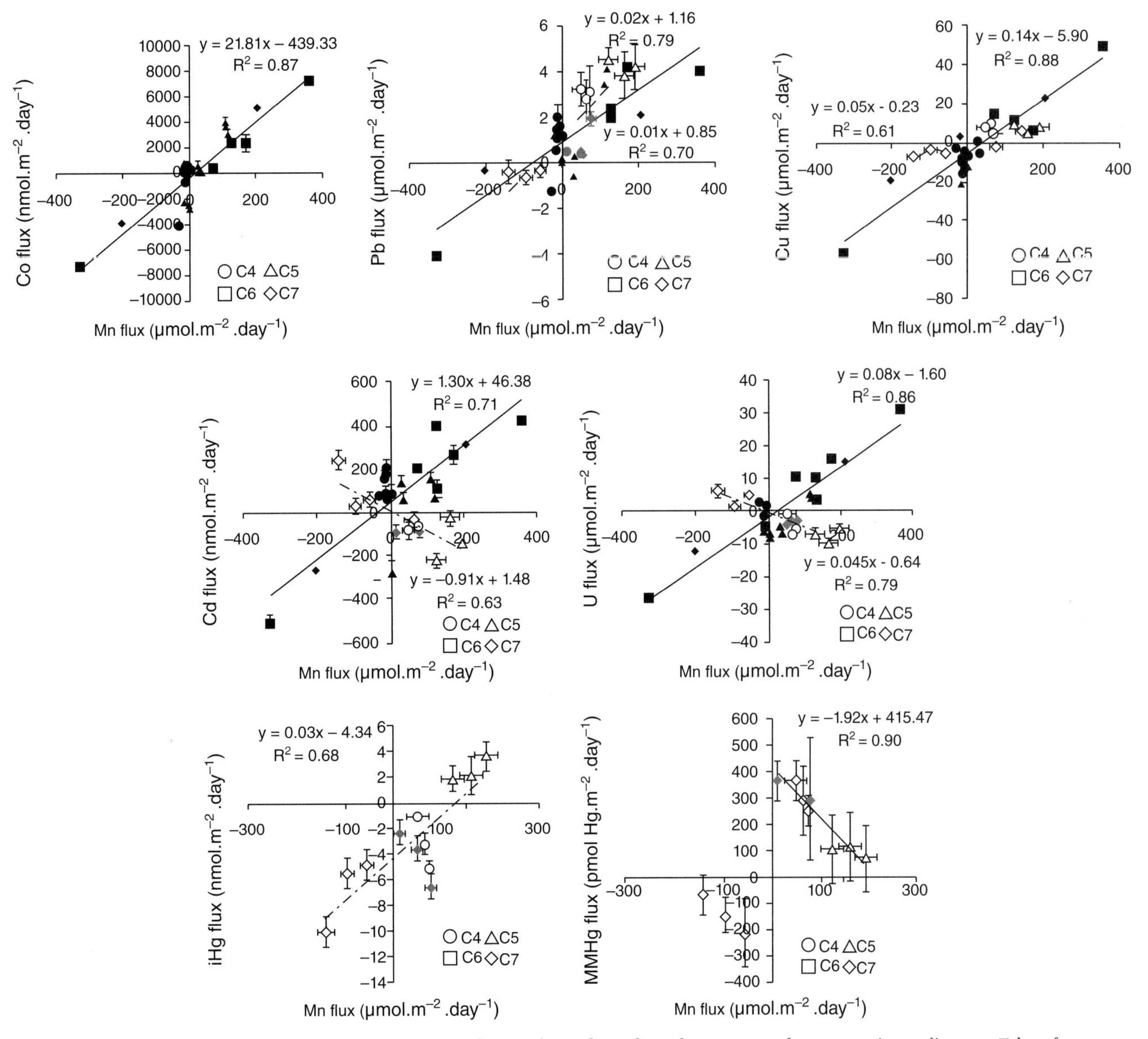

Fig. 6.17 Graphs showing the relationship between the fluxes of metals to that of manganese from estuarine sediments. Taken from Point et al. (2007) *Estuarine Coastal and Shelf Science* **72**: 457–71, reprinted with permission of Elsevier. Note: MMHg = methylmercury.

the flux. The extent of this is dependent on the specific location and the sediment type, the presence or absence of benthic biota, their type and the extent of physical mixing. Alternative methods have been used to estimate the flux to try and take these additional factors into account but each approach has some limitation or potential artifact, and therefore use of more than one method is likely to provide superior information.

6.3.3 Modeling metal cycling in sediments

The transport-reaction mechanisms that control the distribution of metal solutes and solids in the sediment can be derived by assuming, in the first order, that the reactions occur predominantly in the vertical, or in other words, the sediment is assumed to be horizontally homogeneous. This assumption will be discussed more next. The vertical movement of solids or solutes in sediments is determined by advective and diffusive processes, reactions such as redox cycling and precipitation and dissolution, and complexation, and the sources and sinks of the particular element. Equation 6.8 is generally applicable for both solids and solutes [127]:

$$\partial\theta C/\partial t = \partial/\partial x[D\theta(\partial C/\partial x) - \omega\theta C] + \theta\Sigma S + \theta\Sigma R \quad (6.8)$$

where C is the concentration of the metal of interest, x is the depth below the sediment water interface (cm), D is the overall diffusion coefficient (incorporates both molecular and dispersive processes (bulk mixing and hydrodynamic) (units are typically $cm^2 s^{-1}$), θ is the tortuosity and is related to the sediment porosity (φ) and ω is the advection rate (units typically $cm\ s^{-1}$). The various reactions are noted by $\sum R$, and $\sum S$ refers to the non-local transport processes (bio-irrigation, deposit feeding, wave pumping) and both are typically expressed in units of mass $s^{-1} cm^{-3}$. Often the equations are solved for the steady state composition, or under the assumption that $\sum S = 0$. If the equations are to be solved over time, finite difference approaches are usually used to numerically solve the equation. Such an approach is detailed in [127], for example.

In sediment diagenesis, it is often concluded that there is a series of redox reactions that occur during the oxidation of organic matter and that these are the primary processes driving sediment redox status and metal cycling. The reactions are assumed to occur in order of their energy yield (Fig. 6.18) and therefore aerobic oxidation is the initial oxidation pathway until the oxygen is depleted. The remaining reaction scheme is: nitrate reduction (denitrification), followed by Mn and Fe reduction, then sulfate reduction and finally methane formation. For each reaction, as these are primarily mediated by bacteria in marine sediments, there is a limiting concentration below which the pathway is no longer favorable, and in modeling the redox status of the sediment, the model sequentially consumes the organic matter input to the sediment surface through the sequence of reactions [127]. In many cases, it is assumed that the rate of organic carbon oxidation is a first order process and that the rate constant does not vary with depth, and that there is only some fraction of the organic matter that is metabolized. The fraction that is oxidized can also be considered to be linear or some other function of depth.

For example, in modeling the dynamics of Fe and Mn in sediments, it is necessary to consider the organic phase, and the partitioning of Fe and Mn into various inorganic solid phases (oxides, carbonates and sulfidic), as well as the major dissolved constituents involved in the reactions above and their products, such as NH_4^+, HS^-, HCO_3^-, CH_4, and their speciation. The rate expressions for such a model are given in Table 6.6. The first six reactions are those of organic matter degradation by the various pathways (Fig. 6.18) [56]. Additional inorganic reactions involving oxidation-reduction of the products of organic matter degradation and other dissolved constituents are also included, as well as precipitation/dissolution processes. Porewater pH and system buffering is determined by the presence and speciation of the carbonate and sulfide systems, and the calculation of alkalinity (reaction A-20 in Table 6.6) [127].

For the transport-reaction terms in the model, it is assumed that both solid and dissolved species are transported, but through different mechanisms. Typically, random sediment mixing due to bioturbation is included in the model and this impacts both the solid phase and sediment porewater, while there are additional diffusion processes acting on the dissolved constituents. This sediment mixing value is often related to the sediment accumulation rate (ω) (i.e., the rate of supply of fresh organic matter). Overall, a transport-reaction equation is needed for each dissolved and solid phase constituent included in the model.

The results of such model studies for different ocean and freshwater environments are given in Table 6.7 [127]. The major differences highlighted by model simulations are in the pathways for organic matter degradation. Degradation is essentially aerobic for deep sea sediments (~80% of the organic matter degradation is via aerobic oxidation), while coastal systems are often dominated by sulfate reduction (>80% of the total degradation for shelf and estuarine sediments). At the other extreme, in freshwater environments, most of the degradation is through either oxic processes (oligotrophic environments) or methane production (simultaneous oxidation and reduction of organic matter; $2CH_2O \rightarrow CO_2 + CH_4$) in more eutrophic systems. Fe and Mn reduction are of minor importance in NOM remineralization, even in the marine environment. However, the table also shows that organic matter degradation (i.e., biotic processes) is important in the reduction of Mn and Fe in sediments.

The oxygen penetration depth is a function of the relative supply of organic matter compared to the rate of diffusion of oxygen into the sediments. In the deep sea, organic matter supply is low and sediments are permeable, and so oxygen supply is sufficient to support aerobic degradation [128]. In estuarine sediments, sediment type and high organic matter input results in the lowest oxygen penetration and the dominance of sulfate reduction as the organic matter degradation pathway given the relatively high concentrations of sulfate. In freshwater systems, sulfate reduction is often substrate limited. Increased input of sulfate through anthropogenic activities (acid rain) has increased the extent of sulfate reduction in some remote oligotrophic environments with a variety of consequences.

While oxide dissolution is not important in organic matter degradation, it plays a major role in the cycling and transport of Fe and Mn in sediments. Dissimilatory Fe reduction can occur through a variety of different pathways in sediments, with organic and inorganic constituents acting as sources of electrons. Organic reaction pathways include fermentation of sugars, as well as consumption of organic acids (e.g., acetate) and other organics (phenols), with the formation of bicarbonate (C oxidation) as the product. Inorganic sources of electrons are elemental S and hydrogen. The bacteria promoting these reduction reactions obtain energy and as most of the Fe^{III} is present in the environment as solid oxides, it is worth considering the energy released for the reactions with the various oxide phases. In some cases, a

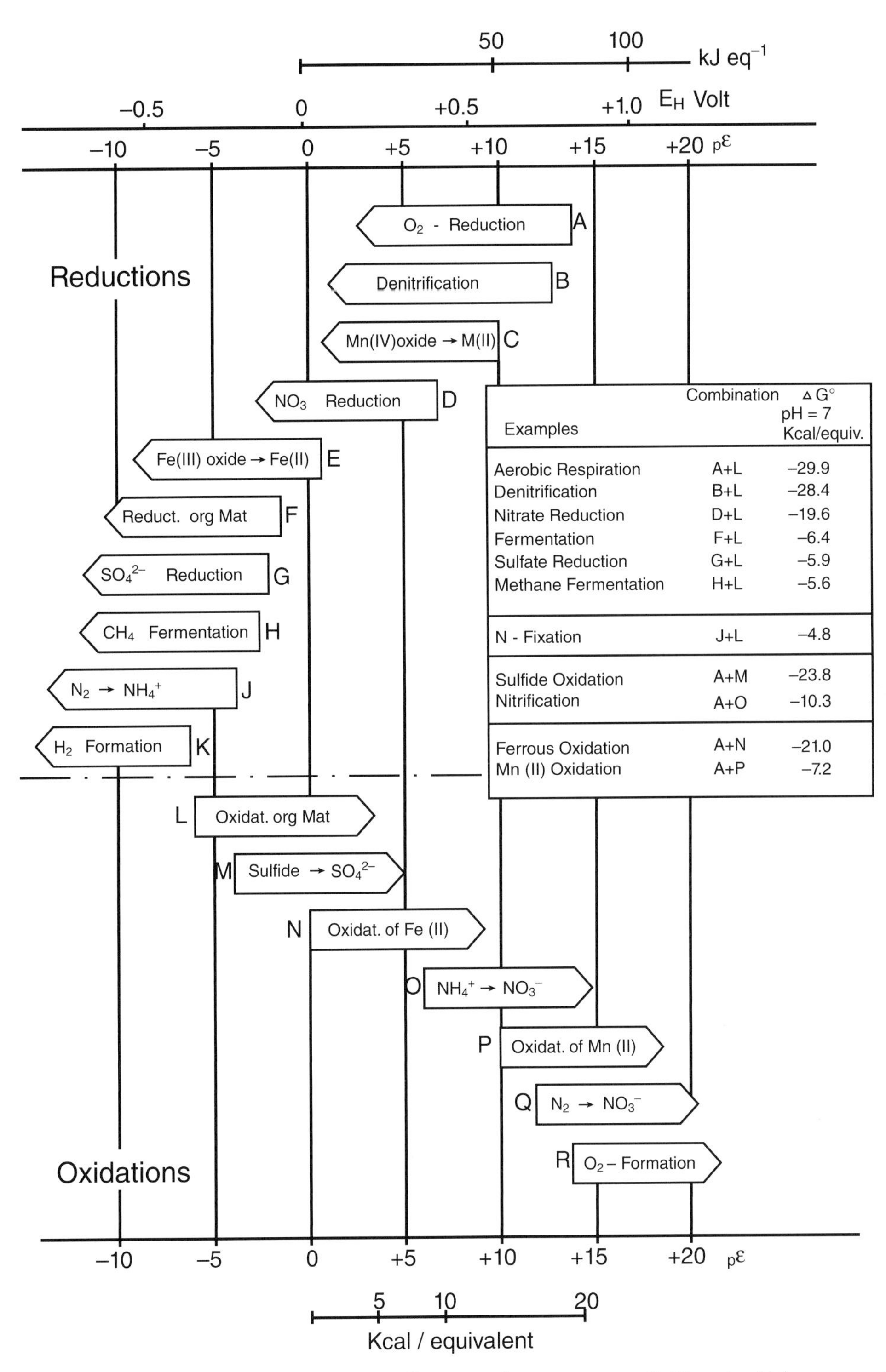

Fig. 6.18 The proposed order of redox diagentic reactions in sediments and water as controlled by microbial processes. Taken from Stumm and Morgan *Aquatic Chemistry* (1996) [56] and reprinted with permission of John Wiley & Sons, Inc.

Table 6.6 Reactions included in the model of Van Cappellen and Wang [127] for the oxic and anoxic degradation of organic matter (Reactions A–1 to A–6) and inorganic redox reactions. Reprinted with permission from Taylor & Francis.

Irreversible Reactions (A–1 to A–19) and Alkalinity Conservation (A–20. Reactions A–1 to A–6 represent the net degradation of organic matter deposited from the water column. Reactions A–7 to A–16 describe the reoxidation of secondary species produced during the oxidation of organic matter. Reactions A–17 to A–19 correspond to the non-reductive precipitation of carbonate and sulfide mineral phases. The irreversible production or consumption of protons is buffered by the dissolved carbonate/sulfide acid-base interconversions (A–20).

$$(CH_2O)_x(NH_3)_y(H_3PO_4)_z + (x+2y)O_2 + (y+2z)HCO_3^- \xrightarrow{R_1} (x+y+2z)CO_2 + yNO_3^- + zHPO_4^{2-} + (x+2y+2z)H_2O \quad \text{A-1}$$

$$(CH_2O)_x(NH_3)_y(H_3PO_4)_z + \left(\frac{4x+3y}{5}\right)NO_3^- \xrightarrow{R_2} \left(\frac{2x+4y}{5}\right)N_2 + \left(\frac{x-3y+10z}{5}\right)CO_2 + \left(\frac{4x+3y-10z}{5}\right)HCO_3^- + zHPO_4^{2-} + \left(\frac{3x+6y+10z}{5}\right)H_2O \quad \text{A-2}$$

$$(CH_2O)_x(NH_3)_y(H_3PO_4)_z + 2xMnO_2 + (3x+y-2z)CO_2 + (x+y-2z)H_2O \xrightarrow{R_3} 2xMn^{2+} + (4x+y-2z)HCO_3^- + yNH_4^+ + 2HPO_4^{2-} \quad \text{A-3}$$

$$(CH_2O)_x(NH_3)_y(H_3PO_4)_z + 4xFe(OH)_3 + (7x+y-2z)CO_2 \xrightarrow{R_4} 4xFe^{2+} + (8x+y-2z)HCO_3^- + yNH_4^+ + zHPO_4^{2-} + (3x-y+2z)H_2O \quad \text{A-4}$$

$$(CH_2O)_x(NH_3)_y(H_3PO_4)_z + \frac{x}{2}SO_4^{2-} + (y-2z)CO_2 + (y-2z)H_2O \xrightarrow{R_5} \frac{x}{2}H_2S + (x+y-2z)HCO_3^- + yNH_4^+ + zHPO_4^{2-} \quad \text{A-5}$$

$$(CH_2O)_x(NH_3)_y(H_3PO_4)_z + (y-2z)H_2O \xrightarrow{R_6} \frac{x}{2}CH_4 + \left(\frac{x-2y+4z}{2}\right)CO_2 + (y-2z)HCO_3^- + yNH_4^+ + zHPO_4^{2-} \quad \text{A-6}$$

$$Mn^{2+} + \frac{1}{2}O_2 + 2HCO_3^- \xrightarrow{R_7} MnO_2 + 2CO_2 + H_2O \quad \text{A-7}$$

$$Fe^{2+} + \frac{1}{4}O_2 + 2HCO_3^- + \frac{1}{2}H_2O \xrightarrow{R_8} Fe(OH)_3 + 2CO_2 \quad \text{A-8}$$

$$2Fe^{2+} + MnO_2 + 2HCO_3^- + 2H_2O \xrightarrow{R_9} 2Fe(OH)_3 + Mn^{2+} + 2CO_2 \quad \text{A-9}$$

$$NH_4^+ + 2O_2 + 2HCO_3^- \xrightarrow{R_{10}} NO_3^- + 2CO_2 + 3H_2O \quad \text{A-10}$$

$$H_2S + 2O_2 + 2HCO_3^- \xrightarrow{R_{11}} SO_4^- + 2CO_2 + 2H_2O \quad \text{A-11}$$

$$H_2S + 2CO_2 + MnO_2 \xrightarrow{R_{12}} Mn^{2+} + S^{o} + 2HCO_3^- \quad \text{A-12}$$

$$H_2S + 4CO_2 + 2Fe(OH)_3 \xrightarrow{R_{13}} 2Fe^{2+} + S^{o} + 4HCO_3^- + 2H_2O \quad \text{A-13}$$

$$FeS + 2O_2 \xrightarrow{R_{14}} Fe^{2+} + SO_4^{2-} \quad \text{A-14}$$

$$CH_4 + 2O_2 \xrightarrow{R_{15}} CO_2 + 2H_2O \quad \text{A-15}$$

$$CH_4 + CO_2 + SO_4^{2-} \xrightarrow{R_{16}} 2HCO_3^{2-} + H_2S \quad \text{A-16}$$

$$Mn^{2} + 2HCO_3^{2-} \xrightarrow{R_{17}} MnCO_3 + CO_2 + H_2O \quad \text{A-17}$$

$$Fe^{2+} + 2HCO_3^{2-} \xrightarrow{R_{18}} FeCO_3 + CO_2 + H_2O \quad \text{A-18}$$

$$Fe^{2+} + 2HCO_3^{2-} + H_2S \xrightarrow{R_{19}} FeS + 2CO_2 + 2H_2O \quad \text{A-19}$$

$$CO_3^{2-} + \delta CO_2 + \delta H_2O + (1-\delta)H_2S \rightleftarrows (1+\delta)HCO_3^- + (1-\delta)HS^- \quad 0 \le \delta \le 1 \quad \text{A-20}$$

Table 6.7 Depth integrated rates of carbon remineralization for different marine and freshwater environments and the fraction of degradation by each process (Reactions A–1 to A–6 in Table 6.6). Taken from [127]. Reprinted with permission from Taylor & Francis.

Process	Deep Sea	Shelf	Coastal Estuarine	Oligotrophic Lake	Eutrophic Lake
Corg oxidation	7	79	981	38	759
% via oxic respiration	80	6.0	4.3	48	3.0
% via nitrate reduction	11	5.8	1.6	9.9	9.3
% via metal reduction	3.5	<1	3.5	0.4	0.4
% via sulfate reduction	6	87	91	4.7	0
% via methanogenesis	0	0	0	37	87

range of values is given as the values depend on the type, age and molar surface area [128, 129].

$$4Fe(OH)_3 + CH_2O + 7H^+ = 4Fe^{2+} + HCO_3^- + 10H_2O$$
$$\Delta G^0 = -376\text{--}-228 \text{ kJ mol}^{-1}$$
$$4FeOOH + CH_2O + 7H^+ = 4Fe^{2+} + HCO_3^- + 6H_2O$$
$$\Delta G^0 = -387\text{--}-239 \text{ kJ mol}^{-1}$$
$$2\alpha\text{-}Fe_2O_3 + CH_2O + 7H^+ = 4Fe^{2+} + HCO_3^- + 10H_2O$$
$$\Delta G^0 = -236 \text{ kJ mol}^{-1}$$
$$2Fe_3O_4 + CH_2O + 11H^+ = 6Fe^{2+} + HCO_3^- + 6H_2O$$
$$\Delta G^0 = -328 \text{ kJ mol}^{-1}$$

The control over which oxide phase is preferentially reduced in the environment depends on a number of factors, with one of the major factors being the surface area of the particular phase. It has been found that amorphous phases are preferentially reduced over crystalline phases as the amorphous phases tend to have much higher surface areas. It has been suggested that one reason for the importance of surface area is the adsorption of the Fe^{II} onto the oxide surface which tends to hinder further reduction [129].

In modeling Fe, and Mn, distributions in sediments, it is necessary to consider both the reduction and oxidation pathways, and the movement of the solid and dissolved phases of the elements. The relative burial of oxide phases is controlled by the rate of sedimentation at the surface, and the rate of bioturbation. For the dissolved constituents, bioirrigation, physical pumping and pressure effects, and diffusive movement are the primary drivers. The depth of these processes relative to the redox boundary will have a large impact on the transport of dissolved and particulate Fe and Mn. Typically, organisms living below the redoxcline will need to bioirrigate their burrows to obtain oxygen while bioturbation is often most prevalent in the surface layers. Mixing of sediment from below the redoxcline will bring sulfidic sediments, and Fe-S phases, into contact with more oxidized compounds and therefore lead to their oxidation, either chemically or biologically. In many cases, it is not oxygen directly that oxidizes the reduced Fe and Mn, but intermediate compounds that have been produced chemically or biologically by these reactions.

The main oxidants for reduced Fe are oxygen, nitrate and oxidized Mn (Mn^{IV}). The reaction with oxygen is rapid and Fe will not therefore escape oxidation by O_2 in sediments with a significant oxic layer. Nitrate oxidation appears to require another metal as a catalyst and it is likely that Fe oxidation by nitrate is mostly microbially mediated in sediments, by nitrate reducing bacteria, which use Fe^{II} as an electron source. Bacterial Fe oxidation by photoautotrophs is also possible in some environments. Fe oxidation by Mn oxides has been shown to be rapid in the laboratory but it appears that the precipitation of Fe oxides can coat the Mn oxide surfaces and reduce the reaction rate.

It should be noted that many models assume that the sediment does not have substantial horizontal gradients in terms of redox. This has been studied by a number of investigators and the role of organism burrows in influencing the penetration of oxygen into sediments, and the resultant sediment geochemistry, is now well documented. Initial studies and modeling were conducted by Bob Aller and Bernie Boudreau, who developed a radial diffusion model approach to estimate oxygen penetration from burrows into the sediment [109, 130–133]. The model assumes that the concentration extends radially from the center of a burrow, and that the burrow size and penetration of oxygen or the chemical of interest from the burrow is small compared to the distance between burrows. The following equation, here assuming constant porosity, is a modification of Equation 6.8:

$$\partial C/\partial t = D\partial^2 C/\partial x^2 + (D/r)\partial/\partial r(r\partial C/\partial r) + \Sigma R(r, x, t) \qquad (6.9)$$

The reaction term in this case is a function of r, x and t, and the exact form of the equation will depend on the specifics of the environment being studied. It is possible to generalize

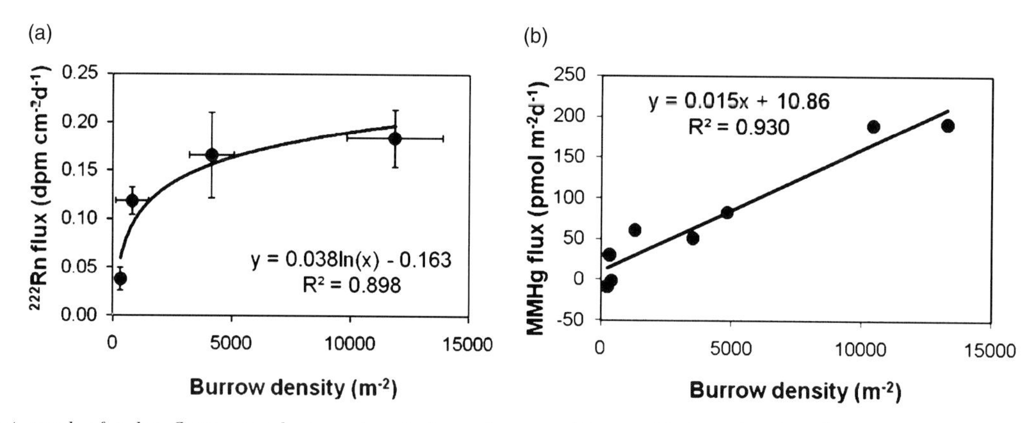

Fig. 6.19 (a) A graph of radon flux versus burrow density for sediments of Boston Harbor, showing the increase in flux with increasing burrow density; and (b) the corresponding flux of methylmercury relative to burrow density. Reprinted with permission from Benoit et al. (2009) *Environmental Science and Technology* **43**: 3669–74 [105]. Copyright (2009) American Chemical Society.

the equation using horizontally averaged values, and this simplifies the equation as it can then be expressed in terms of the average porewater concentration and reaction rates (which is what has likely been measured in the field). Bioirrigation can be included in the model as a function of depth and time [126]. Alternatively, it is possible to define a benthic flux enrichment factor which would generalize the impact of bioirrigation on the movement of dissolved solutes through the sediment porewater. Such a value could be determined from radiotracer studies or measurements and such studies have shown that this factor is related to water depth, and can range from 1–20 for nutrients and non-reactive tracers (^{222}Rn, Br) [129].

Schull et al. [134] have extended the earlier models to include burrow ventilation activities of the organisms in contrast to the assumption that the burrows are fully flushed. The water exchange was rather modeled as a non-local exchange with overlying water and the authors used measurements of ^{222}Rn and compared the measured profiles with the estimated supported ^{222}Rn profile. The study compared the results of models using an averaged burrow ventilation rate, a model that assumed fully flushed burrows, and a model based on molecular diffusion. The results show that the extent of ventilation can have a large impact on fluxes in/from the sediment, and that burrows in the field are not typically fully flushed [134]. Thus, it may not be appropriate to use a constant value for the burrow ventilation as intermittent flushing will allow time for various redox and adsorption/desorption reactions to occur within the burrows that could strongly influence chemical gradients and fluxes of reactive elements and compounds.

Other processes could also result in a heterogeneous distribution in sediments, such as the presence of a large particulate organic entity, burrowing of organisms, or by the deposits at depth of surface feeding organisms. Methane ebullition is another processes that could affect sediment redox on the small scale. These impacts are more transient than the impacts due to irrigated burrows. In such situations it would be also necessary to model the sediment in three dimensions, or to use averaged values as discussed above for bioirrigation.

It has been shown that burrows have a substantial impact on the cycling of Fe and Mn, and other metals [129]. In particular, bioirrigation can have a substantial impact on the extent of Hg methylation as this typically occurs in the low redox zone and this is greatly extended by the presence of burrows [105, 135]. The effect is illustrated in Fig. 6.19. In the absence of burrows, oxygen penetration is small and methylation occurs only in a limited band close to the sediment interface. The presence of burrows leads to a larger redox transition zone, which would extend the methylation zone. However, as shown, it is possible that methylation could become hindered at very high burrow densities, and would then occur deeper in the sediments. Both the flux of CH_3Hg from the sediment and that of ^{222}Rn were related to the burrow density in the sediments of Boston Harbor, USA. Physical processes could also result in a similar effect on methylation, as found for shelf sediments where oxygen penetration, and low sulfide levels resulted in a larger integrated methylation potential compared to estuarine sediments [116].

Models have also been used to investigate the fate of other metals in sediments. One example is the examination of the redox sensitive metals, U, Mo and Re, which have been used in a number of studies as paleoproxies of past redox status, as their incorporation into sediments is highly dependent on redox (Morford et al. [136] and references therein). All these elements are more soluble under oxic conditions, although these metals do adsorb to oxide surfaces – all exist as oxyanions in oxic waters. They are trapped in sediment as redox changes, and are therefore termed *authigenic*, as their zone of concentration depends on the *in situ* redox chemistry [1]. These metals are typically released into porewater in the surface sediments during remineralization of sinking particulate organic material. Deeper in the sediments, these elements partition to the sediment and the zone of partitioning is a function of the metal chemistry. U appears to be removed in the region of Fe and sulfate reduction, and its removal may be microbially-mediated (i.e., U reduction may be used as an energy source), while Re is removed in deeper sediments, and its removal is kinetically slow [136]. It has been suggested that Re is removed with incorporation into Fe-S phases. While Mo is also incorporated under sulfidic conditions into sulfide minerals, this often occurs deeper in the sediments than U and Re.

The observed profiles of these elements in porewater and sediments of Buzzards Bay, in the northeastern USA, were compared to model estimations which incorporated the reactions for the release and/or removal of metals from porewater outlined above. These metals do not partition as strongly to the sediments as the heavy metals, but exhibit large differences in their partitioning across the redox interface. For example, the K_D for Mo, derived from the published profiles, ranged from around 100 at the surface, and at depth, to values of 150–500 in the zone of incorporation. This is a very low K_D, less than that for Cd, for example. For U, the highest K_D values were 6000, with values at the surface around 1000. The partitioning of Re is intermediate to these two elements. Thus, changes in the sediment characteristics can have a large impact on the porewater concentration and on the movement of the elements vertically, especially as the elements are adsorbed primarily through association with oxide phases and/or sulfide phases. For many of the transition and heavy metals, association with organic matter in addition to inorganic phases leads to a stronger partitioning and a more consistent K_D over the redox interface.

It is worth mentioning, while not directly applied to marine systems, that some modelers have taken a fugacity approach to modeling freshwater ecosystems. This will be discussed more in Chapter 8, but basically the aquivalence approach defines a parameter, Q, in units of mol/m^3, which is related to the concentration in different media (i.e., dissolved, particulate) through the use of an aquivalence capacity, Z, which is dimensionless [137]. Calculation of Z values assumes a value for the water phase of $Z_W = 1$, and Z values in other compartments are then calculated using partition coefficients (K_{D12}), the dimensionless ratio of Z values of a chemical in phases 1 and 2. Therefore, as an example, for the particulate phase, $Z_P = Z_w.K_{Dpw}$ (K_{Dpw} is the dimensionless particle-to-dissolved chemical partition coefficient). To account for speciation and the interconversion of multiple species (i.e., different complexes of a metal in solution), the concentration of species j in phase i can be calculated as $C_{ij} = Q_{ij}.Z_{ij}$, where Z_{ij} reflects the species-specific partition coefficients. The total concentration in a particular phase is $C_{iT} = \Sigma(X_{ij} \cdot C_{ij})$. This approach has been applied to metal speciation and flux in lakes through the incorporation of chemical equilibrium speciation modeling into a fate and transport model [137, 138]. However, a number of parameters need to be measured, derived or estimated based on the literature, especially if there is complex speciation of the metal in solution.

6.3.4 Modeling of metal speciation in marine sediment porewaters

In addition to fate and transport modeling, it is possible to model more accurately the speciation and distribution of the various metal complexes in the porewater, and to account for changing partitioning throughout the sediment column. As redox changes, and as depth increases, the amount and nature of the organic matter in the sediment varies. The reduced S content of the sediment increases with depth and appears to be related to the concentration of inorganic sulfide. As the different metals interact with the main sediment binding phases to a differing degree, the relative distribution of metals changes with redox, and this impacts the speciation and mobility of the metals in the sediments.

In modeling metals in porewater, a number of approaches can be taken. Firstly, the interaction with the sediment can be parameterized using a K_D value, and this can be coupled with a detailed thermodynamic modeling of the dissolved species. In this case, the dynamics of the interactions between the dissolved and solid phases are not specifically modeled. However, it would be much better to include a proper formulation of the interaction between the dissolved species and the various sediment phases. This approach has been used in a number of instances to look at the distribution between the dissolved and solid phases in sediments.

To constrain and validate the interactions for Hg in sediments, Hollweg [139] developed a model to estimate the dissolved porewater speciation and to predict the K_Ds assuming the system was at steady state and that there was equilibrium partitioning between the dissolved and solid binding phases (including POM, oxides, FeS, and FeS_2), and any insoluble Hg phases (e.g., HgS). The best model was considered that with the closest fit to the experimental data. The various potential inorganic complexes, including sulfide and polysulfide complexes, and Hg-NOM interactions, which have been characterized to differing degrees in the literature, were constrained based on the concentrations and measured

K_D values for Hg, and the concentrations of ancillary parameters in the sediments and porewater ($\sum S^{2-}$, AVS, CRS, POM, DOM) of the Chesapeake Bay and shelf. In estimating the polysulfide levels it was assumed that there was solid elemental S in the sediment, allowing these complexes to form [49, 140]:

$$HgS\,(s) + HS^- + (n-1)S^0 = HgS_nSH^- \quad \log K = -3.9$$

While the equation is written in terms of the solid HgS, it is not necessary for it to be present for polysulfides to form. Thus, this equation could be written in terms of Hg^{2+} or another species using the relationships discussed previously as:

$$HOHgSH^0 + HS^- + (n-1)S^0 = HgS_nSH^- + H_2O \quad \log K = 6.1$$

So, assuming the elemental sulfide has an activity of 1, the concentration of the polysulfide complexes is greater than the neutral complex concentration around 1 μM HS^-:

$$[HgS_nSH^-]/[HOHgSH^0] = 10^{6.1}[HS^-]$$

Additionally, there is a deprotonated form and it is possible for substitution with OH^- [49]:

$$HgS_nSH^- = HgS_{n+1}^{2-} + H^+$$

$$HgS_nSH^- + OH^- = HgS_nOH^- + SH^-$$

The latter species is only dominant at high pH (>9) and therefore is not relevant for marine and most environmental situations.

These equations suggest that it is possible to have high concentration of dissolved Hg in porewaters but this has not been found in field studies. For example, recent studies in the Chesapeake Bay/shelf the measured concentrations were always <100 pM [108]. Data from one estuarine and one of the shelf stations from this study are shown in Fig. 6.20 for total Hg, reduced sulfide and organic content of bulk sediments and porewater. The constants that were used in the calculations that were performed using MINEQL+ are given in Table 6.8, and their choice is discussed in various publications [108, 141, 142]. As noted earlier in Section 2.3, there is debate in the literature over the value of the constant for the formation reaction of HOHgSH ($Hg^{2+} + H_2O + HS^- = HOHgSH + H^+$) and what this reaction truly represents, and the value used here was chosen to best fit to the data, both partitioning and relationships to Hg methylation. As noted earlier, the larger value could represent the formation of Hg-S nanoparticles, stabilized in solution by organic matter, as this colloidal material would be measured as "dissolved" by the analytical techniques. Comparison of model and empirical data for a number of ecosystems suggest that this is a reasonable simulation of the field situation [108] (Section 2.3). Additionally, the differences between the two constants are of similar order to the differences between the constants for M_xS (s) and their related clusters in Table 6.3, suggesting that Hg is also likely to form clusters in environmental solutions and that these entities are stable in the environment.

Hollweg [139] used the thermodynamic data to predict the K_D based on the dissolved binding constants and the constants for Hg association with the solid phase, and showed that this approach could provide a reasonable estimation of the complexation and competition between binding sites in solids and porewater across a large geochemical gradient (freshwater to ocean salinity, low to micromolar sulfide, large gradient in NOM), except for the sites with high sulfide levels (>50 μM $\sum S^{2-}$) (Fig. 6.21). For these sites, allowing the formation of polysulfides resulted in underestimation of the K_D by several orders of magnitude and this correlates with the discussion above; that is, the lack of correlation was due to the prediction of nM porewater concentrations by the model. In the absence of polysulfides, the dissolved complexation of Hg was dominated by association with NOM and with dissolved sulfide species, even with the model including binding with Cl^- and OH^- [139].

Solid phases that were modeled included the organic matter fraction, with the thiol content estimated based on the measured organic C : S ratio of the bulk organic material and assumptions concerning the fraction of the reduced S that was present as thiols and available for metal complexation [108, 139], which was estimated from the literature, for example, [143]. The ratio of available thiol binding sites to total organic S was 0.06 [108]. Both solid-phase amorphous FeS (estimated based on the AVS concentration) and pyrite (estimated from the CRS concentration) were included in the model as Hg binding sites. Literature constants for the association for Hg binding with FeS were used [144]. From the work of Morse and colleagues, the extent of association of Hg with pyrite (based on Hg DTMP) was derived [145, 146] (Table 6.8). In the model it was assumed that Hg associated with pyrite was unreactive that is, it was not available to partition into the porewater. The model was run to estimate the sensitivity of the results to the various binding phases and to determine which solid phase and dissolved species dominated in the estuarine and coastal environments of the Chesapeake Bay region. To compare the model results and the measured data, partition coefficients were determined by summation of the total Hg in the particulate and dissolved phases (Fig. 6.21).

While the model did not provide an accurate depiction of the observed partitioning between the solid and dissolved phases for all environmental conditions, it is consistent with other recent modeling attempts that have included polysulfide-Hg interactions [142]. Furthermore, lab studies do demonstrate nM filtered Hg concentrations in the presence of HgS solid and elemental sulfur [49]. Therefore,

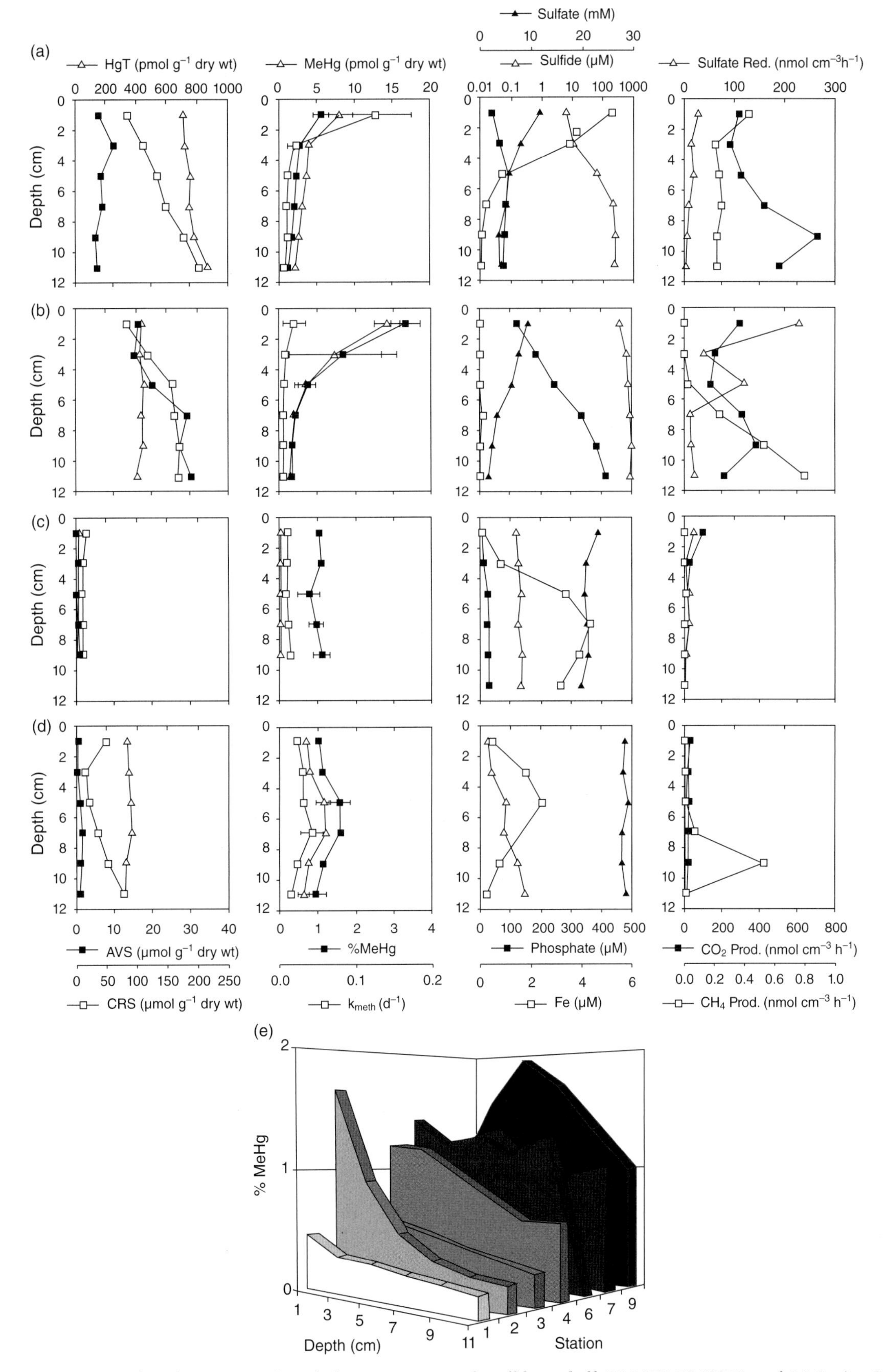

Fig. 6.20 Concentrations of total mercury and methylmercury (MeHg), and % MeHg in sediments with depth, and ancillary parameters (dissolved and solid phase reduced sulfide (AVS and CRS), iron, sulfate and phosphate) and microbial activity (sulfate reduction rate, carbon dioxide and methane production rate) and methylation rate (k_{meth}) for (a) Station 3, and (b) Station 2, Chesapeake Bay site within the mid-Bay region (respectively, 38.56°N, 76.48°W and and 38.56°N, 76.44°W); (c) Station 6, on the offshore shelf (37.09°N, 75.70°W); and (e) Station 9 on the slope (36.33°N, 74.72°W). Also shown in (e) is the vertical distribution in %methylmercury for sites ranging from Station 1 (freshwater) to Stations 6 and 7 on the shelf, and Station 9 on the slope (~600 m). Taken from Hollweg et al. (2010) *Limnology and Oceanography* **55**: 2703–22 [116] and reprinted with permission of Elsevier.

Table 6.8 Equilibrium constants used in the modeling of mercury partitioning in sediment porewater. Summarized from [108, 139], and references therein.

Complex	LogK	Complex	LogK
Dissolved			
$HgOH^+$	10.7	HOHgSH	40.5
$Hg(OH)_2$	22.2	$Hg(SH)_2$	37.7
$Hg(OH)_3^-$	20.9	HgS_2H^-	31.5
$HgCl^+$	7.2	HgS_2^{2-}	23.2
$HgCl_2$	14.0	HgS_x^{2-}	−11.7
$HgCl_3^-$	15.1	$Hg(SR)_2$	42
$HgCl_4^{2-}$	15.4		
HgOHCl	18.1		
Solid		*Solid*	
$RFeSHg^+$	35.1	$Hg(SR)_2$	42
$RFeS_2$-Hg	f(CRS)*		

Notes: # Solid thiol content was estimated using literature information for the fraction of organic matter that is reduced, and the fraction of reduced S as thiol; a ratio of RSH/total OS of 0.06.
**The relationship between Hg incorporation and pyrite content was estimated based on data for the degree of trace metal pyritization. The following relationship was derived: DTMP-Hg = 0.157ln(CRS) + 0.248.*

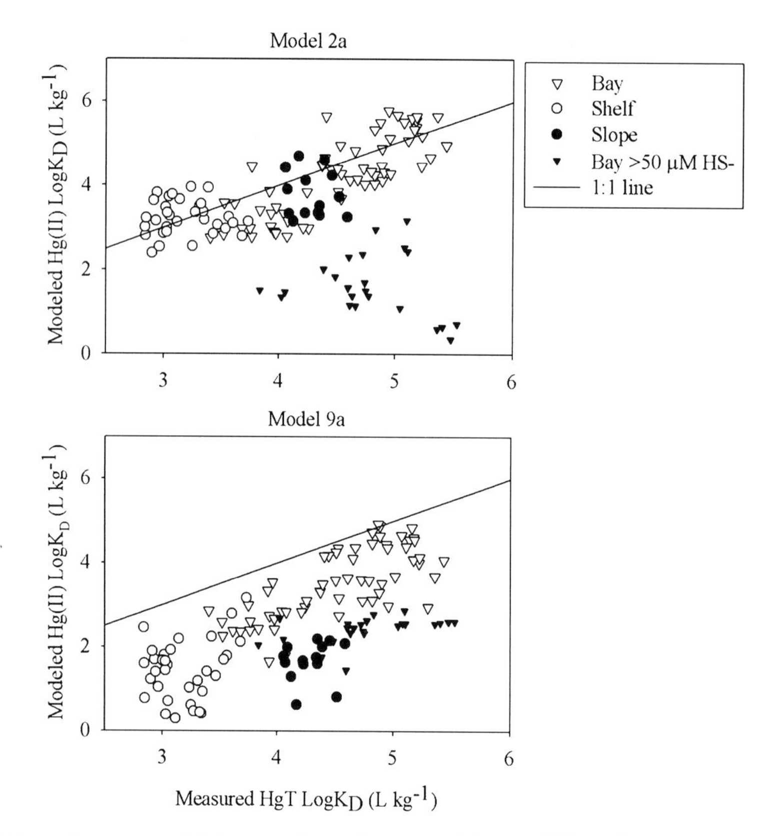

Fig. 6.21 Comparison of measured versus modeled porewater-sediment partition coefficients for mercury across Stations 1 (freshwater end of the Chesapeake Bay), 2–4 within the Chesapeake Bay, 6 and 7 on the shelf and 9 on the slope for all seasons and years of sampling. Two different model results are shown, as discussed in the text. Taken from Hollweg (2010), PhD thesis, used with permission of the University of Connecticut [108].

either the complexes that form in the lab do not form in the field, or they represent the formation of nanopolysulfide entities that don't form in the environment, or the assumption that the activity of S in the sediments is unity is incorrect. The model results suggest that the activity of S is around 0.01 and using this value would significantly improve the relationship. Alternatively, it is possible that at high sulfide concentrations that the filtered fraction is dominated by colloidal material (FeS/NOM and other metal nanoparticles) and that the partitioning is no longer controlled by complexation to surface sites but is determined by the relative concentration of colloidal particulates compared to solid phase material, and the degree of association of Hg to these phases. This idea needs to be investigated further.

While the model could not predict the partitioning across all sites, it could provide a reasonable prediction of the methylation rate constant (k_{meth}). There was a strong correlation between the rate constant and the estimated value for the neutral species composition, using the higher value for the equilibrium constant. The relationship was less strong if the lower value was used for the equilibrium constant. The following relationship was developed for the sediments of the Chesapeake Bay and adjacent shelf and slope, and included both the estimated speciation and the sulfate-reduction rate (SRR) as sulfate-reducing bacteria are considered the primary methylators in these sediments:

$$k_{meth} = HOHgSH{*}0.00121 + SRR{*}0.00159 + 0.004$$
$$(p < 0.0001;\ \text{adjusted } r^2 = 0.58)$$

The formulation of this relationship is similar to others that have been developed to examine the methylation rate in coastal systems [147].

As discussed in Section 6.2.3, there is the potential that the fraction that is modeled using the larger value for the equilibrium constant includes both truly dissolved and colloidal HgS species. Therefore it may be more appropriate to designate this fraction as $\Sigma HOHgSH_{filtered}$ or with some similar notation. It has been recently demonstrated using laboratory culture experiments that Hg is methylated even in the presence of these colloidal Hg-S species, although at a slower rate that in the presence of dissolved inorganic Hg [148]. It is probable that the species being taken up are still dissolved complexes but that there is a rapid partitioning between the dissolved and colloidal phases (colloidal pumping) that supports the methylation observed. Such an idea is consistent with the studies of Slowey and others [61, 62].

6.3.5 The importance of sediment resuspension and extreme events in coastal metal dynamics

The importance of net sedimentation as a sink for metals in estuarine and other ocean environments depends on the degree of return of metals to the water column in the dissolved phase (discussed in Section 6.3.2) or through remobilization of particulate material into the water column. Such mobilization can be driven by wind and tidal effects in shallow ecosystems and macrotidal estuaries, or can be the result of aperiodic events such as seasonal storms, spring runoff and high flow events, and in some locations, due to the impact of hurricanes and other large storm systems (extreme events). Sediment resuspension can also result from anthropogenic activities (e.g., dredging and trawling). Sediment resuspension takes place when the bottom shear stress is sufficient to disrupt the cohesion of the bottom materials and is a function of the properties of bottom sediments, such as grain size, type of sediments, organic content, and water content. Once resuspended, particles tend to resettle by gravity when the shear stress diminishes and the residence time of the particle in the water column depends on its size and density.

The impact of such resuspension events on metal dynamics is driven by a number of physical and chemical parameters such as: (1) resuspension and/or storm intensity and duration, and event frequency; (2) sediment particle size, which will determine its relative transport prior to deposition; (3) particle chemistry as oxidation of phases during transport may result in metal release; and (4) degree of repartitioning of metals between the dissolved and particulate phases. The release of metals during resuspension is difficult to study in situ and therefore a number of approaches have been used to investigate these processes within the laboratory or at the mesocosm scale. Since sediment resuspension in shallow ecosystems controls the movement and redistribution of particles, it can play a major role in the mobility and bioavailability of metals. Additionally, in some locations, and especially in highly dynamic estuarine systems, sediment that has been temporarily deposited can often be resuspended and transported as a result of high flow events (spring runoff events) and storms. This has been documented in the Hudson and Delaware Rivers in the USA [149, 150] and elsewhere. In these locations, sediment resuspension and transport is important even though the site may not be a location of net sediment accumulation.

Some laboratory studies have shown that resuspension of sediments results in the release of organic contaminants and trace metals into overlying water [151–154]. In contrast, others concluded from their small reactor experiment that surficial sediments were not significant sources of trace metals to water column when resuspended [155]. These authors postulated, however, that this might not be applicable to anoxic sediments from deeper layers because of the potential for oxidative release of metals. However, it has been shown that while resuspension and oxidation of sulfidic sediment may initially result in a significant release of Fe, Mn, Cd, Cu, and Pb from the solids, Cu and Pb were rapidly scavenged by the (hydr)oxide phases that form as a result of the oxidation of Fe and Mn. This was not the case, however, for Cd as 50% remained in the dissolved fraction even after this precipitation [156]. This result is consistent with the fact that Cd forms relatively weak complexes with oxides phases and is strongly complexed by chloride in saline waters.

Another study examining the effects of dredging on mobilization of trace metals (Zn, Cu. Cd and Pb) found that dissolved trace metal concentrations were not increased even though there was more metal associated with the suspended material [157]. Studies examining Hg and CH_3Hg in mesocosms in the presence of periodic resuspension found similar results [158]. Overall, the results of these laboratory experiments likely overestimate the impact of resuspension on metal release as they are typically run for short duration, and also the shear stress and mixing and the TSS is often exaggerated compared to the environment.

Additionally, upon resuspension, a steady state is not instantaneously obtained. In a laboratory experiment examining the partitioning of radioactive trace metals between seawater and particulate matter, it was found that for a group of elements (e.g., Na, Zn, Se, Sr, Cd, Sn, Sb, Cs, Ba, Hg, Th, and Pa) constant distribution coefficients were reached after 2–3 days. In contrast, another group of elements (e.g., Be, Mn, Co, and Fe) showed an increasing distribution coefficient over the whole experimental observation time (108 days), indicative of the dynamic interactions between phases [159], as noted in Section 6.2.1.

To further examine the kinetics of these processes, during a mesocosm study with induced tidal resuspension (4 h on/2 h off), investigators added a dissolved spike of a stable Hg isotope (^{199}Hg) to the water column to examine the dynamics of Hg removal from the water column [160]. The spike was added during a period of resuspension and within 12 h most of the added spike was associated with the particulate phase (Fig. 6.22). However, the isotope was still detected in the water column three weeks later (in the particulate), indicating that there is continual interaction between the resuspended particles and the filtered phase due to adsorption/desorption process and colloidal/particle interactions. Interestingly, over time, there was a decrease of the total Hg isotope load (particulate + filtered) in the water column, suggesting its burial in surface sediments and its removal over time from the resuspended layer. The half-life of the Hg spike in the water column was about 1.3 days. These results confirm the notion that while only a small depth of sediment is resuspended (a few mm typically), there is sufficient re-mixing of particles within the upper sediment layers such that it is not the same particles that are resuspended over time.

One approach that mimics reality in terms of resuspension is the use of erosion chambers that apply a range of shear stresses, which can be matched to those typically encountered in coastal environments. The particles and metals released as the shear intensity is increased can therefore be assessed for each shear stress. One such study of sediments in Boston Harbor, MA, USA [161] showed that, at low shear stresses, the initially released sediment particles were enriched up to 50 times in metal content (Ag, Cu, and Pb) compared to the bulk sediment (Fig. 6.23a). As the shear-stress level was increased in the chamber, the total eroded mass increased, but particle metal enrichment decreased. Another study using sediments from the Gulf of Lyons and using a similar erosion device suggested that these differences are due to differences in grain size, and the propensity of metals to be associated with the finer particles [162]. Similar studies using sediments taken from the mesocosm experiments discussed above also showed an enrichment in the surface layers of the sediment, and this reflected the fact that biological material (as indicated by Chl a) was maximal in the top mm of the sediment (Fig. 6.23b). The surface layer sediments had enhanced CH_3Hg concentrations, suggesting that resuspension could result in metal enrichment that could enhance the uptake of metals into the benthic food chain, especially through surface deposit feeders, such as amphipods. In the Boston Harbor study, resuspension fluxes were estimated based on an erosion model, which included the intensity and frequency of different types of resuspension events, and the chamber results (Kalnejais et al. [161]). It was concluded that in this environment, most of the metal flux due to resuspension occurred during relatively low energy events, and the fluxes were higher than those associated with riverine inputs.

Extreme events, such as hurricanes, can also result in metal release and/or metal redistribution within the ecosystem. A number of studies were initiated in the Gulf of Mexico after the passage of the two major hurricanes, Katrina and Rita, in 2006. It was found that post hurricane sediment deposition throughout the impacted region resulted in a storm layer that was up to 20 cm thick [163], as found in previous studies [164]. The storm deposits were generally composed of silty clays with a coarser, somewhat sandy 1–2 cm basal layer. Surface sediments from the storm layer were characterized by relatively high mineral surface areas (SA of 30–50 $m^2 g^{-1}$) and elevated OC contents (%OC of 1.0–2.0%), and marked differences in the SA and %OC in upper sediments was found at offshore locations between these deposited sediments and the prior sediments (SA 5–15 $m^2 g^{-1}$; %OC 0.2–0.6%). Analysis of the sediment characteristics reinforce the idea that the source of the storm deposits was the finer fraction of resuspended seabed sediments, with little evidence for inputs from local land-derived sources or autochthonous algal production. Data from other studies reinforce this conclusion [165]. Overall, the magnitude of sediment and organic matter deposition on the seabed after the storm greatly exceeded the annual

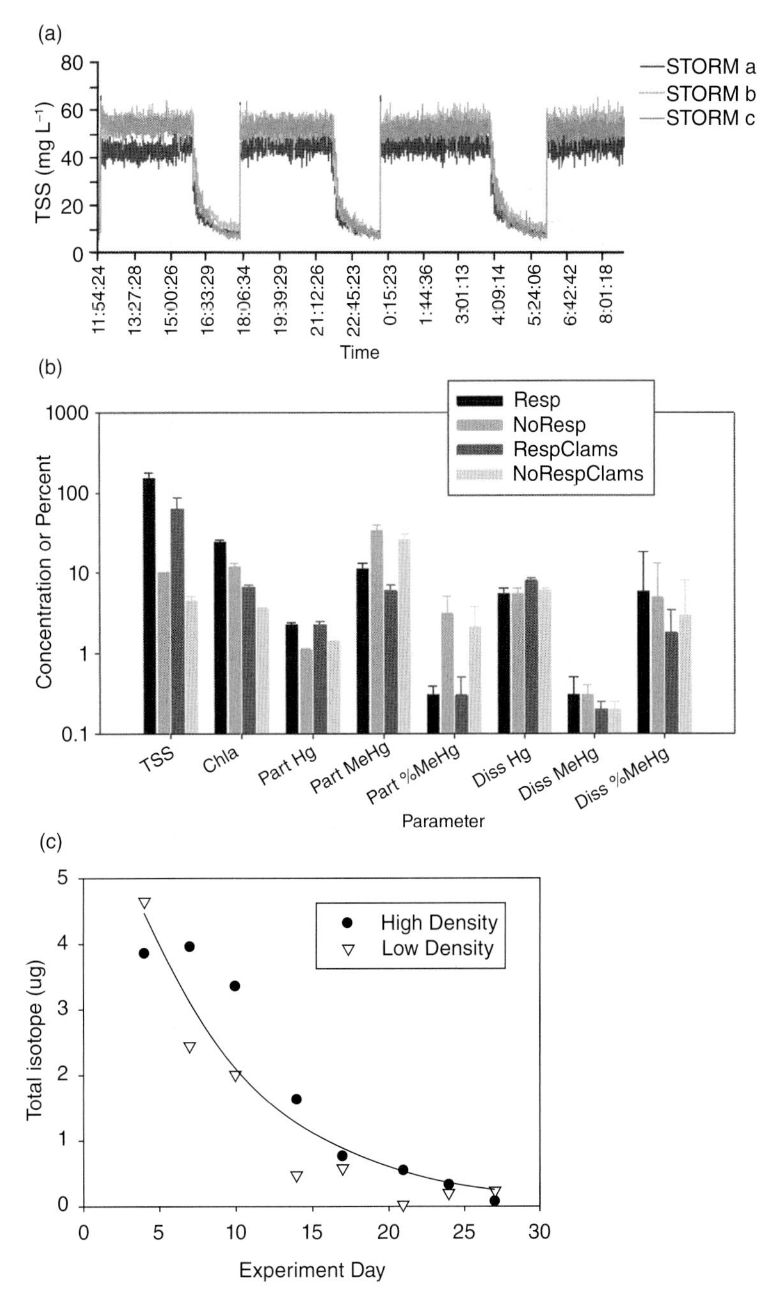

Fig. 6.22 Changes in the concentration of total mercury and methylmercury in the dissolved and particulate phases during mesocosm experiments with simulated tidal resuspension: (a) the change is suspended particulate loading over the simulated tidal cycle; (b), the average concentrations in triplicate mesococms for each treatment with or without resuspension and with or without added clams in the mesocosms; in (c), the rate of change in water column total concentration of a spike Hg addition to the mesocosm is shown for mescosms with resuspension and different densities of clams. Data taken from Kim et al. (2004) *Marine Chemistry* **86**: 121–37 [158] and from Bergeron (2005), MS thesis, University of Maryland [160] and figures created using data from both sources.

inputs from the Atchafalaya River and coastal primary production.

Elevated levels of metals were also observed in post-hurricane measurements of oyster tissue in the Gulf of Mexico when compared to the 20-year historical record for most sites, but concentrations were only significantly higher post-hurricane for Ni and Pb [166]. A comparison of the chemical distribution before and after the hurricanes indicated an overall increase of metal distribution after the storm, with 76% of the locations having metal levels that exceeded the 20 year median concentrations.

A study of Hg distribution as a result of the hurricanes reinforced the data of Goni et al. [163] that these hurricanes resulted in a major redistribution of sediment in the shallow

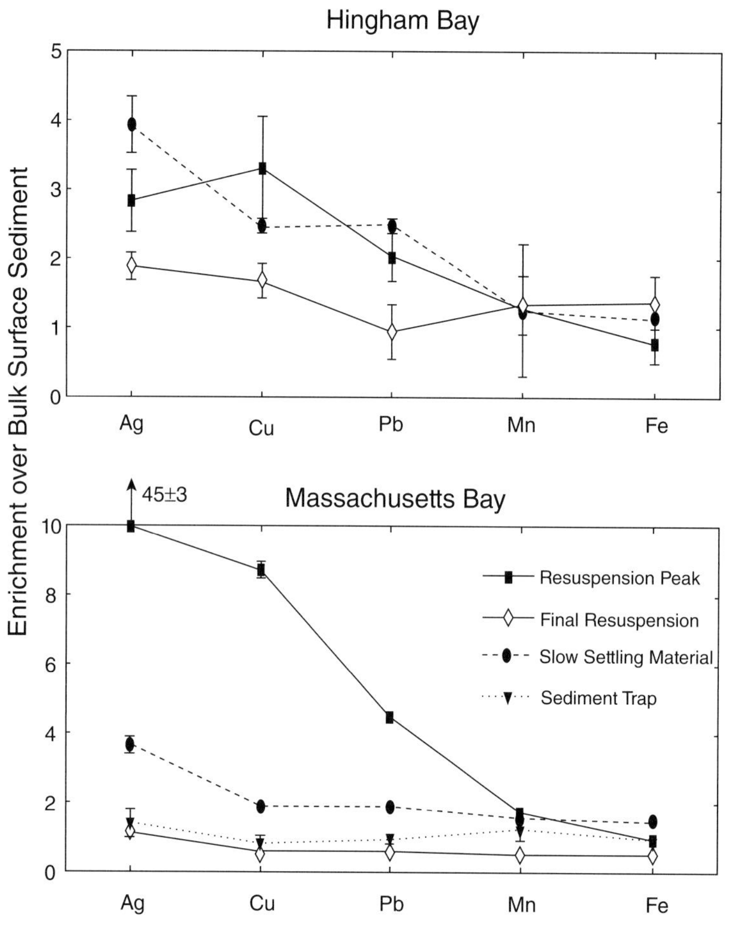

Fig. 6.23 Particle metal contents normalized by the average bulk surface sediment content (Table 6.1) for both Hingham Bay and Massachusetts Bay (MB). The particles shown are as follows: (1) the resuspension peak (solid square) – particles resuspended at or one stress step beyond the erosion threshold, (2) final resuspension (open diamond) – the resuspended particles sampled at the maximum shear stress, (3) slow settling material (solid circle) – eroded particles in suspension after 8 h, and (4) sediment trap (solid triangle) – particles collected in USGS sediment traps moored 16 m above the seafloor, 4 km from the MB sampling site, 2001–03 data. Reprinted with permission from Kalnejais et al. (2007) *Environmental Science and Technology* **41**: 2282–8. Copyright (2007) American Chemical Society.

Gulf region [167]. Firstly, sediment analysis and isotope measurements (^{210}Pb and ^{137}Cs) confirmed substantial differences in the upper sediments before and after the hurricanes for sites directly (Site A'2) or within close proximity of the path of Hurricane Katrina (Sites CB6 and D3) (Fig. 6.24). The differences over time before and after the hurricane were distinct at Site A'2 [167]. Further sediment analysis using X-ray fluorescence (XRF) of a suite of major and minor elements, and principal component analysis of the sediment data confirmed the differences in the upper sediments relative to the deeper sediments at Station A'2 (Fig. 6.24b), and also showed differences at the other stations (data for Station CB6 shown) [167]. However, these data also suggest that there was substantial sediment redistribution after the hurricanes such that the concentrations of Hg and CH_3Hg, and the other metals, were similar in the surface layers at Station A'2 in March and July 2006 compared to that of July 2005, while at the other stations there appeared to be changes in sediment characteristics after the hurricanes. This suggests substantial redistribution of sedimentary material that had been temporarily deposited in these locations by the hurricanes.

Overall, the field and laboratory studies discussed provide substantial evidence that extreme events can have a large impact on metals distributions and fate in the coastal environment. The mesocosm experiments and the data from the Gulf of Mexico suggest that sediment resuspension can enhance the methylation of Hg, and could potentially increase the transfer of metals through the coastal food chain. The overall importance of such events cannot be fully realized as there is insufficient data available at present to accurately determine their potential impact globally.

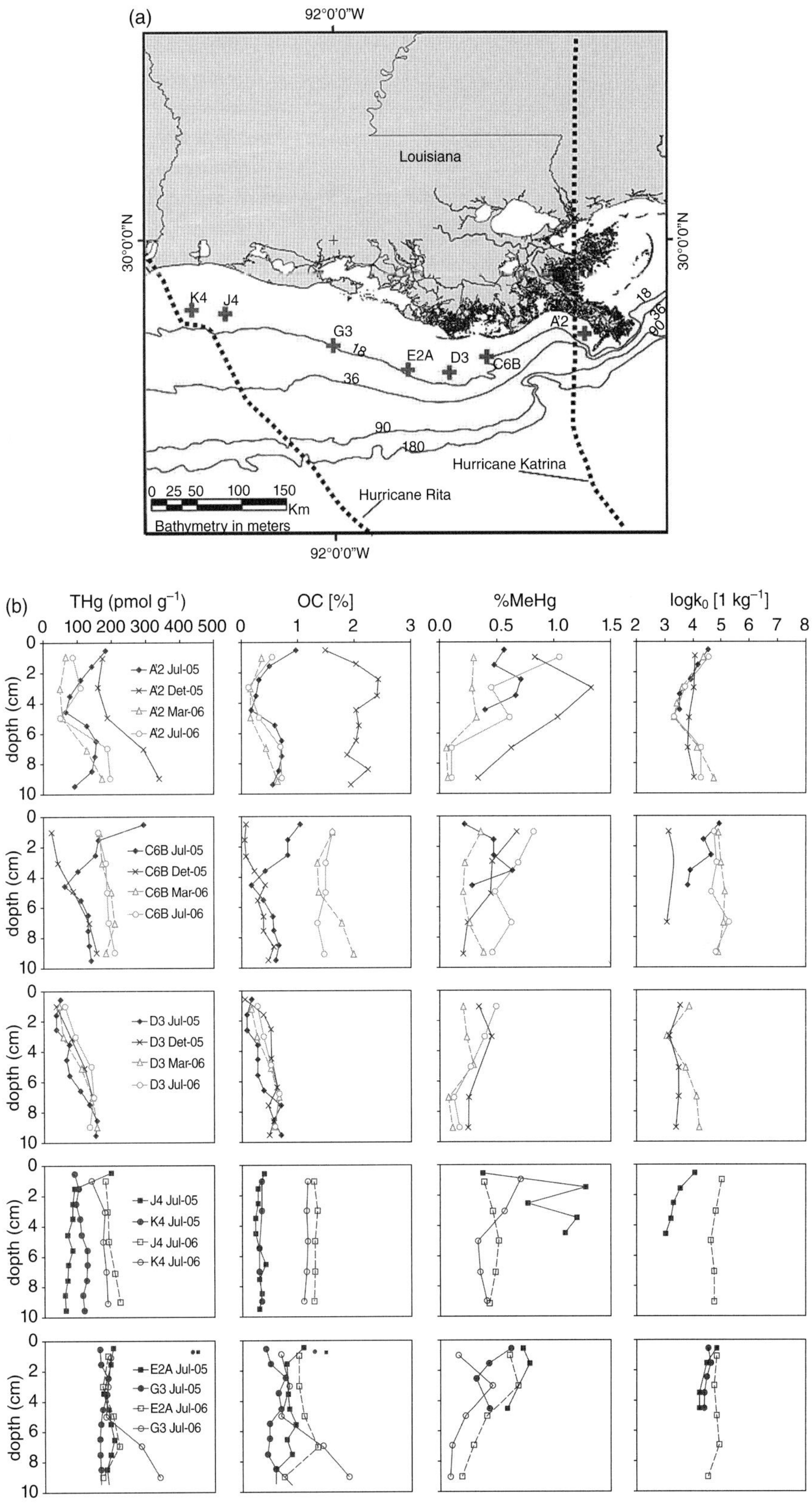

Fig. 6.24 (a) A map showing the locations of the various sampling stations; (b) the concentrations of total organic carbon, total mercury and %methylmercury and log K_D for sediments collected between July 2005 and July 2006, before and after hurricanes Katrina and Rita; (c) principal component analysis showing the relationship between the major component (PC1) and the bulk parameters measured at three sites shown over time. The sites and depths that are outliers from the general trend are highlighted. Figure reprinted from Liu et al. (2009) *JGR-Biogeosciences* **114** [167] with permission of the American Geophysical Union, copyright (2009).

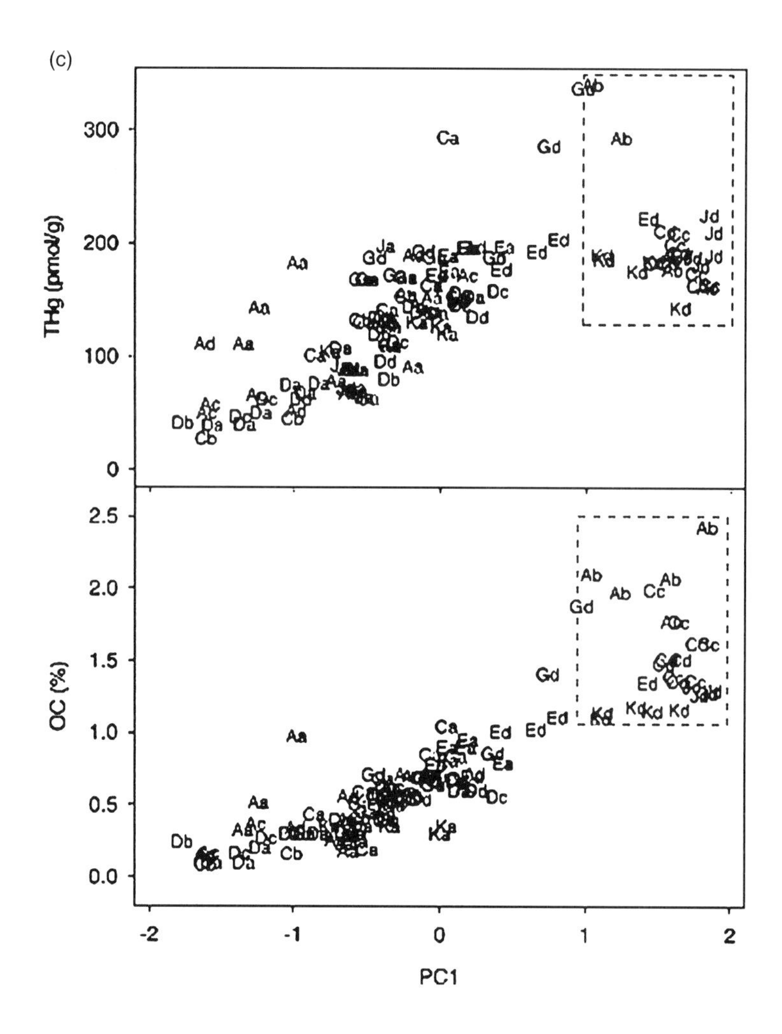

Fig. 6.24 (Continued)

6.3.6 Deep ocean sediments and manganese nodules and accretions

Deep ocean sediments are classified in different ways but range from sediments that are mostly derived from external inputs (*cosmogenic*, atmospheric and terrestrial inputs) to those that are highly modified and dominated by deep ocean inputs from hydrothermal systems [84]. Cosmogenic (extraterrestrial) material is enriched in Fe, Ni, Mg, and Si and is typically a small fraction of the sediment component. *Lithogenic* (terrestrially-derived) material forms the basic component of deep pelagic clays, which are primarily aluminosilicates, in regions where biogenic inputs are not important. In regions underlying waters with high surface productivity, the sediments are either dominated by silicates (siliceous sediments) or carbonates (calcerous sediments). Siliceous sediments dominate in regions with highly productive surface waters such as the equatorial Pacific, the Southern Ocean, and the North Pacific where surface productivity is dominated by diatoms and other Si-forming organisms [84]. Calcerous sediments occur in locations where surface productivity is dominated by coccolithiophores, which precipitate carbonate skeletons, and in regions shallow enough that carbonate dissolution does not occur: $CaCO_3$ behaves anomalously to other solids in that its solubility increases with decreasing temperature. High pressure also increases the solubility. Therefore, while precipitation may occur and be thermodynamically favorable in surface waters, dissolution will occur at sufficiently deep and cold locations. Oceanographers call the depth at which dissolution and deposition are equal as the *calcite compensation depth*. This depth varies for different minerals; calcite (4000–6000 m, depending on ocean basin) or aragonite (around 3000 m).

The major components overall of pelagic marine sediments are aluminosilicates (shale), manganese oxides, biogenic materials (carbonates, silica and barite), zeolites and phosphorites [84]. Phosphorites precipitate mainly on continental shelves and slopes and under highly productive upwelling regions. Volcanic ash and hydrothermal inputs are not major components but can dominate in some

Table 6.9 Compiled average data for different types of deep ocean sediments showing the contrasting concentrations of both major and minor elements. Concentrations given either as a percent value (%mass/mass) or as μmol kg^{-1}. Summarized from [84] and converted to molar units where appropriate.

Element	Pelagic Clay	Fe/Mn Nodule	Basal Sediment	Ridge	Element	Pelagic Clay	Fe/Mn Nodule	Basal Sediment	Ridge
Al	8.40%	2.70%	2.74%	0.50%	*Y*	449.9	1687.3	1439.8	–
As	267.0	1869.2	–	1935.9	*Zr*	1642.9	6133.6	2464.4	–
Bi	2.5	33.5	0.8	–	*Nb*	150.7	538.2	54.9	–
Cd	3.7	89.0	3.6	35.6	*Pd*	0.1	0.1	–	0.2
Co	1256.4	45,840.4	1392.2	1782.7	*La*	302.4	1130.3	705.5	208.8
Cr	1730.8	673.1	288.5	1057.7	*Hf*	23.0	44.8	9.0	–
Cu	3937.0	70,866.1	12,440.9	11,496.1	*Ta*	5.5	55.3	11.6	–
Fe	6.50%	12.50%	20%	18%	*W*	21.8	544.1	–	–
Ga	286.9	92.7	63.0	–	*Re*	0.002	0.005	–	–
Ge	22.0	7.4	30.6	–	*Os*	0.07	0.01	–	–
Hg	0.5	1.4	–	3.7	*Ir*	0.002	0.04	–	0.0
Mn	0.67%	19%	6%	6%	*Pt*	0.03	1.0	–	–
Mo	281.5	4171.0	0.0	312.8	*Ag*	1.0	0.8	1.7	57.5
Ni	3918.2	112,436.1	7836.5	7325.4	*Au*	0.02	0.02	–	0.2
Pb	386.1	4343.6	482.6	733.6	*Lanthanides*				
Sb	8.2	328.4	139.6	–	*Ce*	720.9	3783.0	242.7	60.0
Sc	422.6	222.4	–	–	*Pr*	71.0	255.5	137.0	0.0
Se	2.5	7.6	32.9	–	*Nd*	298.2	1095.7	603.3	159.5
Sn	33.7	16.8	5.1	–	*Pm*	–	–	–	–
Th	56.0	129.3	10.3	–	*Sm*	55.5	199.5	123.7	33.2
Ti	22,504.9	32,778.9	–	1174.2	*Eu*	12.2	59.2	35.5	9.9
U	10.9	21.0	17.6	92.4	*Gd*	52.8	203.4	143.7	38.1
V	2357.6	9823.2	–	8840.9	*Tb*	8.9	34.0	0.0	0.0
Zn	2599.4	18,348.6	7186.5	5810.4	*Dy*	45.5	190.8	127.4	44.9
Group I and II					*Ho*	9.1	42.4	28.5	0.0
Li	8213.3	11,527.4	18,011.5	–	*Er*	24.5	107.6	77.1	33.5
B	21,296.3	27,777.8	11,388.9	46,296.3	*Tm*	3.4	13.6	0.0	0.0
Be	288.6	277.5	743.6	–	*Yb*	22.1	1156	75.1	32.9
Rb	1286.5	198.8	187.1	–	*Lu*	3.1	10.3	12.6	5.0
Cs	45.1	7.5	0.0	–					
Sr	2054.8	9474.9	4006.8	–					
Ba	16,751.6	16,751.6	45,375.1	43,699.9					

locations. Many of the cations (transition metals, heavy metals, Groups I and II) are found associated with Mn phases (Table 6.9). It is thought that Mo forms $MnMoO_4$ which results in its elevated concentration, and W, and other oxyanions, may also be incorporated in a similar fashion. Many cations and anions are associated with Fe oxide phases. Overall, the relative association of elements with these phases can be explained in terms of physicochemical properties and the dynamics of their precipitation and dissolution.

The concentrations of metals and minor elements in ocean sediments vary over many orders of magnitude. Data for pelagic clays, average Mn/Fe nodules, basal sediments and ridge sediments are gathered in Table 6.9 as representative of the range in sediment concentrations. The data are gathered alphabetically for the more common trace metals and metalloids that have been discussed in the book. Additionally, data for the rarer elements from the second and third row of the transition series, and the lanthanides, as well as the heavier elements from Groups I and II are included in the table. The concentrations are given as percentage values for the major elements (Al, Fe, and Mn) and as μmol kg^{-1} for all the other elements. Concentrations range from low values, of <0.1 μmol kg^{-1} for many of the heavier transition metals (e.g., Au, Re, Os, Ir, Pt), to 6.5% for Fe and 8.4% for Al for pelagic clays. Besides Mn (0.7%), most of the remaining first row transition metals exist in pelagic clay sediments in the 1–4 mmol kg^{-1} range.

Overall, there is a reasonable correlation (log-log) between the elemental concentrations in the different types of marine sediment. The correlation between pelagic clays and Mn/Fe nodules is shown in Fig 6.25. It is apparent from this plot

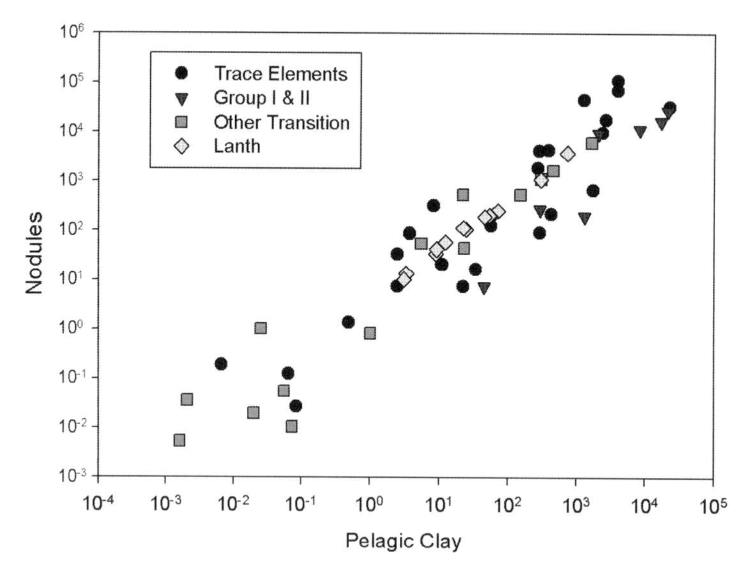

Fig. 6.25 The correlation between the concentrations of elements in pelagic clays versus that of nodules. Data from Lui and Schoonmaker (2004), re-plotted in this original figure.

that concentrations are enriched or depleted to some degree for most metals (molar ratio nodule/clay concentration 0.1–10). Some elements are enriched to a higher degree in nodules (Pb, Ni, Mo, Mn, Cu, Cd, Bi, W, ratio 10–30) with Sb and Co being highly enriched (ratio >30). It is apparent as these elements include elements that exist as both cations and anions in seawater that enrichment is not entirely a function of element chemistry but must result to some degree from differences in source profiles – pelagic clays represent to a large degree external inputs to the ocean whereas nodules are formed due to redox-related processes, and due to the supply of Mn from hydrothermal sources. Enrichment in elements that are enriched in hydrothermal vent solutions may account to some degree for the differences in concentration between these two sediment materials. In summary, while the data is more limited and except for a few anomalies, the molar ratio of elements between ridge and nodules is close to 1, suggesting similar sources and removal processes.

As it is not possible to discuss in detail all the sediment types and the chemistry of the elements in each, discussion that follows is focused on the mechanisms of formation of Mn nodules and the impact of various processes on their metal content. Hydrothermal systems are discussed in more detail in the following section given their unique biogeochemistry and the important reactions that occur. Manganese and Fe oxides both play an important role in ocean trace metal dynamics, especially in regions where the other dominant metal-binding solids (organic matter and sulfide phases) are present at low concentrations. As noted earlier, the degradation of organic matter is primarily through aerobic processes in deep ocean sediments and therefore there is little sulfide mineral formation.

In addition, the destruction of organic matter is essentially complete and ocean deep sediments have a very low organic content. Therefore, oxide minerals provide the main binding phase for many trace metals. The surface area of MnO_2 is ~260 $m^2 g^{-1}$ and its low pH_{zpc} results in it being able to bind cations effectively [168]. Similarly, Fe oxides have a large surface area, as discussed in Chapter 3. However, in the deep ocean, Mn is a more important binding phase because of differences in the redox cycling of Fe and Mn – Mn is reduced more readily, but its rate of oxidation is slower [169]. Therefore, Mn and Fe are separated spatially in environments where diagenetic recycling occurs, or where there is differences in the precipitating phases in deeper sediments. Additionally, ~90% of the Mn added to the deep ocean originates from hydrothermal inputs and the concentration anomalies that results from such inputs can be detected thousands of km from the source. In contrast, most of the Fe that is emitted is rapidly precipitated (within minutes), primarily as sulfide minerals and many of the heavy metals (Pb, Hg, Cd, Zn, Cu) are removed as well close to the source. The remaining Fe is oxidized and removed within the near field of the source, along with many of the oxyanions (e.g., As, Cr, V, P) and the rare earth elements (REE). However, given the slow oxidation rate of Mn, most of the Mn is oxidized and deposited within a few hundred km of the source, but given its enrichment in hydrothermal fluids, the remaining fraction still results in an elevation in concentration at greater distances [168].

The Mn-rich deposits that result from the various processes of precipitation and formation in the deep ocean vary widely in the relative concentrations of the transition metals and in the ratio of Fe/Mn, compared to crustal abundances

Table 6.10 Concentrations of various metals in manganese nodules and related materials compared to those found in pelagic clay and average shale. Data taken from [168, 169].

Element	Average Shale	Pelagic Clay	Mn Nodules	Mn Crusts (Co–Rich)	Baltic Fe/Mn Concretions	Hydrothermal Crusts
% Mn	0.05	0.43	17–33	20–28	8.7–29	54
% Fe	5.2	5.4	7–23	14–17	10–23	0.07
% Co	<0.001	0.011	0.09–0.44	0.67–1.2	~0.01	–
% Ni	0.003	0.021	0.23–1.4	0.24–0.50	<0.01–0.08	0.03
% Cu	0.005	0.023	0.17–1.2	0.03–0.10	<0.01	0.02
Mn/Fe	0.01	0.08	0.73–5.4	1.2–2.0	0.43–2.9	774

(Table 6.10). Red clays, which cover about 30% of the world's ocean floor, are slightly enriched in Mn and some transition metals relative to Fe, when compared to average shale. The enrichment reflects the impacts of diagenesis on Fe and Mn distributions and the enhanced trapping of the transition metals by the oxide phases in these low organic sediments.

The main deposits of Mn in the marine environment are classified as: (1) Mn *nodules* which generally form in deep water (>4 km) where sedimentation is low (<5 mm kyr^{-1}) and are found primarily associated with red clay deposits; (2) Mn *crusts* which form on seamounts and plateaus in >1 km of water, in locations where currents prevent sediment accumulation; and (3) Fe/Mn *accretions* which form in shallow water such as coastal seas. Formation is either of: (1) *hydrogeneous deposits*, which form directly through precipitation from an oxidizing seawater environment; (2) *diagenetic deposits* which form as a result of diagenetic processes in the underlying sediments; and (3) *hydrothermal deposits*.

Hydrogeneous deposits are slow growing (~2 mm every 10^6 yr) and they are mainly formed as a result of the relatively high Mn/Fe ratios of deep waters. Their Mn/Fe ratio is ~1 (Table 6.10). Nodules are found in all ocean basins and their location is controlled to some extent by deep water flow and by the availability of nuclei (e.g., bone fragments, teeth, bits of volcanic rock) around which they form. The content of nodules is heterogeneous and nodules within the same location also vary substantially in composition. In the southwestern Pacific Ocean, nodule abundance is high (>20 kg m^{-2} in some locations) but most are small (<30 mm) and spheroidal, and they are mostly hydrogeneous in origin. The eastern tropical Pacific Ocean, in the vicinity of the Clarion-Clipperton Fracture Zone, is dominated in mass by large nodules up to a maximum of 140 mm. These are flatter "hamburger" type nodules that span the sediment-water interface and have hydrogeneous formation on the upper surface and diagenetic formation within the sediment. In regions of lower abundance (3 kg m^{-2}), the nodules are enriched in Ni and Cu (~3.4%). In other areas, where the abundance is higher (16 kg m^{-2}), the metal content is lower (0.3% Ni + Cu). The tops of the nodules are more enriched in Fe, Co, and Pb, while the bottoms are more enriched in Mn, Cu, Zn, and Mo. This region has been considered for commercial exploitation of the nodules given their metal content and the practicality of their exploitation.

A study of the uptake of elements by ferromanganese nodules examined uptake in terms of speciation and the solid-phase associations [170]. Based on sequential leaching experiments, it was determined that elements fell into two groups in terms of their associations with either the Mn oxide or Fe (hydr)oxide phases. The weakly associated metals such as the alkali and alkali earth metals, and transition metals (Co, Ni, and Zn), Tl^I and Y were overall associated with the Mn phases and this was attributed to the fact that these surfaces are more highly negatively charged. The association was mostly due to a coulombic interaction, as predicted by computer thermodynamic modeling. In contrast, negatively charged metal ion complexes (halides: Cd, Hg, Tl, carbonates: e.g., Cu, U, hydroxides, and oxyanions: most other metals and metalloids) bound preferably to the more positively charged surface of the amorphous Fe oxyhydroxides. For these elements, binding strength is related to their dissolved speciation, and in the deep ocean environments where these nodules form organic matter associations do not appear to be dominant in complexation. For some metals, surface oxidation enhances incorporation. Differences in both the type of nodule formation and the location appear to have a secondary effect on the degree or incorporation of the various elements studied. It was suggested that these differences could be used for paleoceanographic studies.

6.3.7 The biogeochemistry of metals in hydrothermal systems

The importance of hydrothermal processes as a source of metals to deep ocean waters was discussed in some detail in Section 2.2.6, as was basic biogeochemistry related to these inputs. Here, the focus will be more on the overall chemistry and reactions that are important in the fate and transport of trace elements through the hydrothermal systems, and those involved in determining the fate and transport of the trace elements after release into deep ocean seawater. A depiction of the primary chemical reactions involved in hydrothermal systems is depicted in Fig. 6.26(a). Reactions between seawater and the ocean crust can be broken down into two basic interactions, whose products are termed *geothermal* or *hydrothermal* solutions, depending on the reaction conditions. Geothermal solutions are formed during the passage of water through the hotter base material and the elements are either leached or added to the waters during passage but no extensive chemical transfer is occurring. In contrast, hydrothermal solutions are formed due to both chemical and heat transfer and extensive reactions, as indicated in Fig. 6.26b(a). Plumes emitted from hydrothermal systems have been detected in the deep ocean, as illustrated by the data in Fig. 6.26(b).

The relative extent of reaction and enrichment of trace elements in the hydrothermal fluids is dependent on the extent of reaction, the redox status, and the temperature. Factors that impact the type of hydothermal solutions are the rock type (crystalline or amorphous) through which the water is circulating, the temperature and the depth of interaction and the overall water residence time. Additionally, the age of the system is important as rocks become depleted over time, especially of trace constituents. Overall, the reactions that occur in the high temperature, deep reaches below the surface (1–3 km) result in the conversion of sulfate in seawater to sulfide, and the conversion of bicarbonate to CO_2 and/or CH_4, both of which result in the release of H^+ (increased acidity and an overall loss of alkalinity). There is removal of Mg and the addition of Ca and the increase in the concentration of metals, especially Fe and Mn, which are reduced and made soluble, and Cu, and many other minor elements (Fig. 6.25) [56, 171]. The *in situ* pH of hydrothermal solutions is relatively acidic (3.3 ± 0.5).

The relative enrichment of metals has been estimated by comparing the concentration normalized to Cl to that of crustal material. It is worth noting that it is difficult to estimate the flux of metals as given the T and P of the vent solutions many metals are present at concentrations that are saturated relative to STP, and therefore precipitates in the collected samples can be due to sampled particles as well as post-sampling precipitates [172]. It is not always easy to distinguish between the two phases, and to collect dissolved samples directly at these depths. Most trace elements are enriched, as much as 7–8 orders of magnitude relative to seawater, with only a few of the oxyanions (e.g., Mo, U) being removed within the deeper layers, likely through their reduction, in an analogous fashion to sulfate, and precipitation. Other oxyanions are released to the surface waters in conjunction with many metal cations. Overall, the relative composition is highly dependent on the location and many other factors, and therefore estimates of fluxes of metals from hydrothermal systems are relatively crude compared to those for other sources to the ocean (see Table 2.8).

Additionally, as metals are removed as sulfide precipitates in the sub-floor, this will further increase the acidity of the vent fluids, as the predominant form of sulfide at these pH values is the diprotonated acid:

$$Fe^{2+} + H_2S = FeS + 2H^+$$

Alternatively, the acidity is reduced if there is appreciable organic matter present or if the surroundings have been depleted and therefore do not adequately buffer the fluid composition. It is likely that the overall controlling redox couple in the deeper regions is that of H_2/H_2O and vents fluids are enriched in H_2 and CH_4. When the plumes vent into the deep ocean, the rapid change in temperature results into the precipitation of sulfide minerals initially, and potentially $CaSO_4$, while the precipitation of oxide phases occurs less rapidly as this requires oxidation prior to precipitation. Fe and Mn behave differently as Mn does not form highly insoluble sulfide phases and additionally, its rate of oxidation is slower than that of Fe. Even given the precipitation of sulfide minerals in the "chimneys" at hydrothermal vent locations, the plumes of material are still enriched in total Fe and Mn by several orders of magnitude over seawater. Initially the plumes rise vertically due to their relative buoyancy, but after cooling, tend to be transported horizontally away from the source.

It has been shown that there is a diverse and complex biotic community that is living around the hydrothermal system [173] and that some microorganisms are producing organic material (represented as CH_2O) (*chemolithotrophy*, i.e., deriving the energy for carbon fixation from chemical reactions):

$$CO_2 + H_2S + O_2 + H_2O = CH_2O + H_2SO_4$$

Here C (+4 to 0) and O (0 to −2) are reduced while S is oxidized (−2 to +6) for an eight-electron change overall. The organisms have a complex system to carry out these reactions as they involve both oxidized and reduced species that could react inorganically with each other (e.g., O_2 and H_2S), and there is also the potential for sulfide to "poison" the organisms. This microbial production supports a diverse invertebrate community that lives at the vents, such as tube worms and mussels. The tube worms actually lack a mouth or anus and contain in their modified gastrointestinal tract symbiotic sulfur oxidizing bacteria. The worms actually live off the organic waste products of the bacterial symbionts, and provide the bacteria with the required gases for their production. The worms contain blood vessels that transport the oxygen and H_2S needed by the bacteria. The other invertebrates have also developed similar symbiotic relationships to allow their persistence.

Other microbes are able to use H_2 as the source of electrons in fixing CO_2:

$$2CO_2 + 6H_2 = CH_2O + 3H_2O$$

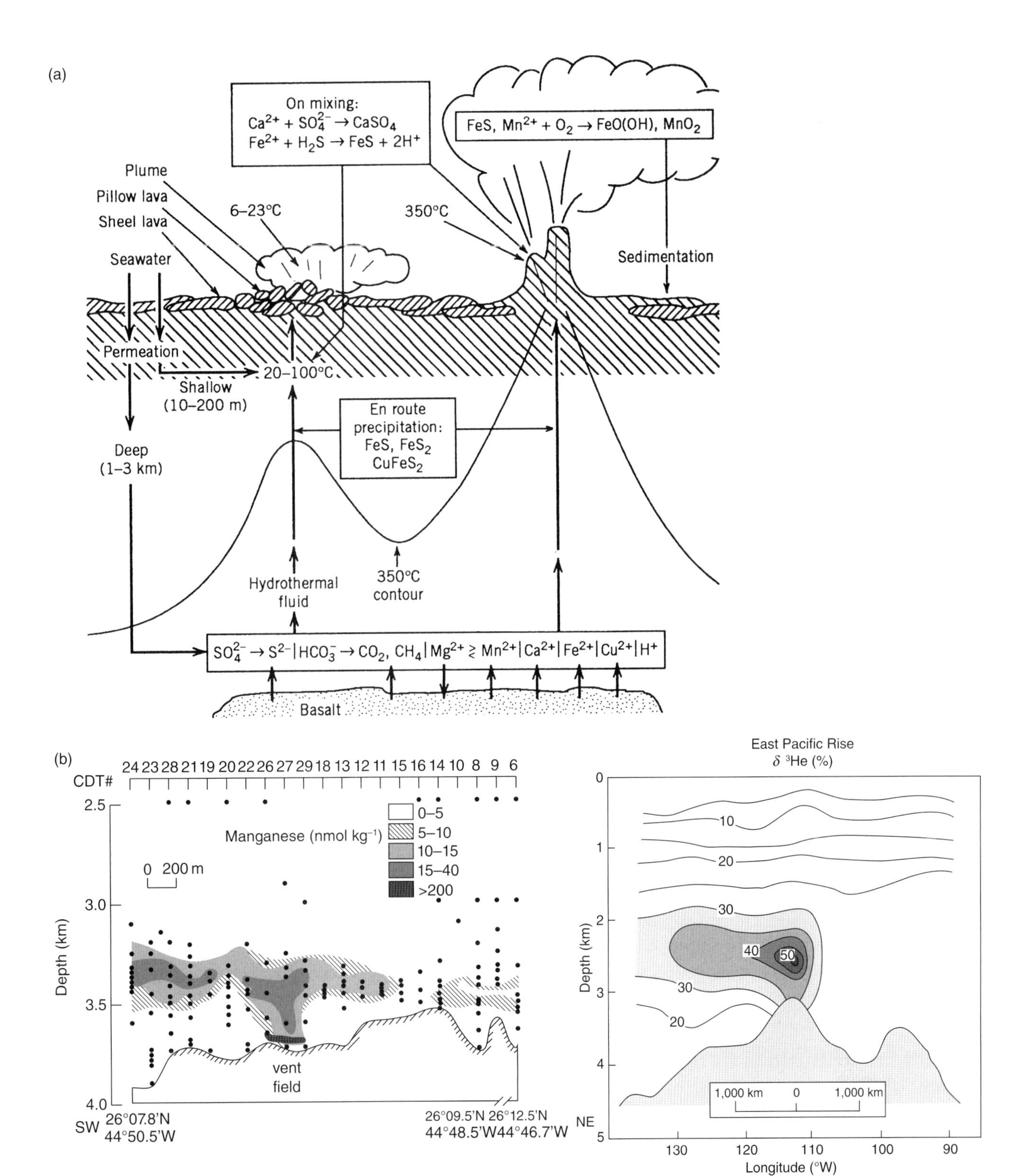

Fig. 6.26 (a) Diagram of hydrothermal systems and main reactions that occur. Figure taken from Stumm and Morgan (1996) *Aquatic Chemistry* but original figure from Jannash and Motti (1985) *Science*, and reprinted with permission of AAAS; (b) depictions of concentration profiles for various plume indicators – ^3He profiles and Mn profiles from the vicinity of hydrothermal systems. Figures taken from Chester (2003) *Marine Geochemistry* (Mn figure) and reprinted with permission of John Wiley & Sons, Ltd, and from German and Van Damm (2004) in The Oceans and Marine Geochemistry, vol. 6, *Treatise of Geochemistry* [171], reprinted with permission from Elsevier.

There is a large suite of organisms that oxidize the reduced species that emanate from the vents and these either rely on oxygen or other oxidized species (e.g., NO_3^-) as the electron acceptor. Sulfide, hydrogen and metal (Fe, Mn) oxidizing bacteria use both, and/or O_2 and NO_3^- (termed denitrifying bacteria if nitrate is used) while sulfur and sulfate reducing bacteria can grow by respiring hydrogen instead of organic matter. Finally, it is likely that bacteria inhabit the sulfide "chimneys" that form at some hydrothermal vents – so-called "black smokers" because of the precipitating sulfide minerals.

The study of hydrothermal systems in terms of metal inputs is still in its infancy [174, 175]. Oxidation rates, and inputs rates from vents, overall determine the extent to which metals are precipitated in the nearfield or transported further before deposition [176–178]. Recent studies have focused on examining the extent of formation of colloidal phases [179], either containing sulfides or oxides, which could be produced either during the precipitation of the sulfide chimneys, and are probably stabilized by the presence of organic material, as discussed in Section 6.2.3. This idea is further supported by the fact that higher fluxes appear to occur in the presence of NOM [174]. Differences in the extent of formation of colloidal material and/or organic complexation could be a major reason for the large differences in concentration found across systems. Also, the coagulation of the colloidal material could contribute to the precipitation post-collection mentioned earlier. Ongoing and future studies should further improve our understanding of the importance of hydrothermal inputs to the ocean budgets of metals, and will inform the extent to which the metals can be transported long distances from their source. For these metal inputs to be relevant to other ocean studies there is the need for the transport of these metals to the upper ocean, and the mechanisms for such transport are slow. Much more study and modeling is needed to better assess the importance of hydrothermal inputs of bioactive metals, such as Fe, Zn, and Co, and of potentially toxic metals, such as Hg and Pb, to the overall biogeochemical cycling of these metals in the ocean.

6.4 Metal distributions in open ocean waters

6.4.1 Vertical distribution of metal(loid)s and controlling factors

Vertical distributions of dissolved constituents in the ocean tend to fall into three major categories reflecting their reactivity and interaction with the major biogeochemical processes in the ocean, and due to their different reactivity and partitioning to solid material (Fig. 2.13). For the major ions, and some trace metals that are relatively unreactive in ocean waters, their vertical distributions vary only slightly with depth, and between ocean basins, and constituents in this class have long residence times (τ) in ocean waters, greater than that of water (~10^3 yrs), and often around 10^5 yrs. These are termed unreactive or *conservative* constituents. They are likely highly soluble, and may be chemically and biologically unreactive, although this is not always so. An example is Mo, which is a required nutrient metal, but its relatively high concentration (105 nM) (Table 6.1) results in little change in its distribution even though it is taken up by microbes in surface waters and released during particle dissolution. It has a long residence time (τ ~8.2×10^5 yrs) as its river concentration is much less (~5 nM) than that in the ocean [2]. It has a much longer residence time than most reactive metals ($\tau < 100$ years). Other trace metals and metalloids that exhibit conservative behavior in the ocean are Sb ($\tau \sim 5.7 \times 10^3$ yrs), W and U ($\tau \sim 5 \times 10^5$ yrs) [2, 3]. Most of these elements also exist as oxyanions in seawater in relatively high concentration (nM–µM) and therefore it is the relative unreactive nature of such compounds, especially in terms of interactions with particles, that leads to their long residence time. Similarly, most major ions (e.g., Na, Ca, K, Cl) have a biochemical role but their high concentration results in this uptake having little impact on their distribution in open ocean waters. The major nutrients and minor elements are also found as oxyanions in their most oxidized forms in seawater (nitrate, phosphate, silicate and sulfate). As discussed in Chapter 8, the potential toxicity of some metalloids relates to the fact that they exist in solution as oxyanions and can therefore be taken up inadvertently by microbes and can interfere with biochemical pathways of the nutrient oxyanions.

Overall, most of the oxyanions exist in relatively high concentrations (>10 nM) in seawater compared to the other trace elements, with As having an average concentration of 23 nM and a relatively long residence time (Table 6.1); similarly with Se. When considering all the metals, there is a general trend that the variability in concentration decreases as the concentration increases for all the elements in Table 6.1, and this reflects the fact that many biologically or chemically reactive elements are those that show the largest temporal and spatial variability, and as a result of their reactivity are those with the lower concentrations overall. Some chemically reactive but biologically toxic elements (e.g., Pb, Hg, Ag) are also found at relatively low concentrations, and as discussed in Section 8.3.4, their toxicity relates in part to the relative low concentration in the prehistoric (anoxic) ocean environment compared to the relative abundance at that time of the nutrient metals (Fe, Mn, Mo, and Cu).

In classifying the metal distributions, a second group of elements is exemplified by most of the micronutrient elements which are required for photosynthetic and other enzymes and biochemicals, and these elements are assimilated into biological material in the surface ocean and their distribution reflects biological particle production and destruction processes in the ocean. They typically exhibit surface water depletion in concentration and have elevated concentrations at depth, due to their release during particle remineralization in the deeper ocean waters. Sinking particles are mostly remineralized in the region of the permanent thermocline (~500–1000 m) and the release of metals results in a rapid increase in concentration in this region. In addition, deep water concentrations tend to increase along the flow path of deep water with concentrations being highest in the central North Pacific. Overall, the circulation of the "deep conveyor belt" begins as surface waters sink in the North Atlantic and around Antarctica, and are ultimately transported to the North Pacific where they are returned to the surface via deep water upwelling [2]. Constituents with this distribution have been called *nutrient-like* or *recycled* elements. Trace elements and metalloids which fall into this class include Zn, Cd, Ge, and Se [180].

The observation of a linear correlation between the concentrations of Zn and Si, and Cd and P, in the ocean (Figs. 2.14 and 6.27) has been attributed to their overall scavenging in surface waters by uptake into biogenic particles and their release at depth due to particulate remineralization. It is known that Zn has multiple roles in cellular metabolism and, most importantly, it is part of an important enzyme, carbonic anhydrase, in marine phytoplankton, which is required to interconvert CO_2 and HCO_3^- within cells [181] (Section 8.3.3). Thus, Zn is actively accumulated into plankton in surface waters. Also, the deep water concentrations of both Si and Zn are higher in the Pacific Ocean compared to the Atlantic Ocean and this reflects their continued buildup during water transport.

The correlation of Cd and P profiles also suggests a biochemical role for this metal (Fig. 6.27). Initially, Cd was thought only to be a toxic metal, even though it exhibits a nutrient profile in the ocean. Its biochemical role was not known until the demonstration of its substitution for Zn in carbonic anhydrase in some marine phytoplankton [181] (Section 8.3.3). The differences in the behaviors of Zn and Cd may reflect the fact that Cd is assimilated more into soft tissue that is more easily remineralized, or may reflect differences in their incorporation into enzymes and cellular organelles. One potential explanation for the differences in the remineralization rate of Si and P is that, to a large degree, Si is not assimilated by zooplankton feeding on phytoplankton and so the passage of Si through the zooplankton gut and incorporation into fecal material likely accounts for its remineralization in deeper waters than P, which is highly assimilated by zooplankton (~70% assimilated) [182], and not found in high concentrations in fecal material. Thus, the remineralization of P likely occurs more from sinking phytoplankton and other more organic-rich material, and the similarity in profiles of Cd and P could be related to the similarity in their sources to deep waters. Overall, the Atlantic/Pacific ratio is much greater for Si than for Zn, suggesting that there is also scavenging of Zn in deep waters. Alternatively, ocean waters are undersaturated with respect to solid silica (opal) and thus the Si is released into solution in deep waters and is not further scavenged.

Copper (Cu) is another element that is both required for some enzymes but has a more complex profile (Fig. 6.27) given its relatively high particle reactivity compared to Zn and Cd. Similarly, Fe is a known micronutrient with a surface depleted profile in the ocean (Fig. 6.28). However, the relative enhancement of deep Pacific Ocean water concentrations seen for Zn and Si is not apparent for Fe [183] because of the high particle reactivity of Fe; it is continually scavenged from the deep waters over time. Therefore, its deep water concentration reflects the short-term history of particulate input to deep waters rather than the longer-term cycling of these deep waters. Additionally, Fe is relatively insoluble in its oxidized form and its complexation to dissolved organic ligands in surface waters tends to enhance its solubility. In deeper waters such ligands are present at much lower concentrations and therefore there is less complexing capacity. This may lead to precipitation and removal of Fe in deeper waters relative to the surface waters.

As more elemental distributions are examined in detail in the ocean, and with the accumulation of more data and more accurate information, the relationships between major nutrients and trace metals are shown to be more qualitative in nature. Overall, given the ever increasing complexity of knowledge about the factors controlling the speciation and fate of trace elements in ocean waters, these distinctions are more difficult to separate for a variety of elements. *Hybrid* distributions are a combination of the major categories, such as found for Fe and Cu. Additionally, both of these metals have redox chemistry such that there is the potential for their release in some low oxygen sub-thermocline waters due to reduction and solubilization of their oxides [184].

The third category is characterized by elements that are particle reactive and/or relatively insoluble, and especially those metals that also have the majority of their input from the atmosphere. Elements in this class are referred to as exhibiting *scavenging* or non-cycling distributional features which are often dominated by their point of entry and removal in the ocean. These metals have short residence times ($<10^3$ yrs). For example, surface water concentrations of Hg and Pb are typically elevated compared to deep waters [185–187], reflecting their dominant atmospheric input pathway to the ocean. Such metals are scavenged and removed from the surface ocean via particle uptake, released back into the water column during the dissolution of the biological particles, but are often re-scavenged from deep ocean waters and therefore do not

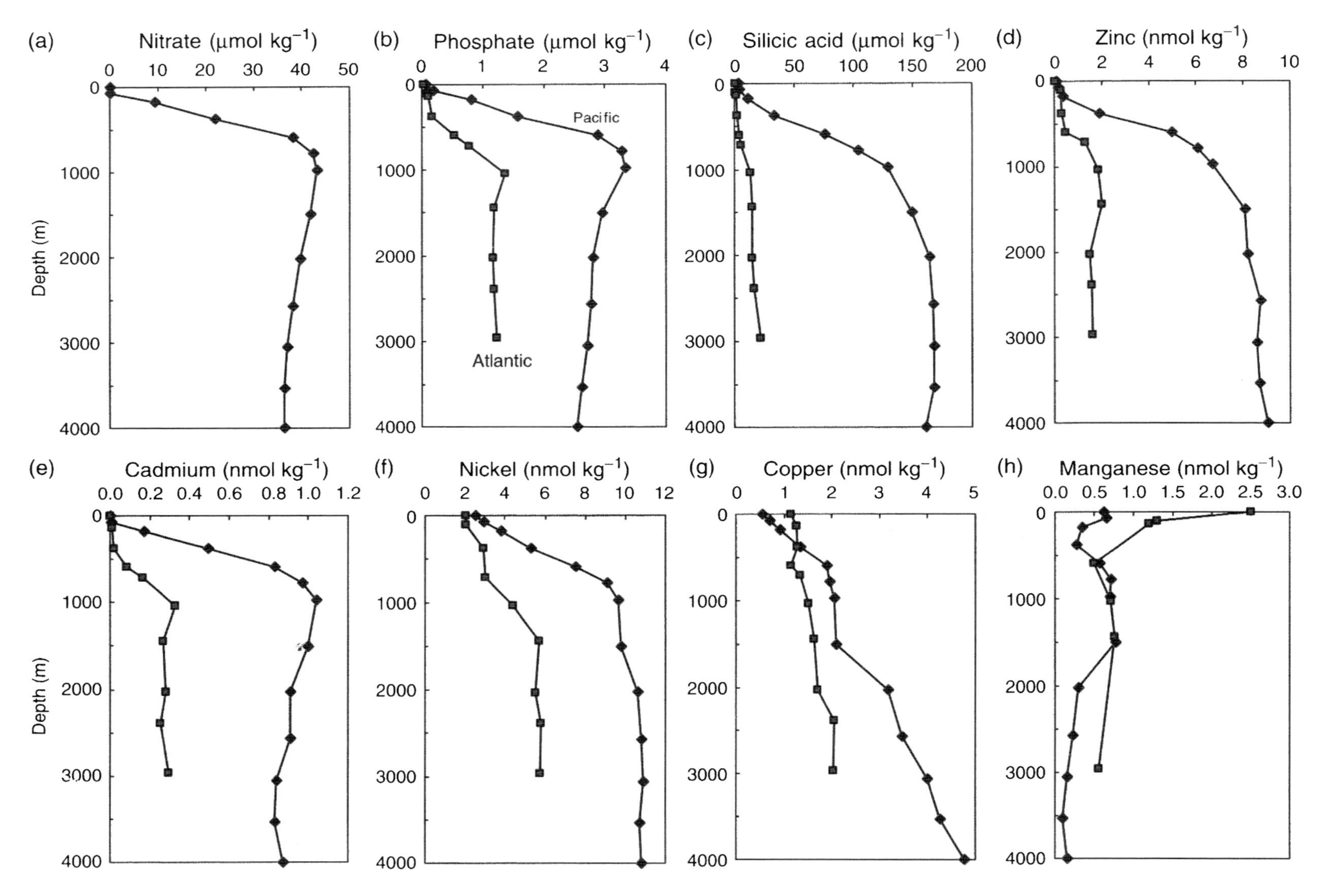

Fig. 6.27 Profiles in the North Atlantic and North Pacific for cadmium, zinc, nickel, copper and manganese, and the nutrients (nitrate, phosphate and silicate) their distributions correlate with most closely. Taken from Sunda (2010) in *Marine Chemistry and Geochemistry* and reprinted with permission of Elsevier.

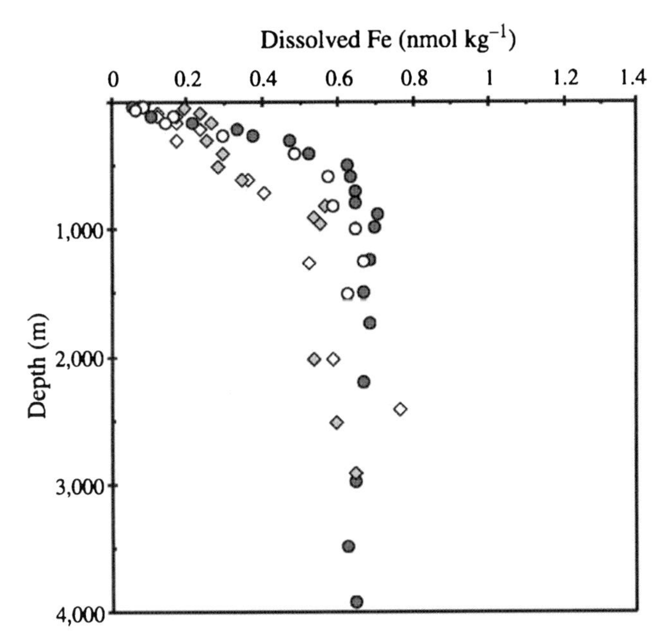

Fig. 6.28 Representative profiles of iron for the North Pacific and the North Atlantic Oceans showing the similarities in the distributions. Taken from Bruland and Lohan (2004) in The Oceans and Marine Geochemistry, vol. 6, *Treatise of Geochemistry* [3], reprinted with permission from Elsevier.

have a continuing increasing concentration with depth, and their concentration is depleted in deep ocean waters, or does not increase with depth, or with time. The distribution of Pb in ocean waters (Fig. 6.29) is illustrative of a scavenged profile and a metal with a strong atmospheric signal [187].

The overall difference between the concentration of Pb in Atlantic Ocean waters compared to the Pacific is however not just a consequence of ocean chemistry and physics [3]. The input of Pb to the environment has been greatly perturbed by human activity and therefore the ocean profile reflects to a large degree the magnitude of the inputs to the different ocean basins over time. The historical industrialization and use of Pb in gasoline in North America and Europe is clearly reflected in the higher inputs of Pb to the North Atlantic and the higher concentrations of Pb in these ocean waters (Fig. 6.29b). Additionally, as Pb is a highly particle reactive metal it is not accumulated in the deep waters, in a similar fashion to Fe, as discussed above. Another metal with a similar ocean chemistry and profile is Al which also has a strong atmospheric signal, and which is actively removed from deep waters via particle scavenging. This leads to a depletion of Al in deep Pacific Ocean waters.

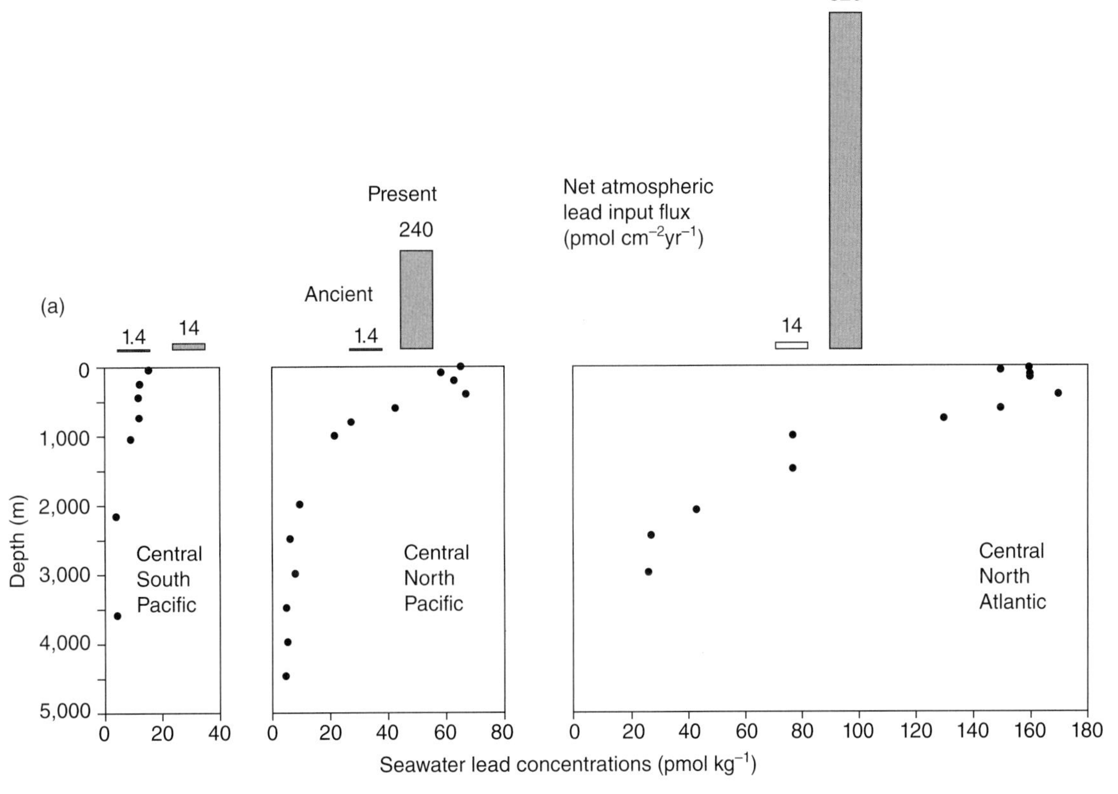

Fig. 6.29 (a) Distributions of lead in various ocean basins showing the changes in concentration between basins in the 1980s; and (b) data from corals near Bermuda showing how the concentrations and lead isotope ratios have changed over time; and (c) comparison of the vertical distribution of lead with that of another particle reactive element, aluminum. Part (a) reprinted from Bruland and Lohan (2004) in The Oceans and Marine Geochemistry, vol. 6, *Treatise of Geochemistry* [3], reprinted with permission from Elsevier; (b) from Reuer et al. (2003) reprinted with permission from Elsevier; and (c) from Chester (2003) *Marine Geochemistry* [1], reprinted with permission of John Wiley & Sons, Inc.

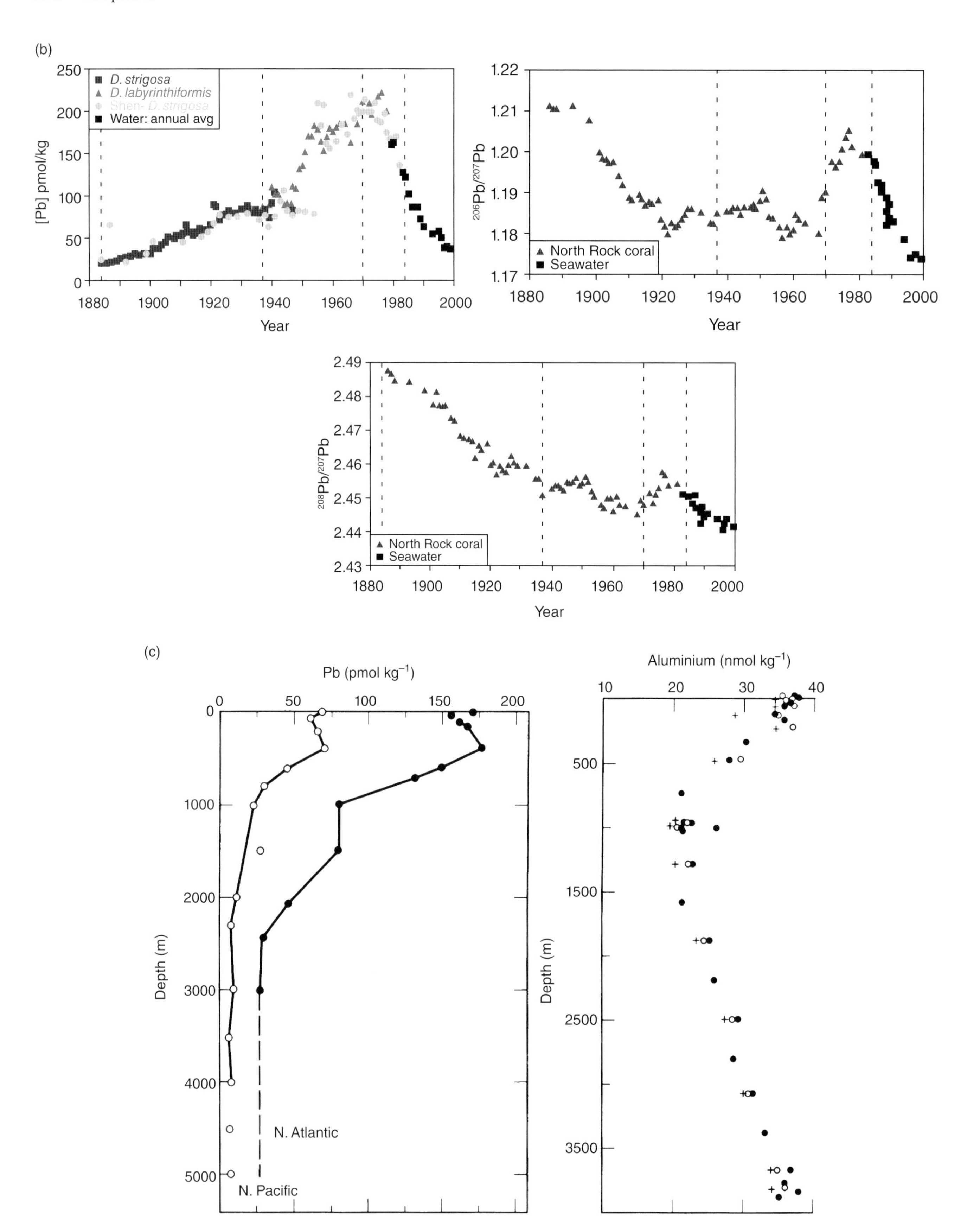

Fig. 6.29 (Continued)

In reality, many metal(loid)s have relatively complex distributions in the ocean as most are taken up to some degree in surface waters due to uptake (purposeful or by accident) into phytoplankton and bacteria, or adsorption onto particles, and are removed from the surface ocean by particle sinking. Particle remineralization releases the metal(loid)s but many are scavenged to deep ocean particulate. Their overall profile therefore reflects their solubility, reactivity and ability to be transformed into other forms. Methylation, for example, of Hg, or by reduction in low oxygen zones (Fe, Mn and Cu) alters their speciation and impacts their overall vertical distribution and their relative concentration in the Pacific and Atlantic Oceans.

Overall, it is difficult to make generalizations about the distribution of metal(loid)s in the ocean as even the biogeochemical cycling of elements with the least anthropogenic impact have been impacted by human activity, which has changed the input from the terrestrial environment through biomass burning and clearing, agricultural activities, urbanization of the coastal zone and similar related activities. Acidification of the atmosphere due to anthropogenic releases has changed the weathering characteristics of precipitation and has altered riverine inputs. Current ongoing acidification of the ocean as a result of changing atmospheric CO_2 levels has changed the precipitation rate and scavenging of dissolved and particulate metals from the surface ocean. Changes in algal composition can have a similar effect. Therefore, given these changes and their varied impact, it is worth briefly discussing the ocean cycling of the metal(loid)s in categories that reflect their major similarities, and contrast their differences.

A compilation of data has been presented elegantly in a figure composed by Yoshiyuki Nozaki (Fig. 6.30). This figure illustrates that data is lacking for some of the elements [188] and provides a valuable synopsis of ocean elemental cycling. The oxyanions (e.g., V, Cr, Mo, W, Re, Os, As) generally have a profile with little vertical variation because of their relatively high solubility and lower particle reactivity. Many elements have a profile with a lower surface water concentration, increasing with depth and these include the "nutrient" metals (e.g., Fe, Cu, Cd, Zn) and also some metals that are primarily from a crustal source (e.g., Sc, Ti, Y, Zr, Nb). The more particle reactive metals whose input is dominated by the atmosphere show the opposite profile (e.g., Pb). The concentration found in the ocean is a function of the relative source strength (i.e., concentration in the Earth) and the element's solubility and reactivity in seawater, and its importance as a nutrient to organisms. Many of the heavy metals are found in low concentrations (pM) due to their relative poor abundance in the crust and to their relatively high partitioning to the solid phase. The unreactive elements are present in seawater at higher concentrations. It is worth reiterating the idea that the relative concentration in seawater is not a direct function of the relative input strength but is more related to the reactivity and solubility of the metal in seawater.

6.4.2 Coordinated ocean studies: GEOTRACERS and prior and related programs

There have been a number of coordinated ocean studies that have resulted in a dramatic increase in the amount of information on trace metals and their cycling. Currently, the GEOTRACERS program, a coordinated global research program to study the biogeochemical cycles of trace elements and their isotopes [189], is being enacted. As noted elsewhere, trace elements and isotopes play important roles in the ocean as nutrients (e.g., Fe, Zn), and impact the ocean biological community's structure and function. Additionally, they can serve as tracers of biogeochemical processes, especially the radioisotopes, as well as indicators of the strength of external inputs, such as the anthropogenic signal. Finally, they can improve our understanding on how the chemistry of the ocean has changed over time. Many heavy elements have been added to the biosphere primarily through human activity and these elements therefore provide a proxy for this insult. Understanding the biogeochemical cycling of nutrient metals informs understanding of other important areas of research, such as the carbon cycle and climate change, and provides insight into how ocean ecosystems have changed over geological time.

The goal of the GEOTRACERS program is to ascertain the global distribution of a suite of elements and isotopes (Table 6.11). The total concentration, speciation and physical form will be determined and used to identify the processes involved and quantify their fluxes, and to establish the sensitivity of these distributions to changing environmental conditions. In addition to ocean distributions, external sources to the ocean, such as atmospheric wet and dry deposition, margin inputs and continental runoff, sediment-water exchange and hydrothermal inputs, will be assessed.

The GEOTRACERS program builds on and extends past global and international programs. The GEOSECS program in the 1970s was a major initiative to establish baseline data for assessing future chemical changes and to examine the large-scale oceanic processes [190]. This program was instituted prior to the development of low level techniques and contamination-free sampling and handling ("clean techniques") for determining trace metal concentrations, as discussed in Chapter 1. Therefore little reliable trace metal data was generated through this study. However, measurements of radiotracers were a large and integral part of the program and yielded a wealth of important information that has driven scientific understanding and ocean research since this time. More recently, a number of ocean expeditions have been completed under the auspices of the International Ocean Commission (IOC), as reported in a number of Special Issues (*Marine Chemistry*, **49**, 1995; **61**(1–2), 1998; *Deep Sea Research II*, **46**(5), 1999; **48**(13), 2001) and while there were

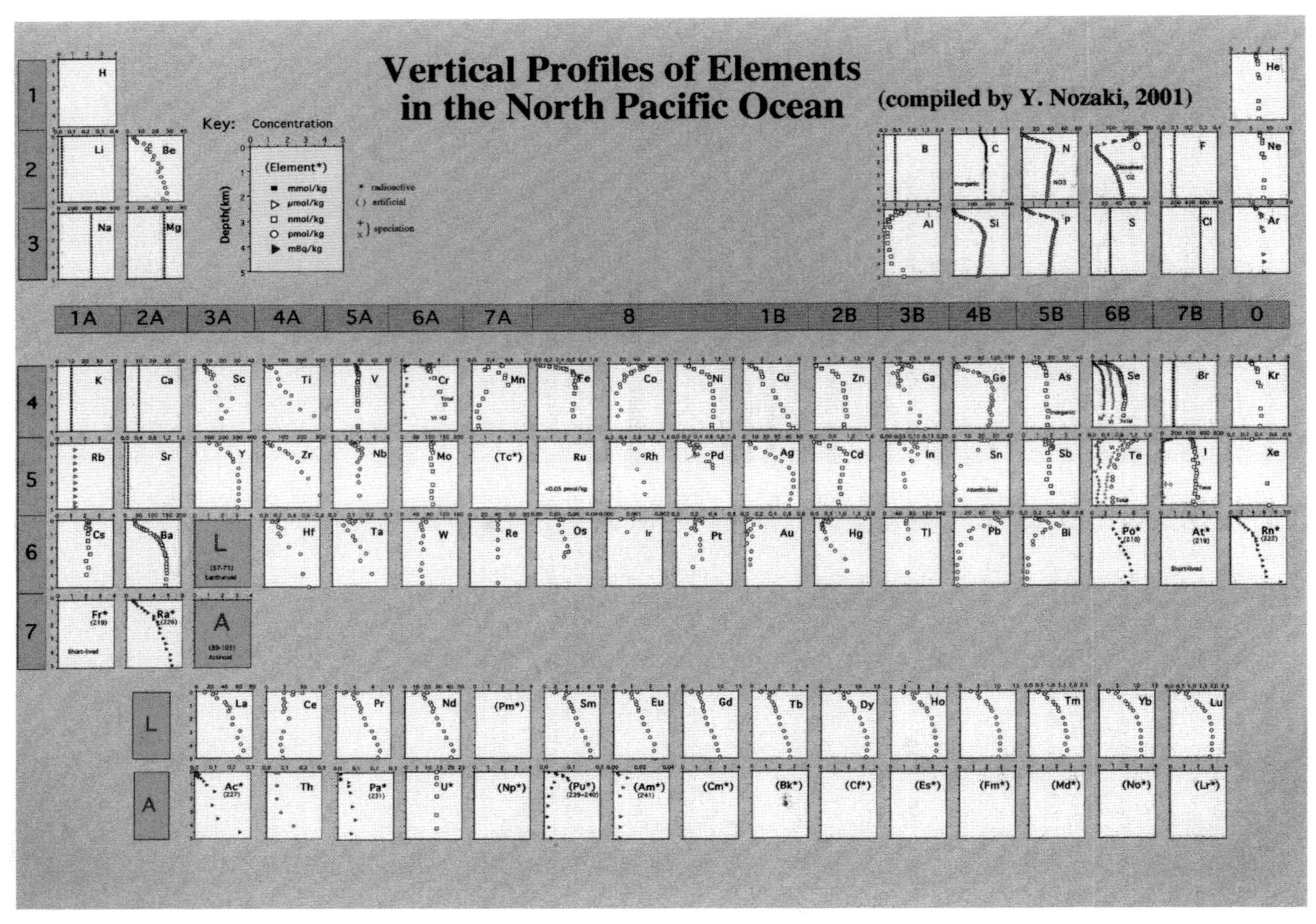

Fig. 6.30 Representative distributions of elements for the North Pacific Ocean. Figure reprinted from Nozaki (1997) *EOS* [188] with permission of the American Geophysical Union, copyright (1997).

Table 6.11 The major trace elements and isotopes that will be included in the Geotracers program studies. Those indicated in italics are not considered key parameters but are likely to be measured on cruises. Data taken from the GEOTRACERS Science Plan [188].

Parameter	Key uses
Fe	Micronutrient
Al	Tracer of atmospheric inputs (mineral dust)
Zn	Micronutrient
Mn	Tracer of Fe input; redox cycling, margin inputs
Cd	Micronutrient; paleoproxy for nutrient
Cu	Micronutrient
Co	*Micronutrient*
Hg, Pb, Ag, Sn	*Anthropogenic source indicators*
$\delta^{15}N(NO_3^-)$	Modern and paleoproxy
$\delta^{13}C$	Modern and paleoproxy for nutrients and circulation
^{230}Th	Flux of particles, scavenging and ocean circulation
^{231}Pa	Paleoproxy of circulation and productivity; particle tracer
Pb isotopes	Tracer of natural and anthropogenic inputs
^{210}Pb	Source tracer
Nd, Hf isotopes	Tracer of natural sources to the ocean
Ra isotopes	*Indicators of groundwater discharge (coastal waters)*
^{3}He	*Hydrothermal inputs*
Metal stable isotopes	*Paleoproxy; redox, chemical and biological processes*
O isotopes	*Nutrient paleoproxy*
Other U-series isotopes	*Particle processes, inputs and removal processes*

issues with contamination for some metals, the data from such studies provide a useful backbone to the GEOTRACERS studies.

More recently, a group of researchers developed a clean rosette/GoFlo sampling system [191] and techniques for high throughput analysis to collect high resolution trace metal profiles in the upper 1000 m. Samples are analyzed on board for dissolved Al and Fe using Flow Injection Analysis, with subsamples collected for analysis of dissolved transition and heavy metals. Results from sampling in the Atlantic Ocean illustrate the wealth of information that such studies provide (Fig. 6.31) [191, 192]. A minimum in the Fe concentration within the 40–80 m depth region between 18–4°N, which coincides with the chlorophyll maximum, suggests that the Fe is being actively accumulated by microorganisms in this zone. Deeper in the water column, in the low oxygen waters, increasing Fe concentration is consistent with remineralization and release of Fe from sinking particles. Nutrient concentrations were also elevated. For Al, elevated concentration in subsurface waters between 30 and 20°N (Fig. 6.31) likely reflect the subduction of subtropical waters that sink in late winter and are advected toward the Equator [192]. The subtropical regions receive enhanced atmospheric dust inputs from North Africa, deposited to the western Atlantic. In contrast, the higher concentrations at depth at 30–45°N probably reflect outflow from the Mediterranean. These two regions of Al enrichment are good demonstrations of the geochemical connection between atmospheric transport and the metal profiles in the North Atlantic.

As noted in Chapter 5, the SEAREX Program (Section 5.2.4) was a large scale initiative that provided a wealth of information on the inputs of metals to the open ocean, and there have been a number of other large scale atmospheric studies since these initial collections. During the studies discussed earlier, aerosol samples were collected using a MOUDI (micro-orifice uniform deposit impactor) cascade impactor to obtain size-fractionated aerosol samples [193]. Samples have been collected across all ocean basins. The results from the basin scale distributions of dissolved and particulate trace elements and the aerosol data will further improve understanding of inputs and global ocean biogeochemical cycles. This information will also inform other important oceanographic questions and help the development of ocean/atmospheric models. The proposed atmospheric measurements during cruises under the GEOTRACERS program will expand and enhance these early measurements and allow for an assessment on how inputs have changed over time.

GEOTRACERS will be global in scope, consisting of ocean transects complemented by regional process studies. Sections and process studies combine fieldwork, laboratory experiments and modeling. Cruises are planned to cross regions of prominent sources and sinks (such as dust plumes, major river discharges, hydrothermal plumes and continental margins), to sample principal water masses, and major biogeographic provinces. Models will be used to interpret the distributions of elements and provide estimates of the principal sources and sinks, as well the dynamics of internal cycling. The primary metals and isotopes (core parameters) that will be measured on all cruises are outlined in Table 6.11. A number of other metals and isotopes will be included where possible (given in italics in Table 6.11).

Additionally, given the current and historic difficulties in making uncontaminated measurements of metals in the ocean at the low levels present, a detailed and thorough intercalibration of methods for use by the international community has been initiated, and is ongoing. These intercalibration draws from the recent success of efforts to improve ocean Fe analysis [189, 194]. In addition to a number of intercalibration exercises and intercalibration cruises, reference materials are being developed to ensure sufficient

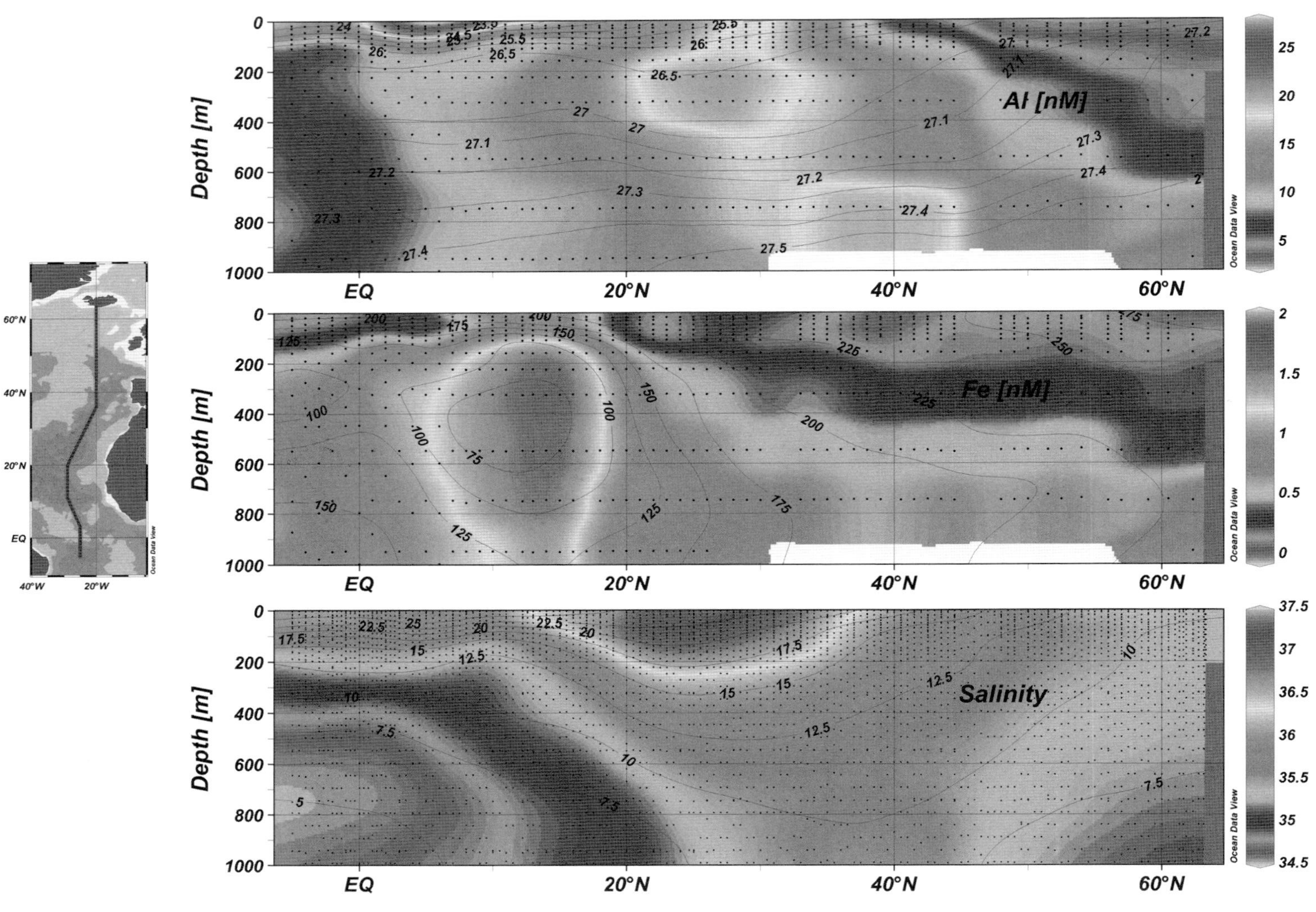

Fig. 6.31 Distributions of aluminum, iron and salinity in the upper 1000 m of the North Atlantic Ocean during the transect indicated on the attached figure. Figure reprinted from Measures et al. (2008) *Global Biogeochemical Cycles* **22** [192] with permission of the American Geophysical Union, copyright (2008).

methods to validate sampling and analytical techniques. Locations in the major ocean basins have also been designated as reference locations and samples have been collected and compared between laboratories to generate a consensus value for field validation of techniques. Validated sampling and analytical protocols have been published to allow the international community to adopt and compare analytical methods. For example, data from the analysis of samples for their Fe content [189] show that in most cases, different laboratories can obtain comparable results.

The results of the large-scale coordinated studies provide a wealth of information that substantially improves on data collected during the last two decades. This has allowed many of the major components of the cycling of metals in the ocean to be understood; their sources, sinks, major uptake and removal pathways and overall biogeochemistry. The current understanding will be detailed in the following sections which will examine different classes of metals and metalloids based on their similarity in chemistry, their requirement to organisms, or their potential toxicity, especially for the metals that have a strong anthropogenic source component.

6.4.3 Iron, manganese, and aluminum cycling in open ocean waters

Atmospheric and riverine inputs are the main sources of these elements to the oceans and much of this input is in the particulate or colloidal fraction. While Fe is primarily present as Fe^{III} complexes and solids in oxic seawater, Mn can be found in its more reduced +2 oxidation state, and there is some evidence for the presence of Mn^{III}, as its stability can be enhanced by complexation [195]. The reason for the persistence of the reduced Mn forms is its relatively slow oxidation rate compared to Fe [196, 197]. While both Mn and Fe can undergo redox cycling in the surface ocean, as discussed in detail in Section 5.5.1 for Fe, the oxidation of Mn is hindered in ocean waters and this allows the reduced form to persist in surface waters.

Aluminum (Al) ranges widely in concentration and is elevated in regions of enhanced atmospheric input, such as the equatorial zones (50–60 nM), and is very low in polar surface waters (<10 nM). It is present as Al^{III} ($Al(OH)_4^-$, $Al(OH)_3$) [198]. Concentrations are highest in the North Atlantic (8–30 nM) and lowest in the North Pacific (<2 nM) due to its depletion via particle scavenging. It has high concentrations in the surface Mediterranean (>100 nM) due to the enhanced atmospheric (dust) inputs in that region. Aluminum is the element with the largest interocean variation.

Riverine inputs dominate for these elements but as the metals are all highly particle reactive/insoluble and most of the riverine input is removed in estuarine and coastal waters. Given their increased insolubility with pH, these elements have low ocean concentrations even though they are some of the most abundant elements on the surface of earth. The relative insolubility of both Fe^{III} and Mn^{IV} leads to the rapid removal of these metals from the surface ocean. In regions of enhanced atmospheric input, this can lead to a surface water maximum in concentration. While both Mn and Fe have a role in the biochemistry of marine organisms (Chapter 8), the concentration of Mn is relatively high given its relatively high concentration in ocean waters compared to its nutrient requirement (Table 6.1) and depletion of its surface concentration due to biological activity is not often found. In contrast, dissolved Fe can show surface ocean depletion in regions of high primary productivity. In seawater, the concentration of dissolved Fe is higher than would be predicted based on its primary solubility in terms of Fe (hyd) oxides due to the presence of Fe-binding ligands in solution [3, 181]. Iron is strongly hydrolyzed in seawater with the dominant inorganic complexes being $Fe(OH)_4^{2-}$, $Fe(OH)_3$ and $Fe(OH)_2^+$. In contrast, Mn forms only weak associations with both inorganic and organic ligands and its main inorganic form in seawater is Mn^{2+}.

Representative profiles for Fe and Mn in the major ocean basins are shown in Figs. 6.27, 6.28 and 6.32. The potential influence of organic complexation on the filtered concentration can be demonstrated by the following calculation. Given the formation constant (log K = 38.8) for the formation of $Fe(OH)_3$ (s), and a pH of 8.2, it can be estimated that the total dissolved Fe concentration at equilibrium (making appropriate ionic strength corrections) is around 0.1 pM. The average measured concentration is 0.5 nM which is three orders of magnitude higher. This is consistent with the electrochemical measurements that have shown that 99% or more of the Fe in surface ocean waters is complexed to organic ligands, as discussed earlier in this chapter. Additionally, given that the measured values are for filtered waters it is likely that this value reflects the presence of both organic and inorganic Fe-containing colloidal phases. For Mn, the dissolved concentration is much lower than the concentration expected from precipitation of Mn^{IV} hydroxide or carbonate phases, and is maintained as a result of the relative rates of oxidation and reduction of Mn in surface and deep ocean waters, due to the slow oxidation kinetics of Mn^{II}.

On area of focus in recent studies is the potential for the persistence of Fe^{II} in surface waters due to its photochemical reduction and stability when complexed to organic matter. While the exact concentration of Fe^{II} is a matter of debate due to potential artifacts related to the current measurement techniques, it is clear that reduction does occur (Section 5.3). As an example, Fe speciation measurements were made on the CLIVAR cruise discussed earlier [199] which allowed for an assessment of the bioavailability of Fe and distribution of Fe^{II} in the upper ocean. These results suggest that Fe^{II} occurs in surface seawater with surface maxima

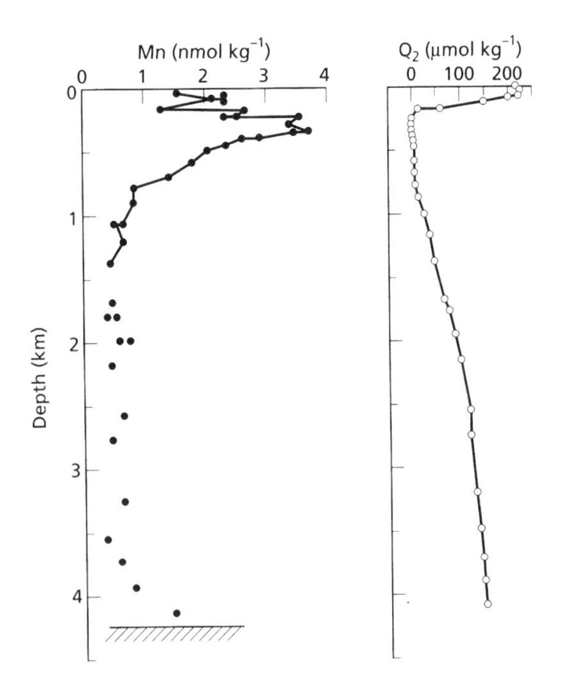

Fig. 6.32 Distribution of manganese in the Pacific Ocean showing the increase in Mn concentration in the low oxygen region. Figure reprinted from Chester (2003) *Marine Geochemistry* [1], with permission from John Wiley & Sons, Inc.

under most conditions and the reduced concentrations are 10–15% of the total dissolved Fe, although these authors suggested that these estimates are conservative. These initial data illustrate how high resolution sampling greatly expands our understanding of the short-term and long-term biogeochemical cycling of metals in the ocean.

The distributions of Fe and Mn in the water column are modified in the regions of low oxygen or anoxic waters. In some ocean water column locations, oxygen concentrations are sufficiently low that both Fe and Mn can be reduced, either through biologically-mediated pathways or abiotically. Given that the reduced forms are substantially more soluble than the oxidized species, decreasing oxygen content can lead to a change in the partitioning and speciation of both metals. This has been observed, for example, in the oxygen minimum zone of the equatorial Pacific Ocean [1].

In addition to these sub-oxic environments, the distributions of Fe and Mn are highly modified in anoxic environments, as discussed in Section 6.3 and in Section 7.2 for freshwater ecosystems. In marine environments, the distribution of dissolved Fe, in particular, is a strong function of the redox state. This results from the fact that while low pε values lead to Fe(III) reduction and an increase in solubility, as sulfide levels increase, Fe(II) is precipitated as Fe-sulfide phases (FeS and FeS_2) and this leads to a maximum in dissolved Fe at low, positive pε values, with removal of Fe, and low dissolved Fe, in the anoxic waters (see Fig. 3.22).

In permanently or seasonally stratified systems, the redox cycling of metals can lead to a large gradient in their concentration. Metals and other constituents that are strongly associated with the oxic particulate phases of Fe and Mn show similar distributions. The profiles of metals in the Black Sea provide one "classic" example of such distributions across the water column redox interface (Fig. 6.33) [200–202]. Other locations where such profiles and distributions have been studied include Framvaren Fjord [203]; the Carioco Trench [204] and the Pettaquamscutt Estuary [205]. The concentration of particulate Fe and Mn, and associated metals, are enriched above the interface due to the diffusion of dissolved, reduced Fe and Mn from below, and their subsequent oxidation in the higher oxygen waters. Sinking of these particulate materials and their dissolution below the interface leads to a peak in dissolved species below the interface. This cycling across the interface has been termed the "ferrous wheel" in analogy to the amusement park ride.

Another location where increased concentrations of dissolved Mn and Fe can be found is in association with hydrothermal vents. While Fe^{II} is rapidly oxidized in oxic waters, the rate of Mn^{II} oxidation is substantially lower and therefore elevated concentrations have been found to persist in the vicinity of hydrothermal vents, in plumes that are enriched in other tracers of such activity (^3He) [1, 206] (Fig. 6.26b). As discussed in Section 6.3.7, the removal of metals with the precipitation of Fe and Mn oxides after emission of fluids from hydrothermal systems results in the local scavenging and removal of many metals and decreases the importance of hydrothermal sources as input to the global ocean. These processes are important in the formation of oxide deposits and also in the formation of nodules, as discussed in Section 6.3.6.

6.4.4 The biogeochemical cycling of zinc and cadmium in the ocean

The distribution of Zn in open ocean waters appears to be strongly controlled by biogeochemical processes, and its distribution is that of a "classic" nutrient metal. Its distribution is consistent with Zn incorporation into microbial tissues in the surface waters [3], as discussed in detail in Section 8.3.3. This is also consistent with the fact that the main source of Zn to the ocean is riverine and terrestrial input, and the atmospheric input is a relatively small component (Table 6.1). Additionally Zn is relatively soluble in seawater in the presence of inorganic ligands and has an average concentration of 5 nM (Table 6.1), greater than that of Fe and Mn, which are much more abundant elements in the terrestrial environment and in riverine inputs. Additionally, there is no important redox chemistry for Zn in ocean waters. There is also evidence for important Zn complexation to dissolved organic ligands (Fig. 6.27) and it is highly likely that these

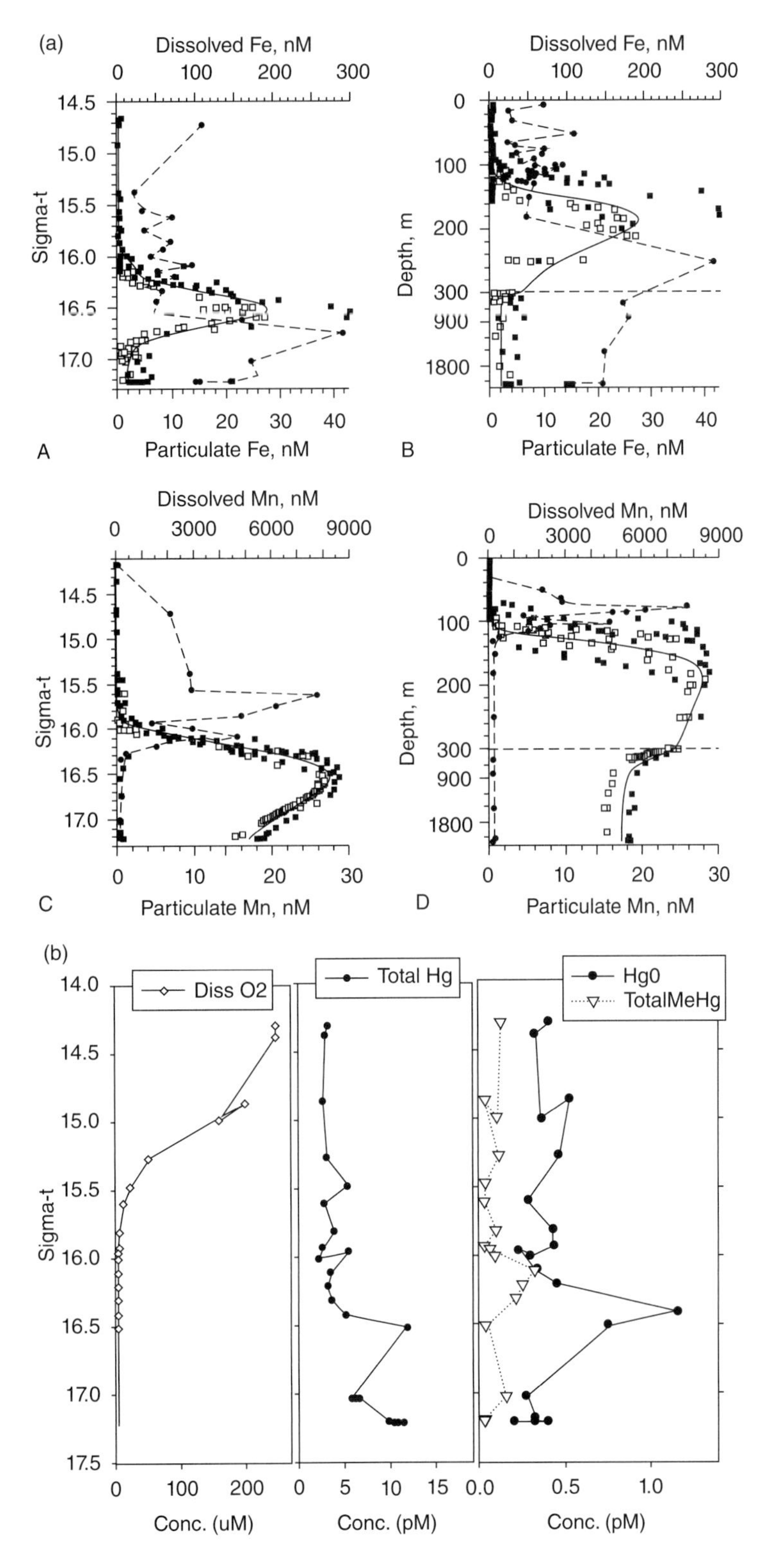

Fig. 6.33 (a) Distributions of (A–D) redox sensitive metals (Mn and Fe) in the Black Sea, focusing on the transition zone between the oxic and anoxic waters; and (b) the distribution of various forms of mercury across the same interface. Taken from (a) Oguz et al. (2001) *Deep-Sea Research* **48**: 761–87 [202]; and (b) re-plotted using the data from Lamborg et al. (2009). Both reprinted with permission of Elsevier. Note: In (A) squares indicate particulate and circle indicates dissolved.

ligands are directly or indirectly produced by microbial organisms, even if they may not be produced specifically for Zn complexation. Even given its direct biological role, Zn has a relatively long residence time compared to the other first row transition metals that exist in solution predominantly as cations.

A number of studies in various ocean regimes have confirmed that >70% of the Zn is complexed to organic ligands [1] and that the complexation can be described by two classes of ligands. As Zn is present in the absence of organic ligands mostly as the free ion in seawater (Table 6.1), it is relatively easily acquired by organisms and therefore complexation with specific organic ligands to enhance uptake is likely not needed. Therefore, complexation of Zn by organic matter does not appear to provide a unique advantage to its uptake in contrast to Fe, which is a demonstrated limiting nutrient and therefore is acquired actively through organic complexation by some microbes (Section 8.3.2). For Cu, which can be toxic to some organisms at higher ocean concentrations, and complexation can result in a reduction in toxicity (Section 6.4.3). It has been suggested that complexation may reduce the rate of scavenging of Zn by sinking particulate, and may buffer the free ion concentration, and this could be a reason for the production of Zn-binding organic ligands by microorganism even though these reduce the free ion concentrations [1]. Alternatively, the complexation may be "by accident" with ligands designed for complexation of other metals binding to Zn that is present at a higher concentration. For example, the formation constants of Co, Ni, Cu, and Zn are relatively similar for many simple organic compounds and it is likely that many ligands produced intentionally for one metal may inadvertently bind to others. Therefore, Zn complexation may not be a result of specific Zn-binding ligands. The lack of a strong relationship between estimated ligand concentration and total Zn concentration supports this contention (Fig. 6.27).

In a similar fashion to Zn, the distribution of Cd, which is from the same group in the Periodic Table, in open ocean waters is strikingly similar to that of phosphate (Fig. 6.27) and it is considered a nutrient metal. There is evidence of its depletion from surface waters in many ocean regions. Similarly to Zn, Cd inputs to the ocean are not dominated by the atmosphere, but in contrast to Zn, the inorganic chemistry of Cd in the ocean is dominated by chloride complexation and the relative fraction of Cd as the free metal ion is a few percent. It is also apparent that Cd is complexed to organic ligands in the ocean but the relative degree of complexation is lower than that of Zn [1], and this probably reflects the greater importance of Cl complexation for Cd rather than its binding strength to organic ligands. For example, the value the complexation constants for Cd and Zn binding to EDTA are very similar. Therefore, the major difference in speciation in seawater is likely not due to differences in the relative binding capacities to organic matter, but the differences in the binding strength to the Cl ion.

The concentration of Cd in the ocean is relatively low compared to other transition and Group 12 metals. Its residence time in the ocean is not well-characterized as there is little information on its distribution through all the ocean basins. Given its sources and biogeochemical cycling, it is probable that it has a similar residence time to Zn and other relatively soluble transition metals.

6.4.5 Copper, cobalt, and other nutrient transition metals in the oceans

Cobalt has received recent attention due to the acknowledged potential importance of Co as a co-factor in cobalamin and other enzymes. The uptake of Co by microbes especially in surface waters could lead to differences in the distributions of Co and Ni, another transition metal with similar chemistry and sources. Copper is also a required nutrient metal and shows a distribution that is a mixture of that expected for a nutrient element, modified because of the tendency of Cu to complex with organic matter. This complexation likely increases the solubility and residence time of Cu in deep waters and accounts for the relatively low K_D of this metal compared to others from the first row transition series.

Many studies have demonstrated the importance of organic complexation for Cu ([1] and references therein) and that this complexation is more important in surface waters than in deeper waters [207]. It has been speculated that these ligands are biologically-derived and that certain microorganisms, such as cyanobacteria which are susceptible to Cu toxicity, are responsible for the production of such ligands. This has been demonstrated in some instances [208, 209]. Concentrations of Cu increase from lower levels in the upper ocean (<1 nM) to around 1.5–2 nM in the deep North Pacific Ocean (Fig. 6.34). Inorganically-complexed Cu is a very small fraction (<1% for surface waters of the North Pacific) and increases to 5–10% of the total in deeper waters. As noted in Table 6.1, inorganic Cu is primarily present as the neutral $CuCO_3$ complex in seawater.

In contrast, both Ni and Co form much weaker complexes with inorganic ligands and are present predominantly as free metal ions in seawater. These differences in speciation likely impact the mechanisms of accumulation of these metals into microorganisms. As will be noted in Table 8.1, all these metals have been shown to have a biochemical role, with Cu being incorporated into many enzymes, while the involvement of Ni is relatively small. The concentration of Ni (~8 nM) is higher than that of Cu, while the concentration of Co is sub-nM (Figs. 6.27 and 6.30). While both Ni and Co are found as organically-bound, the relative fraction is smaller than for the other transition metals discussed, being 30–50%, or greater, in various open ocean locations.

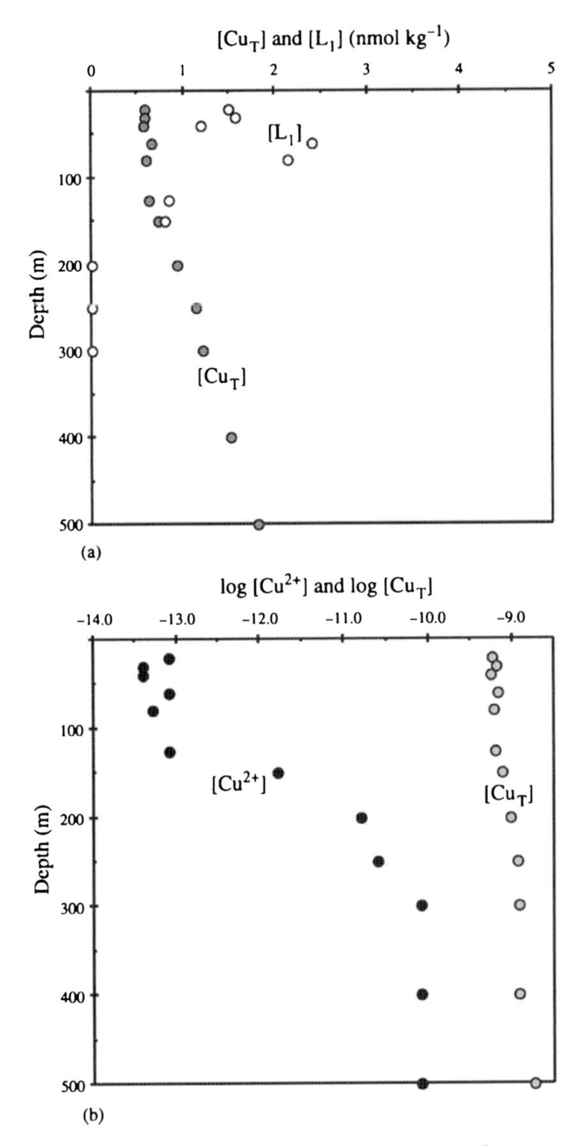

Fig. 6.34 Distribution of copper in the upper ocean showing (a) the concentration of total copper (Cu_T) and estimated total ligand content (L_T), and (b) the estimated concentrations of the free ion (Cu^{2+}) assessed using electrochemical techniques. From Bruland and Lohan (2004) in The Oceans and Marine Geochemistry, vol. 6, *Treatise of Geochemistry* [3], reprinted with permission from Elsevier.

The presence of cobalamin in surface ocean waters has been recently demonstrated although it is present at low pM concentrations [210]. It is possible that most of the organically complexed Co in seawater has a biological origin. The identification of specific biochemicals containing metals raises the possibility that some of the metal, identified as complexed to organic ligands, has been released into the environment as a result of cell leakage, purposeful cellular export (e.g., metals bound to metallothioniens and phytochelatins, as discussed in Section 8.3), cell death, from the release due to grazing or even from fecal material. Alternatively, these ligands could be purposely released into solution to aid in metal assimilation, as occurs for Fe (e.g., release of siderophores), or to reduce metal toxicity, as discussed for Cu previously. Another example is the presence and assimilation of heme (Fe-containing) compounds as it has been shown that microbes can assimilate heme in laboratory cultures [211].

6.4.6 Anthropogenic metals – lead, silver, and mercury

The three main heavy metals that have been examined in detail in the marine environment are Pb, Hg, and Ag. The measurement of Pb was driven by the known input and impact of Pb emitted from its use in leaded gasoline as well as its presence in other anthropogenic sources, as described in Section 2.2. The measurement and examination of Hg is primarily due to the toxicity and bioaccumulation of CH_3Hg in marine food chains and the associated health concerns. Silver is also known to be toxic to organisms (see Chapter 8) but its levels in the open ocean are generally low (<1 nM) [212]. It was first studied in the coastal environment and used as an indicator of local sewage and related inputs [213, 214], primarily as a response to the extensive use of Ag in the photographic industry. As a result of changes in photographic technology (i.e., the emergence of digital cameras), there may have been reductions in local coastal inputs more recently.

All these metals bind strongly to particles and are emitted to the atmosphere from anthropogenic sources although both Pb and Ag will be removed to a degree from stacks by particulate emission control devices. For the open ocean, atmospheric inputs are important but additionally coastal inputs also contribute both Hg and Pb from terrestrial runoff and point source inputs. Mercury is different as is can be present as a dissolved gas (Hg^0) as discussed in Section 5.2.4, and therefore air-sea exchange involves both deposition and gas evasion.

Concentrations of Ag in the ocean typically vary from low pM values in the surface ocean up to 100 pM in deep waters, although there are locations where higher concentrations are found (>100 pM) [215–219] (Fig. 6.35a). Silver has a nutrient-type profile and it has been shown that its vertical distribution mimics that of Si, suggesting its incorporation into the more recalcitrant tissues of microbes and other organisms in the surface ocean and its relatively slow release from sinking particulate material. However, the Ag/Si ratio is much higher in surface waters than at depth suggesting relative differences in the rate of incorporation of these elements in surface waters, or in their degree of release from sinking material [216, 218]. This could also reflect the fact that Ag is likely not readily assimilated by zooplankton and therefore fecal pellets would be enriched in Ag, as they are in Si. Alternatively, the strong binding of Ag to thiols may

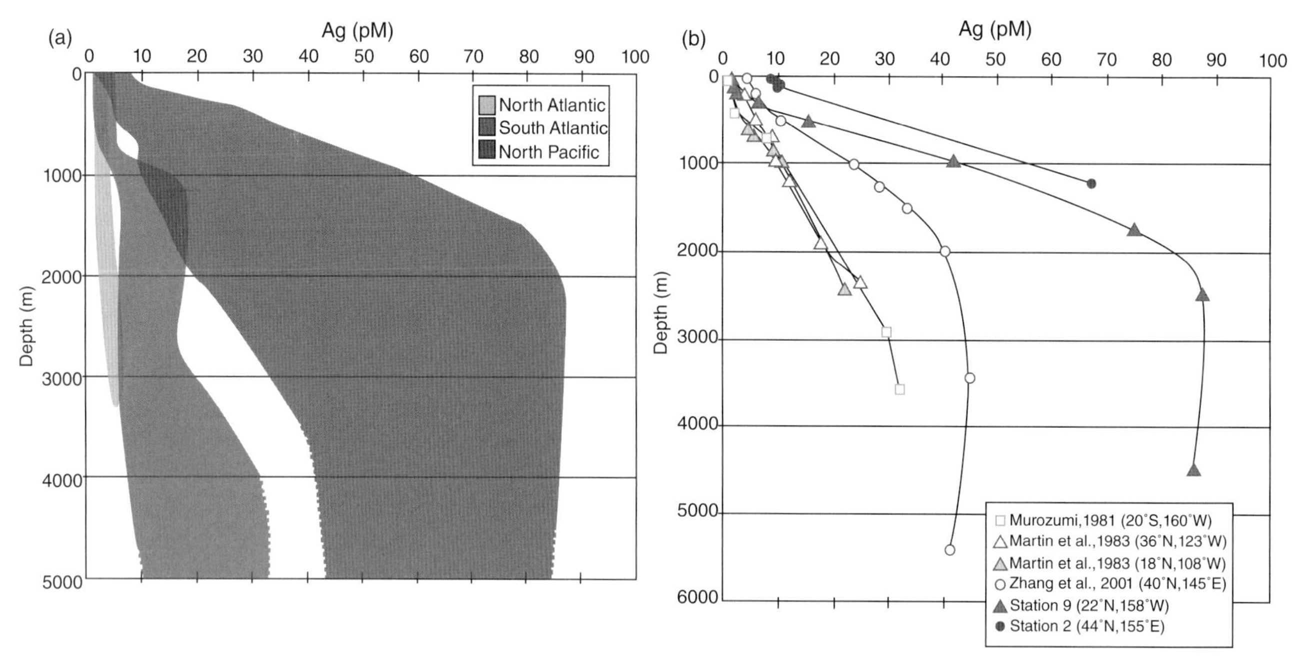

Fig. 6.35 (a) distribution of silver in various ocean basins; and (b) changes in concentration in the North Pacific over time. Figure reprinted from Ranville et al. (2005) *Geochemistry Geophysics Geosystems* **6** [218], with permission of the American Geophysical Union, copyright (2005).

lead to its strong retention in refractive organic matter rather than its association with Si material directly. The difference in the Ag/Si ratios in surface waters between the different oceans reflects differences in the relative inputs of atmospheric Ag to these oceans. Furthermore, the typical vertical distribution can be modified due to the presence of different water masses, as found in the high latitude North Atlantic [215].

Concentrations of Ag appear to change dramatically between ocean basins ([218] and references therein) (Fig. 6.35b). The waters of the high latitude North Atlantic have low Ag concentrations (<10 pM) relative to more surface waters in more impacted regions of the Atlantic, and this likely reflect the relative low Ag inputs and the fact that these are the source waters for deep water formations. The highest deep water concentrations are found in the North Pacific Ocean and this indicates the continual net increase in Ag concentrations as Ag is being continuously released from settling particulate matter but is not being scavenged in deep waters in a similar fashion to other metals.

While Ag is supplied to the ocean mainly from river and coastal sources, direct atmospheric deposition is however not negligible. A recent paper [212] shows, for example, that the atmospheric imprint cannot be ignored in all locations. In the surface waters of the North Pacific Ocean, Ag correlated with Se, which is often used as an indicator of coal burning emissions due to its volatility and relatively high concentration in coal [220] (Section 2.2). The data obtained suggest that atmospheric inputs from anthropogenic sources in Asia are sufficient to impart a signal on North Pacific Ocean surface waters. A similar signal for Al results from the high dust input to this region. Similarly, comparison of intermediate waters in this region suggests that Ag concentrations have increased in the last 20 years, reflecting increased inputs of Ag to the North Pacific [218] (Fig. 6.35b). In contrast, a similar comparison for Hg concluded that there had not been a substantial increase in concentration between 1980 and 2002 [221], although more recent evidence suggests that concentrations have increased recently [222].

The study of Pb in the ocean is couched in the history of the realization, through the demonstration of Clair Patterson in the 1970s, that most measurements of Pb in the ocean were incorrect due to sample contamination during collection and handling, as discussed in Chapter 1. A number of ongoing and detailed studies since then have been able to document in detail the overall contamination of the ocean by anthropogenic Pb, and the role of Pb released from the use of Pb in gasoline. Additionally, these studies have demonstrated the resultant decrease in ocean Pb as a result of the phasing out of Pb additives for gasoline [223]. However, there is still Pb input to the atmosphere from other anthropogenic sources and current studies are evaluating the extent of this input. One unique aspect of examining Pb geochemistry is the fact that it has numerous isotopes, and some of these are daughters of the U-Th decay series. As noted in Section 6.2.1, ^{210}Pb is radioactive and is a much used dating tool. Additionally, ^{206}Pb is the stable product of the ^{238}U decay series, ^{207}Pb the stable product of the ^{235}U decay series and ^{208}Pb is the stable product of the ^{232}Th decay series. Thus, the isotopic ratio of Pb in the environment is altered due to the presence of either U or Th in the medium. It has been demonstrated that Pb from different locations has ratios which are different enough to track the sources of the Pb in the ocean and other environments. Additionally, because of the number of isotopes, plots of different isotope ratios allow for more resolution of the source signals. This is illustrated in Fig. 6.36. Data for the South and Equatorial Atlantic Ocean collected in 1996 is plotted against the various potential source signals and it is concluded that the waters represent a mixture of Pb from a number of natural (e.g., Saharan dust) and anthropogenic signals, including inputs from North America, large rivers and coastal inputs. In contrast, the data for the deep waters of the North Pacific, which represent the "oldest" marine water masses, shows a Pb signal that is interpreted to represent the input of Pb to these deep ocean waters via particle settling and from the anthropogenic enrichment of atmospheric deposition.

As noted previously in Section 6.4.1, much of the Pb in the ocean reflects the different source locations and spatial and temporal distribution of these inputs. This is illustrated in Fig. 6.37 for the North Pacific [223]. The vertical profiles of Pb also reflect differences in the source signal as concentrations are higher in the Atlantic Ocean compared to the Pacific (see Fig. 2.16 and 6.29). For example, the concentrations in the upper waters of the North Atlantic in the vicinity of Bermuda have been decreasing in the last 30 years (see Fig. 2.19) [224], in response to the phasing out of Pb in gasoline in many countries. Concentrations of Pb have decreased from values above 150 pM in the late 1970s to concentrations around 50 pM today. Note that the predicted pre-industrial Pb concentration is 1.4 pM for the deep waters of the North Pacific, and about 14 pM for the Atlantic. These differences in concentration reflect the differences in the natural source signal relative to the basin size as well as the importance of particulate scavenging in the deep ocean in removing Pb from the water column.

It can therefore be concluded that while the surface waters near Bermuda have decreased in concentration they are still substantially elevated above background and reflect the continual input of Pb from combustion and other industrial sources. The recent data from the North Pacific show horizontal differences in concentration that likely reflect the differences in atmospheric inputs to the surface ocean. Highest surface concentrations are found in the mid-latitudes and this reflects the heightened input of Pb to these waters. Air masses from the Asian continent typically track through the mid-latitudes and therefore the concentration reflects this anthropogenic signal. The upper ocean waters are much higher in concentration than the deep waters. Also, the values obtained in this study are similar to previous measurements at this location, suggesting little change in the recent past.

For Hg, both box and numerical models of the global cycle support the notion that anthropogenic releases of mercury (Hg) to the environment have impacted the biosphere substantially, with enhanced deposition (a factor of 3–5) to the open ocean as a result of long range transport and subsequent deposition [225, 226]. The global cycling and air-sea exchange of Hg has been discussed in detail in Sections 5.2.4 and will not be reiterated here. Limited data support the premise that Hg in the open ocean has increased substantially during the past 200 years, primarily as a result of greater atmospheric Hg deposition [227, 228]. Besides atmospheric inputs, local and regional-scale contamination of the coastal zone by Hg has occurred as discussed earlier in this chapter, due to runoff from the terrestrial environment and from point source inputs [229, 230].

Mercury distributions in surface waters reflect the magnitude of the atmospheric deposition source and the strength of local removal process (scavenging and gas evasion) superimposed on water circulation [221, 222, 226] and concentrations change seasonally depending on the variability of atmospheric deposition, evasion and removal of Hg by particulate sinking. A recent modeling output shows the potential variability that could exist for various Hg species in the ocean (Fig. 6.38). Concentrations vary between ocean basins due to differences in the relative impact of anthropogenic Hg in deposition, and these are changing over time. Recently, more of the anthropogenic inputs are from Asia with less from North America and Europe. Surface water concentrations are higher in the Atlantic Ocean and Mediterranean Sea than in the Pacific Ocean [222, 231–238], and there is evidence for recent decreases in concentration in the North Atlantic and the Mediterranean Sea and these are consistent with model predictions [226, 239, 240].

In many locations, the concentration and distribution of Hg appears to be relatively uniform even given its known reactivity. However, the lack of data is to a degree responsible for this observation as it appears that signals of Hg concentration can be transient, and changes are likely to occur on short timescales in the upper ocean [221]. For example, the input of Hg from the atmosphere was demonstrated though an increase in the concentration of Hg in the waters of the North Pacific Ocean seasonal mixed layer during the summer. However, this signal is eliminated as a result of deep water mixing in the fall. Seasonal mixing and latitudinal transport of sinking surface water within the permanent thermocline is a mechanism for the transport of Hg deposited at higher latitudes, which may have been deposited to the ocean a decade or more previously [1, 2], to tropical and other regions [226, 241]. Mercury in the thermocline waters of the North Atlanticis is an example of this mid-depth signal

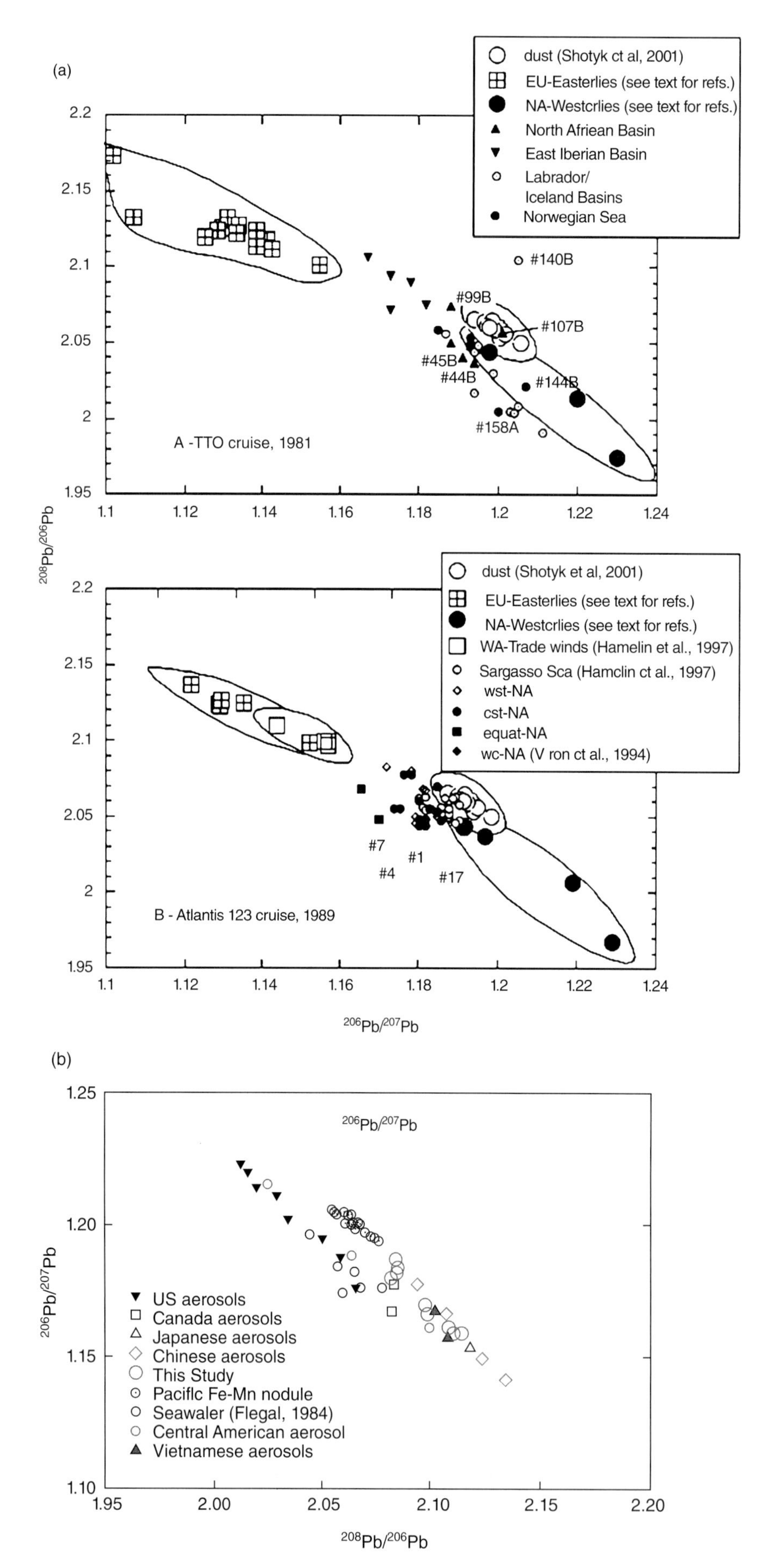

Fig. 6.36 Relationship between the various stable lead isotopes for difference waters, atmospheric aerosols and ocean sediments. Part (a) reprinted from Weiss, D., Boyle, E.A., Wu, J.F. et al. (2003) Spatial and temporal evolution of lead isotope ratios in the North Atlantic Ocean between 1981 and 1989. *Journal of Geophysical Research-Oceans*, **108**(C10), Article Number: 3306, DOI: 10.1029/2000JC000762, 2003, with permission of the American Geophysical Union, copyright (2003) and (b) from Wu et al. (2010) *Geochimica et Cosmochimica Acta* **74**: 46229–38 [224] and reprinted with permission of Elsevier.

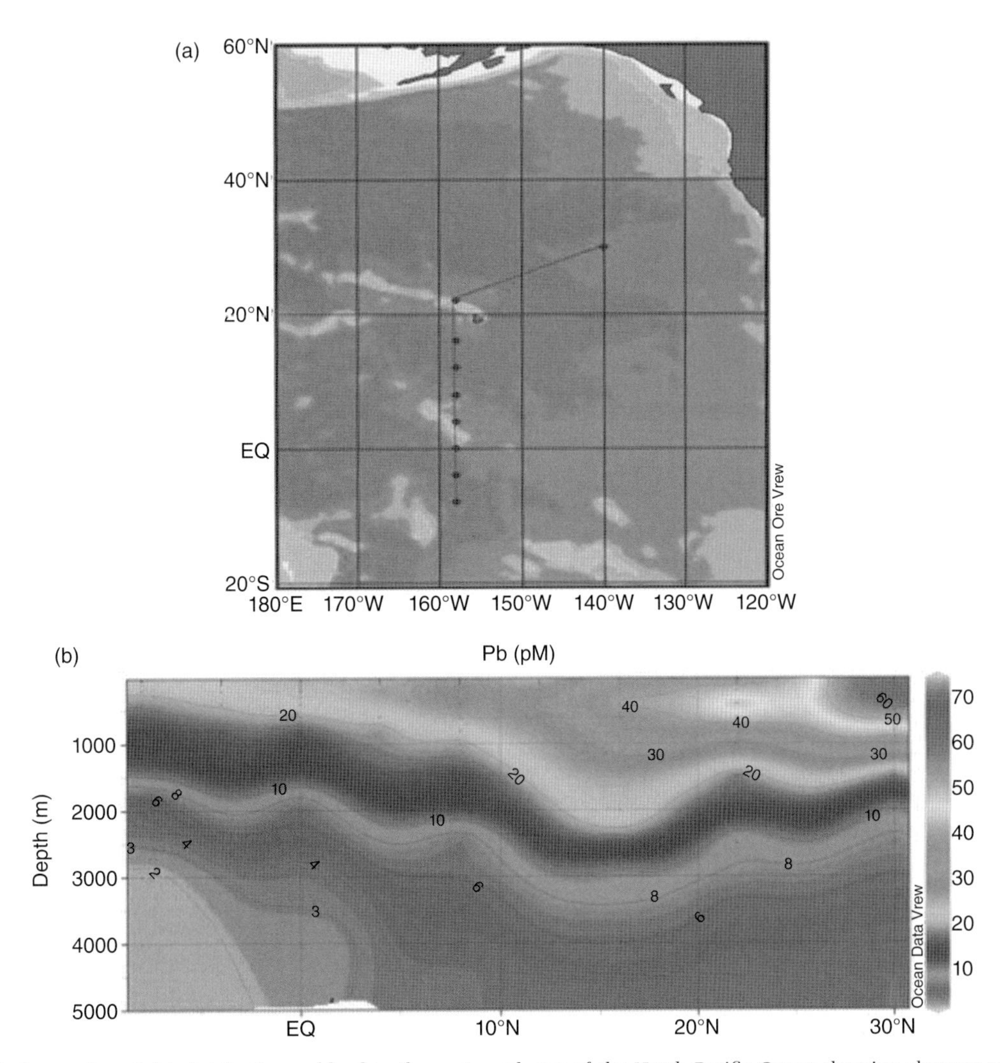

Fig. 6.37 (a) Cruise track and (b) distributions of lead in the water column of the North Pacific Ocean showing changes with latitude and location. Taken from Wu et al. (2010) *Geochimica et Cosmochimica Acta* **74**: 46229–38 [224] and reprinted with permission of Elsevier.

as shown in Fig. 6.39(a), which compares the concentrations of Hg in the North Pacific. The higher historical concentrations in the more recent data suggest the input of Hg to the mid-depth waters of both oceans, and this coincides with an increasing anthropogenic signal from Asia to the North Pacific, and such changes are not evident in the deep ocean waters [221, 222]. Such upper-ocean cycling confounds the understanding of how the Hg concentration in ocean surface waters has changed as a result of increased anthropogenic inputs, and model predictions suggest that the response time of the upper ocean to changes in atmospheric Hg concentrations is decadal or longer [240].

The distribution of Hg in ocean waters reflects the sources and cycling as well as the internal cycling of mercury, demethylation, oxidation and reduction. The biological production and destruction of methylated Hg species, primarily CH_3Hg and $(CH_3)_2Hg$ in the ocean, is discussed in detail in Section 8.4. The photochemically-driven redox chemistry at the air-sea interface, and air-sea exchange, is discussed in Section 5.2.4. Vertical profiles of Hg speciation are shown in Figs. 6.38(b–d), and these profiles illustrate the most important parts of the cycling on Hg speciation in the ocean. In numerous profiles there appears to be an enhancement in the concentration of methylated Hg (CH_3Hg and $(CH_3)_2Hg$) at mid-depth. Note that in some studies samples were acidified to preserve them for later analysis and because of the instability of $(CH_3)_2Hg$, this results in the quantification of total methylated Hg ($\Sigma(CH_3)_xHg$; $x = 1 - 2$). Recent measurements in the North Pacific Ocean demonstrate the recent improvements in analytical capabilities (DL < 10 fM) as well as showing the typical profile of an enhancement in concentration of methylated Hg in the region of the oxygen minimum zone [242]. This has been confirmed by other studies in the North Pacific [222] and Equatorial Pacific [235], and in other ocean basins [243].

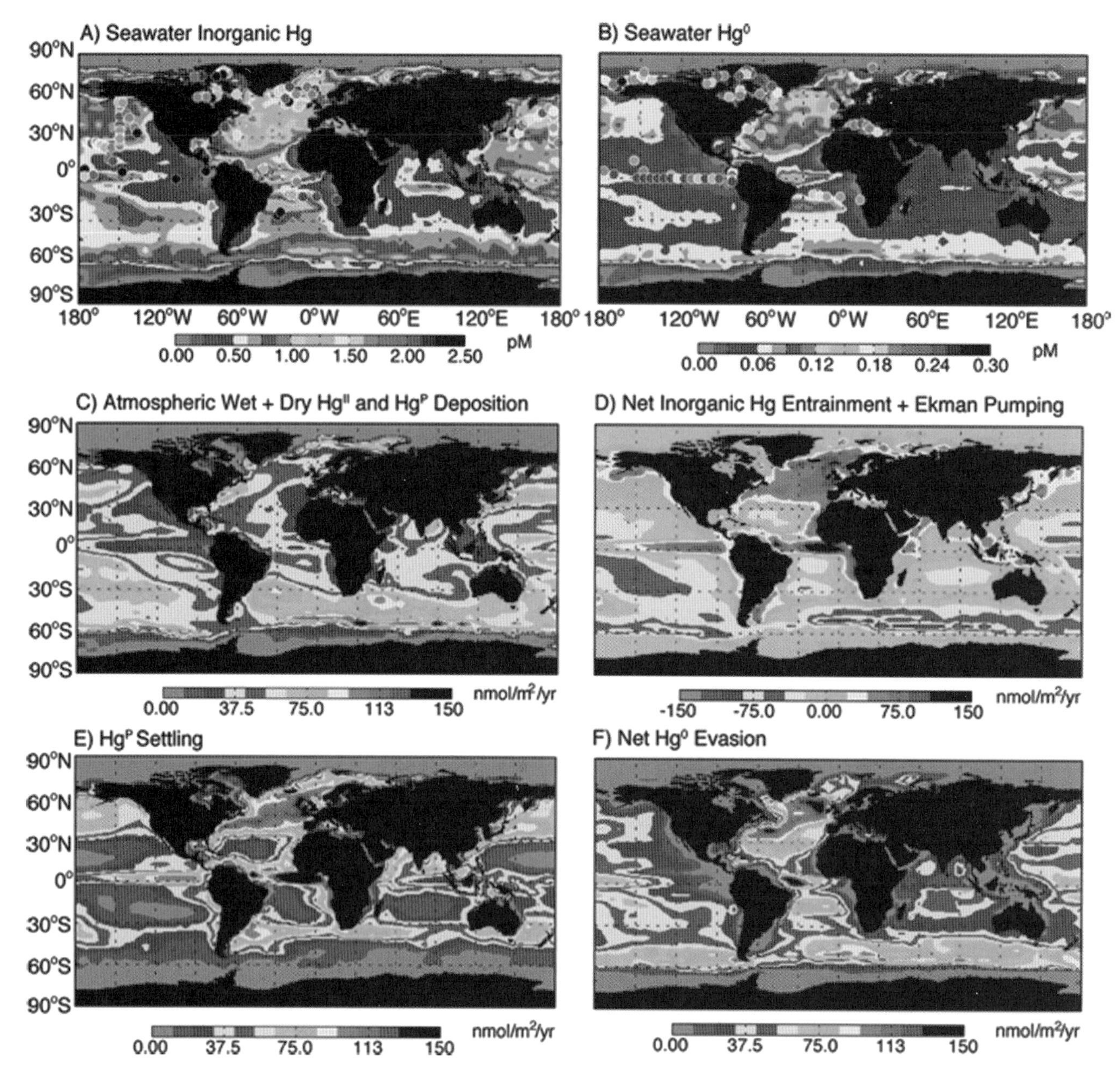

Fig. 6.38 Modeled surface water concentrations of inorganic mercury and elemental mercury, and the estimated net flux of elemental mercury from the ocean to the atmosphere for the global ocean. Shown on the figure (circles) are also the measured values from the literature for comparison. Additional figures show the mercury fluxes into the mixed layer (atmospheric wet and dry deposition and inputs from below (Ekman pumping and entrainment) and the particulate removal flux. (A) total inorganic mercury concentration in seawater (pM) and (B) elemental mercury concentration. (C) Atmospheric wet and dry deposition flux; (D) inputs of mercury from below the mixed layer (entrainment and Ekman pumping); (E) removal of mercury via particle settling; and (F) the evasion of elemental mercury at the ocean surface. All fluxes in nmol $m^{-2} yr^{-1}$. Reprinted with permission from Soerensen et al. (2010) *Environmental Science and Technology* **44**: 8574–80 [240]. Copyright (2010) American Chemical Society.

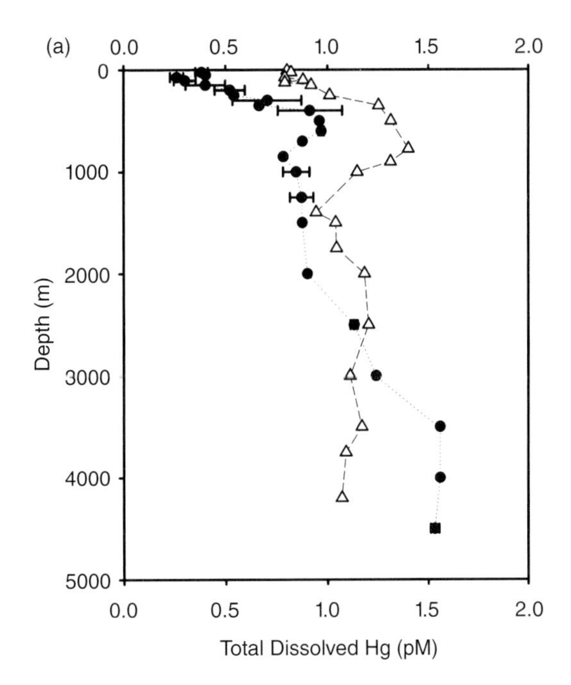

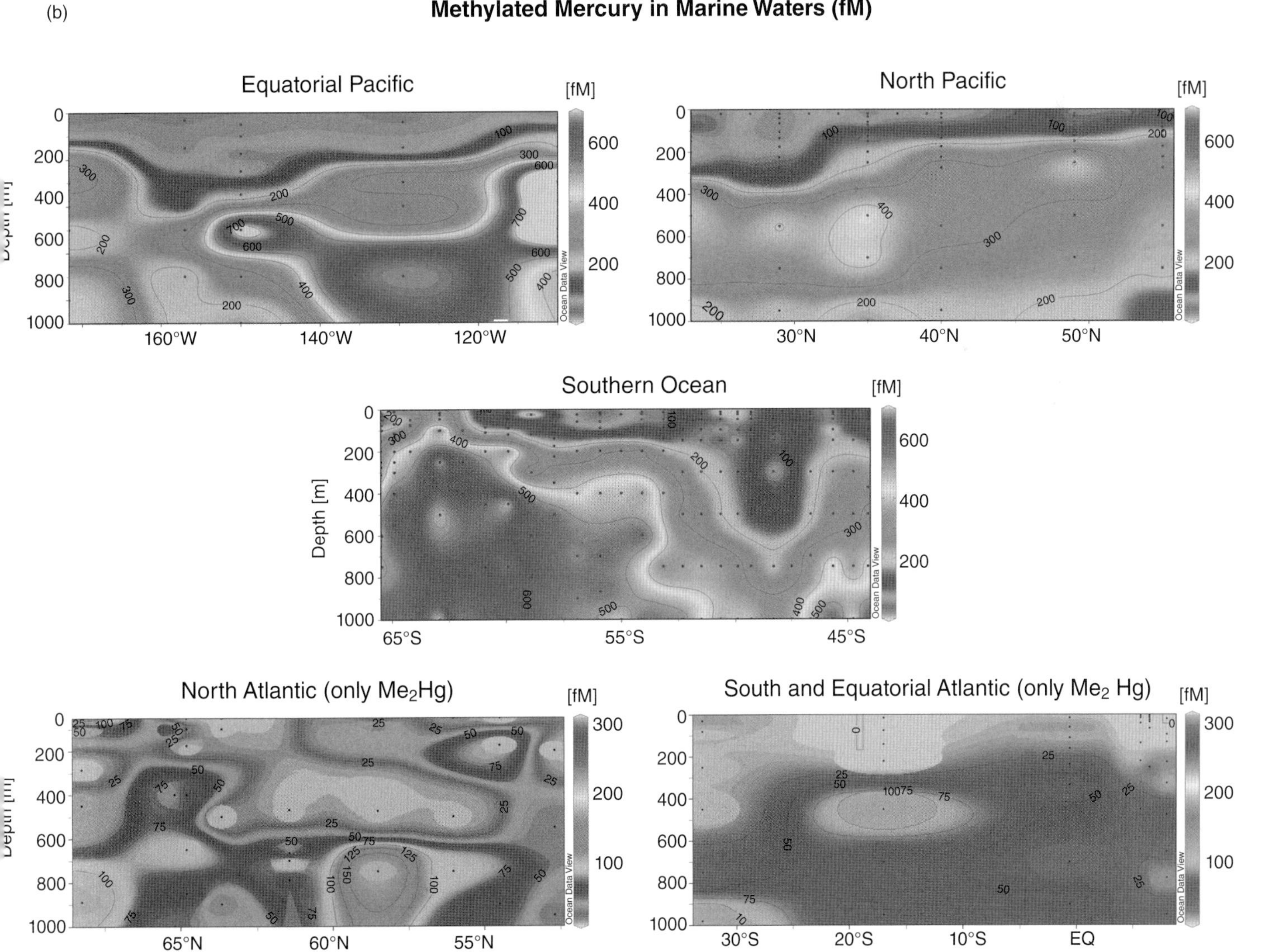

Fig. 6.39 Vertical profiles of (a) total mercury for the North Atlantic △ and North Pacific ● and (b) total methylated mercury for various ocean basins showing the details of the vertical distributions. Part (a) from Lamborg et al. (2012) *Limnology and Oceanography*, Copyright (2012) by the Association for the Sciences of Limnology and Oceanograpy; and (b) from a variety of sources with figure reprinted from Mason et al. (2012) *Environmental Research*, and reprinted with permission of Elsevier.

The analysis of the data and the relationships to environmental parameters suggest that the profiles can be best explained in terms of production of methylated Hg during the decomposition and remineralization of organic matter [222, 243, 244]. This suggests that these species are also being produced below regions of high productivity where there is substantial organic matter recycling. Concentrations of CH_3Hg may not build to high levels in surface waters, as seen in the profiles in Fig. 6.39, even though production may be occurring. For $(CH_3)_2Hg$, and to a lesser extent for CH_3Hg, photochemical reactivity likely results in their depletion in the mixed layer [236, 245]. The more recent studies examining the production of methylated Hg suggests that the presence of methylated Hg in low oxygen waters is more due to the same factors that cause the oxygen depletion – heightened bacterial activity and slow vertical mixing, lack of water ventilation, and particulate scavenging – than to the activity of particular microorganisms [222, 244]. In freshwater and coastal environments and sediments, sulfate and iron reducing bacteria have been demonstrated to be the most important methylating organisms [110, 142]. While little is known about the microorganisms or processes whereby Hg is methylated in the ocean, the fact that methylation occurs in upper ocean waters (Fig. 6.39), and the dominance in many instances of $(CH_3)_2Hg$, suggests that the pathways may be very different from those for freshwater and coastal environments, and in sediments.

The vertical distributions found in the various studies depicted in Fig. 6.39 and found in studies in other ocean waters, are consistent overall. Sunderland et al. [222] used a model that incorporated the extent of organic matter remineralization to demonstrate a correlation between the extent of degradation and the production of methylated Hg. Clearly, some microorganisms, or perhaps even some chemical/enzyme released into the water column during the degradation of biologically-produced organic matter, are responsible for Hg methylation in the ocean [243, 246]. One intriguing possibility is the fact that cobalamin, which can methylate Hg abiotically in the laboratory, is present in the ocean upper waters at pM levels [210], of the same order as the inorganic Hg content. Perhaps, this enzyme and other methyl-donating chemicals (see Section 8.4) are responsible for Hg methylation in the ocean water column.

Finally, the more recent studies also illustrate the fact that a substantial fraction of the Hg in the upper ocean waters can be methylated Hg. The fraction as methylated Hg is as high as 40% in the data shown in Fig. 6.39 and this is consistent with the results of other studies. While methylated Hg is produced during organic matter degradation it is not stable in ocean waters, especially $(CH_3)_2Hg$, and therefore if there is not continued production the concentration will slowly decrease. This accounts for the lack of a strong correlation between methylated Hg and apparent oxygen utilization (AOU), which reflects the integrated amount of degradation since the water mass left the ocean surface. Overall, the concentration of methylated Hg is determined by the complex interactions that occur throughout the water column: production and destruction of methylated Hg by microbial processes and abiotic mechanisms; scavenging and release from particles; and uptake into the food chain. The cycling of methylated Hg in the ocean is distinctly different from that of the coastal and freshwater environment, which was discussed earlier in this chapter and in more detail in Section 7.2.3.

Overall, there are strong similarities between the three major heavy metals in terms of their strong association with organic matter and the particulate phase. Mercury and Ag have similarities in that they are both Class B metals and form strong associations with reduced sulfide, and exist in seawater as chloride complexes in the absence of organic complexes. There are strong similarities in their potential to be assimilated by passive diffusion across membranes of the neutral complexes, as discussed further in Chapter 8. Lead can also be associated with sulfides and forms strong complexes, but to a lesser degree compared to Hg and Ag. Also, Pb does not form strong chloride complexes and therefore its inorganic speciation in the ocean is mostly dominated by the neutral $PbCO_3$ complex (Table 6.1). Therefore there is also the potential for passive accumulation of Pb, although this has not been examined in any detail. Atmospheric sources are all important for these metals to the open ocean, and this is primarily related to the fact that their global cycles have been substantially impacted by anthropogenic activities. Besides Hg, the accumulation and fate of these elements in ocean microorganisms and in the oceanic food chain has been little examined and this could be an important area for future research.

6.4.7 Metalloids and other oxyanion cycling in seawater

Most of the metalloids of interest in marine systems (e.g., As, Se, Sb, Ge) exist as oxyanions, and in a number of oxidation states (See Tables 6.1 and 2.1) [247], although they are also found as methylated compounds, or even as larger metalloid-containing species, such as arsenobetaine and selenoproteins. These organic species and their formation mechanisms are detailed in Chapter 8. For example, As can be found as either As^{III} or As^{V} and as mono-, di- and trimethyl arsenic in marine waters. As noted in Section 8.4, it is thought that the methylation of As is a detoxification/elimination mechanism for As from phytoplankton as As^{V} can be taken up inadvertently by microorganisms in low phosphate waters (both exist as polyprotic acids with similar pK_as). The methylation process is an oxidative methylation process and the As^{V} is initially reduced to As^{III} before being methylated. This is a different mechanism to Hg methylation by sulfate reducing bacteria. There is also often As^{III} present in conjunction with phytoplankton in surface waters, due to this reduction pathway, which is contrary to what is expected based on thermodynamic equilibrium calculations. This is shown, for example, in the upper waters of the North Atlantic (Fig. 6.40). However, As^{III} is a small fraction of the total, as are the methylated species [247]. In estuarine environments and freshwaters, As^{III} and the methylated forms can be a larger fraction of the total dissolved As [248, 249].

Representative profiles of As, Sb, and Se, showing their speciation and vertical distributions are shown in Fig. 6.40. Both methylated Se and Sb compounds exist in natural waters, as well as both the oxidized and reduced inorganic forms, which are all present as oxyanions. The distribution and speciation of Sb is similar to that of As in the ocean water column [1, 250] (Fig. 6.40). The dominant oxidation state is +V, but with the presence of the +III oxidation state

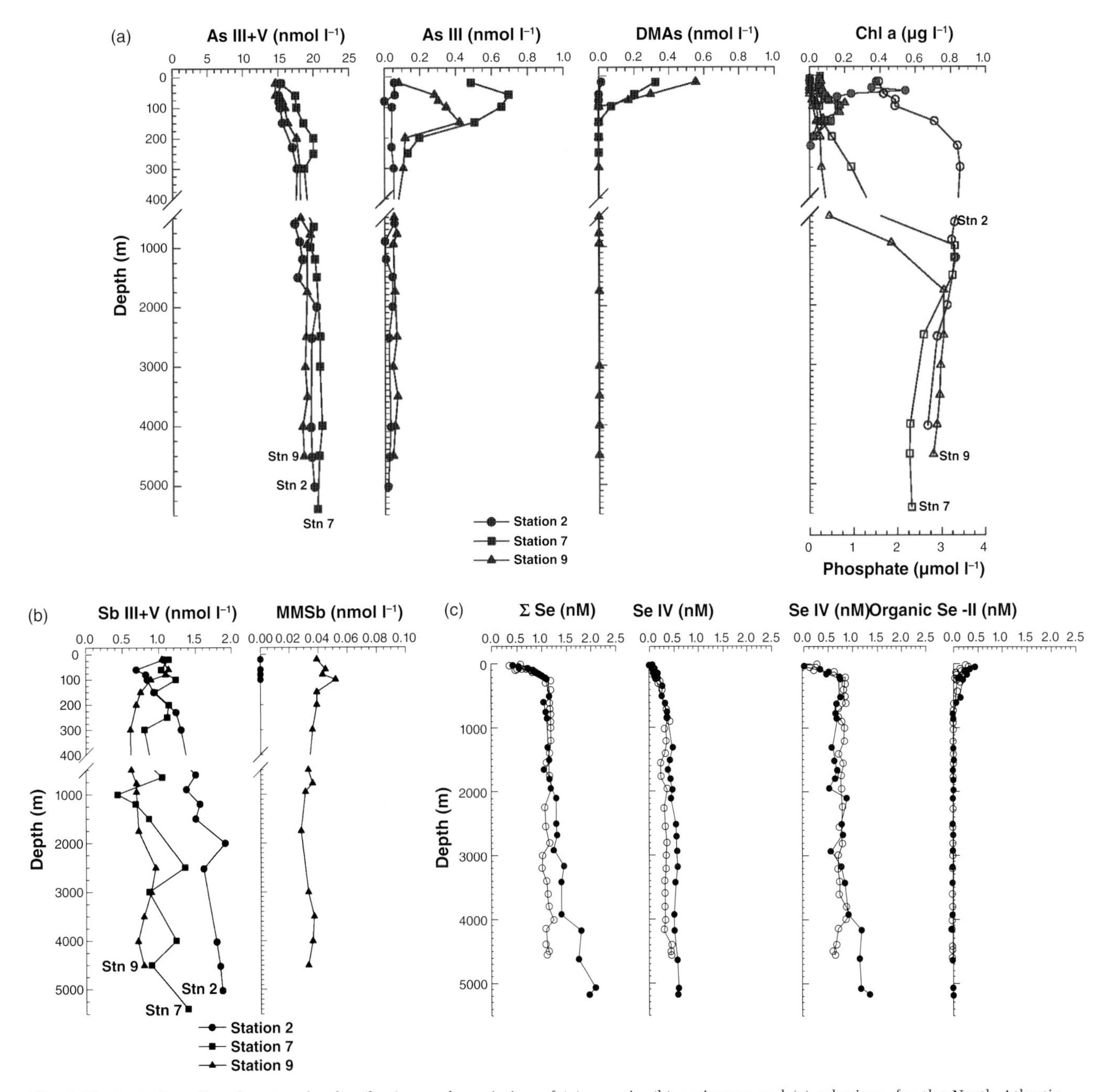

Fig. 6.40 Vertical profiles showing the distribution and speciation of (a) arsenic, (b) antimony and (c) selenium, for the North Atlantic Ocean showing the profiles of the inorganic and organic forms. Taken from Cutter et al. (2001) and reprinted with permission of Elsevier.

in the upper waters and the presence of methylated species, making up about 10% of the total dissolved Sb. In the North Pacific, dissolved As displayed mildly scavenged behavior, as suggested by a correlation with Al, and there was also a suggestion of an anthropogenic atmospheric signal based on higher levels closer to the Asian mainland [250]. The mono-methylated form dominated over the other methylated species and showed relatively conservative behavior.

The two main inorganic redox states of Se appear to cover a similar range in concentration, with deep waters having a ratio of Se^{IV}/Se^{VI} of >0.5, but <1. Both inorganic species appear to be depleted in the surface waters, likely due to their uptake and incorporation into biota. Selenium is an essential element although it is only required at low concentrations. The distribution of organic Se (Se^{-II}) suggests its persistence through the water column, either due to its continual formation and release from microbes and/or from organic matter dissolution, or due to its stability. As most of the bioorganic Se compounds are proteins, it is likely the former (continual production) is occurring. This is also demonstrated by a decrease in organic Se with water mass age for the sinking waters of the North Atlantic, suggesting net destruction. As noted in the atmospheric chapter, volatile Se compounds can be formed (analogs to methylated sulfide species) and the evasion of these compounds could also result in depletion of Se from surface waters. The cycling of Se in the upper ocean is not well understood and needs further study [247].

Germanium (Ge) is a most unusual element in that its ocean distribution is dominated by its methylated forms [247]. Germanium is found as the mono- and dimethylated species with the monomethyl species being the dominant form throughout the water column [1]. It has a conservative distribution which indicates its high stability and little is known about its formation mechanisms. In the South Pacific Ocean, for example, inorganic Ge ranged from 50–100 pM [251]. In the South Pacific, CH_3Ge is about 200 pM and accounts for 50% or more of the total dissolved Ge. The inorganic form shows depletion in surface waters (<10 pM) and highest concentrations at depth (>100 pM), and has a nutrient-type distribution. In the deep South Pacific, the Ge/Si ratio was approximately 0.72×10^{-6}. This is not surprising given its location in the periodic table and its speciation in seawater, which is very similar to that of silicate, and the Si:Ge ratio in plankton skeletal material is relatively constant over the ocean as a result [251, 252].

There is little information on the ocean distribution of tin (Sn). Much of the focus of study of this element has been due to its use as antifouling agents (e.g., tetrabutyltin) and the biogeochemistry of these is discussed in Chapter 8. In the surface ocean its concentration is around 20 pM, and it is lower at depth, showing a scavenged profile. There is similarly in the distribution of Sn and technecium (Tc) (Fig. 6.30) which ranges from 0.5 pM in surface waters to <0.1 pM in deep waters [198]. Bismuth is higher in surface waters (0.25–0.45 pM) and shows a deep water scavenged profile. There is evidence for a mid-depth maximum, likely related to its release with the dissolution of Fe-oxides particles as it is primarily present as an oxyanion (BiO^+, $Bi(OH)_2^+$) [198]. Its estimated residence time is very short (~20 years) but many details of its ocean cycling are poorly understood, and this rapid removal appears anomalous. The concentration of gallium (Ga) (2–70 pM, average 17 pM) increases from low surface concentrations with depth and the concentration in the North Atlantic is higher than the North Pacific. Atmospheric inputs are the main source to the open ocean for Ga and Bi [198]. There is a mid-depth depletion and overall Ga has a similar distribution to that of Al.

Overall, there has been little recent study of the inorganic and organic speciation of the metalloids in the ocean water column. It is probable that new insights and understanding could be gained from the examination of the various fractions in more detail, especially the "organic fraction". It is not clear whether these compounds are derived directly from microorganisms, and other biota, or are primarily produced during organic matter remineralization. It has, however, been shown that microorganisms contain small and large molecular weight metalloid-containing molecules. One aspect discussed in Section 5.2.4 is the formation of volatile Se compounds and the similarity between these compounds and S-containing analogs (e.g., $(CH_3)_2S_ySe_z$; $y = z = 0 - 2$; $y + z = 1$ or 2). These Se-containing compounds could potentially be used as tracers of biological processes given the ability to measure low levels of Se in the environment, and the potential to do experiments using the stable isotopes of Se as a tracer of specific pathways.

6.4.8 Other transition metals, the lanthanides, and actinides

Much less is known about the concentrations and distributions of the elements in the second and third row of the transition series. As can be seen from the plots in Fig. 6.30, the third row transition metals are typically present at less than 1 pM except for W (~80 pM), Re (~40 pM) and Tl (~60 pM), which exist as oxyanions in seawater and are relatively soluble. Their distributions are conservative based on the limited available information [253]. The second row transition metals tend to have somewhat higher concentrations which likely reflect their higher abundance in the Earth's crust. Most of the second row metals have a distribution with a lower surface concentration, overall increasing with depth. For example, Y increases from 100 to 300 pM, and Zr from <50 to 300 pM (Fig. 6.30). The profile for Nb has more structure but this likely reflects the fact that it has been studied in greater detail due to its potential use of Nd isotopes as a tracer of water mass and sources [198], as detailed in a special issue on Nd (*Quaternary Science Reviews*, Vol. 29, 2010).

Concentrations of Zr, Hf, Nb, Ta, Mo, and W have been measured in the western North Pacific Ocean [254]. Concentrations of W are relatively constant (40–50 pM) while Zr is <50 pM in surface waters and up to 300 pM in deep waters. In contrast, Hf, while having low concentrations in surface waters has a relatively constant concentration with depth but concentrations are low (<1 pM); Ta is also <1 pM. Low concentrations of Nb were also found (3–7 pM) [255]. Concentrations of Os, Pt, Au are typically <1 pM while Tl has a concentration around 60 pM. Most of these elements have been little studied as they are neither toxic at these concentrations nor are they thought of as essential metals or valuable tracers. There is some evidence that W can have a biological role, however.

A number of these metals (Zr, Hf, Nb, and Ta) are termed refractory and exist mostly as hydroxide species in solution, and they are particle-reactive and likely this accounts for their low concentration in the ocean. Concentrations of these metals are typically higher in river water and vary from about 5–50 times higher. Similarly, these metals are enriched in seawater relative to their ocean concentrations. These differences provide additional evidence for their rapid removal from seawater and their relative insolubility ([254] and references therein).

Many of the transition metals, especially those at the left of the periodic table also exist in seawater predominantly as oxyanions. The dominant oxidation state of V in surface seawater is V^{+V} (HVO_4^{2-}) but it can be reduced in anoxic zones, and is much less soluble in the V^{+IV} oxidation state (vanadyl ($V^{IV}O^{2+}$) [244]. The vertical distribution of V is relatively conservative and its concentration ranges from 30–36 nM. Vanadyl is a smaller cation, and binds even more strongly to chelating surface groups than the larger anionic vanadate. Experimental results suggest that V can be reduced to V^{III} in sulfidic regions, leading to V enrichments in both sulfate reducing sediments and ore deposits [256, 257]. Vanadium removal from pore waters is thought to occur below the level of either Mn or Fe oxyhydroxide reduction [258] but it is also suggested that V is associated with oxyhydroxide phases. Analysis of particulate material from hydrothermal plumes confirms that V is coprecipitated from seawater with Fe oxide particles [259].

Molybdenum is an important element in the ocean as it is incorporated into enzymes and has a biochemical function. The stable oxidation state in oxic seawater is Mo^{VI}, and the major species is MoO_4^{2-}. Increased Mo concentrations are observed in MnO_2-rich sediments, because of Mo adsorption onto Mn oxyhydroxides [260]. Reduction of Mo^{VI} to Mo^{IV} and its resultant authigenic enrichment happens under sulfidic conditions. Hydrothermal vent fluids are depleted in Mo, which is most likely a reflection of their removal in conjunction with sulfide precipitation, as discussed in Section 6.3.7.

Rhenium is thought to exist in the +VII oxidation state as ReO_4^- in oxygenated seawater [261]. Its detrital concentration is extremely low, making it an ideal tracer for authigenic mineral formation. The redox behavior of Re appears to be less complex than V, Mo, or U because it is not strongly related to Mn or Fe cycling [256]. Sediment data suggests that Re begins to be authigenically enriched in sediments at or just below zones of Fe and U reduction, but before Mo precipitation in suboxic environments [256, 257]. Furthermore, hydrothermal processes play a negligible role in Re geochemistry [253].

Other important oxyanions include some of the actinide and lanthanide species. For example, the stable form of U in oxygenated waters is U^{VI}. Carbonate ions complex with dissolved U in seawater, creating $[UO_2(CO_3)_3]^{4-}$ which dominates the speciation in most natural waters [1]. The dominant source of U to the ocean is fluvial input [262]. There is little difference in the vertical of interocean concentration of U given its long residence time ($2 - 4 \times 10^5$ yrs) and it is around 14 nM [262]. Anoxic basins, organic-rich shelves and hemipelagic sediments are sites for authigenic U deposition [263], as discussed in Section 6.4. It appears that U is scavenged from the water column by particulates to the sediments, where it is fixed through reduction and subsequent adsorption or precipitation, possibly as uranite, ($U^{IV}O_2$ (s)) at the depth of Fe remobilization [256, 263]. Laboratory studies also suggest that U is released from Fe-Mn oxides as they are reduced [263]. Sediment analyses from hydrothermal vent sites indicate that U is strongly enriched in hydrothermal sulfide deposits [264].

The so-called "platinum group elements" (Ru, Rh, Pd, Os, Ir, and Pt) (PGEs) are all found in low concentration, and this is related mostly for their strong removal by processes in the deep earth rather than their insolubility in seawater. Their concentrations range from 0.6 pM (Pd) to 0.5 fM (Ir). Their ratio of their concentration in seawater compared to the earth's crust is much lower than many metals (3×10^{-4} to 2×10^{-6}) [253]. The relatively most soluble trace elements have ratios of 10^{-2} to 10^{-3}. The oxidation states of the various elements in oxic seawater are: Ru^{IV}, Rh^{III}, Pd^{II}, Os^{VI}, Ir^{II},^{I} and Pt^{III}. Most of the elements have a somewhat scavenged profile with depletion in the surface waters and higher concentrations at depth (Rh, Pd) or show depletions due to redox transformations. Osmium shows depletions in low oxygen waters and is suspected to have redox chemistry in seawater, and exists as an oxyanion in oxic waters [253].

Osmium isotopes have potential uses as paleoproxies as they are formed by the decay of radioactive elements; ^{187}Re to ^{187}Os; ^{190}Pt to ^{186}Os. The half-lives of both are in the

40–500 billion years. The ratios show distinct differences in sources to the ocean and differences have been found in sediment material. However, these analyses are difficult and complex due to the low concentrations of both the parent and daughter compounds in seawater. It has been suggested that the Os isotopic signature of marine sediments can be a useful proxy for extraterrestrial material [253]. Finally, given the enhanced usage of PGEs in the recent industrial past, in catalysts, especially in catalytic converters, and other industrial products, these elements can serve as useful tracers of industrial activity and the associated inputs to the ocean. Palladium usage for example, increased by a factor of 30 between 1990 and 2000.

The lanthanides, which with the inclusion of Y and Sc are termed the "rare earth elements" (REEs), typically show a scavenged profile with low concentrations in surface waters and higher concentrations at depth; a "nutrient-type" distribution with higher concentrations in the deep Pacific compared to deep Atlantic waters [265]. Cerium is the exception in that it has a high surface water concentration (~10 pM) and its concentration below the thermocline is relatively constant around 5 pM. The concentration ranges of the other lanthanides are: La 10–60 pM; Nd 10–40 pM; Dy 5–15 pM; Pr, Sm, Gd, Er, and Yb 2–10 pM; and the remainder being <3 pM (Eu, Tb, Ho, Tm, and Lu). Essentially all exist in the +III oxidation state and their chemistries are very similar [265], likely due to the fact that the outer electron configuration (6*s* orbitals) is similar due to the filling of the inner 4*f* orbitals with electrons across the series, which leads to an overall contraction of ionic radius across the series. Of all the REEs, Ce has the most different chemistry due to its redox behavior in seawater, and existence in the +IV oxidation state. The only other exception is Eu, which is found as Eu^{II} in seawater. Overall, given their similarities but differences in properties, such as the increasing complexation with atomic mass, the REE's are considered useful tracers of marine processes [265].

For Ce, the higher oxidation state (+IV), which exists in oxic water, is more particle reactive and this leads to the depletion of Ce relative to the other REEs; the so-called "cerium anomaly". Other uses of REEs as tracers relate to the elemental isotopes formed by radioactive decay: ^{143}Nd from ^{147}Sm ($t_{1/2}$ 10.6×10^{11} yr) is a useful tracer of sources and mixing in the ocean, as is ^{138}Ce formed from ^{138}La ($t_{1/2}$ 2.97×10^{11} yr) [66]. The most dominant sources of most REEs to the ocean is riverine input, with removals during estuarine mixing of 65–75% of the river input. Removal of REEs from the ocean are mainly due to particulate scavenging and their ocean mean residence time is estimated to range from 400–2900 years, except for Ce which has a much lower residence time (50 yrs). This is likely due to the oxidation of Ce in surface seawater (Ce^{3+} to CeO_2, which is very insoluble), which is bacterial-mediated oxidation, and its subsequent efficient removal from the ocean. There is some suggestion that hydrothermal inputs of REE's may be important and that there is also the potential for their remineralization from coastal sediments [265].

Studies of the actinides has essentially been restricted to those elements that have half-lives that make them useful geochemical racers, as discussed elsewhere in this chapter. These elemental concentrations are often reported in radioactive units (disintergrations per unit time). In terms of concentrations, these are generally low except for U, which has a relatively conservative distribution (12–15 nM) as it exists as a relatively soluble oxyanion. In contrast, Th is present at low pM concentrations (<0.2 pM).

Overall, there is little information in the literature for the elements present at trace concentrations unless there is a particular reason for examining their concentration. Some of these trace elements could provide valuable information as tracers, especially if analytical methods continue to improve and reduce the detection limits for these elements. Additionally, studies are needed to examine their interaction with dissolved inorganic and organic ligands and with the particulate surface. Finally, some of the rare elements are now finding use in technological development and this has lead to their rapid increase in environmental concentration. There is always a concern that these increasing concentrations may impact the microorganisms in the ocean and potentially impact the food chain if they are highly bioaccumulative elements.

6.4.9 Particulate metal fluxes to the deep ocean

Determination of the flux of material to the deep ocean is difficult as this flux is not consistent over time and is often dominated by episodic and aperiodic events that transport the majority of the material to the deep ocean. This sinking material is a combination of the remains of biotic material created in the surface zone through primary and secondary production and includes both organic and inorganic material, and remains of the material deposited to the surface via wet and dry deposition (dust and other particulate). All these fractions and phases are processed differently on their journey through the ocean. The flux to the seafloor can be constrained by comparison to sedimentary material and its rate of burial, and to the rate of export of material from the surface ocean, and should fall, for most elements, within the constraints of these two end-members. It is difficult to make measurements of flux and there is the potential for artifacts associated with the dominant methodology used to estimate the vertical particle flux, which is through the deployment of sediment traps followed by analysis of the collected material.

To illustrate this, the estimated particulate fluxes to the deep ocean (>3000 m) are collected in Table 6.12 for a number of trace elements and these data are derived from measurements using sediment trap collected material in both the major ocean basins [1]. For some elements, the flux is correlated with the flux of POC and this suggests that the major vector for the removal of metals to the deep sediment is through their transport and scavenging by sinking biogenic organic material. The flux of the lithogenic elements (e.g., Fe, Mn) may not correlate with the POC flux in regions where atmospheric inputs of terrestrial particulate is important. In such cases, these elements would correlate with Al rather than with POC. Additionally, the concentration of elements in the sinking particles noted in Fig. 6.41 are lower than those for the surface sediment, which illustrates the further enrichment in the concentration that occurs in the sediment due to the further degradation of organic material after settling at the deep sediment surface.

As noted in Section 6.3.6, less than 5% of the material sinking out of the mixed layer of the ocean reaches the deep ocean waters, and the transport of material through sinking material depends on its propensity to adsorb or be taken up into biological material, and its complexation in the dissolved phase. Metals with low K_Ds will be less enriched in this sinking material. Therefore, the metals are not partitioned in this deep collected material in proportion to their concentration in surface water particulate. Chester [1] calculated the relative enrichment factors (relative to Al and crustal ratios) of sinking material and showed that for the cations listed in Table 6.12, Pb, Cd and Zn were enriched, with the Pb enrichment factor being >20 for some locations, and Cd and Zn having a similar range of enrichment (>5 but <20). The degree of relative enrichment is consistent with their relative affinity for organic matter. The crustal metals (Fe, Mn, Co, V) has ratios ~1, indicating little enrichment, while Ni and Cu were enriched in some trap materials but not in others suggesting that the particulate source and the deep water speciation likely play an important role for these metals which have intermediate binding affinity.

As noted in the previous sections of this chapter many metals have a nutrient-type profile and therefore these elements are released from particulate matter during remineralization in the deeper ocean waters, and their extent of readsorption determines the extent of their deep ocean flux. For the "nutrient" metal(loid)s (those actively incorporated into microorganisms), their uptake in surface waters and their readsorption in the deeper ocean depends on a

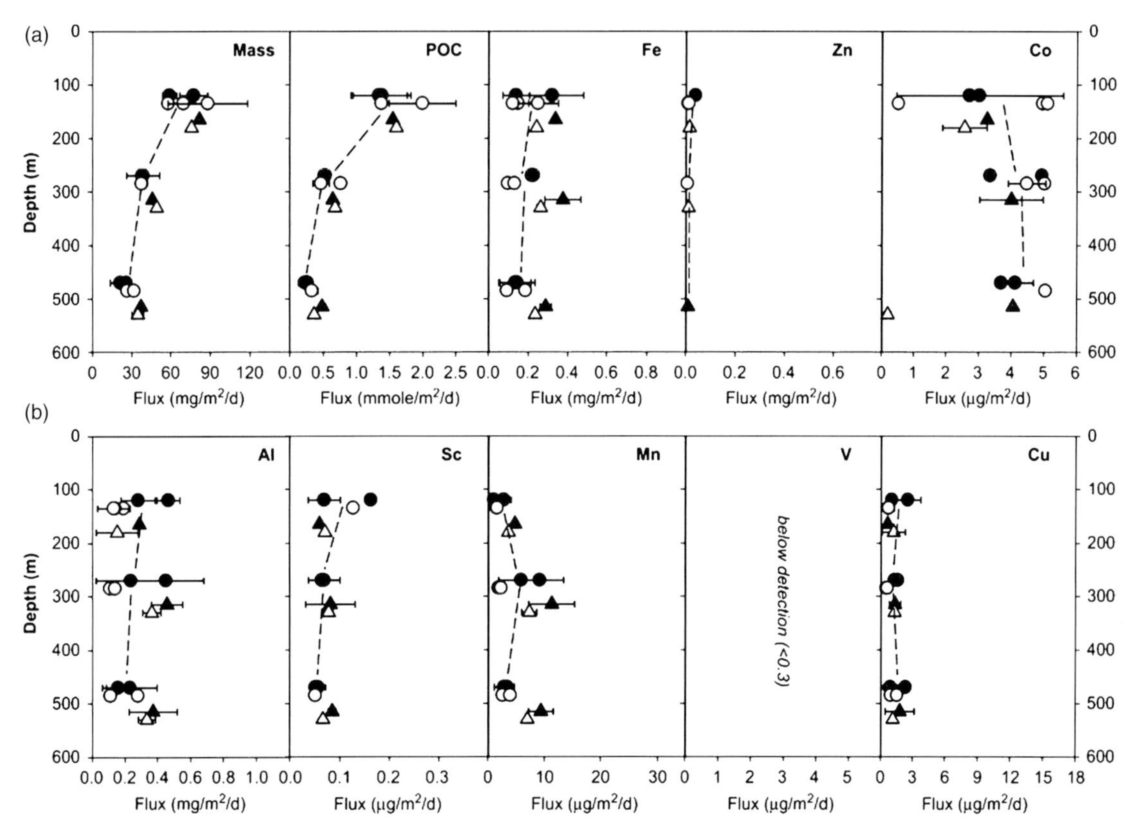

Fig. 6.41 (a) Sediment trap determined fluxes for a variety of metals and for ancillary parameters for a station in the North Pacific Ocean (ALOHA); (a) mass, POC, Fe, Zn; and Co and (b) Al, Sc, Mn, V, and Cu; (c) the corresponding relationships between the metals (Fe, An, Cu, Mn, Co, and Sc) and aluminum and (d) the same metal relationships with carbon illustrating the major controls over metal flux. Parts (a) and (b) taken from Lamborg et al. (2008a and 2008b) *Deep-Sea Research II* **55** [268, 269], reprinted with permission of Elsevier.

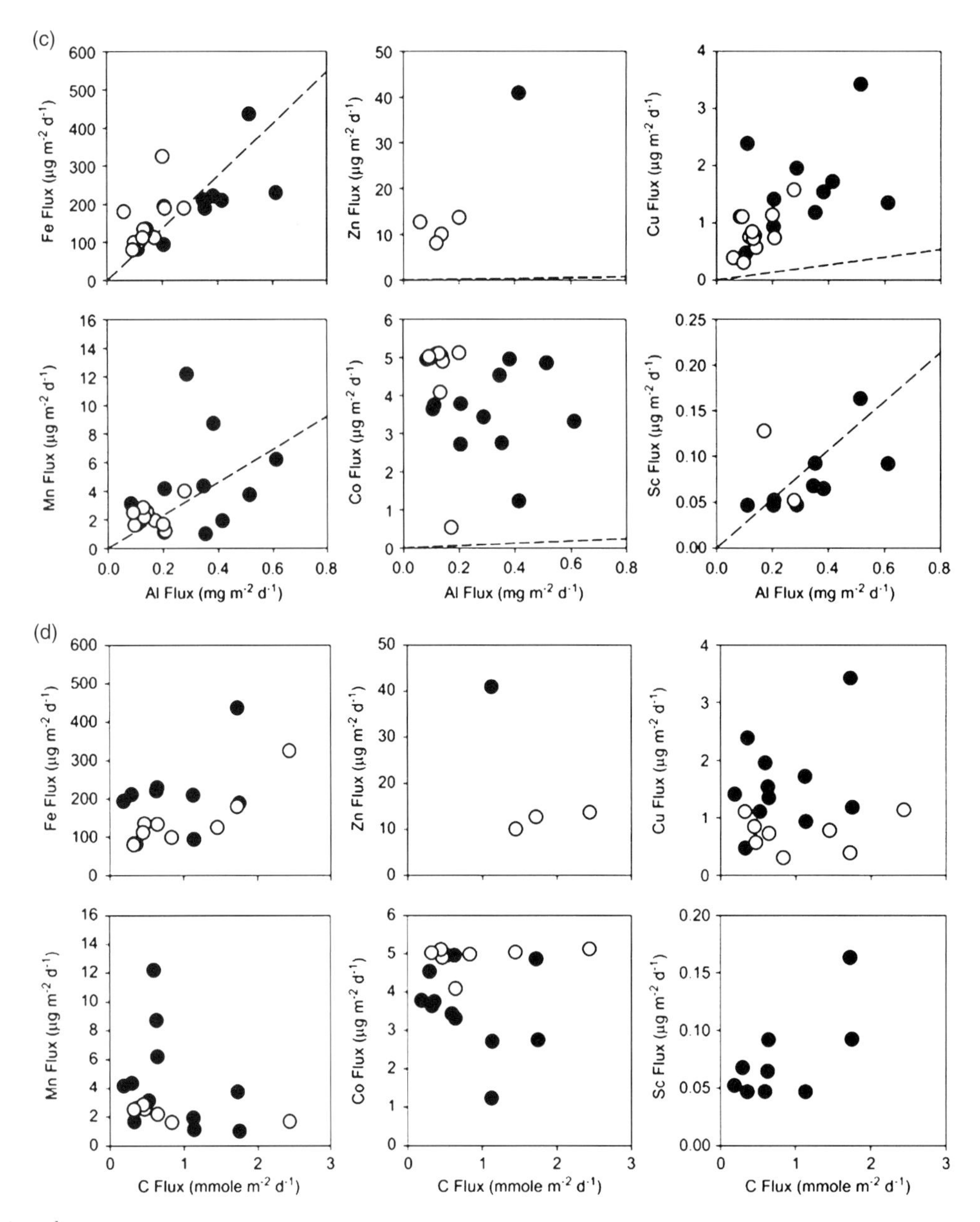

Fig. 6.41 (Continued)

different set of parameters, and therefore their flux ratio may not be similar to the relative incorporation into particulate matter in the mixed layer of the ocean.

While Table 6.12 is a synthesis of data available from earlier studies, and is informative, some more recent studies have begun to examine the flux of material and the export of metal(loid)s in particulate matter in more detail. In particular, the VERTIGO Program [266] measured the concentration of many elements in flux material collected at two locations in the North Pacific, one in the vicinity of Hawaii (ALOHA, 22.75°N, 158°W) and one in a more northern western section of the ocean (47°N, 161°E) [267, 268]. Data from the ALOHA site is shown in Fig. 6.41, as the collected particulate at this site appears more representative of open ocean fluxes (40–60% POM; with lithogenic material being of minor importance). The focus is on the upper ocean – the so-called "twilight" zone. A decreasing mass and POC flux is evident with depth (Fig. 6.41a), and this is also true for the other major biological components (N, P and biogenic Si; data not shown) [267]. Such changes with depth below the mixed layer are consistent with other studies and the literature [269]. However, the data for the major cations (Ba, Ca, Al) is less clear, showing scatter between the fluxes at each depth per deployment and between the two types

Table 6.12 Estimated fluxes of metals into the deep ocean as reported in [1], and taken from references therein. Data converted to molar units.

Metal	Part. Conc. (mmol kg^{-1})		Flux ($\mu mol/m^2/yr$)	
	Min.	Max.	Min.	Max.
Al	685	1780	20,300	32,450
Fe	160	420	1960	8500
Mn	7.3	20	100	265
Ni	0.43	1.0	6.2	22.0
Co	0.09	0.17	1.9	3.7.0
V	0.8	1.4	7.2	29.0
Cu	0.5	1.6	5.7	23.0
Pb	0.1	0.5	1.2	9.0
Zn	1.1	3.0	15	44
Cd	0.004	0.009	–	0.03

of trap used in this study. The ^{234}Th flux is relatively constant over this depth range [270], as is the flux of the "crustal" elements (Fe, Al, and Mn), while the transition metals measured (Zn, Co, Cu) tend to follow the biogenic material more closely. These data suggest that these more refractory elements are being retained in the particulate phase even while the organic matter is being remineralized, and that the fate of the biogenic material and that of the crustal elements are decoupled.

The data in Fig. 6.41(b) compare the fluxes of the elements with that of Al and POC to further examine these notions [68, 268]. There is a relatively good correlation between the fluxes of Al and Fe, Mn and Sc, and these are consistent with known crustal ratios while the flux of the transition metals are substantially higher than expected based on their crustal ratio. However, for these elements, there is also not a strong correlation with the POC flux (Fig. 6.41c), suggesting that the factors controlling their export from the upper ocean are complex. Overall, based on the ALOHA data it can be concluded that the flux of Fe is dominated by lithogenic material while that of Zn is dominated by its association with biogenic material, and specifically its relationship to that of biogenic Si. Overall, the data from the VERTIGO study provide a detailed picture which reinforces the concepts about the controlling factors in the export of trace elements from the upper ocean and the potential for these elements to be released in the deeper ocean waters during particulate remineralization and through deep water scavenging.

While recent studies have refined the notions of the factors controlling the export of trace elements from the surface ocean, the information available is still relatively sparse and this should be a topic for future research. It is not clear if metals are transported mainly through the flux of coagulated microbial biomass or through sinking of fecal pellets; that is, it is not clear how important upper water processing and repackaging are to the metal flux. For most trace elements, there is little formal understanding of the overall budget of the element in the oceanic realm, and the major sources and sinks. For the elements that partition strongly to particulate matter, deep ocean sedimentation is the most important sink, although this is not true for some of the volatile elements such as Hg that have an important sink in oceanic invasion. This is limited to a few metals, however. The particle reactive metals are also likely strongly associated with particles in the coastal environment, and therefore their inputs to the ocean are likely dominated by atmospheric sources, except for the major crustal elements. For the metals whose input to the ocean has been substantially exacerbated by anthropogenic inputs, the decoupling between the rate of deep ocean sedimentation and atmospheric change likely means that the ocean concentration of these elements is increasing. For many of the nutrient metals, however, the anthropogenic component is not sufficient to radically alter the concentrations in the surface ocean and for these elements any changes in bioavailability due to changes in ocean chemistry and climate are likely the main drivers of their ocean concentration and how it may change in the future.

6.5 Chapter summary

1. The concentration of metal(loid)s in ocean waters is a function of the concentration of these elements in rivers and in the atmosphere, and of their solubility, the strength of their complexes with inorganic and organic ligands, especially NOM, and their tendency to interact with solids (microorganisms and inorganic material and detrital phases).

2. Colloidal phases are important in many instances but the current techniques used for filtration do not adequately quantify colloidal material. Colloidal phases often form during the precipitation of metals and these colloids are often stabilized by NOM. The distribution of a metal(loid) between the phases (dissolved, colloidal and particulate) is a steady state situation as there is continual exchange between the phases, and this exchange is often reversible.

3. Metals are partitioned in the particulate phases in such a way that some of the metal is readily bioavailable (reactive) while some of the metal is essentially unreactive. These differences relate to the strength of binding of the metal, and also reflect that the metal may be incorporated into the solid phase rather than bound to the surface.

4. Many metals are complexed primarily to NOM in ocean waters and the extent of complexation influences both the dissolved concentration and the bioavailability of the metal in solution. There is the potential that lower molecular weight biochemicals are important in metal complexation in ocean waters. For most metal(loid)s, concentrations, even in the absence of anthropogenic inputs, are higher in coastal and estuarine waters than in offshore waters.

5. Metal(loid)s interact strongly with iron sulfide phases (FeS and FeS_2) in sediments and the extent of interaction has a strong influence over the reactivity, mobility and bioavailability of the metal. Various methods have been derived to quantify the extent of interaction.

6. Sediment metal(loid) concentrations are typically higher in nearshore waters although concentrations can be elevated in open ocean sediments due to the slow rate of sedimentation and the importance of post-deposition interactions. There is a legacy of past anthropogenic inputs in the deeper layers of many estuarine sediments.

7. Meta(loid)l concentrations and distributions in sediment porewater is controlled by the redox state of the porewater, and by the extent to which the elements complex with dissolved ligands, such as NOM and sulfide, and the extent of interaction with the solid phase (NOM and Fe-S phases). Colloidal phases are likely important in porewaters. Metal(loid)s can exchange across the sediment-water interface both in the particulate and dissolved phase. The magnitude of the dissolved flux depends on the speciation of the meta(loid) in the sediment porewater and overlying water, and on the extent to which the diffusive exchange is enhanced by bioturbation/bioirrigation and physical processes, such as sediment resuspension. Fluxes of metal(loid)s are often correlated and the direction can be predicted from their chemical reactivity.

8. Models have been developed to examine the biogeochemical cycling of metal(loid)s in sediments that incorporate mixing processes, chemical speciation and the impact of benthic invertebrates and microbial transformations. Models have been developed to examine horizontal as well as vertical distributions of metal(loid)s.

9. Deep ocean sediments consist of a variety of different types, and the concentrations of metal(loid)s in the sediments range over many orders of magnitude. One important class of sediments in terms of their metal(loid) content are manganese nodules. Hydothermal systems are a potentially important source of metal(loid)s to the ocean although it is difficult to estimate the overall input given the variability in concentrations across systems, and the complexity of the reactions, which include precipitation of sulfide and oxide phases, during the interaction of the fluids and the overlying seawater.

10. The vertical distributions of metal(loid)s in the ocean reflects their sources and internal biogeochemistry, and whether the elements are taken up into microbes in surface waters in similar fashion to nutrients. Organic matter degradtion at depth releases such metals and their profiles therefore often, but not always, show surface water depletions and higher concentrations at depth. A wide variety of distributions are found and these are discussed in terms of groups of elements with similar chemistry for the transition metals and for the other metals and metalloids. Both essential and potentially toxic elements are discussed. Specific programs that have focused on studying trace elements are discussed.

11. The export of trace elements from the upper ocean to depth is discussed as this has an important impact over the long term fate and deep water concentrations. This is a field in its infancy.

References

1. Chester, R. (2003) *Marine Geochemistry*. Blackwell Science, Malden.
2. Broeker, W.S. and Peng, T.-H. (1982) *Tracers in the Sea*. Eldigio Press, New York.
3. Bruland, K. and Lohan, M.C. (2004) Controls on trace metals in seawater. In: Elderfield, H. (ed.) *The Oceans and Marine Geochemistry*. Vol. 6 in Holland, H.D. and Turekian, K.K. (Exec. eds) *Teatise on Geochemistry*. Elsevier Pergamon, Amsterdam, pp. 23–47.
4. Maybeck, M. (2004) Global occurence of major elements in rivers. In: Drever, J.I. (ed.) *Surface and Ground Water, Weathering and Soils*. Vol. 5 in Holland, H.D. and Turekian, K.K. (Exec. eds) *Teatise on Geochemistry*. Elesevier, Amsterdam, pp. 207–223.
5. Gaillardet, J., Viers, J. and Dupre, B. (2004) Trace elements in river waters. In: Drever, J.I. (ed.) *Surface and Ground Water, Weathering and Soils*. Vol. 5 in *Treatise on Geochemistry*, H. Holland, Turekian, KK, Editor. Elsevier, Amsterdam, pp. 225–272.
6. Conley, D.J., Schelske, C.L. and Stoermer, E.F. (1993) Modification of the biogeochemical cycle of silica with eutrophication. *Marine Ecology Progress Series*, **81**, 121–128.
7. Billen, G. and Garnier, J. (2007) River basin nutrient delivery to the coastal sea: Assessing its potential to sustain new production of non-siliceous algae. *Marine Chemistry*, **106**(1–2), 148–160.
8. Stordal, M.C., Santschi, P.H. and Gill, G.A. (1996) Colloidal pumping: Evidence for the coagulation process using natural colloids tagged with Hg-203. *Environmental Science & Technology*, **30**(11), 3335–3340.
9. Honeyman, B.D. and Santschi, P.H. (1988) Metals in aquatic systems. *Environmental Science & Technology*, **22**, 862–871.
10. Santschi, P. (1988) Factors controlling the biogeochemical cycles of trace elements in fresh and coastal marine waters as revealed by artificial radioisotopes. *Limnology and Oceanography*, **33**, 848–866.
11. Clegg, S. and Whitfield, M. (1990) A generalized model for the scavenging of trace metals in the open ocean – I. Particle cycling. *Deep-Sea Research*, **37**, 809–832.
12. Valiela, I. (1995) *Marine Ecological Processes*. Springer-Verlag, New York, 686.
13. Bianchi, T.S. (2007) *Biogeochemistry of Estuaries*. Oxford University Press, New York, 706 pp.

14. Zhang, Y.Y., Zhang, E.R. and Zhang, J. (2008) Modeling on adsorption-desorption of trace metals to suspended particle matter in the Changjiang estuary. *Environmental Geology*, **53**(8), 1751–1766.
15. Allen, H.E. (1995) *Metal Contaminated Aquatic Sediments*. Ann Arbor Press, Chelsea, MI, p. 292.
16. Wen, L.-S., Santschi, P.H. and Tang, D. (1997) Interactions between radioactively labeled colloids and natural particles: Evidence for colloidal pumping. *Geochimica et Cosmochmica Acta*, **61**, 2867–2878.
17. Anderson, R.F. (2004) Chemical tracers of particle transport. In: Enderfield, H. (ed.) *The Oceans and Marine Geochemistry*. Vol. 6 in Holland, H.D. and Turekian, K.K. (Exec. eds) *Teatise on Geochemistry*. Elsevier, Amsterdam, pp. 247–273.
18. Bruland, K.W. and Coale, K.H. (1986) Surface water 234Th/238U disequilibria: Spatial and temporal variations of scavenging rates within the Pacific Ocean. In: Burton, J.D., Brewer, P.G. and Chester, R. (eds) *Dynamic Processes in the Chemsitry of the Upper Ocean*. Plenum Press, New York, pp. 159–172.
19. Murnane, R., Sarmiento, J. and Bacon, M. (1990) Thorium isotopes, particle cycling models, and inverse calculations of model rate constants. *Journal of Geophysical Research*, **95**(16), 195–116, 206.
20. Coale, K.H. and Bruland, K.W. (1987) Oceanic stratified euphotic zone as elucidated by $^{234}Th/^{38}U$ disequilibria. *Limnology and Oceanography*, **32**, 189–200.
21. van der Loeff, M.M.R. (2010) Uranium-thorium decay series in the oceans. In: Steele, J.H., Thorpe, S.A. and Turekian, K.K. (eds) Overview, in *Marine Chemistry and Geochemistry*. Elsevier, Amsterdam, pp. 203–213.
22. Turner, A., Millward, G.E. and Le Roux, S.M. (2004) Significance of oxides and particulate organic matter in controlling trace metal partitioning in a contaminated estuary. *Marine Chemistry*, **88**(3–4), 179–192.
23. Alvarez, M.B., Malla, M.E. and Batistoni, D.A. (2001) Comparative assessment of two sequential chemical extraction schemes for the fractionation of cadmium, chromium, lead and zinc in surface coastal sediments. *Fresenius' Journal of Analytical Chemistry*, **369**(1), 81–90.
24. Usero, J., Gamero, M., Morillo, J. and Gracia, I. (1998) Comparative study of three sequential extraction procedures for metals in marine sediments. *Environment International*, **24** (4), 487–496.
25. Wen, L.S., Warnken, K.W. and Santschi, P.H. (2008) The role of organic carbon, iron, and aluminium oxyhydroxides as trace metal carriers: Comparison between the trinity river and the trinity river estuary (Galveston Bay, Texas). *Marine Chemistry*, **112**(1–2), 20–37.
26. Tang, D.G., Warnken, K.W. and Santschi, P.H. (2002) Distribution and partitioning of trace metals (Cd, Cu, Ni, Pb, Zn) in Galveston Bay waters. *Marine Chemistry*, **78**(1), 29–45.
27. Wen, L.S., Santschi, P., Gill, G. and Paternostro, C. (1999) Estuarine trace metal distributions in Galveston Bay: Importance of colloidal forms in the speciation of the dissolved phase. *Marine Chemistry*, **63**(3–4), 185–212.
28. Wells, M.L., Kozelka, P.B. and Bruland, K.W. (1998) The complexation of "dissolved" Cu, Zn, Cd and Pb by soluble and colloidal organic matter in Narragansett Bay, RI. *Marine Chemistry*, **62**(3–4), 203–217.
29. Wells, M.L., Smith, G.J. and Bruland, K.W. (2000) The distribution of colloidal and particulate bioactive metals in Narragansett Bay, RI. *Marine Chemistry*, **71**(1–2), 143–163.
30. Jiann, K.T., Wen, L.S. and Santschi, P.H. (2005) Trace metal (Cd, Cu, Ni and Pb) partitioning, affinities and removal in the Danshuei River estuary, a macro-tidal, temporally anoxic estuary in Taiwan. *Marine Chemistry*, **96**(3–4), 293–313.
31. Twiss, M.R. and Moffett, J.W. (2002) Comparison of copper speciation in coastal marine waters measured using analytical voltammetry and diffusion gradient in thin-film techniques. *Environmental Science & Technology*, **36**(5), 1061–1068.
32. Saito, M.A. and Moffett, J.W. (2001) Complexation of cobalt by natural organic ligands in the Sargasso Sea as determined by a new high-sensitivity electrochemical cobalt speciation method suitable for open ocean work. *Marine Chemistry*, **75**(1–2), 49–68.
33. Bruland, K.W., Rue, E.L., Donat, J.R., Skrabal, S.A. and Moffett, J.W. (2000) Intercomparison of voltammetric techniques to determine the chemical speciation of dissolved copper in a coastal seawater sample. *Analytica Chimica Acta*, **405**(1–2), 99–113.
34. Rue, E.L. and Bruland, K.W. (1997) The role of organic complexation on ambient iron chemistry in the equatorial Pacific Ocean and the response of a mesoscale iron addition experiment. *Limnology and Oceanography*, **42**(5), 901–910.
35. Rue, E.L. and Bruland, K.W. (1995) Complexation of iron(III) by natural organic-ligands in the central North Pacific as determined by a new competitive ligand equilibration adsorptive cathodic stripping voltammetric method. *Marine Chemistry*, **50**(1–4), 117–138.
36. Fones, G.R. and Moffett, J.W. (2002) In-situ Cu speciation measurements in Boston Harbor using DGT. *Abstracts of Papers of the American Chemical Society*, **224**, 018-GEOC.
37. Lamborg, C.H., Tseng, C.M., Fitzgerald, W.F., Balcom, P.H. and Hammerschmidt, C.R. (2003) Determination of the mercury complexation characteristics of dissolved organic matter in natural waters with "Reducible Hg" Titrations. *Environmental Science & Technology*, **37**(15), 3316–3322.
38. Zeng, H.H., Thompson, R.B., Maliwal, B.P., Fones, G.R., Moffett, J.W. and Fierke, C.A. (2003) Real-time determination of picomolar free Cu(II) in seawater using a fluorescence based fiber optic biosensor. *Analytical Chemistry*, **75**(24), 6807–6812.
39. Dryden, C.L., Gordon, A.S. and Donat, J.R. (2007) Seasonal survey of copper-complexing ligands and thiol compounds in a heavily utilized, urban estuary: Elizabeth River, Virginia. *Marine Chemistry*, **103**(3–4), 276–288.
40. Xia, K., Skyllberg, U.L., Bleam, W.F., Bloom, P.R., Nater, E.A. and Helmke, P.A. (1999) X-ray absorption spectroscopic evidence for the complexation of Hg(II) by reduced sulfur in soil humic substances. *Environmental Science & Technology*, **33**(2), 257–261.
41. Haitzer, M., Aiken, G.R. and Ryan, J.N. (2002) Binding of mercury(II) to dissolved organic matter: The role of the mercury-to-DOM concentration ratio. *Environmental Science & Technology*, **36**(16), 3564–3570.

42. Cozic, A., Viollier, E., Chiffoleau, J.F., Knoery, J. and Rozuel, E. (2008) Interactions between volatile reduced sulfur compounds and metals in the Seine Estuary (France). *Estuaries and Coasts*, **31**(6), 1063–1071.
43. Cohen-Atiya, M. and Mandler, D. (2003) Studying thiol adsorption on Au, Ag and Hg surfaces by potentiometric measurements. *Journal of Electroanalytical Chemistry*, **550**, 267–276.
44. Yu, M.Q., Sun, D.W., Tian, W., Wang, G.P., Shen, W.B. and Xu, N. (2002) Systematic studies on adsorption of trace elements Pt, Pd, Au, Se, Te, As, Hg, Sb on thiol cotton fiber. *Analytica Chimica Acta*, **456**(1), 147–155.
45. Rickard, D. and Luther, G.W. (2007) Chemistry of iron sulfides. *Chemical Reviews*, **107**(2), 514–562.
46. Bowles, K.C., Ernste, M.J. and Kramer, J.R. (2003) Trace sulfide determination in oxic freshwaters. *Analytica Chimica Acta*, **477**(1), 113–124.
47. Luther, G.W. and Rickard, D.T. (2005) Metal sulfide cluster complexes and their biogeochemical importance in the environment. *Journal of Nanoparticle Research*, **7**(4–5), 389–407.
48. Shea, D. and Helz, G. (1988) The solubility of copper in sulfidic waters: Sulfide and polysulfide complexes in equilibrium with covellite. *Geochimica et Cosmochimica Acta*, **52**, 1815–1825.
49. Jay, J.A., Morel, F.M.M. and Hemond, H.F. (2000) Mercury speciation in the presence of polysulfides. *Environmental Science & Technology*, **34**(11), 2196–2200.
50. Borchardt, L. and Easty, D. (1984) Gas chromatographic determination of elemental and polysulfide sulfur in kraft pulping liquors. *Journal of Chromatography*, **299**, 471–476.
51. Chadwell, S.J., Rickard, D. and Luther, G.W. (2001) Electrochemical evidence for metal polysulfide complexes: Tetrasulfide (S-4(2-)) reactions with Mn2+, Fe2+, Co2+, Ni2+, Cu2+, and Zn2+. *Electroanalysis*, **13**(1), 21–29.
52. Kuganathan, N. and Green, J.C. (2008) Mercury telluride crystals encapsulated within single walled carbon nanotubes: A density functional study. *International Journal of Quantum Chemistry*, **108**, 797–807.
53. Chakraborty, I., Mitra, D. and Moulik, S.P. (2005) Spectroscopic studies on nanodispersions of cds, hgs, their core-shells and composites prepared in micellar medium. *Journal of Nanoparticle Research*, **7**, 227–236.
54. Kristl, M. and Drofenik, M. (2008) Sonochemical synthesis of nanocrystalline mercury sulfide, selenide and telluride in aqueous solutions. *Ultrasonics Sonochemistry*, **15**, 695–699.
55. Mahapatra, A.K. and Dash, A.K. (2006) Alpha-hgs nanocrystals: Synthesis, structure and optical properties. *Physica E*, **35**, 9–15.
56. Stumm, W. and Morgan, J.J. (1996) *Aquatic Chemistry*. John Wiley & Sons, Inc., New York.
57. Luther, G.W., Theberge, S.M. and Rickard, D.T. (1999) Evidence for aqueous clusters as intermediates during zinc sulfide formation. *Geochimica et Cosmochimica Acta*, **63**(19–20), 3159–3169.
58. Ravichandran, M., Aiken, G., Ryan, J. and Reddy, M. (1999) Inhibition of precipitation and aggregation of metacinnabar (mercuric sulfide) by dissolved organic matter isolated from the Florida Everglades. *Environmental Science and Technology*, **33**, 1418–1423.
59. Deonarine, A. and Hsu-Kim, H. (2009) Precipitation of mercuric sulfide nanoparticles in nom-containing water: Implications for the natural environment. *Environmental Science & Technology*, **43**(7), 2368–2373.
60. Lau, B.L.T. and Hsu-Kim, H. (2008) Precipitation and growth of zinc sulfide nanoparticles in the presence of thiol-containing natural organic ligands. *Environmental Science & Technology*, **42** (19), 7236–7241.
61. Slowey, A.J. (2010) Rate of formation and dissolution of mercury sulfide nanoparticles: The dual role of natural organic matter. *Geochimica et Cosmochimica Acta*, **74**(16), 4693–4708.
62. Aiken, G.R., Hsu-Kim, H. and Ryan, J.N. (2011) Influence of dissolved organic matter on the environmental fate of metals, nanoparticles, and colloids. *Environmental Science & Technology*, **45**(8), 3196–3201.
63. Ravichandran, M., Aiken, G.R., Reddy, M.M. and Ryan, J.N. (1998) Enhanced dissolution of cinnabar (mercuric sulfide) by dissolved organic matter isolated from the Florida Everglades. *Environmental Science and Technology*, **32**, 3305–3311.
64. Dyrssen, D. and Wedborg, M. (1991) The sulphur-mercury (II) system in natural waters. *Water, Air, and Soil Pollution*, **56**, 507–519.
65. Tossell, J.A. (2001) Calculation of the structures, stabilities, and properties of mercury sulfide species in aqueous solutions. *The Journal of Physical Chemistry*, **105**, 935–941.
66. Benoit, J.M., Gilmour, C.C., Mason, R.P. and Heyes, A. (1999) Sulfide controls on mercury speciation and bioavailability to methylating bacteria in sediment pore waters. *Environmental Science & Technology*, **33**(6), 951–957.
67. Benoit, J.M., Mason, R.P. and Gilmour, C.C. (1999) Estimation of mercury-sulfide speciation in sediment pore waters using octanol-water partitioning and implications for availability to methylating bacteria. *Environmental Toxicology and Chemistry*, **18**(10), 2138–2141.
68. Biswas, P. and Wu, C.Y. (2005) Critical review: Nanoparticles and the environment. *Journal of the Air & Waste Management Association*, **55**(6), 708–746.
69. Gaiser, B.K., Fernandes, T.F., Jepson, M.A., et al. (2012) Interspecies comparisons on the uptake and toxicity of silver and cerium dioxide nanoparticles. *Environmental Toxicology and Chemistry*, **31**(1), 144–154.
70. Rosenkranz, P., Chaudhry, Q., Stone, V. and Fernandes, T.F. (2009) A comparison of nanoparticle and fine particle uptake by *Daphnia magna*. *Environmental Toxicology and Chemistry*, **28**(10), 2142–2149.
71. Neal, A.L. (2008) What can be inferred from bacterium-nanoparticle interactions about the potential consequences of environmental exposure to nanoparticles? *Ecotoxicology*, **17**, 362–371.
72. Priester, J., Stoimenov, P., Mielke, R., Webb, S., Ehrhardt, C., Zhang, J., Stucky, G. and Holden, P. (2009) Effects of soluble cadmium salts versus CdSe quantum dots on the growth of planktonic *Pseudomonas aeruginosa*. *Environmental Science & Technology*, **43**(7), 2589–2594.
73. Kach, D.J. and Ward, J.E. (2008) The role of marine aggregates in the ingestion of picoplankton-size particles by suspension-feeding molluscs. *Marine Biology*, **153**(5), 797–805.
74. Tueros, I., Rodriguez, J.G., Borja, A., Solaun, O., Valencia, V. and Millan, E. (2008) Dissolved metal background levels in marine waters, for the assessment of the physico-chemical

status, within the European water framework directive. *The Science of the Total Environment*, **407**(1), 40–52.

75. Schaule, B.K. and Paterson, C.C. (1981) Lead concentrations in the Northeast pacific: Evidence for global anthropogenic perturbations. *Earth and Planetary Science Letters*, **54**, 97–116.
76. Mason, R.P. and Lawrence, A.L. (1999) Concentration, distribution and bioavailability of mercury and methylmercury in sediments of Baltimore Harbor and the Chesapeake Bay, Maryland, USA. *Environmental Toxicology and Chemistry*, **18**(11), 2438–2447.
77. Morse, J.W. and Luther, G.W. (1999) Chemical influences on the trace emtal-sulfide interactions in anoxic sediments. *Geochimica et Cosmochimica Acta*, **63**, 3371–3378.
78. Morse, J.W. (1994) Interactions of trace-metals with authigenic sulfide minerals – Implications for their bioavailability. *Marine Chemistry*, **46**(1–2), 1–6.
79. Ankley, G.T. (1996) Evaluation of metal/acid-volatile sulfide relationships in the prediction of metal bioaccumulation by benthic macroinvertebrates. *Environmental Toxicology and Chemistry*, **15**(12), 2138–2146.
80. Mason, R.P. (2000) The bioaccumulation of mercury, methylmercury and other toxic trace metals into pelagic and benthic organisms. In: Newman, M.C. and Hale, R.C. (eds) *Coastal and Estuarine Risk Assessment*. CRC Press, Boca Raton, pp. 127–149.
81. Chen, Z. and Mayer, L.M. (1999) Assessment of sedimentary Cu availability: A comparison of biomimetic and avs approaches. *Environmental Science and Technology*, **33**, 650–652.
82. Mahony, J.D., Di Toro, D.M., Gonzalez, A.M., Curto, M., Dilg, M., De Rosa, L.D. and Sparrow, L.A. (1996) Partitioning of metals to sediment organic carbon. *Environmental Toxicology and Chemistry*, **15**, 2187.
83. Campbell, P. and Tessier, A. (1996) *Ecotoxicology of Metals in the Aquatic Environment: Geochemical Aspects*. CRC Press, Boca Raton, pp. 11–58.
84. Li, Y.-H. and Schoonmaker, J.E. (2004) Chemical composition and mineralogy of marine sediments. In: Mackensie, F.T. (ed.) *Sediments, Diagenesis and Sedimentary Rocks*. Vol. 7. *Treatise on Geochemistry*, H.D. Holland and K.K. Turekian, (Exec. eds). Elsevier, Amsterdam, pp. 1–35.
85. Steinberg, N., Suszkowski, D.J., Clark, L. and Way, J. (2004) *Health of the Harbor*, Hudson River Foundation report, New York, NY.
86. Sañudo-Wilhelmy, S.A. and Gill, G.A. (1999) Impact of the clean water act on the levels of toxic metals in urban estuaries: The Hudson River estuary revisited. *Environmental Science and Technology*, **33**, 3477–3481.
87. Klinkhammer, G. and Bender, M.L. (1981) Trace metal distributions in the Hudson River estuary. *Estuarine, Coastal and Shelf Science*, **12**, 629–643.
88. Balcom, P., Fitzgerald, W.F. and Mason, R.P. (2010) Synthesis and assessment of modern and historic heavy metal contamination in New York/New Jersey Harbor estuary with emphasis on Hg and Cd. Hudson River Foundation Final Report, pp. 75.
89. Hollister, J.W., August, P.V., Paul, J.F. and Walker, H.A. (2008) Predicting estuarine sediment metal concentrations and inferred ecological conditions: An information theoretic approach. *Journal of Environmental Quality*, **37**(1), 234–244.
90. Luther, G.W. (1995) Trace metal chemistry in porewaters. In: Allen, H.E. (ed.) *Metal Contaminated Aquatic Sediments*. Ann Arbor Press, Chelsea, MI, pp. 65–80.
91. Mason, R., Bloom, N., Cappellino, S., Gill, G., Benoit, J. and Dobbs, C. (1998) Investigation of porewater sampling methods for mercury and methylmercury. *Environmental Science & Technology*, **32**(24), 4031–4040.
92. Huerta-Diaz, M.A., Rivera-Duarte, I., Sanudo-Wilhelmy, S.A. and Flegal, A.R. (2007) Comparative distributions of size fractionated metals in pore waters sampled by in situ dialysis and whole-core sediment squeezing: Implications for diffusive flux calculations. *Applied Geochemistry*, **22**(11), 2509–2525.
93. Luther, G.W., Glazer, B.T., Ma, S.F., et al. (2008) Use of voltammetric solid-state (micro)electrodes for studying biogeochemical processes: Laboratory measurements to real time measurements with an in situ electrochemical analyzer (ISEA). *Marine Chemistry*, **108**(3–4), 221–235.
94. Luther, G.W., Brendel, P.J., Lewis, B.L., Sundby, B., Lefrancois, L., Silverberg, N. and Nuzzio, D.B. (1998) Simultaneous measurement of O-2, Mn, Fe, I-, and S(-II) in marine pore waters with a solid-state voltammetric microelectrode. *Limnology and Oceanography*, **43**(2), 325–333.
95. Berg, P., Glud, R.N., Hume, A., Stahl, H., Oguri, K., Meyer, V. and Kitazato, H. (2009) Eddy correlation measurements of oxygen uptake in deep ocean sediments. *Limnology and Oceanography-Methods*, **7**, 576–584.
96. Tengberg, A., Hall, P.O.J., Andersson, U., et al. (2005) Intercalibration of benthic flux chambers II: Hydrodynamic characterization and flux comparisons of 14 different designs. *Marine Chemistry*, **94**(1–4), 147–173.
97. Gill, G.A., Bloom, N.S., Cappellino, S., Driscoll, C.T., Dobbs, C., McShea, L., Mason, R. and Rudd, J.W.M. (1999) Sediment-water fluxes of mercury in Lavaca Bay, Texas. *Environmental Science & Technology*, **33**(5), 663–669.
98. Berg, P., Roy, H., Janssen, F., Meyer, V., Jorgensen, B.B., Huettel, M. and de Beer, D. (2003) Oxygen uptake by aquatic sediments measured with a novel non-invasive eddy-correlation technique. *Marine Ecology Progress Series*, **261**, 75–83.
99. Burdige, D.J., Kline, S.W. and Chen, W.H. (2004) Fluorescent dissolved organic matter in marine sediment pore waters. *Marine Chemistry*, **89**(1–4), 289–311.
100. Jahnke, R., Richards, M., Nelson, J., Robertson, C., Rao, A. and Jahnke, D. (2005) Organic matter remineralization and porewater exchange rates in permeable South Atlantic Bight continental shelf sediments. *Continental Shelf Research*, **25**(12–13), 1433–1452.
101. Thomas, C.J., Blair, N.E., Alperin, M.J., DeMaster, D.J., Jahnke, R.A., Martens, C.S. and Mayer, L. (2002) Organic carbon deposition on the North Carolina continental slope off Cape Hatteras (USA). *Deep-Sea Research Part II*, **49**(20), 4687–4709.
102. Covelli, S., Faganeli, J., Horvat, M. and Brambati, A. (1999) Porewater distribution and benthic flux measurements of mercury and methylmercury in the Gulf of Trieste (northern Adriatic Sea). *Estuarine, Coastal and Shelf Science*, **48**(4), 415–428.
103. Choe, K.Y., Gill, G.A., Lehman, R.D., Han, S., Heim, W.A. and Coale, K.H. (2004) Sediment-water exchange of total mercury

and monomethyl mercury in the San Francisco Bay-Delta. *Limnology and Oceanography*, **49**(5), 1512–1527.

104. Point, D., Monperrus, M., Tessier, E., Amouroux, D., Chauvaud, L., Thouzeau, G., Jean, F., Amice, E., Grall, J., Leynaert, A., Clavier, J. and Donard, O.F.X. (2007) Biological control of trace metal and organometal benthic fluxes in a eutrophic lagoon (Thau Lagoon, Mediterranean Sea, France). *Estuarine, Coastal and Shelf Science*, **72**(3), 457–471.
105. Benoit, J.M., Shull, D.H., Harvey, R.M. and Beal, S.A. (2009) Effect of bioirrigation on sediment-water exchange of methylmercury in Boston Harbor, Massachusetts. *Environmental Science & Technology*, **43**(10), 3669–3674.
106. Riedel, G.F., Sanders, J.G. and Osman, R.W. (1997) Biogeochemical control on the flux of trace elements from estuarine sediments: Water column oxygen concentrations and benthic infauna. *Estuarine, Coastal and Shelf Science*, **44**(1), 23–38.
107. Riedel, G.F., Sanders, J.G. and Osman, R.W. (1999) Biogeochemical control on the flux of trace elements from estuarine sediments: Effects of seasonal and short-term hypoxia. *Marine Environmental Research*, **47**(4), 349–372.
108. Hollweg, T.A. (2010) *Mercury Cycling in Sediments of the Chesapeake Bay and the Mid-atlantic Continental Shelf and Slope, in Marine Sciences.* University of Connecticut, Storrs, p. 221.
109. Boudreau, B.P. (1986) Mathematics of tracer mixing in sediments. 2. Nonlocal mixing and biological conveyor-belt phenomena. *American Journal of Science*, **286**, 199.
110. Hollweg, T., Gilmour, C. and Mason, R. (2009) Factors controlling the methylation of mercury in sediments of the Chesapeake Bay and mid-Atlantic continental shelf. *Marine Chemistry*, **114**, 86–101.
111. Hayduk, W. and Laudie, H. (1974) Prediction of diffusion coefficients for nonelectrolytes in dilute aqueous solutions. *American Institute of Chemical Engineer Journal*, **20**(3), 611–615.
112. Li, Y.-H. and Gregory, S. (1974) Diffusion of ions in sea water and in deep-sea sediments. *Geochimica et Cosmochimica Acta*, **38**, 703–714.
113. Goulet, R.R., Holmes, J., Page, B., Poissant, L., Siciliano, S.D., Lean, D.R.S., Wang, F., Amyot, M. and Tessier, A. (2007) Mercury transformations and fluxes in sediments of a riverine wetland. *Geochimica et Cosmochimica Acta*, **71**(14), 3393–3406.
114. Boudreau, B.P. (1997) *Diagenetic Models and Their Implementation*. Springer, Berlin.
115. Warnken, K.W., Gill, G.A., Santschi, P.H. and Griffin, L.L. (2000) Benthic exchange of nutrients in Galveston Bay, Texas. *Estuaries*, **23**(5), 647–661.
116. Hollweg, T.A., Gilmour, C.C. and Mason, R.P. (2010) Mercury and methylmercury cycling in sediments of the mid-Atlantic continental shelf and slope. *Limnology and Oceanography*, **55**(6), 2703–2722.
117. Schwarzenbach, R.P., Gschwend, P.M. and Imboden, D.M. (1993) *Environmental Organic Chemistry*, 1st edn. John Wiley & Sons, Inc., New York.
118. Benoit, J.M., Gilmour, C.C. and Mason, R.P. (2001) The influence of sulfide on solid phase mercury bioavailability for methylation by pure cultures of *Desulfobulbus propionicus* (1pr3). *Environmental Science & Technology*, **35**(1), 127–132.
119. Mason, R.P., Reinfelder, J.R. and Morel, F.M.M. (1996) Uptake, toxicity, and trophic transfer of mercury in a coastal diatom. *Environmental Science and Technology*, **30**(6), 1835–1845.
120. Hammerschmidt, C.R., Fitzgerald, W.F., Lamborg, C.H., Balcom, P.H. and Visscher, P.T. (2004) Biogeochemistry of methylmercury in sediments of Long Island Sound. *Marine Chemistry*, **90**(1–4), 31–52.
121. Hartmann, M. and Muller, O.J. (1982) Trace metals in interstitial waters from the central Pacific Ocean sediments. In: Fanning, K.A. and Manheim, F. (eds) *The Dynamic Environment of the Ocean Floor*. Lexington Books, Lexington, pp. 285–301.
122. Klinkhammer, G., Heggie, D.T. and Graham, D.W. (1982) Metal diagenesis in oxic marine sediments. *Earth and Planetary Science Letters*, **61**, 211–219.
123. Mason, R.P., Kim, E.H., Cornwell, J. and Heyes, D. (2006) An examination of the factors influencing the flux of mercury, methylmercury and other constituents from estuarine sediment. *Marine Chemistry*, **102**(1–2), 96–110.
124. Shine, J.P., Ika, R. and Ford, T.E. (1998) Relationship between oxygen consumption and sediment-water fluxes of heavy metals in coastal marine sediments. *Environmental Toxicology and Chemistry*, **17**(11), 2325–2337.
125. Shank, G.C., Skrabal, S.A., Whitehead, R.F. and Kieber, R.J. (2004) Fluxes of strong cu-complexing ligands from sediments of an organic-rich estuary. *Estuarine, Coastal and Shelf Science*, **60**(2), 349–358.
126. Skrabal, S.A., Donat, J.R. and Burdige, D.J. (1997) Fluxes of copper-complexing ligands from estuarine sediments. *Limnology and Oceanography*, **42**(5), 992–996.
127. Van Cappellen, P. and Wang, Y. (1995) Metal cycling in surface sediments: Modeling the interplay of transport and reaction. In: Allen, H.E. (ed.) *Metal Contaminated Aquatic Sediments*. Ann Arbor Press, Chelsea, MI, pp. 21–64.
128. Burdige, D.J. (2005) Burial of terrestrial organic matter in marine sediments: A re-assessment. *Global Biogeochemical Cycles*, **19**(4), Article # GB002368.
129. Burdige, D.J. (2006) *Geochemsitry of Marine Sediments*. Princeton University Press, Princeton.
130. Boudreau, B.P. (1987) Mathematics of tracer mixing in sediments. 3: The theory of nonlocal mixing within sediments. *American Journal of Science*, **287**, 693.
131. Aller, R.C. (1980) Tracking particle-associated processes in nearshore environments by use of Th-234-U-238 disequilibrium. *Earth and Planetary Science Letters*, **47**, 161.
132. Aller, R.C. (1984) Estimates of particle-flux and reworking at the deep-sea floor using Th-234 U-238 disequilibrium. *Earth and Planetary Science Letters*, **67**, 308.
133. Boudreau, B.P. (1994) Is burial velocity a master parameter for bioturbation? *Geochimica et Cosmochimica Acta*, **58**(4), 1243–1249.
134. Shull, D.H., Benoit, J.M., Wojcik, C. and Senning, J.R. (2009) Infaunal burrow ventilation and pore-water transport in muddy sediments. *Estuarine, Coastal and Shelf Science*, **83**(3), 277–286.

135. Benoit, J.M., Shull, D.H., Robinson, P. and Ucran, L.R. (2006) Infaunal burrow densities and sediment monomethyl mercury distributions in Boston harbor, Massachusetts. *Marine Chemistry*, **102**(1–2), 124–133.
136. Morford, J.L., Martin, W.R., Francois, R. and Carney, C.M. (2009) A model for uranium, rhenium, and molybdenum diagenesis in marine sediments based on results from coastal locations. *Geochimica et Cosmochimica Acta*, **73**(10), 2938–2960.
137. Bhavsar, S.P., Diamond, M.L., Evans, L.J., Gandhi, N., Nilsen, J. and Antunes, P. (2004) Development of a coupled metal speciation-fate model for surface aquatic systems. *Environmental Toxicology and Chemistry*, **23**(6), 1376–1385.
138. Bhavsar, S.P., Gandhi, N., Diamond, M.L., Lock, A.S., Spiers, G. and De La Torre, M.C.A. (2008) Effects of estimates from different geochemical models on metal fate predicted by coupled speciation-fate models. *Environmental Toxicology and Chemistry*, **27**(5), 1020–1030.
139. Hollweg, T.A., Mason, R.P. and Gilmour, C.C. (2012) Modeling Hg in sediments. *Journal of Geophysical Research-Biogeochemical*, Submitted, see also Ref 108.
140. Paquette, K.E. and Helz, G.R. (1997) Inorganic speciation of mercury in sulfidic waters: The importance of zero-valent sulfur. *Environmental Science & Technology*, **31**(7), 2148–2153.
141. Skyllberg, U. (2008) Competition among thiols and inorganic sulfides and polysulfides for Hg and MeHg in wetland soils and sediments under suboxic conditions: Illumination of controversies and implications for MeHg net production. *Journal of Geophysical Research-Biogeosciences*, **113**, Article # G00C03.
142. Benoit, J.M., Gilmour, C.C., Heyes, A., Mason, R.P. and Miller, C.L. (2003) Geochemical and biological controls over methylmercury production and degradation in aquatic ecosystems. In: Cai, Y. and Braids, O.C. (eds) *Biogeochemistry of Environmentally Important Trace Elements*. ACS Series, Washington, DC, pp. 262–297.
143. Ravichandran, M. (2004) Interactions between mercury and dissolved organic matter – A review. *Chemosphere*, **55**(3), 319–331.
144. Miller, C.L. (2005) The role of organic matter in the dissolved phase speciation and solid phase partitioning of mercury. PhD thesis in Marine, Estuarine and Environmental Sciences, University of Maryland: College Park.
145. Huerta-Diaz, M. and Morse, J. (1992) Pyritization of trace metals in anoxic marine sediments. *Geochimica et Cosmochimica Acta*, **56**, 2681–2702.
146. Morse, J.W. (1995) Dynamics of trace metal interactions with authigenic sulfide minerals in anoxic sediments. In: Allen, H.E. (ed.) *Metal Contaminated Aquatic Sediments*. Ann Arbor Press, Chelsea, MI, pp. 187–199.
147. King, J.K., Kostka, J.E., Frischer, M.E., Saunders, F.M. and Jahnke, R.A. (2001) A quantitative relationship that remonstrates mercury methylation rates in marine sediments are based on the community composition and activity of sulfate-reducing bacteria. *Environmental Science & Technology*, **35**(12), 2491–2496.
148. Zhang, T., Kim, B., Leyard, C., Reinsch, B. C., Lowry, G. V., Deschusses, M. A. and Hsu-Kim, H. (2012) Methylation of mercury by bacteria exposed to dissolved, nanoparticulate, and microparticulate mercuric sulfides. *Environmental Science & Technology*, **46**, 6950–6958.
149. Geyer, W.R., Woodruff, J.D. and Traykovski, P. (2001) Sediment transport and trapping in the Hudson River estuary. *Estuaries*, **24**(5), 670–679.
150. Cook, T.L., Sommerfield, C.K. and Wong, K.C. (2007) Observations of tidal and springtime sediment transport in the upper delaware estuary. *Estuarine, Coastal and Shelf Science*, **72**(1–2), 235–246.
151. Petersen, W., Willer, E. and Willamowski, C. (1997) Remobilization of trace elements from polluted anoxic sediments after resuspension in oxic water. *Water, Air, and Soil Pollution*, **99**(1–4), 515–522.
152. Simpson, S.L., Apte, S.C. and Batley, G.E. (1998) Effect of short term resuspension events on trace metal speciation in polluted anoxic sediments. *Environmental Science & Technology*, **32**(5), 620–625.
153. Laima, M.J.C., Matthiesen, H., Lund-Hansen, L.C. and Christiansen, C. (1998) Resuspension studies in cylindrical microcosms: Effects of stirring velocity on the dynamics of redox sensitive elements in a coastal sediment. *Biogeochemistry*, **43**(3), 293–309.
154. Bloom, N.S. and Lasorsa, B.K. (1999) Changes in mercury speciation and the release of methyl mercury as a result of marine sediment dredging activities. *The Science of the Total Environment*, **237–238**, 379–385.
155. Brassard, P., Kramer, J.R. and Collins, P.V. (1997) Dissolved metal concentrations and suspended sediment in Hamilton Harbour. *Journal of Great Lakes Research*, **23**(1), 86–96.
156. Caetano, M., Madureira, M.J. and Vale, C. (2003) Metal remobilisation during resuspension of anoxic contaminated sediment: Short-term laboratory study. *Water, Air, and Soil Pollution*, **143**(1–4), 23–40.
157. Van Den Berg, G.A., Meijers, G.G.A., Van Der Heijdt, L.M. and Zwolsman, J.J.G. (2001) Dredging-related mobilisation of trace metals: A case study in the Netherlands. *Water Research*, **35**(8), 1979–1986.
158. Kim, E.H., Mason, R.P., Porter, E.T. and Soulen, H.L. (2004) The effect of resuspension on the fate of total mercury and methyl mercury in a shallow estuarine ecosystem: A mesocosm study. *Marine Chemistry*, **86**(3–4), 121–137.
159. Nyffeler, U.P. (1984) A kinetic approach to describing trace element distributions between particles and solution in natural aquatic systems. *Geochimica et Cosmochimica Acta*, **48**, 1513.
160. Bergeron, C.M. (2005) The impact of sediment resuspension on mercury cycling and the bioaccumulation of methylmercury into benthic and pelagic organisms. M.S. thesis, MEES, University of Maryland: College Park. pp. 108.
161. Kalnejais, L.H., Martin, W.R., Signall, R.P. and Bothner, M.H. (2007) Role of sediment resuspension in the remobilization of particulate-phase metals from coastal sediments. *Environmental Science & Technology*, **41**(7), 2282–2288.
162. Law, B.A., Hill, P.S., Milligan, T.G., Curran, K.J., Wiberg, P.L. and Wheatcroft, R.A. (2008) Size sorting of fine-grained sediments during erosion: Results from the western Gulf of Lyons. *Continental Shelf Research*, **28**, 1935–1946.

163. Goñi, M., Alleau, Y., Corbett, R., Walsh, J.P., Mallinson, D., Allison, M.A., Gordon, E., Petsch, S. and Dellapenna, T.M. (2007) The effects of Hurricanes Katrina and Rita on the seabed of the Louisiana Shelf. *Sedimentary Record*, **1**, 4–9.
164. Goni, M.A., Gordon, E.S., Monacci, N.M., Clinton, R., Gisewhite, R., Allison, M.A. and Kineke, G. (2006) The effect of Hurricane Lili on the distribution of organic matter along the inner Louisiana Shelf (Gulf of Mexico, USA). *Continental Shelf Research*, **26**(17–18), 2260–2280.
165. Mitra, S., Lalicata, J.J., Allison, M.A. and Dellapenna, T.M. (2009) The effects of Hurricanes Katrina and Rita on seabed polycyclic aromatic hydrocarbon dynamics in the Gulf of Mexico. *Marine Pollution Bulletin*, **58**(6), 851–857.
166. Johnson, W.E., Kimbrough, K.L., Lauenstein, G.G. and Christensen, J. (2009) Chemical contamination assessment of Gulf of Mexico oysters in response to Hurricanes Katrina and Rita. *Environmental Monitoring and Assessment*, **150**(1–4), 211–225.
167. Liu, B., Schaider, L.A., Mason, R.P., Bank, M.S., Rabalais, N.N., Swarzenski, P.W., Shine, J.P., Hollweg, T. and Senn, D.B. (2009) Disturbance impacts on mercury dynamics in Northern Gulf of Mexico sediments. *Journal of Geophysical Research-Biogeosciences*, **114**, Article # G00C07.
168. Glasby, G.P. (2000) Manganese: Predominat role of nodules and crusts. In: Schulz, H.D. and Zabel, M. (eds) *Marine Geochemistry*. Springer, Berlin, pp. 335–372.
169. Maynard, J.B. (2004) Manganiferous sediments, rocks and ores. In: Mackensie, F.T. (ed.) *Sediments, Diagenesis and Sedimentary Rocks*. Vol. 7 in Holland, H.D. and Turekian, K.K. (Exec. eds) *Teatise on Geochemistry*. Elsevier Pergamon, Amsterdam, pp. 289–307.
170. Koschinsky, A. and Hein, J.R. (2003) Uptake of elements from seawater by ferromanganese nodules: Solid-phase associations and seawater speciation. *Marine Geology*, **198**, 331–351.
171. German, C.R. and Von Damm, K.L. (2004) Hydrothermal processes. In: Elderfield, H. (ed.) *The Oceans and Marine Biogeochemistry*. Vol. 6 in *Treatise on Geochemistry*, H.D. Holland and K.K. Turekian, (Exec. eds). Elsevier, Amsterdam, pp. 181–222.
172. Von Damm, K.L. (2010) Hydrothermal vent fluids, chemistry of. In: Steele, J.H., Thorpe, S.A. and Turekian, K.K. (eds) *Marine Chemistry and Geochemistry*. Elsevier, Amsterdam, pp. 81–88.
173. Brock, T.D. and Madigan, M.T. (1991) *Biology of Microorganisms*. Prentice Hall, Englewood Cliffs, NJ.
174. Sander, S.G. and Koschinsky, A. (2011) Metal flux from hydrothermal vents increased by organic complexation. *Nature Geoscience*, **4**(3), 145–150.
175. Rubin, K. (1997) Degassing of metals and metalloids from erupting seamount and mid-ocean ridge volcanoes: Observations and predictions. *Geochimica et Cosmochimica Acta*, **61**(17), 3525–3542.
176. Statham, P.J., German, C.R. and Connelly, D.P. (2005) Iron(II) distribution and oxidation kinetics in hydrothermal plumes at the Kairei and Edmond vent sites, Indian Ocean. *Earth and Planetary Science Letters*, **236**(3–4), 588–596.
177. German, C.R., Colley, S., Palmer, M.R., Khripounoff, A. and Klinkhammer, G.P. (2002) Hydrothermal plume-particle fluxes at 13 degrees N on the East Pacific Rise. *Deep-Sea Research Part I*, **49**(11), 1921–1940.
178. Gartman, A., Yucel, M., Madison, A.S., Chu, D.W., Ma, S.F., Janzen, C.P., Becker, E.L., Beinart, R.A., Girguis, P.R. and Luther, G.W. (2011) Sulfide oxidation across diffuse flow zones of hydrothermal vents. *Aquatic Geochemistry*, **17**(4–5), 583–601.
179. Yucel, M., Gartman, A., Chan, C.S. and Luther, G.W. (2011) Hydrothermal vents as a kinetically stable source of iron-sulphide-bearing nanoparticles to the ocean. *Nature Geoscience*, **4**(6), 367–371.
180. Donat, J.R. and Bruland, K.W. (1995) Trace elements in the ocean. In: Salbu, B. and Steinnes, E. (eds) *Trace Metals in Natural Waters*. CRC Press, Boca Raton, pp. 247–281.
181. Morel, F.M.M., Milligan, A.J. and Saito, M.A. (2004) Marine bioinorganic chemistry: The role of trace metals in the oceanic cycles of major nutrients. In: Elderfield, H. (ed.) *The Oceans and Marine Geochemistry*. Vol. 6 in Holland, H.D. and Turekian, K.K. (Exec. eds) *Treatise on Geochemistry*. Elsevier Pergamon, Amsterdam, pp. 113–143.
182. Reinfelder, J.R. and Fisher, N.S. (1996) The assimilation of elements ingested by marine copepods. *Science*, **251**, 794.
183. Bruland, K.W., Orians, K.J. and Cowen, J.P. (1994) Reactive trace metals in the stratified central North Pacific. *Geochimica et Cosmochimica Acta*, **58**, 3171–3182.
184. Landing, W. and Bruland, K.H. (1987) The contrasting biogeochemistry of iron and manganese in the Pacific Ocean. *Geochimica et Cosmochimica Acta*, **51**, 29–43.
185. Boyle, E.A., Sherrell, R.M. and Bacon, M.P. (1994) Lead variability in the western North-Atlantic Ocean and central Greenland ice – Implications for the search for decadal trends in anthropogenic emissions. *Geochimica et Cosmochimica Acta*, **58**(15), 3227–3238.
186. Fitzgerald, W.F. and Mason, R.P. (1996) The global mercury cycle: Oceanic and anthropogenic aspects. In: Baeyens, W., et al. Vasiliev, O., and Ebinghaus, R. (eds) *Global and Regional Mercury Cycles: Sources, Fluxes and Mass Balances*. Kluwer Academic, Dordrecht, The Netherlands, pp. 85–108.
187. Schaule, B. and Patterson, C.C. (1983) Perturbations of the natural lead depth profile in the Sargasso Sea by industrial lead. In: Wong, C., Boyle, E., Bruland, K.W., Burton, J.D. and Goldberg, E.D. (eds) *Trace Metals in Seawater*. Plenum Press, New York, pp. 487–503.
188. Nozaki, Y. (1997) A fresh look at element distributions in the North Pacific. Available from: http://www.agu.org/pubs/eos-news/supplements/ (accessed October 22, 2012).
189. GEOTRACERS (2010) The geotracers program website. Available from: http://www.geotraces.org/ (accessed October 22, 2012).
190. Craig, H. and Turekian, K.K. (1980) The GEOSECS Program: 1976–1979. *Earth and Planetary Science Letters*, **49**, 263–265.
191. Measures, C.I., Landing, W.M., Brown, M.T. and Buck, C.S. (2008) A commercially available rosette system for trace metal-clean sampling. *Limnology and Oceanography-Methods*, **6**, 384–394.
192. Measures, C.I., Landing, W.M., Brown, M.T. and Buck, C.S. (2008) High-resolution Al and Fe data from the Atlantic Ocean

CLIVAR-CO2 repeat hydrography A16N transect: Extensive linkages between atmospheric dust and upper ocean geochemistry. *Global Biogeochemical Cycles*, **22**(1), Article # GB003042.

193. Buck, C.S., Landing, W.M., Resing, J.A. and Measures, C.I. (2010) The solubility and deposition of aerosol Fe and other trace elements in the North Atlantic Ocean: Observations from the A16N CLIVAR-CO2 repeat hydrography section. *Marine Chemistry*, **120**(1–4), 57–70.
194. Lohan, M.C., Aguilar-Islas, A.M. and Bruland, K.W. (2006) Direct determination of iron in acidified (ph 1.7) seawater samples by flow injection analysis with catalytic spectrophotometric detection: Application and intercomparison. *Limnology and Oceanography-Methods*, **4**, 164–171.
195. Trouwborst, R.E., Clement, B.G., Tebo, B.M., Glazer, B.T. and Luther, G.W. (2006) Soluble Mn(III) in suboxic zones. *Science*, **313**(5795), 1955–1957.
196. Luther, G.W. (2010) The role of one- and two-electron transfer reactions in forming thermodynamically unstable intermediates as barriers in multi-electron redox reactions. *Aquatic Geochemistry*, **16**(3), 395–420.
197. Luther, G.W. (2005) Manganese(II) oxidation and Mn(IV) reduction in the environment – Two one-electron transfer steps versus a single two-electron step. *Geomicrobiology Journal*, **22**(3–4), 195–203.
198. Orians, K.J. and Merrin, C.L. (2010) Refractory metals. In: Steele, J.H., Thorpe, S.A. and Turekian, K.K. (eds) *Marine Chemsitry and Geochemistry*. Elsevier, Amsterdam, pp. 52–63.
199. Hansard, S.P., Landing, W.M., Measures, C.I. and Voelker, B.M. (2009) Dissolved iron(II) in the Pacific Ocean: Measurements from the PO2 and P16N CLIVAR-CO2 repeat hydrography expeditions. *Deep-Sea Research Part I*, **56**(7), 1117–1129.
200. Yemenicioglu, S., Erdogan, S. and Tugrul, S. (2006) Distribution of dissolved forms of iron and manganese in the Black Sea. *Deep-Sea Research Part II*, **53**(17–19), 1842–1855.
201. Konovalov, S., Samodurov, A., Oguz, T. and Ivanov, L. (2004) Parameterization of iron and manganese cycling in the Black Sea suboxic and anoxic environment. *Deep-Sea Research Part I*, **51**(12), 2027–2045.
202. Oguz, T., Murray, J.W. and Callahan, A.E. (2001) Modeling redox cycling across the suboxic-anoxic interface zone in the Black Sea. *Deep-Sea Research Part I*, **48**(3), 761–787.
203. Swarzenski, P.W., McKee, B.A., Sorensen, K. and Todd, J.F. (1999) Pb-210 and Po-210, manganese and iron cycling across the O-2/H2S interface of a permanently anoxic fjord: Framvaren, Norway. *Marine Chemistry*, **67**(3–4), 199–217.
204. Percy, D., Li, X.N., Taylor, G.T., Astor, Y. and Scranton, M.I. (2008) Controls on iron, manganese and intermediate oxidation state sulfur compounds in the cariaco basin. *Marine Chemistry*, **111**(1–2), 47–62.
205. O'Sullivan, D.W., Hanson, A.K. and Kester, D.R. (1997) The distribution and redox chemistry of iron in the Pettaquamscutt Estuary. *Estuarine, Coastal and Shelf Science*, **45**(6), 769–788.
206. Lupton, J.E. and Craig, H. (1981) The major ^{3}He source on the East Pacific Rise. *Science*, **214**, 13–18.
207. Donat, J. and Dryden, C. (2010) Transition metals and heavy metal speciation. In: Steele, J.H., Thorpe, S.A. and Turekian, K.K. (eds) *Marine Chemistry and Geochemistry*. Elsevier, Amsterdam, pp. 72–80.
208. Dupont, C.L., Nelson, R.K., Bashir, S., Moffett, J.W. and Ahner, B.A. (2004) Novel copper-binding and nitrogen-rich thiols produced and exuded by *Emiliania huxleyi*. *Limnology and Oceanography*, **49**(5), 1754–1762.
209. Mann, E.L., Ahlgren, N., Moffett, J.W. and Chisholm, S.W. (2002) Copper toxicity and cyanobacteria ecology in the Sargasso Sea. *Limnology and Oceanography*, **47**(4), 976–988.
210. Okbamichael, M. and Sanudo-Wilhelmy, S.A. (2005) Direct determination of vitamin B-12 in seawater by solid-phase extraction and high-performance liquid chromatography quantification. *Limnology and Oceanography-Methods*, **3**, 241–246.
211. Hopkinson, B.M., Roe, K.L. and Barbeau, K.A. (2008) Heme uptake by microscilla marina and evidence for heme uptake systems in the genomes of diverse marine bacteria. *Applied and Environmental Microbiology*, **74**(20), 6263–6270.
212. Ranville, M.A., Cutter, G.A., Buck, C.S., Landing, W.M., Cutter, L.S., Resing, J.A. and Flegal, A.R. (2010) Aeolian contamination of Se and Ag in the North Pacific from Asian fossil fuel combustion. *Environmental Science & Technology*, **44**(5), 1587–1593.
213. Sanudo-Wilhelmy, S.A. and Flegal, A.R. (1992) Anthropogenic silver in the Southern California Bight – A new tracer of sewage in coastal waters. *Environmental Science & Technology*, **26**, 2147–2151.
214. Flegal, A.R., Brown, C.L., Squire, S., Ross, J.R.M., Scelfo, G.M. and Hibdon, S. (2007) Spatial and temporal variations in silver contamination and toxicity in San Francisco Bay. *Environmental Research*, **105**, 34–52.
215. Rivera-Duarte, I., Flegal, A.R., Sanudo-Wilhelmy, S.A. and Veron, A.J. (1999) Silver in the far North Atlantic. *Deep-Sea Research Part II*, **46**, 979–990.
216. Zhang, Y., Obata, H. and Nozaki, Y. (2004) Silver in the Pacifc Ocean and the Bering Sea. *Geochemical Journal*, **38**, 623–633.
217. Zhang, Y., Amakawa, H. and Nozaki, Y. (2001) Oceanic profiles of dissolved silver: Precise measurements in the basins of the western North Pacific, Sea of Okhotsk and the Japan Sea. *Marine Chemistry*, **75**, 151–163.
218. Ranville, M.A. and Flegal, A.R. (2005) Silver in the North Pacific Ocean. *Geochemistry Geophysics Geosystems*, **6**, Art. No. Q03M01.
219. Ndung'u, K., Thomas, M.A. and Flegal, A.R. (2001) Silver in the western Equatorial and South Atlantic Ocean. *Deep-Sea Research Part II*, **48**(13), 2933–2945.
220. Ondov, J.M., Choquette, C.E., Zoller, W.H., Gordon, G.E., Biermann, A.H. and Heft, R.E. (1989) Atmospheric behavior of trace elements on particles emitted from a coal-fired power plant. *Atmospheric Environment*, **23**, 2193–2204.
221. Laurier, F.J.G., Mason, R.P., Gill, G.A. and Whalin, L. (2004) Mercury distributions in the North Pacific Ocean – 20 years of observations. *Marine Chemistry*, **90**(1–4), 3–19.
222. Sunderland, E.M., Krabbenhoft, D.P., Moreau, J.W., Strode, S.A. and Landing, W.M. (2009) Mercury sources, distribution, and bioavailability in the North Pacific Ocean: Insights from data and models. *Global Biogeochemical Cycles*, **23**, Article # GB2010.

223. Wu, J.F., Rember, R., Jin, M.B., Boyle, E.A. and Flegal, A.R. (2010) Isotopic evidence for the source of lead in the North Pacific abyssal water. *Geochimica et Cosmochimica Acta*, **74**(16), 4629–4638.
224. Wu, J.F. and Boyle, E.A. (1997) Lead in the western North Atlantic Ocean: Completed response to leaded gasoline phaseout. *Geochimica et Cosmochimica Acta*, **61**(15), 3279–3283.
225. Mason, R.P. and Sheu, G.R. (2002) Role of the ocean in the global mercury cycle. *Global Biogeochemical Cycles*, **16**(4), Article # 1093.
226. Sunderland, E.M. and Mason, R.P. (2007) Human impacts on open ocean mercury concentrations. *Global Biogeochemical Cycles*, **21**, GB4022.
227. Selin, N.E., Jacob, D.J., Yantoscha, R.M., Strode, S., Jaegle, L. and Sunderland, E.M. (2008) Global 3-D land-ocean-atmosphere model for mercury: Present-day vs. Preindustrial cycles and anthropogenic enhancement factors for deposition. *Global Biogeochemical Cycles*, **22**(3), Article # GB2011.
228. Mason, R.P., Fitzgerald, W.F. and Morel, F.M.M. (1994) The biogeochemical cycling of elemental mercury: Anthropogenic influences. *Geochimica et Cosmochimica Acta*, **58**(15), 3191–3198.
229. USEPA (1997) *The Incidence and Severity of Sediment Contamination in Surface Water of the United States*. Three volumes, Office of Science and Technology, Washington, DC.
230. Balcom, P.H., Fitzgerald, W.F., Vandal, G.M., Lamborg, C.H., Rolfllus, K.R., Langer, C.S. and Hammerschmidt, C.R. (2004) Mercury sources and cycling in the Connecticut River and Long Island Sound. *Marine Chemistry*, **90**(1–4), 53–74.
231. Cossa, D., Michel, P., Noel, J. and Auger, D. (1992) Vertical mercury profile in relation to arsenic, cadmium and copper at the eastern North Atlantic ICES reference station. *Oceanologica Acta*, **15**, 603–608.
232. Cossa, D., Cotte-Krief, M.H., Mason, R.P. and Bretaudeau-Sanjuan, J. (2004) Total mercury in the water column near the shelf edge of the European continental margin. *Marine Chemistry*, **90**(1–4), 21–29.
233. Cossa, D., Martin, J.-M., Takayanagi, K. and Sanjuan, J. (1997) The distribution and cycling of mercury species in the western Mediterranean. *Deep-Sea Research*, **44**, 721–740.
234. Mason, R.P. and Sullivan, K.A. (1999) The distribution and speciation of mercury in the South and equatorial Atlantic. *Deep-Sea Research Part II*, **46**(5), 937–956.
235. Mason, R.P. and Fitzgerald, W.F. (1993) The distribution and biogeochemical cycling of mercury in the equatorial Pacific Ocean. *Deep-Sea Research*, **40**(9), 1897–1924.
236. Mason, R.P., Lawson, N.M. and Sheu, G.R. (2001) Mercury in the Atlantic Ocean: Factors controlling air-sea exchange of mercury and its distribution in the upper waters. *Deep-Sea Research Part II*, **48**, 2829–2853.
237. Mason, R.P. and Fitzgerald, W.F. (1996) Sources, sinks and biogeochemical cycling of mercury in the ocean. In: Baeyens, W. et al. (ed.) *Global and Regional Mercury Cycles: Sources, Fluxes and Mass Balances*. Kluwer Academic, The Netherlands, pp. 249–272.
238. Gill, G.A. and Fitzgerald, W.F. (1988) Vertical mercury distributions in the oceans. *Geochimica et Cosmochimica Acta*, **52**, 1719–1728.
239. Soerensen, A.L., Sunderland, E.M., Holmes, C.D., Jacob, D.J., Yantosca, R.M., Skov, H., Christensen, J.H., Strode, S.A. and Mason, R.P. (2010) An improved global model for air-sea exchange of mercury: High concentrations over the North Atlantic. *Environmental Science & Technology*, **44**(22), 8574–8580.
240. Selin, N.E., Sunderland, E.M., Knightes, C.D. and Mason, R.P. (2010) Sources of mercury exposure for us seafood consumers: Implications for policy. *Environmental Health Perspectives*, **118**(1), 137–143.
241. Mason, R.P., O'Donnell, J. and Fitzgerald, W.F. (1994) Elemental mercury cycling within the mixed layer of the equatorial Pacific Ocean. In: Watras, C.J. and Huckabee, J.W. (eds) *Mercury as a Global Pollutant: Towards Integration and Synthesis*. Lewis, Boca Raton, pp. 83–97.
242. Hammerschmidt, C.H. and Bowman, K.L., et al. (2012) Vertical methylmercury distribution in the subtropical North Pacific Ocean, Original Research Article. *Marine Chemistry*, **132–133**, 77–82.
243. Heimburger, L.E., Cossa, D., Marty, J.C., Migon, C., Averty, B., Dufour, A. and Ras, J. (2010) Methyl mercury distributions in relation to the presence of nano- and picophytoplankton in an oceanic water column (Ligurian Sea, north-western Mediterranean). *Geochimica et Cosmochimica Acta*, **74**(19), 5549–5559.
244. Malcolm, E.G., Schaefer, J.K., Ekstrom, E.B., Tuit, C.B., Jayakumar, A., Park, H., Ward, B.B. and Morel, F.M.M. (2010) Mercury methylation in oxygen deficient zones of the oceans: No evidence for the predominance of anaerobes. *Marine Chemistry*, **122**, 11–19.
245. Whalin, L., Kim, E.-H. and Mason, R. (2007) Factros influencing the oxidation, reduction, methylation and demethylation of mercury in coastal waters. *Marine Chemistry*, **107**, 278–294.
246. Cossa, D., Averty, B. and Pirrone, N. (2009) The origin of methylmercury in open mediterranean waters. *Limnology and Oceanography*, **54**(3), 837–844.
247. Cutter, G.A. (2010) Metalloids and oxyanions. In: Steele, J.H., Thorpe, S.A. and Turekian, K.K. (eds) *Marine Chemistry and Geochemistry*. Elsevier, Amsterdam, pp. 64–71.
248. Nice, A.J., Lung, W.S. and Riedel, G.F. (2008) Modeling arsenic in the Patuxent estuary. *Environmental Science & Technology*, **42**(13), 4804–4810.
249. Hellweger, F.L. (2005) Dynamics of arsenic speciation in surface waters: As(iii) production by algae. *Applied Organometallic Chemistry*, **19**(6), 727–735.
250. Cutter, G.A. and Cutter, L.S. (2006) The biogeochemistry of arsenic and antimony in the North Pacific Ocean. *Geochemistry Geophysics Geosystems*, **7**, Article # Q05M08.
251. Santosa, S.J., Wada, S., Mokudai, H. and Tanaka, S. (1997) The contrasting behaviour of arsenic and germanium species in seawater. *Applied Organometallic Chemistry*, **11**(5), 403–414.
252. Sutton, J., Ellwood, M.J., Maher, W.A. and Croot, P.L. (2010) Oceanic distribution of inorganic germanium relative to

silicon: Germanium discrimination by diatoms. *Global Biogeochemical Cycles*, **24**, Article # GB2017.
253. Ravizza, G.E. (2010) Platinum group elements and their isotopes in the ocean. In: Steele, J.H., Thorpe, S.A. and Turekian, K.K. (eds) *Marine Chemistry and Geochemistry*. Elsevier, Amsterdam, pp. 29–28.
254. Firdaus, M.L., Norisuye, K., Nakagawa, Y., Nakatsuka, S. and Sohrin, Y. (2008) Dissolved and labile particulate Zr, Hf, Nb, Ta, Mo and W in the western North Pacific Ocean. *Journal of Oceanography*, **64**, 247–257.
255. Sohrin, Y., Fujishima, Y., Ueda, K., Akiyama, S., Mori, K., Hasegawa, H. and Matsui, M. (1998) Dissolved niobium and tantalum in the North Pacific. *Geophysical Research Letters*, **25**, 999–1002.
256. Wanty, R.B. and Goldhaber, M.B. (1992) Thermodynamics and kinetics of reactions involving vanadium in natural systems – Accumulation of vanadium in sedimentary-rocks. *Geochimica et Cosmochimica Acta*, **56**(4), 1471–1483.
257. Wanty, R.B., Goldhaber, M.B. and Northrop, H.R. (1990) Geochemistry of vanadium in an epigenetic, sandstone-hosted vanadium-uranium deposit, Henry Basin, Utah. *Economic Geology and the Bulletin of the Society of Economic Geologists*, **85**(2), 270–284.
258. Hastings, D.W., Emerson, S.R., Erez, J. and Nelson, B.K. (1996) Vanadium in foraminiferal calcite: Evaluation of a method to determine paleo-seawater vanadium concentrations. *Geochimica et Cosmochimica Acta*, **60**(19), 3701–3715.
259. Trefry, J.H. and Metz, S. (1989) Role of hydrothermal precipitates in the geochemical cycling of vanadium. *Nature*, **342** (6249), 531–533.
260. Crusius, J., Calvert, S., Pedersen, T. and Sage, D. (1996) Rhenium and molybdenum enrichments in sediments as indicators of oxic, suboxic and sulfidic conditions of deposition. *Earth and Planetary Science Letters*, **145**(1–4), 65–78.
261. Crusius, J. and Thomson, J. (2000) Comparative behavior of authigenic Re, U, and Mo during reoxidation and subsequent long-term burial in marine sediments. *Geochimica et Cosmochimica Acta*, **64**(13), 2233–2242.
262. Krishnaswami, S. (2005) Uranium-thorium series isotopes in ocean profiles. In: Steele, J.H., Thorpe, S.A. and Turekian, K.K. (eds) *Marine Chemsitry and Geochemistry*. Elsevier, Amsterdam, pp. 214–224.
263. Barnes, C.E. and Cochran, J.K. (1993) Uranium geochemistry in estuarine sediments – Controls on removal and release processes. *Geochimica et Cosmochimica Acta*, **57**(3), 555–569.
264. Mills, R.A., Thomson, J., Elderfield, H., Hinton, R.W. and Hyslop, E. (1994) Uranium enrichment in metalliferous sediments from the mid-Atlantic ridge. *Earth and Planetary Science Letters*, **124**(1–4), 35–47.
265. Nozaki, Y. (2010) Rare earth elements and their isotopes in the ocean. In: Steele, J.H., Thorpe, S.A. and Turekian, K.K. (eds) *Marine Chemistry and Geochemistry*. Elsevier, Amsterdam, pp. 39–51.
266. Buesseler, K.O., Trull, T.W., Steinber, D.K., et al. (2008) Vertigo (vertical transport in the global ocean): A study of particle sources and flux attenuation in the North Pacific. *Deep-Sea Research Part II*, **55**(14–15), 1522–1539.
267. Lamborg, C.H., Buesseler, K.O., Valdes, J., Bertrand, C.H., Bidigare, R., Manganini, S., Pike, S., Steinberg, D., Trull, T. and Wilson, S. (2008) The flux of bio- and lithogenic material associated with sinking particles in the mesopelagic "Twilight zone" of the northwest and north central Pacific Ocean. *Deep-Sea Research Part II*, **55**(14–15), 1540–1563.
268. Lamborg, C.H., Buesseler, K.O. and Lam, P.J. (2008) Sinking fluxes of minor and trace elements in the north pacific ocean measured during the Vertigo program. *Deep-Sea Research Part II*, **55**(14–15), 1564–1577.
269. Trull, T.W., Bray, S.G., Buesseler, K.O., Lamborg, C.H., Manganini, S., Moy, C. and Valdes, J. (2008) In situ measurement of mesopelagic particle sinking rates and the control of carbon transfer to the ocean interior during the vertical flux in the global ocean (Vertigo) voyages in the North Pacific. *Deep-Sea Research Part II*, **55**(14–15), 1684–1695.
270. Buesseler, K.O., Pike, S., Maiti, K., Lamborg, C.H., Siegel, D.A. and Trull, T.W. (2008) Thorium-234 as a tracer of spatial, temporal and vertical variability in particle flux in the North Pacific. *Deep-Sea Research Part II*, **56**(7), 1143–1167.

Problems

6.1. Many metals are required micronutrients. Speculate how the predicted changes in ocean acidification may have changed or could change the bioavailability of Fe, Zn, and Mo in the surface ocean to microorganisms. Show any equations/calculations that may support your answer.

6.2. The concentrations of metals in seawater are controlled by a number of different factors. In many cases, their concentration is determined by a dominant precipitation reaction. Given the following concentrations for metals in seawater, and the following bulk composition of seawater, determine which of the metals' concentration is controlled by a precipitation reaction. In terms of the carbonate system, assume that the surface ocean is in equilibrium with the atmosphere ($\log P_{CO2} = -3.5$) and that the pH is 8.1. Use equilibrium constants from the literature. If the concentration given in the table is smaller than is predicted based on the dominant precipitation reaction, provide a reasonable explanation why the concentration is not determined by a solubility reaction. Similarly, for those metals whose concentrations exceed that predicted by a solubility reaction, provide an explanation for why these metals persist in solution at a higher than predicted concentration. The relevant information is:

Metal to Consider	Metal Concentration (M)	Major Constituents	Free Ion Concentration (M)*
Ca(II)	1.03×10^{-2}	Na^+	0.46
Fe(III)	5×10^{-10}	K^+	0.01
Mn(III)	3×10^{-10}	Cl^-	0.55
Cu(II)	4×10^{-9}	SO_4^{2-}	0.011
Pb(II)	1×10^{-10}	–	–

Note: **The free ion concentration for the major constituents is calculated by assuming that Na and K are 98% and Cl 100% in the free ion form, but that only 40% of the sulfate is present as a free ion in seawater, the rest being complexed by the major cations.*

6.3. Write a short essay contrasting the biogeochemical cycling of iron and cadmium. In the essay briefly discuss: (1) the relative importance of the atmosphere, terrestrial inputs (rivers and groundwater) and hydrothermal systems as sources to the ocean; (2) the major forms of both elements in the ocean and the role of complexation in controlling their concentration. Discuss (3) their uptake into microorganisms and their biochemical role in the cell. Finally, discuss (4) how changes over time (the recent anthropogenic past (~100 yrs) and over geological time) have changed their concentration in the ocean.

6.4. Using data in Fig. 6.2 and Equation 6.3, estimate the value of the colloidal partition coefficient (K_C) for the various metals. Assume a) that the K_D value at $1\,mg\,l^{-1}$ SPM (TSS) is representative of K_P and b) that the ratio of $M_C/SPM = 0.1$. Use the data for each metal at $100\,mg\,l^{-1}$ SPM to estimate K_C. For which metals is the estimated value of K_C greater than K_P?

6.5. Use the information in Fig. 6.9 to estimate the binding constants for Cd and Cu to organic matter in the colloidal fraction. Assume (a) that all the metals in the <1 kDa fraction are bound to inorganic ligands and b) that the metals in the 1–10 kDa fraction are only bound to organic matter (RCOOH; $pK_a = 4.5$), and that the binding can be approximated by formation of a 1:1 complex ($MOOC^+$ for M^{2+}). Finally assume that 10% of the total carbon in this fraction is as reactive carboxylic acid sites. Do the calculations at salinities of 5 and 30, making the appropriate ionic strength corrections. Are the values of K different? How close are these simplified estimates to values in the literature?

6.6. Estimate binding constants using the data in Fig. 6.27 for Zn and Fig. 6.34 for Cu, both for the 100 m data. Use the ligand concentrations given in the figures and assume the ligand can be represented as RCOOH ($pK_a = 4.5$), and that a 1:1 complex is formed in each case. Adjust the value of the organic matter binding constant to best fit the data. Assume typical concentrations for the major ions in seawater, and make ionic strength corrections.

6.7. You want to do some phytoplankton culture experiments with Cu and Cd and you talk to your professor who gives you this formula for a seawater culture medium: total carbonate = 2.4×10^{-3} M, total $SO_4 = 3 \times 10^{-2}$ M, total Cl = 0.5 M, total Na = 0.5 M, total K = 1×10^{-2} M, total Ca = 1×10^{-2} M and total Mg = 5×10^{-2} M, pH = 8.2 containing $EDTA_T = 1 \times 10^{-5}$ M. The EDTA of course is added to buffer the concentration of the metals in solution. You want to add Fe as an essential nutrient and your professor suggests adding $Fe(III)_T = 1 \times 10^{-7}$ M. For your experiments you wish to have around 10 pM of Cu and Cd as the free metal ions. Your professor does some calculations in his head, mumbling incoherently in a fashion you cannot follow, and then says you should add $Cu_T = 5 \times 10^{-10}$ and $Cd_T = 10^{-10}$ M. You want to check the medium for consistency and also the suggested metal concentrations. You turn to speciation calculations.

a. Firstly you determine the speciation and free ion concentrations of EDTA, Fe, Cu and Cd. What are they? Was the professor right in terms of Cd and Cu? What is the dissolved inorganically-complexed Fe concentration? Is EDTA effective in its role as buffering the metals in solution?

b. You wonder what would be the impact of changing the EDTA concentration over a range of 10^{-3} to 5×10^{-5} M. Discuss the effect on the speciation of Fe, Cd, Cu, Ca, and Mg.

6.8. The preservation of inorganic and organic carbon and metals (Fe, Mn, Mo, and Zn) in sediments depends on a number of competing processes. What would the "typical" profile of each constituent be in:

a. deep sea pelagic sediments, and

b. continental shelf sediments?

Draw such profiles and explain the different regions or factors that impact the profiles. For each parameter, discuss the major controls over these distributions focusing on:

i. processes related to remineralization/dissolution, and

ii. processes enhancing burial/retention in sediments.

Discuss the magnitude of the burial flux relative to the input at the sediment surface.

6.9. The flux of constituents from sediments depends on a number of factors. Discuss the conditions under which the flux from an estuarine sediment may be enhanced for:

a. Fe and Mn;
b. organic carbon;
c. copper;
d. arsenic; and
e. uranium.

6.10. Scientists have done experiments where they have added Fe to seawater to stimulate primary productivity. Using typical seawater concentrations from the literature, predict what would be the primary form of Fe if Fe was added to a total concentration of 10^{-5} M? How much Fe is present in solution? If Fe oxide forms under the conditions given, how would that effect the speciation of a metal such as Zn, which is also potentially an important nutrient metal ($Zn_T = 10^{-9}$ M)? In answering this question one should use the HFO Model of Dzomback and Morel, or a similar formulation to examine the potential for surface complexation of Zn.

6.11. Recent studies have suggested that the upper ocean contains low nM concentrations of small molecular weight thiols such as cysteine and glutathione. Estimate the concentrations of each ligand than would be needed to complex 50% of the following metals in ocean waters, assuming not other organic complexes, and the absence of free sulfide: Hg, Cu, Cd, and Fe.

6.12. In Fig. 6.41, Lamborg et al. give values for the flux of metals out of the surface ocean. a) How do these export fluxes (for the region of 100–200 m) compare with atmospheric inputs to this ocean area of all the metals? (see Chapter 5 or find values in the literature if needed). If the fluxes are not similar (i.e., not within a factor of two) suggest potential reasons for why the export flux is higher or lower than the atmospheric input.

a. For most of the metals there does not appear to be a substantial change in the flux between the upper and lowest traps. Is this consistent with profiles of the concentrations of these metals in the Pacific Ocean? (Again, search the literature for data in not available in Chapter 6.) Explain the instances where there appears to be a discrepancy.
b. If only 10% of the ~500 m flux was buried in deep ocean sediments, how rapidly would the deep ocean concentrations be changing?
c. The metal fluxes seem decoupled from those of carbon and appear more related to that of inorganic material (as represented by Al). Most of these metals are considered nutrient metals and therefore the lack of correlation appears contradictory. Can you give an explanation for the data?

CHAPTER 7
Trace metals in freshwaters

7.1 Overview of metal cycling in freshwaters

The major difference in the chemistry of freshwaters in comparison to marine waters is the ionic strength as the major ions tend to range in concentration from a few mM to 100 μM (Table 7.1) while the trace metals and other dissolved constituents have similar concentrations to the ocean [1]. Additionally, many freshwaters have high particulate loads (TSS) and higher NOM (Table 2.6). The pH of freshwaters covers a large range (4–8.5, typically) depending if the waters are poorly buffered systems that have been acidified by acid rain and other mechanisms or systems which are buffered by interaction with carbonate and other bedrock. If waters are at equilibrium with the atmosphere, then the dissolved ionic carbonate ions are present at high concentration at high pH (>7) and are minor constituents in low pH systems ($pK_{a1} = 6.3$). Alkalinity, as discussed in Chapter 3, is often a useful measure of the system composition. The acidity of unimpacted rainwater is around pH 5.65 (equilibrium with atmospheric CO_2) and unimpacted lakes have higher pH values typically because of the higher concentrations of anions in solution. In "pristine" rainwater, $[HCO_3^-] = [H^+] = 10^{-5.65}$ which is much lower than the range of values shown in Table 7.1. Given the relationships between pH, alkalinity and total carbonate concentrations, it is possible to construct a diagram that relates these three variables (Fig. 7.1) [1], as if two are known, the other is fixed by the thermodynamic relationships discussed in Chapter 3.

As noted in Chapter 2 (Fig. 2.9) there is a broad relationship between the concentration of metal(loid)s in freshwaters and their crustal abundance, although generally cations that bind strongly to solid phases or are highly insoluble fall below the relationship while the anions of metal(loid)s tend to be more soluble on average. However, there are large ranges in concentration across systems which are exemplified by Fig. 7.2 which plots the range in values for a number of major and minor elements in river water [4]. In converting between ppb and molar concentrations, for the first row transition metals and their associated metalloids (atomic mass = 48–78), 10 ppb (10 μg l^{-1}) = 128–208 nM (Table 3.21). In comparing freshwater and saline waters it is worth noting that in freshwaters Ca, and then Mg, are the dominant cations, with bicarbonate as the dominant anion [1]. The major anion concentrations in groundwater are mostly similar to that of surface waters except for sulfate which is often present in high concentration in groundwater. It has been suggested that the major ion concentration of surface waters depends on the type of ecosystem, and whether the inputs are rainfall dominated, dominated by water-rock interactions or resulting from a dominance of evaporation and crystallization, or some combination of these factors [4]. The type of system can often be assessed from an examination of the relative Na concentration to the other major cations and the total dissolved ion concentration (TDS) as both are high for evaporation dominated environments, of which the ocean is the most obvious example. Rain-dominated systems also have a higher ratio but a low TDS while systems dominated by bedrock interactions have a relatively low ratio, as they are dominated by inputs from carbonate weathering and similar processes.

An understanding of the chemistry, reactivity and bioavailability of metals in freshwaters requires knowledge of the speciation of the metals in these environments. A typical speciation for "freshwater" and the range in concentration

Trace Metals in Aquatic Systems, First Edition. Robert P. Mason.

Table 7.1 Concentration ranges for major ions in surface waters. Taken from [1–3].

Major Ion	Log(Conc.) Range	Component	Log(Conc.) Range
Bicarbonate	−2.3—4	pH	4–8.5
Carbonate	−4—6	Alkalinity	−2.3—4
Chloride	−3—5	NOM as C	−3—4
Sulfate	−3—5	Amino acids	−5—7
Sodium	−3—4	Thiols	−7—9
Potassium	−4—5	NOM acidic groups	−4—6
Calcium	−3—4		
Magnesium	−3—4		

for the more important trace metals in different freshwater ecosystems are listed in Table 7.2. This speciation is based on a particular set of conditions, in the absence of NOM, but gives an indication of the dominant forms in solution. Concentrations are given for surface waters and subsurface waters and it is often the case that the concentrations are of the same order [5]. Partitioning to the solid phase is often the most important factor controlling concentration. The concentrations of metals in freshwaters can range widely as there are a number of factors that determine their overall concentration and speciation. In surface waters (rivers and lakes), conditions are typically oxic and therefore concentrations of metals are mostly controlled by their relative solubility and their strength of partitioning to the solid phase.

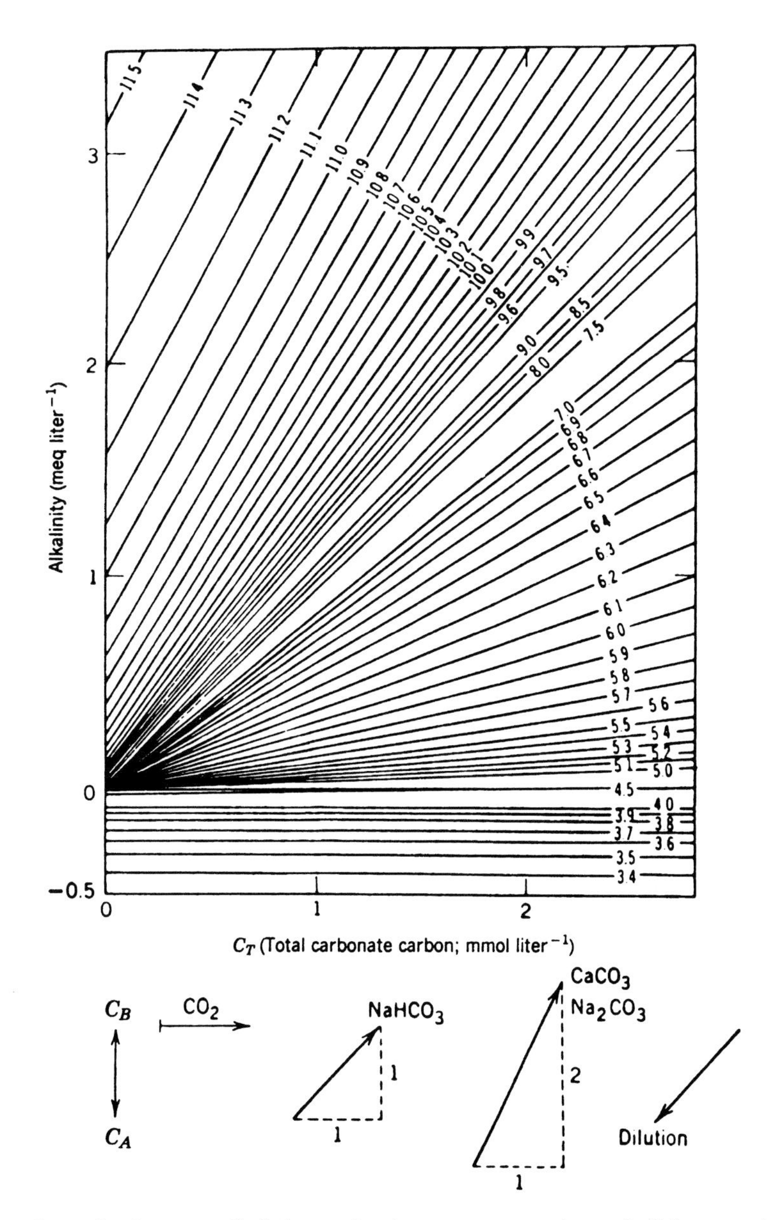

Fig. 7.1 Graph showing the relationship between alkalinity, total carbonate concentration and pH for natural waters. Taken from Stumm and Morgan (1996) *Aquatic Chemistry* [1] and reprinted with permission of John Wiley & Sons, Inc.

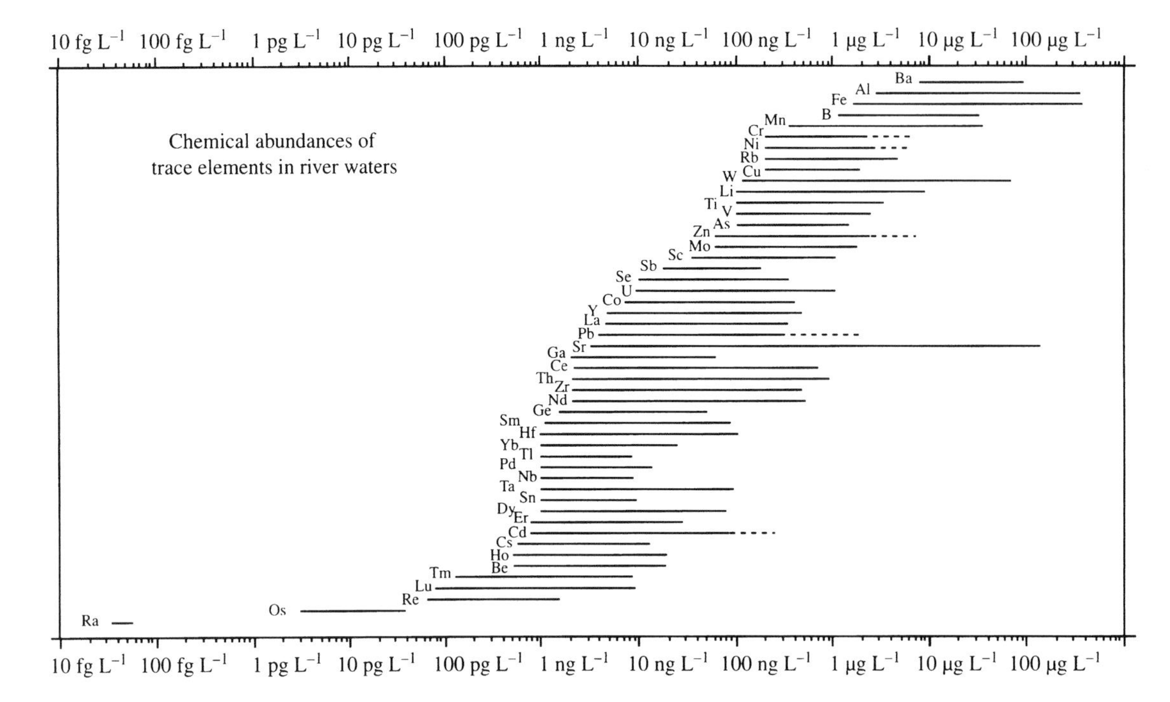

Fig. 7.2 Plot showing the range in the concentrations of elements in river water. Length of bar illustrates the concentration range for each element. Figure reprinted from Gaillardet et al. (2004) in *Treatise of Geochemistry* vol. 5 [6], reprinted with permission of Elsevier.

Table 7.2 The average concentration, and range, of the elements of interest found in uncontaminated surface waters and groundwaters. Data taken from [1, 2, 5, 6].

Element	Av/Conc. Range Surface Waters (nM)	Average logK_D	Groundwater and Soil Solution Conc. (nM)	Major dissolved in species in freshwater*	% as free ion
Al	1850 (1500–8300)	6.3	300	$Al(OH)_x^{3-x}$	<<
V	20	5.2	1–15	HVO_4^{2-}	<<
Cr	20	5.0	1–10	CrO_4^{2-}	<<
Mn	145 (90–360)	5.1	Large range	Mn^{2+}	<<
Fe	720 (90–2240)	5.1	Large range	$Fe(OH)_x^{3-x}$	<<
Co	3.4	5.0	0.4–5	Co^{2+}	50
Ni	8.5 (3.4–34)	5.3	2–15	Ni^{2+}	40
Cu	24 (15–80)	4.8	1–100	$CuCO_3^0$	2
Zn	460 (15–770)	3.9	5–100	Zn^{2+}	40
As	23	3.5	1–20	$HAsO_4^{2-}$	<<
Mo	5.2	3.8	6	MoO_4^{2-}	<<
Ag	2.8	2.4	0.1–0.5	Ag^+, $AgCl^0$	6
Cd	0.18	4.7	0.3–1	Cd^{2+}, $ClCO_3^0$	50
Hg	0.01 (0.005–0.02)	5.5	–	$Hg(OH)_2$	<<
Pb	0.5 (0.2–1.0)	6.0	0.3–3	$PbCO_3^0$	5
Ga	–	–	–	$Ga(OH)_4$	<<
Se	–	–	0.2–3	SeO_4^{2-}	<<
Sb	8.2	3.4	0.5–5	$Sb(OH)_6^-$	<<
W	–	–	–	WO_4^{2-}	<<
U	–	–	–	?UO_4^{2-}	<<

Notes: *Calculations done in the absence of organic matter in solution, and assuming "typical" concentrations of major ions in solution (Table 7.1).

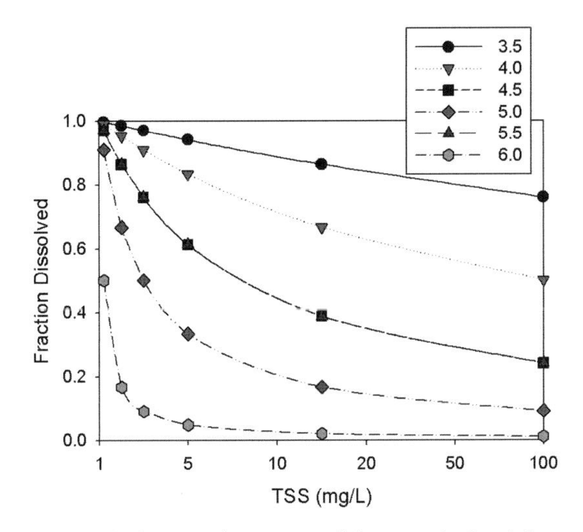

Fig. 7.3 Graph showing the impact of the magnitude of the distribution coefficient (log K_D in figure) on the partitioning of an element between the dissolved and particulate phases over a range of suspended matter (TSS) concentrations.

The partitioning of metal(loid)s differs by several orders of magnitude with the heavy metals often being predominantly bound to the solid phase in higher TSS environments. The highest K_Ds are for Hg, Ag and Pb, and for Al, which is highly insoluble in freshwater (Table 7.2). The crustal elements and some of the first row transition metals also have relatively high K_D's (Fe, Mn, Ni, and Co). While some oxyanions are relatively soluble and have low K_Ds (e.g., As, Mo, Se, Sb), this is not always the case (e.g., Cr, V) [5, 6]. The K_Ds of Cd and Zn are typically lower than that of other transition metals. These differences have a large impact on the total concentration in freshwaters as the elements with high K_Ds will have total concentrations that are a strong function of TSS, while for the others, the TSS has less impact. This is shown in Fig. 7.3 for a range in K_D values. The strength of binding can have a large impact over the fate and transport of metal(loid)s. For an element with a K_D of 10^6, 50% of the element is dissolved at a TSS of $1\,mg\,l^{-1}$ and <10% at $5\,mg\,l^{-1}$, which are typical concentrations for many rivers and lakes. In contrast, for a K_D of 10^4, more than 75% of the element is dissolved at $100\,mg\,l^{-1}$ and essentially all the element is dissolved at TSS < $10\,mg\,l^{-1}$.

The general controls over the cycling of trace elements in lakes will be discussed in detail as there can be strong seasonal cycles in temperate ecosystems that are due to the large temperature differences over the year. Seasonal cycling in large systems is often less important and profiles and concentrations of trace metals in the Great Lakes of North America and other large ecosystems are relatively constant seasonally and over time although they have likely increased in concentration due to anthropogenic inputs. Rivers tend to have a chemistry that is dominated by the solid phase and often also dominated by seasonal cycles, such as strong runoff in spring. This is especially true for locations where there is substantial water input due to snow melt, or where there are often large storms and extreme events (hurricanes, tropical storms). Additionally, many low latitude river systems, particularly those on islands and locations with high mountain ranges and a small coastal plain, have high flow and high particulate loading. Human activity, such as deforestation, agriculture and urbanization has resulted in higher suspended loads in many of these systems. The global input of freshwater and its transport of solids, carbon and other constituents was discussed in detail in Chapter 2 and will not be reiterated here.

This chapter will discuss the general principles pertinent to trace elements in freshwater systems and then will focus on a few specific topics to provide some notion of the complexities of trace element cycling in freshwater systems, be these small streams or rivers, small or large lakes, or the subsurface environment. Issues related to the impact of metal(loid)s on drinking water and human health will be examined in the context of discussing how changes in chemistry and system redox can have a dramatic impact on the concentration and form of toxic elements in freshwaters, and how human activity has exacerbated these changes.

7.2 Trace element cycling in lakes

The cycling of metal(loid)s in lakes is governed by the biogeochemistry of these systems and the relative importance of atmospheric versus fluvial inputs. For most trace elements, it is the fluvial inputs that are the dominant source of metals to lakes and reservoirs because of the strong association of metals with the solid phase. Most lakes have a relative small surface area compared with the area of the watershed and for this reason, direct atmospheric input is not normally an important source of metal(loid)s. As discussed in Section 2.1 and based on Eq. 2.2, the input directly from the atmosphere to that from watershed runoff will be equal when the watershed/lake area ratio (A_{WB}/A_{WS}) is equivalent to the transmission or retention factor (T_{WS}). Retention factors for strongly particle reactive metals (e.g., Pb, Hg) range around 0.1–0.2 while those of weakly binding metals and the more soluble metalloids (e.g., Cd, As, Se) are between 0.3–0.5 [7,24]. Additionally, much of the metal(loid) input is associated with particulate matter and release from the solid phase in such systems is governed by the chemistry of the ecosystem. Many lakes, especially in temperate climates, become stratified in summer due to the strong changes in seasonal temperature, and this stratification results in oxygen depletion in the deeper waters of the lake, and can aid particulate dissolution.

The stratification in lakes is mostly a temperature driven phenomenon as salinity differences are often small between surface and deep waters [8], in contrast to estuaries. These changes have a dramatic impact on the concentrations of metals and other constituents. Most lakes are *holomictic*, which means they have a uniform distribution of temperature at some time during the year even if they are stratified at other times. Lakes that are permanently stratified are termed *meromictic* [8]. Often these lakes have both a temperature and a dissolved solids (density) difference that maintains the stratification in the winter. Many meromictic lakes have bottom waters that are permanently anoxic. Seasonally stratified lakes have an *epilimnion*, the region above the thermocline, and a *hypolimnion* below. If mixing occurs once during the year, in the fall, lakes are termed *monomictic*, and many temperate and tropical lakes fit this category. However, at higher latitudes, ice cover in winter also often results in stratification of the lake and therefore these lakes mix twice a year, in spring and in fall, and are termed *dimictic* lakes [8]. Deeper meromictic lakes such as Lake Pavin in France and the Lower Mystic Lake in Boston, USA can also additionally stratify in summer, forming three layers: the epilimnion, the hypolimnion, which together are called the *mixolimnion*, and the deeper, permanently stratified region, the *monomolimnion*.

Meromictic lakes cover a large range in size and location, from the Rift Valley Lakes in Africa (Lake Tanganyika, 32,900 km^2, maximum depth 1470 m; Lake Kivu, 2700 km^2, max. depth 480 m) and other large lakes where their great depth prevents deep water mixing and organic matter degradation at depth is sufficient to deplete oxygen levels. Many deep lakes are also geologically active and many crater lakes in Africa (e.g., Lake Nyos and Lake Monoun), Europe (Lake Pavin and Lake du Bourget in France) and elsewhere are meromictic. Other meromictic lakes are found in locations where there is deep marine water seepage (e.g., Jellyfish Lake in Palau, Asia), relic seawater (e.g., Lower Mystic Lake, MA, USA) or regions where the surroundings have high salt content (e.g., Great Salt Lake, NV and Green Lake, NY, USA). One intriguing meromictic lake is Lake Vanda in the Antarctica Dry Valleys, which is permanently covered with ice, is 75 m deep and has highly saline bottom waters.

Most of the studies of lakes have been done in the Northern Hemisphere and have focused on temperate ecosystems. Additionally, because of the complications of determining external inputs, many studies have focused on lakes with little external input, except from the atmosphere. *Seepage* lakes do not have appreciable surface water inputs and their hydrologic cycle is dominated by atmospheric inputs, evaporation and seepage of groundwater into and from the lake [8]. In contrast, *drainage* lakes have both an inlet and outlet, and there are intermediate lake types which have no inlet, but have an outlet. Finally, humans have constructed many *artificial* lakes (impoundments, reservoirs and dams) and these have their own biogeochemical cycling, especially immediately after construction. The elemental cycling will be discussed by focusing on a few specific examples rather than attempting to discuss all the studies in the literature. While studies of the transition metals have often involved the examination of numerous elements simultaneously this is often not the case for studies focused on Hg, or on the metalloids. For this reason, separate sections will focus on relevant aspects of the freshwater biogeochemistry of these elements. Mercury lakes studies, again which are focused on temperate lakes, will be highlighted, while the focus on As will be its transport and fate in groundwater.

7.2.1 Processes influencing metal(loid) fate and dissolved speciation in lakes

There are many papers examining the seasonal cycling and changes in constituents for lakes in the temperate regions of North America, Europe and Asia. The basic cycling is described by Wetzel [8] and similar texts and here we will focus our attention on a few studies that examined in detail the distributions of metals and metalloids in such lakes. Three lakes will be discussed in terms of their general chemistry and metal cycling: Lake Hall in North America [9] which is a small, meromictic lake; Lake Pavin, a meromictic lake in France [10]; and Esthwaite Lake in England [11]. The major factors controlling metal distribution seasonally in stratified systems is oxygen distribution, and an example is shown in Fig. 7.4 for a temperate lake in North America [8]. Over winter the waters of the lakes are well-mixed and there is little vertical stratification in any parameter. However, as the surface waters warm in spring and summer, temperature stratification occurs and this prevents the mixing of oxygen from the surface to depth. As the summer progresses, the oxygen content of the bottom waters decreases to undetectable levels (Fig. 7.4a and b) due to the oxidation of sinking organic matter and/or sediment oxygen demand, and the Fe and Mn concentrations simultaneously increase with oxygen depletion due to their reduction dissolution from TSS and the surface sediment (Fig. 7.4c and d). Mixing in the fall leads to a uniform profile in oxygen and to lower levels of Fe and Mn due to precipitation and removal of these metals from the water column. Note that the concentration of Mn increases before Fe in the late spring because of its dissolution at a higher pε than Fe (Fig. 7.4). However, due to the slower oxidation rate of Mn^{II}, it persists in the water column in the fall for a longer period than Fe^{II}.

Profiles of the seasonal cycle of metal concentration and distribution in Esthwaite Lake further illustrates the impact of seasonal stratification on metal cycling (Figs. 7.5 and 7.6 for Esthwaite Lake) [11]. In this lake, anoxia develops and sulfide builds up in bottom waters. The low oxygen conditions and the development of anoxia in the sediments, leads to the reductive dissolution of oxide phases, and the release of metals and metalloids into solution. During the time of low oxygen, dissolved Fe levels, and total Fe levels, increase substantially in these oxygen deficient waters and total Fe concentrations are orders of magnitude higher than in the surface waters (Fig. 7.5b). Most of the Fe and Mn is in the dissolved fraction, and is likely present as reduced Fe (Fe^{II}) and Mn (Mn^{II}) as the oxidized forms are highly insoluble [8, 11]. In the fall, stratification breaks down due to cooling of the surface waters and the water column mixes, often due to a physical disturbance such as a storm, leading to the increase in total Fe and Mn throughout the water column. These higher concentrations persist in the water column for some period of time as the dissolved Fe and Mn is oxidized and precipitated, and removed from the water column by particle settling [1, 8, 11]. Because of the low solubility of oxidized Fe (Fe^{III}) and Mn (Mn^{IV}), dissolved concentrations are low in all cases except for the low oxygen waters where reduced forms are present.

It has been found in some instances that lakes that have ice cover in winter can have a second period of low oxygen as the ice cover restricts oxygen diffusion but photosynthesis and production of organic material is still occurring, due to algae within the ice layer, or due to growth within the lake as the ice often allows sufficient light penetration for photosynthesis [8]. Continued respiration of this newly produced or residual organic material results in the consumption of oxygen and potential oxygen depletion because of the lack of physical mixing of the water column due to ice cover. Oxygen is also potentially consumed by inorganic reduced species, such as the oxidation of reduced metals (Fe and Mn) and other species (ammonia, sulfide, methane) that could flux from sediments to the water column [1, 4]. In many instances, this chemical oxygen consumption, which may not be direct but occurs through a series of coupled reactions, is not considered in studies of oxygen cycling in shallow freshwater ecosystems.

The dissolved concentration of Fe changes dramatically as the redox changes. In oxic waters, Fe is mostly insoluble as Fe^{III}. Upon oxygen depletion, reductive dissolution leads to a buildup of Fe^{II} which is relatively soluble. However, in the presence of increasing sulfide, precipitation of FeS occurs

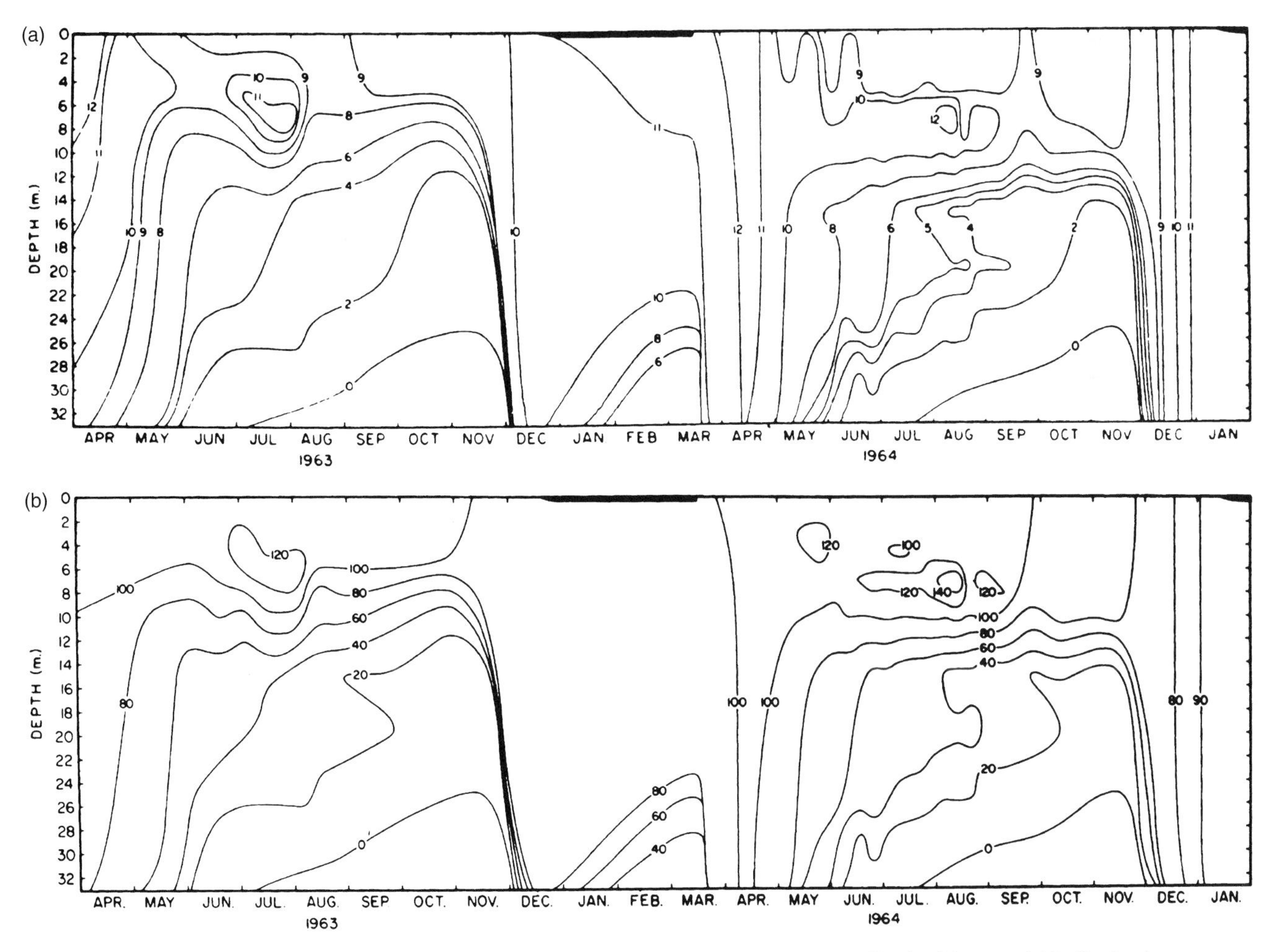

Fig. 7.4 Concentration isopleths for: (a) dissolved oxygen; (b) percent oxygen saturation; (c) dissolved iron; and (d) dissolved manganese in Crooked Lake, Indiana, USA. Taken from Wetzel (2001) *Limnology* [8], reprinted with permission of Elsevier.

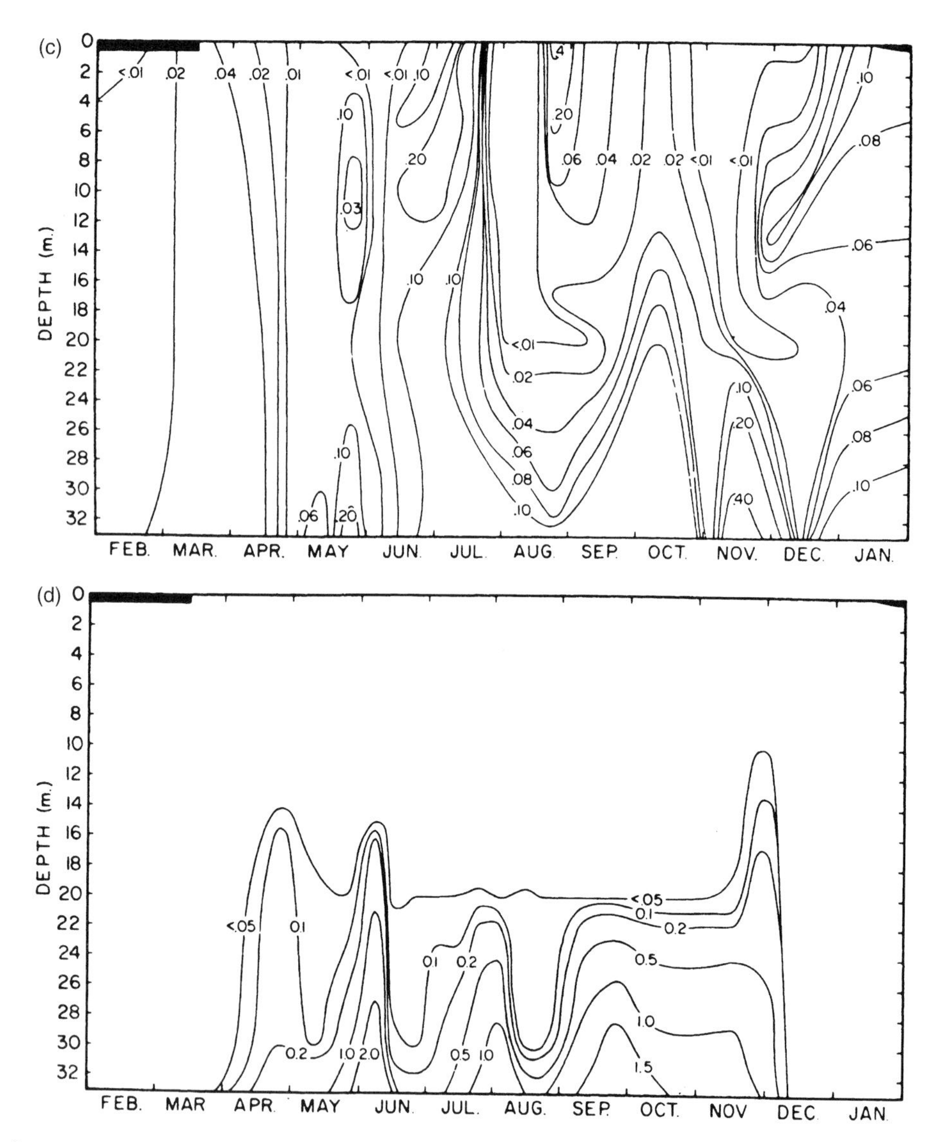

Fig. 7.4 (Continued)

and this decreases the dissolved Fe concentration. The pH-pε diagrams show the distribution between oxidation states and forms under different conditions (Fig. 7.7) [4], and illustrate that solubility and form is a strong function of both parameters.

The cycling of Mn and Fe are similar as Mn is also relatively insoluble in its oxidized state and more soluble when reduced to Mn^{II} (Fig. 7.5b). Other metals and metalloids that are adsorbed onto or included in the oxide phases will be released back to the water column when reductive dissolution occurs (Fig. 7.8) [11]. Thus, the fate and degree of sedimentation of metals in lakes is controlled to a large degree by the lake biogeochemistry, the intensity and degree of stratification, and the overall dynamics of the system. For example, many elements can be transported from surface waters to the deeper waters by particle settling and can be released in low oxygen or anoxic bottom waters and sediments as a result of oxide reductive dissolution. It is important to note that at the near neutral pH conditions of many lakes both cations and anions interact relative ly strongly with the oxide surfaces [1, 4] and so this process is important for transition metals, heavy metals and the metalloids. Additionally, elements taken up biotically in surface waters can be released as this organic matter is remineralized [2].

The profiles of various transition metals in Lake Esthwaite in August, when the waters are stratified and devoid of oxygen illustrate their differences in biogeochemical cycling [11]. Some of the metals (Co, Zn) show higher dissolved deep water concentrations, and given that these metals are not strongly associated with particles (relatively low K_D) (Table 7.2), their total concentrations show a similar distribution. The higher deep water concentration and the depleted surface water concentration for Zn suggests biotic

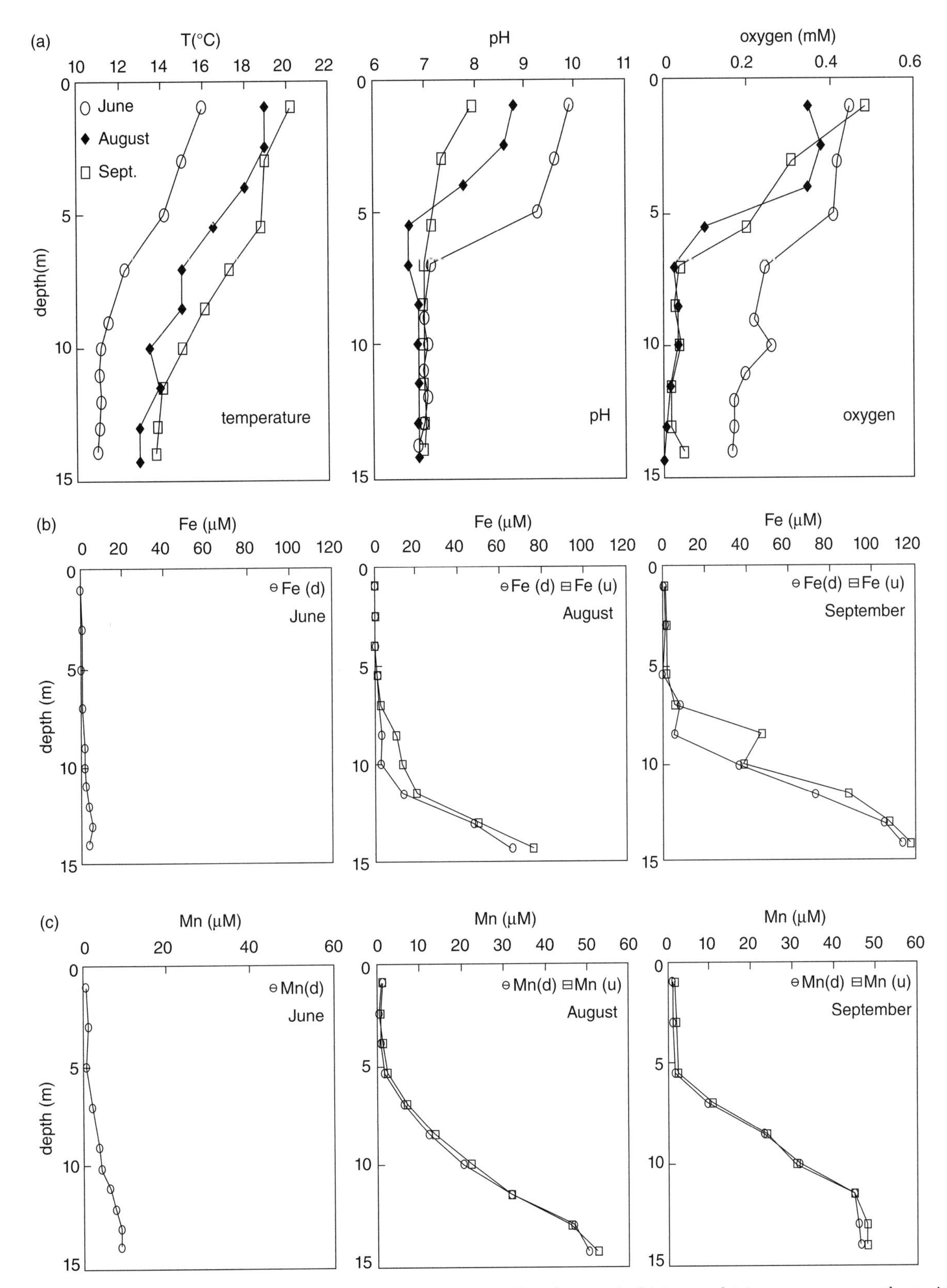

Fig. 7.5 Vertical distributions of (a) ancillary parameters (temperature, pH and oxygen); (b) iron and (c) manganese over the period June to September for Esthwaite Lake in England. For the metals both the dissolved and particulate concentrations are shown. Taken from Achterberg et al. (1997) *Geochimica et Cosmochimica Acta* **61**: 5233–53 [11] and reprinted with permission of Elsevier. Note: (d) = dissolved; (u) = unfiltered.

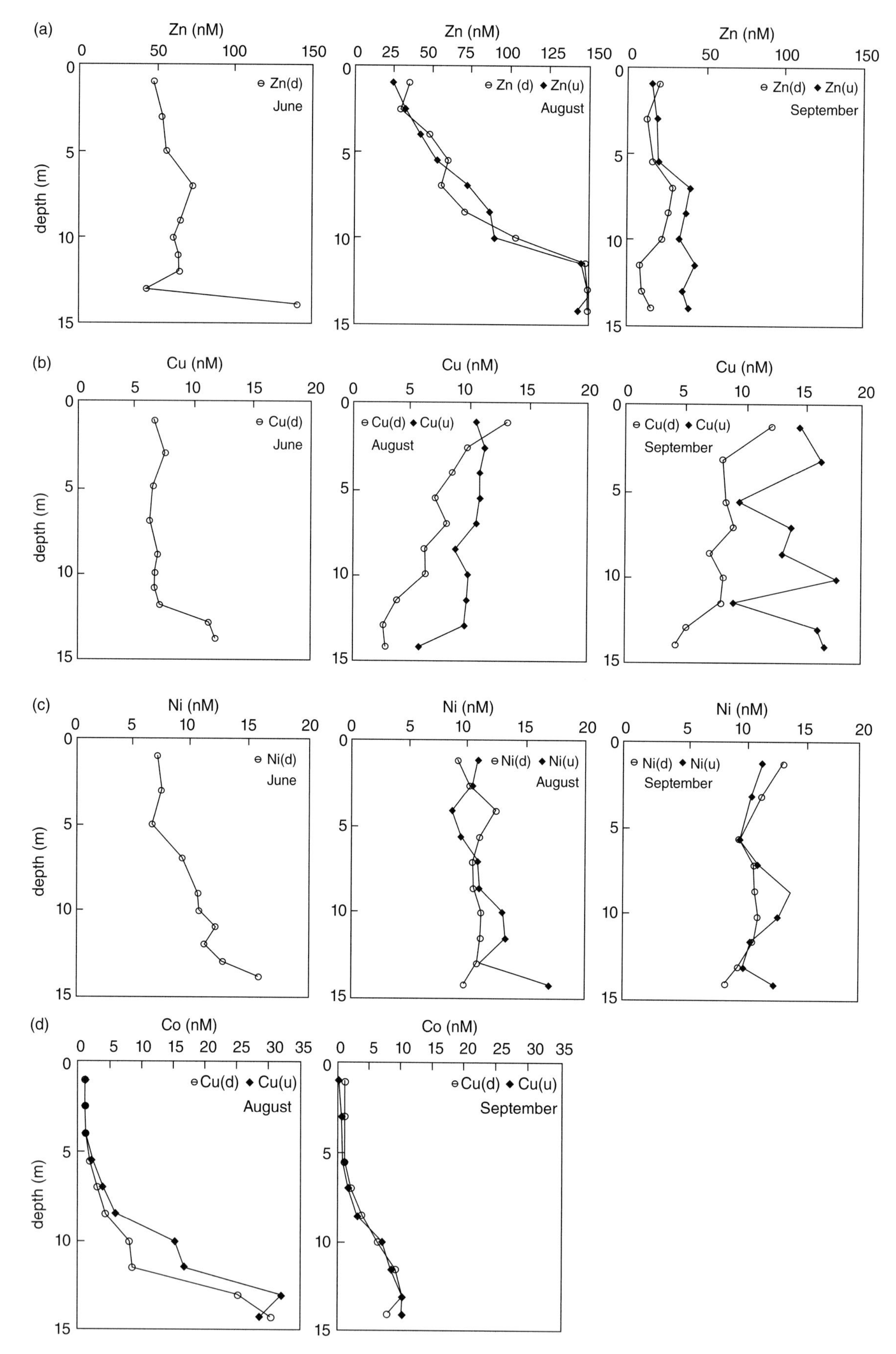

Fig. 7.6 Vertical distributions of dissolved and particulate (a) zinc; (b) copper (c) nickel, and (d) cobalt over the period June to September for Esthwaite Lake in England. Taken from Achterberg et al. (1997) *Geochimica et Cosmochimica Acta* **61**: 5233–53 [11], reprinted with permission of Elsevier.

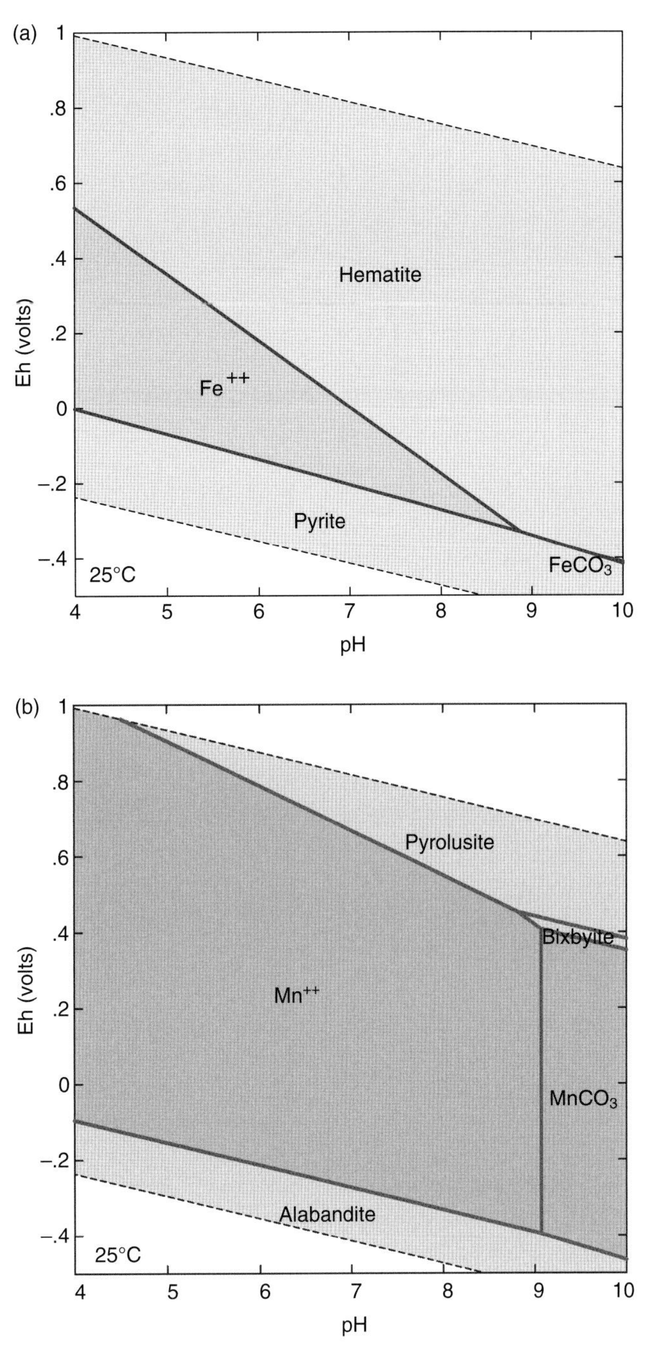

Fig. 7.7 The pH-E_h diagram for (a) iron and (b) manganese (both 1 nM activity) in the presence of atmospheric oxygen and carbon dioxide ($pCO_2 = 10^{-3.5}$ atm) (carbonate species at equilibrium) and total sulfur at 10^{-6} M at 25°C. Figure generated using the computer equilibrium program *The Geochemist's Workbench*, Release 9.

uptake in surface waters and release of Zn at depth. Additionally, Zn could be released from sediments due to organic matter decomposition and oxide dissolution. Similar processes likely account for the Co profiles obtained as both Co and Zn are required "nutrient" metals. In September, when anoxia is more developed, concentrations in the deeper water are lower, but still elevated compared to surface waters, and this change may be attributed to the removal of metals via sulfide precipitation, directly or through adsorption to Fe-S phases.

The Zn data for Hall Lake [9], which is meromictic and had 10 μM sulfide in bottom waters at the time of sampling, higher than that of Esthwaite Lake [11], provides further evidence of the potential importance of sulfide precipitation

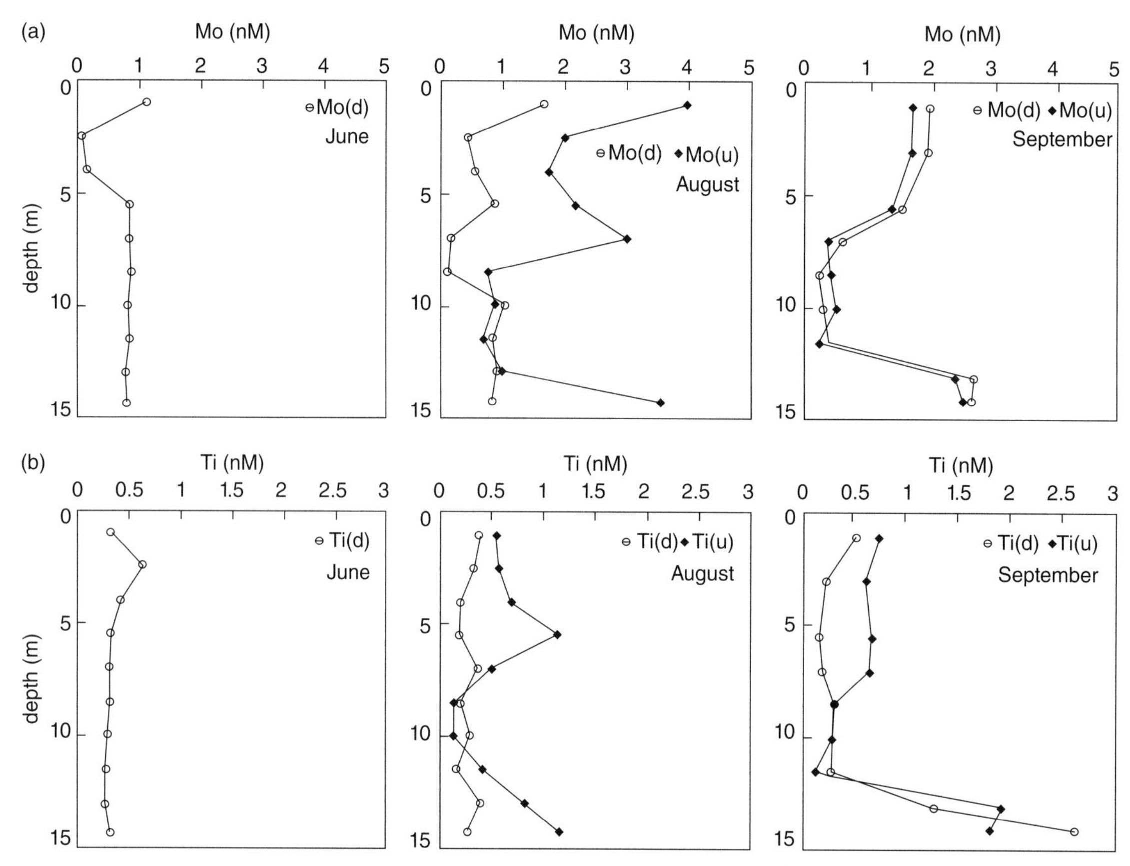

Fig. 7.8 Vertical distributions of dissolved and particulate (a) molybdenum; and (b) titanium over the period June to September for Esthwaite Lake in England. Taken from Achterberg et al. (1997) *Geochimica et Cosmochimica Acta* **61**: 5233–53 [11], reprinted with permission of Elsevier.

in controlling the deep water Zn concentration (Fig. 7.9). This is also apparent for Pb and Cu in Lake Hall, two other metals that forms relatively insoluble sulfides. No Hg or Ag measurements were made. Correlations between reactive S in the particles in the deeper waters of Hall Lake and Zn, Pb, and Cu provide further evidence for their importance in scavenging metals. Computer modeling of the conditions at 12 m (in the anoxic regime) in Hall Lake predicts the formation of ZnS, PbS, NiS, and CoS solid phases, but the presence of dissolved CuS^- due to the reduction of Cu^{II} to Cu^{I} in the presence of sulfide [9].

In comparison to Zn, there is no strong vertical gradient in the concentrations of Cu and Ni in Esthwaite Lake (Fig. 7.6) [11], and this is attributed to the complexation of these metals by organic matter. Modeling of the speciation of the metals in Hall Lake predicts that only Cu is mostly complexed (>80%) to organic matter in surface waters [9], with a minor component of the other transition metals being organically complexed. In Esthwaite Lake, the modeling estimations also predict that Cu is the transition metal with the highest fraction as an organic matter complex in oxic waters [11].

The studies mentioned earlier [9, 11] also measured oxyanions and so these studies provide a basis for an overall discussion of the cycling of oxyanions in lakes. Both redox chemistry and metal oxide dissolution have an impact on the fate of the trace metal(loid)s that exist as anions in solution, so it is also necessary to discuss the impact of seasonal changes and other factors on their distribution and fate in freshwater. Speciation studies show that both oxidation states of Cr were present in Esthwaite Lake but that the concentrations of Cr^{III} were higher in the lower oxygen waters, as expected. The reduced form of Cr is less soluble but given the concentrations of Cr in this lake (low nM levels), precipitation does not occur. It is also likely that the Cr^{III} is associated with the NOM and this stabilizes its concentration in solution. The cycling of As, another important oxyanion which exists in multiple oxidation states in natural waters, is discussed in Section 7.3.4. Typically, its speciation is dominated by As^{V} in oxic waters and As^{III} in anoxic waters and both inorganic forms exist as oxyanions in solution.

Mo concentrations in lakes are much lower than in the ocean, and there is the potential for it to be a limiting nutrient in some freshwaters, as discussed in Chapter 8. In Esthwaite Lake, it is present in low nM concentrations and is mostly in solution in the deeper waters during low oxygen conditions but is partitioned between the dissolved and particulate phase in the surface waters. This could be due to

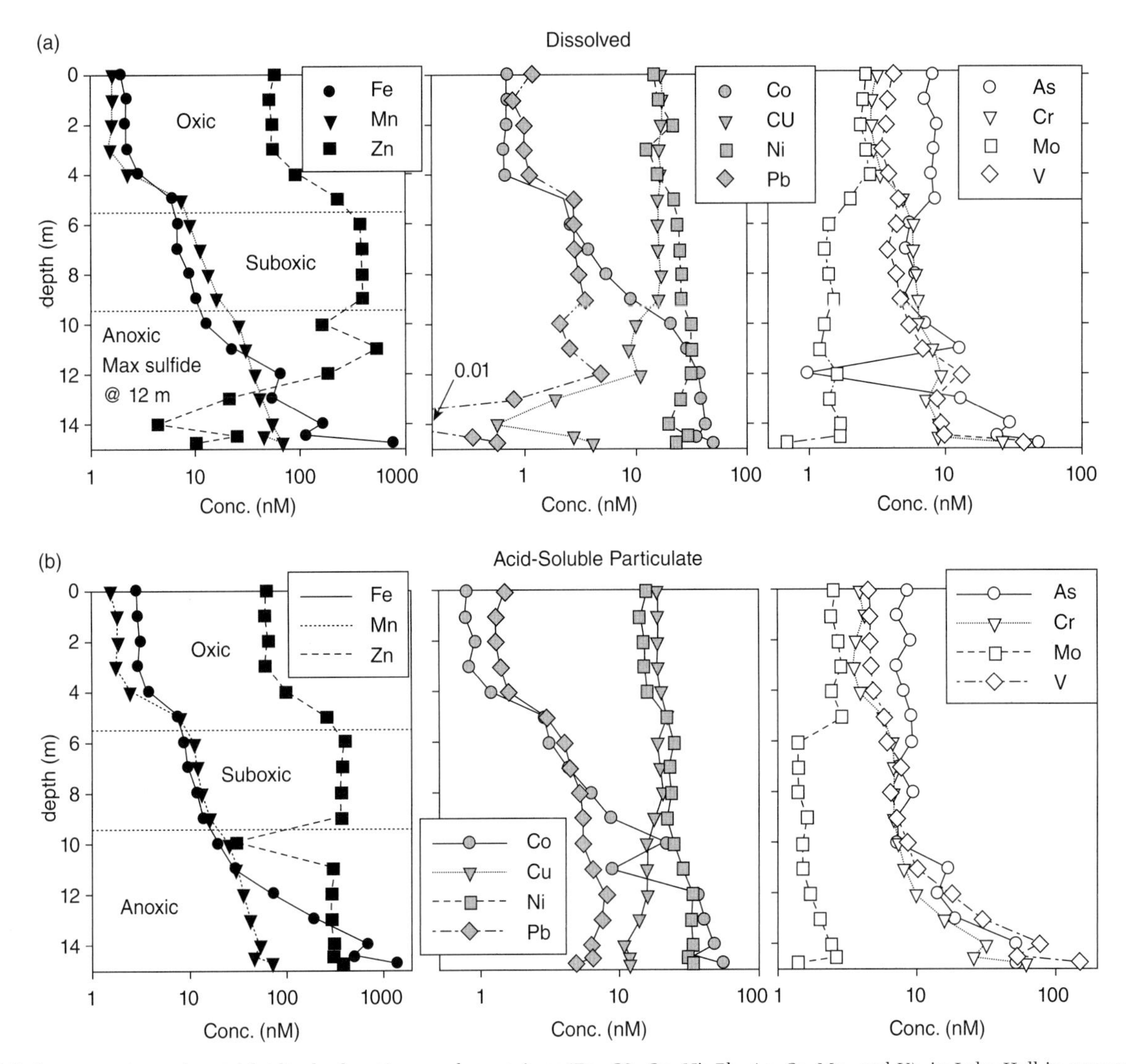

Fig. 7.9 Concentrations of metal(loid)s, both cations and oxyanions (Zn, C0, Cu, Ni, Pb, As, Cr, Mo, and V), in Lake Hall in summer: (a) dissolved concentrations and (b) acid-soluble particulate fraction. From Balistrieri et al. (1994) *Geochimica et Cosmochmica Acta* **58**: 3993–4008 [9], reprinted with permission of Elsevier.

biological uptake but there is also the potential for Mo scavenging by metal oxides, as occurs for the other oxyanions, such as arsenate and phosphate.

A MINEQL model estimation was run for a "typical" lakewater to examine the interaction between oxyanions in terms of adsorption to hydrous ferric oxide (HFO) (Fig. 7.10). In contrast to many other presentations of anion association with HFO, this model study examined which would be the dominant species attached if all were present in solution at the same time. Major ion concentrations in the model were 10^{-4}M for Na, K, Cl, sulfate, 10^{-3}M for Ca, 5×10^{-4}M for Mg and the system was open to the atmosphere. The concentration of total phosphate was 10^{-7}M, As^{V} and Cr^{VI} 20 nM, and 5 nM for Se^{VI}. The total Fe concentration was 5×10^{-6}M (~0.5 mg l^{-1} HFO) and given the formulation of the HFO model [12] this is equivalent to 10^{-6}M weak binding sites on the surface which the anions could associate with. For the minor species, this means that all the oxyanion could potentially adsorb, but for sulfate, as its concentration is two orders of magnitude greater than the number of sites, adsorption will never dominate its speciation. However, the model output does show it is the most important adsorbed species, with two potential complexes (≡Fe(wk)-SO_4^- and ≡Fe(wk)OH-SO_4^{2-}) with the surface complexes decreasing from about $10^{-6.5}$M at pH 4 to $<10^{-9}$M at pH 7. Note that at the higher pHs, Ca adsorption becomes important and competes with the anions for the surface sites (Fig. 7.10a).

Additionally, given the relatively high phosphate concentration, its adsorbed species are also important (=Fe(wk)-HPO_4^- or =Fe(wk)-H_2PO_4), depending on the pH, and the adsorbed species were greater than 50% of the total phosphate at lower pHs. For As, only a small fraction is adsorbed at pH 4 but this rapidly increases with pH until all

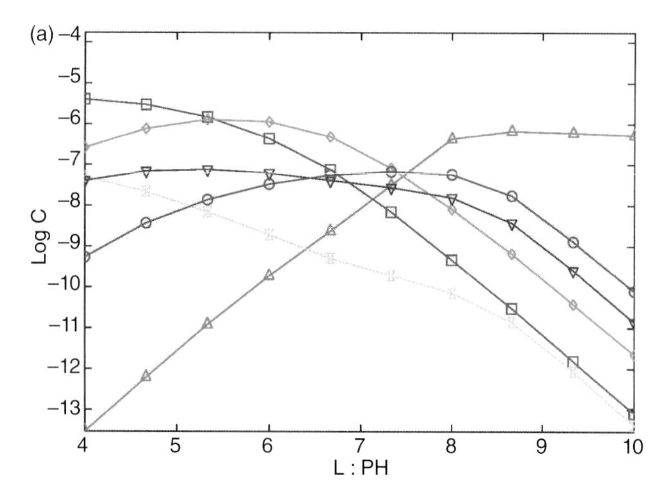

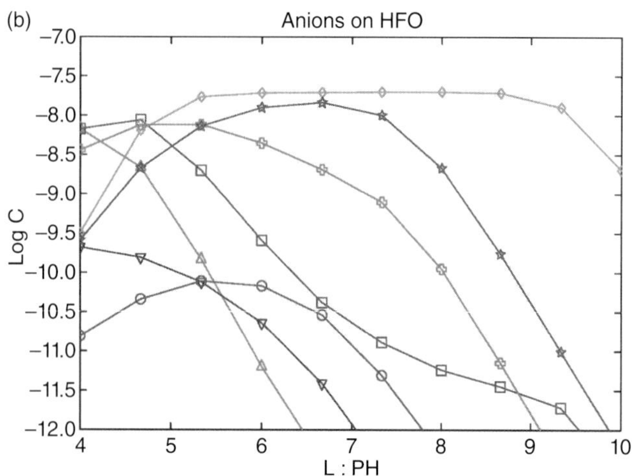

Fig. 7.10 (a) Concentration of adsorbed major anions, and calcium, on the surface of hydrous iron oxide (HFO) under conditions as described in the text. Most anions form more than one type of complex with HFO. The symbols represent: upside triangle: Ca complex; square and diamond: sulfate complexes; star, down triangle and circle: phosphate complexes; (b) Minor anion distributions: upside triangle, square and diamond, arsenic (As(V)) complexes (reduced As was not modeled); down triangle and circle: selenate complexes; cross and star: chromate complexes. Figures generated using the computer program MINEQL+.

As is adsorbed (Fig. 7.10b). Above a pH of 8, desorption occurs. The adsorption of As to solid surfaces is dealt with in more detail in Section 7.3.2. Adsorption of Cr^{VI} was 10% or less of the total Cr for pHs below 7 with the maximum adsorption around pH 5.5. Adsorption of Se was negligible in the presence of other anions although it would adsorb if no other anions were present. Overall, these modeling results support the known importance of Fe precipitation and dissolution on the fate and transport of phosphate and As in natural waters. Fe cycling does not appear to impact Se geochemistry and its impact on Cr^{V} is small.

Lake Pavin in France has been extensively studied and provides an interesting example of trace element cycling in a permanently stratified lake [10, 13]. The *mixomolimnion* is found below 60m and the waters above the permanent *redoxcline* stratify in summer forming three layers, with low oxygen waters developing in the mid-depth waters (Fig. 7.11). Vollier et al. [10] measured a number of metals throughout the water column and examined the role of oxide phases and anoxia in metal cycling. Sulfide levels increase to around 30 μM and NOM levels in the anoxic waters (up to 370 μM) are much higher than in the surface waters (<50 μM) [13]. Similarly, Na concentration is twice as high in the anoxic waters than in surface waters, and Cl 30% higher, but their concentrations are strongly correlated. Other alkali and alkali earth metals (Li, Rb, Cs) were also elevated in the bottom waters, and their concentrations correlated with Na or Li [10]. However, Ba was depleted relative to Li suggesting its removal from bottom waters while Mn, As, Co, and Ni were relatively enriched in the bottom waters, up to 30 times higher than could be predicted from mixing alone. Particulate Fe and Mn are highest just above the redoxcline due to the diffusion upward and precipitation of Fe after oxidation (Fig. 7.11c). The peak in Mn is slightly higher in the water column indicative of its slower rate of oxidation compared to Fe. Profiles of Co, Ni and As suggest that these metals are being transported through association with oxide phases (Fig. 7.12), as found in other systems. Highest bottom water concentrations were around 1.6 nM for Cu, 16 nM for Zn, 80 nM for Co, 120 nM for As, 150 nM for V while levels for Cd and Pb were near the method detection limits (Fig. 7.12) [10]. Mo showed conservative behavior vertically. The concentrations and relative magnitude of the trace elements are very different than found in other systems (Table 7.2) confirming the uniqueness of this permanently stratified system, and the importance of oxide phases in metal transport and their dissolution in the enhancement of deep water concentrations.

Through the determination of oxidation state, it appeared apparent that the As^{V}/As^{III} ratio decreased with depth in the bottom waters, indicative of reduction occurring in the water column of this lake [10]. Precipitation is important for some elements and a peak in particulate Mo coincident with the sulfide peak is explained by precipitation of Mo-S phases. Isolation of organic matter from the bottom waters [14] showed that metals were strongly associated with organic matter in the deeper waters and this explains the high concentrations found compared to other systems. Most of the DOC, U, Mo and V was shown to be colloidal and this association reduces their removal via settling particulate matter.

Mercury was also measured in Lake Pavin and will be discussed briefly here. Further discussion of Hg cycling in lakes is found in Section 7.2.3. Concentrations of inorganic Hg species and methylated Hg (CH_3Hg) showed the importance of Fe and Mn cycling in the fate and distribution of

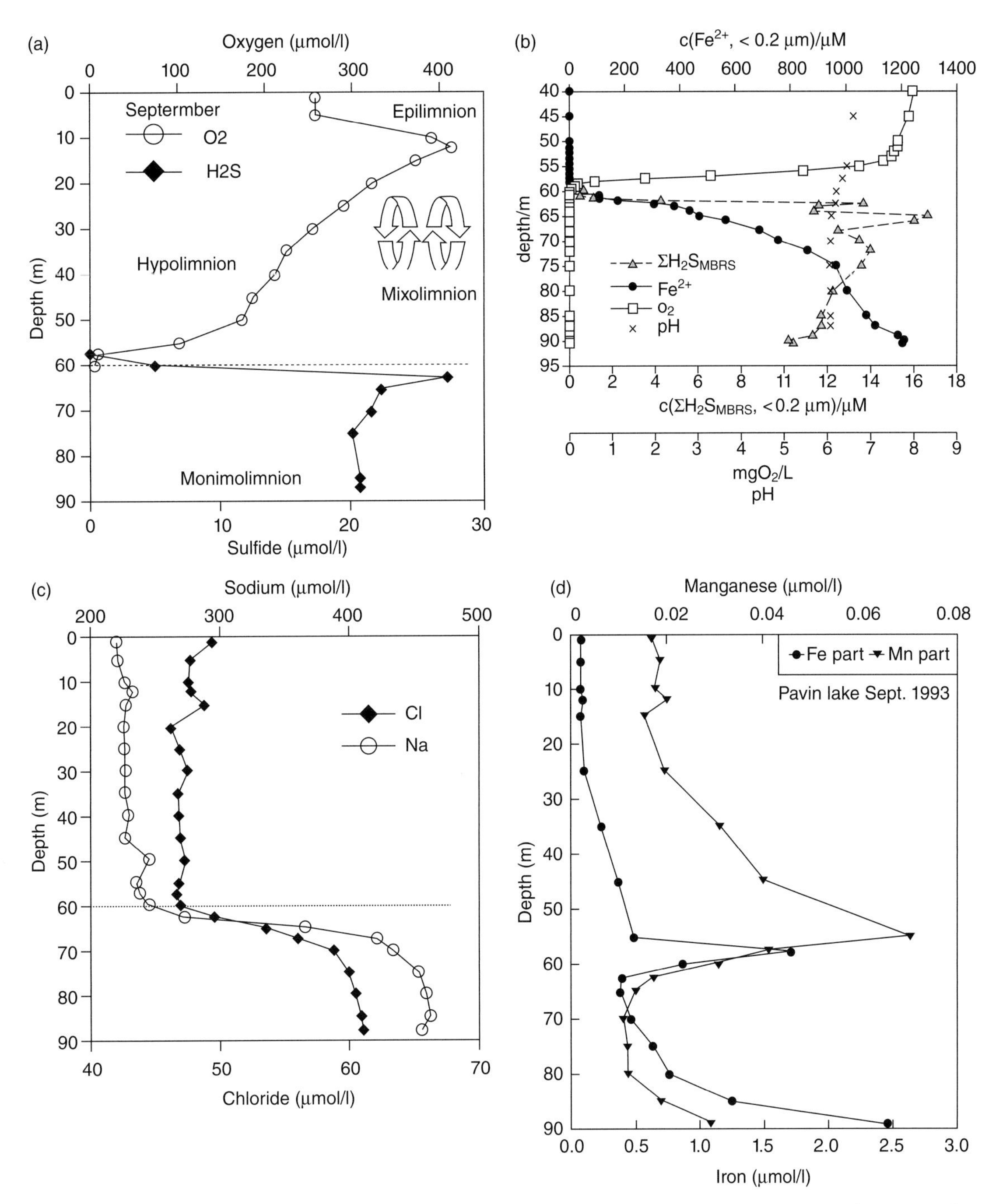

Fig. 7.11 Distributions of (a) oxygen and sulfide through the entire water column; (b) reduced iron, sulfide and pH across the redox interface; (c) vertical distributions of major ions (Na and Cl); (d) particulate iron and manganese in Lake Pavin in France. Parts (a), (b) and (d) taken from Viollier et al. (1995) *Chemical Geology* **125**: 61–72 [10]; Part (c) from Bura-Nakic et al. (2009) *Chemical Geology* **266**: 311–17 [13]. All figures reprinted with permission of Elsevier.

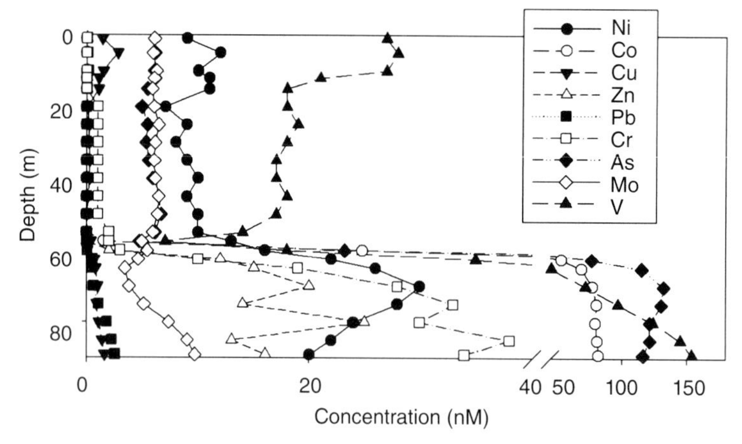

Figure 7.12 The concentration profiles for metal (cations) (Ni, Co, Cu, Zn, and Pb) and metal(loid) oxyanions (Cr, As, Mo, and V) in Lake Pavin. Plot created using data in Viollier et al. (1995) *Chemical Geology* **125**: 61–72 [10] used with permission of Elsevier.

Hg species. Total dissolved Hg was highest in the surface waters (Fig. 7.13) and this reflects the importance of atmospheric sources of Hg to this lake [15]. The dissolved Hg also increased across the chemocline due to the release of Hg during Fe cycling across the interface. Particulate Hg and CH_3Hg concentrations were highest at 60 m, coincident with the particulate Fe peak, and the peak in TSS. Dissolved and total CH_3Hg concentrations were low in surface waters and highest across the redox interface and in the anoxic bottom waters. This is consistent with data from many other lake studies, as discussed further in Section 7.2.3.

The influence of redox conditions can be examined using pH-pε diagrams, as discussed in Chapter 3, and shown previously for Fe. Many of the transition metals exist in one primary oxidation state in freshwater over natural conditions and therefore their distribution and major speciation is determined by the solubility of their major phases, and by their propensity to form dissolved complexes. These diagrams do not include the potential for complexation to NOM, nor do these diagrams take into account the potential for adsorption reactions that modify the relative dissolved concentrations. Even given these caveats, they do present one simple manner of illustrating the biogeochemistry of metals in solution. A few examples will further highlight the major biogeochemical processes discussed earlier.

The pH-pε/E_H diagrams for Cu, Zn, Cd and Pb in the O-CO_2-S system are shown in Figs. 7.14(a–d) [16]. Recall, from Chapter 3, that E_H and pε are related through the equation: $p\varepsilon = (F/2.3RT)E$, where F is the Faraday constant. In these diagrams an activity of 1 nM is assumed for each metal, resulting in a total concentration that is above the measured concentrations in most natural freshwaters. In oxic waters, Cu is typically insoluble at high pH, precipitating out as tenorite or cuprite, depending on the pε, with tenorite being more stable in fully oxygenated water. At low pε, Cu precipitates as a sulfide. The diagrams for Zn, Cd and Pb show strong similarities, with differences in the pH at which precipitation occurs and with Cu and Zn forming oxide phases while Cd and Pb form carbonate solids. The pH-pε diagram for Cu in the absence of sulfide (Fig. 7.14d) shows that, at low pε, the elemental metal is the most stable phase under these conditions. The pH-pε diagram for Cr shows the differences between the transition metals that form cations in freshwaters and that of a metal that exists primarily as an anion in oxic waters. The transition from Cr^{VI} to Cr^{III} occurs at a relatively high pε, especially in low pH waters, and therefore it is likely that both species can be found in the environment, with the reduced metal being dominant in low oxygen or anoxic waters [4, 16]. In contrast to Fe, Cr is much more soluble in its oxidized form and therefore is highly mobile in oxic waters but is relatively insoluble in reduced environments.

Further illustrations of the cycling of oxyanions in freshwaters are the pH-pε/E_h diagrams of As and Se (Fig. 7.15) [16]. The acid–base chemistry of the oxyanions is also shown as As^{III} and As^{V} both are triprotic acids. However, under the conditions of natural waters, As^{V} is typically either as $H_2AsO_4^-$ or $HAsO_4^{2-}$ while As^{III} is mostly in the undissociated form. Arsenic forms a number of sulfide solid phases at low pε and this accounts for its presence often as a trace element in sulfide ores that are extracted and refined for other metals, as discussed in Section 7.3.3.

Selenium is also present in a number of oxidation states, but with only acid-base chemistry for Se^{IV} at the pH of natural waters (Fig. 7.15) [1, 4, 16]. Se, being below S in the periodic table, does not form sulfide complexes but has similar chemistry to S in its most reduced state, as it forms H_2Se and also can dissociate into its acid–base pairs. Given the similarity in the chemistry of S and Se, there is the potential for cations forming complexes in solution with reduced Se. These may be stronger complexes than that with reduced S. It is also possible for Se to be found in NOM,

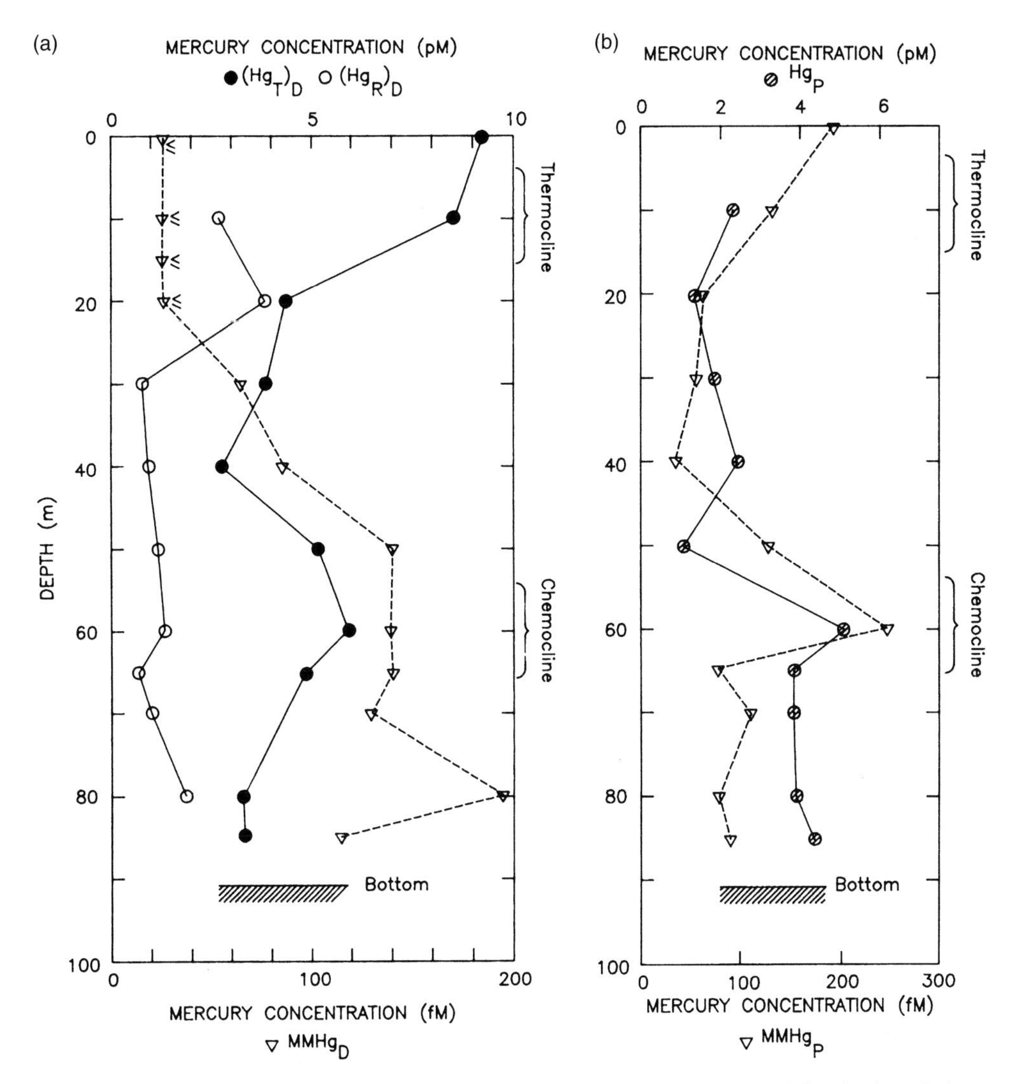

Fig. 7.13 Profiles of mercury speciation in Lake Pavin: (a) for total mercury, reactive mercury and dissolved methylmercury (MMHg in the figure); (b) particulate mercury and methylmercury. From Cossa et al. (1994) in *Mercury Pollution: Integration and Synthesis* [15] and reprinted with permission of Lewis Publishers.

derived from microorganisms, which may contain reduced Se from seleno-proteins, and these so-called "selenothiols" could also be important in metal complexation. However, given the typically low concentrations of Se in the environment, there are likely few situations where Se^{-II} outcompete S^{-II} for metal complexation. However, this has been little studied and there is no substantial evidence in the literature of its importance. One metal that binds substantially more strongly to Se^{-II} is Hg, and there is some suggestion of its potential importance in Hg cycling. The equilibrium constants for the various complexes are given in Table 7.3 [17] for the reactions of the form:

1) $Hg^{2+} + OH^- + HL^- = HOHgLH$, where $L = S^{2-}$ or Se^{2-}
2) Similarly, formation of HgL_2^{2-}
3) Formation of HgL_2H^-
4) $Hg^{2+} + HL^- = HgL_{(s)} + H^+$
5) Acid dissociation constants: $H^+ + HL^- = H_2L$; $H^+ + L^- = HL^-$

For most of the complexes formed, the formation constant for the Se complex is many orders of magnitude larger and so this complex would potentially outcompete sulfide in the complexation of Hg in natural waters. The concentration of reduced Se in the water column is however poorly known but could easily be important in oxic waters where sulfide levels are nM. Furthermore, it is likely that the same differences in binding strength occur for the organic thiols and selenothiols in the environment and therefore it would probably not be surprising to find that reduced organic Se is important in Hg complexation in natural waters.

The discussions in this section have focused on the cycling of metal(loid)s in lakes and the impact of redox on their inorganic speciation and mobility. The importance of adsorption to oxide phases was again demonstrated, and it is clear that the transport of metal(loid)s with oxide phases

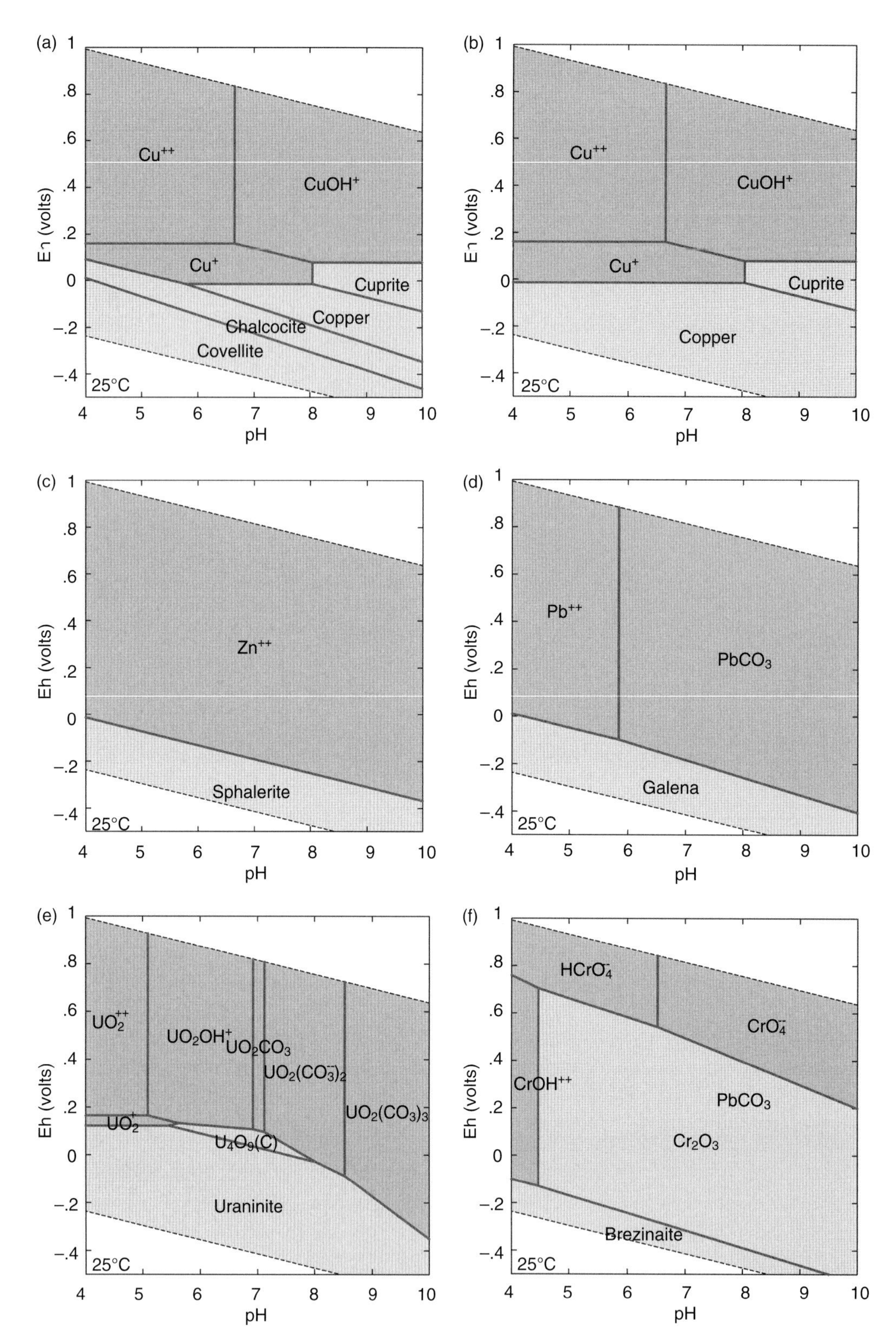

Fig. 7.14 The pH-E_h diagram for various metals (1 nM activity, except Hg at 10 pM) in the presence of atmospheric oxygen and carbon dioxide ($pCO_2 = 10^{-3.5}$ atm) (carbonate species at equilibrium) and total sulfur at 10^{-6} M at 25°C. Figure generated using the computer equilibrium program *The Geochemist's Workbench*, Release 9; (a) copper; (b) copper, but no sulfide; (c) zinc; (d) lead; (e) uranium; and (f) chromium.

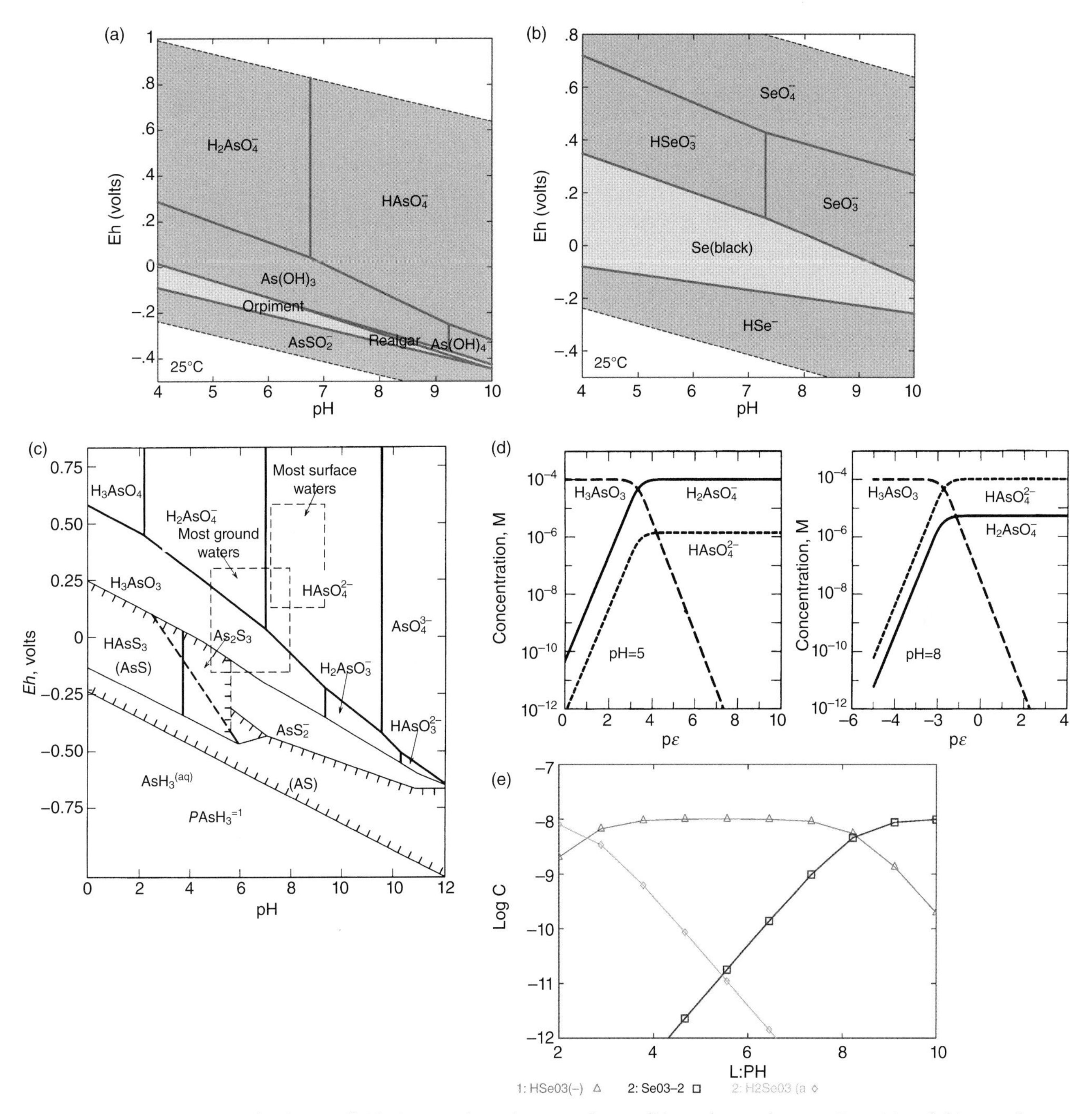

Fig. 7.15 The pH-pε diagrams for the metalloids: (a) arsenic; and (b) selenium under the same conditions to Fig. 7.14; (c) a comparable diagram for arsenic under conditions found in groundwater and at 10^{-5} M arsenic (1 mM sulfur); (d) the acid-base speciation of arsenic across the pε gradient using these conditions; and (e) the acid-base speciation of selenium (10^{-8} M) under conditions of natural waters. Parts (a) and (b) created using the computer equilibrium program *The Geochemist's Workbench*, Release 9; Parts (c) and (d) from Schnoor (1996) *Environmental Modeling* and reprinted with permission of John Wiley & Sons, Inc.; and Part (e) created using the MINEQL+ program.

is important due to metal release in low oxygen or anoxic environments. However, it is also necessary to discuss the importance of organic complexation in trace metal cycling in freshwater systems and this is discussed in more detail in the following section.

7.2.2 Modeling the speciation and association of trace elements in stratified systems

Much of the early focus on the speciation of metals in freshwater considered the inorganic speciation and this is important for some metals in some systems. However, in the majority, for cations, the speciation in freshwater is often dominated by complexation to NOM. The NOM is considered in many models as being dissolved material but in reality it is likely to be colloidal. Organic colloids in rivers and lakes can have a large component that is primarily from soil and terrestrial sources with the remainder being material released from microorganisms. The modeling approaches are based on apparent stability constants derived from studies with NOM, and therefore account to some degree for the presence of colloidal material. However, some models do not account for the potential for these entities to have surface charge which could have some impact on metal binding across pH and ionic strength, as discussed in Chapter 4.

A detailed examination of the modeling approach for examining metal complexation to NOM in natural waters was discussed in Chapters 3 and 4 and will not be reiterated here. Some additional examples will be considered here to further illustrate this concept. As noted previously, the modeling done for Lake Hall and Esthwaite Lake [9, 11] illustrated the importance of organic complexation for Cu, and potentially for some other metals. Modeling results from a study comparing two lakes (Lake Sempach and Lake Greifen) using different model formulations, is shown in Fig. 7.16 for Cu and Cd [3]. The modeled and "measured" free ion concentrations are compared over a range of total concentration, for a "typical" freshwater with 230 μM DOC, and pH 8. These results show that the models apparently underestimate the strength of binding of Cu and Cd to organic ligands as the modeled free ion concentration is higher than that measured using speciation methods based on electrochemical approaches and ligand exchange measurements [18], as described in Section 6.2.3. Sigg and Behra [3] summarize the results of a number of studies which suggest that the binding constants ($logK_{cond}$) for freshwater systems are greatest for Cu ($logK_{cond}$ = 13.5–16), followed by Ni ($logK_{cond}$ = 12–14), Co ($logK_{cond}$ 9.5–11.6), Cd ($logK_{cond}$ = 9.5–10.5) and Zn ($logK_{cond}$ = 7.8–9.5).

The binding of these metals to NOM is typically thought to occur through the association of the metals with the abundant carboxylic acid groups present in organic matter, as discussed in Chapter 4. However, more recent studies, and studies focusing on the Group B "soft" metals, such as Hg and Ag, suggest that binding to organic S (thiols) is potentially important and that these account for the stronger sites estimated in most studies examining binding constants.

Table 7.3 Comparison of the equilibrium constants (log values given) for comparable reactions of mercury with reduced sulfur and selenium. Taken from [16]. Reprinted with kind permission from Springer Science+Business Media.

Anion	HOHgLH	HgL_2^{2-}	HgL_2H^-	HgL(s)	H_2L
S^{2-}	30.3	23.2	31.5	38.9	6.88; 13.48
Se^{2-}	51.2	33.0	38.8	45.0	3.48; 11.60

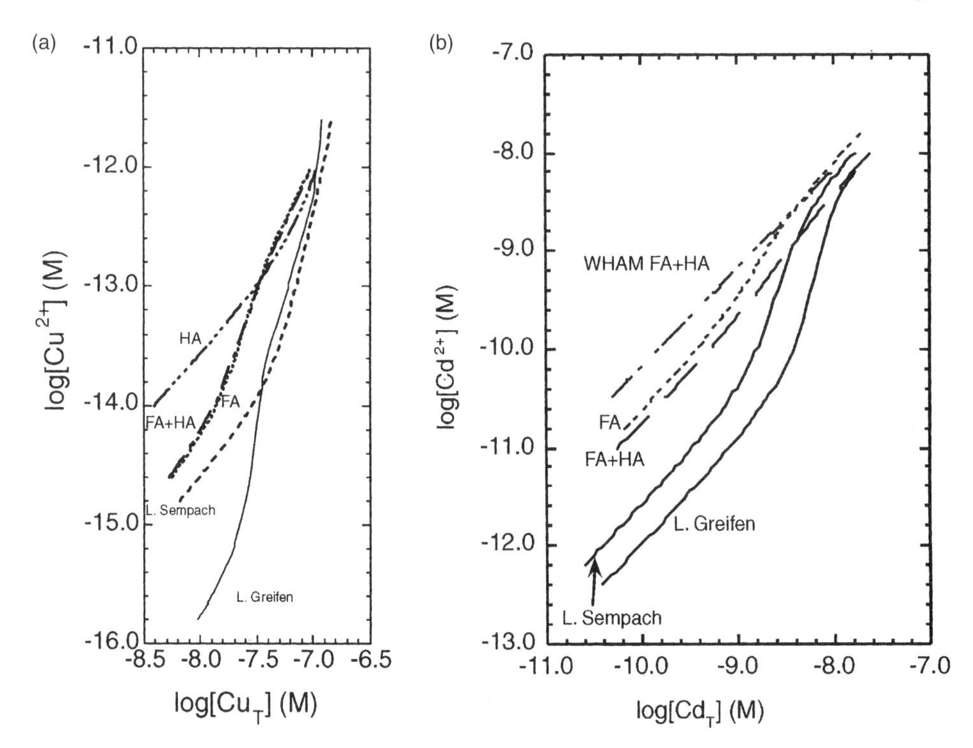

Fig. 7.16 Graphs showing the results of modeling of the speciation of (a) copper and (b) cadmium in different lakes using different model formulations, as discussed in the text. Taken from Sigg and Behra (2005) in *Metal Ions in Biological Systems* vol. 44 [3] and reprinted with permission of Taylor&Francis.

Many of the computer models that include NOM binding constants do not include or assume any association with reduced S groups. Smith et al. [19] reviewed the literature in this regard and estimated the range in concentrations for the various binding sites (concentration of ligands) for carboxylic (RCOOH), amino (RNH_2) and S-containing (both inorganic (sulfide clusters) and organic RSH groups) for natural waters (Table 7.4). Clearly, there are many more carboxylic acid sites but for some metals the binding constants for the RSH groups is substantially higher than that for the other ligands to compensate for the lower ligand concentrations. For example, for an oligotrophic lake the ratio of RCOOH/RSH is about 3×10^5 but the ratio of the binding constants for these ligands range from $\sim 10^4$ for Ni and Co to 10^{10}–10^{11} for Ag, Hg and Pb. Thus, for the "soft" metals, the binding to thiols is more important than the binding to carboxylic acid groups. Overall, Cu is intermediate and is likely bound to thiols in some instances. There is also the potential for reduction of Cu by reduced S groups and the formation of Cu^I [1].

Using the range in published values for the various binding constants for metals and the range of estimated ligand concentrations for the various NOM groups, Smith et al. [19] concluded that the higher values estimated in the literature for Cu complexation in natural waters must reflect the binding of Cu to RSH groups, while the lower values likely represent the binding of Cu predominantly to RCOOH groups in situations where the total ligand concentration was relatively high (10^{-4}–10^{-6} M). Similarly, the higher log K values estimated for Ag binding to NOM reflect the likely association of Ag with RSH groups. These results suggest that speciation models must take into account thiol complexation in estimating the speciation of many metal cations in natural waters.

Hamilton-Taylor et al. [20] applied the WHAM model (Version 6) to data collected in Esthwaite Lake in 1996 of dissolved and particulate metals and experimental data on the dissolution of metals from suspended particles. In contrast to the earlier study, sulfide levels were higher and peaked at mid-depth of the hypolimnion at ~13 μM. Lower sulfide in deeper waters is attributed to precipitation of FeS. Two peaks in labile particulate Fe and Mn, above and below the redox interface were attributed to, respectively, oxide and sulfide phases, and analysis of elemental composition confirmed this. The vertical location was consistent with that of other parameters. Previous studies suggested that the dissolution of Mn oxides, rather than Fe oxides, is the main source of elevated trace metal concentrations at depth in this lake, and this likely stems from the fact that particulate Mn was higher than particulate Fe in the surface waters, which is not the case generally.

Modeling the dissolved concentrations in Esthwaite Lake [20] indicated that Co and Ni were mostly bound in inorganic complexes or present as the free metal, with a small fraction bound in organic matter. Binding to Mn oxides was more significant than binding to Fe oxides, but both accounted for <20% of the Co, depending on the specific model parameters used. This is consistent with other results discussed earlier. In contrast, Pb was predicted to be bound to oxide phases and complexed to organic matter, with a small fraction being present as inorganic complexes. Finally, the model predicted that Cu was mostly associated with organic material and not with the oxide phases, or inorganic complexes to any significant degree. Again, these predictions are consistent with other studies and models.

In the anoxic waters it was concluded that removal of Cu and Pb was occurring because of the adsorption or incorporation of the metals into Fe-S phases rather than from precipitation of the metal sulfide complexes. Additionally, the results suggest that the Fe-S "particles" span a range of sizes, and include colloidal fractions, which would be associated with the filtered fraction. The recent evidence on metal-sulfide nanoparticle formation [21, 22] suggests that this prediction is a valid interpretation of the data.

Overall, these modeling studies are consistent with the interpretation of the field data and knowledge of the important chemistry and interactions between metals and dissolved complexing agents (inorganic anions and NOM) and the degree of adsorption of metal(loid)s to oxide and sulfide phases. The discussion in this section also builds and reflects the discussions in other sections of the book, notably the discussions on speciation and partitioning in the previous chapter. Clearly, the governing principles are similar for all aquatic systems and as long as the ionic strength of the solution and the importance of surface adsorption are correctly parameterized in the modeling effort, the model data appear to provide a consistent interpretation of the factors controlling the speciation, distribution and concentrations of trace species in environmental waters.

Table 7.4 Estimated typical concentrations of metal binding ligands in natural organic matter in freshwater environments. Taken from [2, 3, 18].

Environment	DOC (μM)	RCOOH (μM)	RNH_2 (μM)	RSH (μM)
Seawater	<40	2	0.04	10^{-4}
Groundwater	60	4	0.08	5×10^{-4}
Average River	415	25	0.5	1.5×10^{-2}
Oligotrophic Lake	160	1	0.01	6×10^{-3}
Eutrophic lake	830	5	1	10^{-2}
Dystrophic lake	2500	15	1	0.1
Marsh/Bog	1200–2500	<15	1	0.1

7.2.3 Focus topic: Mercury cycling in lakes

The biogeochemical cycling of Hg in lakes has been the focus of recent research, driven mostly by the elevated levels of CH_3Hg in freshwater fish and the associated human and wildlife health concerns. The factors influencing the formation and degradation of CH_3Hg are discussed in detail in Section 8.4.3 and this is not the focus of the discussion here. This section will discuss in detail the biogeochemical factors that influence the distribution and fate of inorganic Hg and CH_3Hg in lakes [23]. Most of this work has been completed on temperate ecosystems in Europe and North America, but given the underlying principles, the relationships developed for these ecosystems are likely to be generally applicable. Inputs, excluding point source human-derived sources, of Hg to lakes are either from the atmosphere, the watershed or from groundwater. As noted earlier in this chapter, watershed inputs are only important for ecosystems with a relatively large watershed as Hg is typically strongly retained within the watershed soils and sediments. Additionally, Hg in river and stream inputs is often strongly associated with particulate material or bound to NOM [24], and it is therefore likely that such Hg is less bioavailable for biological incorporation or chemical reaction [23]. Atmospheric Hg is much more labile given the composition of NOM in precipitation, and the low particulate content. As noted in the previous chapters, Hg^{II} can be reduced both photochemically and biotically in surface waters. Therefore, a substantial fraction of the Hg deposited from the atmosphere can be returned through reduction and subsequent volatilization of the Hg^0 produced, which often builds up to supersaturated levels in surface waters. Photochemical reduction is mediated by NOM and light levels [25, 26]. Oxidation of Hg^0 is also possible and this is thought to be primarily a photochemical process. Overall, within lakes, the dominant processes are these redox reactions, methylation and demethylation, uptake of Hg and CH_3Hg by microorganisms and abiotic/detrital particles, and these are illustrated in Fig. 7.17 [26].

In lakes, the dominant process whereby Hg is methylated is biotic, with the major product being CH_3Hg and sulfate reducing bacteria are the most important methylators, although there is some evidence for other phylogenetically similar organisms also being important [23, 27, 28]. The

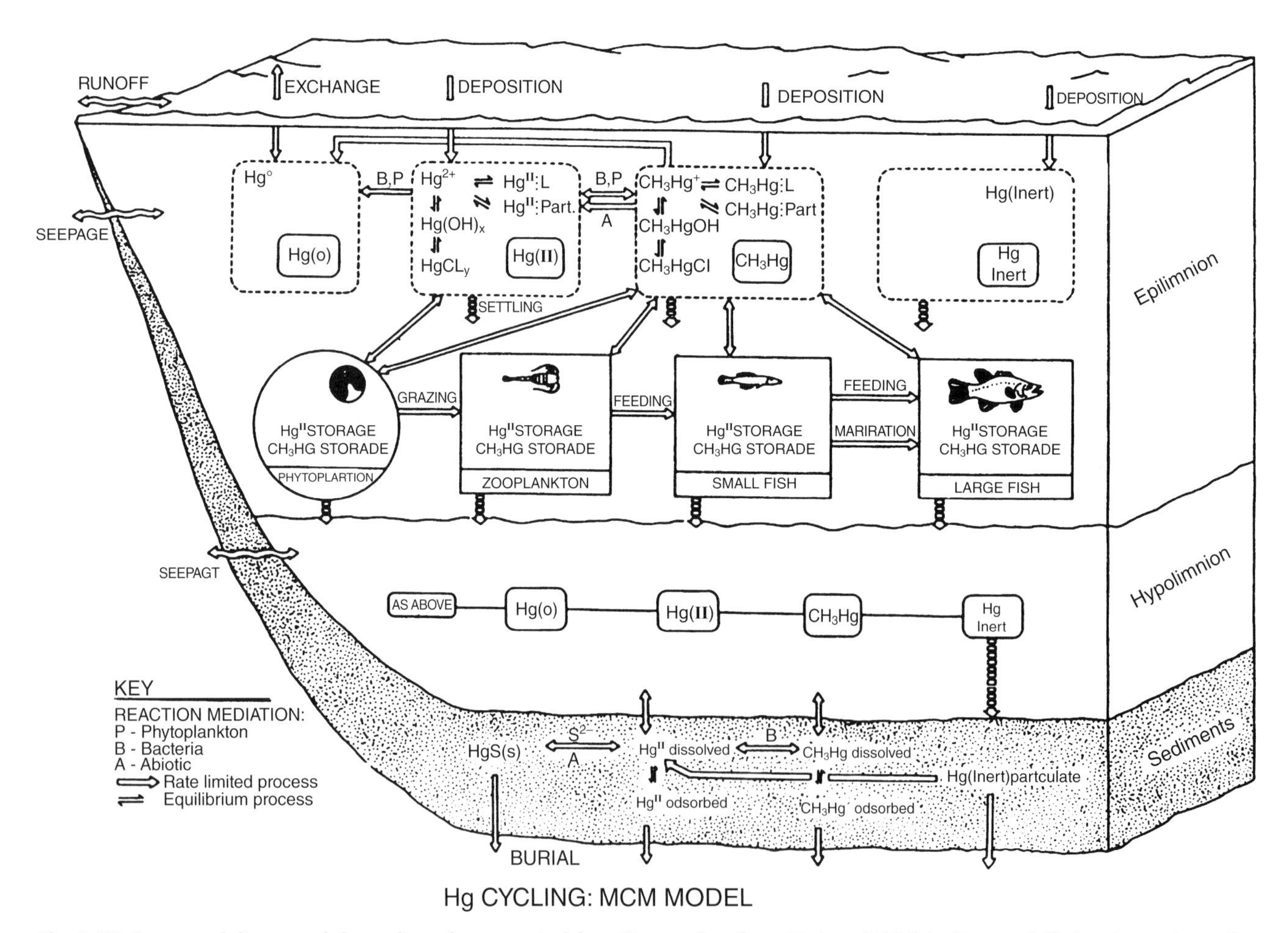

Fig. 7.17 Conceptual diagram of the cycling of mercury in lakes. Figure taken from Hudson (1994) in *Mercury Pollution: Integration and Synthesis* [26] and reprinted with permission of Taylor&Francis.

details of the methylation pathway are not clearly understood, as discussed in detail in Section 8.4.3. Additionally, biotic demethylation is not well understood except for systems with elevated Hg and CH_3Hg content, where the *mer* operons involved in Hg and CH_3Hg detoxification are induced in some bacteria [29]. Abiotic photodemethylation has been also demonstrated and appears to be enhanced by UV radiation in surface freshwaters, especially those with low DOC and/or TSS (i.e., low color and reflectance). Methylation often occurs primarily in the sediments although water column methylation has been shown for some anoxic systems. Therefore, the processes whereby CH_3Hg is transported from the sediments to the water column are of importance, as discussed for marine systems in Section 6.3.2. In stratified systems, recycling of sinking material is driven by the degradation of organic matter and/or the dissolution of oxides phases, in a similar manner to that described for other +II cations earlier in this section.

The redox cycling of Hg is important and it is often the case that the distribution in the water column is far from thermodynamic equilibrium. The pH-pε diagram for inorganic Hg is shown in Fig. 7.18(a) [1]. Elemental Hg (Hg^0) is stable thermodynamically at intermediate pε values and in the presence of sulfide, Hg precipitates as a sulfide (Fig. 7.18b), although as discussed in Section 6.2.3, there are a range of sulfide complexes that form in equilibrium with the solid phase that maintain the dissolved concentration at a much higher level (nM levels) than it would be in the absence of complexation (log $K_{sp} = -42$). The relative importance of the inorganic complexes in freshwater and seawater for Hg over the range of pε of natural waters is shown in Fig. 7.18(b) [1]. In freshwater, at pH 8, the inorganic complexes are present at a concentration that is at least 10 orders of magnitude higher than that of Hg^{2+}, and this difference is even more marked in seawater. The speciation across a range of pH concentrations for oxic seawater and freshwater are shown in Fig. 7.18(c) and this illustrates the importance of chloride speciation in higher chloride environments for both forms of Hg. These diagrams again however neglect the importance of complexation of Hg to NOM, which is the dominating complex in most freshwater systems.

Rather than attempt to summarize and document the many studies that have been completed, a few studies will be the focus of discussion and will be used to demonstrate the major mechanisms and relationships that exist. In oxic waters, in the absence of NOM, Hg forms strong associations

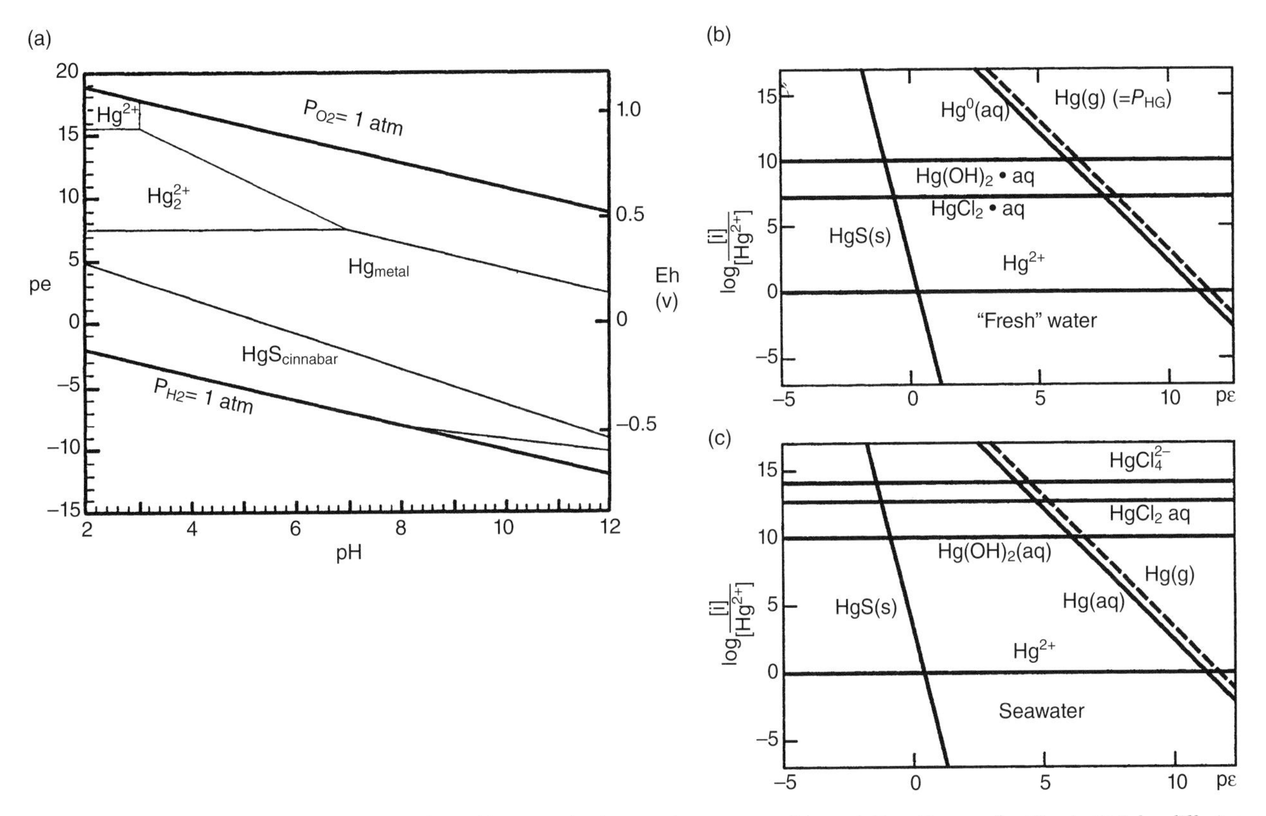

Fig. 7.18 Speciation of mercury across pH and pε. (a) pH-pe for inorganic mercury; (b) speciation diagram for Hg at pH 8 for differing pe in freshwater; and (c) similar speciation-pe diagram for seawater. Part (a) from Schnoor (1996) *Environmental Modeling* and used with the permission of John Wiley & Sons, Inc.; Parts (b) and (c) from Stumm and Morgan (1996) *Aquatic Chemistry* [1] and used with permission of John Wiley & Sons, Inc.

with Fe and Mn oxides and the dissolved speciation is dominated by complexation to chlorides in systems with $>10^{-3}$ M Cl, or by formation of hydroxide complexes [26]. In most ecosystems, the free metal ion concentration is typically a very small fraction of the total dissolved Hg^{II}, being $<2 \times 10^{-6}$ % of the total even in pure water and in contrast to other +2 cations, it is unlikely that uptake of Hg^{II} from solution into microorganisms is through the direct accumulation of the free metal ion. The same is not true for CH_3Hg^+, which forms weaker complexes with inorganic and organic ligands than Hg^{II} and therefore it is possible for the free ion to be an important component in solution. As has been speculated in a number of publications, it is thought that the neutrally charged inorganic complexes are those that cross the membranes of microorganisms readily and therefore it is clear that pH and salinity have an important impact on the bioavailability of Hg and CH_3Hg.

There have been a number of studies of large lakes as these are important in terms of their fisheries and their potential contamination. Most of these studies have shown that concentrations in the water column are not substantially elevated compared to the ocean, and are typically lower than that of smaller freshwater systems. This likely reflects the lower surface area to volume ratio and the relatively small importance of river inputs to these systems. Studies in the North American Great Lakes suggest that elevated levels of CH_3Hg in fish reflect the more oligotrophic nature of these systems and the longer food chains rather than an inherent higher level of Hg input or methylation. For example, a comparison found higher levels of total Hg in Lake Michigan (~ 2 pM) compared to Lake Superior (~1 pM) [30]. Levels of CH_3Hg were <0.05 pM. These studies confirmed that gas evasion and sedimentation were the major loss processes in such systems, as they are for the ocean.

Studies have been completed in lakes that are dimictic as well as in meromictic lakes, such as Lake Pavin. As shown in Fig. 7.14 [15], the data from Lake Pavin show that there are elevated levels of Hg and CH_3Hg in the upper levels of the chemocline, in conjunction with the maxima in particulate Fe, but not Mn, suggesting their active scavenging by Fe precipitation processes. Dissolved Hg and CH_3Hg were also relatively elevated in this region which may just reflect equilibrium partitioning, or may also be due to an enhanced content of colloidal material in the filtered fraction at these depths. Data from studies in other lakes provide similar records of the impact of seasonal stratification on Hg cycling and CH_3Hg production. One set of studies were conducted in Wisconsin and were some of the first in the USA to use clean sampling techniques [31]. These studies were focused on an experimentally divided seepage lake, Little Rock Lake, (one side was acidified), and surrounding lakes, and examined the changes in the lake with acidification.

The early results from these studies in Little Rock Lake [32–35] and studies in other ecosystems have shown: (1) the importance of air-water exchange of Hg^0 in the cycling of Hg in lakes, especially lakes without substantial watershed contributions; (2) the importance of seasonal stratification in the buildup of CH_3Hg in bottom waters and in its supply to the food chain; (3) the lack of $(CH_3)_2Hg$, in contrast to the oceans; and (4) the importance of *in situ* production of methylated Hg compared to external inputs. The Wisconsin studies also provided the first insights into why fish in low acid lakes had higher CH_3Hg concentrations. Studies in Sweden and other Scandinavian countries conducted during the same period came to similar conclusions about Hg cycling in temperate lakes [36, 37].

The mass balance for Hg and CH_3Hg in the acidified portion of Little Rock Lake is instructive of the processes occurring (Fig. 7.19) [26]. This mass balance clearly demonstrates the importance of in-lake methylation – the amount of CH_3Hg in the food chain is about 20 times that of the atmospheric input. Additionally, the estimated annual bioaccumulative flux ($0.3\,mmol\,yr^{-1}$) is six times the atmospheric input. Further studies in lakes have suggested that most of the CH_3Hg production is within the sediments, with release to the water column during stratification and changes in sediment redox. There is also evidence that water column methylation can occur, especially in low oxygen or anoxic waters [26, 38]. Details of the processes of methylation and demethylation in environmental waters are discussed in Section 8.4.3.

The mass balance [26] also shows the importance of particulate transport between the surface water and sediments and this is the main vector for the transport of inor-

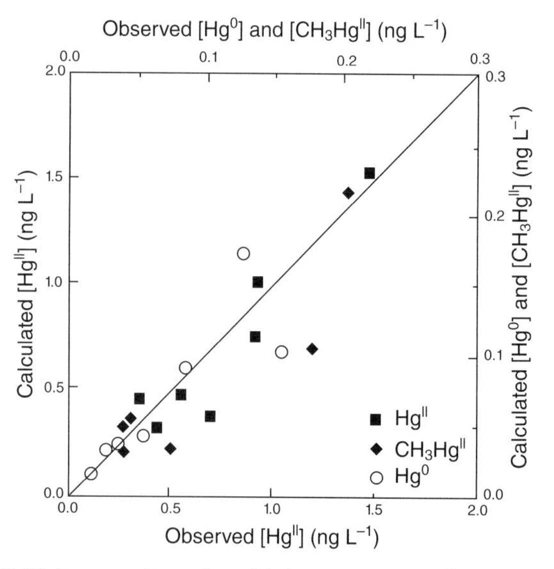

Fig. 7.19 A comparison of modeled versus measured concentrations for various forms of mercury in lakes within northern Wisconsin, USA. Figure taken from Hudson et al. (2004) in *Mercury Pollution: Integration and Synthesis* [26] is reprinted with permission of Lewis Publishers.

ganic Hg input from the atmosphere to the sediment where methylation is occurring. These earlier studies did not examine demethylation in detail and recent work has demonstrated that in the sediments of many ecosystems there is a constant and rapid cycling between inorganic Hg and CH_3Hg, and this is mostly microbially-mediated [23]. This cycling, and the flux of CH_3Hg from sediments to the water column, is similar to the processes occurring in coastal sediments, as discussed in detail in Section 6.2.

It has often been difficult in studies of Hg and CH_3Hg cycling in lakes to separate the various potentially competing mechanisms and the importance of each is clearly a function of a number of parameters – pH, NOM content, ionic strength, degree of stratification and anoxia, TSS, and biotic activity. One innovative study approach [39, 40] was the use of stable Hg isotopes that were deliberately added to a lake (Lake 658) in the Experimental Lakes Area of Canada. This multi-investigator effort was the METAALICUS project [39] aimed to simulate the impact of increased anthropogenic deposition on Hg cycling in the lake, and also to assess its recovery after years of enhanced additions. The input was increased to correspond to the average Hg deposition rate of the eastern USA, or about five times the pre-addition level. Additionally, by adding different isotopes to the upland watershed, an associated wetland and the lake surface, it was possible to track the importance of the various pathways in contributing to the CH_3Hg accumulating in the aquatic food chain [40, 41]. Prior to the addition of the isotopes, the inputs of Hg to the lake were dominated by watershed inputs with atmospheric deposition being a relatively minor component.

These studies showed that the system responded rapidly to the Hg isotope added directly to the lake surface with the documentation of enriched isotopic CH_3Hg in surface waters shortly after the addition of the Hg^{II} isotope to the lake. Its presence in sediment trap material (sinking particulate) was documented within weeks [40]. Furthermore, evidence of its presence in the surface sediment and the zooplankton and benthos was found within a month (Fig. 7.20) [40]. These results suggest that Hg input from the atmosphere is rapidly cycled through the system and is actively methylated and transported within the ecosystem and bioaccumulated into the food chain.

The major sinks for Hg added to a lake are outflow, evasion to the atmosphere, and accumulation of Hg in sediments. Studies of Hg loss during the Hg isotope spike additions to lakes or other freshwaters suggest that a significant portion of the isotope (20% or greater) added is lost to the atmosphere through net Hg^{II} reduction and subsequent evasion [40, 42]. Various studies confirm the magnitude of the evasional loss [42–45]. Thus, re-emission is an important process in Hg cycling in lakes, as it is for the ocean [46]. Sedimentation is the other major loss and is important especially for lakes with substantial terrestrial input as the particulate material

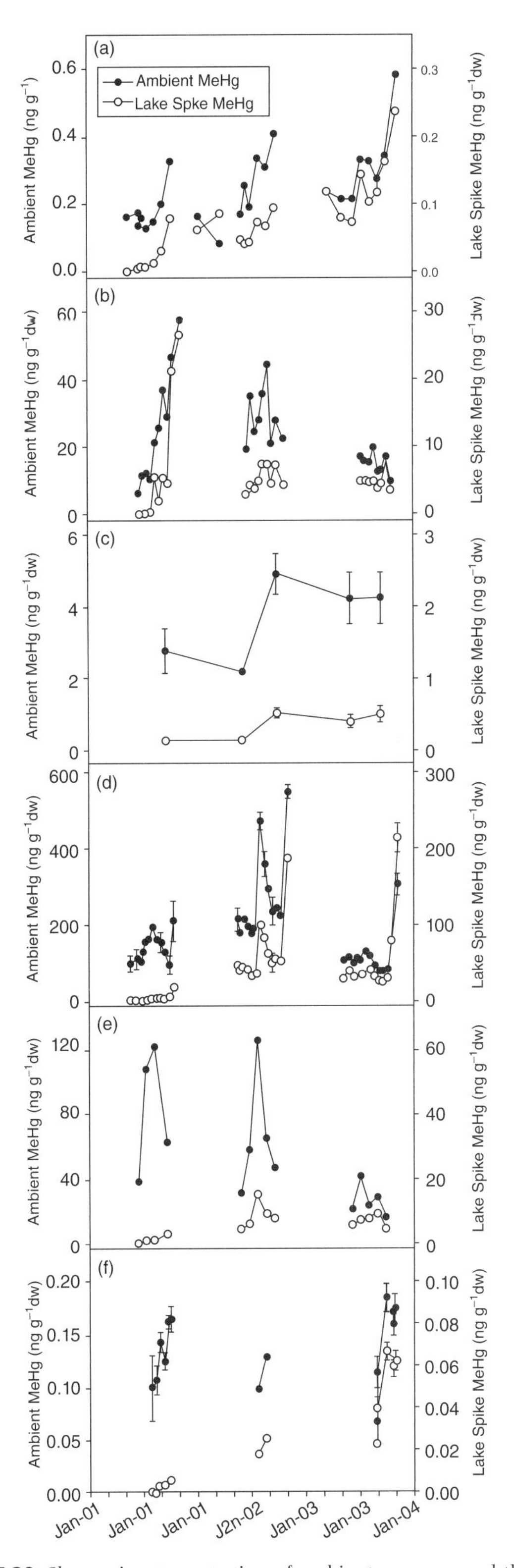

Fig. 7.20 Change in concentration of ambient mercury and the added mercury spike in various compartments (water, sediment and biota) after the spike addition began in Lake 658 in the Experimental lakes Area, Canada. Figure reprinted from Harris et al. (2007) *PNAS* **104**: 16586–91 [40], and used with permission. Copyright (2007) National Academy of Sciences, USA.

supplied through runoff increases the relative sedimentation rate [36]. Studies have shown that the relative removal of Hg via sedimentation is a function of the residence time of water in the lake for drainage lakes, and removal increases with residence time [45]. Additionally this study found, as have a number of studies, that the relative amount of dissolved Hg transported from the watershed to a lake is a function of the NOM content, which is often related to the amount of wetlands in the watershed. This is mostly related to the strong association of Hg with colloidal NOM and its transport, as discussed later for Hg transport in rivers (Section 7.3.1).

The internal cycling and processes governing the exchange of Hg and CH_3Hg between surface waters and sites of methylation/demethylation are highlighted in a detailed study of the distribution of Hg and CH_3Hg in the METAALICUS lake at the sediment-water interface [47] (Fig. 7.21a). While previous studies had demonstrated the importance of particulate transport to the sediment, the details of the sediment-water cycling had not been examined in detail. Profiles were collected using a specifically designed sampler that allowed small interval sampling within 1m above the sediment. These data illuminate the processes responsible for cycling of the redox sensitive metals. There was a clear increase in filtered Fe and Mn at the sediment-water interface, indicative of its release from surface sediment or from recently settled particulate matter. These increases coincided with the depletion of oxygen and the presence of sulfide. Dissolved organic matter content also increased, suggesting its release in concert with the Fe and Mn, and indicating its transport from the surface via particle setting. These changes were even more apparent in the graph focused on the bottom meter (Fig. 7.21b) [47], and highlight the cycling of Fe, with the particulate (oxidized) Fe peak being above that of the dissolved component. There was also evidence for colloidal Fe species.

The concentrations of dissolved Hg and CH_3Hg, for both the ambient and the spiked Hg, show a peak just above the sediment water interface, which suggests that they are also released in concert with particulate dissolution and are likely transported along with the organic matter attached to the settling inorganic phases. These results do not suggest that there is significant release of CH_3Hg from the sediment and indicate that remineralization may be an important source of CH_3Hg in stratified waters of lakes, and that the CH_3Hg buildup is not entirely due to *in situ* production in these waters and sediments. Also, the reduction of Fe and its potential for mediating dissolved sulfide levels may have a role in mediating Hg methylation, as discussed in Chapters 6 and 8. Finally, the fraction of the isotope that is methylated is higher than that of the ambient Hg, providing a further indication that the recently added Hg to the lake is more bioavailable on average than the ambient Hg within the system [47].

A focus of the METAALICUS project was the response time and the rate at which the newly added Hg is transported through the system. The results of the study have shown that the isotope added to the surface of the lake was methylated much more efficiently than the total amount of Hg already in the system, and especially compared with the Hg being added to the lake from the watershed [40]. This demonstrates that the bioavailability of Hg being added to a lake or aquatic system changes over time. This makes sense, especially as inorganic Hg added to the lake from the atmosphere through wet deposition is only weakly complexed to inorganic and organic ligands, and the dry deposited gaseous inorganic Hg is also labile (see Section 5.5.2),. Over time the Hg will become incorporated into NOM and particulate material, bound to inorganic sulfides and form other strong associations that will reduce its availability to the methylating organisms, and thus decrease its bioavailability. While such a processing was thought to occur prior to the METAALICUS study, it was clearly demonstrated by the results and preferential bioaccumulation of isotope CH_3Hg in the food chain (Fig. 7.20) [40]. As noted above, there is rapid recycling in the system between the inorganic Hg and CH_3Hg but clearly through each cycling some of the Hg is being sequestered in forms or locations where its bioavailability is less. This result has an important management outcome as it suggests that the ecosystems should respond relatively rapidly to any decrease in Hg input from the atmosphere. This is currently being cataloged during the "recovery" stage of this lake after cessation of Hg additions.

The results presented by Harris et al. [40] showed that little of the isotope added to the watershed was transported to the lake over the initial years of the study. This confirms that while there is a yield of Hg and CH_3Hg from the watershed to a lake, this is not recently deposited Hg but that which has cycled through the watershed for an extended period, perhaps for many years to decades. Given this retention in the watershed, it is also apparent that any lake with a significant input of Hg from the watershed will respond more slowly to changes in atmospheric deposition. Recent attempts to examine such differences using models suggest that there would be a rapid initial decrease in concentration, that would be greatest for a seepage lake or river, within 5–10 years, and a slower decrease to steady state over decades to a century [48].

One important way that human activity has exacerbated CH_3Hg levels in fish is through the impact of reservoir flooding/new impoundment construction. For example, a study in Finland of 18 newly formed reservoirs showed that fish concentrations increased compared to natural lakes by 30% on average, and that these elevated concentrations remained for an extended period, with fish concentrations not returned to pre-inundation levels after 20–30 y [49]. Fish concentrations were related to reservoir age, as well as

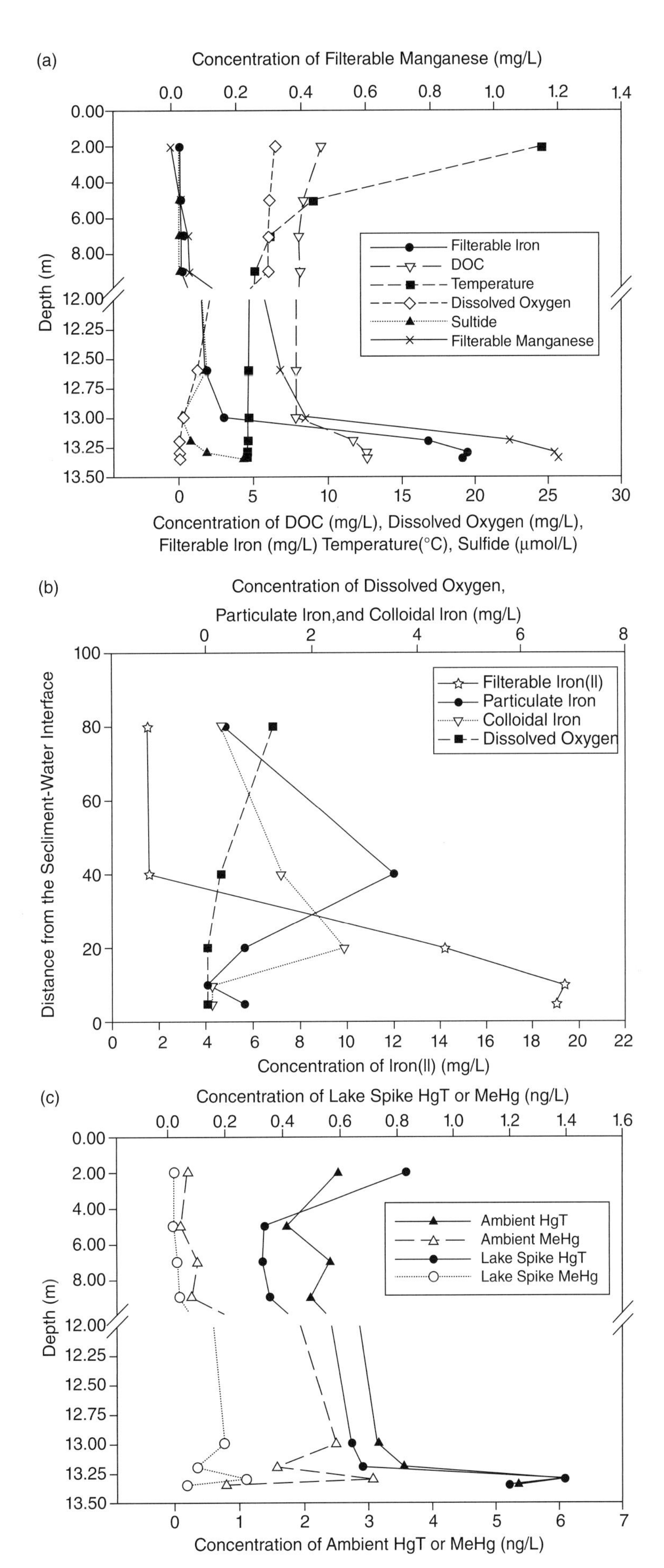

Fig. 7.21 Cycling of mercury forms and (a) the distribution of ancillary parameters (oxygen, iron and manganese); and (b) a closeup of their changes near the bottom; and (c) the distribution of mercury and methylmercury with depth in Lake 658 in the Experimental Lakes Area of Canada within a few meters of the bottom showing the impact of redox cycling and sediment water interactions of the distribution of the various species. Taken from Chadwick et al. (2006) *Science of the Total Environment* **368**: 177–88 [47], reprinted with permission of Elsevier.

the pH and NOM content of the reservoir waters. Other studies have also found similar correlations [50], and similar timeframes of impact [51]. Studies in tropical environments [52] have confirmed the results found in more temperate regions. Impacts were also related to the extent of flooding that occurred. Concentration increases in piscivorous pelagic fish were greater than those of benthivores [51]. While flooding of soils with higher carbon levels lead to more CH_3Hg formation, the presence of higher levels of DOM in the water tended to counter bioaccumulation [50, 53]. While such effects are dramatically demonstrated with new reservoir formation, there is also evidence that seasonal draw-down and replenishing of reservoirs, and other water bodies, can also enhance the production of CH_3Hg within the system.

However, most of the studies discussed so far have all been focused on temperate ecosystems and the METAALICUS study is focused on a location in a remote forested region with a watershed that is dominated by granite outcrops and has a thin soil layer. It is difficult to assess the degree to which these studies are transferable to tropical ecosystems. One study in a tropical lake demonstrated an important location where methylation can occur, within the roots of macrophytes [54]. Methylation was also influenced by temperature. It was demonstrated that the microbial consortium associated with these roots was responsible for the methylation, which was inhibited by Mo, and stimulated by sulfate, indicating methylation by sulfate reducing bacteria.

Another study focused on a reservoir in French Guiana, which was created by flooding the surrounding forest and which still contains anoxic bottom waters. The concentration of Hg and CH_3Hg were 2–3 times higher in these bottom waters compared to the surface [44]. Vertical profiles showed maxima in dissolved CH_3Hg just above the redox interface and at the sediment-water interface. These results again suggest that the sinking and remineralization of particulate matter has an important role in Hg and CH_3Hg transport. A mass balance for the reservoir indicated that there was more production of CH_3Hg within the lake (60%) than was contributed via watershed inputs (34%) and atmospheric deposition (6%). Additionally, it was evident that the overall methylation rate was substantially higher than that of temperate systems. This could be a result of either the temperature differences or due to the impact of reservoir creation on Hg methylation. Finally, given the high %CH_3Hg in the bottom waters and the release of this water downstream, it was concluded that this reservoir system was both an important source of CH_3Hg to the fish within the reservoir and downstream [44]. These studies also suggested that the reservoir may be a more important contributor to elevated CH_3Hg levels in watershed fish than that related to elemental Hg contamination as a result of gold mining using Hg amalgamation [45]. Indeed, while total Hg was 5–10 times higher in upstream sites where mining activity was occurring, fish CH_3Hg was about eight times higher in sites downstream of the reservoir.

7.3 Trace elements in rivers and groundwater

Much of the study of metal(liod)s in freshwater environments have focused on lakes as these are easier to study and are recreationally and often economically more important than rivers. In many instances studies of rivers have been driven by concerns derived from point source contamination and while these are important, the associated chemistry is likely not representative of remote, uncontaminated systems. Even less is known about metal cycling in groundwater given the difficulties in cleanly collecting samples of groundwater without metal contamination. As shown in Fig. 7.22(b), there is a large variation in the concentrations of the constituents of groundwater, and similar data for surface waters is given in Fig. 7.22(a). The following sections will discuss the general characteristics of the fate and transport of metal(loid)s in rivers and groundwater and the discussions on rivers compliments the discussions in Chapter 6 which dealt with cycling of metal(loid)s in the coastal zone. After discussing biogeochemistry in general terms, a few specific examples will be discussed in detail to illustrate the complexity of freshwater ecosystems. These sections will focus on As fate and transport in groundwater and potential groundwater fluxes to the ocean.

7.3.1 Trace elements in rivers

Gaillardet et al. [6] compiled a database of dissolved (filtered) concentrations of trace elements in rivers in the major continents (Africa, Europe, North and South America and Asia) and these are shown in Table 7.5 for the more important and abundant elements and in Table 7.6 for other more rare elements, from the second and third row of the transition metals, and including the actinides [6]. Some of the more minor elements from Groups I and II of the periodic Table are also listed. All concentrations for the trace elements are given in nM. The estimated fluxes to the coastal zone for each metal (kmol yr^{-1}) are also included in Tables 7.5 and 7.6, and these are the fluxes for dissolved constituents. In rivers, Al and Fe are present in the highest amounts as their average concentration is above 1 μM, although these values likely represent to some degree the presence of colloidal material in the filtrate. All the data reported in the tables is for filtered waters but reflects either filtration through 0.2 or 0.4 μm filters and this can make an important difference for some metals, especially if these elements form insoluble oxide phases or attach to organic matter, both of which form the majority of the colloidal phases in freshwater. All the other elements listed have nM or sub-nM dissolved concentrations. The total concentration of a metal in

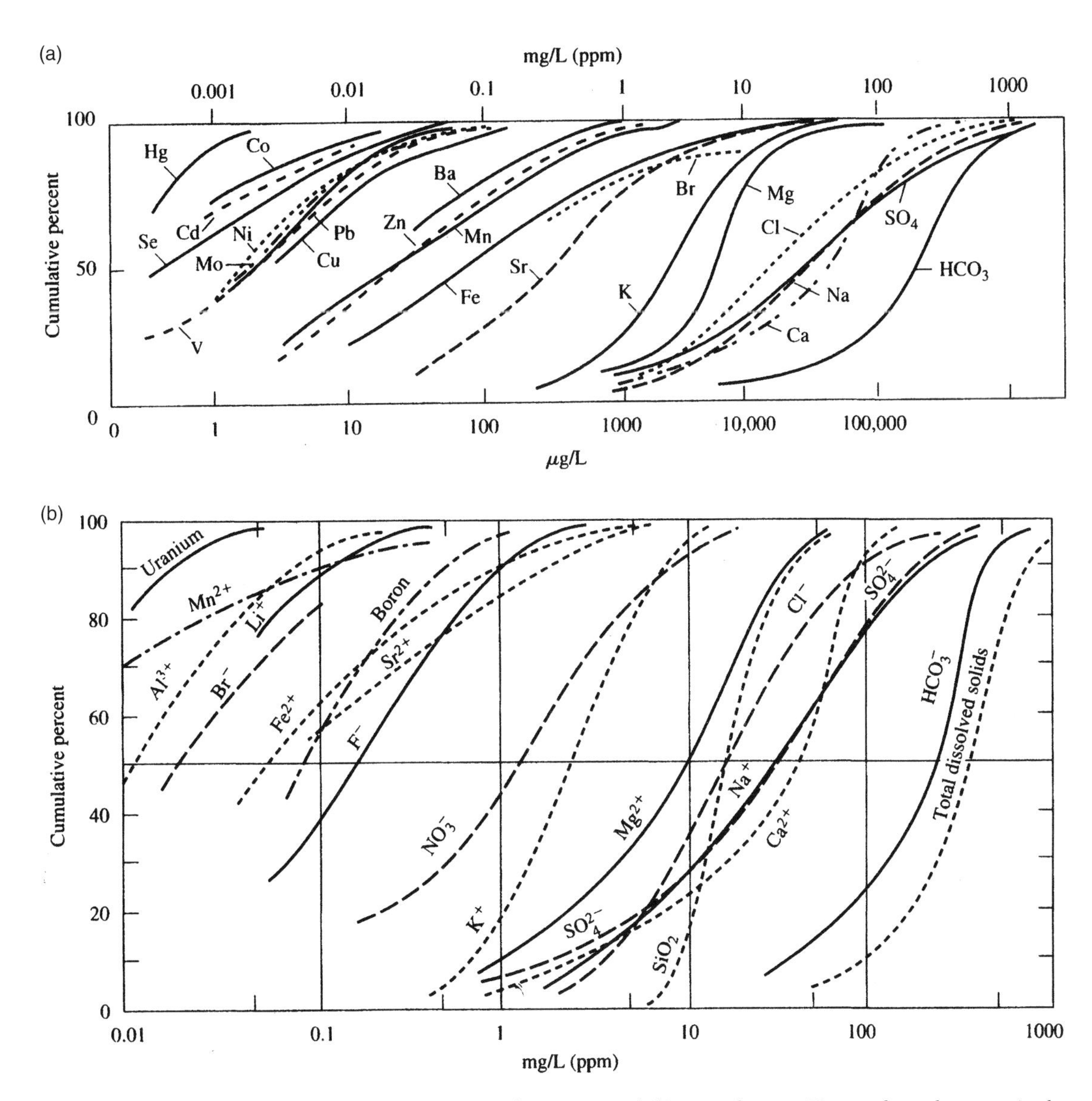

Fig. 7.22 Relative concentrations of various elements in (a) surface water and (b) ground water. Figures show the range in the concentrations, on a mass basis, for each element. Reprinted from Lagmuir (1997) *Aqueous Environmental Chemistry* [2] and used with permission of Prentice Hall/Pearson.

a river is mostly controlled by the suspended load given the typically higher TSS concentrations in rivers compared to lakes and the ocean. Values of TSS range over several orders of magnitude from low mg l^{-1} for small rivers and streams under base flow to over 1 $g^{-} l^{-1}$ for some large Asian rivers and large rivers in other continents. Dissolved organic carbon concentrations in rivers show less variability and range from 120–3000 μM. Most unimpacted rivers are slightly acid or slightly basic, depending on the terrain being drained (pH 5.5–8.1).

The relative mobility of elements can be estimated by comparing their average dissolved concentration in rivers (C_W) to that of average crust (C_c). This is shown in Fig. 7.23 [6], which shows the relative normalized concentration to that of Na ($C_W/C_C = 1$), which is a highly soluble and therefore mobile metal. The mobile elements (ratio >0.1) compromise the major cations and anions, and many of the elements that form oxyanions in solution. As noted in Section 7.1, this includes most of the Group I and II metals, and some of the more mobile trace metal cations (Cd, Cu, Mn, Co, and Ni). The relatively immobile elements include transition metals (e.g., Zn, Cr, Y, V) and the rare earth elements, some heavy metals that bind strongly to particulate (e.g., Hg, Pb), and some metalloids (Ge, Ga) and Th. The most immobile metals are those that are the most insoluble (e.g., Nb, Zr, Ti, Ta) and include Fe and Al, although the

Table 7.5 Typical range in dissolved concentrations (minimum, maximum and average value) of the more abundant metals in rivers and their estimate flux as dissolved constituents to the coastal zone. Taken primarily from Gaillardet et al. [6]. Reprinted with permission of Elsevier.

Metal	Min (nM)	Max (nM)	Ave (nM)	Flux (kmol/yr)	Metal	Min (nM)	Max (nM)	Ave (nM)	Flux (kmol/yr)
Ag	–	–	–	–	*Mn*	7.29	2077	619	23133
Al	370	18519	1185	44,444	*Mo*	1.15	23.98	4.38	167
As	1.34	33.38	8.28	307	*Ni*	2.04	177.17	13.65	511
Au	–	–	–	–	*Pb*	0.02	18.34	0.38	14.48
Bi	–	–	–	–	*Sb*	0.16	2.22	0.57	21.35
Cd	0.04	3.56	0.71	26.69	*Sc*	1.22	39.37	26.69	1001
Co	0.08	4.41	2.51	93.38	*Se*	0.38	2.91	–	–
Cr	3.85	221	13.46	500	*Sn*	–	–	–	–
Cu	3.15	55.1	23.31	866	*Th*	0.01	0.08	0.18	64.66
Fe	179	13262	1183	44 265	*Ti*	0.49	23.97	2.39	88.06
Ga	0.01	1.72	0.43	15.78	*U*	0.04	20.58	1.56	58.80
Ge	0.06	1.10	0.09	3.44	*V*	1.96	56.97	13.95	530
Hg	0.001	0.03	0.01	0.40	*Zn*	1.53	96.33	9.17	352

Table 7.6 Typical average concentrations of the less abundant metals in rivers and their estimate flux as dissolved constituents to the coastal zone. Taken primarily from Gaillardet et al. [6]. Reprinted with permission of Elsevier.

Main Group	Average (nM)	Flux (kmol/yr)	Trans. III	Average (nM)	Flux (kmol/yr)	Lanth.	Average (nM)	Flux (kmol/yr)
Li	265	9940	*Y*	0.45	16.9	*Ce*	1.86	70
B	945	35,185	*Zr*	0.43	16.4	*Pr*	0.28	10.6
Be	1.11	36.63	*Nb*	0.02	0.68	*Nd*	1.05	39.5
Rb	19.1	712	*Pd*	0.26	9.87	*Pm*	–	–
Cs	0.08	3.0	*La*	0.86	32.4	*Sm*	0.24	8.64
Sr	685	25,571	*Hf*	0.03	1.23	*Eu*	0.07	2.43
Ba	16.75	6265	*Ta*	0.06	0.22	*Gd*	0.25	9.54
Trans. II			*W*	0.54	20.1	*Tb*	0.03	1.26
Y	0.45	16.9	*Re*	0.00215	0.08	*Dy*	0.18	6.77
Zr	0.43	16.4	*Os*	0.00005	0.002	*Ho*	0.04	1.64
Nb	0.02	0.68	*Ir*	–	–	*Er*	0.12	4.48
Ru	–	–	*Pt*	–	–	*Tm*	0.02	0.71
Rh	–	–	*Au*	–	–	*Yb*	0.10	3.47
Pd	0.26	9.9				*Lu*	0.01	0.51

relative concentration of these depend strongly on factors such as pH and redox status [6]. The relative mobility is obviously a first order approximation as there is high variability in concentration across rivers due to factors such as NOM content, pH, colloidal matter and other factors.

The mean flux of each metal is shown in the tables and this estimate is based on the concentrations and the total river flow ($3.74 \times 10^{13} m^3 yr^{-1}$), which is essentially equivalent to that of the largest rivers. In making this estimate there has been an effort to remove the impact of pollution by not considering obviously polluted rivers but the impact of human activity on these estimates is still apparent because of many other factors, such as changes in land use within a river's watershed. Also, given the lack of data for some elements, these estimates are likely biased based on the heterogeneous distribution of sampling. For many metals, especially those that have a high log K_D value (>5; Table 7.2), the dissolved flux is a small fraction of the total flux as the solid loading is the major component. This is illustrated in Fig. 7.3 which relates the fraction of the metal in the dissolved fraction to the overall K_D. As much of the TSS is removed due to sedimentation in estuaries, the dissolved flux is likely of the same order as the input from estuaries to the coastal zone. Of course, the degree to which this is true depends on the relative removal of the element in the estuary as some metals, such as Fe, Al and the heavy metals are

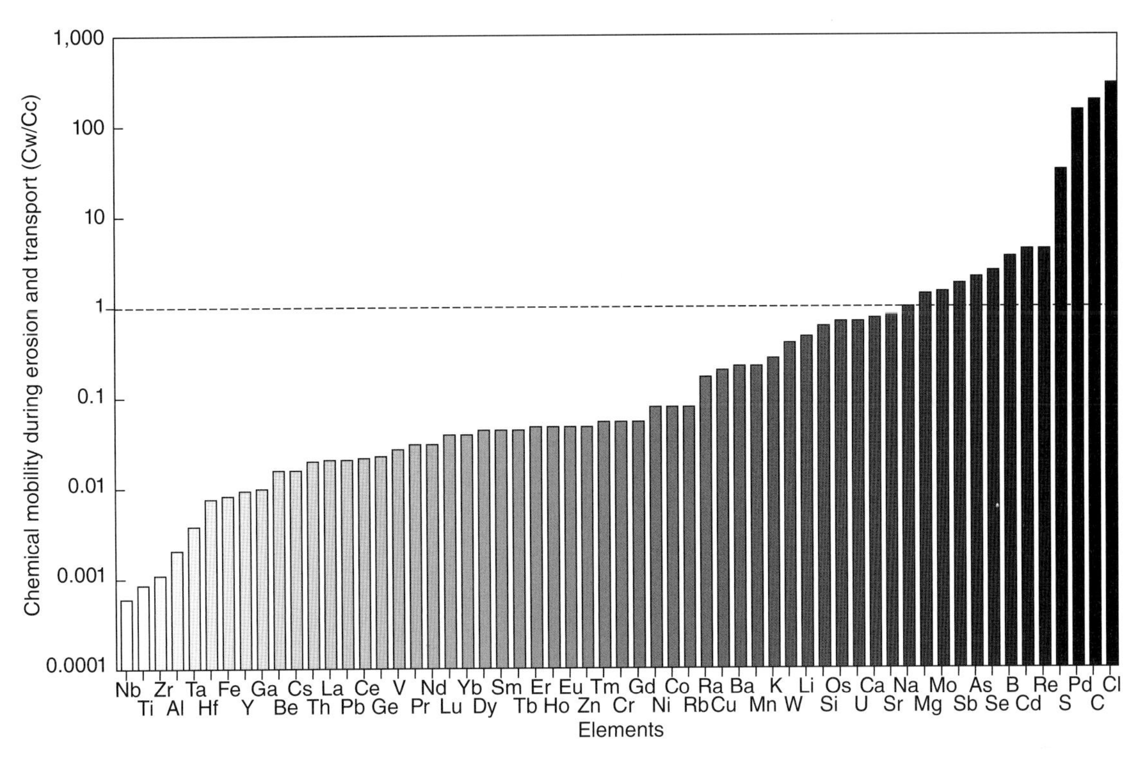

Fig. 7.23 Relative concentration for various elements to sodium for river waters. Taken from Gaillardet et al. (2004) in *Treatise of Geochemistry* vol. 5 [6], reprinted with permission of Elsevier.

strongly retained (>90%; Section 2.2.4), while some metals are transferred to a larger degree (50–90% retained) (e.g., Zn, Cd) [2].

Correlations between elements have been documented in rivers and this reflects to a large degree the similarities in the geochemistry of the various elements [6]. Globally, alkali and alkali earth metals track each other and this reflects their relative solubility. Trace elements that exist as oxyanions often track each other and the major anions. Overall, however, it is difficult to generalize across all rivers as the degree of correlation is strongly related to the overall ionic strength and composition of the waters, and the mineral composition of the drainage basin. Additionally, concentrations of trace elements likely vary on a seasonal basis and there is some indication that variations may be on a shorter timescale, such as diurnal, for some elements. Such diurnal changes are driven by photochemistry or by changes in the redox status of sediments and the impact of this on trace element fluxes from sediments. Additionally, changes in the relative distribution of metals between truly dissolved, colloidal and particulate phases may impact the distribution, estimated based on typical filtration.

Speciation has an important impact on the fate and transport of trace elements in rivers and this can be illustrated by the so-called "Born plot" which relates the metal first hydrolysis stability constant ($\log\alpha_{MOH}$ for $M^{n+} + H_2O = MOH^{(n-1)+} + H^+$) to the charge (Z) and to the cation radius (r) (Fig. 7.24). Highly hydrolyzed metal(loid)s, including the oxyanions, have a value of $\log(Z/r^2)$ of >1.3. Intermediate species can exist as cations in solution and form complexes with a variety of inorganic ligands.

To examine these relationships more closely, it is useful to focus on some specific examples. As an example of a regional study, consider studies done to examine the concentration, speciation and flux of metals in mid-west USA rivers, including the upper portions of the Mississippi [55]. This study examined both headwater streams and receiving waters in forested (non-calcareous) and agriculturally-dominated (calcareous) watersheds, which ranged in flow from 0.5 to $1980\,m^3\,s^{-1}$ (16 to 70 000 cfs) (Mississippi sites). Suspended load ranged from 2 to $50\,mg\,l^{-1}$ and DOC from 150 to $1500\,\mu M$. The metals examined were Al, Cd, Cu, Pb, and Zn. The investigators examined the major factors controlling concentration and concluded that stream order alone was an unsatisfactory predictor. Watershed type was important as filterable levels of Al, Pb, and Zn were significantly lower, and total levels and K_Ds higher, in calcareous (agriculturally-dominated) watersheds. Headwater streams tended to have a higher K_D than associated receiving waters. The fraction of Al, Cd, Pb, and Zn in the particulate phase was greater in agriculturally-dominated waters. Seasonal effects were also apparent. These relationships appeared to reflect, to a large

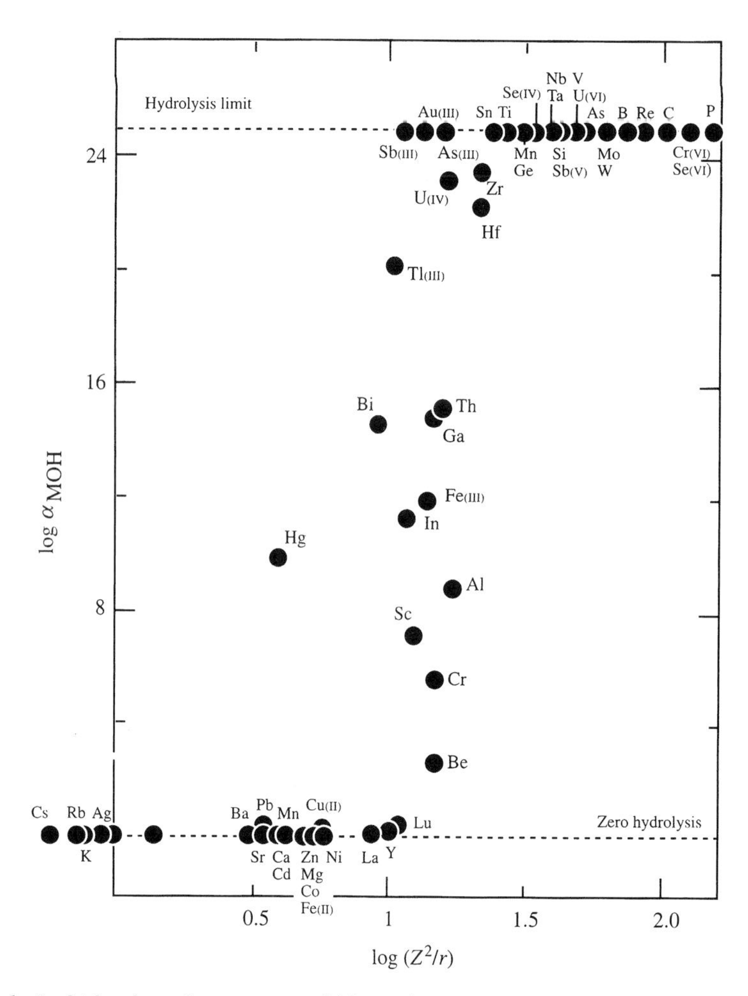

Fig. 7.24 A so-called "Born plot" which relates the extent to which an element is hydrolyzed in water (log α_{MOH} is a measure of the affinity for OH^- ions) relative to its polarizing power, which is reflected by the ratio of the charge squared (Z^2) to the ionic radius. Taken from Gaillardet et al. (2004) in *Treatise of Geochemistry* vol. 5 [6], reprinted with permission of Elsevier.

degree, variations in DOC across the different watersheds for most of the metals, and the relationships between K_D and DOC are shown for Zn and Pb in Fig. 7.24. Concentrations and watershed yields of Cu were not predicted by DOC levels, and were more a function of discharge and basin area.

Gundersen and Steinnes [56] looked at the speciation of some metals in a variety of river systems with a range of pHs and differing NOM content. The degree of adsorption of Cu, Zn, and Cd to particles and colloids was pH-dependent, being low (<10%) for low pH (<5.1) rivers and higher (20–40%) for circumneutral rivers. These results agreed with models based on the adsorption to oxide surfaces, although the model results tended to over predict adsorption. The results suggested the importance of NOM was either its enhancing the adsorption to the solids or increasing the concentration in the dissolved fraction, and therefore that the role of NOM is related to the strength of binding to the individual metals. This reflects the limited number of binding sites on the solid phase and therefore, as the NOM content increases, it will have more impact by binding metals in solution and reducing adsorption for those metals that form strong associations. These experimental results [56] are consistent with the discussions in Chapters 3 and 4 of this book. The importance of the colloidal fraction was also investigated by Hill and Aplin [57], who examined the fractionation between the dissolved, colloidal and particulate fractions for a variety of metals. The colloidal fraction was >50% of the filtered metal for Fe, Al, Zn, Ni, Cu, and Pb while it was less important for Mn, and Group I and II metals, and this is consistent with the relative strength of complexation of the metals to oxide surfaces and to NOM.

Dupre et al. [58] compared measured values (using ultrafiltration and other fractionation techniques) with model estimates for the partitioning of metals between inorganic and organic ligands in river waters. This examination indicated that for Al, Fe, Th, and Cu, the model and

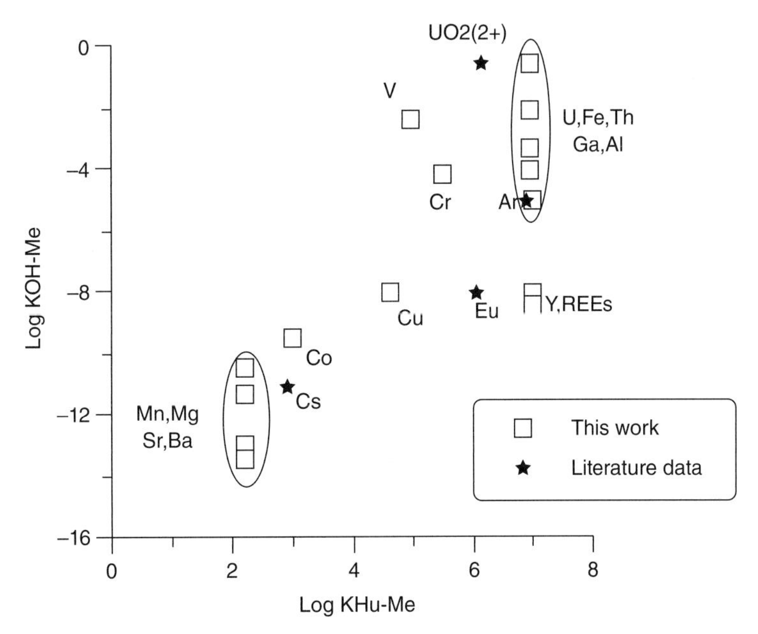

Fig. 7.25 Relationship between stability constants for various metals with humic substances relative to that with hydroxide. Taken from Dupre et al. (1999) *Chemical Geology* **160**: 63–80 [58], reprinted with permission of Elsevier.

measurements were in agreement. For these metals, the majority of the dissolved metal was associated with the organic fraction, in agreement with the other studies. Estimated stability constants for the interaction of metals with organic matter were well correlated with the first hydrolysis constant for the respective metal (Fig. 7.25) [58] which is consistent with the knowledge that the majority of these metals are binding to the carboxylic acids groups on the NOM.

In a study of the South Fork Coeur d'Alene River, Balistrieri and Blank [59] examined the dissolved speciation of metals using two approaches – filtration and DGT (diffusion gradient in thin film) gels. These devices record the concentration of the permeable species that penetrate the gel, which is interpreted to represent the concentration of dissolved inorganic low molecular weight complexes plus the free ion (the labile ion concentration). Overall, these measurements showed that for this system there were differences between the measured dissolved concentration and the DGT value for Cd (32–100% of dissolved Cd), Zn (62–100% of dissolved Zn), Cu (2–43% of the dissolved concentrations) and Pb (16–100% of the dissolved Pb) [59]. The observation that the difference is greatest for Cu, and least for Cd and Zn, is consistent with their known binding strengths to NOM and oxides phases, which are likely a large fraction of the non-labile fraction in the filtered waters.

To further examine these results, Balistrieri and Blank [59] modeled their data using a variety of models (WHAM: Windermere Humic Aqueous Model version 6; the NICA-Donnan formulation, discussed in Chapter 4, and the SHM: the Stockholm Humic Model) that incorporate binding to NOM using a number of different formulations and approaches. The results of these modeling studies (Fig. 7.26) using the default parameters in each model show some differences but all lead to similar conclusions that are consistent with the dissolved/DGT data for Cd and Zn; that most of the metal is present as the free ion or as inorganic complexes (typically >80%). This is not the case for Cu and Pb as there are substantial differences between the model predictions, with the SHM model predicting the highest inorganic labile fraction for Pb. Conversely, for Cu, the WHAM and SHM models appear to have the best agreement. Clearly, these results suggest that the various model parameterizations still need to be better calibrated and developed so that predictions are improved compared to measurement. However, even given these differences, there are consistent trends in prediction for these metals that have important environmental implications for metal uptake and bioaccumulation – most of the Cd and Zn is present as a free ion or labile in solution, while very little of the Pb (<20%) and Cu (<2%) is labile, considering all models. Overall, these results are consistent with what has been found in other ecosystems, as is discussed in other sections of this chapter, and elsewhere in the book.

The association of cations with NOM and the relationships that exist likely reflect both the direct association of cations with "dissolved" (colloidal) NOM and also the potential for NOM to be associated with colloidal inorganic phases that would pass through traditional filtration devices. The importance of colloids has been discussed on numerous occasions

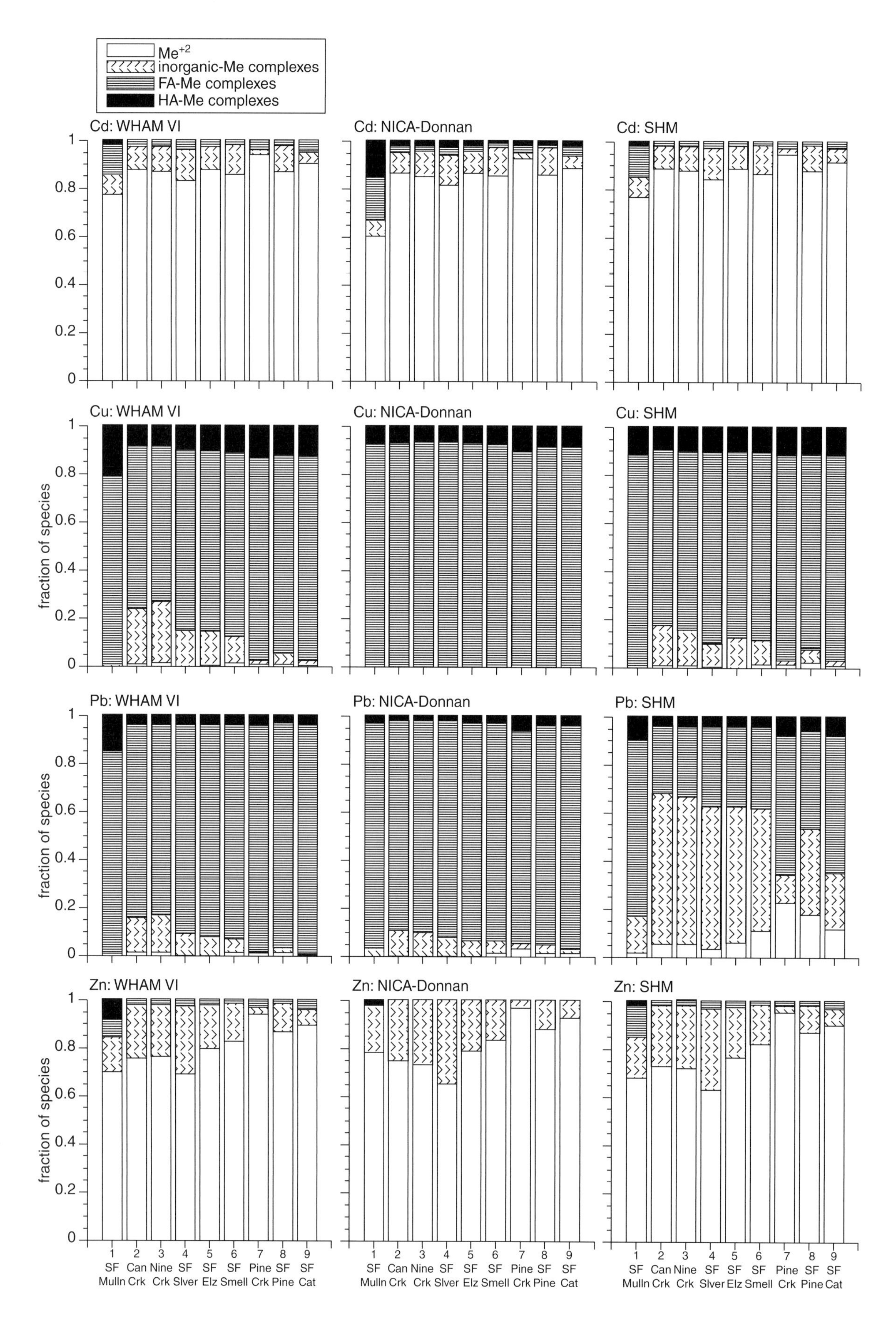

Fig. 7.26 A comparison of the results obtained for the speciation of metals (Cd, Cu, Pb, and Zn) using three different modeling approaches (WHAM, NICA-Donnan and SHM models (described in Chapter 4) for study sites within the Coeur d'Alene River Basin, Idaho, USA. Figure from Balistrieri et al. (2008) *Applied Geochemistry* **23**: 3355–71 [59], reprinted with permission of Elsevier.

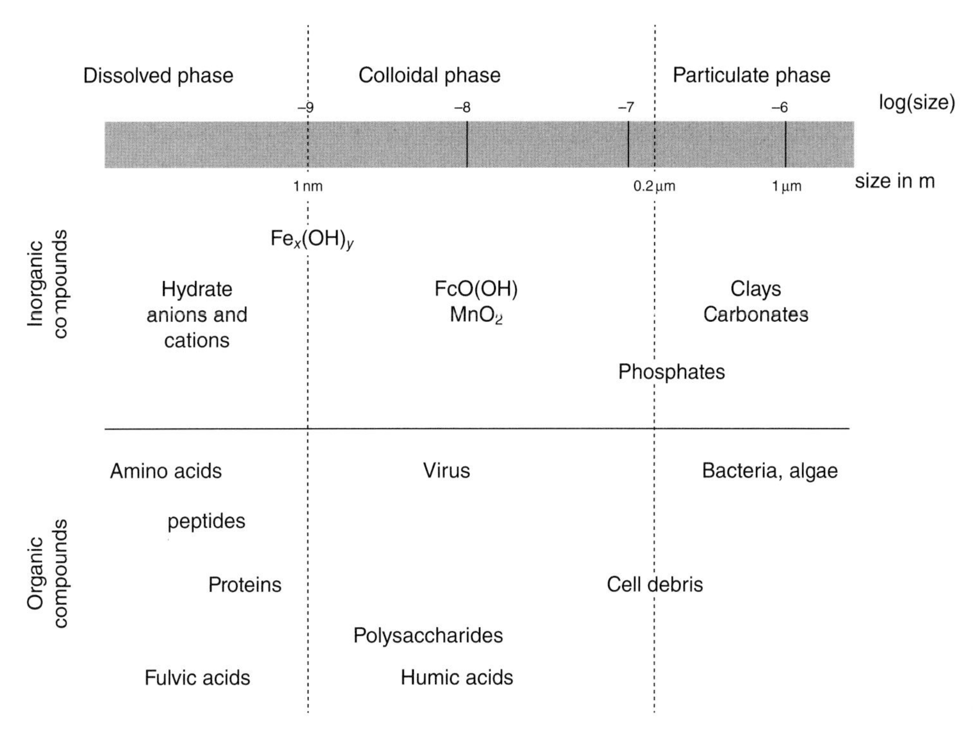

Fig. 7.27 Distribution of particle sizes and the colloidal fraction. Taken from Stumm (1992) *Chemistry of the Solid-Water Interface* and used with permission of John Wiley & Sons, Inc.

throughout this book and it is worthwhile to discuss their role in more detail, especially considering the relatively high concentrations of NOM in natural freshwaters. The division between the various fractions is ill-defined and this is shown in Fig. 7.27 [1], which shows the traditional division between dissolved and particulate phases based on filtration devices and the typical range in size for colloids, and inorganic and biotic particles. The smallest microbes are >0.1 μm, while colloids range above that size but are mostly in the <0.2 μm fraction. Colloids consist of both inorganic and organic phases; soil-derived entities (e.g., silicates, kaolinite and other clays), Fe, Mn oxides and biotic debris, including exudates and polymeric substances, and humic material. Colloids are also a mixture of agglomerates of smaller particles and therefore are both organic and inorganic. Some small microorganisms and viruses fall into the colloidal fraction. The dual inorganic/organic nature has been demonstrated by analysis of field samples where both oxide phases and organic matter contribute to the colloidal fraction and to the binding of metals to colloids [6]. In anoxic environments, if sufficient S is present, polysulfides and other metal-S multinuclear entities are found.

Colloids have different roles in trace metal biogeochemistry [1]. Physical aggregation and coagulation of materials can enhance removal of metal(loid)s from the dissolved phase and/or can allow for metal transport over substantial distances, as discussed further later in the groundwater section (Section 7.3.2). Colloids provide surfaces for the adsorption of metal(loid)s, especially as they contain hydroxide (oxide surfaces), carboxylic (NOM) or other acidic surface sites. The can act as either Lewis acids or bases depending on their surface charge. Furthermore, they can act as a catalyst for redox reactions by adsorption and mediation of oxidants and reductants, or by donating or accepting electrons directly, especially oxide and sulfide phases. Depending on their composition, they may also have a role in photochemical reactions, mediating the reactions discussed, for example, in Section 5.3. Biologically-derived colloids may contain enzymes and other reactive molecules that could enhance extracellular reactions and breakdown of organic matter.

The relative importance of colloidal organic surfaces in metal complexation is a function of location and of the size fraction of the NOM (i.e., its relative age and degree of degradation). In organic-rich waters, organic colloids will obviously dominate while the opposite will be true for temperate, clear water and low TSS systems. In Fig. 7.27, the relative size and fraction of various inorganic and organic compounds are shown and this illustrates the complexity of the situation. No separation method provides an absolute distinction between these two fractions. Indeed, much of the discussion in this and other chapters on the association of metal(loid)s with NOM in filtered field samples could easily be due to the association with colloids rather than truly dissolved NOM. However, there is increasing evidence for the presence of small molecular weight NOM with high binding capacity, such as thiols (cysteine, glutathione, phytochelatins), many of which are derived from biological processes and are either released during cell lysis or excreted into the environment. The relative importance of these compounds in binding cations is likely a function of the relative amount of high molecular weight humic material which also contains substantial reduced S content, and has available thiol sites for reaction with metals.

Additionally, as metals are often associated with NOM in rivers, as in other aquatic systems, any biogeochemical changes in NOM can have an important impact on mobility and speciation, especially through the impact of photochemical processes on NOM lability and its binding strength to metals. A number of investigators have examined the importance of photochemistry using both field experiments [60] and laboratory studies [61, 62], in concert to similar studies that have been done in marine systems. Studies in rivers impacted by acid mine drainage have been used as a natural laboratory to study the interaction of light, Fe and NOM as these systems have relatively high Fe content in suspended particles. While solubility of Fe is enhanced by pH, it is likely that photochemical production of oxidants through degradation of NOM is the driver of these reactions as the peak concentration of reduced Fe is found during the middle of the day, when light intensity is the greatest. However, direct reduction of Fe is also possible. The following light induced reactions likely occur in such a system [60, 61]:

$$FeOH^{2+} \rightarrow Fe^{2+} + OH^{\bullet}$$

$$NOM + O_2 \rightarrow DOM\ radical + O_2^{\bullet -}$$

$$HO_2^{\bullet} + HO_2^{\bullet} \rightarrow H_2O_2 + O_2$$

$$H_2O_2 + 2Fe^{2+} \rightarrow 2Fe^{3+} + 2OH^-$$

The hydroxyl radical ($OH^{\bullet}$), the superoxide anion ($O_2^{\bullet -}$) and hydrogen peroxide are directly produced by the photochemical reactions and these reactive species could all be involved in reactions with other metals, in addition to the redox cycling of Fe. In addition to abiotic oxidation of Fe, which is slow, there is the potential for the reaction to be microbially-mediated. McNight and Duren [60] discuss previous studies that have shown a diurnal cycle in dissolved Fe concentration, where Fe^{II} concentrations, in the range of 8–24 μM, comprised 40–90% of the total dissolved Fe. The system was shown to be at pseudo-equilibrium at midday with the rates of photochemical reduction being balanced by chemical and biological oxidation reactions. The estimated photoreduction rates, accounting for all the reactions occurring, including microbial oxidation of Fe^{II}, ranged from 0.2 to $1.9 \times 10^{-4}\,\mu M\,s^{-1}$. These rates are consistent with reduction of oxide phases within the streambed as there is evidence for much higher rates of reduction for dissolved Fe^{III} and for reduction of colloidal Fe. While not discussed in the paper, it is likely that there is release and scavenging of metals associated with the photochemical dissolution during the day, and with net precipitation that occurs at night [6]. Dissolution of Mn oxides could also occur via similar photochemical mechanisms.

The type of NOM has an influence on the rate and extent of reaction as it has been shown that lower molecular weight NOM lead to the production of higher steady state Fe^{II} concentrations, and a higher fraction of the total Fe as Fe^{II}, in both freshwater and seawater samples [63]. However, the rate of hydrogen peroxide formation was greater for the high molecular weight fractions in seawater, and lower in freshwater. It was hypothesized from these data that Fe^{II} formation mostly involved dissolved Fe^{III} species and that complexation stabilized the Fe^{II} in the freshwaters compared to the seawater samples, leading to the higher steady state concentrations. It was also concluded that terrestrial NOM is more photochemically reactive than marine NOM that is typically microbially-derived. These results are also consistent with the reaction pathways shown previously.

The production of the reactive oxygen species will also influence the redox chemistry of Cu, Cr and Hg, and other metals. There is substantial evidence in the literature for the presence of Hg^0 in surface waters and the role of NOM in Hg^{II} reduction has been demonstrated (e.g., Tseng et al., [64]), as discussed in detail in Section 7.2.3. More studies are needed to examine the impact of NOM photochemical transformations on metal(loid) cycling. In one set of laboratory experiments looking at the longer-term impact of NOM photochemical degradation [62], a 20% decrease in DOC content was found over three weeks during natural light incubations. Dissolved Fe (<20 nm filtration) also decreased which was thought to be due to its release from DOC during degradation and its subsequent precipitation. Decreases were also found for Ce, Cu, Cr, Pb, V, and U, but not for Mo, Mn, Co, Zn and the Group I and II metals. These results are consistent with the release of these metals through DOC degradation and their scavenging by the precipitating Fe. The metals that showed little change in concentration are those that do not associate strongly with NOM, and which are also not strongly adsorbed to oxide phases.

While photochemistry can have a diurnal impact on concentrations of metal(loid)s in rivers, changes in NOM over a seasonal cycle can also impact the filtered concentration as many elements in freshwater are dominantly complexed to NOM. This has been demonstrated in a number of cases [6]. For example, in the Mississippi, concentrations of Fe and Mn vary substantially with flow and many trace metals showed concentrations fluctuations that correlated with the Fe and Mn changes, but with less variability. As discussed by Gaillardet et al. [6], metals, such as Zn, Pb, and the REEs, were highest at high flow, while oxyanions (V, Mo, U) and some cations (Cu, Ni, Cd) were highest under low flow conditions. However, these patterns were not found in the Amazon so there appears to be little consistent trend across systems on the impact of large scale factors such as flow on metal concentration. It is likely that the variability is related to the changes in NOM content, levels of TSS, the POC fraction in the TSS, and the degree of sediment resuspension as these all impact the K_D and therefore the dissolved content of the waters. Additionally, not to belabor a point, the presence of colloidal material in the filtered fraction will be a strong function of flow and TSS, and may be partially responsible for the differences seen seasonally in these large rivers.

Overall, the cycling of trace elements in rivers has been less studied than that of lakes, and many of the studies have only focused on specific locations that have been contaminated. Studies have focused on measurements to ascertain inputs and fluxes to the coastal zone or to other aquatic systems. Additionally, much of the details of the interactions of metals with suspended solids and the mobility and interaction of trace metal(loid)s with colloidal material is similar to that found in other freshwaters discussed in this chapter and elsewhere in the book.

7.3.2 Trace elements in groundwater

The composition of elements in groundwater can be characterized as the steady state composition resulting from the reactions involved in mineral dissolution and precipitation and the related incorporation of elements into these phases. Metal(loid)s are also adsorbed onto/released from the surfaces of the particles/minerals through which the water is percolating. Therefore, it can be considered that the major ion composition of the water is a reflection of its surroundings [65]:

$$\text{Final water} = \text{initial water} + \text{dissolving minerals} - \text{precipitating minerals}.$$

For metals, redox reactions and changes within the aquifer can also lead to substantial changes in concentration, and these reactions are often mediated by microorganisms, as discussed later, for example for acid mine drainage (Section 7.3.3) and As cycling (Section 7.3.4). Therefore, it is possible that these redox reactions are not at equilibrium and perhaps in these instances it would be better to consider the changes in terms of the kinetics of the mediated reactions, and the hierarchy of the various reactions in terms of their energy yield. Overall, the pH and redox state of groundwaters is between that of oxic surface waters and anoxic environments (Fig. 7.28) and therefore, as will be discussed later, small changes in redox state or microbial activity can have a large impact on the fate of trace elements that are relatively easily transformed between oxidation states, or which are strongly adsorbed to oxide phases. Many metals are highly particle reactive and therefore are not likely to be transported through the subsurface as dissolved ions. There is the potential for enhancement of the dissolved concentration due to the presence of NOM in the filtered fraction or due to the formation of colloidal precipitates and/or sulfide clusters and other nanomaterials.

Additionally, while important in many aquatic systems, colloidal transport can be an important mechanism for trace metal(loid) transport in groundwater. In soils and aquifers, mobile colloidal particles originate mostly from *in situ* generation of submicron-sized mineral and organic matter particles, which are naturally present, and which often contain both inorganic and organic fractions, and may be a coagulation of smaller nanomaterials. Such particles can be composed of clays and (hydr)oxides (e.g., Fe, Al, Mn, or Si), carbonates, phosphates and other precipitates, and humic and fulvic acids and other NOM, including biotically-derived polymers, and viruses, bacteria, and other microorganisms [66, 67]. The release of colloidal particles in soils and aquifers is driven by hydrology, and fluctuations in water saturation and velocity, and changes in solution composition (i.e., pH, ionic strength, redox potential). These impacts are much more important in the unsaturated zone and fast-flowing systems. Rapid changes in water flow will cause mobilization and the associated changing chemical conditions will enhance colloid release and transport. Other factors, including human-derived perturbations, can also cause chemical and physical perturbations that result in strong gradients in chemical composition and redox potential and enhance colloidal transport.

Colloidal inorganic phases, such as Fe oxides, often result in enhanced transport of metal(loid)s that are associated with these phases through surface adsorption. Additionally, coating of these materials with organic molecules can also increase metal adsorption and transport. The specific surface area and surface charge help define the colloidal reactivity, and its adsorption capacity for trace metal(loid)s [66, 67]. The smallest particles contribute most to surface area although they do not contribute significantly to the total colloidal mass, and therefore are mostly responsible for colloid-facilitated transport of trace substances. The surface charge of colloids is an important parameter and most natural colloidal particles have a net negative surface charge, although the exact charge will differ with pH, especially for organic matter and clay minerals. Oxides are positively charged above their point of zero charge, which is around pH 8–9, but most are coated with NOM, or have adsorbed major anions, which results in a more negative charge overall. The transport of colloidal particles is mostly related to the tendency of the colloids to aggregate or attach to stationary surfaces, both of which depend on electrostatic interactions and van der Waals forces, although other processes are also important (gravitational settling and straining) [66].

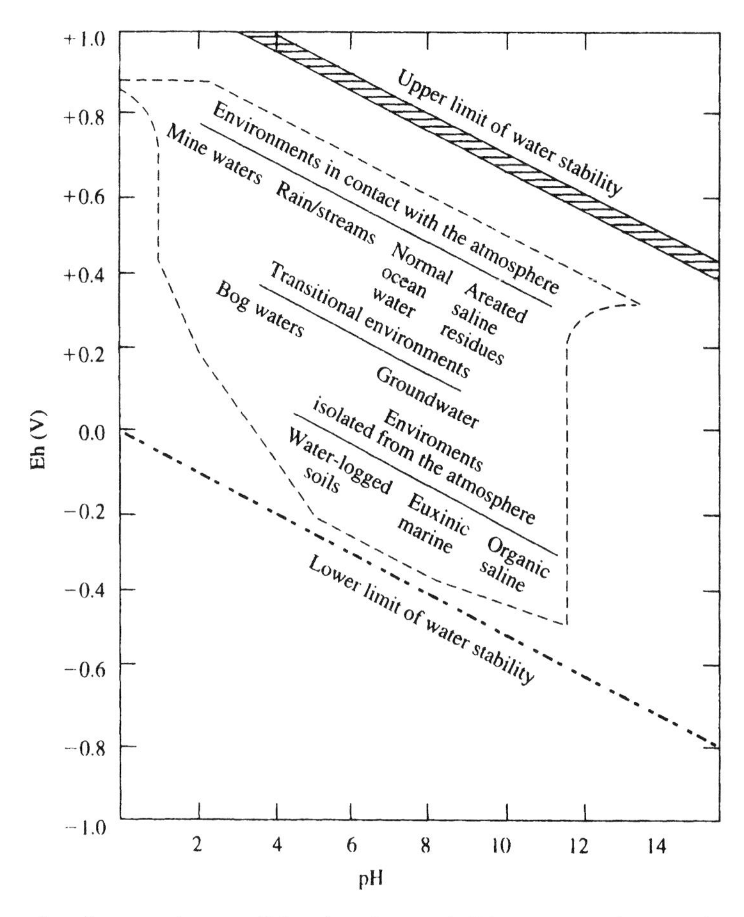

Fig. 7.28 Range in pH and pε values for natural waters. Taken from Langmuir [2] *Aqueous Environmental Geochemistry* and reprinted with permission of Prentice Hall/Pearson.

Transport can be extensive because colloidal settling velocities are low and filtration by straining is ineffective.

Ionic strength and the degree of surface adsorption both can impact the overall surface charge and typically these factors enhance coagulation and colloidal adsorption to surfaces [68]. Laboratory studies have shown that colloid-facilitated transport can become the major transport mechanism of strongly adsorbing trace metals, such as Pb and Hg [66]. Other metals that bind less strongly to oxides and organic matter, such as Zn and Cd, are less affected. Many examples of enhanced transport due to colloidal material in both natural and anthropogenically-impacted groundwater systems are in the literature [66, 67]; and references therein. Both laboratory column elution studies and field investigations have shown that colloidal transport can substantially enhance the rate of migration, up to 20 times faster than that found in the absence of colloids. Studies have examined the impact of colloids on the transport of Pb, for example, where it was shown that the majority of the Pb being transported was colloidal [67]. Other studies have looked at Ni, As, and Zn transport. One factor that was found in some studies was that the rate of enhancement was a function of concentration, with less enhancement of transport at higher concentration, suggesting that there is increased interaction with the solid phase under these conditions.

Many studies have examined the transport of Hg and CH_3Hg in groundwater as this can be important for contamination of lakes close to industrial sites where Hg was used (e.g., in the chlor-alkali industry; in industry where Hg compounds were used as catalysts) [69] or in urbanized locations [70]. Studies have shown that groundwater input can be important in these situations as well as in pristine seepage lakes where groundwater input may be an important source of Hg to the lake. The flow of groundwater across the *hyporeic* zone can be an important region for Hg methylation and other redox transformations [71, 72]. The hyporeic zone is the section of the sediments, spoils, gravel or sands that form the interface between the deeper sediments/soils and groundwater and the flowing surface water, where much water cycling occurs and where organic matter degradation and other microbial processing is enhanced. This region can

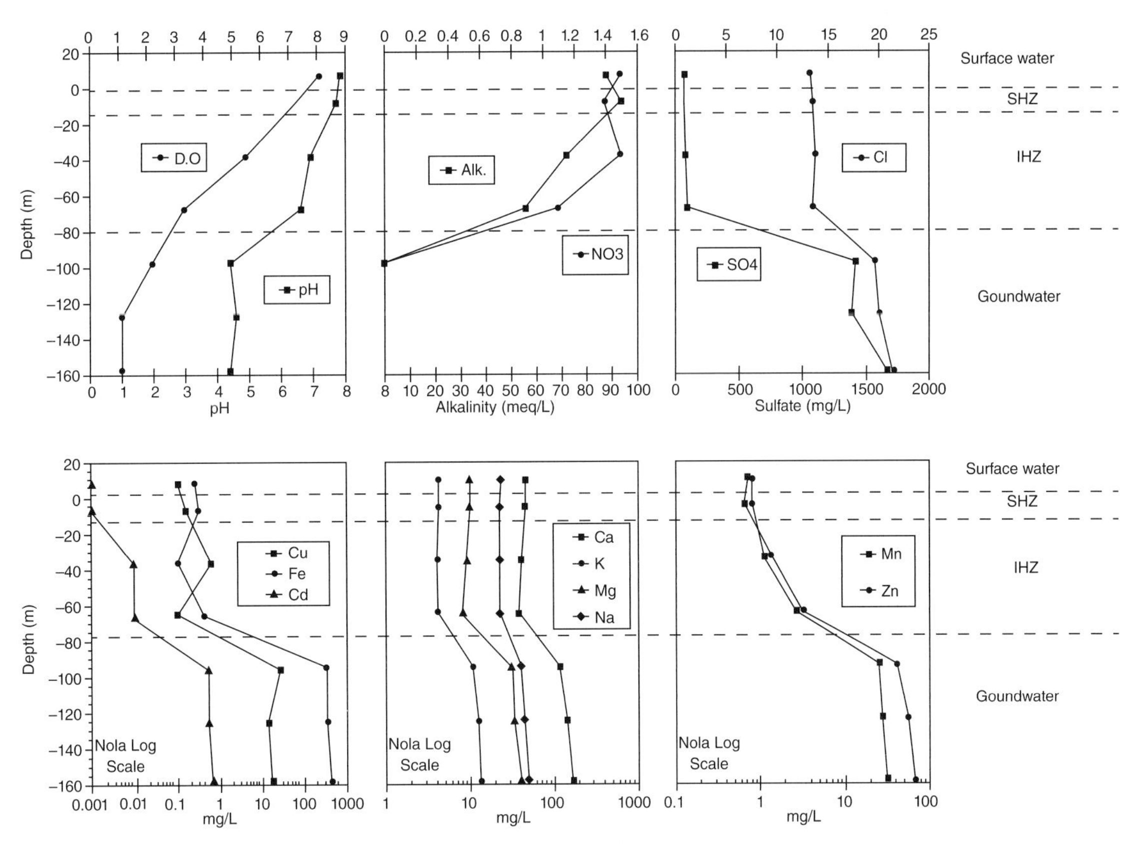

Fig. 7.29 Profiles of various elements and constituents across the hyporeic zone of a small stream. Reprinted with permission from Ren and Packman (2005) *Environmental Science and Technology* **39**: 6387–94 [73]. Copyright (2005) American Chemical Society.

be important for other metal cycling as well as it was shown that at the stream-subsurface interface the mobility of hematite colloids, and that of the associated trace metals was enhanced [73].

A study examining the distribution of metals across the steam/sediment interface [74] showed that the hyporeic zone is the major region of mixing between the groundwater, which is often higher in concentration of trace elements and other constituents, and the surface water. In this study the mixing region was broken down into the surface hyporeic zone and the interactive hyporeic zone, where much of the mixing occurs. The distributions (Fig. 7.29) [73] show a relatively constant concentration for the more conserved elements (alkali and alkali earth elements), but not for the other metals, suggesting that there is a removal mechanism for the metals in this zone. The distribution of Fe, and observations of oxide precipitation at the sampling site, indicate that Fe, which is mostly reduced in the groundwater is precipitated during travel through the transition zone, and is therefore removing metals from solution. Indeed, analysis of the solid phase confirms the presence of most of the metals which were more abundant in the dissolved phase in the groundwater.

Other regions where groundwater transport and colloidal enhancement is important is the movement of trace elements from waste disposal sites and other contaminated point locations into the surrounding environment. Many studies have examined the transport of radionuclides given the groundwater contamination that has been found in various locations. These studies have examined the transport of ^{137}Cs, and many of the actinides (U, Th, Pu) [67]. Some small enhancement of Cs transport has been noted but as Cs is not strongly adsorbed to particles, its transport is mainly in the dissolved phase. However, enhanced transport of the actinides has been found and one suggested reason for this is their tendency to be hydrolyzed in solution and their potential to precipitate as colloidal material. In one study it was observed that transport of U was enhanced when the ionic strength of the flowing solution was decreased and it was suggested that this could result in the release of U into the groundwater during periods of heavy rain and its infiltration into the subsurface [74]. Modeling of metal transport by colloids in groundwater has been developed to examine the fate and transport of a variety of contaminants, including radionuclides, organic contaminants and trace metals [67]. Most early models considered three phases (mobile and attached colloids and solids) and assumed equilibrium partitioning of the colloids and the associated contaminants. More recently, a four-phase model, with the inclusion of the liquid phase and contaminant partitioning within all phases has been developed, and which does not assume steady state. For example, in this model, the unsteady mass balance for a contaminant associated with the suspended colloidal particles is given as [67]:

$$\varepsilon\frac{\partial(X_1C)}{\partial t}=\varepsilon\frac{D_B\partial^2(X_1C)}{\partial x^2}+\frac{\upsilon_0\partial(X_1C)}{\partial x}+r_rX_3-r_cX_1+ b_1\varepsilon K_aC_c+\varepsilon K_dCX_1 \quad (7.1)$$

where ε is the porosity, C is the concentration of colloids, X_1 is the mass fraction of the contaminant in the colloids and X_3 the fraction adsorbed to solids , D_B is the Brownian diffusivity, υ_0 is the velocity, r_r is the rate of release and r_c the rate of capture of colloids by the solid phase, C_c is the concentration of contaminant in the aqueous phase, K_a and K_d are the rate constants for sorption and for desorption from the colloidal particles, and b_l represents the fraction of total adsorption that takes place. The left-hand side of the equation represents the rate of accumulation of contaminant onto the colloids. Similar equations can be written for the other fractions in the model. More recent models [67] have been expanded to include non-constant parameters, such as changing porosity due to plugging phenomenon, hydrodynamically-driven colloidal release and a threshold concentration below which colloidal capture is negligibly small. These changes reflect experimental results and make the overall model more realistic.

Such models have been used to examine the most important processes responsible for the mobility of adsorbed species via colloidal transport, and both field and laboratory studies have shown this to be an important phenomenon [67], especially for metals that are strongly bound to colloids. Both facilitation of transport and retardation can occur, especially through entrapment and/or plugging conditions. Such retardation occurs when there is low initial porosity, high colloidal concentration, and high velocity.

Overall, trace metal(loid) transport in the subsurface has received relatively little attention except for specific cases that have warranted further study and examination, and these are mostly associated with contamination or transport of toxic metals. As noted, there has been substantial study of the mobility of the radionuclides and of metal(loid)s such as As, Pb and Hg, but there has not been much systematic study of metal transport in general. In some of the following sections there will be further discussion of groundwater mobility, such as that associated with As contamination of drinking water (Section 7.3.4) and the exchange of trace species across the land-sea interface (Section 7.3.5).

7.3.3 Focus topic: Mining impacts and acid mine drainage

Mining has left a large legacy on the Earth's surface due to the interaction of water with mining wastes. Mining causes environmental impacts both at the site of ore extraction and during its refining and processing [75]. Both processes can result in waste products that can leach trace metals into the environment and are important sources of pollution in many areas of the world. The impact of mining on freshwaters will be discussed to highlight these impacts and to discuss the environmental chemistry that is associated with mining and refining.

Most metals in minerals and ores that are commercially extracted are sulfide or oxide minerals, although some carbonate minerals are exploited [75]. Metals such as Zn, Cu, Hg, and Pb are mostly extracted from their primary ore while the metalloids and some heavy metals are by-products in other ores, especially if they are precious metals or relatively rare in the environment. For example, Ag ores are chloragyrite (AgCl) and acanthite (AgS) but it is also extracted from ores primarily mined for Cu and Zn. Arsenic is extracted as arsenopyrite (FeAsS) or arsenolite (As_2O_3). Zn and Cd are both found in sulfide and carbonate ores. Additionally, some metals are found in the environment in their elemental state, principally Au, Ag, and Hg.

The most pervasive impact of mining is its role in the generation of acid mine drainage, which is mostly associated with abandoned mines [76]. Reactions of water with the exposed mine face or the associated mine waste piles from coal mining or mining of sulfide ores can lead to the generation of low pH runoff, which can be as low as −1. The low pH waters often have a high metal load which results in extensive impacts on the environment. Many of the acid mine drainage sources are close to or connected with streams and rivers leading to the wide dispersal of the released metals. While the focus of these processes is largely related to mining sites and mine wastes, these processes can also occur naturally, and there is a large and diverse microbial community that is associated with these redox reactions [75, 76]. Ore recovery often involves smelting of the sulfide ores and the waste streams associated with these processes can also be a major source for metals and acid mine drainage. These wastes include waste rock (of too low grade to be processed) and mill tailings, and for coal, ash and slag residues. Coal can contain up to many hundred μmol kg^{-1} of heavy metals and the metalloids and these are further concentrated (by a factor of 5–50), in the slag produced by coal incineration, except for the more volatile elements such as Se and Hg.

Base metal mining, primarily for Cu, Pb, and Zn, relies mostly on sulfide ores such as chalcopyrite ($CuFeS_2$), sphalerite [(Zn,Fe)S] and galena (PbS), and all these ores contain many other trace metal(loid)s such as Cd, As, Ge, and In. Platinum group metals are mostly found as sulfide ores, or as ores such as $PtAs_2$ (sperrylite) or $PdTe_2$ (moncheite) [76]. Precious metal mining can also rely on sulfide minerals although the processing of ores containing elemental Au does not often involve smelting but uses either mercury amalgamation or cyanide solubilization for extraction. Both these methods of Au extraction result in their own environmental insult [77]. Recovery through Hg-Au amalgamation impacts both the aquatic environment and the atmosphere as the amalgam is mostly heated to release the Hg to the atmosphere and recover the Au.

Cyanide leaching relies of a number of approaches to recover elemental Au from the Au-CN complexes in solution. The reactions involved in cyanidation can be written as:

$$4Au + 8NaCN + O_2 + 2H_2O \rightarrow 4Na[Au(CN)_2] + 4NaOH$$

Gold is oxidized to the +I oxidation state in the presence of oxygen. Other precious (Ag) and heavy metals (Cu, Hg, Zn) which are often present in the ores, especially if sulfidic, are also solubilized by the cyanide solution. The "traditional" method of recovering the Au from solution was addition of elemental Zn, which was oxidized and the Au reduced. Other methods are now employed as the Zn-CN solutions produced were highly toxic. However, given traces of metals in the initial ore, mining wastes from Au and Ag extraction typically contain a cocktail of many elements [77].

Pyrite and other sulfide ore oxidation depends on a number of factors, including the presence of oxygen, T, pH, and electrode potential, as well as the presence of microorganisms [76]. It has been demonstrated that the oxidation rate is enhanced by several orders of magnitude if biotically-mediated. There are a number of different pathways and these will be discussed briefly. The reaction for pyrite with oxygen is:

$$2FeS_2 + 14O_2 + 2H_2O = 2Fe^{2+} + 4SO_4^{2-} + 4H^+$$

The reduced Fe^{II} is oxidized and precipitates and this generates further acidity:

$$4Fe^{2+} + O_2 + 4H^+ = 4Fe^{3+} + 2H_2O$$

$$4Fe^{3+} + 12H_2O = 4Fe(OH)_3 + 12H^+$$

And overall:

$$4FeS_2 + 15O_2 + 14H_2O = 8SO_4^{2-} + 4Fe(OH)_3 + 16H^+$$

or 4 mol of H^+ per mole of pyrite. Oxidation can be enhanced by the presence of Fe^{3+}. At low pH, dissolution can occur without oxidation, releasing reduced Fe and sulfide. Similar reaction schemes occur with other Fe-S minerals, and for other metal sulfide minerals. However, for ores of metals such as Zn, which is relatively soluble, the generation of acidity is less than that with pyrite. Oxidation of PbS, by either oxygen or Fe^{3+}, leads to the precipitation of $PbSO_4$.

In many locations, mine tailings, especially those generated historically, contain the legacy of incomplete extraction of elements from their ores and therefore these locations can be an important local source of metal(loid) input to the environment. Two important examples are the extraction of As from its sulfide ore, arsenopyrite (FeAsS), and mercury from HgS (cinnabar). As noted in Section 2.2.1, the Aberjona watershed, and other environments, have been contaminated with As and other metals as a result of the use of sulfide minerals in the production of sulfuric acid for industrial uses. For Hg, cinnabar is resistant to oxidation and can persist in the environment and therefore the extraction of Hg from the ore was primarily done by roasting the ore and distilling the elemental Hg produced. This process was inefficient leaving behind roasting wastes that were both able to produce acid mine drainage as well as leach substantial quantities of Hg into the environment. One location where these problems are acute is California, especially given that Hg mined in one region of the state was then used for Au extraction in other locations leading to Hg contamination from both Hg mining and its recovery and in Hg usage in Au extraction. Additionally, the regions around the major Hg mines in Slovenia (Idrija) and Spain (Almaden), there has been large scale and intensive contamination that has been studied in detail by numerous investigators [78–80].

Numerous bacteria have derived the ability to exist on the oxidation of sulfide minerals for energy production. These are both Fe and S oxidizing bacteria. Acidophilic sulfide oxidizing bacteria, such as *Acidothiobacillus thiooxidans*, are common in acid mine drainage ecosystems as are Fe reducing bacteria such as *Acidothiobacillus ferrooxidans*. Most sulfide oxidizing microbes can exist on a variety of substrates such as H_2S, S^0, $S_2O_3^{2-}$, and $S_4O_6^{2-}$ as well as in the presence of sulfide minerals [76]. It is thought that solubilization is necessary for microbial utilization by either S- or Fe-oxidizing bacteria, but this can be by either direct processes – that is, the organism releases chemicals to dissolve the solid – or through indirect means. Bacterial oxidation can enhance the rate of reaction by a factor of 10 to >100.

Overall, the examination of waters impacted by acid mine drainage and other mining-related insults provides "classic" examples of the changes that can occur when material is transported from regions of anoxia to the oxygenated environment. As most oxidation-reduction reactions produce or consume hydrogen ions, such changes can have a large impact on the pH of the waters being impacted. Furthermore, the dramatic differences in the solubility of Fe, Mn, and other metals with changing oxidation state can lead to dramatic reversals in the concentration of metals that may be associated with these solid phases. Finally, as most minerals that are extracted and processed are not pure minerals and contain many trace constituents, the extraction and refining for one element in particular often leads to the contamination of the environment with these trace constituents given the large volume of material that is often processed, and the fact that concentrations of most metals in freshwaters is in the nM or lower range.

7.3.4 Arsenic in surface water and groundwater

Concentrations of As in freshwaters are typically in the nM range (<50 nM and typically 10–20 nM for uncontaminated or impacted systems) and the dominant form is As^{V} in oxic surface waters [81]. However, as discussed in Chapter 8, As^{V} is an oxyanion and has a very similar acid–base chemistry to phosphate and is therefore often taken up inadvertently by microbes in aquatic surface waters. Many organisms have devised mechanisms to deal with this inadvertent uptake, and often reduce and/or methylate the As into forms that can be excreted. There have been a number of studies [81–83] that have documented that surface waters have a substantial fraction of the total As as As^{III}, which also exists in solution as an oxyanion, contrary to what would be predicted at chemical equilibrium. Additionally, these studies have measured elevated levels of the methylated forms. The predicted dominant forms of As over the pH and pε (E_H) range of natural waters is shown in Fig. 7.15. The reduced form only becomes dominant under relatively reducing conditions and a near-neutral pH.

However, converse to the thermodynamic predictions, in low oxygen environments in lakes, much of the As can be present as Av^{V} and this has been attributed to the release of As in conjunction with the dissolution of sinking Fe-oxide particles from the surface. In bottom waters where low oxygen persists, but there is no buildup of sulfide, the rate of reduction of As^{V} is slow, and it has been shown that As^{V} can also persist bound to oxide colloids. However, in the presence of sulfide, most of the As is found as As^{III} (Fig. 7.30) [83]. These differences may be partially attributed to the importance of microbially-mediated As reduction. One location where As cycling has been investigated in detail in a system of lakes outside Boston, MA that were contaminated by As from industrial activity [83–86]. Studies in this system also showed that As-reducing bacteria existed in the

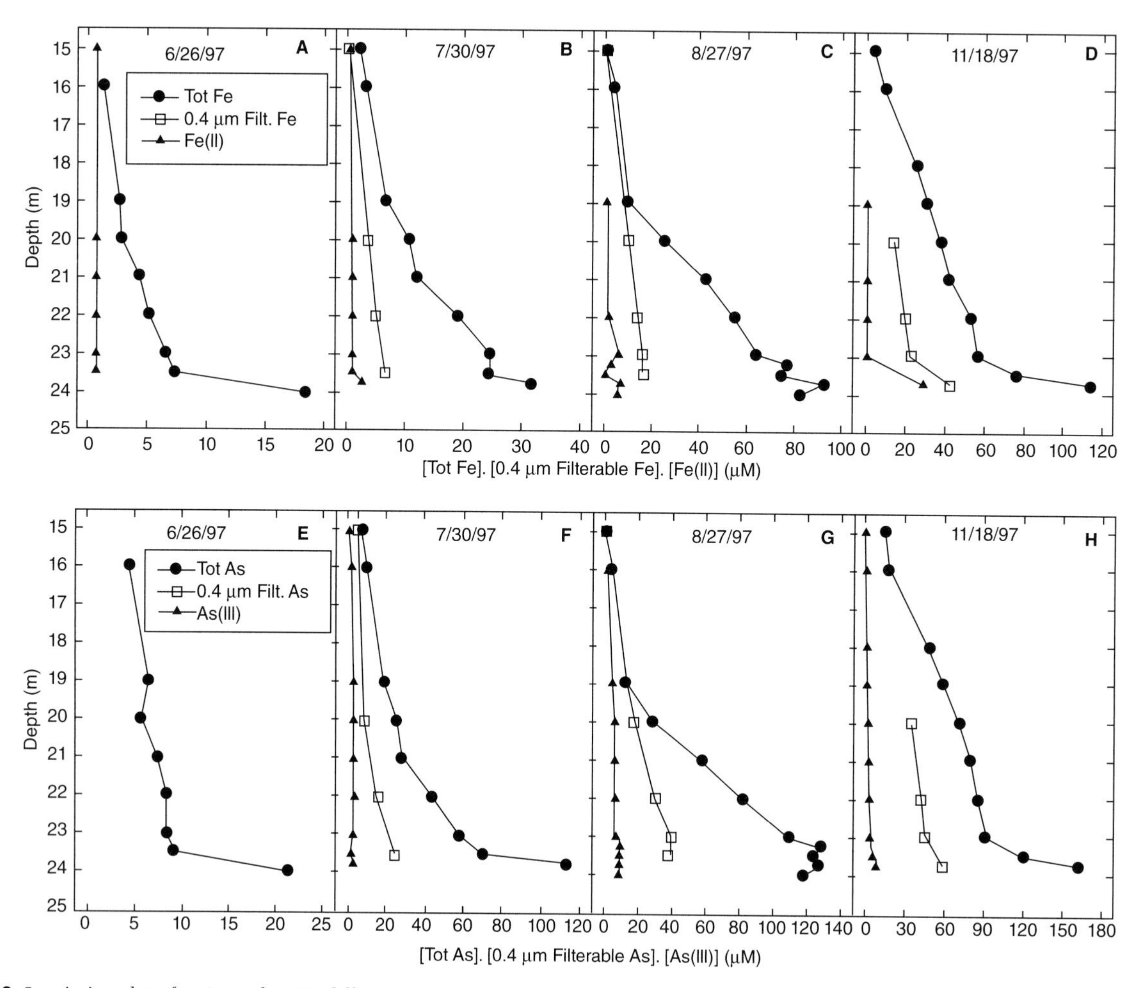

Fig. 7.30 Speciation data for As and Fe in different size fractions for the waters of the Upper Mystic Lake, MA, USA. Taken from Senn and Hemond (2004) *Environmental Toxicology and Chemistry* **23**: 1610–16 [83] and used with permission from John Wiley & Sons, Inc.

sediments of the high As locations [87, 88] and as discussed further below, there have been subsequent studies showing that As-reducing bacteria are relatively common.

It has been suggested that differences in the binding capacity of the oxidized and reduced forms can enhance As mobility. However, if one considers the binding of either form to oxides, using, for example, the Fe-oxide model [12], then this conclusion is not universal. For example, considering the species that are dominant at pH 7, the following equilibrium equations can be defined [83]:

For As^{III},

$$\equiv FeOH + H_3AsO_3 = \equiv FeH_2AsO_3 + H_2O \quad \log K = 5.4$$

For As^{V}

$$\equiv FeOH + HAsO_4^{2-} + H^+ = \equiv FeHAsO_4^- + H_2O \quad \log K = 12.0$$

$$\equiv FeOH + H_2AsO_4^- + H^+ = \equiv FeH_2AsO_4 + H_2O \quad \log K = 10.8$$

From these equations one can conclude that at pH 7, As^{III} will absorb more strongly than As^{V}, and this dominance will increase with pH [89]. However, at pH values below 6, the oxidized form will dominate according to these equations. Therefore it is likely not universally true that the oxidized form interacts more strongly with oxide phases than reduced As, and therefore it is not always apparent whether oxide dissolution or As reduction are more important processes in the release of As from the solid phase. Additionally, there is also the potential that As^{III}, in addition to binding to the surface of the oxides, can also be oxidized on mineral Mn surfaces. This does not occur with Fe oxides, and is due to the reductive dissolution of Mn oxides [1]:

$$As^{III} + Mn^{IV} = As^{V} + Mn^{II}$$

It was noted that the pH has an important impact on the reaction rate as at higher pHs, the Mn^{II} remains absorbed to the mineral surface and this can hinder the reactions. Additionally, the presence of other cations also likely impacts the reaction rate.

Studies in the Mystic Lakes near Boston have shown that As^{V} persists in bottom waters even after extended periods of low oxygen after stratification [83–85] (Fig. 7.30). However, examination of the partitioning provided evidence for the importance of colloidal phases in the persistence of the oxidized As [83]. These authors found that while 25 to 60% of total As was in particulate, most of the filtered As was in a "colloidal" size fraction (50 nm – 0.4 μm), which was similar to the Fe distribution. This was also counter to the expectation that most of the Fe would be reduced and therefore soluble (Fe^{2+}). These results again highlight the need to determine what is truly dissolved when doing environmental studies as "classical" filtration (0.2 or 0.4 μm filtration) often includes colloidal material. The authors concluded from model output and their field results that most of the As in the lake's hypolimnion was present as As sorbed on Fe(III) oxides, even after several months of anoxia.

These discussions are also pertinent to the observed enhanced mobility of As is groundwater, which has become an important problem in many regions of the world, especially Bangladesh and other Asian countries, and is a topic worthy of discussion as it demonstrates the complexity of the geochemical cycling of As in aquatic systems [81]. The concentration of As that can pose a cancer risk is very close to that found in some environments. The value set for drinking water varies around the globe (130–665 nM), but is close to the value found in some uncontaminated groundwaters in contact with high As-containing soils and minerals. It is difficult to define a "typical" concentration for As in the surface earth as it varies widely. A survey in the USA and Canada [90] found As levels to range from <10 μmol kg^{-1} to 3 mmol kg^{-1} in parts of Alaska. Highest concentrations are found in pyrite-containing rocks as As is bound up in various sulfide minerals, especially as arsenopyrite (FeAsS), realgar (AsS) and orpiment (As_2S_3), as well as As co-precipitated with Fe-S minerals, as discussed in Section 7.3.3 on acid mine drainage.

In the oxic environment, concentrations of As in solution in contact with solids is most often controlled by adsorption rather than precipitation of a specific As solid, and the degree of adsorption is strongly pH dependent. The adsorption isotherm for As^{V} to Fe oxide is shown in Fig. 7.10(b). As the pH increases, As can be released from solids and it is possible that NOM can enhance the dissolution as it forms associations with As, and especially with As^{III}, probably as a result of its neutral charge. Given the similarity in structure and the dominant forms in solution, a number of oxyanions (e.g., sulfate, silicate and phosphate) can compete with As for binding sites on the surface of oxide minerals (Fig. 7.10).

While changes in pH can cause As to be released into solution, changes in redox potential and oxide dissolution are often more dominant processes for its release [81, 91], driven by changes in hydrology and organic content of groundwater. This is thought to be the main processes whereby groundwaters in East Asia and other locations have become "contaminated" with As. Concentrations in the groundwater range from low levels, even in regions of Asia (<10 nM) to extremely elevated levels (>4 μM) (Fig. 7.31a). In contrast, concentrations in groundwater in North America are <700 nM and mostly <100 nM (Fig. 7.31b) [92].

None of the soils in the region of high As in groundwater appear to be highly elevated in As in Bangladesh, for example, and the minerals are mostly not sulfide ores. Rather the processes accounting for the release of As are due the dissolution of oxides, which is likely microbially-mediated, as well as the potential production and release of As^{III} by As-reducing bacteria [93, 94]. The increases in As are a recent phenomenon because the population has switched from using surface waters, which have become

(a)

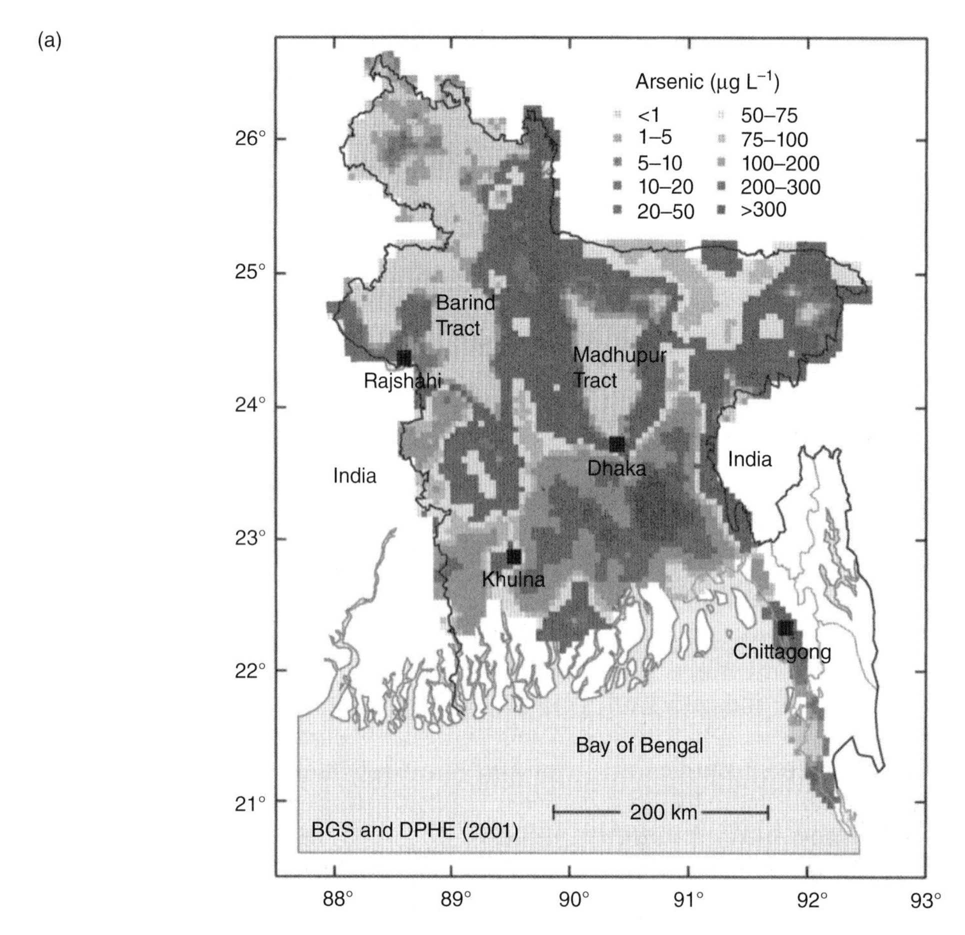

(b)

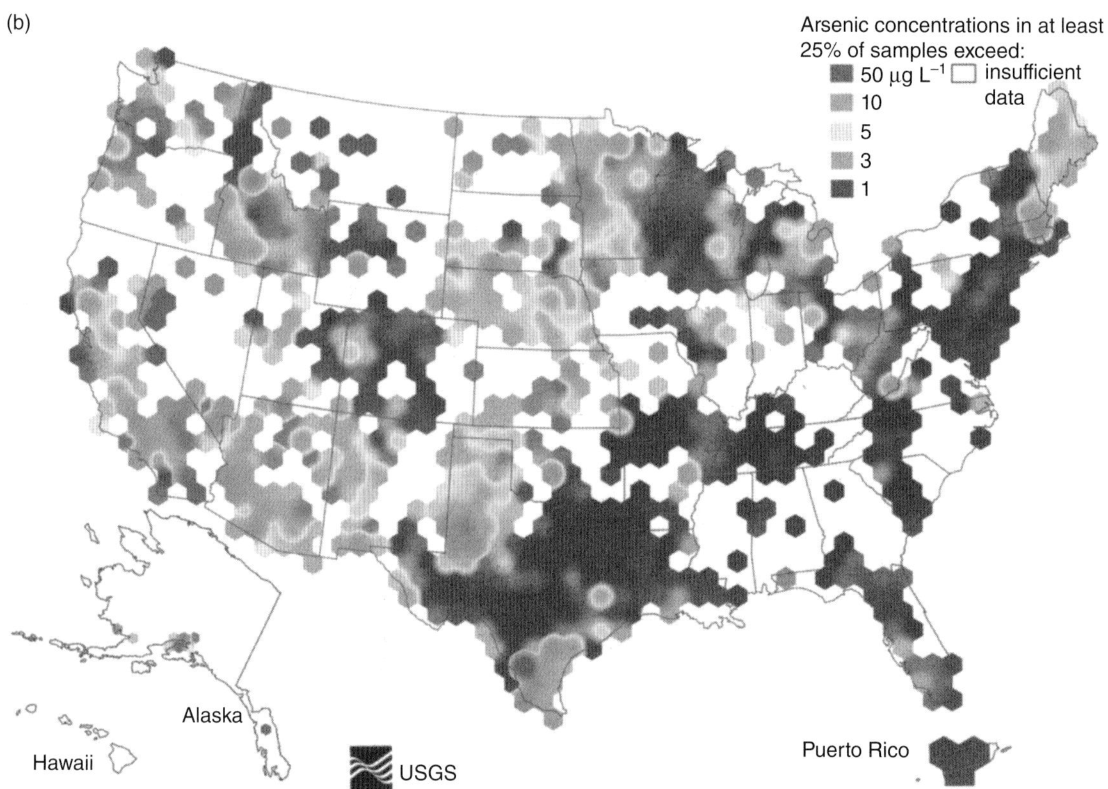

Fig. 7.31 Arsenic concentrations and distributions in groundwater (a) in Bangladesh and (b) in North America. Figure reprinted from Plant et al. (2004) in *Treatise on Geochemistry* vol. 9, reprinted with permission of Elsevier. Note: 50 μg l^{-1} = 0.67 nM.

contaminated with fecal and other bacteria, to using well water. Additionally, due to a rapid increase in the use of water in agriculture, and specifically for rice growing, which has increased rapidly in recent years in Bangladesh, much groundwater is being extracted. The increase in water use and recycling of water through the upper soil layers appears to have enhanced the bacterial populations and activities in the subsurface due to the increased percolation of organic matter and/or nitrate into the deeper layers where oxic respiration is replaced by denitrification, oxide reduction and even arsenic and sulfate reduction, although sulfate is often low in these systems.

The processes and reactions that are likely occurring can be exemplified by the following, where the Fe reduction reaction here is shown as a mediated bacterial reaction in conjunction with organic matter oxidation (represented here as CH_2O). Such a reaction would release As^V into the groundwater as the minerals are dissolved. Alternatively, As reduction would release As^{III} into solution.

$$CH_2O + 4Fe(OH)_3 + 7H^+ = HCO_3^- + 4Fe^{2+} + 10H_2O$$

$$CH_2O + 2H_2AsO_4^- + H^+ = HCO_3^- + 2H_2AsO_3$$

It is not possible to determine directly the relative magnitude of the various reactions as it is possible for secondary reactions to occur. For example, organic matter may enhance the oxidation of Fe^{II} and it is also possible that As^{III} can be oxidized through the further reduction of manganese oxides [5], as noted earlier.

A model was developed to try and validate these mechanisms for a groundwater system which included a mobile water phase, attached (immobile) bacteria and a solid phase [95]. In this model, organic matter input is derived from organic matter release from the solid phase and from the bacteria. In addition to the bacterially-mediated reactions, the model also included inorganic partitioning and other pertinent reactions. Concentrations of the most important species in the sediment column and in the influent water are gathered in Table 7.7. The model included aerobic respiration, denitrification, two pathways for Fe reduction – one which released As^V to account for reduction in locations where As levels are high – and As^V reduction. A flow-through column experiment (30 cm column, 23 day experiment) was designed to mimic the percolation of water vertically through the soil. Steady state was achieved within the model timeframe and the results show oxygen depletion within the upper 12 cm, with most of the organic matter degradation occurring in this region. Overall, the model was able to simulate the measured concentrations of As, Fe, nitrate, and NOM in the effluent over the time period, and showed that after an initial depletion in dissolved NOM, the growth, respiration and decay of bacteria maintained the concentration of NOM in the system at a reasonable level.

Table 7.7 Concentrations in the soils in the region modeled by Razzak et al. [95] showing the relatively low organic content of the sediments and the differences in concentration of the column influent and effluent waters.

Constituent	Soil Conc. (mmol kg^{-1})	Influent Water (nM)	Effluent (day 23)
Fe	700	28	172 μM
Mn	21	42	NR
As	0.12	29	35 nM
Nitrate-N	1.6	NR	DL
pH	6.2	7.9	NR
POC (%)	1.95%	–	1.1 mM

Notes: NR = not reported; DL = detection limit.

This suggests that in the environment, while NOM levels may be low, the rapid biogeochemical cycling can maintain the system through these coupled cycles.

Another study focusing on deep sediments in West Bengal (10–20 m) looked at As release in lower oxygen environments, where sulfate reduction could be occurring [94]. These studies focused on identifying the organisms present based on microbial assays (total bacterial and archaeal communities by 16S rRNA gene analysis) and specifically looking for the genes associated with As and sulfate reduction (respectively, genes *arrA* and *dsrB*). Slurry incubation experiments using the sediments demonstrated that microbial activity was occurring even though the sediments had low POC (0.02% at 15 m, for example). Three experimental conditions were used: (1) sterile soils; (2) unamended; and (3) acetate amended (10 mM) and soils were handled under anoxic conditions throughout the preparation and incubations. An increase in As^{III} was initially found in the non-sterile treatments but the concentration decreased over time. A decrease in sulfate suggested that active sulfate reduction was occurring and over time it is likely that the formation of sulfide in the system and precipitation of sulfide minerals accounted for the decrease in As^{III}. Iron reduction was maximal early in the incubation and then decreased in concert with the decrease in As. These results suggest that while Fe reduction may lead to a release of As into the system, which initially occurred in the incubations, sulfate reduction can have the opposite effect, removing reduced As through precipitation of sulfide minerals [94, 96]. However, another modeling study [89] suggests that the formation of thioarsenite species under conditions of high sulfide, low Fe could be a mechanism for enhanced As mobility. This only occurs in regions where sulfate reduction is actively occurring.

It was concluded from the West Bengal study that the main As reducing bacteria were closely related to a number of *Geobacter* species, which are known to be important Fe-reducers in the environment. Sulfate reduction was likely due to the presence of *Desulfotomaculum* species, although the techniques used in this study were not sensitive enough to detect the presence of sulfide or S-minerals in the slurries. The work of others [88] have identified other organisms (e.g., *Desulfotomaculumn auripigementum, sulfurospirillum* species and *Chrysiogenes arsenatis*) that grow (respire) on As^V, and point out the difference between As reduction as a detoxification pathway and As reduction as an energy yielding process. These authors showed that As reduction could be, depending on conditions, an higher yielding and more favorable reaction than sulfate reduction.

The results of one specific set of investigations provide convincing evidence for the role of human changes on the release of As into the groundwater [93]. These studies examined the importance of various surface activities on the content and release of As into deeper groundwaters. The peak in As tended to coincide with the depth level of most groundwater wells (Fig. 7.32a). Using concentrations of various conservative and reactive elements and methane and dating of water masses using ^{14}C and tritium (^3He), these authors concluded that the groundwaters were relatively young and were derived mostly from percolation of water through organic rich, anoxic sediments of impoundment ponds, which leached reactive NOM, which during its degradation lead to the release of As. The pumping of water from groundwater wells helps speed up the transport of these surface waters to depth and results in a more rapid recycling (Fig. 7.32b–d) [93] and more sediment reduction that would have occurred prior to this changed hydrologic regime. While much of the groundwater being pumped is used to irrigate rice fields, it does not appear that the rice fields are directly related to the As release and buildup in the groundwater as most of the NOM in the sediments of these fields is recalcitrant, and not available to the microbial community.

The understanding of the factors controlling As mobilization in soils and into groundwater is rapidly evolving and many papers are being published that will refine the overlying principles discussed here. This will lead to the necessary level of understanding that is required to control this important problem. Overall, this is a complex problem, as is the other metal concerns in freshwaters highlighted already. Additionally, as discussed in the next section, there is the potential for movement of chemicals from the freshwater to the ocean via groundwater flow across the land-sea interface.

7.3.5 Metal inputs from groundwater and margin exchange processes

The interaction of the terrestrial and coastal environment is complex, and much of the focus has been on inputs from surface waters – rivers and anthropogenic sources. However, groundwater exchange is also an important process whereby terrestrial metal(loid)s are delivered to the coastal ocean, and these inputs can be greater than those due to inputs from coastal sediments. It is therefore worthwhile to examine the larger scale processes involved in groundwater inputs of metal(loid)s to the coastal and open ocean. Groundwater inputs are a significant source of water and its constituents to the coastal and open ocean, but the exact magnitude of the water input, and the fluxes of dissolved constituents is still poorly known. This input has been termed *submarine groundwater discharge* (SGD), and by definition includes all flow of water to the ocean from the continental margins, regardless of the composition or driving force [97]. Input and mixing of freshwater and seawater within the continental margin aquifer, due to convection, and wave and tidal pumping greatly increase the overall water flux. There have been many efforts to estimate the extent of this input, and its impact on the hydrological cycle [97]. The magnitude of global SGD is within the errors of the estimates of the major water fluxes at the Earth's surface (precipitation and evapotranspiration) and is estimated to be substantially less than $20 \times 10^{13}\,m^3\,yr^{-1}$ [97], but could be significant compared to global river discharge ($35–40 \times 10^{13}\,m^3\,yr^{-1}$). In contrast, by making assumptions concerning the relative magnitude of groundwater discharge to rivers compared to the associated coastal regions, Zekster et al. [98] estimated that SGD was about 5–6% of river input.

A number of other approaches have been used to estimate SGD [97, 99]. The details of these methods will not be discussed here but include the use of thermal gradients, electromagnetic techniques to measure conductivity (salinity) and/or electrical resistivity, seepage meter measurements, and tracer techniques. The geochemistry of Ra, which is produced by decay of Th *in situ*, makes it a useful tracer of SGD. As summarized by Moore [97], tracer studies using radioisotopes, particularly ^{228}Ra ($t½ = 1600\,yr$) and ^{226}Ra ($t½ = 5.7\,yr$) have provided local and regional estimates of SGD input up to $10^6\,m^3\,km^{-1}\,d^{-1}$. The short-lived Ra isotopes ^{223}Ra ($t½ = 11\,d$) and ^{224}Ra ($t½ = 3.66\,d$,) are useful for assessing mixing within the subterranean estuary. Radon (^{222}Rn; $t½ = 3.8\,d$) can also be used. It should be noted that measurements based on Ra provide an estimate of the total water input and not just the freshwater discharge.

There are different types of SGD, ranging from freshwater input that is driven purely by a hydraulic head to SGD with a strong marine component. The physical processes at the coastal margin have a large impact on the hydraulic flow and can enhance the flow due to mixing of the freshwater and seawater, driven by tidal action, changes in sediment permeability, thermal gradients, tidal differences and sea level change, and groundwater removal by humans. It has been estimated that the amount of seawater cycling through the coastal permeable sediments due to ocean physical forcing is equivalent to a few percent of the

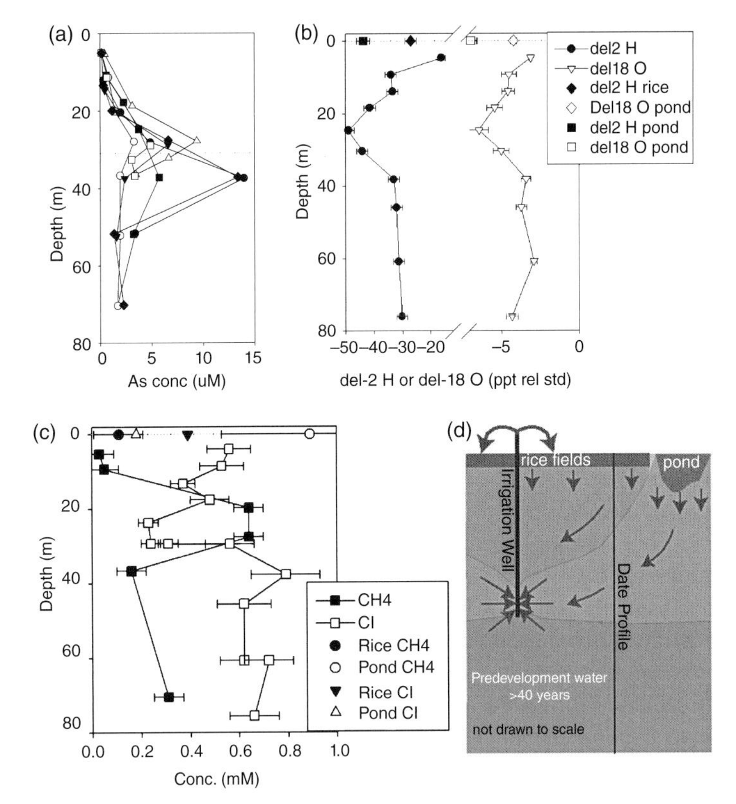

Fig. 7.32 (a) Arsenic concentrations in seven closely located well clusters; (b) the measured concentrations of isotopes (deuterium (del-2 H) and oxygen (del-18 O), both referenced against standards and plotted as ppt; (c) the measured concentrations of chloride and methane (mM for each); and (d) a physical description of recharge layering in the aquifer. In (b) and (c) the concentrations of the various species in surface pond waters and waters from rice fields is also shown. Error bars represent analytical uncertainty. More details provided in the reference. Data taken and redrawn from Neumann et al. (2010) *Nature Geoscience* **3**: 46–52 [93] with permission of Nature Publishing Group. (Original version in colour.)

freshwater input. Offshore, pressure gradients can still drive SGD through permeable shelf sediments, in waters many hundreds of meters deep. These inputs are either driven by hydraulic gradients and have a strong freshwater component or they can involve recycling of marine waters driven by thermal gradients due to geothermal heating. Deep submarine springs can be important sources of metals, such as Ba and Sr, radioisotopes (Ra) and other constituents [100]. The release of Ba is substantial and is ascribed to its release due to ion exchange reactions and/or release due to oxide dissolution [97].

As noted earlier, the water discharged to the ocean is not fresh, but a mixture of fresh and saline waters, whose composition has been altered by the mixing and biogeochemical processing of the two water end-members. Abiotic and biotic processes, such as chemical reactions, bacterially-mediated redox processes, adsorption/desorption, dissolution/precipitation, and remineralization of organic matter all occur during mixing in the subsurface aquifers, and therefore the coastal regions where these processes occurs have been termed *subterranean estuaries*. Early work focused mostly on carbonate chemistry and nutrient fluxes but more recent studies have focused on how these interactions modify the flux of metals to the coastal ocean. Such processing is important for metals, due to their redox chemistry, and their propensity to associated with organic matter and attach to particles [97].

For example, a study conducted on the shores of Long Island Sound using a series of wells along a flowpath showed that metal and other constituent concentrations changed during transport and mixing of fresh groundwater with marine waters [101]. The extent of the changes depended on season and the element or constituent. Overall, while Si and DOC appeared to behave conservatively, the concentrations of some elements, such as Co and Ni, increased by 2–10 fold, indicating sources of these metals along the flowpath.

In contrast, the nutrients (phosphate and nitrate) appeared to be removed during transit. These results indicate that much care and careful measurement is needed, and likely seasonal data is required, if representative estimates of groundwater inputs are to be generated.

Other studies of trace metal fluxes associated with SGD have also shown that many metals exhibit non-conservative behavior along the flow path from land to sea [102], reinforcing the fact that realistic estimates of metal fluxes cannot be obtained from only a few measurements of groundwater concentrations. Even with this caveat, most studies on SGD as a source of trace metals to coastal waters have concluded that the SGD source is comparable in magnitude to river or atmospheric fluxes. Additionally, it has been suggested that the degradation of NOM in coastal aquifers could contribute to the export of dissolved rare earth elements and that this flux was comparable to river inputs along the southeastern USA coastline [103].

The input of Fe from SGD is an important research area as such inputs may supply this limiting nutrient and support primary production in Fe limited regions. Studies on the southeastern Atlantic coast of Brazil [104, 105] demonstrate a SGD flux to the coastal ocean and, based on Ra tracers, suggest that the cross-shelf Fe flux was comparable to 10% of the atmospheric flux to the entire South Atlantic Ocean. While the data is limited, these results suggest that the SGD of Fe could rival its other inputs. Charette and Sholkovitz [102] however suggest that the oxidation of Fe within the subterranean estuary could scavenge metals and P in the oxic reaches before discharge to the coastal water. Recent work has examined Fe isotopes within the subterranean estuary and concluded that these were consistent with redox processing within the sediments, and that the Fe isotope signature may provide another tracer of SGD inputs [106].

Bone et al. [107] found that As, a metalloid that is often strongly associated with Fe oxide phases, was released in SGD in their study in Waquot Bay in Massachusetts, USA. Dissolved As concentration correlated with that of dissolved Fe, Mn, and P, and similarly for the solid phase, suggesting that it was being released in concert with Fe oxide dissolution [107]. Measurements of sediment concentrations showed that there were regions of substantial Fe and Mn deposition. The overall vertical distribution, with the maximum Mn content closer to the surface than that of Fe was consistent with their known redox cycling (see Fig. 7.33) [108]. Elevated concentrations of Th, As, and P coincided with the high Fe sediments while the Ba distribution was more closely aligned with that of Mn. These are consistent with their known binding characteristics to Fe and Mn oxides. It is suggested that the Fe-coated sediments act as an "iron curtain", trapping various constituents in this zone of the sediment [109]. It is likely that such regions are also important in trapping heavy metals, with the likelihood of them being trapped or released being controlled by oxygen content, pH and the organic content of the groundwater as the presence of significant binding strength in solution will result in transport of metals that associate strongly with organic matter through this zone.

Investigations of the Hg input into Waquoit Bay, Massachusetts, concluded that this flux was substantial, and was an order of magnitude greater than atmospheric fluxes to this coastal region [110]. Similarly, Laurier et al. [111] compared Hg input by SGD on the southern coast of the English Channel (the Pays de Cayx) with the input to this region from the Seine estuary. Higher concentrations of Hg in mussel tissue in the Pays de Cayx had suggested an additional input, and their study confirmed that fluxes in groundwater to the Pays de Cayx were higher than inputs from the Seine estuary [111].

Conversely, some metals will be removed under anoxic conditions and therefore the subterranean estuary could be an important sink. Uranium is an example of such a metal and Windom and Niencheski [104] examined U removal in the freshwater–sea water mixing zone of permeable sediments along the coast of southern Brazil. The depletion of U in the mouths of rivers on the southeast coast of North America has also been used to estimate the amount of sea water entering anoxic portions of the subterranean estuary, and suggested that this subterranean removal process should be considered when estimating a global budget for U and other redox-sensitive metals [112].

While the number of studies of metal inputs via SGD is still limited, it is evident that this flux is important in many cases and cannot be ignored when considering ocean budgets. More studies are clearly needed across a wide range of ecosystem types to further constrain the magnitude of the fluxes, and to obtain a clearer picture of the biogeochemical processes that alter the transport of metals from the terrestrial environment through the subterranean estuary.

7.4 Human activities and their impact on trace metal(loid) concentrations in drinking water and receiving waters

The discussions previously have provided some examples of the impact of human activity on water quality, both for surface waters and for groundwater. For example, acid mine drainage impacts water quality worldwide with the source of the acidic waters being active or abandoned mines. Additionally, the previous section discussed the impact of changes in hydrology on the concentration of As in groundwater and its impact on the population reliant on this water for drinking and irrigation. Groundwater contamination with heavy metal(loid)s and radioactive elements is again a global problem, and is not purely due to inorganic contaminants, with much impact due to the extraction of coal and other hydrocarbon fuels.

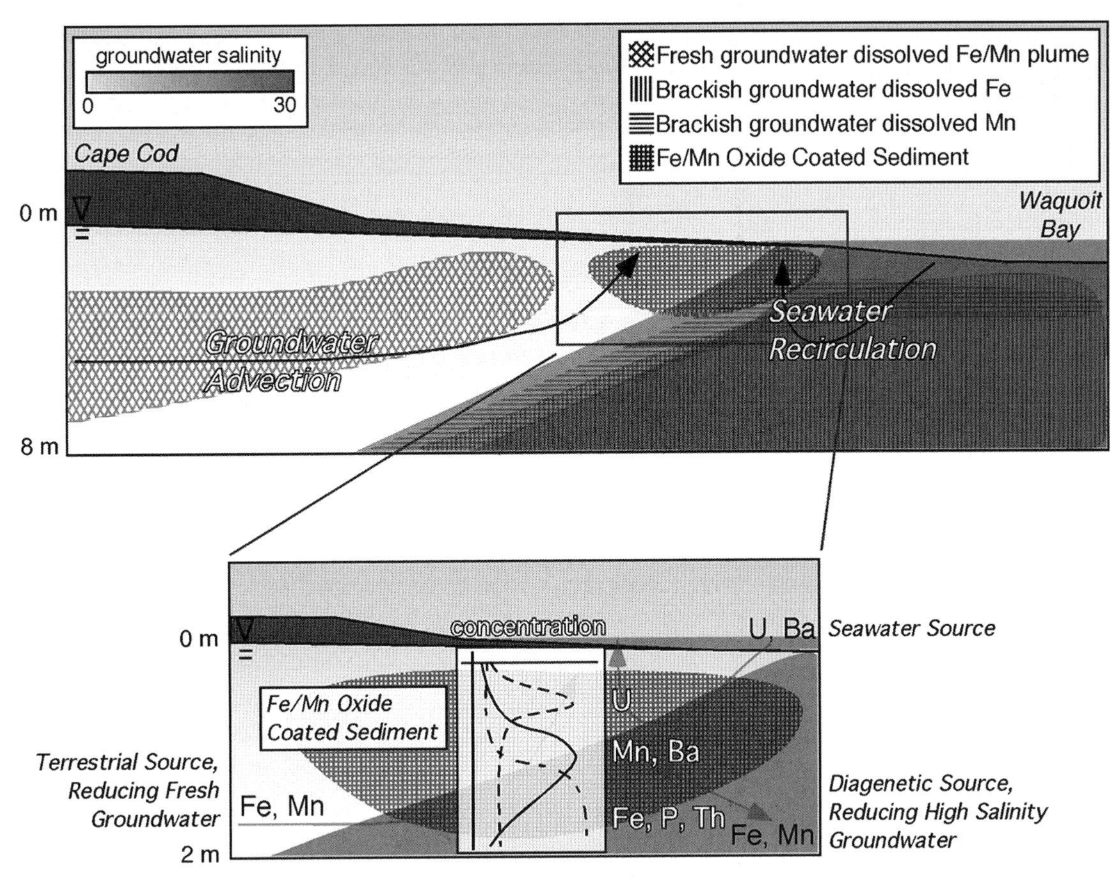

Fig. 7.33 Schematic depiction of the cycling of redox sensitive elements in the submarine estuary. Figure reprinted from Charette et al. (2005) *Geochimica et Cosmochimica Acta* **69**: 2095–109 [108], reprinted with permission of Elsevier.

In the following sections, to further discuss the impact of human activity on the aquatic environment, two other examples will be discussed. Firstly, the impact of various practices on municipal drinking water will be highlighted by the discussion of the impact of changing water chemistry on Pb in municipal drinking water. To round out the section, the impact of waste water treatment discharges on surrounding waters will be discussed through the examination of the chemistry of some heavy metals and the impact of factors in the plant on their fate and transport in the receiving waters.

7.4.1 Lead in drinking water

There have been ongoing concerns in many cities about the contamination of drinking water with Pb ([113] and references therein). This is especially prevalent in older cities which may have some Pb pipes, or have used Pb-containing solder, or have brass fittings containing Pb (often up to 2% Pb in brass). The problems seem to be partially related to changes in the disinfectants used in drinking water [113]. One recent well documented case was the high levels of Pb discovered in drinking water in Washington, DC in early 2000, which was then attributed to a change from the use of chlorine to the use of chloramines (typically NH_2Cl) as the final disinfectant for the water supply [114]. A number of studies originated from this discovery, which was manifest by high Pb levels in children's blood in the city, and was traced to the water supply through sampling. Initially, in the 1990s, many water supplies increased the water pH as a means to reduce corrosion and used free chlorine to do this. However, concern about the potentially harmful by-products of chlorine's reaction with other constituents lead to the proposed use of chloramines instead, which is less aggressive and also does not appear to enhance corrosion.

While Pb exists in the +II oxidation state in most natural waters, its most oxidized form is Pb^{IV}. Initially it was thought that the Pb-containing "scale" forming in water pipes was primarily Pb^{II}, which precipitates either as carbonate or hydroxide solid at the concentrations, pH and alkalinity of drinking waters [113]. However, further investigation, and the realization that the concentration of Pb in drinking waters was much lower than that expected at equilibrium with solid Pb^{II} phases, suggested that the Pb was likely present as the more insoluble PbO_2. Subsequently, it has been shown that the highly oxidized, low pH environments with residual oxidant that often occur in the drinking water (free chlorine or chloramines at 20–50 μM levels) result in the long term formation of PbO_2 in pipes. A number of factors appear to influence the rate of oxidation including the alkalinity, pH and presence of ligands that may stabilize Pb^{II} in solution [1, 4]. The rate of oxidation is slower at higher alkalinity and it has been shown, for example, that the presence of orthophosphate inhibits the oxidation of Pb^{II} to Pb^{IV}.

However, there are a number of mechanisms whereby the oxidized Pb^{IV} can be reduced and solubilized and this appears to be why subtle changes is water chemistry can lead to dramatic changes in Pb concentration. Natural organic matter (NOM) is known to assist the solubilization of Pb [113]. It is possible that the Pb particles on the surfaces, which have been shown to exist mostly as microcrystals (nanoparticles; <100 nm in diameter) in these environments, are coated and mobilized by the presence of the NOM, in the same way that NOM stabilizes other nanoparticles, as discussed in Section 6.2.3. Reductive dissolution can also occur in the presence of reduced Fe and Mn, but oxidation of the Fe, and precipitation of Fe^{III} (hydr)oxides can lead to a coating of the surface and inhibition of the reaction [115]. This is not the case with Mn oxidation.

Furthermore, it was shown in one study [116] that while the rate of dissolution of PbO_2 was slower in the presence of 40 μM of OCl^- or NH_2Cl than in their absence, the rate of dissolution was much slower in the presence of the free chlorine. Thus, these results suggest that changes in the disinfectant used could have a large impact on the rate of Pb release into solution, especially in water in stagnant pipes. Indeed, these authors also demonstrated that the rate of oxidation increased as the OCl^- concentration decreased, suggesting that its presence acts as an inhibitor of the dissolution of the oxide. These authors noted that the overall redox potential of the system is controlled by the oxidation reactions that are occurring:

$$2Pb^{2+} + 4H_2O = 2PbO_2(s) + 8H^+ + 4e^- \quad \log K_{Pb} = -99,\ E_H^0 = -1.46\ V$$

$$O_2(aq) + 4H^+ + 4e^- = 2H_2O \quad \log K_{O2} = 86,\ E_H^0 = 1.27\ V$$

$$2NH_2Cl + 4H^+ + 4e^- = 2Cl^- + 2NH_4^+ \quad \log K_{NH2Cl} = 95,\ E_H^0 = 1.40\ V$$

$$2HOCl + 2H^+ + 4e^- = 2Cl^- + 2H_2O \quad \log K_{HOCl} = 100,\ E_H^0 = 1.48\ V$$

From these equations, and the E_H^0 values, which refer to standard conditions, it is apparent that small changes in chemistry and the binding of Pb in solution, or the pH, chloride concentration and other variables could have a large impact on which reaction is favorable and likely provides the reasoning why changes in the disinfectant type, from OCl^- to NH_2Cl, could result in large changes in the dissolved Pb concentration. It is also possible that in the environment of water pipes the system could be far from equilibrium and microenvironments could develop in the absence of flowing water to enhance the reactions.

It has been further suggested that fluoride treatment of the water can also have a negative impact [117]. These authors state that the addition of fluosilicic acid, a common additive to drinking water in place of F^-, can exacerbate the dissolution of Pb by forming a stable Pb complex and that this process is enhanced when chloramines are used, through the enhancement of reaction by traces of ammonia. These authors also claim that the presence of traces of ammonia can enhance the dissolution of copper from fittings and therefore also enhance the Cu contamination of the water supply.

One additional pathway that has been discussed in the literature is the formation of galvanic cells in the presence of free chlorine or other oxidants in solution, and the presence of Pb-containing brass fittings [118]. These studies have shown it is possible for the elemental Pb to be oxidized and be solubilized, acting as the cathode, with the copper pipe acting as an anode, and that the formation of dissolved Pb hydroxides can lead to a decrease in pH that can overall exacerbate the reaction. However, research suggests that passivation of surfaces due to precipitation can reduce the extent of this process and overall the relative importance of electrochemical processes in this environment is not well characterized. It is possible that these reactions are very localized and can occur close to the source at which the drinking water is collected due to differences in the materials used in the construction between the pipes and the household fittings.

Overall, this example further demonstrates the complexities that exist in natural waters and in other aqueous media, and that when dealing with elements that are present in trace quantities there are often very small changes in the overall chemistry of the medium that can have a large impact on the concentration and form of the metal(loid) in solution, and can have a large environmental impact. The discussion in this section also further cautions that thermodynamic predictions, while valuable, are not often a reasonable representation of the actual environment and that studies are needed to evaluate and confirm the results of such model predictions.

7.4.2 Metal inputs from waste water treatment facilities and other industrial discharges

Studies of the cycling of metals in sewage treatment plant facilities have focused on various aspects, such as the relative removal of metals across the plant, the locations within the plant where removal occurs, and the factors controlling the speciation and concentration of metals in the effluent. Studies have also examined the role of organic complexation and the potential for effluent metal association with sulfide colloids (nanoparticles) for metals in effluents. We will examine these factors in detail in the following sections. Again, these studies illustrate the chemical reactions and processes involved in the fate and transport of metal(loid)s in freshwater ecosystems.

Sewage treatment facilities or waste water treatment plants (WWTP's) are ubiquitous throughout the world and are necessary for the removal of nutrients, inorganic and organic contaminants, and bacteria and other microbes that can contaminate surface waters and impact humans and other wildlife. Many facilities must deal with a large range in influent concentrations and often the capability of a facility to remove a particular element or constituent between influent and effluent is a function of the input concentrations. A comprehensive study by Shafer et al. [119] provides a good illustration of the ability of WWTPs to remove metals. Highest removal appears to be for those elements that bind strongly to solids (Fig. 7.34a) (e.g., Ag, Fe, Al, Cu, Pb) and there is a strong correlation between the extent of removal and the K_D of the metal within the systems examined (Fig. 7.34b). Removal of Hg is also typically >90% [120]. Interestingly, Cr has a high removal compared to the other oxyanions (V, As, Mo) which are not strongly retained by the plant. This may be due to changes in oxidation state within WWTPs as Cr^{III} is much less soluble than Cr^{VI}. The weakly binding metals, such as Cd and Zn, are not retained to a large degree by WWTPs. Overall, Cr, Cu, and Pb are retained to a higher degree than predicted based on K_D, while Mo, V, and As are retained to a lower degree compared to expectation (Fig. 7.34b). Silver appears to be an enigma as its retention is less than that predicted based on K_D.

Studies of Ag binding in effluents reveal that the conditional binding constants ($K = [ML^{n-1}]/[M^{n+}][L^-]$) are on the order of 10^{12} for these systems [121]. A value of this magnitude is high for Ag relative to its binding to inorganic ligands and other organic ligands, except those containing thiol groups. Similarly, the study of Van Veen et al. [122] found a relatively high binding constant for Cu for WWTP effluents. The results of Hsu and Sedlak [123] indicate a binding strength for Hg of greater than 10^{25}. Lamborg et al. [124] estimated the binding constant for one effluent to be around 10^{23}.

The study of Hsu-Kim and Sedlak [120] showed that the Hg binding ligands formed in WWTPs were stable to oxidation and were similar in behavior to compounds formed between Hg and S-containing organic molecules. However, they were not as labile as Hg complexes formed with NOM, suggesting that the associations were not just those between Hg and the thiol groups of NOM. The strongly Hg-bound complexes, in similar fashion to metal-sulfide complexes and clusters, were destroyed by chlorine disinfection, but

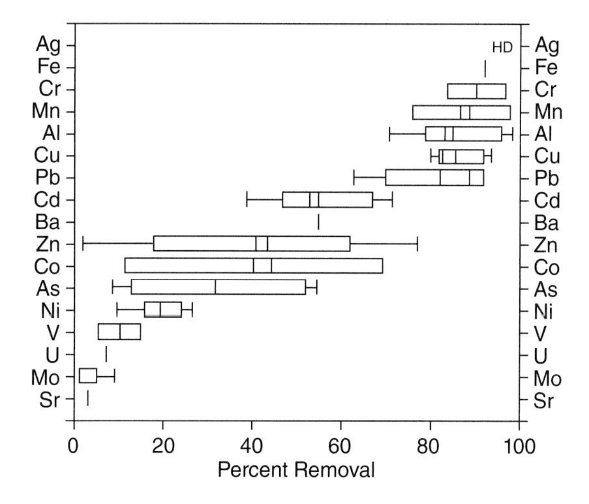

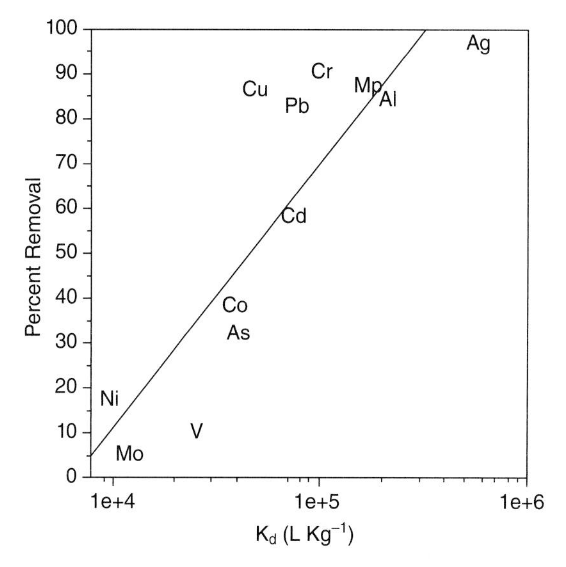

Fig. 7.34 Relationship between removal of elements over a sewage treatment plant and the degree of partitioning to the metals to suspended solids. Figure taken from Shafer et al. (1998) *Environmental Toxicology and Chemistry* **17**: 630–41 [119] with permission of John Wiley & Sons, Inc.

were relatively stable in oxygenated solution showing loss of up to 75% of the ligands over a two week period. A study of Van Veen et al. [122] similarly showed that the copper (Cu) complexing capacity of effluent changed slowly over a 20-day period and that this complexing ability still remained in some effluents after this time period (Fig. 7.35). In addition, Van Veen's study showed that mixing of the effluent and receiving waters in various dilution ratios did not eliminate the Cu complexing capacity – to the contrary, the amount of complexation after mixing was greater than would be predicted based on conservative mixing, suggesting that there is excess complexation capacity in effluents above that of the metals present. Overall, even though 10–70% of the Cu in the different effluents was in the dissolved

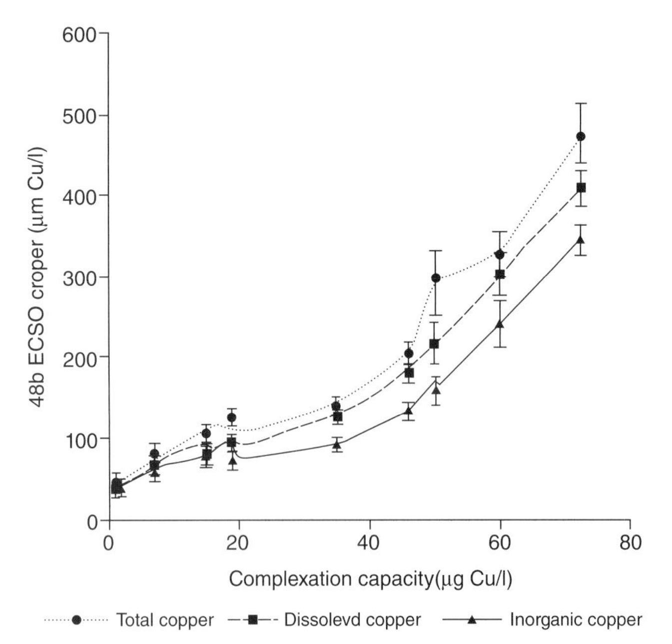

Fig. 7.35 Toxicity data for experiments with *Daphnia* and effluents and their mixture with receiving waters showing the impact of mixing on toxicity. Figure reprinted from Van Veen et al. (2002) *Environmental Toxicology and Chemistry* **21**: 275–80 [122], with permission of John Wiley & Sons, Inc.

fraction, calculations indicate that very little of the filterable Cu was present as inorganic Cu complexes.

These results suggest that the metal complexing ligands in WWTP effluents are reacting through S-linkages, and likely involving NOM as well, and are forming complexes or associations including metal, S and NOM that are relatively stable against degradation or ligand exchange reactions. The most logical conclusion is that these entities involve metal-sulfide clusters stabilized by NOM as such agglomerates would be stable to oxidation and persist in receiving waters, as discussed in Section 6.2.3. It is not clear whether metals in such associations are bioavailable.

The study of Shafer et al. [119] also investigated the fate of Ag upon mixing of the effluent with the receiving waters. In this instance, the effluent had a conductivity of about three times the receiving waters, pH and NOM were similar, but the dilution ratio was about 10 : 1. Therefore the mixing should not have changed the chemistry of the receiving waters in any substantial way. However, they still found non-conservative behavior and a rapid removal of the Ag immediately downstream of the outlet (within 100 m), with losses in a range of 30–60%. High Ag sediment concentrations confirmed that removal of Ag occurred, most probably due to coagulation and sinking. Measurements in the water column indicated a rapid increase in particulate Ag and a decrease in filterable Ag, consistent with the other data. Thus, their study appears to provide a case where differences between the effluent and receiving waters, although not apparent in the measurements made, lead to a rapid coagulation and removal of Ag to the sediment downstream of the mixing zone. Clearly, more investigation of the possible reasons for coagulation or flocculation, or the opposite, particulate disintegration, need to be examined in order to fully understand the fate of Ag and other particulate-reactive metals in effluents and in receiving water. Data is needed for a variety of different mixing regimes if the complexities of these interactions are to be fully understood.

While Shafer et al. [119] found non-conservative behavior, the experiments of Van Veen et al. [122] found conservative behavior in their laboratory-based mixing experiments. These authors studied the fate of Cu in WWTP effluents upon mixing with receiving waters, the complexing capacity of the effluents for Cu, and that of mixtures of effluent and receiving waters. They also conducted toxicity assays with the effluents themselves and with mixtures of the effluents and receiving waters. As shown in Figure 7.35, the toxicity (EC50) increased with increasing complexing capacity of the waters and mixtures so it was evident that the complexed Cu was not bioavailable to the test organisms – *Daphnia*. These results confirm the idea that the metals bound up in colloidal fractions, or bound to NOM are not as bioavailable as the inorganically bound metal or the free metal ion. This study also showed an "excess" complexing capacity in the effluents. Thus it is possible that the mixing of the effluents and the receiving waters could lead to an overall decrease in the bioavailability of the mixture compared to even that of the receiving waters.

7.5 Metal stable isotopes and their use

Radioisotopes, as discussed in Section 6.2, have long been used as tracers of a variety of processes, such as the removal of particles from the surface ocean, ocean circulation patterns, fluxes at the sediment-water interface and inputs to the ocean from groundwater and other sources. They have also been widely used for age dating of sediments and other environmental matrices. In contrast, stable isotopes of heavier elements have only been used more recently as tracers and for study of environmental processes. The stable isotopes discussed here are those that do not occur as a result of radiochemical decay on the Earth. The stable isotope ratios of the lighter biochemical elements (C, N, O, S) have been used to examine chemical and biological processes and the fate and transport and food chain dynamics of ecosystems for many decades. However, the heavier elements, such as Fe, Mo, and Hg, have only recently been used for such investigations primarily because of the prior lack of instrumentation with sufficient sensitivity and resolving power to resolve the small isotopic differences found in nature. The recent advances in instrumentation, primarily through the development and rigorous evaluation of multicollector ICP-MS (MC-ICP-MS) for isotope analysis, is now allowing analysis of heavy element isotopes with sufficient precision [125] that the relatively small changes in fractionation can be relatively easily detected on a small enough sample that detailed experiments and extensive field measurements are now possible. The development and testing of the techniques for isotope measurement is crucial as the processes involved in ICP-MS, such as ionization of the sample, can lead to substantial fractionation. Therefore, in most instances, the extent of fractionation is referenced to that of a standard so that the instrument effects are effectively eliminated.

In reporting the isotope fractionation, it is customary to report values relative to a reference material whose isotope fractionation is well characterized. The values are reported as "per mil" (because the difference is multiplied by 1000) and the notation used is $\delta^{xxx}E$ where xxx is the atomic number of the element, E. The formulation, using Fe as an example, is:

$$\delta^{56}Fe = [((^{56}Fe/^{54}Fe)_{sample}/(^{56}Fe/^{54}Fe)_{standard}) - 1] \times 1000 \qquad (7.2)$$

Many trace metals have now been shown to fractionate during chemical and biological processing and there is an explosion of studies examining the extent of the fractionation for all the potential processes involved. Such metal fractionation can potentially be used as a "biosignature" of its incorporation into biological tissue, and they are also indicators of chemical and biological processing. The formation and destruction of chemical bonds during redox reactions, precipitation and dissolution, adsorption and desorption, and the impacts of mass differences on rates of transport in aquatic media and across membranes, all result in fractionation [125, 126]. The overall reaction pathway can have an impact on the degree of fractionation and the type of complexes formed in solution will also influence any exchange reactions.

It can be shown that the strength of the bonds in molecules is dependent on the isotope mass and therefore the resultant rates of reactions are mass dependent. *Kinetic isotope effects* result when reactions are unidirectional, but mass effects also impact the magnitude of the equilibrium constant in reversible reactions, and this is termed the *equilibrium isotope effect* [125]. Isotope fractionation will occur in the environment even if equilibrium is not achieved. For physical processes, such as molecular diffusion, the impact of isotope mass is more apparent.

Stable isotope fractionation is also a potential paleoproxie. While traditional stable isotopes such as Rb/Sr and Sm/Nd - daughters of radioactive decay – and those of the U isotope system (the U/Th/Ra/Rn/Pb decay series) have been used as proxies for ocean temperature, carbon cycling, ocean circulation, inputs, weathering, and other processes, it is now apparent that isotope ratios of single elements may also be used. This is especially so for the "bio-essential" metals which are incorporated into biomolecules and enzymes. In oceanography, Fe and Mo are currently the elements of choice for paleo studies as their fractionation can be used for tracking oxygen conditions and metal transformations in the "paleo-ocean" [125], given the importance of trace metals in the evolution of life, as discussed in Chapter 8. The increasing use of many essential metals (e.g., Fe, Mo, Mn, and Cu), and now their isotopes, to address paleo questions, such as Earth redox changes over geological time, is obvious given that their concentrations are markedly different in oxic versus anoxic environments [126–31].

Iron and Mo are a good comparative pair of metals to examine as they have very different chemistry in seawater, and their residence time in water is very different [125]. While the ocean concentration of Fe is very low, Mo is abundant and has a uniform distribution in the ocean, but can be present in low concentrations in freshwaters, and is removed in anoxic environments. Fe has four isotopes (see Table 7.8) and is fractionated by a wide variety of processes while the processes contributing to Mo fractionation relate mostly to its adsorption/desorption from Mn oxides in ocean sediments, and its incorporation into solids in the presence of sulfides.

It is likely that the isotope ratios of a number of other biometals will also be used in an oceanographic context and in freshwater studies, and there is already study of Cu and Zn. In each case, it is necessary to examine the processes important in their fractionation in detail as the extent of fractionation cannot be yet accurately predicted ([125] and references therein). For Zn, the measurements and the results of laboratory studies examining uptake into phytoplankton and adsorption reactions appear inconsistent with theory and more work is needed to ascertain what processes are most important. Overall, it appears that biological fractionation is more important than chemical processes for Zn. The fractionation of Cu during hydrothermal processing has received substantial study to date and it appears that fractionation is greatest under conditions where processes are occurring at elevated temperatures. Redox transformations of Cu can also result in relatively large fractionations and these are greater than found during biological uptake or ligand exchange and other chemical processes.

In most cases discussed so far, fraction is *mass dependent* (MDF) in that the extent and direction of the fractionation can be predicted based on the known mass differences between the isotopes. For example, reactions or membrane transport processes that are diffusion or kinetically controlled will lead to the depletion of the light isotope relative to the heavier isotope in the remaining medium, and the product will be enriched in the lighter isotope. However, it has been found that for reactions that involve radicals, and mostly those that form a reactive intermediate complex, can

Table 7.8 Typical extent of mass dependent fractionation of various trace elements for various processes in aquatic systems. Compiled from the literature, primarily [125, 130].

Element	Isotope Mass (Abundance)	Sources of Fractionation	Fractionation Range (‰)
Fe	54 (5.84%), 56 (91.7%), 57 (212%), and 58 (0.28%)	Chemical and biological redox processes; kinetic, equilibrium effects; exchange reactions	Widely found; ~5 ‰ for δ^{56}Fe
Mo	92 (14.84%), 94 (9.25%), 95 (15.92%), 96(16.68%), 97 (9.55%), 98 (24.13%), and 100 (9.63%)	Adsorption/ desorption to oxides; redox fractionation is small	~1 ‰ for $\delta^{97/95}$Mo
Zn	64 (48.63%), 66 (27.90%), 67 (4.10%), 68 (18.75%), and 70 (0.62%)	Mechanisms uncertain; biological processing mainly; little redox chemistry	~1 ‰ for δ^{66}Zn
Cu	63 (69.17%) and 65 (30.83%).		~10 ‰ for δ^{65}Cu
Hg		Chemical and biological reduction, volatilization	1–3‰ for δ^{202}Hg
Tl	203Tl (29.5%) and 205Tl (70.5%)		

result in *mass independent fractionation* (MIF). In this case it is mostly some aspect of the atom's nuclear chemistry, due to the relative proton to neutron mass that determines the extent of fractionation by enhancing or decreasing the energy and stability of the reactive intermediate. Therefore the products can become enriched in the isotopes that form the more stable intermediates [133]. Mass independent fractionation was first discovered for O and S, and has been used in studies of cosmochemistry, paleo studies, atmospheric chemistry and biochemistry. To date, the only heavy element for which MIF is known and studied is for Hg. The main effect thought to contribute to the MIF for Hg is the *magnetic isotope effect*, which results from the fact that the odd isotopes have non zero nuclear spin and nuclear magnetic moments, which interact with the electrons and affects the stability of the excited intermediate state in radical reactions. While the overall effects of these interactions are not clearly understood, this is one mechanism whereby MIF could occur during a photochemical reaction. Another effect, termed the *nuclear volume effect*, results from the fact that the nuclear charge radius and volume are not directly related to mass, and varies between odd and even isotopes. The odd numbered Hg isotopes (199 and 201) behave differently from the even isotopes (196, 198, 200, 202, and 204) [133]. Thus, reactions involving Hg can lead to both mass dependent and mass independent fractionation and this provides more diagnostic information, and can enhance the ability to differentiate between processes. The reporting of MIF is done also as a ratio but in this case it is referenced relative to the theoretical value that would have occurred purely due to MDF:

$$\Delta^{xxx}\text{Hg} = \delta^{xxx}\text{Hg} - (\delta^{202}\text{Hg} \cdot \beta_{xxx}) \tag{7.3}$$

where xxx is the mass of the Hg isotope and β_{xxx} is the kinetic or equilibrium fractionation for that isotope. For MIF, errors are smaller than for MDF as the errors are independent of the instrument bias as this is primarily related to MDF fractionation [132].

For Hg, the extent of fractionation for various processes is shown in Fig. 7.36. As with the other metals, most kinetically-controlled processes, such as physical exchange (volatilization, evaporation) or chemical (photo) or biological reduction, follow MDF and the Hg^0 product is kinetically lighter. There is some evidence for equilibrium controlled fractionation and for impact of speciation on the degree of fractionation. Fraction is typically 1–3 per mil (Table 7.8). In contrast, MIF has only been demonstrated for a few processes with the most investigated being photochemical reduction. High fractionation factors can occur, especially if a large fraction of the ionic Hg is reduced. As the product Hg^0 can be lost from the system, and most laboratory studies have used this experimental mode of stripping the product from solution, the extent of fractionation is much greater, especially compared to natural systems where photochemical oxidation reactions may also be occurring. In addition, MIF has been observed during the photodegradation of CH_3Hg.

Recent studies have shown that Fe isotopes are useful tracers of biogeochemical redox cycling of Fe in both the modern ocean, as well as being a paleoproxie of changes in ancient marine environments. This is because both biotic and abiotic redox processes fractionate the Fe isotopes. Measurable fractionation has been demonstrated during dissimilatory Fe^{III} reduction, anaerobic photosynthetic and abiotic Fe^{II} oxidation, and precipitation of Fe oxides and sorption of aqueous Fe(II) onto ferric hydroxides ([125] and references therein). Fractionation can also occur during the precipitation and dissolution of other Fe minerals, and during the interaction and solubilization of Fe^{III} by organic ligands.

Investigations of different Fe deposits have been used for paleo studies. Fe in ocean Fe/Mn crusts can be used to investigate the changes in ocean chemistry through the Cenozoic. It is assumed that when Fe precipitates there is little fractionation and thus the signature in the solids

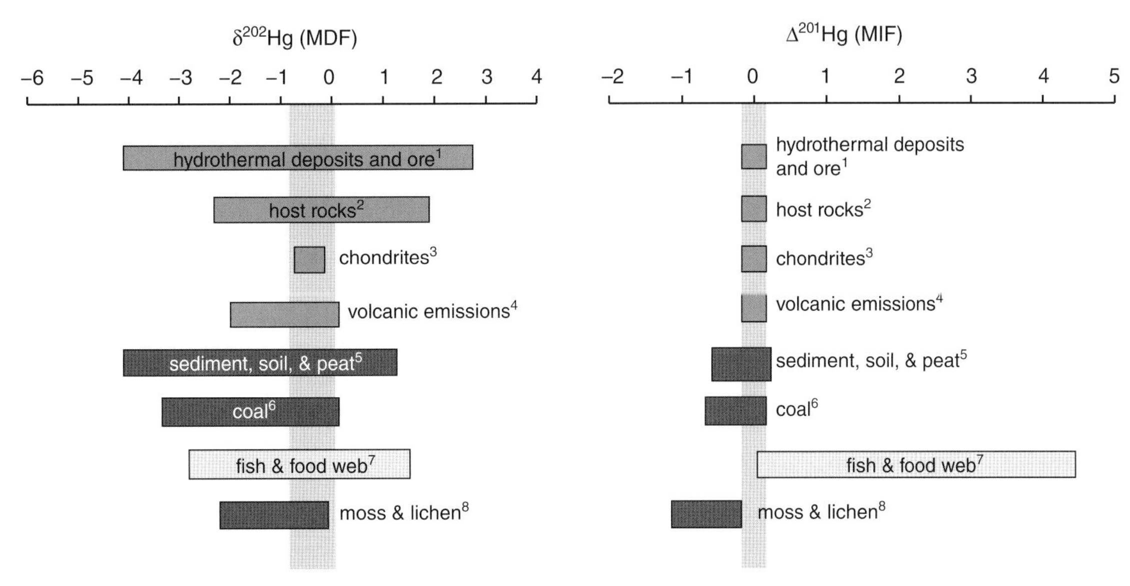

Fig. 7.36 Typical ranges in the fractionation of mercury isotopes during various biogeochemical processes for both mass dependent and mass independent fractionation. Figure taken from Bergquist and Blum (2009) *Elements* **5**: 353–7 [132] and used with permission of the Mineralogical Society of America.

reflects the historic signal. To track further back in time, use is made of shale, which has been shown to accurately reflect the three major stages of the Earth's ocean, which evolved from a Fe-rich environment through a more S-rich stage to the current oxygenated ocean, as described in more detail in Section 8.3.4. Finally, banded Fe deposits are thought to accurately reflect changes in ocean biogeochemistry over time. Similarly, the isotopic signature of Mo has been investigated in these Fe deposits and generally reflects a similar signature and interpretation to that of the Fe isotopes.

7.6 Chapter summary

1. Metal(loid) concentrations in freshwaters can vary widely depending on the type of system and the watershed that is associated with the water body. Overall, the dissolved concentrations of metal(loid)s that are strongly associated with particulate material are depleted in waters relative to those that form oxyanions or partition weakly, and are relative to their crustal abundance.

2. Metal(loid) cycling in lakes can be strongly influenced by the season changes in bulk water chemistry due to physical stratification and oxygen depletion. Metal(loid)s that form strong associations with oxide phases can be transported and released in conjunction with the precipitation and dissolution of these phases. Similarly, association with reduced sulfur phases and organic matter can strongly influence the fate and transport of metal(loid)s in freshwater.

3. The modeling approaches discussed in earlier chapters have been used to successfully describe the distribution and cycling of metals in freshwater systems.

4. The cycling of mercury in lakes is an important topic given the bioaccumulation of CH_3Hg in fish in these systems. The major factors influencing the formation and fate of CH_3Hg were discussed as well as the cycling and sources of Hg to these systems.

5. The controls over metal(loid)s in rivers and groundwater were highlighted and the role of colloids and organic matter in the transport of these elements was discussed. The importance of the watershed characteristics and the flow regime in determining metal(loid) concentrations in these systems was discussed. The role of photochemistry in metal(loid) cycling was also highlighted.

6. The importance of the regions of exchange of groundwater with the surface waters, either within rivers or at the land-sea interface, was highlighted with focus on the role of redox conditions and other factors on their transport or retention within these zones. Methods of tracking the mixing and exchange at the land-sea interface were discussed as well as the potential importance of this process as a source of trace elements to coastal waters.

7. The chapter focused on a variety of topics including the importance of acid mine drainage and associated processes on stream and river chemistry and the importance of human influences on the concentration of arsenic in groundwater wells in Asia. The complexities of the As situation highlights the manner in which subtle changes in organic matter supply and redox dynamics can have a large impact on the concentration of trace elements in aquatic ecosystems.

8. A focus topic was the impact of changes in water treatment procedures on the concentrations of trace metal(loid)s in freshwaters. The problem of elevated Pb in drinking water due to its release with redox changes was highlighted. In addition, there was a discussion of the impact of waste water treatment plants on the receiving waters.

9. The final section of the chapter discussed the emerging discipline of using the stable isotopes of trace metal(loid)s to unravel the chemistry of the past and to understand the sources, sinks and cycling of trace elements in environmental systems.

References

1. Stumm, W. and Morgan, J. (1996) *Aquatic Chemistry*. John Wiley & Sons, Inc., New York.
2. Morel, F.M.M. and Hering, J. (19963) *Principles and Applications of Aquatic Chemistry*. John Wiley & Sons, Inc., New York.
3. Sigg, L. and Behra, R. (2005) Speciation and bioavailability of trace metals in freshwater environments. In: Siegel, A., Siegel, H. and Sigel, R.K.O. (eds) Metal Ions in Biological Systems. *Biogeochemistry, Availability and Transport of Metals in the Environment*. Vol. 44. Taylor & Francis, Boca Raton, pp. 47–73.
4. Langmuir, D. (1997) *Aqueous Environmental Geochemistry*. Prentice Hall, Upper Saddle River, NJ.
5. Adriano, D.C. (2001) *Trace Elements in Terrestrial Environments*. Springer-Verlag, New York.
6. Gaillardet, J., Viers, J. and Dupre, B. (2004) Trace elements in river water. In: Drever, J.I. (ed.) *Surface and Ground Water, Weathering and Soils*. Vol. 5. Treatise on Geochemistry, H.D. Holland and K.K. Turekian (Exec. eds). Elsevier, Amsterdam, pp. 225–271.
7. Scudlark, J.R. and Church, T.M. (1997) Atmospheric deposition of trace elements to the Mid-Atlantic Bight. In: Baker, J.E. (ed.) *Atmospheric Deposition of Contaminants to the Great Lakes and Coastal Waters*. SETAC Press, Pensacola, Florida, USA, pp. 195–208.
8. Wetzel, R.G. (2001) *Limnology*. Academic Press, San Diego.
9. Balistrieri, L., Murray, J.W. and Paul, B. (1994) The geochemical cycling of trace elements in a biogenic meromictic lake. *Geochimica et Cosmochimica Acta*, **58**(19), 3993–4008.
10. Viollier, E., Jezequel, D., Michard, G., Pepe, M., Sarazin, G. and Alberic, P. (1995) Geochemical study of a crater lake (Pavin Lake, France): Trace-element behavior in the monimollimnion. *Chemical Geology*, **125**, 61–72.
11. Achterberg, E., van der Berg, C.M.G., Boussemart, M. and Davison, W. (1997) Speciation and cycling of trace metals in Esthwaite water: A productive English lake with seasonal deep-water anoxia. *Geochmica et Cosmochimica Acta*, **61**(24), 5233–5253.
12. Dzomback, D.A. and Morel, F.M.M. (1990) *Surface Complexation Modeling: Hydrous Ferric Oxide*. John Wiley & Sons, Inc., New York.
13. Bura-Nakic, E., Viollier, E., Jezequel, D., Thiam, A. and Ciglenecki, I. (2009) Reduced sulfur and iron species in anoxic water column of meromictic crater Lake Pavin (Massif Central, France). *Chemical Geology*, **266**, 311–317.
14. Alberic, P., Viollier, E., Jezequel, D., Grosbois, C. and Michard, G. (2000) Interactions between trace elements and dissolved organic matter in the stagnant anoxic deep layer of a meromictic lake. *Limnology and Oceanography*, **45**(5), 1088–1096.
15. Cossa, D., Mason, R.P. and Fitzgerald, W.F. (1994) Chemical speciation of mercury in a meromictic lake. In: Watras, C.J. and Huckabee, J.W. (eds) *Mercury Pollution: Integration and Synthesis*. Lewis Publishers, Boca Raton, pp. 57–67.
16. Dyrssen, D. and Wedborg, M. (1991) The sulfur-mercury(II) system in natural-waters. *Water, Air, and Soil Pollution*, **56**, 507–519.
17. Xue, H.B. and Sigg, L. (1999) Comparison of the complexation of Cu and Cd by humic or fulvic acids and by ligands observed in lake waters. *Aquatic Geochemistry*, **5**, 313–335.
18. Smith, D.S., Bell, R.A. and Kramer, J.R. (2002) Metal speciation in natural waters with emphasis on reduced sulfur groups as strong metal binding sites. *Comparative Biochemistry and Physiology C-Toxicology & Pharmacology*, **133**, 65–74.
19. Hamilton-Taylor, J., Postill, A.S., Tipping, E. and Harper, M.P. (2002) Laboratory measurements and modeling of metal-humic interactions under estuarine conditions. *Geochimica et Cosmochimica Acta*, **66**, 403–415.
20. Rickard, D. and Luther, G.W. (2007) Chemistry of iron sulfides. *Chemical Reviews*, **107**(2), 514–562.
21. Lau, B.L.T. and Hsu-Kim, H. (2008) Precipitation and growth of zinc sulfide nanoparticles in the presence of thiol-containing natural organic ligands. *Environmental Science & Technology*, **42**(19), 7236–7241.
22. Deonarine, A. and Hsu-Kim, H. (2009) Precipitation of mercuric sulfide nanoparticles in NOM-containing water: Implications for the natural environment. *Environmental Science & Technology*, **43**(7), 2368–2373.
23. Benoit, J.M., Gilmour, C.C., Heyes, A., Mason, R.P. and Miller, C.L. (2003) Geochemical and biological controls over methylmercury production and degradation in aquatic systems. In: Cai, Y. and Braids, O.C. (eds) *Biogeochemistry of Environmentally Important Trace Elements*. Vol. 835. ACS Symposium Series. American Chemical Society, Washington, D.C., pp. 262–297.
24. Lawson, N.M. and Mason, R.P. (2001) Concentration of mercury, methylmercury, cadmium, lead, arsenic, and selenium in the rain and stream water of two contrasting watersheds in Western Maryland. *Water Research*, **35**(17), 4039–4052.
25. Jardim, W.F., Bisinoti, M.C., Fadini, P.S., et al. (2010) Mercury redox chemistry in the Negro River Basin, Amazon: The role of organic matter and solar light. *Aquatic Geochemistry*, **16**(2), 267–278.
26. Hudson, R.J.M., Gherini, S., Watras, C.J., et al. (1994) Modeling the biogeochemical cycling of mercury in lakes. In: Watras, C.J. and Huckabee, J.W. (eds) *Mercury as a Global Pollutant: Towards Integration and Synthesis*. Lewis, Boca Raton, pp. 473–526.
27. Fleming, E.J., Mack, E.E., Green, P.G. and Nelson, D.C. (2006) Mercury methylation from unexpected sources: Molybdate-inhibited freshwater sediments and an iron-reducing bacterium. *Applied and Environmental Microbiology*, **72**(1), 457–464.
28. Kerin, E.J., Gilmour, C.C., Roden, E., Suzuki, M.T., Coates, J.D. and Mason, R.P. (2006) Mercury methylation by dissimilatory iron-reducing bacteria. *Applied and Environmental Microbiology*, **72**(12), 7919–7921.

29. Barkay, T., Miller, S.M., Summers, A.O., et al. (2003) Bacterial mercury resistance from atoms to ecosystems. *FEMS Microbiology Reviews*, **27**(2–3), 355–384.
30. Jeremiason, J.D., Kanne, L.A., Lacoe, T.A., et al. (2009) A comparison of mercury cycling in Lakes Michigan and Superior. *Journal of Great Lakes Research*, **35**(3), 329–336.
31. Fitzgerald, W. and Watras, C.J. (1989) Mercury in surficial waters of rural Wisconsin lakes. *The Science of the Total Environment*, **87/88**, 223–232.
32. Fitzgerald, W.F., Mason, R.P., and Vandal, G.M. et al.(1991) Atmospheric cycling and air-water exchange of mercury over mid-continental lacustrine regions. *Water, Air, and Soil Pollution*, **56**, 745–767.
33. Hurley, J.P., Watras, C.J., and Bloom, N.S. et al.(1991) Mercury cycling in northern Wisconsin seepage lakes: The role of particulate matter in vertical transport. *Water, Air, and Soil Pollution*, **56**, 543–551.
34. Hurley, J.P., Watras, C.J., and et al.Bloom, N.S. (1994) Distribution and flux of particulate mercury in four stratified seepage lakes. In: Watras, C.J. and Huckabee, J.W. (eds) *Mercury Pollution: Integration and Synthesis*. Lewis Publishers, Boca Raton, pp. 69–82.
35. Watras, C.J., Bloom, N.S., Hudon, R.J.M., et al. (1994) Sources and fates of mercury and methylmercury in Wisconsin lakes. In: Watras, C.J. and Huckabee, J.W. (eds) *Mercury Pollution: Integration and Synthesis*. Lewis Publications, Boca Raton, pp. 153–177.
36. Lindqvist, O., Johansson, K., and Astrup, M., et al. (1991) Mercury in the Swedish environment-Recent research on causes, consequences and corrective methods. *Water, Air, and Soil Pollution*, **55**, 1–261.
37. Verta, M. and Matilainen, T. (1995) Methylmercury distribution and partitioning in stratified Finnish forest lakes. *Water, Air, and Soil Pollution*, **80**, 585–588.
38. Watras, C.J., Bloom, N.S., Claas, S.A., et al. (1995) Methylmercury production in the anoxic hypolimnion of a dimictic seepage lake. *Water, Air, and Soil Pollution*, **80**, 735–745.
39. METAALICUS (2010) *The METAALICUS project*. Available at http://wi.water.usgs.gov/mercury/metaalicus-project.html (accessed October 11, 2012).
40. Harris, R.C., Rudd, J.W.M., Amyot, M., et al. (2007) Whole-ecosystem study shows rapid fish-mercury response to changes in mercury deposition. *Proceedings of the National Academy of Sciences of the United States of America*, **104**(42), 16586–16591.
41. Hintelmann, H., Harris, R., Heyes, A., et al. (2002) Reactivity and mobility of new and old mercury deposition in a Boreal forest ecosystem during the first year of the METAALICUS study. *Environmental Science & Technology*, **36**(23), 5034–5040.
42. Orihel, D.M., Paterson, M.J., Gilmour, C.C., et al. (2006) Effect of loading rate on the fate of mercury in littoral mesocosms. *Environmental Science & Technology*, **40**(19), 5992–6000.
43. Poulain, A.J., Orihel, D.M., Amyot, M., et al. (2006) Relationship to aquatic between the loading rate of inorganic mercury ecosystems and dissolved gaseous mercury production and evasion. *Chemosphere*, **65**(11), 2199–2207.
44. Muresan, B., Cossa, D., Richard, S., et al. (2007) Mercury speciation and exchanges at the air-water interface of a tropical artificial reservoir, French Guiana. *The Science of the Total Environment*, **385**(1–3), 132–145.
45. Selvendiran, P., Driscoll, C.T., and Montesdeoca, M.R.et al. (2009) Mercury dynamics and transport in two Adirondack lakes. *Limnology and Oceanography*, **54**(2), 413–427.
46. Mason, R.P. (2009) Mercury emissions from natural processes and their importance in the global mercury cycle. In: Pirrone, N. and Mason, R.P. (eds) *Mercury Fate and Transport in the Global Atmosphere*. Springer, Dordrecht, pp. 173–191.
47. Chadwick, S.P., Babiarz, C.L., Hurley, J.P., and Armstrong D.E. (2006) Influences of iron, manganese, and dissolved organic carbon on the hypolimnetic cycling of amended mercury. *The Science of the Total Environment*, **368**(1), 177–188.
48. Knightes, C.D., Sunderland, E.M., Barber, M.C., et al. (2009) Application of ecosystem-scale fate and bioaccumulation models to predict fish mercury response times to changes in atmospheric deposition. *Environmental Toxicology and Chemistry*, **28**(4), 881–893.
49. Porvari, P. (2005) Development of fish mercury concentrations in Finnish reservoirs from 1979 to 1994. *The Science of the Total Environment*, **213**, 279–290.
50. Hall, B.D., Cherewyk, K.A., Paterson, M.J. and Bodaly, R.A. (2009) Changes in methyl mercury concentrations in zooplankton from four experimental reservoirs with differing amounts of carbon in the flooded catchments. *Canadian Journal of Fisheries and Aquatic Sciences*, **66**, 1910–1919.
51. Bodaly, R.A., Jansen, W.A., Majewski, A.R., Fudge, R.J.P., Strange, N.E., Derksen, A.J. and Green, D.J. (2007) Postimpoundment time course of increased mercury concentrations in fish in hydroelectric reservoirs of northern Manitoba, Canada. *Archives of Environmental Contamination and Toxicology*, **53**, 379–389.
52. Hylander, L.D., Grohn, J., Tropp, M., Vikstrom, A., Wolper, H., Silva, E.D.E., Meili, M. and Oliveira, L.J. (2006) Fish mercury increase in Lago Manso, a new hydroelectric reservoir in tropical Brazil. *Journal of Environmental Quality*, **81**(S1), 155–166.
53. Hall, B.D., St. Louis, V.L., Rolfhus, K.R., Bodaly, R.A., Beaty, K.G., Paterson, M.J. and Cherewyk, K.A. (2005) Impacts of reservoir creation on the biogeochemical cycling of methyl mercury and total mercury in boreal upland systems. *Ecosystems*, **8**, 248–266.
54. Mauro, J.B.N., Guimaraes, J.R.D., and Melamed, R.et al. (2001) Mercury methylation in macrophyte roots of a tropical lake. *Water, Air, and Soil Pollution*, **127**(1–4), 271–280.
55. Shafer, M., Overdier, J., Phillips, H., et al. (1999) Trace metal levels and partitioning in Wisconsin rivers. *Water, Air, and Soil Pollution*, **110**, 273–311.
56. Gundersen, P. and Steinnes, E. (2003) Influence of pH and TOC concentration on Cu, Zn, Cd, and Al speciation in rivers. *Water Research*, **37**(2), 307–318.
57. Hill, D.M. and Aplin, A.C. (2001) Role of colloids and fine particles in the transport of metals in rivers draining carbonate and silicate terrains. *Limnology and Oceanography*, **46**(2), 331–344.
58. Dupre, B., Viers, J., Dandurand, J.-L., et al. (1999) Major and trace elements associated with colloids in organic-rich river waters: Ultrafiltration of natural and spiked solutions. *Chemical Geology*, **160**(1–2), 63–80.

59. Balistrieri, L.S. and Blank, R.G. (2008) Dissolved and labile concentrations of Cd, Cu, Pb, and Zn in the South Fork Coeur d'Alene River, Idaho: Comparisons among chemical equilibrium models and implications for biotic ligand models. *Applied Geochemistry*, **23**(12), 3355–3371.
60. McKnight, D.M. and Duren, S.M. (2004) Biogeochemical processes controlling midday ferrous iron maxima in stream waters affected by acid rock drainage. *Applied Geochemistry*, **19**(7), 1075–1084.
61. White, E.M., Vaughan, P.P., and Zepp, R.G., et al. (2003) Role of the photo-Fenton reaction in the production of hydroxyl radicals and photobleaching of colored dissolved organic matter in a coastal river of the southeastern United States. *Aquatic Sciences*, **65**(4), 402–414.
62. Shiller, A.M., Duan, S.W., van Erp, P., and Bianchi, T.S. et al. (2006) Photo-oxidation of dissovled organic matter in river watrer and its effect on trace elements speciation. *Limnology and Oceanography*, **51**, 1716–1728.
63. Tseng, C.M., Lamborg, C., Fitzgerald, W.F., and Engstrom, D.R. (2004) Cycling of dissolved elemental mercury in Arctic Alaskan lakes. *Geochimica Et Cosmochimica Acta*, **68,** 1173–1184.
64. Meunier, L., Laubscher, H., Hugg, S.J., et al. (2005) Effects of size and origin of natural dissolved organic matter compounds on the redox cycling of iron in sunlit surface waters. *Aquatic Sciences*, **67**(3), 292–307.
65. Chapelle, F.H. (2004) Geochemistry of groundwater. In: Dreve, J.I. (ed.) *Surface and Ground Water, Weathering and Soils*. Vol. 5. Treatise on Geochemistry, H.D. Holland and K.K. Turekian (Eexc. eds). Elesevier, Amsterdam, pp. 425–449.
66. Kretzschmar, R. and Schafer, T. (2005) Metal retention and transport on colloidal particles in the environment. *Elements*, **1**(4), 205–210.
67. Sen, T.K. and Khilar, K.C. (2006) Review on subsurface colloids and colloid-associated contaminant transport in saturated porous media. *Advances in Colloid and Interface Science*, **119**(2–3), 71–96.
68. Roy, S.B. and Dzombak, D.A. (1997) Chemical factors influencing colloid-facilitated transport of contaminants in porous media. *Environmental Science & Technology*, **31**(3), 656–664.
69. Henry, E.A., Dodgemurphy, L.J., Bigham, G.N., et al. (1995) Total mercury and methylmercury mass-balance in an alkaline, hypereutrophic urban lake (Onondaga Lake, NY). *Water, Air, and Soil Pollution*, **80**(1–4), 509–517.
70. Barringer, J.L., Riskin, M.L., Szabo, Z., et al. (2010) Mercury and methylmercury dynamics in a Coastal Plain Watershed, New Jersey, USA. *Water, Air, and Soil Pollution*, **212**(1–4), 251–273.
71. Krabbenhoft, D.P., Benoit, J.M., Babiarz, C.L., et al. (1995) Mercury cycling in the Allequash creek watershed, northern Wisconsin. *Water, Air, and Soil Pollution*, **80**(1–4), 425–433.
72. Stoor, R.W., Hurley, J.P., Babiarz, C.L., et al. (2006) Subsurface sources of methyl mercury to Lake Superior from a wetland-forested watershed. *The Science of the Total Environment*, **368**(1), 99–110.
73. Ren, J.H. and Packman, A.I. (2005) Coupled stream-subsurface exchange of colloidal hematite and dissolved zinc, copper, and phosphate. *Environmental Science & Technology*, **39**(17), 6387–6394.
74. Baumann, T., Fruhstorfer, P., Klein, T., et al. (2006) Colloid and heavy metal transport at landfill sites in direct contact with groundwater. *Water Research*, **40**(14), 2776–2786.
75. Fuge, R. (2005) Anthropogenic sources. In: Seleinus, O. (ed.) *Fundamentals of Medical Geology*. Elsevier, Amsterdam, pp. 43–59.
76. Blowes, D., Ptacek, C.J., Jambor, J.L. and Weisener, C.G. (2005) The geochemistry of acid mine drainage. In: Lollar, B. (ed.) *Environmental Geochemistry*. Vol. 5. Treatise on Geochemistry, Holland, H.D. and Turekian, K.K. (Exec. eds). Elsevier, Amsterdam, pp. 149–203.
77. Marsden, J.O. and House, C.I. (2006) *The Chemistry of Gold Extraction*, 2nd ed. Society of Mining, Metallurgy and Exploration, Englewood, CO, 619 pp.
78. Hines, M.E., Horvat, M., and Fageneli, J., et al. (2000) Mercury biogeochemistry in the Idrija River, Slovenia, from above the mine into the Gulf of Trieste. *Environmental Research*, **83**(2), 129–139.
79. Berzas Nevado, J.J., Garcia Bermejo, L.F., and Rodriguez Martin-Domimeadios, R.C., et al. (2003) Distribution of mercury in the aquatic environment at Almaden, Spain. *Environmental Pollution*, **122**(2), 261–271.
80. Munthe, J., Bodaly, R.A., Branfireun, B.A., et al. (2007) Recovery of mercury-contaminated fisheries. *Ambio*, **36**(1), 33–44.
81. Harvey, C.F. and Beckie, R.D. (2005) Arsenic: Its biogeochemistry and transport in groundwater. *Metal Ions in Biological Systems*, **44**, 145–169.
82. Aurillo, A.C., Mason, R.P. and Hemond, H.F. (1994) Speciation and fate of arsenic in 3 lakes of the Aberjona watershed. *Environmental Science & Technology*, **28**(4), 577–585.
83. Senn, D.B. and Hemond, H.F. (2004) Particulate arsenic and iron during anoxia in a eutrophic, urban lake. *Environmental Toxicology and Chemistry*, **23**(7), 1610–1616.
84. Aurilio, A.C., Durant, J.L., Knox, M.L. and Hemond, H.F. (1995) Sources and distribution of arsenic in the Aberjona watershed, eastern Massachusetts. *Water, Air, and Soil Pollution*, **81**(3–4), 265–282.
85. Spliethoff, H.M. and Hemond, H.F. (1996) History of toxic metal discharge to surface waters of the Aberjona Watershed. *Environmental Science & Technology*, **30**(1), 121–128.
86. Spliethoff, H.M., Mason, R.P. and Hemond, H.F. (1995) Interannual variability in the speciation and mobility of arsenic in a dimictic lake. *Environmental Science & Technology*, **29**(8), 2157–2161.
87. Ahmann, D., Krumholz, L.R., Hemond, H.F.. (1997) Microbial mobilization of arsenic from sediments of the Aberjona Watershed. *Environmental Science & Technology*, **31**(10), 2923–2930.
88. Newman, D.K., Ahmann, D. and Morel, F.M.M. (1998) A brief review of microbial arsenate respiration. *Geomicrobiology Journal*, **15**(4), 255–268.
89. Lee, M.K., Saunders, J.A., Wilkin, R.T., et al. (2005) Geochemical modeling of arsenic speciation and mobilization: Implications for bioremediation. In: Oday, P.A., Vlassopoulos, D., Meng, Z. and Benning, L.G. (eds) *Advances in Arsenic Research - Integration of Experimental and Observational Studies and Implications for Mitigation*. Vol. 915. American Chemical Society, Washington, D.C., pp. 398–413.

90. Grosz, A.E., Grossman, J.N., Garrett, R., et al. (2004) A preliminary geochemical map for arsenic in surficial materials of Canada and the United States. *Applied Geochemistry*, **19**(2), 257–260.
91. Reza, A., Jean, J.S., Lee, M.-K., et al. (2010) Arsenic enrichment and mobilization in the Holocene alluvial aquifers of the Chapai-Nawabganj district, Bangladesh: A geochemical and statistical study. *Applied Geochemistry*, **25**(8), 1280–1289.
92. Plant, J., Kinniburgh, D.G., Smedley, P.L. and Fordyce, F.M. (2005) Arsenic and selenium. In: Lollar, B. (ed.) *Environmental Geochemistry*. Vol. 9. Treatise on Geochemistry, H.D. Holland and K.K. Turekian (Exec. eds). Elesevier, Amsterdam, pp. 17–65.
93. Neumann, R.B., Ashfaque, K.N., Badruzzaman, A.B.M., et al. (2010) Anthropogenic influences on groundwater arsenic concentrations in Bangladesh. *Nature Geoscience*, **3**(1), 46–52.
94. Hery, M., van Dongen, B.E., Gill, F., et al. (2010) Arsenic release and attenuation in low organic carbon aquifer sediments from West Bengal. *Geobiology*, **8**(2), 155–168.
95. Razzak, A., Jinno, K., Yoshinari, Y., and Oda, K. (2009) Mathematical modeling of biologically mediated redox processes of iron and arsenic release in groundwater. *Environmental Geology*, **58**(3), 459–469.
96. Kirk, M.F., Roden, E.E., Crossey, L.J., et al. (2010) Experimental analysis of arsenic precipitation during microbial sulfate and iron reduction in model aquifer sediment reactors. *Geochimica et Cosmochimica Acta*, **74**(9), 2538–2555.
97. Moore, W.S. (2010) The effect of submarine groundwater discharge on the ocean. *Annual Review of Marine Science*, **2**, 59–88.
98. Zekster, I.S., Everett, L.G. and Dzhamalov, R.G. (2007) *Submarine Groundwater*. CRC Press, Boca Raton, 466 pp.
99. Burnett, W.C., Aggarwal, P.K., Aureli, A., et al. (2006) Quantifying submarine groundwater discharge in the coastal zone via multiple methods. *The Science of the Total Environment*, **367**(2–3), 498–543.
100. Charette, M.A. and Sholkovitz, E.R. (2006) Trace element cycling in a subterranean estuary: Part 2. Geochemistry of the pore water. *Geochimica et Cosmochimica Acta*, **70**(4), 811–826.
101. Beck, A.J., Rapaglia, J.P., Cochran, J.K. and Bokuniewicz, H.B. (2007) Radium mass-balance in Jamaica Bay, NY: Evidence for a substantial flux of submarine groundwater. *Marine Chemistry*, **106**, 419–441.
102. Charette, M.A. and Sholkovitz, E.R. (2002) Oxidative precipitation of groundwater-derived ferrous iron in the subterranean estuary of a coastal bay. *Geophysical Research Letters*, **29**(10), 1444–1448.
103. Duncan, T. and Shaw, T.J. (2003) The mobility of Rare Earth Elements and Redox Sensitive Elements in the groundwater/seawater mixing zone of a shallow coastal aquifer. *Aquatic Geochemistry*, **9**(3), 233–255.
104. Windom, H. and Niencheski, F. (2003) Biogeochemical processes in a freshwater-seawater mixing zone in permeable sediments along the coast of Southern Brazil. *Marine Chemistry*, **83**(3–4), 121–130.
105. Windom, H.L., Niencheski, F., Moore, W.S. and Jahnke, R.A. (2006) Submarine groundwater discharge: A large, previously unrecognized source of dissolved iron to the South Atlantic Ocean. *Marine Chemistry*, **102**(3–4), 252–266.
106. Rouxel, O., Sholkovitz, E., Charette, M. and Edwards, K.J. (2008) Iron isotope fractionation in subterranean estuaries. *Geochimica et Cosmochimica Acta*, **72**(14), 3413–3430.
107. Bone, S.E., Gonneea, M.E. and Charette, M.A. (2006) Geochemical cycling of arsenic in a coastal aquifer. *Environmental Science & Technology*, **40**(10), 3273–3278.
108. Charette, M.A., Sholkovitz, E.R. and Hansell, C.M. (2005) Trace element cycling in a subterranean estuary: Part 1. Geochemistry of the permeable sediments. *Geochimica et Cosmochimica Acta*, **69**(8), 2095–2109.
109. Spiteri, C., Regnier, P., Slomp, C.P., et al. (2006) pH-Dependent iron oxide precipitation in a subterranean estuary, *Journal of Geochemical Exploration*, **88**(1–3), 399–403.
110. Bone, S.E., Charette, M.A., Lamborg, C.H. and Gonneea, M.E. (2007) Has submarine groundwater discharge been overlooked as a source of mercury to coastal waters? *Environmental Science & Technology*, **41**(9), 3090–3095.
111. Laurier, F.J.G., Cossa, D., Beucher, C. and Breviere, E. (2007) The impact of groundwater discharges on mercury partitioning, speciation and bioavailability to mussels in a coastal zone. *Marine Chemistry*, **104**, 143–155.
112. Moore, W.S. and Shaw, T.J. (2008) Fluxes and behavior of radium isotopes, barium, and uranium in seven Southeastern US rivers and estuaries. *Marine Chemistry*, **108**(3–4), 236–254.
113. Boyd, G.R., Dewis, K.M., Korshin, G.V., et al. (2008) Effects of changing disinfectants on lead and copper release. *American Water Works Association Journal*, **100**, 75–87.
114. Adler, T. (2004) Washington's water woes. *Environmental Health Perspectives*, **112**, A735.
115. Shi, Z. and Stone, A.T. (2009) PbO2 (s, Plattnerite) reductive dissolution by aqueous manganous and ferrous ions. *Environmental Science & Technology*, **43**, 3596–3603.
116. Xie, Y., Wang, Y. and Giammar, D.E. (2010) Impact of chlorine disinfectants on dissolution of the lead corrosion product PbO_2. *Environmental Science & Technology*, **44**, 7082–7088.
117. Maas, R.P., Patch, S.C. and Christian, A.-M. (2007) Effects of fluoridation and disinfection agent combinations on lead leaching from leaded-brass parts, •*Neurotoxicology*, **28**(5), 1023–1031.
118. Nguyen, C.K., Stone, K.R., Dudi, A. and Edwards, M. (2010) Corrosive microenvironments at lead solder surfaces arising from galvanic corrosion with copper pipe. *Environmental Science & Technology*, **44**, 7076–7081.
119. Shafer, M.M., Overdier, J.T. and Armstong, D.E. (1998) Removal, partitioning, and fate of silver and other metals in wastewater treatment plants and effluent-receiving streams. *Environmental Toxicology and Chemistry*, **17**(4), 630–641.
120. Hsu-Kim, H. and Sedlak, D.L. (2005) Similarities between inorganic sulfide and the strong Hg(II) - Complexing ligands in municipal wastewater effluent. *Environmental Science & Technology*, **39**, 4035–4041.
121. Herrin, R.T., Andren, A.W., Shafer, M.M. and Armstrong, D.E. (2001) Determination of silver speciation in natural waters. 2. Binding strength of silver ligands in surface freshwaters. *Environmental Science & Technology*, **35**, 1959–1966.
122. Van Veen, E., Burton, N., Comber, S. and Gardner, M. (2002) Speciation of copper in sewage effluents and its toxicity to

Daphnia magna. *Environmental Toxicology and Chemistry*, **21**(2), 275–280.

123. Hsu, H. and Sedlak, D.L. (2003) Strong Hg(II) complexation in municipal wastewater effluent and surface waters. *Environmental Science & Technology*, **37**, 2743–2749.
124. Lamborg, C.H., Fitzgerald, W.F., Skoog, A. and Visscher, P.T. (2004) The abundance and source of mercury-binding organic ligands in Long Island Sound. *Marine Chemistry*, **90**, 151–163.
125. Anbar, A.D. and Rouxel, O. (2007) Metal stable isotopes in paleooceanography. *Annual Review of Earth and Planetary Sciences*, **35**, 717–746.
126. Anbar, A.D. and Knoll, A.H. (2002) Proterozoic ocean chemistry and evolution: A bioinorganic bridge? *Science*, **297**, 1137–1142.
127. Falkowski, P.G., Katz, M.E., Knoll, A.H., Quigg, A., Raven, J.A., et al. (2004) The evolution of modern eukaryotic phytoplankton. *Science*, **305**, 354–360.
128. Kirschvink, J.L., Gaidos, E.J., Bertani, L.E., Beukes, N.J., Gutzmer, J., et al. (2000) Paleoproterozoic snowball Earth: Extreme climatic and geochemical global change and its biological consequences. *Proceedings of the National Academy of Sciences of the United States of America*, **97**, 1400–1405.
129. Quigg, A., Finkel, Z.V., Irwin, A.J., Rosenthal, Y., Ho, T.Y. et al. (2003) The evolutionary inheritance of elemental stoichiometry in marine phytoplankton. *Nature*, **425**, 291–294.
130. Saito, M.A., Sigman, D.M. and Morel, F.M.M. (2003) The bioinorganic chemistry of the ancient ocean: The co-evolution of cyanobacterial metal requirements and biogeochemical cycles at the Archean-Proterozoic boundary? *Inorganica Chimica Acta*, **356**, 308–318.
131. Zerkle, A.L., House, C.H. and Brantley, S.L. (2005) Biogeochemical signatures through time as inferred from whole microbial genomes. *American Journal of Science*, **305**, 467–502.
132. Bergquist, R.A. and Blum, J.D. (2009) The odds and evens of mercury isotopes: Applications of mass-dependent and mass-independent isotope fractionation. *Elements*, **5**(6), 353–357.
133. Bergquist, B.A. and Blum, J.D. (2007) Mass-dependent and -independent fractionation of Hg isotopes by photoreduction in aquatic systems. *Science*, **318**(5849), 417–420.

Problems

7.1. A lake with the simple composition $Na_T = Ca_T = K_T = Cl_T = SO4_T = CO3_T = 10^{-3}$ M contains an organic complexing agent, Y, characterized by:

$$HY = H^+ + Y^- \quad pKa = 6.0$$
$$CuY^+ = Cu^{2+} + Y^- \quad pK = 7.0$$
$$CaY^+ = Ca^{2+} + Y^- \quad pK = 2.0$$
$$MgY^+ = Mg^{2+} + Y^- \quad pK = 2.0$$

a. Consider all the relevant inorganic complexes but no solid phases. Calculate the Cu speciation at pH = 7.0 for $Cu_T = 10^{-7}$ M, $Y_T = 10^{-6}$ M. How will Cu speciation change as a function of pH?

b. Consider now that $Cu_T = 10^{-6}$ M and that both carbonate and/or hydroxide solid phases may form and precipitate with everything else the same as Part (7.1a). Would solids form in the presence of the organic ligand? If the ligand was not present, what would be the critical pH for precipitation of the Cu carbonate and hydroxide solids?

7.2. The drinking water standard for dissolved lead (Pb) is $50\,\mu g\,l^{-1}$. If water is sitting in contact with Pb pipes, there is the tendency for Pb to solubilize and to increase the dissolved Pb concentration. Calculate the pε at which the total dissolved Pb concentration would be equivalent to the water quality standard under the following conditions: pH = 7 and the bicarbonate concentration is 3 mM. What is the concentration of Pb^{2+}?
The electrode potential, E^0 for $Pb^{2+} + 2e^- = Pb$ is −0.12 V.

7.3. The question is based wholly, or partially, on Achterberg et al. [11], which discusses the speciation of metals in a lake that has seasonally anoxic bottom waters. The paper and the calculations presented in table 2 of the paper show that Cu^+ is the dominant form of copper in the anoxic bottom waters in September while Cu^{2+} dominates at the other depths. Construct a pε-pH diagram for copper for this system, ignoring the potential for organic complexation of the copper. Restrict your diagram to the following limits – pH from 3–10, pε from +10–-5. Consider the total carbonate concentration and concentrations of other components, if required, as in the paper, p. 5235. Use the following total concentrations: $Cu_T = 10^{-8}$ M; $S_T = 2 \times 10^{-6}$ M in your calculations. What is the pε value at which the reduced and oxidized forms of Cu are present at the same concentration at a pH of 7?

7.4. Manganite is one of the minerals in desert varnish, a dark coating on the surfaces of rocks in deserts. Petroglyphs in desert environments were sometimes made by scratching off desert varnish. (http://minerals.gps.caltech.edu/FILES/VARNISH/Index.htm).

Construct the log C versus pH diagram for the Mn(II) species in equilibrium with manganite (a Mn(III) oxyhydroxide mineral) for the pH range of 2–10. Show your work, and clearly identify each of the equations that you derived to describe the lines on the log C versus pH diagram.

Use the stability constants from Table A3.1 for the hydrolysis of Mn^{2+}. (Note: only include Mn^{2+}, $Mn(OH)^+$, and $Mn(OH)_2(aq)$ . . . the other species are negligible.)

Use the following solubility constant for manganite.

$$MnOOH(s) + 3H^+ + e^- = Mn^{2+} + 2H_2O \quad \log K = 25.3$$

Note: This dissolution mechanism involves the reduction of Mn(III) to Mn(II). Assume that the water is in equilibrium with oxygen in the atmosphere ($P(O_2) = 0.21$ atm) and use the $O_2(g) - H_2O$ half reaction to get an expression for $\{e^-\}$ as a function of pH that you can substitute into the mass law above.

7.5. a. In the manipulation of information on trace metal speciation use is often made of the "Scatchard plot" where the ratio of the bound to unbound metal concentration is plotted against the bound metal concentration for a range of exposure concentrations. Discuss what this is and why it is useful. Use as an example the following complexation equation:$Cd^{2+} + L^- = CdL^+$.
What assumptions are inherent in the Scatchard formulation?

b. A metal titration using Fe(III) was done using a lake sample at pH 6.5 and T = 10°C. Using the data from this titration, determine the total concentration of the ligand and the conditional stability constant.

Fe(III) added	Peak Current	Fe(III) added	Peak Current	Fe(III) added	Peak Current
1.0	19.2	11.0	668.1	21.0	1635.0
2.0	45.0	12.0	760.2	22.0	1722.9
3.0	79.4	13.0	855.7	23.0	1820.9
4.0	123.9	14.0	944.0	24.0	1927.3
5.0	178.7	15.0	1042.3	25.0	2021.7
6.0	245.5	16.0	1137.0	26.0	2121.8
7.0	318.8	17.0	1242.5	27.0	2226.5
8.0	398.7	18.0	1328.4	28.0	2314.2
9.0	484.9	19.0	1431.9	29.0	2416.4
10.0	572.6	20.0	1523.5	30.0	2519.3

7.6. Explain qualitatively why the stability constant (i.e., Log K) for reaction with ED is so much larger than that with ammonia even though the bond strengths for each of the N ligand atoms to the Cu(II) atom are nearly the same for the free NH_3 and the amine functional groups on the organic molecule.

$$Cu^{2+}(aq) + 4NH_3(aq) = Cu(NH_3)_4^{2+} \quad \text{Log } K = 11.8$$

$$Cu^{2+}(aq) + ED\,(aq) = Cu(ED)_4^{2+} \quad \text{Log } K = 19.6$$

Where ED => Ethyldiamine or $H_2NCH_2CH_2NH_2$

7.7. Calculate the adsorption of Zn onto hydrous ferrous oxide (HFO) at pHs of 5, 6, 7 and 9 given the following information. Amount of HFO is $90\,mg\,l^{-1}$ ($1\,mmol\,l^{-1}$ Fe) and the ionic strength is 10^{-2} M. Assume that the Zn adsorbs only to the high affinity sites. The specific surface area of the HFO is $600\,m^2\,g^{-1}$ and the density of sites is: 0.2 mol/mol Fe for low affinity sites and 0.005 mol/mol Fe for the high-affinity sites.

CHAPTER 8

Trace metals and organisms: Essential and toxic metals, organometallics, microbial processes, and metal bioaccumulation

8.1 Introduction

The focus of this chapter is the interaction between trace metal(loid)s and organisms in aquatic environments. Such interactions fall into two basic categories. Firstly, the interaction can lead to the adsorption or uptake of an element from solution, and the subsequent retention or elimination of the metal(loid) may then occur. Such accumulation or uptake may be intentional or by accident. Secondly, the interaction may lead a transformation of the form and speciation, and the most important processes in aquatic environments is the *methylation* (*alkylation*) of an inorganic metal to its mono, di or poly methylated form (e.g., $(CH_3)_nM$, n = 1–4, typically, M = metal) or *demethylation* (*dealkylation*) of an organic metal (*organometallic*) compound, or alternatively its reduction and elimination. These reactions mostly occur within the organism, and therefore uptake is required for these reactions to occur. However, it is also possible that redox reactions can occur on the surface of organisms, and therefore while these reactions are biologically-mediated, no uptake of the metal is required. The air–water exchange of such alkylated compounds, and their presence in environmental waters, was discussed in Chapters 5–7.

In terms of the accumulation of metal(loid)s into food webs, the primary and most important accumulation step involves the phytoplankton in the pelagic realm, and bacteria and benthic algae in the benthic realm. For most metal(loid)s, accumulation into higher trophic level organisms (primary and higher level consumers) is from food rather than from the water directly, and therefore the crucial step in bioaccumulation is that between the water and the primary producers. This consideration defines the focus of the chapter and while accumulation and trophic transfer to higher trophic organisms is an important and complex process, it will not be discussed in detail here.

Overall, research on metal(loid)s in the aquatic environment is driven by: (1) concern about their impact on aquatic organisms, and their consumers; and (2) examination of the uptake of essential elements by organisms. The impact of an accumulated metal can therefore be positive if the metal is a required (*essential*) nutrient or has some functionality in the cell due to its increasing concentration enhancing growth rate or cellular metabolism, especially for microbes living within the aquatic water column. Essential elements for aquatic organisms include Fe, Mn, Co and Zn (Table 8.1). Mostly, essential metals form part of enzymes and other biochemicals needed for specific cellular reactions. One definition of an essential element is [1]:

> *An element is considered essential to an organism when a decrease of its accumulation below a certain limit results consistently in a reduction in a physiologically important function, or when the element is an integral part of the organic structure performing a vital function in the organism.*

Alternatively, a metal(loid) may be *toxic* to the organisms and therefore increasing concentration will have an adverse effect. The definition of toxicity could be analogous in form to the definition above for essential elements. Also, it is possible for a single element to be both essential and toxic to an organism if it is required at low concentration for some specific enzyme or biological function but at high concentration interferes with other functions and causes toxicity. Copper and Se are two prime examples of such elements, which are required in low concentration but are toxic at high concentration [1]. Overall, most elements will

Trace Metals in Aquatic Systems, First Edition. Robert P. Mason.

Table 8.1 Major roles of metals in enzymes and cellular biochemistry. Taken from [1–4].

Metal	Biochemical Role
V	Found in ascidians; various oxidation states allows it to play a role in cell defense (e.g., peroxidase activity)
Mn	Important in many functions. Found in chloroplasts, mitochondria (with Fe in superoxide dismutase). Part of Photosynthesis System II (enhances the oxidation of the oxygen in water to O_2)
Fe	Has many roles in heme proteins/porphyrins, part of the chlorophylls of plants, the photosynthetic pigments (cytochromes b, c and f), ferredoxin, nitrogenase, nitrate reductase, catalase, peroxidase, and other enzymes
Co	Cofactor in Vitamin B_{12} (Cobalamin). Can substitute for Zn in carbonic anhydrase of some marine phytoplankton. Also found in various other enzymes and proteins in algae and bacteria
Ni	Ni is found in some important enzymes such as superoxide dismutase and in urease.
Cu	Acts as important cofactors due to its redox characteristics. Found in many enzymes and proteins involved in photosynthesis, and mitochondrial electron transport (cytochrome c)
Zn	No redox chemistry but is a Lewis acid that binds strongly to O and N containing compounds. Principal metal in carbonic anhydrase. Part of metallothioneins. In DNA and RNA polymerases
Se	Found in selenoproteins
Mo	In enzymes associated with redox reactions of C, N and S
W	Found in some enzymes
Cd	Can substitute for Zn in carbonic anhydrase in phytoplankton

be toxic to organisms at some level, but this is most apparent for trace nutrients. In such cases, a "bell-shaped" dose-response curve reflects this dichotomy (see Fig. 1.2). A specific example is the effect of the Zn concentration in solution on the growth of a marine diatom (Fig. 8.1) [1, 2]. This metal is a required nutrient and phytoplankton growth is inhibited at low concentration, but high levels of Zn can also be toxic. Many organisms can regulate the concentration of the required metals to a certain point and maintain cellular levels relatively constant as the external medium concentration increases. This ability is termed *homeostasis*. However, at some point, as metals are mostly accumulated by transport processes that are not directly energy-dependent but do depend on the solution concentration, the ability of an organism to counter the increased concentration in the medium fails, and toxicity occurs.

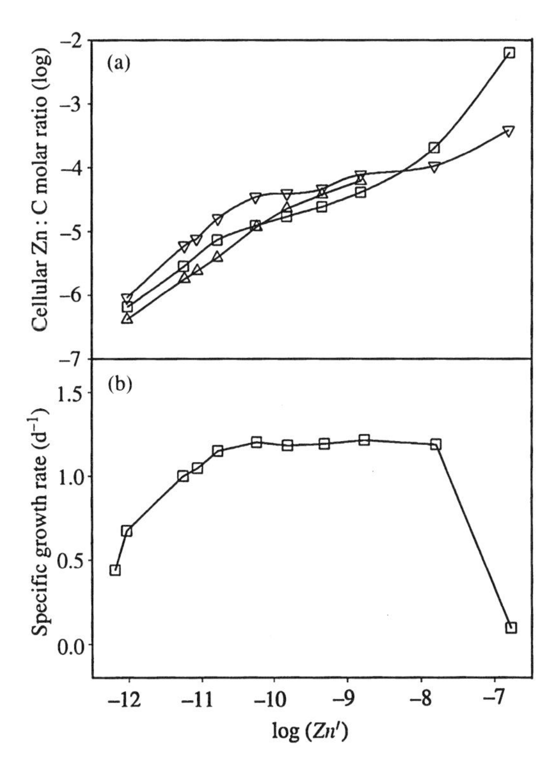

Fig. 8.1 The relationship between Zn concentration in solution and growth rate of the marine diatom *Thalassiosira weissflogii* and two clones of *Emiliana hucleyi* (triangles) showing the effect of both low and high Zn concentrations (labile Zn, Zn′) on the growth rate, respectively due to limitation and toxicity. Figure taken from Morel et al. (2004) [2] in *Treatise on Geochemistry* vol. 6 [1], and was created using information from Sunda and Huntsman (1992) *Limnol Oceanogr.* **37**: 25–40. Reprinted with permission of Elsevier.

The major ions, sodium (Na^+), potassium (K^+) and chloride (Cl^-) all play an important role in cell function and physiology, as do the major alkaline earth metals, Ca^{2+} and Mg^{2+}, and they are taken up via specific channels or pores within the membrane [3]. All these metals are almost entirely present in solution as free metal ions, especially in freshwater as they all form very weak complexes with other charged counter ions, and thus their dissolved concentration is a good representation of their activity in solution and their bioavailable concentration. This is not so for most of the metals and metalloids discussed in this book [2, 4]. Therefore, the total concentration in solution is not a good measure of the bioavailable fraction and two systems, with similar total concentration but with different solution chemistry, may have very different concentrations of bioavailable metal, and the effect will be different for different trace elements.

Many of the elements of the first transition series of the periodic table have some biochemical role and these are outlined briefly in Table 8.1 [1, 2]. Iron is an essential element as it has many varied functions within the cell, for both photosynthetic and non-photosynthetic organisms, but its role in the photosynthetic mechanism is vital. There has been ample demonstration that many oceanic phytoplankton are limited by the lack of sufficient Fe in seawater, as discussed in Chapter 6. The reason for this is the chemistry of Fe^{III}, a relatively insoluble ion in both saline and freshwaters, due to its tendency to be hydrolyzed to insoluble (hydr) oxide species, as discussed in Chapter 3. Of the transition metals, the most important elements appear to be Fe, Mn, Co, and Zn, as discussed further in Section 8.3.

To understand in more detail the role of different metal ions either as essential or toxic and bioaccumulative elements, it is necessary to understand their speciation in solution, as not all forms of the metal can be assimilated into organisms. In addition, it is necessary to clarify the mechanisms of uptake and the potential interferences and competition that may occur during uptake across cellular membranes. Also, there is an evolving understanding of the role of metals in cellular processes, with the realization that some metals, that were previously considered to be toxic, can have a biochemical role – Cd is the prime recent example. There are two complimentary approaches to understanding the interaction of trace metals with organisms. The first is to take a detailed mechanistic approach and examine the processes of uptake, regulation and binding of metals within cells. The second is to bypass the details and examine the overall relationship between the metal concentration in solution and the metal concentration in the organism, and the factors that may influence the cellular concentration. While the former approach has been mostly applied to the study of the essential and potentially limiting "nutrient" metals, and the latter to the more bioaccumulative and toxic metals, it is obvious that there is overlap in these approaches and both yield useful knowledge and understanding for different applications.

The realization that the total concentration of the metal in the aqueous system of interest is not a good predictor of the concentration of the metal that is available to be taken up by the organism has been known by scientists for a while, but such notions have only been slowly incorporated into regulation and management practice. As will become evident below, this is partially a result of the complexity of defining the bioavailable concentration within a particular system. For single cell organisms, the metal must be in solution to be available for uptake and therefore the uptake rate is related to the truly dissolved concentration, which is often a small fraction of the total metal concentration in an unfiltered water sample. It is known, and this was discussed in earlier chapters, that metals bind strongly with particulate matter, through association with oxide and sulfide phases, and with particulate organic matter. The strength of the association depends on whether the "particles" are living or dead, and non-living particles span a range from inorganic solids to detritus derived from decaying biological organisms. Additionally, it is often the case that the particle is a mixture of such phases and the partitioning of the metal is a complex interaction. However, the fractionation of the metal between the dissolved and particulate phases can be measured and the amount of metal in the dissolved fraction estimated, as discussed in earlier chapters. When coupled with modeling of the solution speciation, it is possible in many instances to estimate the bioavailable fraction in solution if the system is at steady state, and if the mechanism of uptake is known; that is, the species that are taken up and the details of the major complexation of the metal in solution, especially in the presence of NOM. Therefore, it is possible, in principle, to understand and model the uptake of metals by organisms.

8.2 Mechanisms of metal accumulation by microorganisms

8.2.1 The transport of metals across membranes

This section is primarily focused on the mechanisms in volved in the uptake of metals across membranes, as this is the primary way that aquatic microorganisms accumulate solutes from their aquatic surroundings. Therefore, the discussion is also pertinent to the transport of metals across the cell membranes of higher organisms. There has been increasing recognition that the membranes of organisms are not purely lipid-bilayers but that they contain proteins which are involved in the uptake of ions and molecules across the membrane [2, 3]. It has been long known that while there is a relationship between the ability of a molecule to partition into an organic phase and its permeability across cell membranes, there are other mechanisms for uptake besides a *passive diffusive* process as many ions are taken up by organisms much more rapidly than can be accounted for by passive processes [3, 5, 6]. The uptake of elements across biological membranes can therefore be divided into a number of different mechanisms that can be broadly defined as: *passive diffusion* (transport), *facilitated transport*, and *active uptake* (Fig. 8.2). *Passive diffusion* occurs due to the partitioning of the molecules, mostly neutrally charged, into the lipid fraction of the cell membrane, while the other uptake pathways involve proteins which are imbedded into the lipid bi-layer matrix and form channels for ion flow. *Facilitated transport* involves transport of the metal entity through a channel in the membrane designed to allow passage of the specific metal and this uptake process is passive in that no energy is required to directly facilitate the process. *Active uptake* also involves channels, but there is an energy requirement associated with the process. The membrane proteins, which are involved in ion and other molecular transport, can account for up to 50% of the mass of the membrane

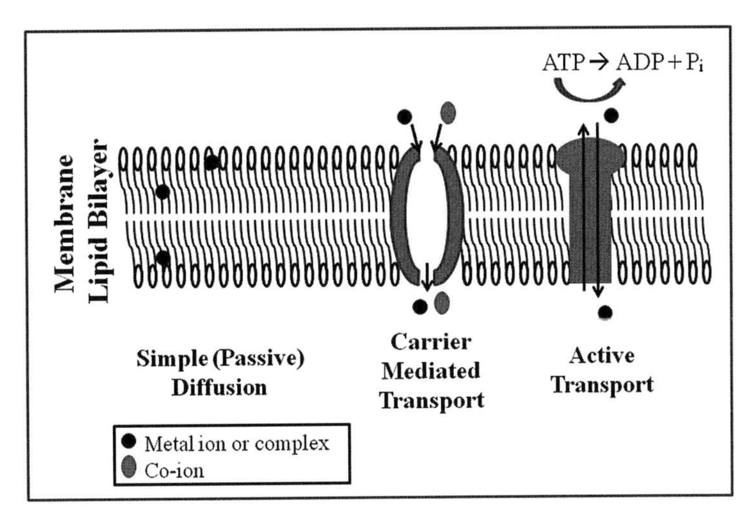

Fig. 8.2 Representation of the major uptake pathways for elements and molecules across the cell membrane, showing the difference between passive (simple) diffusion, carrier-mediated transport and active uptake which requires the expenditure of ATP.

[3]. Overall, the cell surface is negatively charged and therefore there is an electric double layer associated with it and there is an ionic association in the immediate vicinity of the surface, as discussed previously in terms of particles in Chapters 3 and 4. The charge of the surface is a function of the solution pH. Given the charged surface, positively charged ions are more easily assimilated and for most metals, it is the positively charged ions – the free metal or small molecular weight ions – that are mostly taken up. Recall that for most transition metals, they are present in solution as hydrated ions and therefore their free ions are essentially positively charged complexes with water. In some instances, it has been demonstrated that uptake involves the removal of the associated water molecules during transport through the membrane pores – this is the case for Ca^{2+} and Mn^{2+} transport [3].

In terms of uptake, both passive and facilitated transport are processes that do not require direct energy input, while active transport covers those processes that involve the active use of energy to transport the metal across the membrane. These different transport mechanisms are dealt with next, with specific examples for each mechanism, and some description of their relative importance for metal transport into microorganisms.

8.2.2 Passive transport (diffusion)

Passive diffusion or transport across the cell membrane occurs if there is a concentration gradient and the molecule or entity that is diffusing is more soluble in the membrane (lipid) and can therefore passively partition from the solution into the membrane of the organism (Fig. 8.2). The rate of transport (*uptake flux*) (V, in mol $cm^{-2} s^{-1}$) for passive diffusion is given by:

$$V = -KD_i(C_2 - C_1)/l \quad (8.1)$$

where C refers to the concentration (in the medium and in the membrane) (mol cm^{-3}); l is the thickness of the membrane (cm); D_i is the diffusion coefficient of the molecule of interest within the membrane ($cm^2 s^{-1}$); and K is the *partition coefficient* for the diffusing substance between the lipid phase and water. The value of D_i is often not known, and depends on the shape and size of the molecule of interest, although it is related to the molecular diffusion coefficient in water, as discussed in Chapter 4. The *permeability coefficient*, P, or uptake velocity (units: cm s^{-1}), is a parameter that can be measured and this is related to the other parameters by [6]:

$$P = KD_i/l \quad (8.2)$$

If $C_1 = C_2$, passive diffusion will not occur, regardless of the compound, as there is no concentration gradient. If K is small, passive diffusion is small. For charged ions, K is typically very small. Some metals, however, form neutral complexes in solution and these compounds are more hydrophobic and will partition into the membrane (have a larger K). The magnitude of K is a measure of the propensity of the molecule to partition into the organic phase and these values are often estimated from experiments where the partitioning of the compound of interest between water and the organic phase is measured. In organic chemistry, and particularly with the studies looking at the hydrophobicity of organic contaminants, the *octanol-water partition coefficient* (K_{ow}) is the parameter that is most often measured, rather than a specific lipid-water partition coefficient, and this is used as a measure of the potential for a molecule to partition passively into a cell's membrane. The K_{ow} is thought to be a better surrogate for a cellular membrane than a particular lipid as the membrane contains both lipids and proteins, which affects its overall hydrophobicity and polarity [6]. However, it has been shown that for a certain class of compounds (e.g., alkanes) there is a linear relationship between the K_{ow} and the K for partitioning into another organic phase, and so the K for any organic liquid can be determined if the relationship is known. Overall, for comparison of the potential importance of passive diffusion for a particular set of compounds, the K_{ow} provides a relative indication. The K_{ow} has been measured for many organic molecules and there is a reasonable theory concerning the molecular factors that influence its value [7]. The trace metal community has also adopted and used K_{ow} values as a measure of the ability of a metal to partition passively across membranes. The K_{ow} for a number of neutral metal complexes, both inorganic and organic, are given in Table 8.2. For the inorganic complexes, it appears that Hg complexes have the highest values, suggesting that they would partition into cellular membranes more than the other metals. The higher propensity of the inorganic Hg complexes to partition likely results from its tendency to form more "covalent" complexes compared to Cd and Ag, which is related to its inorganic chemistry, such as its electronegativity (Chapter 3). Another complementary measure of the tendency to passively partition has been devised, based on the *"covalent index"*, which was estimated from the Pauling electronegativity for a metal ion and its ionic radius [17]. This index was highest for Hg^{II} and Ag^{I}, with the value for Cd^{II} being much lower, which is overall consistent with the differences in the values for the K_{ow} for the neutral metal chloride complexes in Table 8.2.

For metals in solution it is possible that more than one form or complex of the metal will be sufficiently neutral to partition into octanol, or another organic phase. Therefore, it is necessary to relate and define the *overall octanol-water partition coefficient* (D_{ow}) in terms of the individual K_{ow}s of each complex. Overall, the total partitioning will be a sum of the relative partitioning of all the contributing fractions, which depends on the speciation and the respective coefficients (K_{ow}) of individual species. Thus:

$$D_{ow} = \sum f_i \cdot (K_{ow})_i \tag{8.3}$$

where f_i is the mole fraction of the total metal in solution present as the species i [9]. Thus, all species that can partition will contribute to the overall passive transport, and the importance and rate of uptake will therefore be a function of the solution chemistry that can influence the uptake rate. For example, considering the inorganic Hg complexes with Cl^- and OH^-, one would predict based on the relative K_{ow} values (Table 8.2) that the uptake rate or permeability of $HgCl_2$ should be about three times that of HgClOH, assuming that both complexes have a similar size and D_i. Thus,

Table 8.2 Estimated octanol-water partition coefficients (K_{ow}) for neutrally charged inorganic and organic complexes of various metals. The data taken from references as indicated below the table.

Metal	Inorganic Complexes	K_{ow}	Ref.	Organic Complexes	K_{ow}	Ref.
Hg	$HgCl_2$	3.3	[1]	Hg(cysteine)$_2$	3.7	[2]
	HgOHCl	1.2	[1]	Hg(thiourea)$_2$	4.6	[2]
	$Hg(OH)_2$	0.05	[1]			
	HgSHOH, $Hg(SH)_2$	26	[3]			
	$Hg(S)_6$	26	[3]			
	Hg^0	~10^3	[7]			
		4.2	[1]			
CH_3Hg	CH_3HgCl	1.7	[1]	CH_3Hg(cysteine)	50	[2]
	CH_3HgOH	0.07	[1]	CH_3Hg(thiourea)	630	[2]
	CH_3HgSH	28	[2]			
	$(CH_3)_2Hg$	180	[1]			
Cd	$CdCl_2$	0.002	[8]	Cd(dithiocarbamate)$_2$	1000	[4]
Ag	AgCl	0.09	[5]	–		
Cu	—			Cu(oxine)$_2$	400	[4]
				Cu(oxine)$_2$	70	[6]
				Cu(chloroxine)$_2$	325	[6]
				Cu(dichloroxine)$_2$	690	[6]
				Cu(dibromoxine)$_2$	3715	[6]
				Cu(dithiocarbamate)$_2$	630	[4]
Pb	–			Pb(dithiocarbamate)$_2$	10^4	[4]

References: 1: Mason et al. [9] and Morel and Hering [10] and references therein; 2: Lawson and Mason [8]; 3: Benoit et al. [11]; 4: Phinney and Bruland [12]; 5: Reinfelder and Chang [13]; 6: Kaiser and Escher [14]; 7: Jay et al. [15]; 8: Gutknecht [16].

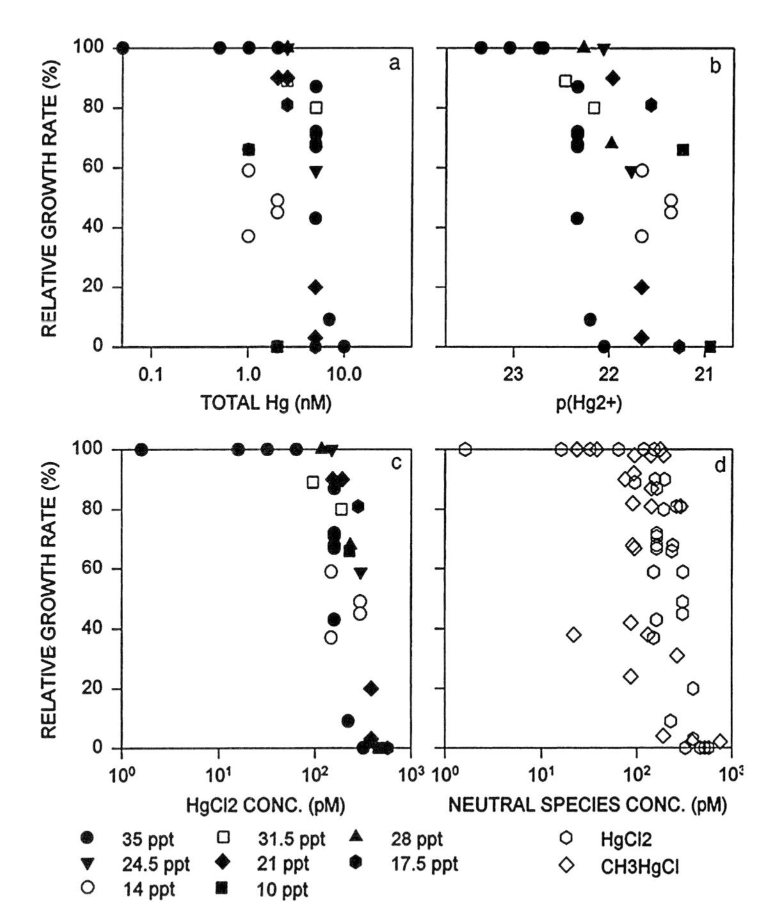

Fig. 8.3 The accumulation of mercury and methylmercury by the marine diatom *Thalassiosira weissflogii* showing the relationship between relative growth rate and (a) the total concentration of inorganic mercury; (b) the free ion ($[Hg^{2+}]$) concentration; (c) the neutrally-charged mercury ($[HgCl_2]$) content; (d) the relationship between relative growth rate and the concentration of neutrally-charged mercury or methylmercury ($[HgCl_2]$ or $[CH_3HgCl]$). Reprinted with permission from Mason et al. (1996) *Environ. Sci. Technol.* **30**: 1835–45 [9]. Copyright (1996) American Chemical Society.

one would predict that uptake would be highest under conditions where $HgCl_2$ was the dominant species in solution, and this has been demonstrated [9, 18]. Equation 8.3 is predicated on the assumption that the individual species in solution can rapidly repartition between the various complexes and maintain equilibrium speciation so that the equilibrium model, which would be used to determine the value of *f*, provides an adequate prediction of the actual solution speciation. This would depend on the ligand exchange characteristics of the metal, which are relatively rapid for all the metals listed in Table 8.2 [10] (see Chapter 4), and therefore this is a reasonable assumption. If this is not the case, which may be so for short extraction times in the presence of NOM, which may form complexes with metals that are kinetically slow to dissociate, then a more complex formulation is needed to calculate the overall partitioning [19].

Passive uptake of Hg and other metals was first proposed based on measurements made with artificial membranes [20]. Passive uptake of Hg and CH_3Hg complexes was then demonstrated with experiments using the estuarine diatom, *Thalassiosira weisflogii* [9]. Uptake was most efficient for the neutral chloride complexes, $HgCl_2$ and CH_3HgCl, compared to the neutral hydroxide complexes, in agreement with the differences in their K_{ow}s (Table 8.2). The results of experiments done at varying concentrations and salinities, which impacts the overall relative $HgCl_2$ content, is shown in Fig. 8.3. The relationship between toxicity and concentration is obviously more related to the $HgCl_2$ concentration than to the calculated and very low free metal ion concentration, or the total concentration. Some of the variability is due to the fact that HgOHCl is also present at small but variable concentrations. Additionally, the similarity in the toxicity based on $HgCl_2$ or CH_3HgCl concentration suggests it is the rate of passive diffusive uptake that is determining the overall toxicity. Further studies with both diatoms [8] and sulfate-reducing bacteria [11] have similarly shown that neutral complexes with sulfide – HgS and CH_3HgSH – and organic thiols, such as cysteine [21, 22] are also efficiently taken up and have K_{ow} values that are higher than that of the chloride complexes (Table 8.2). These results suggest that in the presence of neutral inorganic or simple organic complexes, passive accumulation of Hg and CH_3Hg occurs via partitioning of these complexes into the cell membrane.

Passive accumulation of neutral inorganic complexes has also been demonstrated for other metals. For Ag, it has been shown that the complex AgCl has a higher K_{ow} than the free metal, and that it is taken up by phytoplankton more rapidly than the free metal ion [13]. The uptake rates of AgCl, when normalized to neutral exposure concentration, were similar to those for $Hg(OH)_2$, and both are less than that for $HgCl_2$, consistent with the differences in their K_{ow}s. For Cd, $CdCl_2$ has a low K_{ow} compared to $HgCl_2$ and uptake of $CdCl_2$ across artificial membranes was orders of magnitude less than that for $HgCl_2$ under the same conditions. This is in accordance with the measured differences in K_{ow}. Uptake was much greater than if Cd^{2+} was the primary species in solution. Similarly, it was shown with artificial membranes that uptake of $TlCl_2$ was much greater than that of the free ion [16]. These results confirm that there is the potential for uptake of neutral inorganic metal complexes by passive diffusion into phytoplankton and bacteria, and these metal complexes could also passively cross the membranes of higher organisms. However, for complexes with substantial "ionic" character, other mechanisms dominate as the passive uptake rates are relatively slow.

The accumulation of neutral organic complexes of metals has also been shown. For example, it was shown that while Cu, Cd, and Pb were not accumulated in the presence of EDTA, which forms charged complexes with these metals, all were taken up when complexed to other organic ligands (e.g., oxine and dithiocarbamate) that form neutral complexes with substantial K_{ow}s (Table 8.2) [12, 23]. In all cases, initial accumulation rates were much higher for the neutrally-complexed metal than they were in the absence of the ligand. These studies with Cu-organic complexes have been repeated by others [14, 24]. In these experiments with larger organic complexes, it was found that the observed uptake velocity of the complexes was similar to that of $HgCl_2$ even though the K_{ow}s of the complexes are orders of magnitude higher. As demonstrated by others [6], the size of the molecule is an important consideration as diffusion through the cell membrane limits the accumulation rate. Recall from Section 6.3.2, that the diffusion coefficient is a function of molecular volume. Therefore, for large compounds, uptake is slow even if they are highly lipophillic molecules. The permeability depends on both K_{ow} and D_i, the diffusion rate of the molecule within the membrane (Equation 8.2), and the rate of diffusion is a strong function of the size and shape of the molecule.

In understanding and interpreting the importance of hydrophobicity and molecule size on accumulation rates, it is worth discussing a general linear relationship between P and K_{ow} values, which has been derived and applied to low molecular weight neutrally charged organic compounds diffusing across red blood cell membranes [6]. A relationship exists between the *corrected membrane permeability*, which can be thought of as the potential permeability in the absence of size effects (cm s^{-1}), P* and the molar volume of the molecule:

$$\log P^* = \log P + mv \tag{8.4}$$

where m is a proportionality constant ($m = 0.0546\,\mathrm{mol\,cm^{-3}}$) and v is the *van der Waals volume* (in $\mathrm{cm^3\,mol^{-1}}$) of each species. This empirical relationship predicts the actual (measured) permeability coefficient based on the effects of molecular size on diffusion rate through the membrane. The impact of differences in molecular size is large given the log relationship. When the data obtained for the corrected passive diffusive uptake of a number of metal complexes (log P*) is plotted against the log K_{ow}, and including data for small organic compounds, there appears to be a reasonable relationship, suggesting that Equation 8.3 provides an adequate prediction of the uptake of neutral molecules of the metals by passive diffusion [1, 9].

The molecular size effect explains why the uptake rates for large hydrophobic compounds (large K_{ow}) are low, and data for organic contaminants (e.g., PCBs which have K_{ow} values of $>10^6$) [7] provide an example of this as their uptake rate into microbes is of the same order as Hg. This also explains why, for example, in the presence of polysulfides, there is no increase in the methylation rate of Hg even though a large hydrophobic compound (HgS_6) is predicted to be present in solution [15]. Again, the large molar volume of the complex prevents its rapid uptake across the membrane, and uptake of the smaller neutrally-charged HOHgSH and $Hg(SH)_2$ complexes is what is driving the Hg methylation [25].

One final fact needs to be considered for the uptake of neutral complexes by passive diffusion. As the values for the K_{ow} are small for most metal inorganic complexes, equilibrium would be rapidly reached if there was not a sink in the cell for the accumulated metal. As uptake proceeds, the concentration in the membrane increases, and the complex will therefore diffuse into the cytosol of the cell given the concentration gradient until equilibrium is reached. However, if the metal complex reacts with ligands or sites within the cell, then this "sink" will result in further accumulation. This was shown to be the case for both Hg and CH_3Hg as their complexes rapidly bind to intracellular ligands, or even ligands within the membrane, and therefore continued uptake occurs [9].

The steady state concentration that accumulates in the cell is a function of the uptake rate and the average growth rate, as slower growing cells will accumulate more metal prior to cell division, and for microbes, uptake can be parameterized as:

$$V_C = A \cdot P[M] \quad (8.5)$$

where A is the surface area of the cell (cm^2), P is the permeability ($cm\ s^{-1}$) and [M] is the solution concentration of the passively diffusing compound (M or mol cm^{-3}), and the units of V_C, the *cellular uptake rate,* are mol $cell^{-1}\ s^{-1}$. The cell steady state concentration, or *cell quota,* Q, is then:

$$Q = V_C/g \quad (8.6)$$

where g is the specific growth rate (d^{-1} or s^{-1}) so that the units of Q are mol $cell^{-1}$. While this parameterization is discussed here, it is applicable to all modes of accumulation.

8.2.3 Facilitated transport (accelerated diffusion)

The enhancement of uptake of ions by processes that do not directly involve energy is termed *facilitated transport,* or carrier mediated transport (Fig. 8.2). As most ions have very low permeability in lipids membranes they would not be transported to any significant degree if membranes were entirely composed of a lipid bilayer. However, there are ion channels or pores within the membrane and these can rapidly transport ions from the exterior solution to the interior of the cell [2, 3]. Additionally, often there is transport of ions in both directions so that charge balance is maintained, such as the exchange of Cl^- for HCO_3^-, which involves specific proteins embedded in the membrane with a pore size and shape that is specific in terms of its ability to allow the transfer of HCO_3^- in exchange for Cl^-, but not other negatively charged ions [3].

The process of transporting ions through specific channels results in a much higher uptake rate than would be possible by passive diffusion. However, as the uptake is dependent on the number of transport sites in the membrane, uptake can become limited as the number of sites is finite. In contrast, the rate of passive diffusion is a linear function of the concentration difference between the external solution and the membrane. It has been shown that the facilitated transport processes conform to *Michaelis-Menten kinetics.* As the metal solution (*substrate*) concentration increases, uptake rate initially increases in a linear fashion, but at higher concentrations, decreases in rate to a *maximum uptake rate* (V_{max}):

$$V = V_{max} \cdot [M]/(K_M + [M]) \quad (8.7)$$

where [M] is the concentration of the ion or molecule on interest in the external medium and K_M is the *half-saturation constant,* which is equivalent to $V_{max}/2$. This was discussed in Section 4.6.3.

For trace metal ions in solution, it is reasonable to discuss and explain the uptake in terms of the equilibrium partitioning of the one or more forms or complexes of the metal ion in solution (M_{out}) to a surface transport site on the membrane (X) followed by the internalization of the ion to the cell interior (M_{in}):

$$M_{out} + X = MX \rightarrow M_{in}$$

The relevant equation is derived by considering the binding of the ion to a surface transport site, or interacting with the transporting pore, to be a reversible process (k_L for the forward reaction, k_{-L} for the reverse reaction), and where the internalization of the molecule or ion across the membrane (k_{in}) is irreversible. As k_{in} is the rate limiting step, the concentration of [MX] reaches a steady state value and as the surface sites are regenerated for further interaction after each ion is transported, the total concentration of X (X_T) remains constant. As shown in detail in Morel and Hering [10], such a situation is analogous to a catalytic reaction which proceeds through the formation of a reactive intermediate, and this is why it can be parameterized using the Michaelis–Menten equation, with the following relationships:

$$V_{max} = k_{in}X_T \quad (8.8)$$

and

$$K_M = k_{-L} + k_{in}/k_L \quad (8.9)$$

At high concentrations in the external medium, when [M] >> K_M, then $V \sim V_{max}$ (Equation 8.7) and the uptake rate becomes independent of the external concentration due to saturation of the uptake sites. Alternatively, at low concentration of [M] in the external medium, $K_M >> [M]$ and $V \sim V_{max}.[M]/K_M$, and the uptake rate is linearly related to concentration. As an example, the uptake kinetics of Fe in solution is shown in Fig. 8.4. The uptake rate was measured in a number of different media containing different chelating agents with different binding strengths to Fe (Fig. 8.4a) [10]. In each case, the uptake rate increases as the total concentration in the medium increases, but the intercept of the various relatively parallel lines is different. When the same data is plotted against the calculated free metal ion concentration, then a consistent relationship is found (Fig. 8.4b). This indicates that it is only some fraction of the Fe in solution is bioavailable and while the data in Fig. 8.4(b) suggest that this is the free metal ion, more extensive research has determined that it is the inorganic (labile) fraction in solution (Fe^{3+} plus Fe bound to inorganic ligands (primarily $Fe(OH)_x^{3-x}$)) that controls the uptake rate [10]. Additionally, Fig. 8.4(c) also illustrates that the uptake rate is dependent on the number of uptake sites on the surface of the cell, and that the organisms can regulate these to some extent [26]. The maximum number of transport proteins is determined by

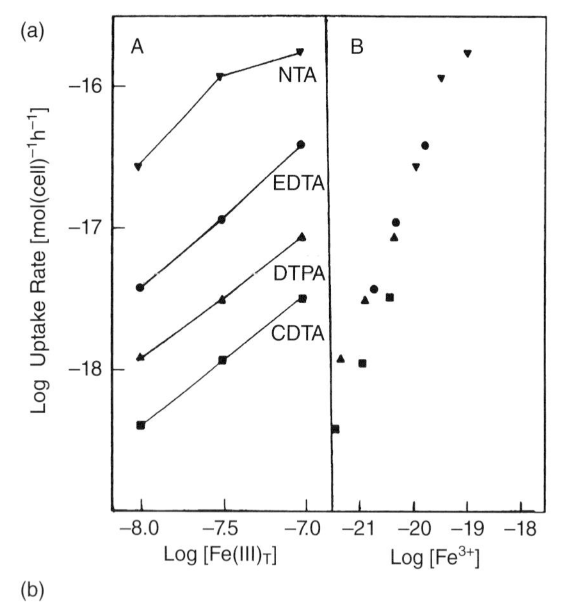

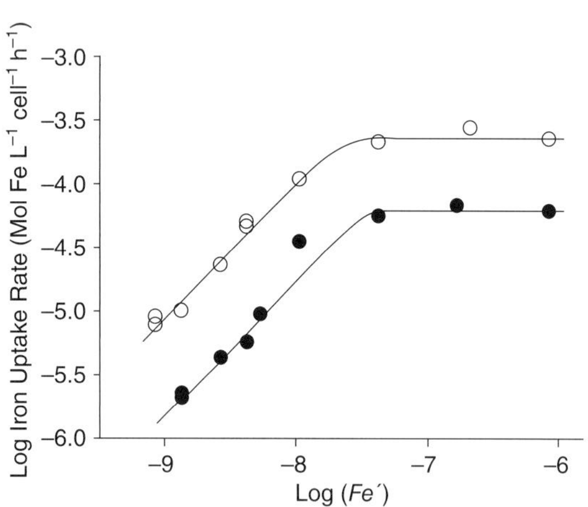

Fig. 8.4 Examples of facilitated transport of metals: (a) the relationship between the uptake rate and total iron concentration in the presence of different organic ligands (*left panel*) and the same data plotted in terms of the free metal ion concentration ([Fe^{3+}]) showing a linear relationship (*right panel*). Figure taken from Sunda (2001) in *The Biogeochemistry of Iron in Seawater* [26] and used with permission from John Wiley & Sons, Inc.; and (b) demonstration of saturation of uptake at high concentrations (Fe′: the labile fraction). Note that phytoplankton grown under iron limitation (open circles) have a higher relative uptake rate that those grown under replete conditions (solid circles). Taken from Morel and Hering (1993) *Principles and Applications of Aquatic Chemistry* [11] and used with permission from John Wiley & Sons, Inc.

cell size (surface area) and the organisms have the ability to increase the number of transport sites if stressed, and this is what occurs to the phytoplankton previously exposed to low Fe levels. This increase in the number of sites under stressed conditions then results in the initial more rapid uptake of Fe by the cells that had been grown in Fe-deficient media.

As with passive diffusion, and shown in Fig. 8.4, it is not the total concentration in the water that defines the rate of uptake but the concentration of the metal ions and complexes that can kinetically react with the surface site. This has been termed the *kinetically-labile* or *reactive* fraction. In many situations, and especially in natural waters, this pool is essentially equivalent to the total dissolved "inorganic" fraction (i.e., [M^{n+}] and the sum of the concentration of all the inorganic complexes), often referred to as [M′] in the literature [10], as discussed earlier for Fe. The dissociation, and therefore the reactivity, of metal-NOM complexes is relatively slow compared to the inorganic complexes, and therefore it is assumed that these complexes are not kinetically-labile, especially for Fe^{3+}. For many of the transition metals, the kinetics of their dissociation reactions are relatively slow, but the same may not be true for some of the heavy metals, whose complexes dissociate much more rapidly (see Section 4.6.3). The differences in reactivity can be illustrated by comparing the values of the rate constants for water exchange [10], which vary from very large values (>10^9 s^{-1} for Pb^{2+}, Hg^{2+} and Cu^{2+}) to very small values (<10^4 s^{-1} for Fe^{3+}, Al^{3+}, and Cr^{3+}). Generally, metal ions that are more highly charged are less kinetically-reactive. This is problematic for Fe especially as it is a required nutrient and present in solution in the ocean mostly as organically-complexed compounds. This is discussed further in Section 8.3.1. Not all organic molecules are less reactive. For example, the fact that the EDTA complexes of most transition metals are relatively reactive kinetically, but not bioavailable, is used in the formulation of culture media for experimental exposures of phytoplankton under laboratory conditions [10], where the labile concentration in solution is a small fraction of the total concentration, but it is buffered by the addition of EDTA to the medium. In agreement with the above discussion, it has been demonstrated that Hg-EDTA are kinetically reactive (readily reducible), and can be reduced on a similar timescale to inorganic complexes. This is not true for Hg-NOM complexes [27].

In many early studies it was found that the free metal ion concentration was a suitable surrogate measure of the labile fraction and that the free metal ion concentration provided the best measure of the rate of accumulation of a metal by phytoplankton. This has been erroneously concluded to be equivalent to concluding that the free metal ion is the form of the metal that passes across the membrane. However, this does not need to be so. Also, it has been shown that there are instances where this model does not apply directly, which is apparent from the discussions about the importance of the kinetics of the reactions. It is also possible that the model of uptake, which is represented above as a ligand exchange reaction between the uptake site and the complex or ion in solution is not directly applicable as there are other scenarios that would lead to metal transfer to the uptake site. These include the formation of a tertiary complex between the active binding location on the transporter and the metal-ligand complex, which has diffused to

Table 8.3 The relative range in concentration of nutrient metals in marine phytoplankton in comparison to open ocean water concentrations and the predicted dissolved inorganic metal (free ion plus inorganic complexes) concentration ([M′]), based on thermodynamic modeling and measured ocean metal-organic matter complexation. The ratio of the cell quota to [M′] for one particular organism (*Emiliana huxleyi*, an open ocean coccolithophorid) (noted as EH in the table) is also given. Data taken from [2, 32] and references therein.

Metal	Range (mmol mol^{-1} P)	Dissolved Open Ocean Concentration (nM)	Predicted M′ concentration (pM)	Ratio EH
Fe	<1–16	0.5	0.1	~40
Mn	1–8	0.3	100–1000	<0.07
Co	0.01–0.5	0.02	≤10	<0.03
Cu	0.1–1.5	4	1–100	<0.02
Zn	0.2–2.4	5	10	0.04
Cd	0.01–0.8	0.6	1	0.4

the membrane surface [2]. Also, there is evidence that some bacteria release Fe-binding ligands (*siderophores*) into solution and that these are then taken up through a transport mechanism, then the rate of accumulation would be different to that described above (see Section 8.3.1). Additionally, it has been suggested that in natural waters with elevated levels of NOM, there is the potential for adsorption of the NOM to the cell surface and this could also facilitate uptake of metals either through enhancing the facilitated transport by forming tertiary interactions with the transport site [28–30] or by changing the membrane permeability [31].

The uptake constants (V_{max}) has been determined for a number of marine phytoplankton species and these range from around 2–100 μmol mol C^{-1} hr^{-1} for Fe, Mn and Zn. Cell quotas (Q) range from 20 to >100 μmol metal mol C^{-1} for Fe, and are less than 50 for Mn and Zn [2, 4]. The value of Q relative to the P content of marine phytoplankton cells, for the major required metals, is shown in Table 8.3. This illustrates the importance of the requirements for the metals Fe and Mn relative to other metals. Also shown in Table 8.3 is the open ocean concentrations of these metals, as well as the predicted inorganically-complexed (readily bioavailable) fraction of the metal in solution ([M′]), which is estimated based on the degree of complexation of the metals to organic ligands in surface ocean waters [2]. Using these data, it is possible to estimate the cellular requirement, in this case for *Emiliana huxleyi*, an open ocean coccolithophorid, relative to [M′]. The larger the value of this ratio, the more likely the metal is a limiting nutrient in ocean seawater. The likelihood of Fe being a limiting nutrient is obvious from this estimation, with the metals Zn and Cd also being potential limiting nutrients. Additionally, the relative requirements for the various metals depend on the degree to which substitution of metals in enzymes are possible.

As shown in Fig. 8.5, for most of the metals, the relative quotas vary by about an order of magnitude or more for different organisms, and thus there is the potential for limitation of, or competitive advantage for, different species depending on the circumstances. For most aquatic systems, Fe is likely to be present in low abundance in a bioavailable form, and given its major requirement, organisms have developed novel ways of obtaining sufficient Fe for cellular growth. This will be discussed in more detail in Section 8.3.1. In some cases, the different requirements can be related to differences between genera of organisms, while in other cases, phytoplankton that inhabit remote open ocean locations have lower requirements. Additionally, the potential for various metals to be substituted within enzymes and proteins can lead to different requirements depending on whether the organisms have this ability, or not.

An important role for Cd was initially suggested by its vertical distribution in seawater, as its distribution has been correlated with that of phosphate [32], as discussed in Chapter 6. Initially the reason for such a correlation was not known. However, there is now evidence that Cd can have a biological role in phytoplankton, in that it can substitute for Zn in the enzyme, *carbonic anhydrase* [33, 34]. Phytoplankton require Zn as it forms the active part of the carbonic anhydrase enzyme, which is a crucial part of carbon sequestration prior to its incorporation into organic matter during photosynthesis. However, under conditions where the concentration of Zn may be limiting it appears that some phytoplankton can produce a form of this enzyme that contains a Cd atom in the place of Zn. Thus, it is necessary to consider Cd as both a beneficial and a potentially toxic metal to aquatic organisms.

As noted in Equation 8.5, the cellular concentration is related to the growth rate and the uptake rate. If the metal accumulation leads to toxic effects, or interferes with the cellular processes such that the growth rate decreases, then the cell quota will rise rapidly as the growth rate decreases. Such a feedback leads to a sharp "edge" in most toxicity (growth rate) versus concentration curves for compounds that are accumulated without energy, as illustrated both in Figs. 8.1 and 8.3. While the facilitated transport sites are relatively specific there is the potential for the uptake of

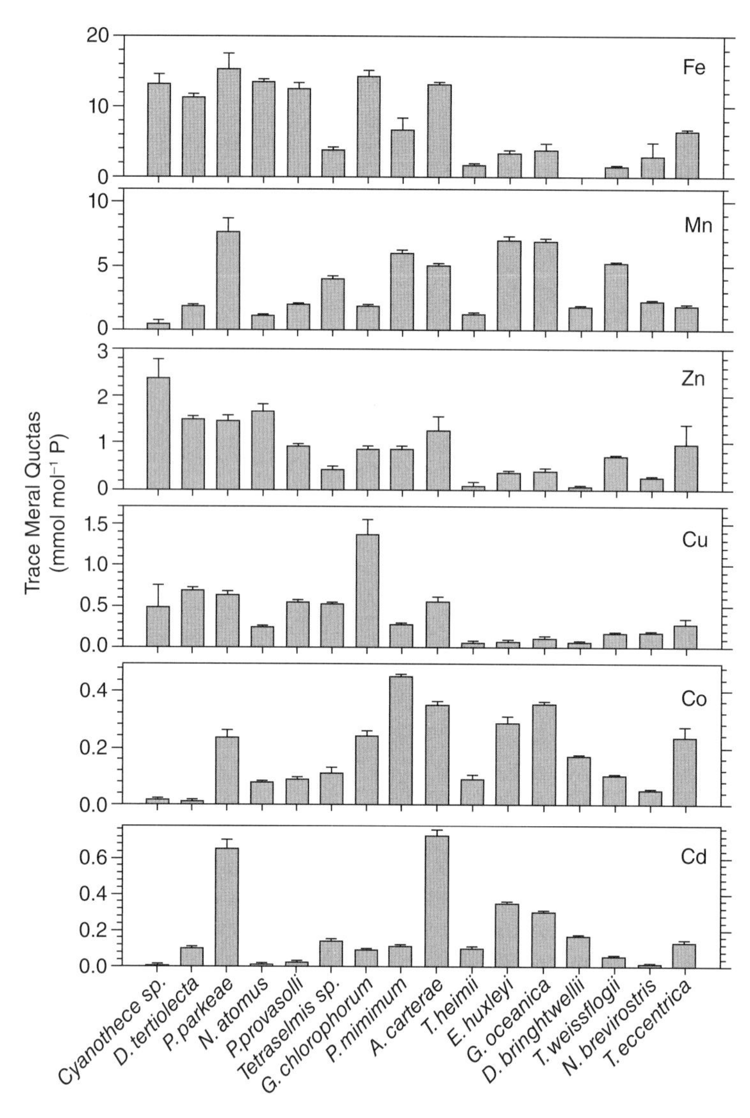

Fig. 8.5 Average cell quota values relative to phosphorous for a number of phytoplankton as summarized by Morel et al. (2004) in *Treatise on Geochemistry* vol. 6 [2], based on information in Ho et al. (2003) *J. Phycology* **39**: 1145–59. Reprinted with permission from Elsevier.

other metals "by accident" and this competitive uptake can lead to potential limitation of the accumulation of an important nutrient. For example, the antagonism of Cu and Cd on Mn uptake has been shown for phytoplankton [2, 4]. Such accidental uptake will occur for metal ions or complexes that have similar size, as most transition metal ions do. Finally, it is worth noting that there is the potential for more than one mode of uptake for a metal. For example, uptake of positively charged complexes has been suggested for example, $CuOH^+$ in addition to uptake of the free metal ion, and facilitated uptake occurs for Cd^{2+} and Ag^+, which can also partition passively as the neutral chloride complexes. The same is true for Hg, as there is evidence that its rate of uptake is not directly related to the neutral complex concentration in all instances [18, 35] and some bacteria have transporters for its active uptake and elimination via detoxification (reduction to elemental Hg) [36].

8.2.4 Active uptake

It is possible to transport molecules against the gradient in concentration, and this is the process of *active transport*, which is not particularly important in terms of trace metal uptake by microbes, but may be important in some situations and will be briefly discussed here. Active uptake can involve the transport of a single molecule or ion (*uniport* transport) or can involve the exchange of ions or molecules across the membrane -*symport* transport of two entities in the same direction, or *antiport* transport of entities in opposite directions. Such transport involves energy and transport systems are therefore termed ATPases, and a well-known example is the Na^+, K^+-ATPase. It has been shown in some toxicity and uptake studies that heavy metal ions can be transported across the membrane by this transport system, or can interact with this transport system in some fashion. Mostly, energy is required to transport ions across the membrane against a concentration gradient. *P-type transporters* are involved in the transport of cations and while these are mostly involved in transport of major cations such as Na^+, K^+, Ca^{2+} and H^+, specific transporters have been identified in bacteria that are able to transport Hg^{2+}, Cd^{2+} and Cu^{2+} ions [1]. These transporters are linked to a detoxification mechanism and the metals are taken up to be converted into less toxic forms. There is a body of literature that suggests for higher level organisms, uptake of many metals that can be toxic occur through channels designed for the active transport of major cations. As noted in Chapter 3, the ionic radii of many metal ions are relatively similar to that of Ca^{2+} and Mg^{2+}, and thus, depending on the selectivity of the uptake channels, these ions could be taken up inadvertently. This has been demonstrated in a number of instances. For example, Cd^{2+} and Pb^{2+} have been shown to be taken up through the Ca^{2+} channels in membranes [5]. For the macro nutrients, it is well known that there is antagonism in terms of phosphate uptake in the presence of As. Both exist as oxyanions in seawater, and As concentrations are sufficiently elevated that uptake of As can occur in low phosphate environments. Indeed, it has been suggested that marine phytoplankton have developed a mechanism of reducing and/or methylating the As^V as a means of detoxification.

While the details of the accumulation of the metalloids, As and Se, and the oxyanions, Mo and W, has not been a focus of research, it is expected that these elements are taken up through transport pathways and mechanisms similar to those used to transport the minor element oxyanions, sulfate and phosphate, and that because this requires the transport of a negatively charged species across the membrane, is likely an active process, involving energy and the transport of a counter ion. For most of the transition metals, uptake involves transport of the free metal ion, which as a positively charged entity is more easily transported by facilitated transport mechanisms across the cell membrane, or through the interaction of the free ion or a complex in solution with the surface transporter. Thus, there is a strong difference in the mode of transport of the cations and the oxyanions across membranes.

8.3 Essential trace metals

8.3.1 An overview of essential trace metals

As noted in Table 8.1, there are a number of essential trace metals and metalloids that are required for a wide variety of biochemical processes. Most metals form part of essential enzymes, especially the metals, besides Zn and Ni, which have active redox chemistry as part of their cellular function (catalysis or electron transfer). For a number of these metals, their concentration in environmental waters is not low enough to be a limiting nutrient and therefore these metals have received less study than the metals, in particularly Fe, and to a lesser extent, Zn, which have low ocean concentrations and therefore are potentially limiting nutrients (Table 8.3). It may seem apparent, and the discussion so far may support this supposition, that trace element limitation is mostly a concern of open ocean environments where concentrations are lower. This is not the case, however, as there is now a large number of studies that have shown micronutrient limitation in lakes [37]. Additionally, the accepted paradigm that "lakes are mostly phosphate limited while marine environments are mostly nitrate limited" is being challenged in both realms. As discussed in Chapter 6, and further in Section 8.3.1, Fe limitation has been shown to be a very important consideration in terms of ocean nutrient limitation, and it has been shown that Fe, and other metals, are also important limiting nutrients in freshwater environments ([37, 38] and cited references in both), and that the statement that "phosphate is essentially always the limiting nutrient" is not valid as both observational and experimental data indicates that most lakes are co-limited by N and P as well as Fe and micro-constituents [37, 38].

Some of the demonstrated limiting nutrients in freshwater environments are relatively easy to appreciate. For example, Mo, which is one of the few elements with a lower concentration in freshwater (Table 3.4; river/ocean ratio ~0.05), has been shown to be a limiting nutrient in freshwater ([37] and references therein), and this is not surprising. Also, concentrations of many elements in surface soils can vary widely depending on their geological origin and so the ratios shown in Table 3.4 do not reflect the large variability in concentration that can occur. Therefore, there are instances of Fe limitation in lakes where Fe concentrations are within the same order as their open ocean concentrations. In the review of the literature by Downs et al. [37], the most prevalent micronutrient limitations of those examined were with Fe and/or Mo (~75% of the studies that looked for either Fe or Mo limitation, demonstrated this). However, they were not always limiting for the same environments. Other limitations were found for Co (~70% of the studies) and Cu (one of five studies found it to be limiting). There is probably bias in these data as studies that either did not find limitation, or environments where limitation is not likely, would not be studied. Nonetheless, the results do clearly demonstrate the importance of considering micronutrient limitation in freshwater ecosystems.

In addition, while nutrient limitation could be considered more likely in large freshwater bodies, where surface water inputs may be more limiting relative to volume, and in oligotrophic systems, where overall nutrient inputs are low, there does not appear to be equivocal evidence to support either of these statements [37]. One reason for this may be that in freshwater environments algae tend to be larger, and to grow in colonies or in filamentous forms, compared to their open ocean counterparts. Green algae, diatoms and other microplankton are more important than the nano and picoplankton of the open ocean. Therefore, nutrient limitation can occur at higher dissolved metal concentrations. However, there is also the potential that there can be shifts in species composition in relation to changes in micronutrient loadings and thus longer term changes may counter shorter term impacts of a specific nutrient. Also, many freshwater ecosystems have higher NOM concentrations, both from in situ sources and terrestrial inputs, and it is possible, as a result, that a smaller fraction of the dissolved metal is present in a bioavailable form. Overall, the role of NOM in metal bioavailability to microbes is complex as increased NOM will tend to increase the dissolved concentration of a complexing metal while, at the same time, decreasing the relative bioavailability of the metal. In many cases, an increase in NOM will result in a decrease in the bioavailable fraction.

Besides Fe, which has received the most attention in terms of micronutrient limitation and bioavailability to photosynthetic and other microbes, Zn has also received attention due to its vital role in the essential enzyme, carbonic anhydrase [2]. The metals Cd and Co have also been the focus of recent study due to their potential to substitute for Zn in this enzyme. Besides being essential for carbonic anhydrase, Zn has a number of other biochemical functions. There is also a cellular requirement of Co due to its presence as a co-factor in the important cellular vitamin B_{12} (cobalamin), which is essential within the cellular machinery; in the transfer of methyl groups between compounds. Other first row transition metals that have a biochemical role are Mn, Cu, Ni, and to some degree V and perhaps Cr. According to [1], Mo and W, metals below Cr in the Periodic Table, which exist in solution as oxyanions, have biochemical roles within cells, although there is the possibility that these oxyanions can also be toxic at high concentration due to their interference with the uptake and cellular pathways associated with sulfate or phosphate.

Other elemental substitution in biochemicals has also been found. There is evidence for a role of Se in replacing S in amino acids and proteins and it is suggested that Se has a role in combating oxidative stress in organisms. Additionally, As is also found in a number of biochemicals, especially in marine microbes and organisms, and while the role of these As-containing compounds is not clearly understood, they appear to play a cellular role. The metalloids, As and Se, will be discussed in Section 8.4 as these are found in a number of different compounds and these appear to be produced by pathways that are similar to those that lead to alkylation of metals, both essential and toxic.

Overall, much of the attention on uptake by microbes has focused on the marine environment and this is mostly driven by the much lower metal concentrations in these waters, and by the importance of carbon uptake and sequestration in the surface ocean, especially given the potential for carbon removal to the deep ocean to be a mechanism to counter current increases in atmospheric CO_2. In these organisms, the trace elements are present in biological tissue at small concentrations, as illustrated by the data for marine phytoplankton in Table 8.3. Most of the essential transition metals appear to be taken up as free metal ions or through specific transport sites in the cell membrane, and therefore their mode of accumulation is described in Section 8.2.2. Overall, the steady state concentration, or quota, Q, is the steady state situation that occurs during exponential microbial growth and is dependent on the bioavailable metal ion concentration in solution, the rate of uptake, which depends on the number of transport sites in the cellular membrane, the ability of the organism to store the required metal internally, and the overall growth rate of the organism. As noted previously, slower growth will lead to a higher Q, all else being equal, while larger cells are often more easily limited by nutrient concentrations as uptake is dependent on membrane surface area, and concentration is related to cellular volume.

Most organisms have developed methods to store metals inside cells so that the free metal ion concentration in the cytoplasm is typically low. Storage of beneficial metals allows for survival during periods of low metals concentration in the environment while storage and binding of toxic metals alleviates toxicity. Often the mechanisms of storage are similar for metals regardless of their necessity or toxicity, as demonstrated for some marine phytoplankton that produce Cd binding proteins at both high and low Cd levels in the medium [2]. In phytoplankton, as in plants, the storage molecules are polypeptides, which are descriptively classified as *phytochelatins*, and which are made up of the small chain polymers of glutathione, cysteine and glycine ($[\gamma$-glutamyl-cysteinyl$]_n$-glycine), where n, the number of repeating units, is 2–11. Studies have shown that a wide variety of marine and freshwater algae produce phytochelatins [39, 40] and their formation in plants growing hydroponically in the presence of metals has also been demonstrated [41]. Terrestrial plants also produce a range of metal binding proteins and peptides [42, 43]. In addition to plants, phytochelatins are found in cyanobacteria and fungi. In invertebrates and higher level organisms, the metal binding proteins are often referred to as *metallothioneins*. These have a similar structure in terms of the active metal binding center being a cysteine-containing moiety. Besides phytochelatins, there is evidence for specific proteins for Fe accumulation and storage (*ferritins*) in bacteria and other organisms [2]. More research in this area is likely to further expand the type and diversity of the metal storage and detoxifying molecules in microorganisms, plants and animals.

Phytochelatins bind very strongly with Class B metals and for example those isolated from marine algae appear to bind most strongly with Cd, and to a lesser extent with Zn, Cu, and other metals. Interestingly, studies looking at a variety of metals showed that Hg, Ag, and Pb did not induce substantial phytochelatin production at environmentally relevant concentrations in the medium [44]. This probably reflects the fact that these metals bind strongly to thiol group in general and as a result bind mostly with compounds in the membrane of microbes, and therefore do not penetrate into the cytoplasm except under high metal loadings. Induction of phytochelatin production is related to increasing metal concentration in the medium, and initially metals appear to be bound by glutathione present in the cells. While increasing metal loading will promote phytochelatin production in marine algae, and increases in the production of cysteine and glycine, it does not result in elevated glutathione concentrations [45]. In contrast, for a freshwater green algae, while increasing concentrations of metals (Zn, Pb, and Ag) and metalloids induced phytochelatin production, these metals had differing impacts on cellular glutathione, with the presence of Pb and Ag increasing its production, while it decreased in concentration in the presence of Zn. These differences may relate to differences in the mechanism of binding of the different metals across organisms. As glutathione has many roles in the cell, the minimal impact of metal uptake reflects the fact that its role in phytochelatin formation is not its major cellular function.

As the focus is trace metals in aquatic systems, the accumulation of metals in terrestrial plants is beyond the scope of the book. However, there are many papers in the literature discussing the uptake of metals by rooted plants and their induction of phytochelatins, and recent research has also focused on using such plants as metal accumulators for the removal of metals from contaminated environments. The interested reader should consult recent reviews [40, 42, 43] and the extensive literature on this topic.

In terms of the assimilation of metals by microbes, much of the focus of environmental studies of essential metal acquisition by eukaryotic microbes has focused on the requirements and nutrient limitations of Fe and Zn. While there has been some focus on the other transition metals, specific sections will detail the wealth of information on Fe, Zn, and a few other metals. Some aspects of the accumulation of other metals have already been discussed. For example, while Mn concentrations in ocean waters appear to be sufficient based on the calculations in Table 8.3, there is a specific interaction that can hinder the accumulation of Mn, due to the competitive uptake of Cu through the same transport site. This has been demonstrated in laboratory culture experiments and is illustrated in Fig. 8.6 [4]. The growth rate in culture is increased by either increasing the Mn concentration or decreasing the Cu concentration (Fig. 8.6a), and the resultant cellular Mn concentration is a function of the relative Mn : Cu concentration (Fig. 8.6b). Overall, the growth rate is a strong function of the intracellular Mn (Fig. 8.6c), indicating its essential need by the organism, and therefore the toxicity and impact of Cu in this instance is through its hindrance of Mn uptake, and the resultant cellular Mn deficiency that results ([4] and references therein) rather than through the direct toxicity of Cu. Current research is demonstrating other instances where there may be such a synergistic impact of one metal, likely through its competitive uptake.

The uptake of oxyanions of As and Mo through channels designed for phosphate and sulfate, and the noted inhibition of sulfate reduction by addition of Mo, is another example of the complexity and potential interactions between elements during uptake and within cells. Such interactions are not always negative, as discussed in Section 8.2.2 where substitution of Cd or Co for Zn in carbonic anhydrase results in the formation of an enzyme that can still function, and thus the presence of Cd in solution at low Zn concentrations does not hinder but enhances growth rate.

8.3.2 Focus topic: Iron uptake by microbes in marine waters

The uptake of Fe by microbes is thought to be primarily through its transport by specific transporter proteins embedded in the cell membrane, and thus that uptake is a facilitated transport mechanism [26]. It has been further demonstrated that a variety of labile Fe complexes can react with these cell surface sites and therefore that the uptake is related to the concentration of reactive Fe species in solution, which is denoted as [Fe′] ([26] and references therein) (Fig. 8.7). Different organisms have somewhat different requirements and thus maximum growth rate is obtained at different concentrations of Fe′ in solution (Fig. 8.7a). Given the relatively slow rate of reaction of Fe^{III} complexes, it has been found that the rate of uptake is often limited by the rate of the exchange reaction between the labile Fe^{III} complexes in solution and the surface site of the transporter on the membrane. Thus, surface normalized uptake rates are often relatively similar for different species of phytoplankton [2, 26] (Fig. 8.7d). As a result, Fe limitation is most severe for large organisms given their lower surface area : volume ratio. As shown in Fig. 8.7(a), the highest growth rate at any Fe′ concentration is for the smallest phytoplankton, *Thalassiosira pseudonana*, an open ocean diatom. Its coastal relative, *T. weissflogii*, whose diameter is around three times larger (volume, 27 times larger) does not grow as fast at any Fe′ concentration even though both organisms have a similar cellular quota, normalized to carbon (Fig. 8.7b). There is a limit in terms of the maximum relative uptake rate, normalized to the cell surface area, in the region of $1–2 \times 10^3\,nmol\,m^{-2}\,d^{-1}$.

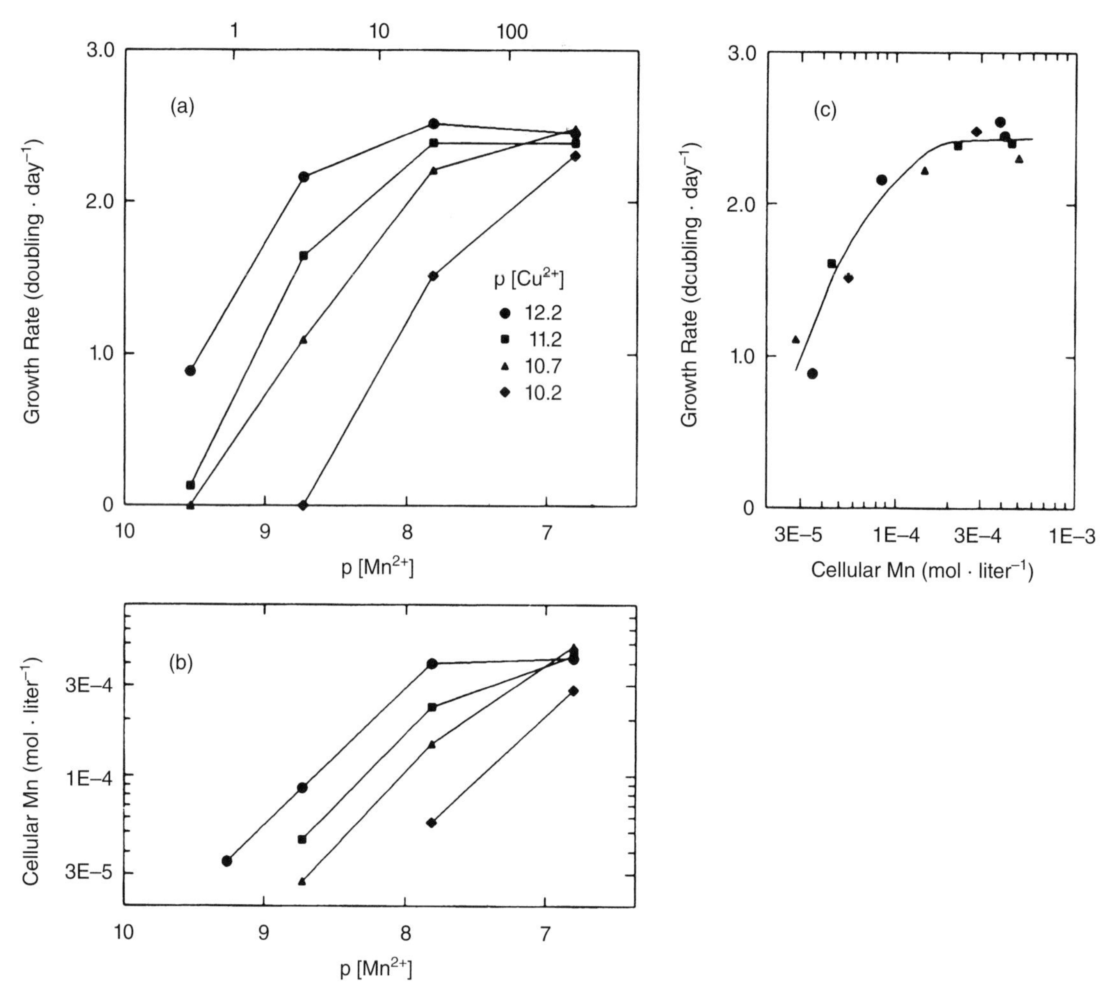

Fig. 8.6 Impact of copper (Cu) on the uptake of manganese (Mn) into phytoplankton. Figures show the relative growth rate against the free Mn ion content and the cellular Mn, as well as the relationship between cellular Mn and the exposure concentration. Figure reprinted from Sunda (2001) in *The Biogeochemistry of Iron in Seawater* [26] and used with permission from John Wiley & Sons, Inc.

These figures focus on the mechanism of uptake but, additionally, there is the need for the Fe species to diffuse to the cell surface and the size (width) of the unstirred (diffusion) layer around a cell is a function of cell volume. It can be shown that for small cells (< ~60 μm diameter) diffusion limitation is not important (which relates to all the cells in Fig. 8.7), and that cells < ~10 μm in size have substantially higher specific Fe uptake rates than larger cells. Not surprisingly, most open ocean phytoplankton are therefore small (<10 μm). In cases when the ocean is rapidly "fertilized" with Fe, either due to dust input or experimentally, there is often an algal bloom which is dominated by larger algal species, which have a competitive advantage at higher Fe concentrations, as shown in Fig. 8.8. Here, the addition of Fe to waters from the equatorial Pacific Ocean results in rapid growth of diatoms, at the expense of *Prochlorococcus*, the most abundant but small phytoplankton (<1 μm diameter) typically found in these ocean waters. *Synechoccocus*, a larger cyanobacteria, does not appear to be affected to the same degree. This pattern has been seen in other studies, where Fe addition has resulted in a shift in species composition. Additionally, there are studies in freshwater environments showing the impact of Fe limitation on algal growth and composition [38].

Alternatively, as larger organisms cannot physically increase their uptake rates, they appear to have adapted through evolution by reducing their cellular Fe requirement, as demonstrated by comparison of the cellular quota of coastal versus marine diatoms [26], and shown in Fig. 8.7(b). As noted previously, the uptake mechanics follow Michaelis–Menten kinetics and therefore the primary mechanism for increasing the rate of uptake is to increase the number of transport sites within the cell membrane, but this increase is finite. An alternative strategy is to reduce the cellular requirement by decreasing the number of Fe-containing molecules, or by rapidly recycling the Fe from molecules if no longer required, or by storing Fe internally under conditions when Fe concentrations in the medium are high.

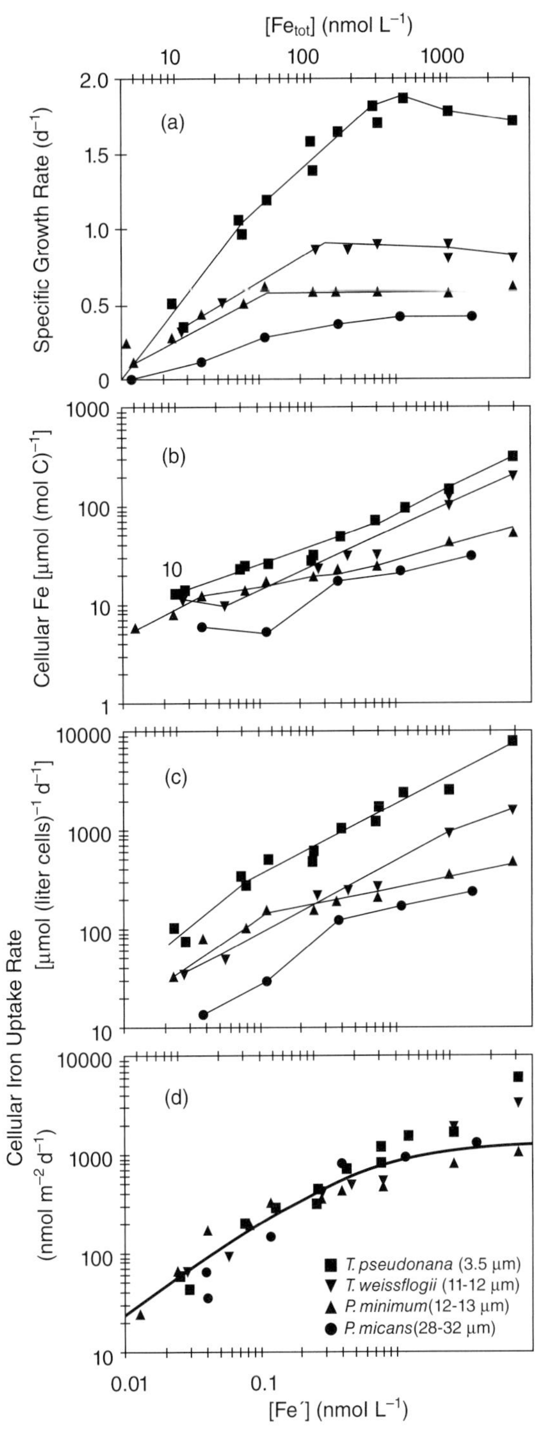

Fig. 8.7 The uptake rate parameters for iron for a number of marine phytoplankton plotted in terms of the labile iron ([Fe′]) concentration: (a) the specific uptake rate; (b) the cellular quota; (c) the cellular uptake rate per cell; and (d) the surface area normalized uptake rate. Taken from Sunda (2001) in *The Biogeochemistry of Iron in Seawater* [6] and used with permission from John Wiley & Sons, Inc.

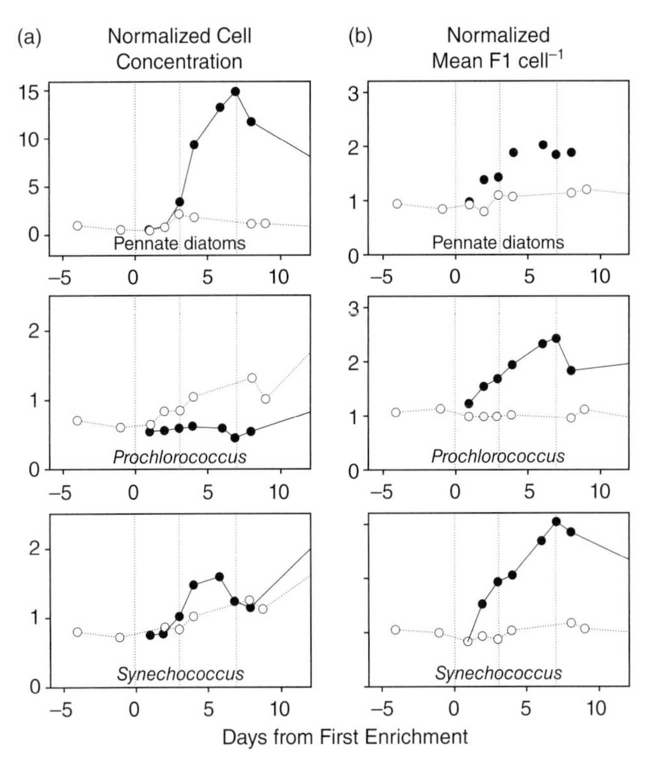

Fig. 8.8 The effect of iron (Fe) addition to the ocean on the growth of different phytoplankton groups. Graphs show the change in cell concentration with time after the Fe addition. The response of different phytoplankton groups in the iron fertilized patch (closed symbols) relative to outside (open symbols) during the equatorial Pacific Ocean IronEx II experiment. Response shown as (a) normalized cell concentration (to initial values outside) and (b) as normalized mean pigment fluorescence (Fl) per cell. Fluorescence is orange fluorescence for *Synechococcus* and red for *Prochlorococcus* and pennate diatoms. Figure drawn using information from the IronEx II experiment in the equatorial Pacific Ocean, and reprinted from Morel et al. (2004) in *Treatise on Geochemistry* vol. 6 [2], reprinted with permission from Elsevier.

In terms of accumulation, the discussion so far has focused on Fe^{III}. It has also been shown that Fe^{II} is readily taken up across the cell membrane. As Fe^{II} forms weaker complexes with dissolved ligands (Chapter 3), any reduced Fe is more bioavailable than Fe that is oxidized. Also, the rate of reaction of Fe^{II} complexes is more rapid than those of Fe^{III} (based on the values of their water exchange coefficients which differ by four orders of magnitude for Fe^{2+} compared to Fe^{3+} [10]), and therefore the Fe^{II} complexes are kinetically more labile. As a result, the abiotic or biotic reduction of Fe at the cell surface or in the immediate environment can lead to a substantial enhancement in its uptake rate. As discussed in Chapter 5, there is the possibility of charge transfer reduction reactions occurring photochemically, whereby a Fe^{III}-organic ligand complex absorbs radiation and the complex is decomposed, releasing Fe^{II} into solution and an oxidized form of the organic compound. This Fe^{II} is in a highly reactive (bioavailable) state and can be assimilated if it is not first oxidized. Complexation may enhance the stability of the reduced form (Fe^{II}), and its uptake, as the exchange kinetics of the Fe^{II} complex reaction with the surface transporter ligand are likely to be sufficiently rapid [26]. For example, model estimates for a Fe limited culture of a coastal diatom under typical light conditions showed that the photochemically produced Fe^{II} was a small direct source of Fe for uptake (~4% of the uptake) [10]. However, even though the majority of the Fe^{II} was re-oxidized, the labile Fe^{III} complexes formed contributed substantially to the labile Fe pool that was then assimilated by the diatom. Thus, the importance of photochemistry cannot be ignored in terms of Fe bioavailability to microbes in environmental waters.

In addition to photochemical reduction of dissolved complexes, it is also possible for Fe^{II} to be released into solution from Fe (hydr)oxide solids due to reductive dissolution mechanisms (Chapter 5). Much of the Fe in surface seawater is present as particulate (>0.4 μm) and while some fraction of this Fe is undoubtedly biological material, there is also likely crystalline and amorphous Fe (hyd)oxide phases present in suspension. Whether these particles are coated with NOM or exist as uncoated surfaces will have a large impact on their potential and rate of photo-dissolution. Additionally, it is possible that Fe-containing particles could be found in micro-reduced zones where the reduction to Fe^{II} would be favored [26]. Larger organisms, and even some phytoplankton, may be able to acquire Fe by "phagocytosis" of particulate material. Overall, the uptake of Fe by aquatic eukaryotes is through interaction of the labile Fe with the surface transport sites and the resultant uptake, and therefore is highly dependent on the speciation and bioavailability of the Fe in solution.

In contrast to algae, bacteria are known to release Fe-binding organic compounds (*siderophores*) into solution to enhance and mediate Fe uptake. Two mechanisms have been identified: (1) the Fe-siderophore complex is taken up, and the Fe released inside the cell, with the recycling of the chelator (siderophore) to the external medium for further Fe assimilation; or (2) the Fe-siderophore complex interacts with the surface transporter and the Fe is exchanged, leaving the unbound chelator in the external medium. It has been suggested that these siderophores can complex and solubilize Fe from both the dissolved and particulate phase [26]. Therefore, bacteria may solubilize Fe after direct attachment to the particulate surface and the release of Fe-dissolving organic molecules into the immediate environment. Also, it appears that some compounds that are released by microbes bind Fe and are then photochemically labile, and thus their primary role may be in producing labile Fe for accumulation, rather than the resultant complex being taken up by the microbe [46]. Thus, there is a potential dual role for siderophores in that they may actively transport the Fe^{III} as a complex to the surface transport site, or may be involved in the production of labile Fe′ in solution that is then assimilated.

The siderophores identified and isolated from marine bacteria are low molecular weight compounds, typically <1 kDa. Their production is typically mediated by gene expression under low Fe conditions in the medium. In both the ocean and in freshwaters, most siderophores identified are hydroxamates and catecholates and are produced by a variety, but certainly not all, bacteria and cyanobacteria. There is no clear evidence to support their production by eukaryotic phytoplankton [2]. Therefore there is a distinct difference in the uptake mechanisms for Fe for bacteria compared to eukaryotes.

Furthermore, in terms of uptake and bioavailability, there is the possibility of surface reductases associated with the cell being involved in Fe reduction, and this may involve specific complexes of Fe. Biological reduction of Fe^{III} at the cell surface has been demonstrated in oceanic as well as terrestrial environments [10, 26]. These processes are active and are driven by cellular energy usage (NADPH/NADP) and do not require direct association of the Fe with the cellular surface, and may actually occur through a series of electron transfer reactions. Such mechanisms have been identified in the presence of marine algae, and may involve pathways such as those associated with nitrate reduction.

A final comment concerns "luxury accumulation" of Fe. At high Fe concentrations it would be advantageous to accumulate and stored Fe internally, and this has been demonstrated for marine algae and bacteria [26]. Studies have shown that Fe quotas for both coastal and marine algae are often 2–3 times the required levels for maximum growth in conditions of high concentrations in the medium, but there is documentation of even higher levels of Fe accumulation, more than 2 orders of magnitude higher than required levels. However, details of the specific storage mechanisms for Fe are not well known for microbes.

Given the discussion above, the impact of organic molecules, either released intentionally or due to microbe decomposition and other mechanisms, on the speciation of Fe and other metals in surface ocean waters is substantial, and a high fraction of most metals are bound to organic ligands in natural waters. This has been well documented and was discussed in Chapter 6. The impact of organic complexation is shown in Fig. 8.9 for Fe, Cd, and Cu [2]. The concentrations of the complexed Fe, presumably to organic ligands, and the total ligand concentration are derived from experimental titrations using electrochemical detection techniques [32], which allow for an estimation of the overall binding constant of the metal to the ligands and their associated concentration. The data in Fig. 8.9 show that most of the Fe is complexed in surface waters as the Fe′ concentration is ~0.1 pM, compared to a total concentration several orders of magnitude higher, and that the ligand concentration is higher than the total Fe, but appears to have a similar

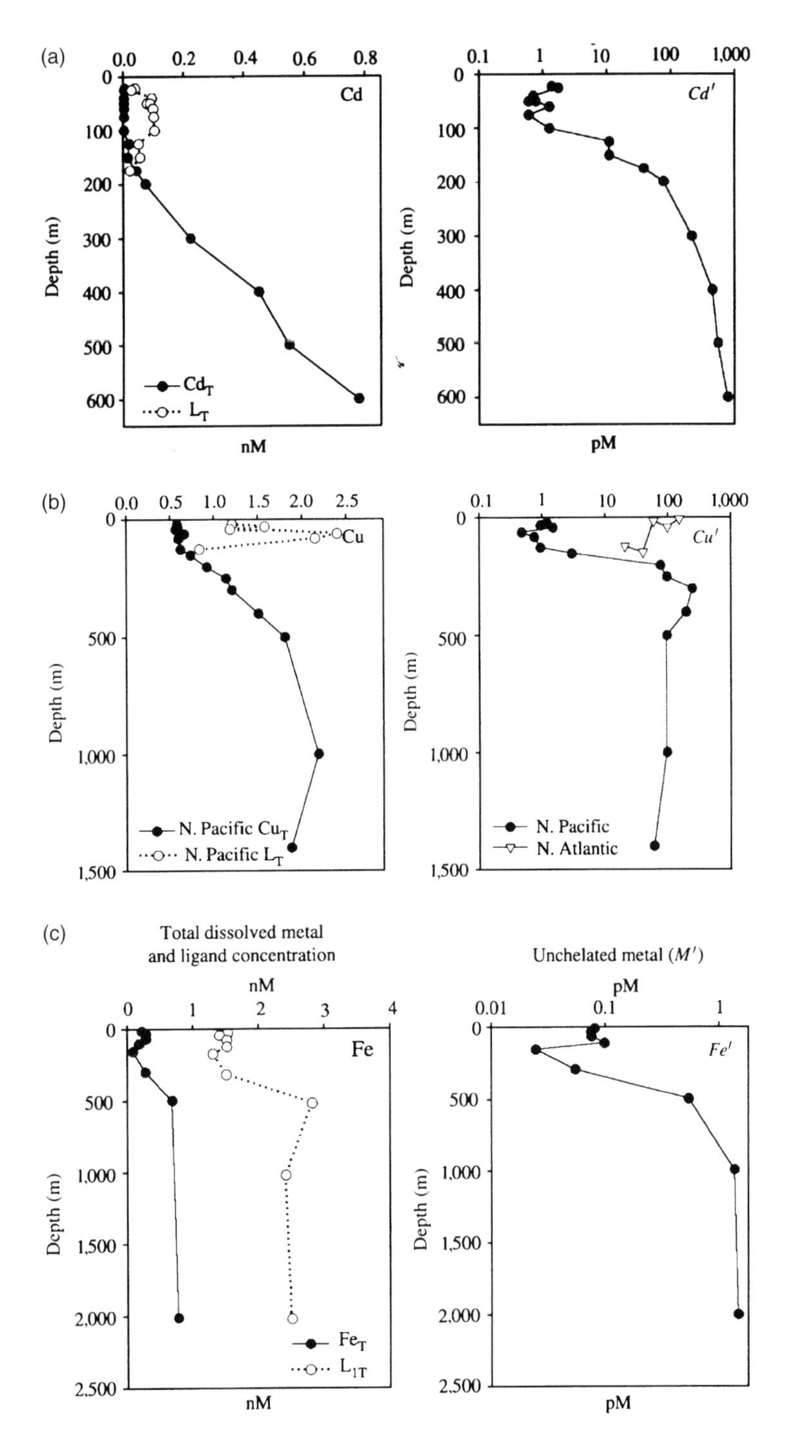

Fig. 8.9 Vertical profiles of the total concentration (M_T), total ligand concentration (L_T) and labile concentration (M') for various metals (M), determined using various speciation techniques: (a) Cd; (b) Cu, and (c) Fe. Data from the North Pacific, with additional data for the North Atlantic for Cu included. From Morel et al. (2004) in *Treatise on Geochemistry* vol. 6 [2], reprinted with permission from Elsevier. Data in figures from numerous sources as detailed in Morel et al. (2004).

vertical profile. For Fe, the ligands appear to occur throughout the measurement area (upper 2000 m) in contrast to Cd and Cu, where the binding ligands are confined to the surface waters. It has been suggested that the Fe-binding ligands are a mixture of siderophores released by bacteria, other biochemicals, and Fe-S clusters, and there may be a substantial "colloidal" component to this unreactive fraction [2]. Overall, it is clear that the concentration and speciation of Fe in ocean waters is strongly impacted by the presence of organic ligands, and by the activity of microbes in these waters. The concentration of Fe^{III} in seawater in the presence of the solid phase (i.e., at saturation) is less than 0.1 nM and therefore the presence of the binding ligands enhances substantially the total dissolved Fe concentration, as shown in Fig. 8.9 and discussed in Chapter 6.

8.3.3 Focus topic: Transition metals and carbon fixation by microbes

Photosynthetic organisms need to acquire CO_2 from their environment for its conversion into organic matter. In aquatic systems, the most important forms of inorganic carbon are dissolved CO_2 and HCO_3^-. Organisms have devised methods of accumulating both species, with CO_2 uptake mostly relying on the passive diffusion across the cell membrane, and HCO_3^- uptake involving co-transport of the anion with a metal co-ion, such as Na^+, in an active uptake process. Overall, the degree to which the carbon acquisition is by uptake of CO_2 or HCO_3^- depends on the photosynthetic organism, and the conditions of the environment (seawater typically has higher HCO_3^- concentrations than freshwater). To allow for continual CO_2 uptake, there is the need to use or store the CO_2 to reduce the internal concentration, and allow for continual passive diffusion. The conversion of CO_2 internally to HCO_3^- is the most obvious and common method for inorganic carbon storage in the cell.

The most important enzyme for CO_2 storage in microbes is *carbonic anhydrase*, an enzyme that catalyzes the conversion of CO_2 to HCO_3^-, and *vice versa*, as CO_2 is the form of carbon used in the initial stages of organic matter formation (*carbon fixation*). However, there appears to be numerous other roles for carbonic anhydrase within cells and the enzyme is found in both prokaryotic and eukaryotic organisms, including non-photosynthetic animals [47]. A number of different forms of the enzyme have evolved, and some organisms have more than one form of carbonic anhydrase. Carbonic anhydrase is a metal-containing enzyme, with the most common metal being Zn, although there is evidence that other metals such as Cd, which is in the same group of the periodic table, and Co can substitute for Zn in forms of the enzyme that are similar and functional [2].

It has been shown that Zn is part of many proteins within cells and that these Zn-containing enzymes perform a variety of functions. However, Zn is not always present in sufficient concentrations in aquatic environments, and this is especially true for open ocean surface waters. Organisms have therefore devised methods to accommodate for low Zn levels. This is in contrast to Fe, discussed in Section 8.3.1, where no substitution has been found by other metals for Fe in the various enzymes. This is likely related to the fact that these enzymes rely on the redox capacity of Fe, and therefore substitution of another metal, with similar redox characteristics and size is not easily accomplished. Substitution of Co for Zn in carbonic anhydrase is found more often in cyanobacteria, and especially the small, open ocean cyanobacteria, such as *Prochlorococcus* and *Synechococcus* [2]. For these organisms, Co is required even in the presence of elevated Zn concentrations for other enzymatic functions. In contrast, open ocean coccolithophorids appear to be able to substitute Cd for Zn in carbonic anhydrase, and possibly other enzymes. For open ocean diatoms, both Co and Cd can substitute for Zn. The substitution for Zn has also been demonstrated for freshwater algae and it was shown that addition of both Co and Cd to the medium can enhance the growth of freshwater microalgae, although the effectiveness of Cd was less than Co [48].

The impact of the various metals on the growth of the coastal diatom, *Thalassiosira weissflogii*, is shown in Fig. 8.10(a). It is apparent that growth can be increased by increasing the CO_2 concentration, which increases its rate of uptake, or by increasing the levels of either Cd or Zn. This impact is also illustrated by isolation and characterization of the amount and activity of the carbonic anhydrase in the cells (Fig. 8.10b). Decreased CO_2 leads to an increase in the amount of storage (more carbonic anhydrase), and in its activity. Addition of either Co or Cd to low Zn cultures results in enhanced activity but there is a clear difference in their impact. While it is apparent that Co will substitute directly for Zn in carbonic anhydrase, in the presence of Cd, the organism produces a different enzyme. Thus, the overall mechanisms of substitution are different. An examination of the ionic radii of the +2 cations shows that Co^{2+} and Zn^{2+} have very similar ionic radii, but that of Cd^{2+} is larger (Tables 3.3 and 3.7) and this provides an explanation for the observations.

The utilization of Zn in surface ocean waters is confirmed by oceanic vertical profiles (Fig. 8.11). In the upper ocean (top 250 m), Zn is strongly depleted and its vertical profile is similar to that of the major nutrients, suggesting its uptake in surface waters and its regeneration at depth due to the release of Zn during decomposition of sinking organic matter. There are strong binding ligands in the surface ocean for Zn, but the profile shows little structure suggesting that these ligands are not necessarily released by organisms primarily to bind and transport Zn, as occurs with Fe (Section 8.3.1). Additionally, while there is substantial evidence for the limitation of primary productivity by Fe, there is no conclusive experimental evidence for Zn, Cd, or Co limitation in both marine and freshwater environments. This is

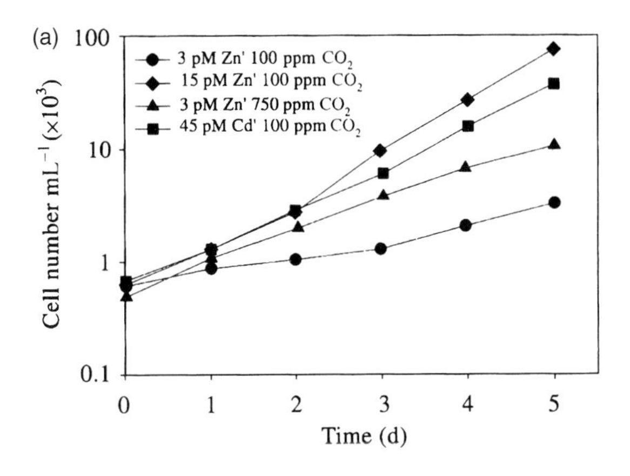

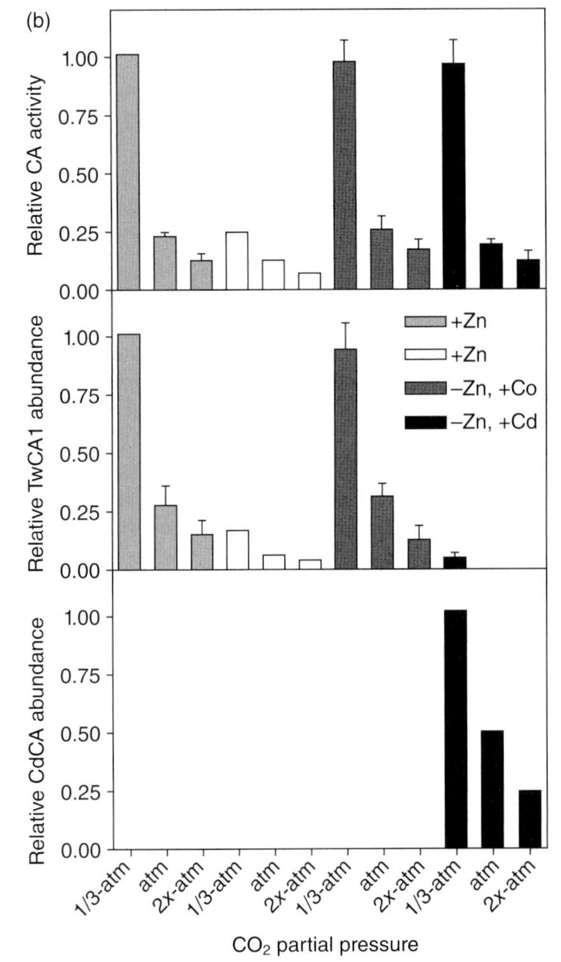

Fig. 8.10 (a) Graph showing the effect of different concentrations of Zn, Cd, and carbon dioxide (CO_2) on the growth rate of the marine diatom, *Thalassiosira weisflogii*. The labile metal concentrations is given in figure; (b) the carbonic anhydrase (CA) activity of the diatom under various treatments showing the amount of the Zn-containing (TwCA1) and the Cd-containing (CdCA) forms of the enzyme expressed. Figures reprinted from Morel et al. (2004) in *Treatise on Geochemistry* vol. 6 [2], reprinted with permission from Elsevier. Data in figures from sources as detailed in Morel et al. (2004).

likely due to the ability of organisms to use a variety of metals in the specific enzymes. Substitution is possible as the major role of the metal is providing the conformation framework and structure to the enzyme.

Besides carbonic anhydrase, a number of other enzymes are needed for carbon fixation, and biomolecule production. In terms of carbon fixation, Fe and Mn are important metals in enzymes involved in photosynthesis. The two parts of the photosynthetic pathway involved in light harvesting and electron transport contain Fe centers. The number of Fe atoms required for the overall photosynthetic molecular apparatus is greater than 20 [2], with the actual number depending on the relative proportion of the various fractions of the overall complex molecular system. This is the largest Fe requirement for a photosynthetic cell. Coupled with the light harvesting complex are the enzymes involved in the oxidation of water, and these are Mn-containing enzymes. These enzymes represent an essential requirement of a photosynthetic organism for Mn [2]. The relatively high concentration of Mn in ocean waters results is little evidence of its depletion in surface relative to deeper waters (Fig. 8.11). Finally, while Fe-containing molecules are involved in photosynthesis and carbon fixation, Fe is also present in the enzymes associated with organic matter respiration. Overall, Fe is an essential part of the cellular apparatus of microorganisms, especially those that photosynthesize.

In a related way, Fe is also important in carbon fixation in that it is needed for the enzymes associated with nitrogen assimilation, and therefore in the formation of proteins and other N-containing biochemicals. Regardless of the way that nitrogen is acquired by photosynthetic microbes, it is assimilated into organic material in its reduced form (NH_4^+) and the enzyme associated with this assimilation of N is Fe-containing. Many phytoplankton assimilate N as nitrate or nitrite, and thus reduction is required prior to their incorporation into biochemicals. The various enzymes involved in the myriad of potential transformations of N, which has many oxidation states (+5 in nitrate, +3 in nitrite, +1 in N_2O, 0 in N_2, −3 in NH_3 and a variety of oxidation states in small organic molecules, such as amines), mostly contain Fe, Cu and Mo, with Fe being the most widespread and important metal. Specifically, for example, nitrate needs to be reduced and the nitrate reductase enzyme ($NO_3^- \rightarrow NO_2^-$) contains both Fe and Mo, while the nitrite reductase enzyme ($NO_2^- \rightarrow NH_3$) has a large Fe requirement as the necessary reduction requires six electrons per molecule [2]. Thus, algae growing while assimilating nitrate have a larger Fe requirement than those utilizing more reduced forms of N. Studies have shown that such limitation of N uptake at low Fe levels occurs in both marine and freshwaters [2, 49]. As an example, in Lake Superior, both P and Fe limitation result in low utilization of N, which has increased as a result of human activity [49]. Other enzymes involved in N metabolism are: (1) the enzyme involved in nitrogen fixation ($N_2 \rightarrow NH_3$) which contains Fe

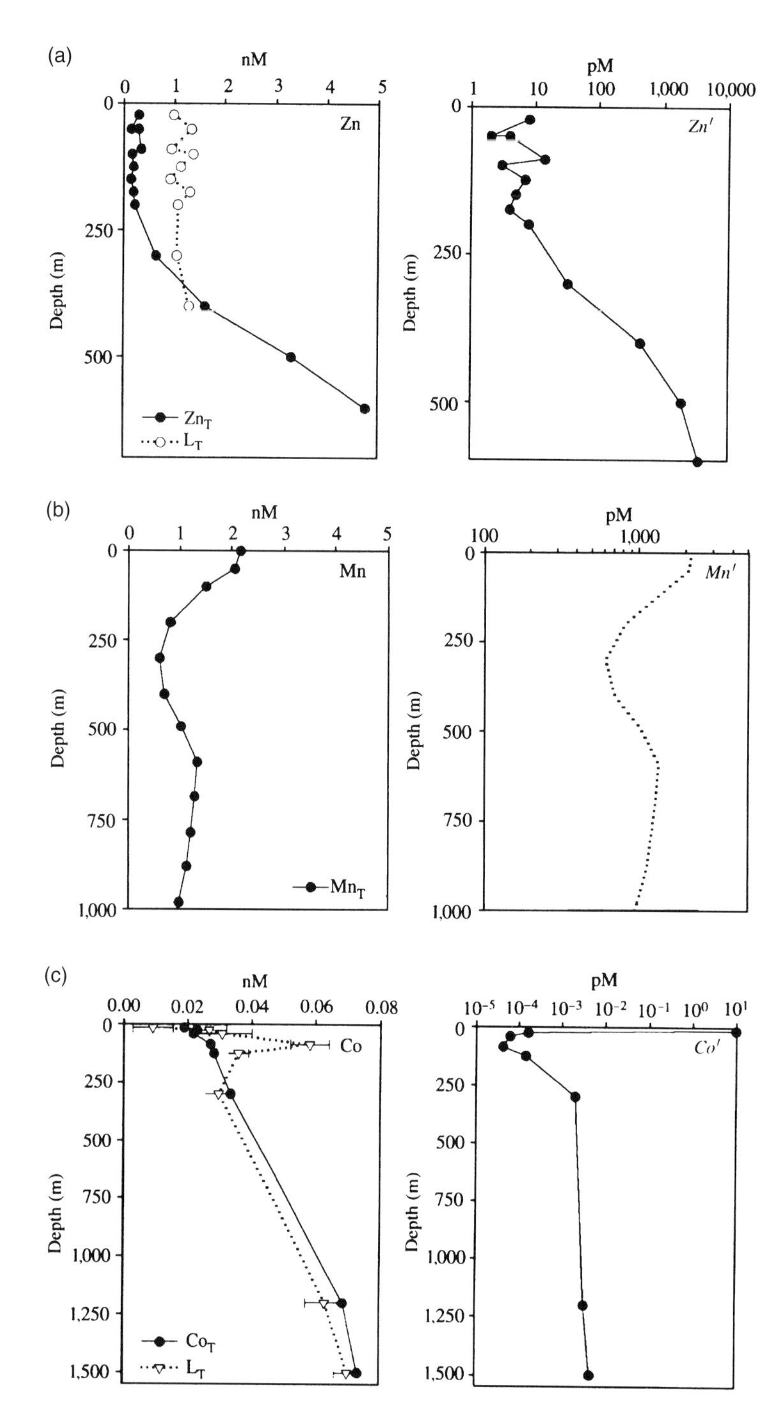

Fig. 8.11 See Fig. 8.9 caption for details. Vertical profiles of M_T, L_T and M′ for (a) Zn, (b) Mn and (c) Co. Reprinted from Morel et al. (2004) in *Treatise on Geochemistry* vol. 6 [2], reprinted with permission from Elsevier. Data in figures from numerous sources as detailed in Morel et al. (2004).

and Mo; and (2) urease, which converts urea to ammonia, and is a Ni-containing enzyme.

Therefore there is a clear role for transition metals, especially those in the first transition series, in the molecular machinery of cells and especially in the enzymes associated with photosynthesis and the production of organic matter from inorganic carbon. This observation leads one to the question why organisms may have utilized metals in their biochemistry that appear to be present in environmental waters at low concentrations. This question has puzzled bioinorganic chemists and ecologists [50] and the answer likely lies in the fact that the evolution of the cellular mechanisms and enzymes is an extremely slow process and that many of these enzymes were developed early in Earth's history. These early microbes lived in a very different environment, one that lacked an oxygenated atmosphere. As discussed in the next section, the incorporation of Fe, Mn, and other trace metals into molecular enzymes makes sense when examined in this context.

8.3.4 The biochemistry of metals and their presence in the early biosphere

The hypothesis has been put forward that the metals incorporated into enzymes and biochemicals were determined by their being the dominant metals present in solution in the environment in the early stages of Earth's development, particularly during the period prior to the oxygenation of the biosphere. This idea can be paraphrased by stating that in the selection of chemical elements for incorporation into cellular function, evolution responded by incorporating those metals whose particular chemistry and abundance allowed their use with less energetic cost required for uptake, given the function for which they are required (electron transfer and catalysis of carbon polymer formation) [50]. Given the absence of atmospheric oxygen in the ancient biosphere, the metals that are relatively soluble in sulfidic or oxygen-free waters, such as Fe, Mn, Co, and Ni, are those that were originally incorporated into biochemicals rather than those that were highly insoluble, such as Hg, Pb, Ag, and the metalloids under these conditions [50–52]. As many early biomolecules contained reduced thiols, the propensity of the heavy metals to bind to thiols and S-containing groups more strongly that the biometals (Fe, Mn, Zn) accounts to some degree for their potential toxicity. Elements such as Zn, Cd and Cu are intermediate in terms of their concentration in the presence of sulfide, and were incorporated into enzymes, but only later in evolution, and are mostly absent from the biomolecules of primitive organisms [50].

In considering the early Earth, compounds containing Fe and S were likely abundant in the low oxygen early environment. The evolution of oxygen on Earth compared to the emergence of various organisms is detailed in Table 8.4. The presence of both inorganic and organic sulfides probably had an important impact on the evolution of Earth. Reactions with metals and sulfide that could have occurred and triggered catalysis and other reactions, forming organic molecules, are [50]:

$$FeS + H_2S = FeS_2 + H\ (bound) + energy$$

$$Ni_4S_3 + H_2S = 4NiS + H\ (bound) + energy$$

$$Co_9S_8 + H_2S = 9CoS + H\ (bound) + energy$$

where H(bound) refers to the incorporation of H with C into the formation of an organic C compound. Such reactions could also occur with Cu and Mo sulfides and these compounds are especially electron-rich. Also, the change in electron configuration from high spin in FeS to low spin in FeS_2 could conceivably have been one major energy source for the origin of life [50].

Table 8.4 Comparison of the timing of major developments in the evolution of organisms on Earth and the corresponding oxygen level in the lower atmosphere. Compiled from [50].

Approx. Time (YBP)	Oxygen Level (relative to Present)	Event/Dominant Organisms
$4–3.5 \times 10^9$	$10^{-4}–10^{-3}$	Prokaryotes
$3–2.5 \times 10^9$	$<10^{-2}$	Aerobic prokaryotes
$2.5–2 \times 10^9$	$\leq 10^{-2}$	Single cell eukaryotes
$\sim 10^9$	~0.2	Multi cell eukaryotes
0.5×10^9	~1	Modern Plants, Animals
10^5	~1	Mankind

Note: YBP: Years before present.

In examining the potential functions of various metals, it can be concluded that given their chemistry and rates of reaction that Fe, Co, Ni and their companion metals in the second and third transition series (Rh, Pd, Pt, and Ir), in combination with S and Se, form a catalytic pathway for the transfer of H and C-containing fragments, as suggested in the equations earlier, as they can form M–H and M–C bonds readily in their lower oxidation states [50]. In contrast, O and N transfer reactions involve metals that readily form oxides and these are many of the first row transition metals (Ti, V, Cr, Mn, Fe), as well as W and Mo, and these do so in their higher oxidation states [52].

In the formation of organic matter (i.e., C-containing polymers), it is necessary for the C to be reduced, and thus on early Earth, as now, the reduction of C, present in the environment primarily in an oxidized form, was necessary. Of all the potential compounds that could catalyze such reactions, Fe/S-containing entities were likely the most abundant. Therefore, most primitive electron transfer systems contained these elements [50]. Overall, Fe-S clusters, later associated with amino acids or proteins, and mostly with the thiol entities of cysteine, have some unique properties that probably allowed them to function as enzymes in early biomolecular systems [53, 54]. The chemistry of S, which can exist in oxidation states −2 to +6 and which can readily form and break bonds, and the multiple oxidation states of Fe, contribute to the functionality of the Fe-S clusters. These metal-amino acid associations were able to transfer electrons and thus mediate redox reactions, and also had a cationic center that was capable of binding strongly to O- and N-containing compounds and ligands. Finally, the Fe-S clusters provided rigidity necessary for the enzyme function. Examples of such early functional molecules were the ferrodoxins, which are found in primitive archea, while evidence suggests the later development of the NAD/NADP system [53]. An examination of the most primitive enzymes indicates that these enzymes, such as Acetyl-Co-A synthetase, hydrogenases and dehydrogenases, and the ferrodoxins, contained mostly the trace elements Fe, Ni and Co [50].

The oxygenation of the atmosphere that occurred as a result of photosynthesis resulted in a dilemma for organisms that still persists today. The sulfidic compounds, and other enzymes with reduced metal centers, are susceptible to degradation in an oxygenated environment. Additionally, and especially for Fe, oxygenation resulted in a strong decrease in the Fe concentration in solution and the surviving organisms therefore needed to devise mechanisms to assimilate a diminishing but fundamentally important resource [51, 54]. Thus, organisms either needed to adapt to the new environment or retreat to anoxic regions, and clearly both pathways were followed through evolutionary change. Many organisms, however, are still today vulnerable to oxidative stress as a result of their evolutionary history that is tied to changes in the electrode potential of the biosphere [53, 54].

The result of models developed to examine this notion in more detail are shown in Fig. 8.12 [51]. These geochemical predictions compare the concentration of a variety of metals

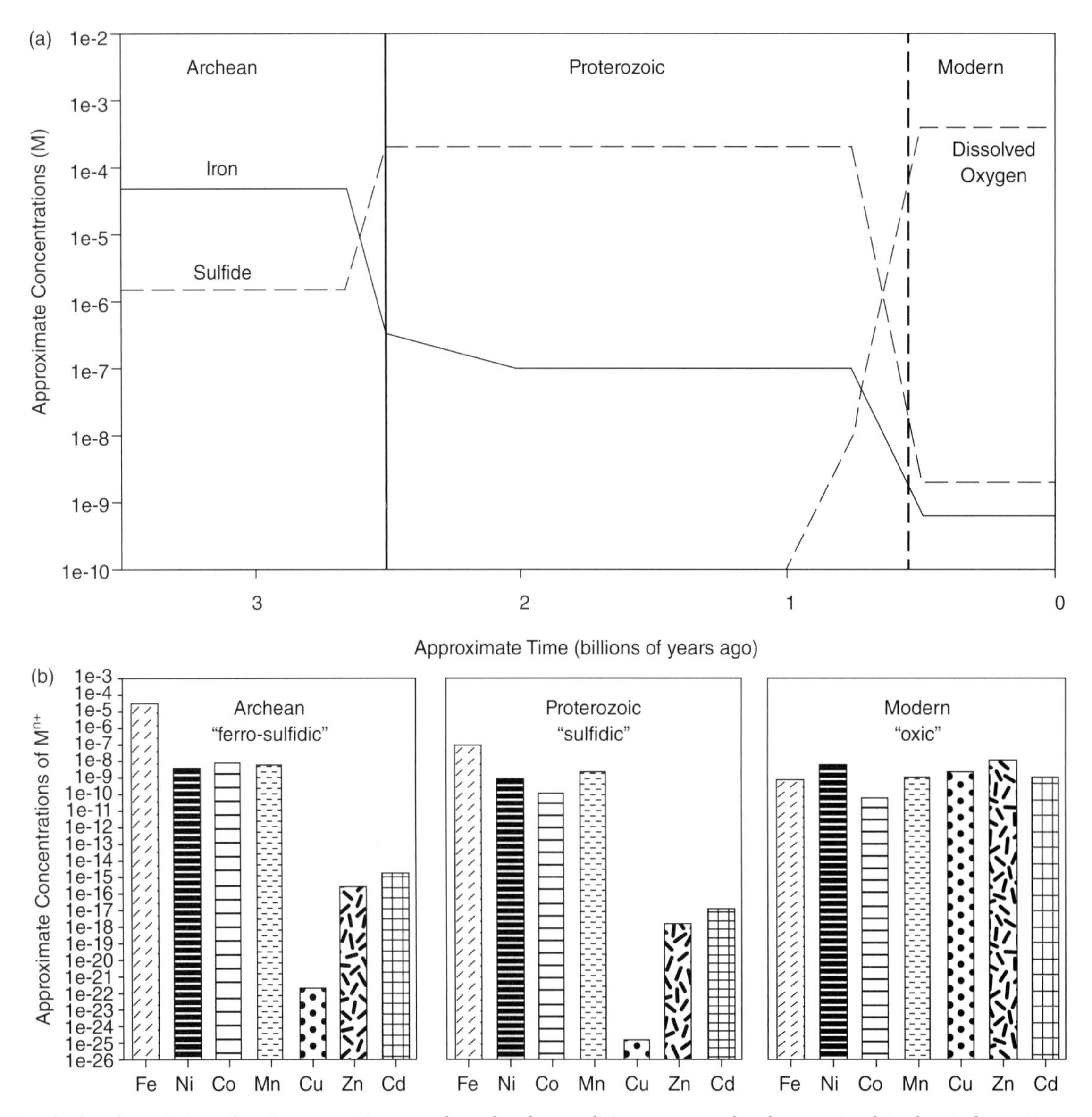

Fig. 8.12 Calculated speciation of various transition metals under the conditions presumed to have existed in the Archean Ocean, in the ocean during the Proterozoic and in the Modern Ocean. (a) Concentrations plotted over geological time; and (b) representative concentrations of the free ion for each major time period. Taken from Saito et al. (2003) *Inorganica Chimica Acta* **356**: 1–11 [51], reprinted with permission from Elsevier.

in ocean waters under Archean "ferro-sulfidic" conditions (>2.5 billion years before present (YBP)), with the "sulfidic" Proterozoic (0.5–2.5 billion YBP), and the current oxic system. The model results and considerations of geochemical cycling are based on the conclusion that the Archean ocean was a high Fe and low sulfide environment, and thus this was the period of highest Fe concentrations. As the production of sulfide increased, the system switched to a more sulfidic environment, with lower overall Fe and other metal concentrations. However, the relative ratios of the metals remained similar and Fe, Mn, Co, and Ni were still the metals in highest concentration (Fig. 8.12) [51]. Consistent with this metal hierarchy is the dominance of Fe and S containing enzymes, as discussed previously, as well as the origin of enzymes such as the Co-containing vitamin B_{12} (cobalamin), which is thought to have been produced 2.7–3.5 billion years ago. This enzyme is found in prokaryotes and not eukaryotes; for example, in ocean cyanobacteria, such as *Prochlorococcus*, but not in eukaryotic phytoplankton. Additionally, cyanobacteria are much more sensitive to elevated Cd and Cu concentrations compared to eukaryotes, and such sensitivity could have an evolutionary context. Conversely, the eukaryotes have many more Zn-containing enzymes than the cyanobacteria, as do higher organisms [52]. All these differences are consistent with the model results that suggest that in the early ocean the concentrations of metals followed the following series: Fe > Mn, Ni, Co >> Cd, Zn, Cu [51].

Overall, these model predictions suggest that Fe, Mn, Ni and Co were abundant in the early environment and therefore the inclusion of the redox active metals, Fe and Mn, in enzymes involved in electron transport is compelling [50–52]. These arguments and modeling also helps explain the current paradox discussed in Section 8.3.1 where ocean microbes have to spend considerable time and energy obtaining Fe from their surroundings. Clearly, development of the specific enzymes under a different chemical environment provides the rationale for the current situation having developed. Therefore, one may conclude that the incorporation of metals into biochemicals and enzymes is both a reflection of their specific chemistry of the metal and their presence in the environment at sufficient levels during the earliest part of the evolution of the Earth and the biosphere. Conversely, toxicity is highest for those elements that were present at very low concentrations in the early sulfidic biosphere. Organisms have derived a number of mechanisms to counter toxicity of metals, including mechanisms of storage within the cells, excretion, and transformation into less toxic or bioaccumulative forms. Conversion of metals through methylation, reduction or incorporation into biomolecules are all possible fates of incorporated toxic and/or required elements. The chapter will focus on the transformations of metal(loid)s within microbial organisms, with a primary focus of methylation processes.

8.4 Organometallic compounds and microbial transformations of metals

8.4.1 Mechanisms of metal methylation

In this book the term *organometallic* or organic metal(loid) compound is used to refer specifically to those compounds found in natural and impacted waters and in which the attachment of the organic moiety to the metal ion is directly through a carbon-metal bound. Most of these M–C bonds are covalent, especially for the metals and metalloids which have filled d and f orbitals [55]. In considering the periodic table, the elements involved in the formation of organometallic compounds of interest here are mostly those in Groups 12 (Cd and Hg), 13 (mainly Tl), 14 (Ge, Sn and Pb), 15 (As, Sb and Bi) and 16 (Se and Te). In most cases, the bond is polarized and the metal(loid) has a δ^+ charge. Most of the compounds that will be considered are relatively small and simple compounds present either in the environment due to human-related inputs, or due to the formation of these compounds in situ. In addition, a number of metalloids (As and Se, in particular) are incorporated into biological molecules and these will be briefly discussed. There is an ever-growing field of organometallic chemistry related to the use of transition metal compounds as catalysts or in organic production synthesis. Such compounds will not be discussed in detail in this book, but many transition metals form carbonyl compounds and while these can be manufactured in the laboratory, their presence has also been detected in the environment; for example, $Mo(CO)_6$ and $W(CO)_6$ in landfill gases [56]. It is not known whether these compounds are formed biotically or abiotically in the environment. Other transition metal carbonyls have been detected in some instances. The range of organometallic compounds found in the environment was discussed briefly in Chapter 5, in connection with the air-sea exchange of the volatile forms, and additionally in the discussions in Chapters 6 and 7. Some of this information will be reiterated here.

Most of the compounds that will be discussed contain one or more methyl groups attached to the metal or metalloid atom. Compounds with other alkyl groups (ethylated, butylated or phenyl-metal compounds) in the environment are mostly added as a result of human activity. Many organometallic compounds are formed biotically in the environment by microorganisms through the process of methylation within the cell. Methylation within cells is a fundamental biochemical process and can be carried out by a number of pathways. It appears, however, that the mechanisms of methylation of metals and metalloids are carried out by one of three pathways, involving either: S-adenosylmethionine, methylcobalamin or N-methyltetrahydrofolate, whose structures are shown in Fig. 8.13.

Methylcobalamin itself contains a metal center, Co, which is bound to five nitrogen atoms within the molecule. Its sixth bond is to the methyl group and the conversion

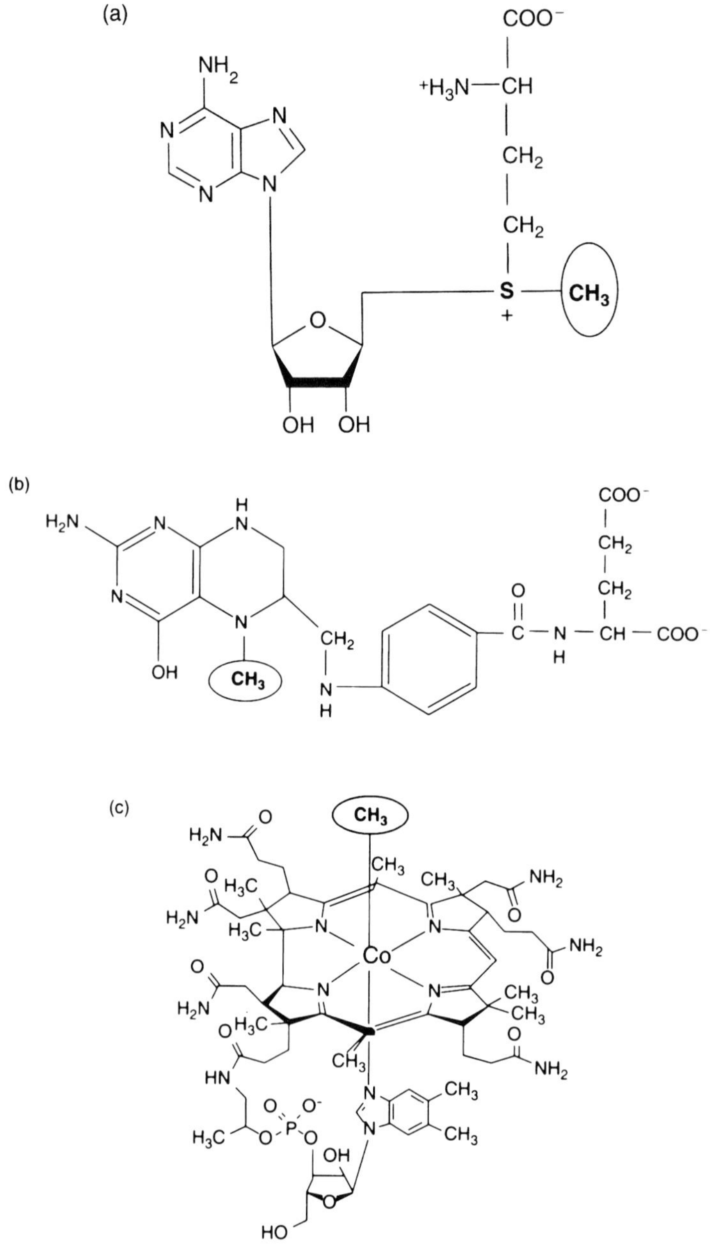

Fig. 8.13 The three major biochemicals responsible for methylation within cells: (a) adenosylmethionine; (b) N-methyltetrahydrofolate; and (c) methylcobalamin. The methyl group that is donated is circled in each case.

between methylcobalamin and cobalamin involves the removal of the methyl group and the reduction of the central Co atom, and thus this process donates a methyl *carbanion* (CH_3^-) via a SN_2 reaction. This is the only known biochemical pathway for such a reaction [57]. In contrast, the methyl group given up by S-adenosylmethionine (SAM) and N-methyltetrahydrofolate is a *carbocation* (CH_3^+) and thus there is a fundamental difference in the way these two compounds react with metals and metalloids. For example, the so-called "Challenger" pathway of methylation of As by SAM requires that initially the As(V) is reduced to As(III) and then methylated [58]. The methylated product is then oxidized to the As(V) state and must be further reduced before addition of more methyl groups (Fig. 8.14). This is also the pathway that leads to the formation of larger biomolecules containing As. In contrast, the methylation of Hg, Sn, and other cations is thought to involve mainly the cobalamin pathway with these elements being methylated while in their most oxidized form. There is no concrete evidence for the methylation of reduced Hg (Hg^I or Hg^0)

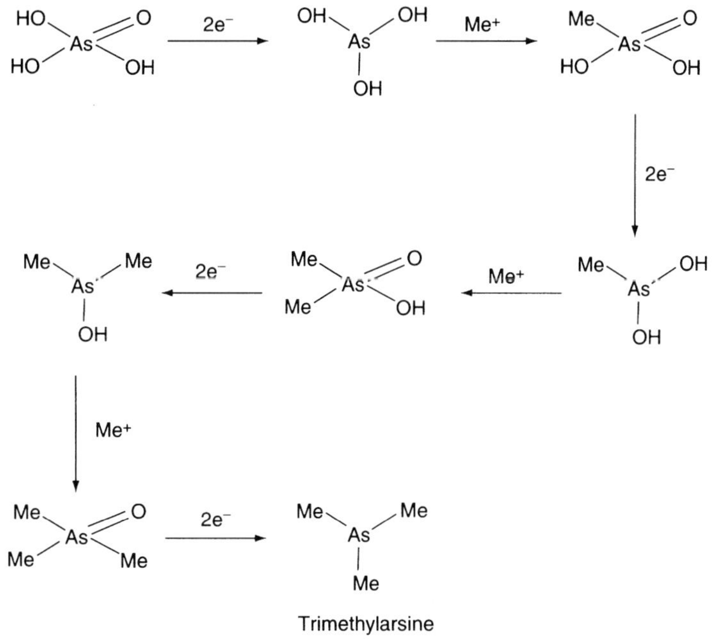

Trimethylarsine

Fig. 8.14 A representation of the Challenger mechanism for the methylation of arsenic showing the oxidation, reduction and methylation steps involved in the production of trimethylated arsenic compounds and the intermediate compounds. Figure reprinted from Feldman et al. (2003) in *Organometallic Compounds in the Environment* [57] with permission from John Wiley & Sons, Inc.

Table 8.5 The acid dissociation constants for the trace elements that form oxyanions and/or acids in solution. For comparison, constants for comparable major elements from the same group of the Periodic Table are included. For oxyanions, formula is HyXOz and the oxidation state of the element X is determined as 2z-y. The different number of significant figures indicates the certainty associated with each value. Reprinted with permission from Perrin [60]. Copyright (1982) American Chemical Society.

Trace Element	pK_{a1}	pK_{a2}	pK_{a3}	Major Element	pK_{a1}	pK_{a2}	pK_{a3}
Cr(VI), H_2CrO_4	0.74	6.49	–	–	–	–	–
As(V), H_3AsO_4	2.26	6.76	11.29	P(V), H_3PO_4	2.16	7.2	12.4
As(III), H_3AsO_3	9.29	–	–	P(III), H_3PO_3	1.3	6.70	–
Ge(IV), H_2GeO_4	9.01	12.3	–	Si(IV), H_4SiO_4	9.9	11.8	12
Se(VI), H_2SeO_4	–	1.7	–	S(VI), H_2SO_4	–	1.99	–
Se(IV), H_2SeO_3	2.62	8.32	–	–	–	–	–
Se(-II), H_2Se	3.89	11.0	–	S(-II), H_2S	7.05	19	
Te(VI), H_2TeO_4	7.68	11.0	–	–	–	–	–
Te(IV), H_2TeO_3	6.27	8.43	–	–	–	–	–
Te(-II), H_2Te	2.6	11.0	–	–	–	–	–

[59]. This is probably a result of the unstable nature of Hg^I in the environment and the non-polar nature of Hg^0 and the fact that as a dissolved gas in most environmental aquatic systems it is not accumulated to any significant degree by microorganisms [9], or other organisms, unless it is oxidized. So, therefore while Hg^0 could likely be methylated by SAM, methylation of Hg^{II} by other pathways appears to be more efficient. Other metal(loid)s present in the environment in an oxidized form that are thought to be methylated only by methylcobalamin are Sn^{2+} and Bi^{3+}, and there is also the potential for methylation of Pb^{2+}, although the majority of the alkylated Pb compounds in the environment are due to releases from human activity, and specifically from the use of Pb compounds in gasoline.

Thus, the various pathways of methylation are entirely distinct and which pathway dominates is a function of the differences in the speciation and oxidation state of the metal and metalloid in aquatic systems. The metalloids As, Se and Sb are present in environmental solutions mostly as an oxyanion of a weak acid (Table 8.5), and therefore their form depends on pH, while most metals exist in solution as complexes of oxidized ions, and the specific complexation of the metal likely controls its uptake. As discussed in Section 8.2, the neutral complexes of some metals, such as Ag, Hg, Tl, can passively cross membranes [13, 16, 20, 21], and these forms of the metal are the most "biologically available", although they react or change form once inside the cell. It is thought that Hg is taken up as a neutral Hg-sulfide complex prior to methylation by sulfate-reducing bacteria [61]. However, there is also the potential for active transport of metals, especially through their uptake via channels designed for major ion transport, or for the acquisition of required metals, such as Fe and Zn, or when combined with low molecular weight thiols.

In addition to biotic methylation of metals and metalloids by microbes, there is also the possibility of abiotic methylation by methyl donor reactions within the aquatic system. It is also possible for methyl transfer reactions to occur from a methylated metal complex to another metal. For example, the methylation of Hg by methylated Sn species has been shown with the reaction rate being dependent on the pH and the chloride concentration as the reaction appears to be hindered by high chloride conditions [62]. There is evidence for the methylation of Hg, Sn, and Pb by methyl donor compounds, including methyl iodide (CH_3I) ([62, 63] and references therein). It appears that while CH_3I will methylate Sn and Pb, the same does not occur for Hg, and Hg appears to catalyze the hydrolysis of CH_3I. The importance of these reactions under environmentally relevant conditions is less certain however and there is conflicting evidence in the literature as many of these studies have been done at high concentrations, and their relevance at the typically low nM to pM concentrations in the environment requires further demonstration.

There is also the possibility of the transfer of a methyl group from dissolved organic matter to a metal bound to NOM, although there is little evidence for this in the literature except for Hg. Gardfeldt et al. [64] showed that there is the potential for CH_3Hg formation due to the decomposition of Hg-acetate complexes, under conditions typical of rainwater/atmospheric solutions, and it is suggested that such reactions could account for some fraction of the CH_3Hg found in rainwater [64, 65]. Also, the methylation of ionic Hg has been shown under conditions of elevated temperature in the presence of NOM. Experimentally, there is the potential for artifactual formation of CH_3Hg during the analytical separation and concentration of CH_3Hg from sediment matrices, a process that is often done by acidification and distillation of the CH_3HgCl formed in the analytical solution [66–68].

Finally, the use of alkylated metal compounds in industry and other uses, such as for their toxicological properties, has led to the dispersion of man-made forms of these compounds in the environment (Chapters 5–7). A well-known example of this use is the Pb compounds – tetraethyllead and tetramethyllead – that were added to gasoline as an *"anti-knock" agent*. The contamination of the environment by alkylated Pb compounds over time is illustrated by the historical record of these compounds in ice and snow, collected in the French Alps (Fig. 8.15) [69]. Their use is now banned

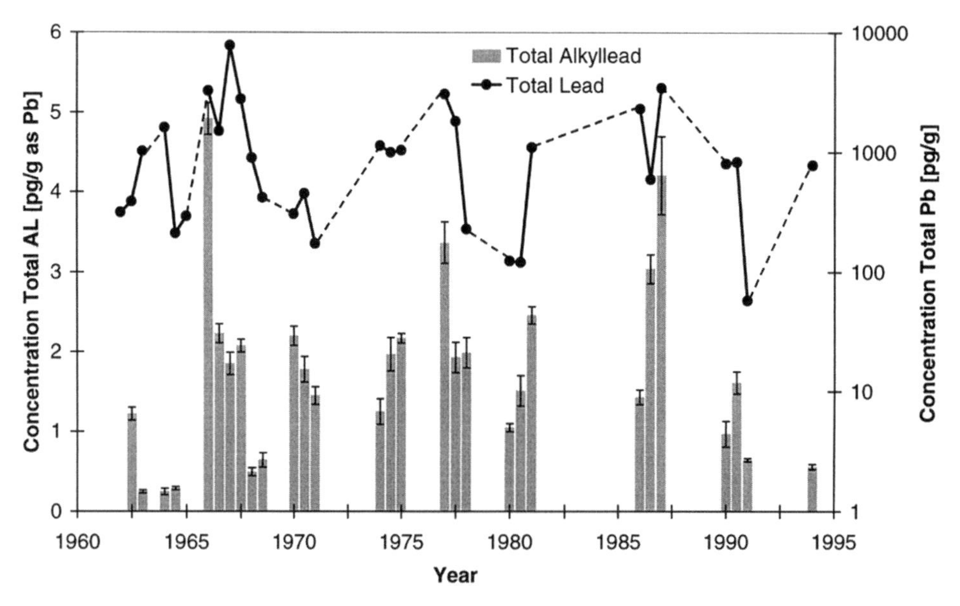

Fig. 8.15 Concentration of total lead (Pb) and alkyllead in high altitude alpine snow showing the changes in concentration over time. Reprinted with permission from Heisterkamp et al. (1999) *Environmental Science and Technology* **33**: 4416–41 [68]. Copyright 1999 American Chemical Society.

in many parts of the world. In addition, tributyltin and other alkyl Sn compounds has been used extensively as an anti-fouling agent for aquatic craft and this has resulted in substantial contamination in many locations.

The thermodynamic stability of the various alkylmetal species varies and this gives some indication of the potential for their existence in the environment [55]. However, it should also be noted that the reaction kinetics are often important, and in some instances mediated by bacteria, and thus the likelihood of the existence of a particular compound in the environment cannot be made based purely on equilibrium calculations. Most organometallic compounds are susceptible to oxidation if one considers the final products to be the metal oxide, CO_2 and water. However, in the environment, it is possible that the oxidation will be only partial to an intermediate, and that, while the overall oxidation reaction may be favored, the formation of the intermediate may result in partial, rather than complete, oxidation, or may hinder the overall kinetics of the reaction. Many of these reactions have relative high activation energy and thus are not likely to proceed at room temperature [55].

Photochemical decomposition, however, is an important mechanism for the destruction of many organometallics in aquatic surface waters as the energy supplied is sufficient to overcome the required activation energy for the decomposition. Additionally, there are many reactive species produced by photochemical interactions, such as ozone and the OH radical, and these can lead to enhanced decomposition in the atmosphere and in surface waters. This is especially true for elements which have volatile compounds, such as Hg, Se, and other metals and metalloids, as their propensity for decomposition will be much higher in the atmosphere compared to surface waters [55, 70]. In the atmosphere, many organometallics have lifetimes on the order of hours to days, rather than days to weeks. The pathways for the decomposition of organometallics often occur in a stepwise fashion with the removal of successive methyl (or other alkyl) groups from the central metal(loid) atom. Examples include the decomposition of $(CH_3)_2Hg$ to CH_3Hg^+ and $(CH_3)_4Sn$ or $(CH_3)_4Pb$ to the tri-, di- and monomethyl forms.

As summarized by Craig et al. [55], the toxicity of the organometallic compounds is usually greatest for the +1 cations and these are often much more toxic than the ionic forms of the metal(loid). Compounds such as CH_3Hg^+ and $(CH_3)_3Sn^+$ are therefore highly toxic as they are bioaccumulative even though they are charged. Arsenic is an exception as the methylated forms appear to be less toxic. For Sn, the triethyl and trimethyl compounds are the most toxic to mammals while the higher alkyl compounds are more toxic to bacteria and microorganisms [71]. This explains the use of tributyltin as an "anti-fouling" agent for the hulls of vessels. Some of the more highly methylated and neutral compounds such as $(CH_3)_4Pb$ and $(CH_3)_2Hg$ likely derive their toxicity from their ability to readily cross biological membranes by passive diffusion as there are neutrally charged and non-polar. After uptake, their subsequent conversion into reactive ions through partial demethylation likely occurs within tissues. Indeed, overall, it is the increased hydrophobicity of the alkyl compounds relative to the metal ion complex that allows for their greater accumulation into tissue relative to the ionic forms.

Mostly, organometallic compounds cause toxicity by interfering with enzyme function and for many of the compounds, by their strong binding to thiol groups, mostly those on proteins, which inhibit the function of the protein or allow for the transport of the metal complex through amino acid channels designed for the transport of peptides and amino acids. Thus, the compounds are taken up as they "mimic" required proteins and other molecules [72]. Other impacts of organometallics are the disruption of major ion (Ca^{2+}, Na^+, K^+) transport across membranes and the general interference with membrane function. These interactions likely account for the higher toxicity of the +1 charged ions compared to other organometallic forms. The impact of such compounds on higher organisms and humans is not within the scope of this book and are detailed in many other texts.

The most common, well-known, and studied organometallic compounds are those of Hg, Pb, As, Se, and Sn. These organometallic compounds have been studied because of their relative abundance but also because of their toxicity to humans and wildlife, as well as microorganisms. The toxicity of the various compounds varies for different organisms exposed to these compounds and also between the different compounds with the same central metal(loid). The well-known example is the difference between the highly toxic inorganic As^{III} and that of the organic compound, arsenobetaine, which is not toxic and is found in many marine organisms (Section 8.4.5). For a more complete discussion of the topic the reader is referred to literature. There are abundant papers on the subject, and specific books such as the text edited by Craig, *Organometallics in the Environment*. The following sections will focus as appropriate on the most important and/or most studied metal(loid)s and their alkylation reactions, and their fate in the aquatic environment.

8.4.2 Less common organometallic compounds

A number of metal(loids) are known to form organometallic compounds although there is little information on the formation, stability and toxicity of many of them. Because of the ubiquity and toxicity of CH_3Hg complexes in the environment, and its bioaccumulative potential, the formation and cycling of methylated Hg compounds will be dealt with in detail next (Section 8.4.3). Zn is known to have an important biochemical role as the constituent of various required enzymes, as discussed in Section 8.3.2 [2]. It is not known to form any stable small alkylated compounds, however. In contrast, Cd, above Hg in the periodic table, has been isolated from the environment as methylated compounds [56], although $(CH_3)_2Cd$ is relatively unstable in water. There have been few studies of organic Cd compounds in the environment in contrast to the inorganic chemistry of Cd that has been well-studied. Overall, Cd, because of its overall weak binding to the solid phase, it is relatively mobile in the environment. Initial evidence for the formation of CH_3Cd^+ was by bacterial cultures isolated from polar waters [73], and these same cultures also produced methylated Pb and Hg compounds [74]. In the environment, evidence suggests that the peak concentration of these methylated species coincided with that of chlorophyll a, suggesting a microbial role in the formation of these compounds in polar waters [73]. It should be noted that the general consensus for Hg is contrary to the findings of these authors, as most studies looking at Hg methylation in oxic waters have not found strong evidence for its production and/or for this being an important environmental pathway [59]. Therefore, these studies of the formation of methylated Cd compounds in ocean waters require more confirmation. However, it is known that Cd can be abiotically methylated by methylcobalamin so the potential for its biological methylation exists, through a similar pathway to the formation of methylated Sn and Hg compounds.

For the Group 13 elements, there is little evidence for the formation of alkylated compounds except for Tl. This is a potentially toxic and bioaccumulative metal and the chemistry of inorganic Tl has received some attention [74]. As noted in Section 8.2.1, it has been shown that TlCl can passively diffuse across membranes and therefore this is one potential form of Tl that would be methylated. It appears that Tl^I is oxidized and methylated at the same time (i.e., via the Challenger mechanism) producing a mono-, di- or trimethylated product of, respectively, +2, +1 and 0 charge. This process occurs under anaerobic conditions with little evidence of aerobic methylation [57]. None of these compounds are highly stable but there is some initial evidence for the presence of these compounds at low pM (<20 pM) concentrations in the Atlantic Ocean, mostly as $(CH_3)_2Tl^+$. There is no evidence for alkylated In and Ga complexes in the environment.

Alkylation of Group 14 elements occurs for the heavy elements, Sn and Pb, and to a lesser degree for Ge. As noted earlier, alkyl Pb and Sn compounds have been widely produced for use in industrial and other applications. Their chemistry will be discussed in Section 8.4.4. For Ge, the mono-, di-, tri- and tetramethylated compounds have been found in the environment. In the ocean, Ge^{IV} exhibits "nutrient-like" behavior and has its highest concentrations in deep ocean waters (~0.1 nM) ([56] and references therein). The methylated form of Ge (CH_3Ge^{3+}), however, appears to be more "conservatively" distributed and is present at somewhat higher concentration (~0.2 nM). Its distribution appears unrelated to that of inorganic Ge. The other methylated Ge species ($(CH_3)_2Ge^{2+}$ and $(CH_3)_3Ge^+$) occur in seawater at similar concentrations (0.1–0.2 nM). It is apparent that methylated Ge compounds can be produced under anaerobic conditions and that these species are not produced in oxic waters in the presence of algae. These compounds can also be made in the laboratory through reactions with CH_3I and methylcobalamin, suggesting that this is the pathway for methylation in the environment, although it has also been suggested that these compounds are formed via the Challenger mechanism [74].

The element of Group 15 that forms extensive organometallic compounds is As, and the other two elements of higher atomic number, Sb and Bi, also have important organometallic chemistry. These compounds and their aquatic chemistry are discussed in Section 8.4.5. Selenium, an element in Group 16, has important and complex organometallic chemistry, and is methylated. In organisms, Se has a large biochemical role as it is incorporated into a number of enzymes [52]. The methylation of Se was first investigated by Challenger and since these early studies it has been shown that Se has an extensive biochemistry and that compounds, such as selenoproteins are important constituents of organisms as they act as antioxidants and have other roles in the cellular mechanism [75, 76]. Compounds such as selenomethionine and Se-adenosylselenomethionine (SeSAM), are found in cells, where Se has replaced S, as could be expected as these elements are from the same group of the periodic table. The methylating ability of SeSAM has been shown to be greater than that of SAM [74]. As noted elsewhere, Se is an element that has a very small range between its toxic concentration and its required intake for many organisms and therefore many instances of Se deficiency or Se toxicity have been reported in the literature [52].

It has been suggested that CH_3SeH and the cation CH_3Se^+ are some of the compounds responsible for the toxicity of Se. As noted in Chapter 5, analogs to the methylated S-containing compounds $(CH_3)_2S$ (DMS) and $(CH_3)_2S_2$ have been identified ($(CH_3)_2Se$, $(CH_3)_2SSe$ and $(CH_3)_2Se_2$). It appears likely that these compounds are the degradation products of certain microorganisms that exist in ocean waters, being formed from the decomposition of 3-dimethylselenopropionate ($(CH_3)_2Se^+CH_2CH_2COO^-$) [74]. Similar processes likely account for their formation in terrestrial waters and other environments. Overall, the formation of methylated Se compounds is relatively ubiquitous as they are apparently produced by both bacteria and algae, and there is a clear role for Se in biochemistry as an analog or substitute for S.

There are a limited number of studies on microbial Se methylation and demethylation [76]. In freshwater environments, γ-Proteobacteria, such as *Pseudomonas* spp. have been identified as Se methylators. Demethylation of $(CH_3)_2Se$ is found in anoxic sediments and it is speculated that methanogens are using this compound for growth in an analogous fashion to their utilization of DMS. Additionally, it is possible that Se(VI) can be used as an electron acceptor by some bacteria, in an analogous way to sulfate reduction [77], and it has also been found that some microbes reduce selenate to S(0). The acid dissociation constants for selenate are very similar to that of sulfate (Table 8.5) and therefore they are present in the same form, as the −2 anion, in most natural waters. This selenate reduction is in contrast to the assimilatory selenate reduction where Se is incorporated into organic matter. The biogeochemical cycle of Se suggests that inorganic Se introduced to the environment is mostly converted to methylated Se compounds in sediments under anaerobic conditions. However, these methylated compounds are also produced in combination with primary production in the water column. It is likely that in sediments, direct methylation is occurring while in the water column the production of the methylated Se compounds results from decomposition of Se-containing biomolecules.

There is evidence for the formation of organo-Te compounds and the mechanisms for their formation appear to be similar to the mechanisms for the formation of organic Se-containing compounds [56]. Certain bacteria have been shown to methylate Te and form the dimethylated compounds, and there is a complex interaction between the ability of these organisms to methylate and the Se concentration. This suggests that Se and Te, which are electronically similar, behave similarly in this regard. Both exist in solution as oxyacids although the pK_as of the selenic and selenous acid are much lower than the corresponding values for telluric and tellurous acid (See Table 8.5) [60]. There has been some suggestion that Po can be methylated but the conditions under which this occurred suggest that these compounds are unlikely in natural environments [74]. The mechanism of methylation is not known and as Po exists in various oxidation states, there are a number of potential methylation pathways. More research is needed to examine in more detail the methylation and cycling of Po and the other less-studied heavy metals and metalloids.

8.4.3 Alkylated mercury compounds

Methylation of Hg is well-documented and studied as research and interest is primarily due to the toxicity of the monomethylated form (CH_3Hg) and its propensity to accumulate to high levels in top level aquatic predators, such as *piscivorous* (fish-eating) fish, birds and mammals, including humans (Table 8.6) [84]. In the environment, levels of CH_3Hg in media increase from low concentrations in the aquatic environment (typically <2% CH_3Hg in sediments; <10% in water) to increasing higher relative concentrations in the food chain (<20% in phytoplankton; >90% in piscivorous fish). Levels in some freshwater environments are high enough to impact fish reproduction and cause other impacts [85], but the primary impact of CH_3Hg accumulation in food chains is on the predators that consume upper trophic level fish. For most humans, consumption of ocean fish is the dominant exposure pathway [86, 87], while wildlife, such as mink, otter and loons, are often exposed due to freshwater and coastal fish consumption [88].

Table 8.6 Representative bioaccumulation factors (BAFs) (log values) for a variety of elements for both phytoplankton and fish.

Element	Log(BAF) Algae*	Log(BAF) Fish*
Zn	4.7	2.5–5.0
Cd	3.7	<2.0–3.5
Ag	5.0	2.7
Hg	4.5	3.0–3.8
CH_3Hg	5.0	6.3
As	4.3	3.0–3.5
Se	4.1	3.8
Pb	5.0	2.3–4.0
Cu	–	1.5–3.5
Co	3.7	2.8
Ni	–	<2.5

*Values estimated from a variety of sources including [78–83].

Acute exposure incidents in Japan and elsewhere in the 1960s provided the initial impetus for study and also provided the first evidence that inorganic Hg could be converted into the more toxic and bioaccumulative CH_3Hg. Methylation of Hg in sediments by the resident microorganisms was demonstrated and many of the early studies examining the role of microbes in methylation were in coastal and estuarine sediments. Selective microbial inhibition studies indicated that methylation was primarily mediated by sulfate-reducing bacteria (SRBs) [89, 90]. The role of SRBs has been confirmed in many studies since then [59] but more recently, iron-reducing bacteria in the genera *Geobacter* have also been observed to methylate Hg [91, 92].

Subsequently, in the late 1980s, methylated Hg compounds were detected in the water column of the remote ocean [93, 94] and it was suggested that the processes of formation were likely biological. Studies in the same period also showed that methylation occurred in lakes, and that this process was ubiquitous in aquatic systems. In ocean waters, and in some freshwater systems, dimethylmercury ($(CH_3)_2Hg$) has been measured and its presence is mostly found in the water column, although there is some evidence for $(CH_3)_2Hg$ in sediment porewaters [70]. Typically, it is found in low oxygen but not anoxic environments and therefore there is some discussion concerning its formation pathway, and whether SRBs are the main microbes involved in its production. Some early studies did suggest that $(CH_3)_2Hg$ is produced in culture [70], and there has been demonstration of its abiotic formation through a disproportion reaction of an intermediate compound, $(CH_3Hg)_2S$, formed in the presence of elevated CH_3Hg and sulfide. More studies are required to fully understand the sources and cycling of methylated Hg, and especially $(CH_3)_2Hg$, in the coastal and open ocean environment, given that most people are exposed to the potentially harmful effects of CH_3Hg primarily through ocean fish consumption [95]. The relative importance of different bacteria in CH_3Hg and $(CH_3)_2Hg$ production in diverse natural aquatic environments is not clearly understood, although it has been a topic of recent investigation. Besides biotic pathways, abiotic mercury methylation is possible by a variety of methylating agents in sediments, including acetate, organic acids, and humic substances [62, 96] (Section 8.4.1). However, biotic pathways of MeHg production are believed to be more important than abiotic pathways [59].

In early studies in culture there was some disagreement about the pathway of Hg methylation but it has since been shown that the methylating agent is cobalamin, and that Hg is methylated as the oxidized ion, Hg^{II} [59]. This is consistent with the pathways of methylation of other cationic metals in solution. Although production varies with the total dissolved Hg concentration (Fig. 8.16), the correlation only explains 41% of the variability, and a number of other variables, such as the speciation and bioavailability of Hg to the methylating organisms, as well as the presence and activity of methylating bacteria, also impact the degree of conversion [59, 96]. As can be seen from Fig. 8.16, for any total Hg concentration in sediments, there is at least two orders of magnitude variability in CH_3Hg.

In addition to Hg methylation, demethylation of CH_3Hg is also biologically-mediated and microbes appear to be important in this conversion as well, although the mechanisms are not as well understood [97, 98]. In uncontaminated environments, the major products appear to be Hg and CO_2 and therefore this pathway has been termed *oxidative demethylation*, in contrast to *reductive demethylation*, where CH_4 is the major carbon product [98]. The mechanism of oxidative demethylation may be analogous to monomethylamine degradation by methanogens or to acetate oxidation by SRBs. Reductive demethylation appears to be prevalent in more contaminated environments and at high CH_3Hg and Hg concentrations it has been shown that a series of inducible genes can be activated (the *mer* operon) that can aid in detoxification of CH_3Hg via demethylation (the *mer B* gene which encodes for organomercury lyase), and reduction of Hg^{II} to Hg^0 (the *mer A* gene which produces mercury reductase) (Fig. 8.17). The *mer B* gene can decompose a variety of organomercury compounds, such as ethylmercury and phenylmercury, and these are found in the some environments due to their manufacture and input. There is also a regulatory (promoter/activator) gene (*mer R*), as well as transport genes and transport proteins in the cell membrane [97]. The overall operon is contained on a plasmid and is readily transferred between bacteria in the environment. However, while membrane Hg^{II} transport proteins are present in bacteria with the *mer* operon, they are absent in SRB and therefore are not involved in the transport of Hg associated with methylation. It is thought that such uptake involves passive diffusion of neutral complexes of Hg across the lipid membrane [59].

In a number of ecosystems, CH_3Hg production has been shown to be strongly related to Hg interactions with sulfide [95, 99–103] (Fig. 8.18) as neutrally charged dissolved Hg–S species (HOHgSH and $Hg(SH)_2$) are believed to be much more bioavailable to sulfate-reducing bacteria than charged complexes [61, 104, 105]. They have a relatively high K_{ow} (Table 8.2) and dominate at low environmental sulfide concentrations. At higher sulfide concentrations, negatively charged Hg-S species and Hg polysulfide species dominate [15]. Therefore, methylation is most prevalent in environ-

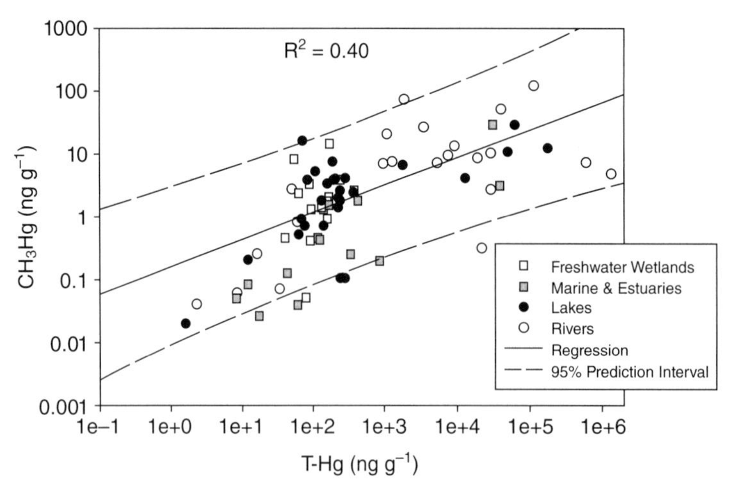

Fig. 8.16 The relationship between total mercury concentration in sediments and the methylmercury concentration for a number of different aquatic ecosystems. Reprinted with permission from Benoit et al. (2003) in *Biogeochemistry of Environmentally Important Trace Elements* [59]. Copyright (2003) American Chemical Society.

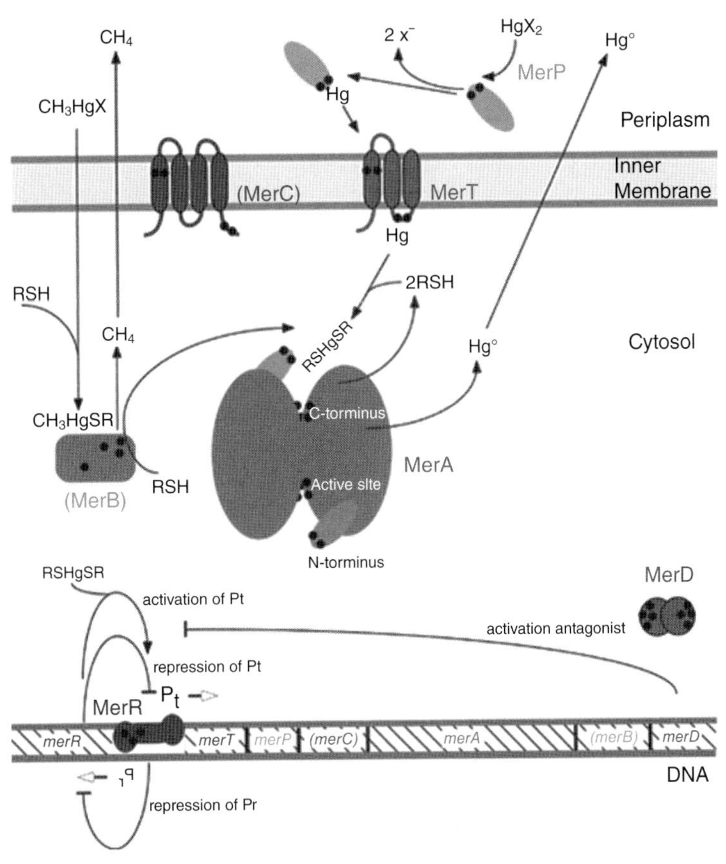

Fig. 8.17 A diagrammatic representation of the pathways and major *mer* genes involved in the detoxification of mercury and methylmercury. Figure reprinted from Barkay et al. (2003) *FEMS Microbial Reviews* **27**: 355–84 [97] and used with permission of John Wiley & Sons, Inc.

ments where sulfate reduction is high, but sulfide accumulation is low, as shown, for example, in Fig. 8.18 [59]. These locations are often the upper layers of sediments, as inputs of fresh organic matter and oxygen diffusion into sediments result in high rates of Fe and sulfate reduction in regions of relatively low sulfide content.

For example, in the upper reaches of the Everglades, sulfate concentrations and sulfate reduction are high and it would therefore be expected that Hg methylation may be elevated. However, both the methylation rate constant, and the $\%CH_3Hg$ in sediments is low, and this correlates with the low fraction of dissolved Hg as HOHgSH. As an opposite contrast, in the lower reaches of the Everglades, sulfate and sulfate reduction rate is much lower, but bioavailable Hg is higher, but the combination still results in low overall methylation. In the middle of the region, high sulfate reduction rates combine with relatively high bioavailability to result in the highest methylation rate constants, and the highest $\%CH_3Hg$ in sediments [59]. This example provides clear evidence of the role of Hg bioavailability to bacteria in controlling Hg methylation. As noted earlier, other complexes of Hg are potentially bioavailable and complexes with small organic ligands can have a relatively high K_{ow}. This is demonstrated by the fact that a bacterial culture of *Geobacter sulfurreducens* can readily accumulate and methylate Hg^{II} when present in the medium as a cysteine complex at a much faster rate than as either sulfide or chloride complexes [22].

The site of methylation within cells has not been clearly identified and not all SRB's methylate Hg, and even organisms from the same genus has different abilities to methylate Hg [59]. While initial studies suggested that methyaltion was associated with the Acetyl-CoA pathway and cobalamin in one organism (*Desulfovibrio desulfuricans* LS), this link does not hold across a variety of organisms, as organisms without this pathway methylate Hg, and *vice versa*. For SRBs, both complete and incomplete organic carbon oxidizers are methylators of Hg [106]. These results appear to confirm the notion that methylation is accidental and not a detoxifying mechanism in these organisms.

In conclusion, therefore, the importance of speciation has been well demonstrated in the uptake and methylation of inorganic CH_3Hg and the same impact is probable on the rate of demethylation of Hg, and on the uptake and methylation of other cations. To date, there has been little study of other metals that are methylated, and so the supposition has little

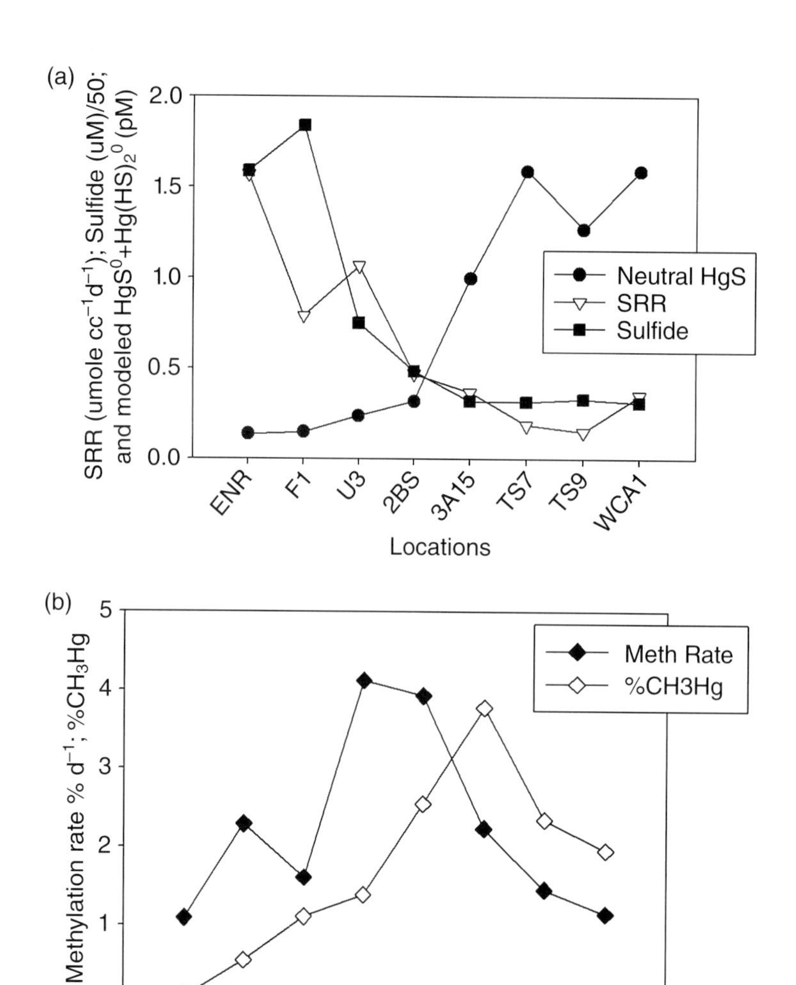

Fig. 8.18 The changes in (a, *Top*) sulfate reduction rate (SRR) in sediments, the sulfide concentration in sediment porewater and the estimated concentration of neutrally-charged mercury-sulfide complexes; and (b *Bottom*) the methylmercury production rate and the percent methylmercury in sediments of the Florida Everglades across the nutrient/sulfate gradient (sites plotted left to right in decreasing concentration). Redrawn with permission from Benoit et al. (2003) in *Biogeochemistry of Environmentally Important Trace Elements* [59]. Copyright (2003) American Chemical Society.

concrete supporting scientific evidence. Of the species that exist in solution as cations, Hg is the element that is methylated to the highest extent, and has been the most studied, primarily as a result of concerns of the human and environmental impact of the accumulation of CH_3Hg in fish, and their consumption. While the results of these extensive studies could be extrapolated to other cations that are methylated, more study is also required to investigate the fate of other cations, which are also methylated and bioaccumulated, or which have been added to the environment as a result of human activity.

8.4.4 Lead and tin compounds

In contrast to the other metal(loid)s, the sources of alkylated Pb and Sn compounds to the environment are primarily due to their use by humans and their releases from anthropogenic sources. Therefore, these two metals are considered together, as well as their being in the same group of the Periodic Table. Additionally, while most of the organic compounds of the other metal(loid)s of environmental relevance are the methylated species, for Pb and Sn, the ethyl, butyl and higher alkylated species are more important. Organo-Pb compounds were first produced in the 1850s and have found a wide variety of uses since then [107]. However, in the 20th century – tetramethyllead and tetraethyllead and mixed compounds (i.e., $(CH_3)_x(CH_3CH_2)_{4-x}Pb$ ($x = 0–4$)) in five different combinations were produced commercially and collectively these compounds are referred to as tetra-alkyl Pb, abbreviated as TAL. These compounds were primarily used as a gasoline additive, as an "anti-knock" agent (an additive added to gasoline to promote consistent ignition within the cylinders and to improve the fuel octane rating). The organic compounds are mostly destroyed during the combustion of the gasoline but the process resulted in the release of Pb to the environment in the fine particulate exhaust of the vehicle. While these compounds are now mostly banned, it was estimated that 65–85% of the Pb use in the early 1980s was related to the production of Pb-containing anti-knock agents, and the contamination of the environment as a result of their use has been mentioned previously (Fig. 8.15).

While most of the TALs in gasoline are decomposed during combustion, studies show that a small fraction (estimates vary from 1 to <0.1% [107]) is emitted from the tailpipe. Other sources such as evaporation and spillage have resulted in detectable levels of these compounds in the environment. Given the amount of additives used in the 1970–1980s in gasoline in the Western Hemisphere, the estimates of annual emissions were in the region of 10^3 tons yr^{-1}. Measurements of Pb in the atmosphere during the period of their usage show that levels of Pb were elevated, with substantial gaseous TAL concentrations of up to 1 nmol m^{-3} (0.5–5% of the total atmospheric Pb) in urban areas, with rural values being 1–2 orders of magnitude lower, and remote ocean air values of <5 pmol m^{-3} ([107] and references therein). While most of the TALs are in the atmosphere in the gaseous phase, inorganic Pb is mostly in the particulate phase. As noted in Section 5.2.4, recent measurements in urban environments still detect TALs although the concentrations are much lower (~10^{-2} nmol m^{-3}).

Strong gradients in concentration suggest that TALs are either efficiently removed from the atmosphere or are relatively unstable to decomposition. Measurement of organic Pb compounds in rainwater shows that the dominant species are the ionic trialkylated forms, and this is consistent with the relatively rapid decomposition of TALs in the atmosphere [107]. The amount of the various TAL compounds in rainwater during the 1970–1980s was relatively low, being a small fraction of the total. Concentrations ranged from <10^{-2} nM for remote locations to 0.5 nM for impacted areas. Concentrations have declined in concert with the phasing out of TAL in gasoline, but the rate of decline has been slower than the rate of decrease in usage. This suggests that there is a legacy of TALs in soils and terrestrial locations that is providing a continued input of these compounds to the atmosphere.

Studies have shown that TALs are subject to photochemical decomposition and are unstable in the presence of the hydroxyl radical, ozone and other oxidants, with their rate of decomposition being highest (>10% h^{-1}) in the presence of $OH^{\bullet}$ [107]. Overall, $(CH_3CH_2)_4Pb$ is much less stable than $(CH_3)_4Pb$, with the differences in the rate of decomposition being a factor of 3–5. No appreciable decomposition occurs in the dark. The products are the trialkylated species which remain in the gas phase but which are highly soluble and scavenged by wet deposition, or are dry deposited. These compounds are also unstable to further decomposition but at rates ten times slower than their respective TAL. These relative decomposition rates are therefore consistent with the measurement of the highest concentrations of the trialkylated species in precipitation.

In surface waters, levels of TAL are low, and were low even during times of their usage, as they are relatively insoluble and unstable and are likely degraded photochemically. Decomposition occurs, especially in the presence of other metals, or in the presence of oxide surfaces, as shown, for example, in Fig. 8.19. Tri- and dialkylated forms are dominant as a result of the stability of the compounds, which is inverse to their degree of alkylation, that is, TALs are the most rapidly degraded, while dialkylated compounds are substantially more stable [107]. Their presence in waters is also consistent with their relative stability with the trialkylated species dominating, and this is confirmed by their detection in aquatic organisms, including fish. Alkylated Pb compounds are highly bioaccumulative and TALs are readily accumulated across membranes.

It is still evident that most of the methylated Pb compounds are from anthropogenic sources given little evidence for the biotic methylation of Pb in the environment. However, there is evidence for bacterial mediation of their decomposition, and it appears that they are also degraded by invertebrates and other organisms. In terms of methylation by organisms, a number of early studies found conflicting results but more careful examination of methylation with either cobalamin, S-adenosylmethionine or methyltetrahydrofolate showed that none of these typical biological methylators methylated Pb. Some studies suggest that methylation can occur in the presence of CH_3I, a compound

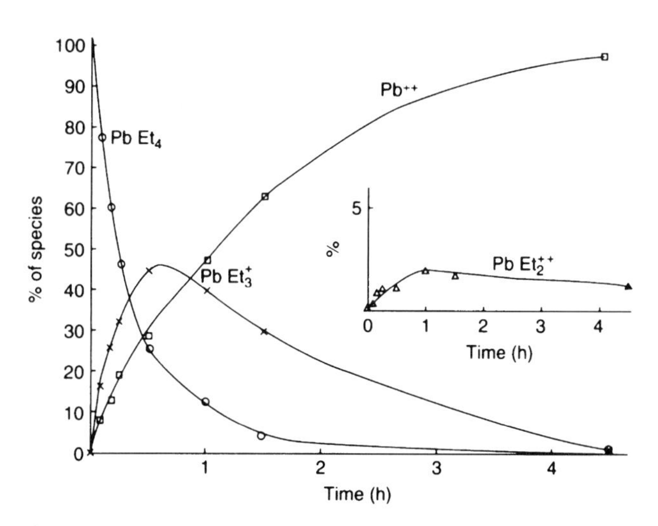

Fig. 8.19 Degradation of alkyllead compounds over time in water in the presence of UV light. Figure reprinted from Yoshinaga (2003) in *Organometallic Compounds in the Environment* [107] with permission from John Wiley & Sons, Inc.

which is found in the marine environment, and has a biological source [107]. Therefore, methylation of Pb in the environment is possible but that it does not readily occur under typical environmental conditions and is therefore not ubiquitous as is the methylation of other metal(loid)s.

Tin has also been added as a variety of organic compounds to the environment primarily through their usage in "anti-fouling" paints and similar applications. TrialkylSn compounds are the most dominant forms and it has been estimated that production is in the region of 50,000 tonnes per year [71]. Tributyltin (often denoted as TBT) is the most common and well-known compound used as a wood preservative, biocide and anti-fouling paint. These compounds are very stable in the biosphere due to the stability of the Sn–C bond and their solubility depends on the degree of alkylation and the associated ligand. Their general formula is R_nSnX_{4-n} where R is the alkyl group and X is a complexing ligand, usually OH^- or acetate, or a halogen anion. An extensive variety of compounds have been produced and these compounds will not be discussed individually here [71]. These compounds have or are being highly regulated and their usage has decreased as a result in many regions of the world.

In the aquatic environment, TBT is highly particle reactive and given its relative stability, is mostly sequestered in the sediments close to its source of emissions/usage. Thus, in the absence of continual input, water column concentrations decrease relatively rapidly. The TBT complexes are relatively soluble being ionic but they also have a relatively high K_{ow} ($\log K_{ow} = 3.7$) relative to other metal compounds and complexes (see Table 8.2), and therefore do bioaccumulate through passive partitioning. Uptake into higher trophic organisms is from water and through the food chain, and filter feeding organisms such as bilavers can accumulate high concentrations. Bioaccumulation factors are in the range of 10^3–10^4, which are comparable to non-alkylated metal(loid)s (Table 8.6). Bioaccumulation and toxicity is highest at the lower levels of the food chain, and this reflects the design and use of these compounds as a biocide to stop growth on ships and other human structures [108]. Metabolism of accumulated TBTs is slow for most higher trophic level organisms.

In the sediments, TBT is slowly degraded to the di- and monobutyl forms, and has a half-life on the order of months to years in sediments, with its half-life in waters being somewhat more rapid. The degradation of these compounds is likely both by biotic and abiotic pathways but the pathways have been little studied. Bacteria have been shown to degrade the TBTs and debutylation leads to the formation of less toxic forms. The Sn–C carbon bond is also susceptible to cleavage in the presence of UV radiation (<290 nm) and likely this degradation pathway accounts for the shorter half-lives of TBT compounds in surface waters [71]. Additionally, it is possible that the TBT compounds and their degradation products can be converted into hydrides, and these compounds are more volatile and have been detected in the environment (Section 5.2.6) [109].

In contrast to organo-Pb compounds, it appears that the mono- and dibutyl compounds can be methylated producing a variety of compounds of the general formula: $(CH_3CH_2CH_2CH_2)_xSn(CH_3)_{4-x}$, x = 1–3. Such compounds have been detected in the environment in waters and sediments. Ethyl-methyl compounds have been found in landfill gas ($(CH_3CH_2)_xSn(CH_3)_{4-x}$, x = 1–3). Such methylation produces more volatile and neutrally charged compounds that are much more mobile in the environment, and can be transported in water and lost to the atmosphere [56, 71]. The details of the methylation processes in the environment are not well understood but methylation has been demonstrated in laboratory studies. It appears likely that methylcobalamin is the main agent for methyl transfer, as expected given the positive charge of the alkyl complexes.

Overall, therefore, the fate, transport and impact of alkylated Pb and Sn compounds is primarily determined by their anthropogenic input into the environment. Both types of compounds are now highly regulated in many parts of the world and therefore their continued impact is more as a legacy pollutant than as a continual input. However, there is evidence that these compounds, and especially TBTs, are relatively persistent in the environment, and therefore will remain a concern in the future. Finally, because some microbes can methylate the alkyl Sn compounds, there is a further mechanisms for their mobilization and redistribution from their primary reservoir, which is coastal and freshwater sediments close to their region of initial application.

8.4.5 Organometallics of arsenic, antimony, and bismuth

The methylation of As (Fig. 8.14) is thought to be a detoxifying mechanism for bacteria and phytoplankton, especially those in the marine environment where the concentration of As is relatively high compared to the required nutrient phosphate. Both exist in solution as oxyanions. However, the first indication of As methylation was the demonstration of the release of volatile As-containing compounds from wallpaper in the 1800s in Europe, from the use of As-containing pigments in their coloration [58]. It was further realized that mold and damp enhanced their formation and therefore a biological role for the production of these volatile compounds was concluded, mostly based on work done by an Italian physician, Gosio. This early work was followed by Challenger and his colleagues in the early 20th century, who developed the original mechanism for As methylation that still bears his name.

It is now known that that methylation is also carried out by a wide variety of fungi, yeasts and bacteria, and eukaryotes [58, 75, 110]. It is thought that As is mainly taken up by phytoplankton through phosphate uptake channels and that

As methylation is a detoxifying mechanism. Inorganic As is present in environmental waters as an oxyacid in both oxidation states: arsenic acid (arsenate) $AsO(OH)_3$ or H_3AsO_4 ($pK_{a1} = 2.2$; $pK_{a2} = 7.0$; $pK_{a3} = 11.5$) and arsenous acid (arsenite) $As(OH)_3$ ($pK_{a1} = 9.3$). The pKas of arsenic acid are very similar to those of phosphate (respectively, 2.1, 7.2, and 12.4) (Table 8.5) and therefore the speciation of As^V and phosphate in most environmental waters is analogous and the As will be in a similar form to phosphate, which is taken up actively through specific channels. At the typical pH of environmental waters, the major species will be $H_2AsO_4^-$ and $HAsO_4^{2-}$. It generally appears that phosphate uptake is active and in conjunction with Na^+ co-transport and therefore a similar mechanism is likely for As. In open ocean surface waters, phosphate concentrations are often very low (<0.5 μM) and As concentrations are typically around 10–20 nM [111, 112]. Therefore, given this relatively small difference in concentration, inadvertent uptake of As is possible [110]. The methylated forms of As can be released from the cells into the environment or they can be incorporated into larger molecules and their toxicity is reduced as a result. Overall, it appears that the methylated forms of As are less toxic than the inorganic forms, which is in contrast to most other organometallics.

The so-called "Challenger" mechanism of As, and other metalloid, methylation involves a series of reductive methylation steps where the addition of the methyl group is via a carbocation addition reaction (Fig. 8.14). The details of the mechanism are relatively well-known although some aspects of the biochemistry within the cell are not entirely elucidated. It is thought that cellular thiols, such as glutathione, are involved in the reduction steps and that S-adenosylmethionine (SAM) is the main methylating agent. This mechanisms has been identified in a variety of microbes [58]. The products are the simple reduced methylated species ($(CH_3)_xAsO(OH)_{3-x}$, $x = 1–3$) as well as oxidized forms, such as trimethylarsine oxide ($(CH_3)_3AsO^-$). As noted in Section 8.4, volatile forms are produced by some organisms, and these are the methylated As hydrides, such as ($(CH_3)_xAsH_{3-x}$; $x = 1–3$) [56, 113]. In marine algae, As is found in a variety of organic compounds, such as arsenosugars and As-containing carbohydrates, with the most common compound being arsenobetaine (trimethylarsinio acetate) ($(CH_3)_3AsCH_2COO^-$) [110]. A variety of As compounds are found in marine and terrestrial animals as well [114]. It is thought that these products are formed from further reactions of the methylated derivatives and that they are ultimately generated by processes similar to that invoked in the Challenger mechanism. The methylated species are also found in measurable concentrations in many environmental waters, both freshwater and marine, and their presence has mostly been correlated with phytoplankton activity although their production via bacteria present in conjunction with the algae is also possible [114]. It is suggested that these compounds are actively exported by the phytoplankton.

The methylation pathway in most aerobic microbes appears to be as described above but there is some speculation that the pathway of methylation may be different for anaerobic organisms. In this case, it has been suggested that methylation may involve cobalamin and therefore involve a different mechanism whereby a carbanion or a radical from methylcobalamin is added to As^{III} in the presence of mediating enzymes [58]. However, while this mechanism is discussed, there appears to be little concrete evidence for this pathway.

Based on the pK_a values (Table 8.5), it is evident that the reduced form of As is present as an undissociated acid at physiological pHs and in most environmental solutions, and therefore that there is the possibility of its loss via passive diffusion from the cells into the surrounding media. There are a number of instances where measurements in the environment have shown that the concentration of reduced inorganic As is greater than that of As^V, and release of As^{III} after microbial reduction is the most likely explanation [76]. There have been a number of As reductases indentified in microbes and these processes appear to be separate from the methylation pathways, and therefore involve different reductants, such as thioredoxin and glutaredoxin [58]. Similar to other metal reductases, the operons are attached to plasmids in bacteria.

Antimony (Sb) has a similar chemistry to As and therefore it is expected to behave in a similar manner to As in terms of uptake and methylation [74]. Its compounds have been widely used in industry, medicine, as a poison, and as cosmetics. The methylated forms, $(CH_3)_xSb_{3-x}$ ($x = 1–3$), are well-known, as are many other organo-Sb compounds, and many have been synthesized for industrial purposes. Volatile Sb compounds were suggested as a possible cause for Sudden Infant Death Syndrome, as Sb compounds were used as flame retardants in mattresses, and this lead to the examination of the mechanisms of their formation [58, 74]. A number of fungi and bacteria have been identified that can methylate Sb, and methylation is much higher in the presence of Sb^{III} and it is apparent that many organisms are not able to reduce and methylate Sb^V. Therefore, the methylation of Sb may be less ubiquitous than that of As, and there appears to be complex interactions if both As and Sb are present in the same culture in terms of relative methylation [58, 74]. Overall, little is known of the speciation and form of Sb in natural waters, and the formula is given either as $Sb(OH)_5$ or $HSb(OH)_6$, which dissociates in water to form the anion, $Sb(OH)_6^-$ (pK_{a1} ~2.2). There is evidence from measurements in seawater and in the presence of marine algae that reduction to Sb^{III} occurs and that the monomethylated form exists in environmental waters [112]. It is also present in freshwaters, and evidence for its formation in environments such as landfills [56]. Overall, however, it appears that Sb^V is the major form in environmental solutions. The presence of volatile Sb compounds in landfill gases and other methanogenic environments confirms that methylation of Sb is microbially mediated [56, 58].

Bismuth exists under some conditions as methylated compounds [57]. The compounds of Bi are used in industrial and pharmaceutical applications. Given the wide use of Bi in many daily-used products, such as Pepto-Bismol, its presence in sewage treatment plants, and in municipal waste deposits, and the loss of volatile forms of Bi from these environments has been demonstrated [56]. Bismuth, in contrast to As and Sb, exists in environmental media as a +3 ion rather than as an oxyanion. It also can be found in the mono-, di- and tri- methylated form, and $(CH_3)_3Bi$ is non-polar while the other forms are ions, which exist as complexes in solution. It has been shown that methylated Bi compounds are produced by methanogens in culture [74]. The trimethylated form is less stable than its As and Sb analogs [57]. In addition to methylated forms, the hydride (BiH_3) has also been isolated from bacterial cultures. In the biotic formation of methylated Bi compounds, it is possible that the methyl group is donated by methylcobalamin, which is consistent with its form in solution as ionic complexes [57]. This contrasts the methylation of As and Sb that occurs through the S-adenosylmethionine pathway, as discussed above. However, there is little detailed information available about the exact nature of the methylation process, and it has been shown to occur in relatively few organisms.

8.5 Bioavailability and bioaccumulation

8.5.1 Introduction

The discussions in Section 8.3 suggest that there is the potential for microbes and other organisms to regulate the concentration of some metals internally, to a degree, and to also potentially store required (nutrient) metal(loid)s for use under conditions where their concentration in the environment may be limiting. The details of these storage processes are not clearly understood for many metals, but this is an area of active research. If an organism is actively storing a metal within its cell, then the external medium metal ion concentration may not be in any direct way related to that of the organism. Thus, as noted in Fig. 8.5, for many metals the intracellular concentration is relatively constant across species and therefore the concentration in the tissue is not a reflection of the exterior insult, but of the ability of the organism to control its cell quota. Of course, under high exposure concentrations, control of the intracellular metal pool becomes compromised and toxicity results. For non-essential metals, toxicity can also be avoided if the organism is capable of binding the metal intracellularly in a non-toxic or reactive complex. Metal regulating or storage proteins or compounds have been identified for Zn, Cd, and some other metals, and are referred to as *metallothioneins* for the molecules that are found in a variety of organisms and *phytochelatins* for the molecules found in phytoplankton and higher plants [42, 43, 115, 116] (See Section 8.3.1). The Fe-storing molecules are called *ferritins* and these have been shown to be extremely effective in storing Fe intracellularly, and are found in most organisms. There is a large amount of information in the literature on the accumulation and storage mechanisms for metals in plants, but the focus here will be mostly on the regulation of metals in unicellular aquatic organisms.

As noted in Section 8.3.1, phytochelatins (PCs) are synthesized in response to exposure to various required and toxic metal ions [2, 42]. They are peptides with the general structure of $(\gamma\text{-Glu–Cys})_n\text{–Gly}$ ($n = 2–11$) (subunits of γ-glutamyl-cysteine bound to glycine) and were discovered first in yeast and plants, and later in marine and freshwater algae [42]. They are synthesized from glutathione (GSH) in a reaction catalyzed by phytochelatin synthesases, and the expression and amount of these enzymes produced is in direct response to increasing metal(loid) concentration. It appears that Cd^{2+} is the strongest activator in a variety of plants and algae, but other cations (Pb, Zn, Cu, Sb, Ag, and Hg), as well as As, can activate PC synthesis [2, 42].

Rather than focusing on sub-cellular processes, many investigators have related cell quotas and tissue concentrations to the concentration in the aquatic medium. Such examination of bioavailability has been mostly derived to provide tools and approaches for prediction of organic contaminant accumulation; for example [78, 117–119]. While the importance of metal speciation and form in solution in assessing bioavailability is recognized (i.e., that total concentration is a poor metric), application of this knowledge to regulation has lagged because of the difficulty in making accurate predictions of the bioavailable metal concentrations in aquatic environments. A consequence of this is the development, for example, of water quality criteria for CH_3Hg in the USA that are based on concentrations in fish rather than the concentration of dissolved CH_3Hg, the most relevant phase. While this approach can provide an easily measured parameter for regulation, complications arise when there is an exceedance and a fish concentration must be related to Hg and CH_3Hg concentrations in solution, and therefore methods to do this need to be developed. Mostly, approaches are based on empirical determination or estimation of a *bioaccumulation factor* (BAF) which relates the tissue concentration to that of the surrounding medium [118, 119].

The impact of speciation of the dissolved metal on its rate of uptake has been discussed above (Section 8.2), and the impact of the dissolved speciation was alluded to in much of the discussion of transport mechanisms. Metal speciation and toxicity in natural waters depends on a complex mixture of ubiquitous inorganic and organic ligands that can alter bioavailability and solubility of metals [21, 120]. The solution pH, suspended solids load, dissolved and particulate organic carbon content (DOC and POC), and the presence of sulfides, all can alter the metal speciation. It is worth reiterating that even if uptake of a metal across a cellular membrane is related to the free metal ion activity, this does not necessarily indicate that this is the form that is accumulated. As long as the form taken up is linearly related to the free ion concentration over the conditions used within the experimental design, or in the field, it is not possible to determine the exact form or forms of the metal that are assimilated from such experiments.

Different approaches have been developed in an attempt to provide accurate prediction of the bioavailable fraction for the metal of interest in the environment of concern [78], and these will be discussed briefly. Metal speciation can vary with changes in water chemistry, and depending on the exchange kinetics associated with complex formation, there may be a fraction of the metal in solution that is "potentially bioavailable". Such fractions have been termed the "reactive" or "labile" metal fraction. Such approaches have been used, for example, in examining the bioavailability of Fe.

For clarity, a definition of terms is required. The US EPA website lists a number of definitions of *bioavailability* and selects the following definition: "A measure of the physicochemical access that a toxicant has to the biological processes of an organism. The less the bioavailability of a toxicant, the less its toxic effect on an organism". For trace metal(loid)s, however, this definition does not encompass the entire spectrum of bioavailability as toxicity is not the only concern, or even the primary issue in some instances as some metals are required nutrients and are actively taken up. In many situations, strong complexing ligands in solution, or the association of metal(loid)s with suspended solids in an exchangeable fashion allow "buffering" of the solution concentration of the available form. Uptake leads to momentary depletion of the bioavailable metal form in solution, but this deficit is removed due to the re-establishment of equilibrium conditions through dissociation of some of the metal(loid) from the "buffering" complexes. Thus, a constant concentration of the available form persists in solution. This is the case as long as the kinetics of the re-equilibration are sufficiently rapid.

Overall, therefore, the following definitions appear to be reasonable for the consideration of bioavailability of metals in aqueous solutions:

The bioavailable fraction of a metal(loid) in solution consists of those forms of the metal(loid), both inorganic and organic, that can be readily transported across biological membranes, either by active or passive processes, and subsequently accumulated, and/or biotransformed to a more toxic and bioaccumulative species.

In addition, a definition is needed for the potentially available fraction:

The potentially bioavailable metal(loid) fraction includes any form of the element in solution or associated with the solid phase that can be converted into a bioavailable form in a time span that is rapid relative to the rate uptake by the organism(s) under consideration.

If bioavailable, a metal can be bioaccumulated through the food chain. The concept of *bioaccumulation* can be defined as [120]:

The net accumulation of an element or compound in a tissue of interest, or a whole organism, that results from exposure.

The bioaccumulation of metals arises from uptake from all environmental sources - potentially, metals in air, dissolved in water and from the solid phase (organic and inorganic phases in soil and sediment), and through food consumption. Bioaccumulation that occurs under steady-state conditions (i.e., where tissue concentration remains relatively constant because uptake is offset by elimination and/or growth) is often the primary concern for environmental regulators and scientists.

If bioaccumulation reaches steady state, then it is possible to define the overall *bioconcentration factor* (BCF) which is the ratio of the concentration in the organism, mostly on a wet weigh basis (C_{org}), to the dissolved concentration in the surroundings [78]. If the relationship is defined relative to the truly dissolved concentration in the water (C_W), and as the organism concentration is usually expressed per kg, and the dissolved concentration per liter, the units of BCF are l/kg, and:

$$BCF = C_{org}/C_w \tag{8.10}$$

It is often not easy to determine the truly dissolved metal concentration because of colloidal material measured in the filtered fraction. Colloidal metal(loid)s are likely not bioavailable for uptake into organisms. Also, it may be that not all of the dissolved fractions are bioavailable for uptake (e.g., metals bound to NOM are often not accumulated) and thus this value is an underestimate for the true BCF. Actually, BCF should be based only on the bioavailable concentration (C_{WB}) which is often impossible to measure and difficult to estimate for metal(loid)s in natural waters. Furthermore, the concept of BCF is also usually applied to the assimilation of the metals by the organisms (i.e., uptake into tissues) and therefore should not include either metal(loid) that is externally adsorbed to the organism (unless it is in the process of being taken up) or present in the gut and other "external association". Such differentiation is not always possible. Again, for higher trophic organisms it is possible to determine the relative distribution of element between the various tissues and to depurate the organisms to remove that associated with food and other external solid materials prior to measurement. For microorganisms, various investigators have devised approaches and techniques to "wash" cells to remove any surface bound metal(loid) prior to analysis.

In addition to BCF, the *bioaccumulation factor* (BAF) has been defined where the bioaccumulation is considered not just from the dissolved phase but from all sources, such as metal(loid)s in food or sediment. If the main source of accumulation is food or sediment, then the BAF can be related to the concentration in this phase. Thus, the difference between BCF and BAF is that the latter takes into account the accumulation from all sources, even if, for example the BAF is still defined as the organism concentration relative to the water concentration.

The concept of relating the organism's burden to the sediment is often used if the organism of interest resides and feeds in/on the sediment [78]. The *sediment bioaccumulation factor* (SBAF) is defined as the concentration in the organism to that of the sediment (C_{sed}):

$$SBAF = C_{org}/C_{sed} \quad (8.11)$$

Such normalization does not always have merit, especially if the metal(loid) is being taken up from the dissolved fraction (porewater), and it also implicitly assumes that if uptake is from the porewater, then the system is at steady state, and does not change substantially over the time of measurement. Given the many physical, chemical and biological factors that may influence the porewater concentration, such an assumption is rarely met and it is therefore less reasonable to determine the SBAF in many instances.

If bioaccumulation occurs at each level of the food chain, and the overall bioaccumulation factor increases with trophic level, then this is referred to as *bioconcentration* through the food web. Thus, while many metals are bioaccumulated, they are not necessarily bioconcentrated up the food chain to piscivorous fish. This is shown, for example, by the BAF values in Table 8.6 for a variety of metals and metalloids. These are representative values and there is substantial variation depending on the metal and the relative size or the organism, especially for microbes (surface area/volume ratios), and depending on the organisms level in the food chain.

In examining the impact of elements or compounds on organisms, it is necessary to consider the factors that influence the rate of uptake, redistribution within the cell or organism, the impact through its interaction with cellular components and/or its storage and complexation, and its elimination [79, 121]. In microorganisms, the cellular quota is determined by the rate of uptake relative to the rate of growth (Section 8.2.2). Often one step in the overall process is rate limiting and this step therefore determines the overall extent of impact of the metal in solution. Mechanisms of uptake, of mediation of cellular free ion concentration, and of conversion or elimination of unwanted elements have been discussed in detail in the proceeding sections of this chapter. Therefore, in terms of unicellular organisms, the approaches and models already discussed provide the approach to determining the impact of a particular metal(loid) on a particular organism. For higher level organisms, the issue is more complex as there are multiple routes of uptake – accumulation directly from water across the body exterior or through the gills or uptake from food directly. These mechanisms are discussed in Section 8.5.2.

8.5.2 Trophic transfer of metals

Although most metal(loid)s enter phytoplankton cells, more efficient trophic transfer of some constituents leads to their enhanced bioaccumulation upon grazing by primary consumers, and their further bioconcentration through the food chain. Of all the toxic metal(loid)s, CH_3Hg and Se transfer from diatoms to copepods is the greatest and this coincides with the relatively greater sequestration of these elements in the diatom's cytoplasm compared to binding to constituents in the cellular membranes and other particulate material (the assay is based on centrifugal separation) [80, 122] (Fig. 8.20). The *assimilation efficiency* (AE) of the element is the relative assimilation from food and is estimated using one of the following the following approaches:

$$AE = A_{ass}/A_{food} = (A_{food} - A_{fec})/A_{food} = \Delta A_{org}/A_{food} \quad (8.12)$$

where A_{ass} is the amount of the element assimilated during feeding, A_{food} is the amount in the food, A_{fec} is the amount in the feces and A_{org} is the amount in the organism. It is often difficult to measure the change in concentration in the organism after feeding, especially if only one dose of food is administered, and so the most used approach is to collect and measure the amount in the feces and compare it to the initial amount in the food. In most instances, to alleviate the

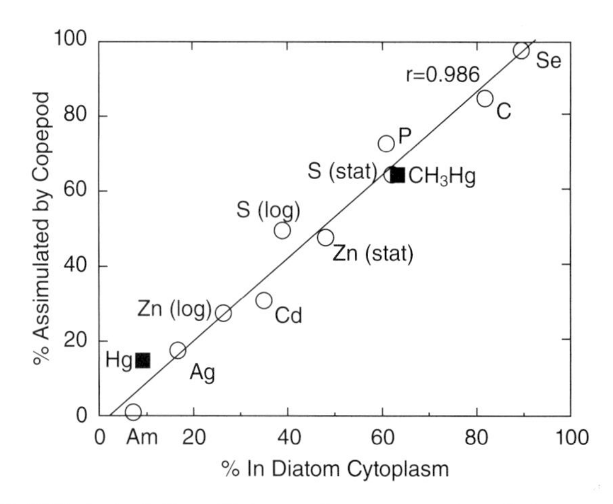

Fig. 8.20 Assimilation efficiency of mercury and methylmercury, as well as other elements, by copepods feeding on phytoplankton. Data besides Hg and CH_3Hg compiled by Reinfelder and Fisher (1991) *Science* **251**: 794 [80]. Figure reprinted from Mason et al. (1995) *Water Air Soil Pollution* **80**: 915–21 [122], reprinted with permission of Elsevier.

need to collect all the feces and know exactly the amount of food used in the experiment, the ratio of the element of interest to one that is not assimilated is used in the calculation. For diatoms, Si can be used as an unassimilated tracer, and in radioactive tracer studies, an element like americium (Am), which has a very low AE (<10%) can be used [9, 80].

Overall, Ag, Cd, and Hg behave similarly with relatively low assimilation efficiencies (<30%) during trophic transfer between phytoplankton and zooplankton. Transfer efficiency is somewhat higher for bivalves feeding on algae [80]. Recent studies suggest assimilation of Cu is similar to that of Zn and Cd [123], and that AE values for freshwater invertebrates are similar to those of marine organisms. For invertebrates and fish, uptake from food (algae or sediment) is much more important than direct uptake from water for most elements of interest [21, 79, 123, 124] and therefore diet, trophic structure and food chain length determine the total metal concentration. For metal(loid)s such as As, Se, Pb, and Cd, bioaccumulation does not occur at all levels of the food chain (Table 8.6) and this may be due to detoxification mechanisms, or may reflect other means of maintaining and preserving cellular metal concentrations [78].

In most cases, the highest BAF values are found in algae, with a reduction in concentration of about an order of magnitude between algae and primary consumers. In one examination of food chain accumulation [125], the AE for Cd decreased from 21–33% for a zooplankton (*Daphnia pulex*) feeding on phytoplankton to <16% for killifish (*Fundulus heteroctitus*) consuming the zooplankton to 38–56% for striped bass (*Morone saxatilis*) consuming the killifish. The AEs for CH_3Hg were >75% for all trophic transfers, while the corresponding values for Po were: 69–87%, 21–41% and 9–21%. Thus, while Cd was relatively strongly assimilated by the striped bass, Po was not, while the opposite was found for the killifish, reflecting the numerous factors that can influence AE. Bioconcentration did not occur at all trophic levels for Po and Cd, but did for CH_3Hg. When considering all trace metal(loid)s, the tissue concentration of elements which are incorporated into cellular biochemicals and/or are required for enzymes, such as As, Se, Zn, Co, are more regulated (homeostasis is maintained) and their concentration will reflect this unless exposure is at high and toxic concentrations.

An examination of the relationship between BAF and exposure concentration demonstrates the ability of organisms to regulate metal levels to some extent [120]. For Zn, examining data in the literature for various levels of the food web, it was apparent that increasing exposure concentration (over several orders of magnitude) lead to a relative small change in tissue concentration, and therefore to a decrease in the BAF by orders of magnitude. This suggests that the Zn concentration is being actively regulated by the organisms and thus the BAF approach provides little information on exposure or impact. A similar effect was found for Cd and Cu, although the magnitude of the change in BAF with exposure was less pronounced. The overall BAFs for Cd and Cu are also less than Zn for most trophic level organisms. For Hg, data on inorganic Hg is similar to that for the other +2 cations, but for CH_3Hg, the evidence strongly supports the conclusion that the BAF is little impacted by the exposure concentration, and therefore organisms have little ability to regulate their CH_3Hg concentration. This is similar to what is found for hydrophobic organic contaminants [120].

In many instances of assessing bioaccumulation, it is apparent that factors not directly related to the metal concentration in the food or water, and/or water column speciation, can have a large impact on the bioaccumulation. Growth rate and system eutrophication have been shown to be two very important variables, for example [18, 126]. Studies examining the impact of "growth dilution" for CH_3Hg have shown that if there is an increase in primary productivity, but not a concomitant increase in CH_3Hg production or input, then the concentration of CH_3Hg in phytoplankton is decreased. This can also impact the overall food chain accumulation if productivity is enhanced at other trophic levels [127]. Increased or decreased growth rate can also strongly impact accumulation for a metal(loid) that is strongly retained and not eliminated, and which is mostly obtained from food. In this case, growth is the primary dilution mechanism and therefore can influence the overall tissue burden. Such effects have been demonstrated in the laboratory and there is an indication that these are also important in natural environments, although this is more difficult to demonstrate.

The extent of toxicity and bioaccumulation into primary consumers is a function of food (sediment or biotic material) chemistry and exposure regime; for example, [129–131]. It is also necessary to consider the impact of the environment, especially for benthic organisms. In terms of benthic invertebrates living in or feeding on sediment, meaningful estimations of toxicity must consider the sediment POC, the oxide and the acid volatile sulfide (AVS) content, all of which strongly bind metals, as changes in these ancillary parameters impacts metal mobility, bioavailability, and ultimate fate. The role of sulfide is most pertinent for the coastal zone and other sulfide-rich environments, which also may be heavily polluted by industrialization and human activity. Several approaches have been developed to assess metal bioavailability and toxicity to organisms residing in sediments. Ankley et al. [132] proposed that transition metal toxicity in sediments (for Cd, Ni, Pb, Zn, and Cu) is related to the presence or absence of AVS, an important sediment metal chelator. Potential toxicity could therefore be assessed by measuring the relative amounts of metal and AVS in the sediments. Since AVS content is defined as that fraction of the sulfide extracted by acid leaching of sediments, the authors proposed that the metal recovered during extraction (the simultaneously extractable metal or SEM fraction) provided a measurement of the bioavailable metal fraction [132, 133]. The paradigm put forward was that if AVS/SEM was >1, the sediments were not toxic because the metals were not bioavailable. This initial approach has been refined because AVS is not the only important binding agent found in sediments, even in sulfidic sediments [129, 134]. The initial studies were done in the laboratory but the approach was extended to examine toxicity under field conditions [135].

This approach has limited predictability, however, in terms of assessing trophic transfer and bioaccumulation into primary benthic consumers [129]. This is probably because porewater is not the only source from which metals accumulate into benthic invertebrates [136, 137]. The accepted paradigm of bioaccumulation during digestion is that only soluble compounds are absorbed across the gut lining and it has been shown that there is a strong relationship between the degree of solubilization of a metal by intestinal fluids during digestion and the overall bioaccumulation into the organism [131, 138]. The extent of metal release from sediments is a function of an organism's intestinal fluid chemistry and overall physiology, as well as the chemistry of the sediments ingested. From a chemical perspective, the degree of solubilization of metals during digestion for most invertebrates appears to be a competitive ligand exchange process between the ligands in the sediment (POC and AVS phases) and the ligands in the intestinal fluids (primarily amino acids). The degree of solubilization of metals from the same sediment by the intestinal fluids of different organisms showed a difference that was relative to amino acid content of the individual organism's intestinal fluid [131, 138]. Therefore it is not just the sediment or particle chemistry that determines the extent of extraction and accumulation of metals by deposit and suspension feeding organisms.

As a result, metal bound to AVS and POC may become bioavailable when the metals are solubilized during invertebrate digestion of sediment [130, 131, 138, 139]. In Baltimore Harbor, for example, where AVS > SEM for many metals (Hg, CH_3Hg, Ag, Cu, Pb, and Cd), accumulation into bivalves still occurred [21]. Chen and Mayer [138] also concluded that SEM-AVS estimations were not a good measure of the bioavailability of Cu to benthic invertebrates. Lee et al. [124] showed that even in the presence of high AVS, bioaccumulation into benthos was related to the extractable sediment metal (Cd, Zn, or Ni) concentration and not the porewater concentration. Overall, it appears that benthic invertebrates accumulate most of their metal burden from food and ingested sediment, and not the porewater or overlying water [140]. Sediment POC affects the sediment sorption properties and metal bioavailability, as has been found for DOC in porewater and overlying water [141, 142]. Organism physiology is also important.

The discussion of the limitation of bioavailability during intestinal solubilization assumes that the membrane transport is not the fundamental determinant of bioaccumulation rate. This has been demonstrated in a number of instances where the uptake of metal complexes across membranes has been investigated using laboratory exposures and perfusion studies using isolated tissues [143–145]. These studies suggest that transition metal ions are taken up through similar mechanisms but that CH_3Hg is not, and may be assimilated while complexed to cysteine and other polypeptides through channels for amino acid assimilation [143, 146]. The oxyanions are likely accumulated in a similar fashion to their inorganic analogs. Overall, whether uptake from water or from food dominates will depend on the metal, the environment and the organism. Elements that are found in solution as oxyanions are mostly accumulated from food while there are specific instances for some metals where uptake from water can be important. However, for most metal(loid)s, uptake from food is the dominant route [122].

Most metal(loid)s accumulated by invertebrates and higher trophic level organisms can be efficiency and relatively rapidly depurated and thus the concentration in tissue in a stable exposure environment will reflect the relative rate of uptake and depuration [122]. If these processes can be considered to be pseudo first order, then the concentration in the tissue at any time can be determined. Over a short timeframe, growth is clearly not an important consideration but for metals that are strongly retained in tissues, it is necessary to take into account changes in concentration due to growth – so-called growth dilution, as discussed earlier.

In understanding the accumulation and trophic transfer of metals between organisms, it is possible to parameterize and model the processes at different levels of complexity. The empirical approach of estimating or measuring a BAF in the laboratory or field is the simplest approach and there are other approaches of varying complexity that will be discussed briefly in Section 8.5.3.

8.5.3 Exposure and bioaccumulation models

Models have been developed to examine in detail the interaction of dissolved metals with the membranes of organisms as well as the uptake and depuration of metals by organisms, and the model approaches will be briefly discussed. The two approaches can be combined but they are typically used to examine different aspects of metal interaction with organisms. In considering the uptake and accumulation of a contaminant by an organism, the following general framework includes all the processes that impact tissue concentration. For a particular metal(loid) [78]:

$$dC_{org}/dt = k_U C_{WB} - k_{-U} C_{org} + \sum AE_i \cdot I_i \cdot C_i - [k_E + k_M + g] C_{org} \tag{8.13}$$

where concentrations are as defined above (C_{WB} here represents the bioavailable dissolved fraction and not the total dissolved concentration (C_W)); k_U and k_{-U} are the direct uptake and elimination rate constants for accumulation directly from water, and loss directly to water (e.g., across gills or other exposed membranes) (all rate constants are assumed pseudo first order (units: d^{-1})); AE is the assimilation efficiency for food type i; I is the weight normalized consumption rate for prey i (g prey (g predator)$^{-1}$ d^{-1}); C_i is the concentration of the metal in the prey (mol g^{-1}); k_E is the elimination rate constant after assimilation into the tissues; k_M is the metabolic transformation rate (which would be applicable in only some instances, such as degradation of methylated species); and g is the growth rate coefficient. At steady state, this equation can be simplified for a variety of situations. For example, if uptake from food is negligible, then Equation 8.12 simplifies to, at steady state:

$$C_{org}/C_{WB} = k_U/(k_{-U} + k_E + k_M + g) \tag{8.14}$$

If steady state is obtained on a short timescale relative to growth and elimination is slow, then, as $k_M = 0$ for most metal(loid)s, the equation simplifies to:

$$C_{org}/C_{WB} = k_U/k_{-U} = BCF \tag{8.15}$$

For some metals, such as Ag, direct interaction with exposed tissues of higher organisms, such as the gills of fish, can result in substantial accumulation, but this has been shown to be dependent on the water column speciation. Therefore, a complexation modeling approach has been derived to allow prediction of the extent of these interactions. This model, termed the *biotic ligand model* (BLM; [147]), calculates the interaction of a dissolved metal with a biological surface by considering the complexation of the metal in solution and by assuming that interaction of the metal with the biological surface can be simulated by a metal-ligand interaction, which can be characterized by an effective binding constant (Fig. 8.21). In the model, the free metal ion is the species that interacts with the membrane surface. This approach is essentially a thermodynamic equilibrium model with one of the "ligands" being the site on the biological surface with which the metal interacts (Fig. 8.21). Based on knowledge of the membrane physiology and the accumulation pathway, the number of binding sites on the surface can be estimated, or this value can be determined empirically from a set of experiments at different known exposure concentrations. The model therefore assumes that the rate of uptake is determined/related to the extent of interaction and that differences in toxicity or bioaccumulation across systems is driven by differences in the dissolved speciation of the metal that is related to differences in pH, and the concentration of inorganic and organic ligands, including NOM. Additionally, it is possible that different metals (and H^+) could compete for the uptake sites on the biological surface, which are finite, as shown in the example for Ag in Fig. 8.21(b) [148]. This modeling approach has been used mostly to examine the impact of Ag and other cations with fish gills, but other applications are clearly possible.

Many of the early studies that formed the basis of the BLM approach were done by Playle and his colleagues [147], who investigated metal–gill interactions and showed that there was competition between major cations and trace ions for a limited number of binding sites on the gill surface. Additionally, this competition affected the degree of metal accumulation at the gill surface. These experiments were used to evaluate gill binding site densities and conditional stability constants, parameters that could then be incorporated into chemical equilibrium models and used to predict the interactive effects of concentration, competition and complexation on metal accumulation. Initial studies focused on Cu, Cd, and Ag [148–150] and demonstrated the importance of speciation in controlling the interaction with the gill surface and the mechanisms of toxicity [151–153].

The BLM approach has been widely applied and used in environmental decision making and regulation [154] and in studies of uptake by fish and invertebrates, including the impact of multiple metals in solution [155]. As a confirmation of the BLM approach, one study demonstrated that the impact of NOM on measured Cu accumulation in trout correlated with measurements of the free ion concentration, using an ion selective electrode, and Cu uptake measured using diffusive gradient in thin films approaches. The study also showed that both quantity and quality of NOM were important factors controlling Cu uptake [156]. Recent examples of the application of the BLM approach include investigations of the influence of major ion concentration and organic matter content on the interaction of Cd, Cu, and Pb with fish gills [157, 158], on the uptake of Ni [159] and Ag by *Daphnia* in the presence of dissolved anions, and in examination of the accumulation mechanisms for a variety of invertebrates. Metal cations apparently alter the functioning of major ion ATPases and intracellular ion concentrations rather than just interacting at the surface of gills and other external membranes [160, 161]. Finally, the BLM approach has been extended to the sediment environment and used in combination with the SEM/AVS analysis, and a sediment partitioning model, to examine toxicity of metals to invertebrates [162].

Kinetic models have been developed to examine the interactions and accumulation of metal(loid)s at the organism level. Factors that are incorporated into such models are the AE from food, uptake from water directly, the depuration (loss) rate, and the growth rate of the organism [125] (Equation 8.12). If uptake from water is small, a *trophic transfer potential* (TTP) can be defined as:

$$TTP = IR \cdot AE/(k_E + g) \tag{8.16}$$

(a)

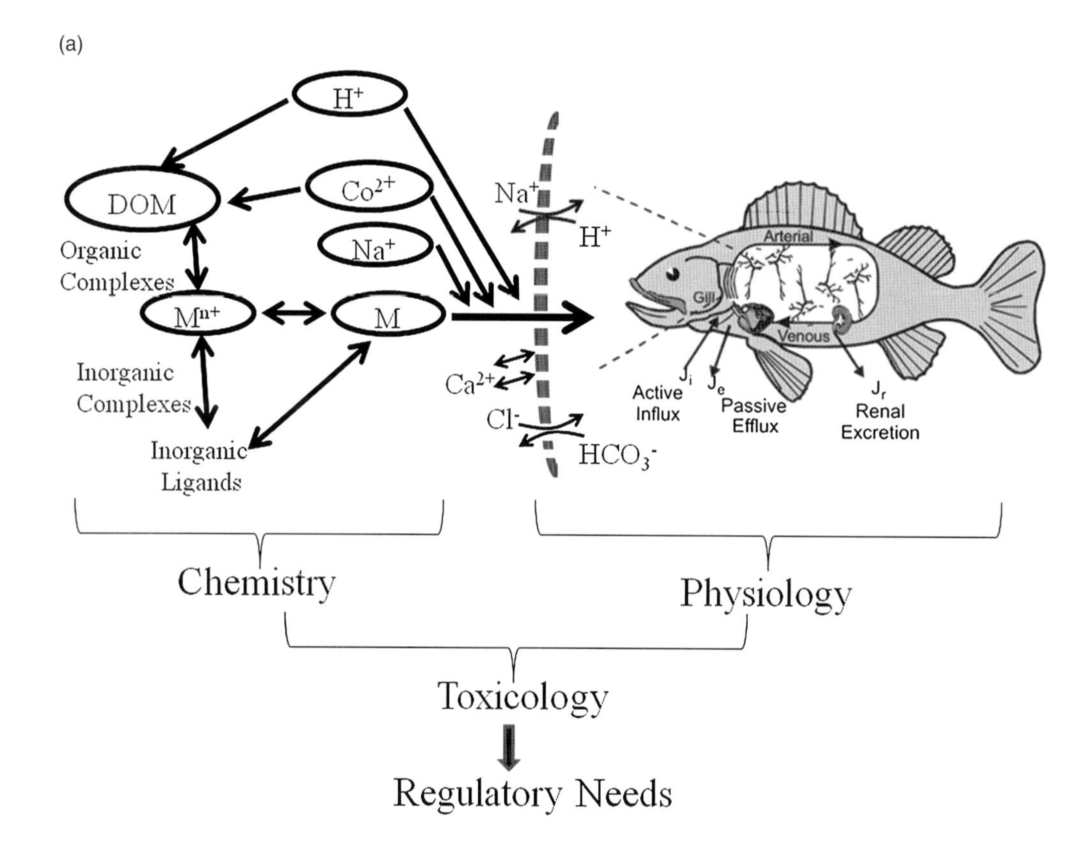

(b)

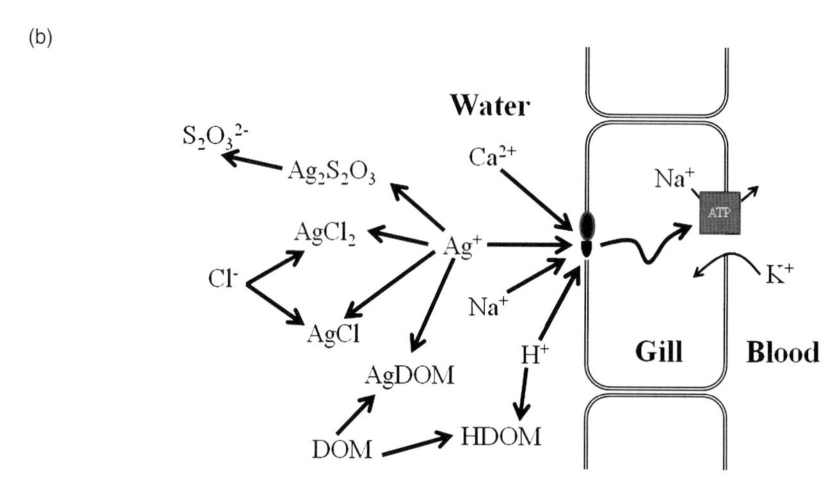

Fig. 8.21 (a) A diagrammatic conceptualization of the Biotic Ligand Model (BLM) for metal (M) uptake by fish across their gills. Figure redrawn from Paquin et al. (2002) *Comparative Biochemistry and Physiology C* **133**: 3–35 [147], reprinted with permission of Elsevier; (b) Schematic depiction of the interactions included in the BLM, with silver (Ag) as the example. Redrawn from McGreer et al. (2000) *Environmental Science and Technology* **34**: 4199–207 [148]. Copyright (2000) American Chemical Society.

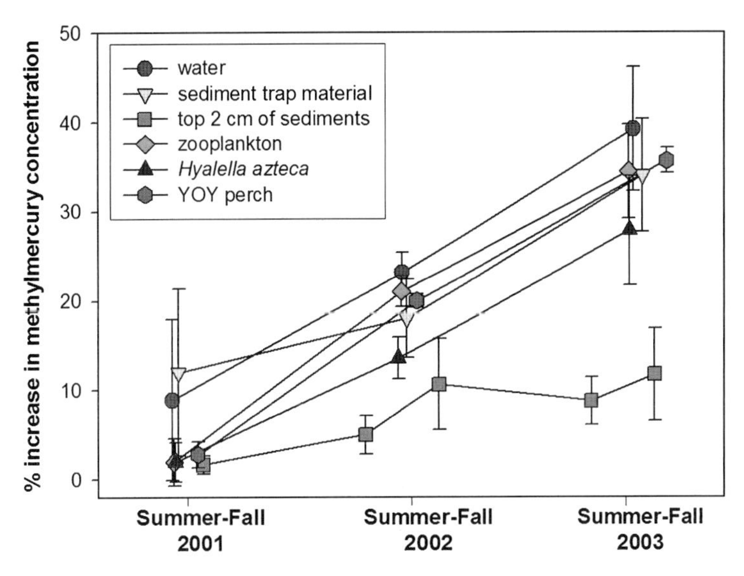

Fig. 8.22 The response of the lake ecosystem and the biota to the increase in inorganic mercury added to the lake surface as a stable isotope beginning in summer 2001. The overall increase in methylmercury in water, sediment and biota as a result of this addition is shown over time. Taken from Harris et al. (2008) *PNAS* **104**: 16586–91, and used with permission. Copyright (2008) National Academy of Sciences, USA.

where IR is the weight specific ingestion rate of food [121]. This value represents the steady state concentration in a consumer that would be derived from prey consumption, relative to that of the prey. Such a model is suitable for examining primary and secondary consumers but may not be suitable for long-lived species. Based on Equation 8.16, a lack of bioconcentration at a trophic step can be due to either a low AE or a high k_E.

If more detailed modeling is required, a bioenergetics model can be developed to examine in detail all the sources of an element to a particular organism, and the importance of the various processes involved in uptake, trophic transfer and elimination, and the impact of environmental conditions. Such modeling is an extension of the approach described earlier and is based on Equation 8.12, and was initially used to examine contaminant bioaccumulation in the 1980s, with much of the focus on the accumulation of organic contaminants [78, 163–165]. If the system is assumed to be at steady state, the BAF can be written in terms of the BCF as defined in Equation 8.15 and the remaining parameters [78]. However, Equation 8.13 is usually used in numerical models to examine the dynamic accumulation of a metal into an organism over time, and to simulate complex interactions of a typical food chain. Also, for many metals, as accumulation from food is the most important pathway, the model approach can rely on a bioenergetics approach. In many instances, in examining metal uptake, the following simplification therefore can be used:

$$dC_{org}/dt = \sum AE_i \cdot I_i \cdot C_i - [k_E + g]C_{org} \tag{8.17}$$

A discussion of such models is beyond the scope of this book but this approach has been used extensively to model the accumulation of CH_3Hg in fish in aquatic food chains [166–168]. The Mercury Cycling Model [169, 170] is one framework that has been used extensively to study remote and contaminated lakes [171–173]. An example of the bioaccumulation of Hg, added to a lake in the Canadian Experimental Lakes Area as a specific isotope, is shown in Fig. 8.22. The measured and modeled increase in CH_3Hg over time in different organisms that constitute the food chain of the lake is shown. For other metals, models have examined the cycling and accumulation of Cd in Lake Erie organisms [174]; Cd in rainbow trout [175]; and the accumulation of various metals in marine bivalves [176]. A similar approach was used to examine the CH_3Hg dynamics in a shallow estuarine system with sediment resupsension [177] (Fig. 8.23). In the figure, both the carbon and the CH_3Hg movement through the ecosystem are shown.

Dynamic modeling approaches have been used to examine the accumulation of CH_3Hg in piscivorous birds [178, 179], and in the examination of the uptake of other metal(loid)s [180, 181]. Additionally, for larger organisms it is possible for internal redistribution of the metals within the various tissues and therefore modeling of the internal redistribution of the metal could enhance the model capabilities. This is also important as elimination is likely based on the concentration in a specific organ (liver, kidney, gill) rather than the overall average body burden. Such approaches have been used, for example, in the examination of Cd accumulation in trout [175] and CH_3Hg accumulation in fish [146, 182].

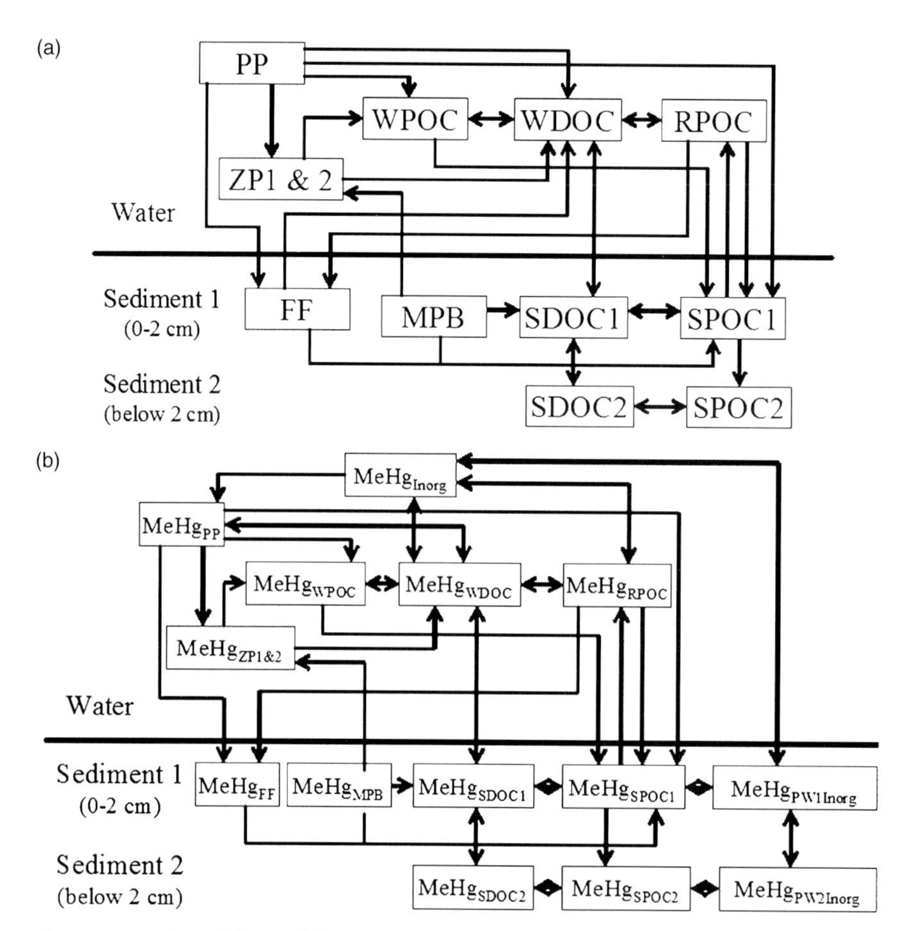

Fig. 8.23 A diagrammatic representation of the modeling framework illustrating (a) the carbon flow within the mesocosm with sediment resuspension and (b) the flow of methylmercury in the various compartments. Taken from Kim et al. (2008) *Ecological Modeling* **218**: 267–89, reprinted with permission of Elsevier.

8.6 Chapter Summary

1. Many metal(loid)s are required by organisms as they form an important part of enzymes and other biomolecules. However, even required metals can be toxic if present in high enough concentrations. Metals are acquired by microbes by using mechanisms that may or not require energy. Passive diffusion and facilitated transport are non-energy requiring processes. Neutral complexes that can partition into the membrane are taken up by passive diffusion. The other uptake processes are more complex and rely on a number of mechanisms.

2. The concentrations of metal(loid)s in microbes reflects both their necessity and they availability in the aquatic environment. Small cells are less easily limited given that these cells have a relatively higher surface-area-to-volume ratio.

3. One of the most important elements for phytoplankton and many other microbes is Fe and this trace element has been shown to be limiting to a greater extent than any other trace element. Fe has an important role in the photosynthetic machinery as well as in other essential enzymes for the transformation of nitrogen and other compounds. Zinc is an important element as it is a key part of the enzyme carbonic anhydrase. Other metals (Co and Cd) can substitute for Zn and therefore potentially remove Zn limitation of phytoplankton growth. Trace element limitation occurs in both freshwater and marine environments.

4. As Fe is often present in low concentrations organisms have devised many ways to increase its availability and uptake. Photochemical reactions in surface waters can also increase its bioavailability. The low level of Fe in the current world, and the inclusion of specific metals in biomolecules, is thought to reflect the relative abundance of these trace elements in the early, oxygen-free aquatic environments on Earth. On the other hand, many of the toxic elements were present in low concentrations in the prehistoric ocean.

5. There are three main mechanisms for methylation of metal(loid)s by microbes. Most are methylated either by a reductive methylation pathway (the Challenger mechanism) or by reaction involving cobalamin. These reactions differ in the form the methyl group is donated. Most of the metalloids are methylated by the Challenger process while Hg, and other Group 12 and other cations are mostly methylated by cobalamin. In addition to methylation in the environment, there are metals that have been added to the environment from anthropogenic sources, most specifically alkylated Pb and Sn compounds. The fate and transport and degradation of Pb and Sn compounds wre discussed in some detail in the chapter.

6. The methylation of Hg has been extensively studied, particularly in freshwater and coastal environments, and it is known that sulfate reducing bacteria and other similar organisms are mostly responsible for the methylation in sediments. Methylation in the ocean is less well understood.

7. The methylation of the metalloids was discussed as this pathway is important and it is also likely that the processes whereby As and Se are incorporated into larger biomolecules first involves their methylation. Methylation of As is thought to be a detoxifying mechanism as As uptake can interfere with the uptake of P by microbes.

8. The concepts of bioaccumulation and biomagnification and the related parameters such as the bioconcentration factor were discussed. The factors that influence the magnitude and variability in bioaccumulation were highlighted, and there was discussion of the factors influencing trophic transfer of metals from phytoplankton to higher organisms. The chapter also briefly discussed the models that are available to examine and understand the accumulation of metal(loid)s into aquatic food chains.

References

1. Lindh, U. (2004) Biological function of the elements. In: Seleinus, O., Alloway, B., Centeno, J.A. et al. (eds) *Essentials of Medical Geology*. Elsevier, Amsterdam, pp. 87–144.
2. Morel, F., Milligan, A.J. and Saito, M.A. (2004) Marine bioinorganic chemistry: The role of trace metals in the oceanic cycles of major nutrients. In: Elderfield, H. (ed.) *The Oceans and Marine Geochemistry*. In Holland, H.D. and Turekian, K.K. (Exec. eds) Treatise on Geochemistry. Elsevier, Amsterdam, pp. 113–143.
3. Simkiss, K. and Taylor, M.G. (1995) Transport of metals across membranes. In: Tessier, A. and Turner, D.R. (eds) *Metal Speciation and Bioavailability in Aquatic Systems*. John Wiley & Sons, Ltd, Chichester, pp. 1–44.
4. Sunda, W. (1989) Trace metal intgeractions with marine phytoplankton. *Biological Oceanography*, **6**, 411–442.
5. Campbell, P. and Tessier, A. (1996) *Ecotoxicology of Metals in the Aquatic Environment: Geochemical Aspects*. CRC Press, Boca Raton, FL, pp. 11–58.
6. Stein, W.D. and Lieb, W.R. (1986) *Transport and Diffusion across Cell Membranes*. Academic Press, Orlando.
7. Schwarzenbach, R., Gschwend, P.M. and Imboden, D.M. (1993) *Environmental Organic Chemistry*. John Wiley & Sons, Inc., New York.
8. Lawson, N.M. and Mason, R.P. (1998) Accumulation of mercury in estuarine food chains. *Biogeochemistry*, **40**(2–3), 235–247.
9. Mason, R.P., Reinfelder, J.R. and Morel, F.M.M. (1996) Uptake, toxicity, and trophic transfer of mercury in a coastal diatom. *Environmental Science & Technology*, **30**(6), 1835–1845.
10. Morel, F.M.M. and Hering, J.G. (1993) *Principals and Applications of Aquatic Chemistry*. John Wiley & Sons, Inc., New York.
11. Benoit, J.M., Mason, R.P. and Gilmour, C.C. (1999) Estimation of mercury-sulfide speciation in sediment pore waters using octanol-water partitioning and implications for availability to methylating bacteria. *Environmental Toxicology and Chemistry*, **18**(10), 2138–2141.
12. Phinney, J.T. and Bruland, K.W. (1994) Uptake of lipophilic organic Cu, Cd, and Pb complexes in the coastal diatom *Thalassiosira-weissflogii*. *Environmental Science & Technology*, **28**(11), 1781–1790.
13. Reinfelder, J.R. and Chang, S.I. (1999) Speciation and microalgal bioavailability of inorganic silver. *Environmental Science & Technology*, **33**(11), 1860–1863.
14. Kaiser, S.M. and Escher, B.I. (2006) The evaluation of liposome-water partitioning of 8-hydroxyquinolines and their copper complexes. *Environmental Science & Technology*, **40**(6), 1784–1791.
15. Jay, J.A., Morel, F.M.M. and Hemond, H.F. (2000) Mercury speciation in the presence of polysulfides. *Environmental Science & Technology*, **34**(11), 2196–2200.
16. Gutknecht, J.J. (1983) Cadmium and thallous ion permeabilities through lipid bilayer membranes. *Biochimica et Biophysica Acta*, **735**, 185.
17. Veltman, K., Huijbregts, M.A.J., Van Kolck, M., Wang, W.X. and Hendriks, A.J. (2008) Metal bioaccumulation in aquatic species: Quantification of uptake and elimination rate constants using physicochemical properties of metals and physiological characteristics of species. *Environmental Science & Technology*, **42**(3), 852–858.
18. Najera, I., Lin, C.C., Kohbodi, G.A. and Jay, J.A. (2005) Effect of chemical speciation on toxicity of mercury to Escherichia coli biofilms and planktonic cells. *Environmental Science & Technology*, **39**(9), 3116–3120.
19. Turner, A. and Williamson, I. (2005) On the relationship between D-OW and K-OW in natural waters. *Environmental Science & Technology*, **39**(22), 8719–8727.
20. Gutknecht, J. (1981) Inorganic mercury (Hg2+) transport through lipid bilayer membranes. *The Journal of Membrane Biology*, **61**, 61.
21. Mason, R.P. (2002) The bioaccumulation of mercury, methylmercury and other toxic elements into pelagic and benthic organisms. In: Newman, M.C., Roberts, M.H., Jr. and Hale, R.C. (eds) *Coastal and Estuarine Risk Assessment*. Lewis Publisher, Boca Raton, pp. 127–149.
22. Schaefer, J.K. and Morel, F.M.M. (2009) High methylation rates of mercury bound to cysteine by Geobacter sulfurreducens. *Nature Geoscience*, **2**(2), 123–126.
23. Phinney, J.T. and Bruland, K.W. (1997) Trace metal exchange in solution by the fungicides Ziram and Maneb (dithiocarbamates) and subsequent uptake of lipophilic organic zinc, copper and lead complexes into phytoplankton cells. *Environmental Toxicology and Chemistry*, **16**(10), 2046–2053.
24. Croot, P.L., Karlson, B., Van Elteren, J.T. and Kroon, J.J. (1999) Uptake of Cu-64-oxine by marine phytoplankton. *Environmental Science & Technology*, **33**(20), 3615–3621.

25. Jay, J.A., Murray, K.J., Gilmour, C.G., et al. (2002) Mercury methylation by *Desulfovibrio desulfuricans* ND132 in the presence of polysulfides. *Applied and Environmental Microbiology*, **68**(11), 5741–5745.
26. Sunda, W.G. (2001) Bioavailability and bioaccumulation of iron in the sea. In: Turner, D.R. and Hunter, K.A. (eds) *The Biogeochemistry of Iron in Seawater*. John Wiley & Sons, Ltd, Chichester, pp. 41–84.
27. Lamborg, C.H., Fitzgerald, W.F., Skoog, A. and Visscher, P.T. (2004) The abundance and source of mercury-binding organic ligands in Long Island Sound. *Marine Chemistry*, **90**(1–4), 151–163.
28. Lamelas, C., Wilkinson, K.J. and Slaveykova, V.I. (2005) Influence of the composition of natural organic matter on Pb bioavailability to microalgae. *Environmental Science & Technology*, **39**(16), 6109–6116.
29. Parent, L., Twiss, M.R. and Campbell, P.G.C. (1996) Influences of natural dissolved organic matter on the interaction of aluminum with the microalga Chlorella: A test of the free-ion model of trace metal toxicity. *Environmental Science & Technology*, **30**(5), 1713–1720.
30. Quigg, A., Reinfelder, J.R. and Fisher, N.S. (2006) Copper uptake kinetics in diverse marine phytoplankton. *Limnology and Oceanography*, **51**(2), 893–899.
31. Vigneault, B., Percot, A., Lafleur, M. and Campbell, P.G.C. (2000) Permeability changes in model and phytoplankton membranes in the presence of aquatic humic substances. *Environmental Science & Technology*, **34**(18), 3907–3913.
32. Bruland, K. and Lohan, M.C. (2004) Controls of trace metals in seawater. In: Elderfield, H. (ed.) *The Oceans and Marine Geochemistry*. In Holland, H.D. and Turekian, K.K. (Exec. eds), Treatise on Geochemistry. Elsevier, Amsterdam, pp. 23–47.
33. Lane, T.W. and Morel, F.M.M. (2000) Regulation of carbonic anhydrase expression by zinc, cobalt, and carbon dioxide in the marine diatom *Thalassiosira weissflogii*. *Plant Physiology*, **123**(1), 345–352.
34. Lee, J.G., Roberts, S.B. and Morel, F.M.M. (1995) Cadmium – a nutrient for the marine diatom *Thalassiosira-weissflogii*. *Limnology and Oceanography*, **40**(6), 1056–1063.
35. Golding, G.R. et al. (2002) Evidence for facilitated uptake of Hg(II) by *Vibrio anguillarum* and *Escherichia coli* under anaerobic and aerobic conditions. *Limnology and Oceanography*, **47**(4), 967–975.
36. Yamaguchi, A., Tamang, D.G. and Saier, M.H. (2007) Mercury transport in bacteria. *Water, Air, and Soil Pollution*, **182**(1–4), 219–234.
37. Downs, T., Schallenberg, M. and Burns, C. (2008) Responses of lake phytoplankton to micronutrient enrichment: A study in two New Zealand lakes and an analysis of published data. *Aquatic Sciences*, **70**(4), 347–360.
38. Sterner, R.W. (2008) On the phosphorus limitation paradigm for Lakes. *International Review of Hydrobiology*, **93**(4–5), 433–445.
39. Ahner, B.A., Price, N.M. and Morel, F.M.M. (1994) Phytochelatin production by marine-phytoplankton at low free metal-ion concentrations - laboratory studies and field data from Massachusetts Bay. *Proceedings of the National Academy of Sciences of the United States of America*, **91**(18), 8433–8436.
40. Kramer, U. (2005) Phytoremediation: Novel approaches to cleaning up polluted soils. *Current Opinion in Biotechnology*, **16**(2), 133–141.
41. Maier, E.A., Matthews, R.D., McDowell, J.A., Walden, R.R. and Ahner, B.A. (2003) Environmental cadmium levels increase phytochelatin and glutathione in lettuce grown in a chelator-buffered nutrient solution. *Journal of Environmental Quality*, **32**(4), 1356–1364.
42. Clemens, S. (2006) Toxic metal accumulation, responses to exposure and mechanisms of tolerance in plants. *Biochimie*, **88**(11), 1707–1719.
43. Ernst, W.H.O., Krauss, G.J., Verkleij, J.A.C. and Wesenberg, D. (2008) Interaction of heavy metals with the sulphur metabolism in angiosperms from an ecological point of view. *Plant, Cell & Environment*, **31**(1), 123–143.
44. Ahner, B.A., Kong, S. and Morel, F.M.M. (1995) Phytochelatin production in marine-algae 1. An interspecies comparison. *Limnology and Oceanography*, **40**(4), 649–657.
45. Ahner, B.A., Wei, L.P., Oleson, J.R. and Ogura, N. (2002) Glutathione and other low molecular weight thiols in marine phytoplankton under metal stress. *Marine Ecology Progress Series*, **232**, 93–103.
46. Barbeau, K., Rue, E.L., Bruland, K.W. and Butler, A. (2001) Photochemical cycling of iron in the surface ocean mediated by microbial iron(III)-binding ligands. *Nature*, **413**, 409–413.
47. Smith, K.S. and Ferry, J.G. (2000) Prokaryotic carbonic anhydrase. *FEMS Microbiology Reviews*, **24**, 335–366.
48. Intwala, A., Patey, T.D., Plolet, D.M. and Twiss, M.R. (2008) Nutritive substitution of zinc by cadmium and cobalt in phytoplankton isolated from the lower Great Lakes. *Journal of Great Lakes Research*, **34**, 1–11.
49. Ivanikova, N.V., McKay, R.M.L., Bullerjahn, G.S. and Sterner, R.W. (2007) Nitrate utilization by phytoplankton in Lake Superior is impaired by low nutrient (P, Fe) availability and seasonal light limitation – A cyanobacterial bioreporter study. *Journal of Phycology*, **43**, 475–484.
50. Williams, R.J.P. and Frausto Da Silva, J.J.R. (2006) *The Chemistry of Evolution: The Development of our Ecosystem*. Elsevier, Amsterdam.
51. Saito, M., Sigman, D.M. and Morel, F.M.M. (2003) The bioinorganic chemistry of the ancient ocean: The co-evolution of cyanobacterial metal requirements and biogeochemical cycles at the Archean/Proterozoic boundary? *Inorganica Chimica Acta*, **356**, 1–11.
52. Williams, R.J.P. (2004) Uptaker of elements from a chemical point of view. In: Selenius, O. (ed.) *Essentials of Medical Geology*. Elsevier, Amsterdam, pp. 61–86.
53. Beinert, H. (2000) Iron-sulfur proteins: Ancient structures, still full of surprises. *Journal of Biological Inorganic Chemistry*, **5**(1), 2–15.
54. Imlay, J.A. (2006) Iron-sulphur clusters and the problem with oxygen. *Molecular Microbiology*, **59**(4), 1073–1082.
55. Craig, P.J., Eng, G. and Jenkins, R.O. (2003) Occurrence and pathways of organometallic compounds in the environment – General considerations. In: Craig, P. (ed.) *Organometallic Compounds in the Environment*. John Wiley & Sons, Ltd, Chichester, pp. 1–56.
56. Feldman, J. (2003) Volatilization of metals from a landfill site. In: Cai, Y. and Braids, O.C. (eds) *Biogeochemistry of*

Environmentally Important Trace Elements. ACS, Washington, DC, pp. 128–140.

57. Feldman, J. (2003) Other organometallic compounds in the environment. In: Craig, P. (ed.) *Organometallic Compounds in the Environment*. John Wiley & Sons, Ltd, Chichester, pp. 353–390.
58. Bentley, R. and Chasteen, T.G. (2002) Microbial methylation of metalloids: Arsenic, antimony, and bismuth. *Microbiology and Molecular Biology Reviews*, **66**(2), 250.
59. Benoit, J.M., Gilmour, C.C., Heyes, A., Mason, R.P. and Miller, C.L. (2003) Geochemical and biological controls over methylmercury production and degradation in aquatic systems. In: Cai, Y. and Braids, O.C. (eds) *Biogeochemistry of Environmentally Important Trace Elements*. Vol. 835. ACS Symposium Series. ACS, Washington, DC, pp. 262–297.
60. Perrin, D.D. (1982) *Ionization Constants of Inorganic Acids and Bases in Aqeous Solution*. Pergamon Press, Oxford.
61. Benoit, J.M., Gilmour, C.C. and Mason, R.P. (2001) Aspects of bioavailability of mercury for methylation in pure cultures of Desulfobulbus propionicus (1pr3). *Applied and Environmental Microbiology*, **67**(1), 51–58.
62. Celo, V., Lean, D.R.S. and Scott, S.L. (2006) Abiotic methylation of mercury in the aquatic environment. *Science of the Total Environment*, **368**(1), 126–137.
63. Hamasaki, T., Nagase, H., Yoshioka, Y. and Sato, T. (1995) Formation, distribution, and ecotoxicity of methylmetals of tin, mercury, and arsenic in the environment. *Critical Reviews in Environmental Science and Technology*, 1064–3389, **25**(1), 45.
64. Gardfeldt, K., Munthe, J., Stromberg, D. and Lindqvist, O. (2003) A kinetic study on the abiotic methylation of divalent mercury in the aqueous phase. *Science of the Total Environment*, **304**(1–3), 127–136.
65. Hammerschmidt, C.R., Lamborg, C.H. and Fitzgerald, W.F. (2007) Aqueous phase methylation as a potential source of methylmercury in wet deposition. *Atmospheric Environment*, **41**(8), 1663–1668.
66. Falter, R. (1999) Experimental study on the unintentional abiotic methylation of inorganic mercury during analysis: Part 2: Controlled laboratory experiments to elucidate the mechanism and critical discussion of the species specific isotope addition correction method. *Chemosphere*, **39**(7), 1075–1091.
67. Hintelmann, H. (1999) Comparison of different extraction techniques used for methylmercury analysis with respect to accidental formation of methylmercury during sample preparation. *Chemosphere*, **39**(7), 1093–1105.
68. Liang, L. et al. (2004) Re-evaluation of distillation and comparison with HNO3 leaching/solvent extraction for isolation of methylmercury compounds from sediment/soil samples. *Applied Organometallic Chemistry*, **18**(6), 264–270.
69. Heisterkamp, M., Van De Velde, K., Ferrari, C., Boutron, C.F. and Adams, F.C. (1999) Present century record of organolead pollution in high altitude alpine snow. *Environmental Science & Technology*, **33**(24), 4416–4421.
70. Mason, R. and Benoit, J.M. (2003) Organomercury compounds in the environment. In: Craig, P.J. (ed.) *Organometallic Compounds in the Environment*. John Wiley & Sons, Ltd, Chichester, pp. 57–99.
71. Cima, F., Craig, P.J. and Harrington, C. (2003) Organotin compounds in the environment. In: Craig, P.J. (ed.) *Organometallic Compounds in the Environment*. John Wiley & Sons, Ltd, Chichester, pp. 101–150.
72. Bridges, C.C. and Zalups, R.K. (2005) Molecular and ionic mimicry and the transport of toxic metals. *Toxicology and Applied Pharmacology*, **204**(3), 274–308.
73. Pongrantz, R. and Heumann, K.G. (1999) Production of methylated mercury, lead, and cadmium by marine bacteria as a significant natural source for atmospheric heavy metals in polar regions. *Chemosphere*, **39**, 89–102.
74. Thayer, J.S. (2002) Biological methylation of less-studied elements. *Applied Organometallic Chemistry*, **16**(12), 677–691.
75. Plant, J., Kinniburgh, D.G., Smedley, P.L. and Fordyce, F.M. (2005) Arsenic and selenium. In: Lollar, B. (ed.) *Environmental Geochemistry*. In Holland, H.D. and Turekian, K.K. (Exec. eds), Treatise on Geochemistry. Elsevier, Amsterdam, pp. 17–65.
76. Stolz, J.E., Basu, P., Santini, J.M. and Oremland, R.S. (2006) Arsenic and selenium in microbial metabolism. *Annual Review of Microbiology*, **60**, 107–130.
77. Fox, P.M., LeDuc, D.L., Hussein, H., Lin, Z.-Q. and Terry, N. (2003) Selenium speciation in soils and plants. In: Cai, Y. and Braids, O.C. (eds) *Biogeochemistry of Environmentally Important Trace Elements*. American Chemical Society, Washington, DC, pp. 339–354.
78. Paquin, P.R. , Farley, K.J., Santore, R.C., et al. (2003) *Metals in Aquatic Systems: A Review of Exposure, Bioaccumulation and Toxicity Models*. SETAC, Pensacola, FL, 168 pp.
79. Fisher, N.S. and Reinfelder, J.R. (1995) The trophic transfer of metals in marine systems. In: Tessier, A. and Turner, D.R. (eds) *Metal Speciation and bioavailability in Aquatic Sytems*. John Wiley & Sons, Inc., New York, p. 363.
80. Reinfelder, J.R. and Fisher, N.S. (1991) The assimilation of elements ingested by marine copepods. *Science*, **251**, 794.
81. Chen, C.Y., Stemberger, R.S., Klaue, B., et al. (2000) Accumulation of heavy metals in food web components across a gradient of lakes. *Limnology and Oceanography*, **45**(7), 1525–1536.
82. Mason, R.P., Laporte, J.M. and Andres, S. (2000) Factors controlling the bioaccumulation of mercury, methylmercury, arsenic, selenium, and cadmium by freshwater invertebrates and fish. *Archives of Environmental Contamination and Toxicology*, **38**(3), 283–297.
83. Watras, C.J. and Bloom, N.S. (1992) Mercury and methylmercury in individual zooplankton: Implications for bioaccumulation. *Limnology and Oceanography*, **37**, 1313.
84. Clarkson, T.W. and Magos, L. (2006) The toxicology of mercury and its chemical compounds. *Critical Reviews in Toxicology*, **36**(8), 609–662.
85. Wiener, J.G., Krabbenhoft, D.P., Heinz, G.H. and Scheuhammer, A.M. (2003) Ecotoxicology of mercury. In: Hoffman, D.J., Rattner, B.A., Burton, G.A. and Cairns, J.J. (eds) *Handbook of Ecotoxicology*. CRC Press, Boca Raton, pp. 409–463.
86. Chen, C.Y., Serrell, N., Evers, D.C., et al. (2008) Meeting report: Methylmercury in marine ecosystems-from sources to seafood consumers. *Environmental Health Perspectives*, **116**(12), 1706–1712.
87. Sunderland, E.M. (2007) Mercury exposure from domestic and imported estuarine and marine fish in the US seafood market. *Environmental Health Perspectives*, **115**(2), 235–242.

88. Evers, D.C., Mason, R.P., Kamman, N.C., et al. (2008) Integrated mercury monitoring program for temperate estuarine and marine ecosystems on the North American Atlantic Coast. *EcoHealth*, **5**(4), 426–441.
89. Compeau, G. and Bartha, R. (1985) Sulfate-reducing bacteria: Principal methylators of mercury in anoxic estuarine sediment. *Applied and Environmental Microbiology*, **50**(2), 498–502.
90. Gilmour, C.C., Henry, E.A. and Mitchell, R. (1992) Sulfate stimulation of mercury methylation in fresh-water sediments. *Environmental Science & Technology*, **26**(11), 2281–2287.
91. Fleming, E.J., Mack, E.E., Green, P.G. and Nelson, D.C. (2006) Mercury methylation from unexpected sources: Molybdate-inhibited freshwater sediments and an iron-reducing bacterium. *Applied and Environmental Microbiology*, **72**(1), 457–464.
92. Kerin, E.J., Gilmour, C.C., Roden, E., et al. (2006) Mercury methylation by dissimilatory iron-reducing bacteria. *Applied and Environmental Microbiology*, **72**(12), 7919–7921.
93. Kim, J.P. and Fitzgerald, W.F. (1988) Gaseous mercury profiles in the tropical Pacific Ocean. *Geophysical Research Letters*, **15**, 40–43.
94. Mason, R.P. and Fitzgerald, W.F. (1990) Alkylmercury species in the Equatorial Pacific Ocean. *Nature*, **347**, 457–459.
95. Sunderland, E.M., Gobas, F., Branfireun, B.A. and Heyes, A. (2006) Environmental controls on the speciation and distribution of mercury in coastal sediments. *Marine Chemistry*, **102**(1–2), 111–123.
96. Fitzgerald, W.F., Lamborg, C.H. and Hammerschmidt, C.R. (2007) Marine biogeochemical cycling of mercury. *Chemical Reviews*, **107**(2), 641–662.
97. Barkay, T., Miller, S.B. and Summers, A.O. (2003) Bacterial mercury resistance from atoms to ecosystems. *FEMS Microbiology Reviews*, **27**, 355–384.
98. Marvin-DiPasquale, M., Agee, J., McGowan, C. et al. (2000) Methyl-mercury degradation pathways: A comparison among three mercury-impacted ecosystems. *Environmental Science & Technology*, **34**(23), 4908–4916.
99. Benoit, J.M., Gilmour, C.C., Mason, R.P., Riedel, G.S. and Riedel, G.F. (1998) Behavior of mercury in the Patuxent River estuary. *Biogeochemistry*, **40**(2–3), 249–265.
100. Conaway, C.H., Squire, S., Mason, R.P. and Flegal, A.R. (2003) Mercury speciation in the San Francisco Bay estuary. *Marine Chemistry*, **80**(2–3), 199–225.
101. Hammerschmidt, C.R., Fitzgerald, W.F., Balcom, P.H. and Visscher, P.T. (2008) Organic matter and sulfide inhibit methylmercury production in sediments of New York/New Jersey Harbor. *Marine Chemistry*, **109**, 165–182.
102. Hammerschmidt, C.R., Fitzgerald, W.F., Lamborg, C.H., Balcom, P.H. and Visscher, P.T. (2004) Biogeochemistry of methylmercury in sediments of Long Island Sound. *Marine Chemistry*, **90**(1–4), 31–52.
103. Hollweg, T.A., Gilmour, C.C., and Mason, R.P. (2010) Mercury and methylmercury cycling in sediments of the mid-Atlantic continental shelf and slope. *Limnology and Oceanography*, **55**(6), 2703–2722.
104. Benoit, J.M., Gilmour, C.C., Mason, R.P. and Heyes, A. (1999) Sulfide controls on mercury speciation and bioavailability to methylating bacteria in sediment pore waters. *Environmental Science & Technology*, **33**(6), 951–957.
105. Benoit, J.M., Gilmour, C.C. and Mason, R.P. (2001) The influence of sulfide on solid phase mercury bioavailability for methylation by pure cultures of *Desulfobulbus propionicus* (1pr3). *Environmental Science & Technology*, **35**(1), 127–132.
106. Ekstrom, E.B., Morel, F.M.M. and Benoit, J.M. (2003) Mercury methylation independent of the acetyl-coenzyme a pathway in sulfate-reducing bacteria. *Applied and Environmental Microbiology*, **69**(9), 5414–5422.
107. Yoshinaga, J. (2003) Organolead compounds in the environment. In: Craig, P.J. (ed.) *Organometallic Compounds in the Environment*. John Wiley & Sons, Ltd, Chichester, pp. 151–194.
108. White, J.S., Tobin, J.M. and Cooney, J.J. (1999) Organotin compounds and their interactions with microorganisms. *Canadian Journal of Microbiology*, **45**(7), 541–554.
109. Tessier, E., Amouroux, D. and Donard, O.F.X. (2003) Biogenic volatilization of trace elements from European estuaries. In: Cai, Y. and Braids, O.C. (eds) *Biogeochemistry of Environmentally Important Trace Elements*. ACS, Washington, DC, pp. 151–165.
110. Edmonds, J.S. and Francesconi, K.A. (2003) Organoarsenic compounds in the marine environment. In: Craig, P. (ed.) *Organometallic Compounds in the Environment*. John Wiley & Sons, Ltd, Chichester, pp. 194–275.
111. Cutter, G.A. and Cutter, L.S. (2006) Biogeochemistry of arsenic and antimony in the North Pacific Ocean. *Geochemistry Geophysics Geosystems*, **7**(5), doi:10.1029/2005GC001159.
112. Cutter, G.A., Cutter, L.S., Featherstone, A.M. and Lohrenz, S.E. (2001) Antimony and arsenic biogeochemistry in the western Atlantic Ocean. *Deep-Sea Research Part II*, **48**(13), 2895–2915.
113. Planer-Friedrich, B., Lehr, C., Matschullat, J., et al. (2006) Speciation of volatile arsenic at geothermal features in Yellowstone National Park. *Geochimica et Cosmochimica Acta*, **70** (10), 2480–2491.
114. Kuehnelt, D. and Goessler, W. (2003) Organoarsenic compounds in the terrestrial environment. In: Craig, P. (ed.) *Organometallic Compounds in the Environment*. John Wiley & Sons, Ltd, Chichester, pp. 223–275.
115. Knauer, K., Ahner, B., Xue, H.B. and Sigg, L. (1998) Metal and phytochelatin content in phytoplankton from freshwater lakes with different metal concentrations. *Environmental Toxicology and Chemistry*, **17**(12), 2444–2452.
116. Le Faucheur, S., Schildknecht, F., Behra, R. and Sigg, L. (2006) Thiols in Scenedesmus vacuolatus upon exposure to metals and metalloids. *Aquatic Toxicology*, **80**(4), 355–361.
117. Burkhard, L.P., Cook, P.M. and Mount, D.R. (2003) The relationship of bioaccumulative chemicals in water and sediment to residues in fish: A visualization approach. *Environmental Toxicology and Chemistry*, **22**, 2822–2830.
118. Burkhard, L.P., Endicott, D.D., Cook, P.M., Sappingtonm, K.G. and Winchester, E.L. (2003) Evaluation of two methods for prediction of bioaccumulation factors. *Environmental Science and Technology*, **37**, 4626–4634.
119. EPA (2003) *Methodology for deriving ambient water quality criteria for the protection of human health*. Technical support document Volume 2: Development of national bioaccumulation factors. In: U.E. Office of Water, Washington, DC (ed.).

120. McGeer, J.C., Brix, K.V., Skeaff, J.M., et al. (2003) Inverse relationship between bioconcentration factor and exposure concentration for metals: Implications for hazard assessment of metals in the aquatic environment. *Environmental Toxicology and Chemistry*, **22**(5), 1017–1037.
121. Reinfelder, J.R., Fisher, N.S., Luoma, S.N., Nichols, J.W. and Wang, W.X. (1998) Trace element trophic transfer in aquatic organisms: A critique of the kinetic model approach. *Science of the Total Environment*, **219**(2–3), 117–135.
122. Mason, R.P., Reinfelder, J.R. and Morel, F.M.M. (1995) Bioaccumulation of mercury and methylmercury. *Water, Air, and Soil Pollution*, **80**, 915–921.
123. Croteau, M.N. and Luoma, S.N. (2005) Delineating copper accumulation pathways for the freshwater bivalve Corbicula using stable copper isotopes. *Environmental Toxicology and Chemistry*, **24**(11), 2871–2878.
124. Lee, B.G., Griscom, S.B., Lee, J.-S., et al. (2000) Influences of dietary uptake and reactive sulfides on metal bioavailability from aquatic sediments. *Science*, **287**(5451), 282–284.
125. Mathews, T. and Fisher, N.S. (2008) Evaluating the trophic transfer of cadmium, polonium, and methylmercury in an estuarine food chain. *Environmental Toxicology and Chemistry*, **27**(5), 1093–1101.
126. Chen, C.Y. and Folt, C.L. (2005) High plankton densities reduce mercury biomagnification. *Environmental Science & Technology*, **39**(1), 115–121.
127. Pickhardt, P.C., Folt, C.L., Chen, C.Y., Klaue, B. and Blum, J.D. (2005) Impacts of zooplankton composition and algal enrichment on the accumulation of mercury in an experimental freshwater food web. *Science of the Total Environment*, **339**(1–3), 89–101.
128. Karimi, R., Chen, C.Y., Pickhardt, P.C., Fisher, N.S. and Folt, C.L. (2007) Stoichiometric controls of mercury dilution by growth. *Proceedings of the National Academy of Sciences of the United States of America*, **104**(18), 7477–7482.
129. Ankley, G.T. (1996) Evaluation of metal/acid-volatile sulfide relationships in the prediction of metal bioaccumulation by benthic macroinvertebrates. *Environmental Toxicology and Chemistry*, **15**(12), 2138–2146.
130. Lawrence, A.L., McAloon, K.M., Mason, R.P. and Mayer, L.M. (1999) Intestinal solubilization of particle-associated organic and inorganic mercury as a measure of bioavailability to benthic invertebrates. *Environmental Science & Technology*, **33**(11), 1871–1876.
131. McAloon, K.M. and Mason, R.P. (2003) Investigations into the bioavailability and bioaccumulation of mercury and other trace metals to the sea cucumber, Sclerodactyla briareus, using in vitro solubilization. *Marine Pollution Bulletin*, **46**(12), 1600–1608.
132. Ankley, G.T., DiToro, D.M., Hansen, D.J. and Berry, W.J. (1996) Technical basis and proposal for deriving sediment quality criteria for metals. *Environmental Toxicology and Chemistry*, **15**(12), 2056–2066.
133. Berry, W.J., Hansen, D.J., Boothman, W.S., et al. (1996) Predicting the toxicity of metal-spiked laboratory sediments using acid-volatile sulfide and interstitial water normalizations. *Environmental Toxicology and Chemistry*, **15**(12), 2067–2079.
134. Mahony, J.D., Di Toro, D.M., Gonzalez, A., et al. (1996) Partitioning of metals to sediment organic carbon. *Environmental Toxicology and Chemistry*, **15**, 2187.
135. Hansen, D.J., Berry, W.J., Mahony, J.D., et al. (1996) Predicting the toxicity of metal-contaminated field sediments using interstitial concentration of metals and acid-volatile sulfide normalizations. *Environmental Toxicology and Chemistry*, **15**(12), 2080–2094.
136. Lawrence, A.L. and Mason, R.P. (2001) Factors controlling the bioaccumulation of mercury and methylmercury by the estuarine amphipod Leptocheirus plumulosus. *Environmental Pollution*, **111**(2), 217–231.
137. Lee, B.G., Lee, J.S., Luoma, S.N., Choi, H.J. and Koh, C.H. (2000) Influence of acid volatile sulfide and metal concentrations on metal bioavailability to marine invertebrates in contaminated sediments. *Environmental Science & Technology*, **34**(21), 4517–4523.
138. Chen, Z. and Mayer, L.M. (1999) Sedimentary metal bioavailability determined by the digestive constraints of marine deposit feeders: Gut retention time and dissolved amino acids. *Marine Ecology Progress Series*, **176**, 139–151.
139. Weston, D.P., Judd, J.R. and Mayer, L.M. (2004) Effect of extraction conditions on trace element solubilization in deposit feeder digestive fluid. *Environmental Toxicology and Chemistry*, **23**(8), 1834–1841.
140. King, C.K., Gale, S.A. and Stauber, J.L. (2006) Acute toxicity and bioaccumulation of aqueous and sediment-bound metals in the estuarine amphipod *Melita plumulosa*. *Environmental Toxicology*, **21**(5), 489–504.
141. Besser, J.M., Brumbaugh, W.G., May, T.W. and Ingersoll, C.G. (2003) Effects of organic amendments on the toxicity and bioavailability of cadmium and copper in spiked formulated sediments. *Environmental Toxicology and Chemistry*, **22**(4), 805–815.
142. Lu, X.Q., Werner, I. and Young, T.M. (2005) Geochemistry and bioavailability of metals in sediments from northern San Francisco Bay. *Environment International*, **31**(4), 593–602.
143. Laporte, J.M., Andres, S. and Mason, R.P. (2002) Effect of ligands and other metals on the uptake of mercury and methylmercury across the gills and the intestine of the blue crab (*Callinectes sapidus*). *Comparative Biochemistry and Physiology C-Toxicology & Pharmacology*, **131**(2), 185–196.
144. Ojo, A.A., Nadella, S.R. and Wood, C.M. (2009) In vitro examination of interactions between copper and zinc uptake via the gastrointestinal tract of the rainbow trout (*Oncorhynchus mykiss*). *Archives of Environmental Contamination and Toxicology*, **56**(2), 244–252.
145. Ojo, A.A. and Wood, C.M. (2007) In vitro analysis of the bioavailability of six metals via the gastro-intestinal tract of the rainbow trout (*Oncorhynchus mykiss*). *Aquatic Toxicology*, **83**(1), 10–23.
146. Leaner, J.J. and Mason, R.P. (2002) Methylmercury accumulation and fluxes across the intestine of channel catfish, Ictalurus punctatus. *Comparative Biochemistry and Physiology C-Toxicology & Pharmacology*, **132**(2), 247–259.
147. Paquin, P.R., Gorsuch, J.W., Apte, S., et al. (2002) The biotic ligand model: A historical overview. *Comparative Biochemistry and Physiology C-Toxicology & Pharmacology*, **133**(1–2), 3–35.
148. McGeer, J.C., Playle, R.C., Wood, C.M. and Galvez, F. (2000) A physiologically based biotic ligand model for predicting the acute toxicity of waterborne silver to rainbow trout in

freshwaters. *Environmental Science & Technology*, **34**(19), 4199–4207.
149. Playle, R.C., Dixon, D.G. and Burnison, K. (1993) Copper and cadmium-binding to fish gills – estimates of metal gill stability-constants and modeling of metal accumulation. *Canadian Journal of Fisheries and Aquatic Sciences*, **50**(12), 2678–2687.
150. Playle, R.C., Dixon, D.G. and Burnison, K. (1993) Copper and cadmium-binding to fish gills – modification by dissolved organic-carbon and synthetic ligands. *Canadian Journal of Fisheries and Aquatic Sciences*, **50**(12), 2667–2677.
151. Janes, N. and Playle, R.C. (1995) Modeling silver-binding to gills of rainbow-trout (*Oncorhynchus-mykiss*). *Environmental Toxicology and Chemistry*, **14**(11), 1847–1858.
152. Morgan, T.P., Grosell, M., Gilmour, K.M., Playle, R.C. and Wood, C.M. (2004) Time course analysis of the mechanism by which silver inhibits active Na+ and Cl– uptake in gills of rainbow trout. *American Journal of Physiology-Regulatory, Integrative and Comparative Physiology*, **287**(1), R234–R242.
153. Wood, C.M., Playle, R.C. and Hogstrand, C. (1999) Physiology and modeling of mechanisms of silver uptake and toxicity in fish. *Environmental Toxicology and Chemistry*, **16**, 71–83.
154. Wood, C.M. (2005) Putting fish gill and gut physiology into environmental regulations. *Comparative Biochemistry and Physiology A-Molecular & Integrative Physiology*, **141**(3), S101–S101.
155. Playle, R.C. (2004) Using multiple metal-gill binding models and the toxic unit concept to help reconcile multiple-metal toxicity results. *Aquatic Toxicology*, **67**(4), 359–370.
156. Luider, C.D., Crusius, J., Playle, R.C. and Curtis, P.J. (2004) Influence of natural organic matter source on copper speciation as demonstrated by Cu binding to fish gills, by ion selective electrode, and by DGT gel sampler. *Environmental Science & Technology*, **38**(10), 2865–2872.
157. Niyogi, S., Kent, R. and Wood, C.M. (2008) Effects of water chemistry variables on gill binding and acute toxicity of cadmium in rainbow trout (*Oncorhynchus mykiss*): A biotic ligand model (BLM) approach. *Comparative Biochemistry and Physiology C-Toxicology & Pharmacology*, **148**(4), 305–314.
158. Schwartz, M.L., Curtis, P.J. and Playle, R.C. (2004) Influence of natural organic matter source on acute copper, lead, and cadmium toxicity to rainbow trout (*Oncorhynchus mykiss*). *Environmental Toxicology and Chemistry*, **23**(12), 2889–2899.
159. Kozlova, T., Wood, C.M. and McGeer, J.C. (2009) The effect of water chemistry on the acute toxicity of nickel to the cladoceran Daphnia pulex and the development of a biotic ligand model. *Aquatic Toxicology*, **91**(3), 221–228.
160. Bianchini, A., Playle, R.C., Wood, C.M. and Walsh, P.J. (2005) Mechanism of acute silver toxicity in marine invertebrates. *Aquatic Toxicology*, **72**(1–2), 67–82.
161. Bianchini, A., Rouleau, C. and Wood, C.M. (2005) Silver accumulation in Daphnia magna in the presence of reactive sulfide. *Aquatic Toxicology*, **72**(4), 339–349.
162. Di Toro, D.M., McGrath, J.A., Hansen, D.J., et al. (2005) Predicting sediment metal toxicity using a sediment biotic ligand model: Methodology and initial application. *Environmental Toxicology and Chemistry*, **24**(10), 2410–2427.
163. Hewitt, S.H. and Johnson, B.L. (1987) *A Generalized Bioenergetics Model of Fish Growth for Microcomputers*. University of Wisconsin, Madison, WI.
164. Thomann, R.V. (1981) Equilibrium model of fate of microcontaminants in diverse aquatic food chains. *Canadian Journal of Fisheries and Aquatic Sciences*, **38**, 280.
165. Thomann, R.V. (1989) Bioaccumulation model of fate of organic chemical distribution in aquatic food chainss. *Environmental Science and Technology*, **23**, 699.
166. Harris, R.C., Rudd, J.W.M., Amyot, M. et al. (2007) Whole-ecosystem study shows rapid fish-mercury response to changes in mercury deposition. *Proceedings of the National Academy of Sciences of the United States of America*, **104**(42), 16586–16591.
167. Henry, E.A., Dodgemurphy, L.J., Bigham, G.N. and Klein, S.M. (1995) Modeling the transport and fate of mercury in an urban lake (Onondaga Lake, NY). *Water, Air, and Soil Pollution*, **80**(1–4), 489–498.
168. Rogers, D.W. (1994) You are what you eat and a little bit more: Bioenergetics-based models of methylmercury accumulation in fish revisted. In: Watras, C.J. and Huckabee, J.W. (eds) *Mercury Pollution: Integration and Synthesis*. Lewis Publishers, Boca Raton, pp. 427–440.
169. Hudson, R.J.M., Gherini, S.A., Watras, C.J. and Porcella, D.B. (1994) Modeling the biogeochemical cycle of mercury in lakes: The Mercury Cycling Model (MCM) and its applications in the MTL study lakes. In: Watras, C.J. and Huckabee, J.W. (eds) *Mercury Pollution: Integration and Synthesis*. Lewis Publishers, Boca Raton, pp. 473–523.
170. TetraTech (2007) *Mercury in the Environment*. Brochure, Palo Alto.
171. Driscoll, C.T., Blette, V. , Yan, C., et al. (1995) The role of dissolved organic-carbon in the chemistry and bioavailability of mercury in remote Adirondack lakes. *Water, Air, and Soil Pollution*, **80**(1–4), 499–508.
172. Gbono-Tugbawa, S. and Driscoll, C.T. (1998) Application of the Regional Mercury Cycling Model (RMCM) to predict the fate and remediation of mercury in Onondaga Lake, New York. *Water, Air, and Soil Pollution*, **105**(1–2), 417–426.
173. Leonard, D., Reash, R., Porcella, D., et al. (1995) Use of the mercury cycling model (MCM) to predict the fate of mercury in the Great-Lakes. *Water, Air, and Soil Pollution*, **80**(1–4), 519–528.
174. Thomann, R.V., Szumski, D.S., Di Toro, D.M. and O'Connor, D.J. (1974) A food chain model of cadmium in Western Lake Erie. *Water Research*, **8**, 841–849.
175. Thomann, R.V., Shkreli, F. and Harrison, S. (1997) A pharmacokinetic model of cadmium in rainbow trout. *Environmental Toxicology and Chemistry*, **16**(11), 2268–2274.
176. Thomann, R.V., Mahony, J.D. and Mueller, R. (1995) Steady-state model of biota sediment accumulation factor for metals in 2 marine bivalves. *Environmental Toxicology and Chemistry*, **14**(11), 1989–1998.
177. Kim, E., Mason, R.P. and Bergeron, C.M. (2008) A modeling study on methylmercury bioaccumulation and its controlling factors. *Ecological Modelling*, **218**(3–4), 267–289.
178. Karasov, W.H., Kenow, K.P., Meyer, M.W. and Fournier, F. (2007) Bioenergetic and pharmacokinetic model for exposure of common loon (*Gavia immer*) chicks to methylmercury. *Environmental Toxicology and Chemistry*, **26**(4), 677–685.
179. Monteiro, L.R. and Furness, R.W. (2001) Kinetics, dose-response, and excretion of methylmercury in free-living adult

Cory's shearwaters. *Environmental Science & Technology*, **35**(4), 739–746.

180. Baines, S.B. and Fisher, N.S. (2008) Modeling the effect of temperature on bioaccumulation of metals by a marine bioindicator organism, Mytilus edulis. *Environmental Science & Technology*, **42**(9), 3277–3282.
181. Ke, C.H. and Wang, W.X. (2001) Bioaccumulation of Cd, Se, and Zn in an estuarine oyster (*Crassostrea rivularis*) and a coastal oyster (*Saccostrea glomerata*). *Aquatic Toxicology*, **56**(1), 33–51.
182. Leaner, J.J. and Mason, R.P. (2004) Methylmercury uptake and distribution kinetics in sheeps-head minnows, *Cyprinodon variegatus*, after exposure to CH(3)Hg-spiked food. *Environmental Toxicology and Chemistry*, **23**(9), 2138–2146.

Problems

8.1. It has been suggested that the uptake of silver by phytoplankton can occur through the passive diffusion of the neutral chloride complex (AgCl) across the membrane. Discuss how the bioavailability of Ag (10^{-7} M) would change going from a freshwater environment (Cl = 10^{-4} M; total sulfate = 10^{-4} M; Ca = Mg = 10^{-3} M; Na = K = 10^{-4} M; pH = 8) to that of seawater (Cl = 0.5 M; total sulfate = $10^{-1.5}$ M; Ca = Mg = K = 10^{-2} M; Na = 0.5 M; pH = 8), assuming equilibrium with the atmosphere throughout in terms of carbonate chemistry (i.e., $[H_2CO_3] = 10^{-5}$ M). Do calculations for freshwater and Cl concentrations of 0.01, 0.1 and 0.5 M, and estimate the salinity where the concentration of AgCl is maximum. Assume conservative mixing to obtain the concentration of the other major ions at the intermediate chloride concentrations. How important do you think passive diffusion of AgCl will be to Ag uptake in freshwater? In seawater?

8.2. Sulfate reduction is an important process in coastal sediments for the degradation of organic matter. But it is possible that other compounds could also be involved in organic matter degradation. For example, organisms have been isolated that are arsenic-reducing bacteria that is, they reduce arsenate to arsenite in the process of respiring organic matter to CO_2.

a. Would this reaction release more or less energy at a pH of 8.2 in coastal seawater sediments per mole of organic matter degraded to CO_2 than that involving sulfate reduction? Assume that the total arsenic concentration is 2×10^{-8} M and that the sulfate concentration is that of "typical" seawater. Relevant reactions are given below. Note that the arsenic species are present as triprotic acids in water and so the relative magnitude of each species will be a function of pH.

b. Consider the possibility of selenate reducing bacteria. Determine whether the reaction is more favorable than either As or S-reduction under the same conditions.

Potentially relevant equations are:

$$H_2AsO_4^- + 3H^+ + 2e^- = H_3AsO_3 + H_2O \qquad \log K = 21.7$$

$$H_2AsO_4^- = H^+ + HAsO_4^- \qquad \log K = -6.8$$

$$\frac{1}{2}SeO_4^{2-} + 2H^+ + e^- = \frac{1}{2}H_2SeO_3 + \frac{1}{2}H_2O \qquad \log K = 19.4$$

$$SeO_4^{2-} + H^+ = HSeO_4^- \qquad \log K = 1.7$$

$$SeO_3^{2-} + H^+ = HSeO_3^- \qquad \log K = 7.9$$

$$HSeO_3^- + H^+ = H_2SeO_3 \qquad \log K = 2.4$$

8.3. The data shown in Fig. 8.9 show that there are large differences in the speciation of copper between surface and subsurface waters. Estimate what the binding constant for an organic ligand complex would need to be to account for the speciation in the surface waters. Assume that the ligand is a carboxylic acid molecule of the formula RCOOH with an acid dissociation constant, $pK_a = 5.8$ and a concentration of 2 nM. For the deep waters, what would the binding constant need to be to account for the speciation at 400 m, assuming that the ligand is present at 1 nM in concentration?

8.4. Recent studies have suggested that the upper ocean contains low nM concentrations of small molecular weight thiols such as cysteine and glutathione. Estimate the concentrations of cysteine that would be needed to complex 50% of the following metals in surface ocean waters, assuming no other organic complexes, and in the absence of free sulfide: Hg, Pb, and Zn.

8.5. Recreate the data for Co, Zn, and Cd for the "sulfidic" Proterozoic Ocean as shown in Fig. 8.12. Do you agree with the model as presented in the figure? Also estimate the dissolved concentrations for Hg and Pb under these conditions.

8.6. Determine the rate constant for the reaction leading to the decomposition of tetraethyllead as shown in Fig. 8.19. Discuss the reaction sequence and write out reactions for the various reactions that could lead to the results shown in the figure.

8.7. Phytoplankton are exposed to a solution containing 2 nM Cd, 10 pM inorganic Hg and 1 pM CH_3Hg. Estimate the relative amounts of each that would be assimilated into zooplankton feeding on these exposed algae using the information in Fig. 8.20 and the data in Table 8.6.

Index

Page numbers in *italics* refer to figures, those in **bold** refer to tables.

Trace Metals in Aquatic Systems, First Edition. Robert P. Mason.